The Nariokotome *Homo erectus* Skeleton

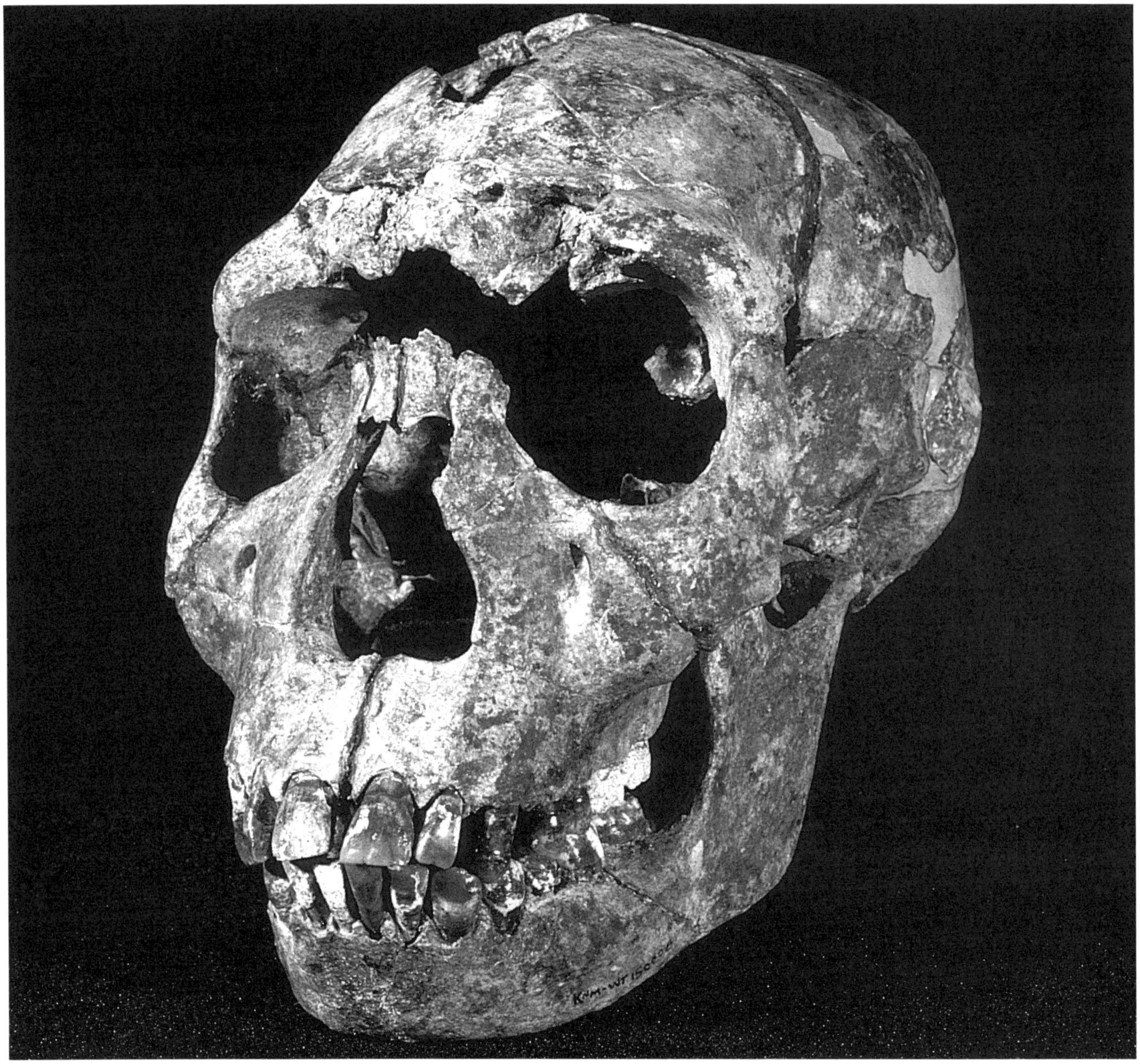

The cranium and mandible (KNM-WT 15000 A and B) of the Nariokotome Homo erectus *skeleton, seen in left three-quarter view.*

The Nariokotome *Homo erectus* Skeleton

Edited by

Alan Walker and Richard Leakey

Harvard University Press
Cambridge, Massachusetts
1993

Printed in the United States of America

This book is printed on acid-free paper,
and its binding materials have been chosen for
strength and durability.

Library of Congress Cataloging in Publication Data

The Nariokotome Homo erectus skeleton / edited by Alan Walker and
Richard Leakey.
p. cm.
Includes bibliographical references and index.
ISBN 0-674-60075-4 (alk. paper)
1. Homo erectus—Kenya. I. Walker, Alan, 1938– .
II. Leakey, Richard E.
GN284.N37 1993
573.3—dc20
92-32985
CIP

CONTENTS

The Nariokotome *Homo erectus* Skeleton

CHAPTER 1

INTRODUCTION

RICHARD LEAKEY AND ALAN WALKER

The most complete early hominid skeleton ever found was excavated from sediments 1.5 million years of age near the Nariokotome sand river in northern Kenya. The discovery of the skull along with the skeleton of an individual who lived before purposeful burial was practiced is exceedingly rare. Designated KNM-WT 15000, this specimen is a male early Homo erectus; *at the time of his death he was about eleven years old in terms of human development. It is a great find in itself, but it also gives us new insights into other hominid fossils and into the life and times of our early human ancestors.*

Over fifty years have gone by since A. M. Champion reported the presence of fossil mammals on the west side of Lake Turkana in northern Kenya (Champion, 1937). He recorded areas "strewn with the remains of fish, saurians and mammals." Since then, major fossil discoveries have been made in the Plio-Pleistocene sediments of the lower Omo Valley to the north and the Miocene-Pliocene sediments of Moruarot, Lothidok, Loperot, Lothagam, and Ekora to the south (see Coppens et al., 1976, and Howell and Coppens, 1983, for historical summaries). But, curiously, the thin strip of Plio-Pleistocene sediments that extends eastward from the Murua Rith and Labur hills (which border the western and northwestern shores of Lake Turkana) to the lake itself escaped notice by any paleontologist.

Our recent paleontological explorations on the west side of Lake Turkana, Kenya, began as a natural offshoot of the Koobi Fora Research Project which was established on the east side of the lake in 1968 (Leakey and Leakey, 1978). The search for Plio-Pleistocene faunas at East Lake Turkana resulted in the accumulation of a very complete fossil record for some time periods. This record has been documented in numerous papers and two volumes (Harris, 1983, 1991). The latter cover the families Bovidae, Camelidae, Deinotheriidae, Equidae, Giraffidae, Elephantidae, Rhinocerotidae, and Suidae.

Attempts to find local sediment sources for the formations to the east of the lake were unsuccessful. Partly for this reason, a controversy developed between members of the Koobi Fora Research Project and others who knew the much better-dated Omo geological sequence and its faunas. The differences were resolved by a series of potassium-argon dates of the East Turkana tuffs. By that means, the deposits on the eastern shore were eventually brought into concordance with the stratigraphic and faunal data from nearby related deposits of the Omo Valley of Ethiopia (see Lewin, 1987, for a historical review). A regional picture of the continuous development of the whole basin has now been published by Brown and Feibel (1991), and an account of the paleoenvironments is given in some detail by Feibel et al. (1991).

Because of the controversy which developed over the ages of certain hominid fossils while the apparent conflict between the Omo and East Turkana data was unresolved, it was decided not to carry out serious prospecting on the west side of the lake until the local stratigraphy had been established and all major dating problems had been settled. This was in large part accomplished by the time the 1984 prospecting season began, and many tuffs had already been correlated geochemically with those exposed on the east side of the lake or in the Omo Valley to the north (Harris and Brown, 1985; Brown et al., 1985). The deposits on the west side of the lake were also important because of their age; some were about 2.5–2.0 million years (Ma) old, and samples from this period are not well represented at East Turkana.

The Discovery of KNM-WT 15000

In August 1984, Kamoya Kimeu found a small piece of hominid cranial vault in the surface pebble lag on the slopes of a small tributary of the Nariokotome sand river about 5 km inland from Lake Turkana (Leakey and Walker, 1985). This site is designated Nariokotome III (Brown et al., 1985; Harris et al., 1988a,b), abbreviated NK3. Sieving operations were begun at the site on 26 August and much of the rest of the cranium was recovered in pieces from the slope wash. On 30 August the facial skeleton was found embedded in the roots of a small *Acacia* tree. Subsequent excavations into the hillside at this level resulted in the discovery of many more parts of a single, juvenile skeleton, which was given the accession number KNM-WT 15000. The excavations in 1984 were curtailed on 21 September for lack of funds.

An announcement about the find was made in October 1984 by the sponsors of the research, The National Geographic Society of Washington, D.C., in order to preempt the release of the story by an international news magazine, which had been told of the discovery by a visitor to the camp. Excavations had resumed by the time a brief initial note had appeared in *Nature* the next year (Brown et al., 1985). Soon two popular accounts of the discovery had been published (Harris, 1986; Leakey and Walker, 1985).

In the second season more bones were recovered, together with all the missing teeth except for the upper right third molar germ. Excavations in the third year produced a few more pieces, in the fourth year yet two more pieces, and in the fifth year none. Work at the site stopped in late 1988 after this last attempt to find more of the skeleton failed. During the five seasons of excavation over 1,500 cubic meters of overburden was removed by hand and wheelbarrow. At the end of our work a roughly L-shaped area of about 425 square meters was exposed. By comparison, the FLK ("Zinjanthropus") excavation at Olduvai Gorge covered about 350 square meters (Leakey, 1971).

We knew in 1985 that more bones and teeth were probably going to be found, and we also knew that these might influence our interpretations of the bones that we already had in hand. We decided, therefore, to wait until we had finished the excavations before carrying out analyses. Instead, we concentrated our analytical efforts on new Miocene hominoids as well as on important new *Australopithecus* fossils from other localities. We did, however, publish a brief account in the *Proceedings of the Second International Congress of Human Paleontology* (Leakey and Walker, 1989), which included a reconstruction of the skeleton in which practically all the parts now known were shown.

THE SCOPE OF THIS VOLUME

We considered several alternative ways of proceeding with the study and publication of the skeleton KNM-WT 15000. Since the specimen is the most complete early hominid ever found, we decided to try to publish this material fairly quickly (by paleoanthropological standards!). In that way other researchers could begin their own studies of this very important fossil. But this approach has meant that we cannot give now as complete and detailed an account as the specimen ultimately deserves, especially in the matter of comparisons with other hominid fossils. Given the intermediate position of *Homo erectus* between the earlier *Homo habilis* on the one hand and *Homo sapiens* on the other, an appropriate treatment would entail comparing the specimen with all known hominid species. Because practically all parts of the skeleton were found, a thorough comparison would mean examining nearly the whole hominid fossil record. We have undertaken here, then, to present descriptions and photographs of the parts of the skeleton without a thorough analysis of many of them. We have also tried to establish an account of the population of early *H. erectus* from which this individual came by considering other specimens of this species from Lake Turkana. Since no single person is an expert on all parts of the early hominid skeleton, we decided to enlist the aid of some of our colleagues with expertise in certain aspects of the preliminary research. As a result, some topics have now been dealt with in some detail, but others have only been mentioned.

This volume is divided into three parts. Part I (Chapters 2–5) concerns the geology, dating, paleoenvironments, and taphonomy of the NK3 site. Part II (Chapters 6–8) is a description of the specimen and a review of other *H. erectus* specimens from the Turkana region. And Part III (Chapters 9–17) brings together a variety of analytical papers on the biology of the KNM-WT 15000 individual in particular and of *H. erectus* specimens found at other sites.

It is important to note that since these chapters were written, Phillip Tobias's account of the early hominids from Beds I and II at Olduvai Gorge has appeared (Tobias, 1991). This work, which gives new meaning to the word *monumental,* is an exhaustive account of some Olduvai fossils, most of which were collected in the 1960s. We have decided, in the interests of getting our own work published promptly, not to send the relevant chapters back to their authors for revisions in the light of Tobias's work. Occasionally, however, comments have been added to the text when a major point of difference occurred.

Is the name *Homo erectus* appropriate?

Up until 1976, there was still a group of scholars who believed that there only was one early hominid species at any one time in the past. A cranium was found at Koobi Fora in 1975 that helped to falsify that hypothesis (Leakey and Walker, 1976). At that time we claimed that two hominid species—*Australopithecus boisei* and *Homo erectus*—lived contemporaneously in eastern Africa from about 1.6 to 1.3 Ma. We called this specimen, KNM-ER 3733, *H. erectus* purely on phenetic grounds and claimed that it was unequivocally in the same species as similar specimens from Java and China. There has been a lot of support for this position (e.g., Coppens, 1983; Conroy, 1990; Howell, 1978; Rightmire, 1979, 1991; Tobias, 1978, 1991; Wolpoff, 1980).

Recently, however, as paleoanthropologists try to apply the principles of cladistics to the problems of hominid taxonomy, there are those who feel that we overstated the case of the resemblances between early African forms and later Chinese ones. Groves (1989), for instance, says that the comparison "would seem to have been greatly overdone." Andrews (1984) claimed that the African *erectus* specimens had none of the uniquely derived features of the Asian ones and wanted to disbar the Asian specimens from a role in later human ancestry. Stringer agreed in 1984 that a cladistic concept of the species *erectus* could not be contructed so as to include the early specimens from Koobi Fora, Olduvai, and the Far East, but by 1987 his more rigorous analyses did support a clade containing the early African and the later Asian *erectus* groups.

Thus, although it seems that there are two opposing positions about the classification of early African *H. erectus* specimens, practically every authority agrees that they are phenetically close to the Asian members. When it comes to cladistic methods, only some practitioners feel that the African specimens should be included in the clade which includes *H. erectus.* We feel, however, that in some of these first cladistic attempts the degree of variability in the early sample has been overlooked. We hope that this has been pointed out properly in Chapters 6 to 8. In any case, we believe that future research will show that these early forms do belong cladistically with those from Asia and that they do not need their own, new, name.

Preliminary interpretations of the specimen

Many interesting findings have come out of the analysis of this fairly complete skeleton. These are dealt with fully in the relevant chapters and also again in the summary. But the reader will find it useful to consider some points that can be made beforehand.

One of the most important results of the discovery of this specimen is that we can now be certain of the association of skull and teeth with the postcranial skeleton for one individual. That means that we can now reexamine previously collected unassociated postcranial specimens with a view to assigning them to species. Most postcranial bones from earlier collections are not associated with skulls or teeth, so the finding of KNM-WT 15000 will enhance their analytical value. We also have, for the first time, the chance of determining cranial capacity and body size in a single early hominid individual.

The specimen also has cleared up some problems that we had with interpreting the early Pleistocene record in Africa. We could not ourselves agree on the affinities of one mandible (KNM-ER 992): one of us believed it belonged with crania such as KNM-ER 1813, and the other thought it more likely that it belonged with crania like KNM-ER 3733. We both now agree that it belongs with *H. erectus,* as Wood (1976) suggested long ago. To give one example of unassociated postcranial bones that now can be assigned with more certainty, the ulna (OH 36) from Olduvai, previously attributed to *Homo erectus* (Day, 1986), is almost certainly that of *Australopithecus boisei.*

Now that we have KNM-WT 15000 we can examine for the first time the physiological consequences of early hominid body proportions. Ruff (1991; see also Chapter 11) has developed his ideas of body proportion and climate to the stage where the climatic model of human body proportions can be almost fully understood. If we use the Nariokotome skeleton's proportions to predict the mean annual temperature, the answer is distinctly equatorial. The finding that early African *H. erectus* were already physiologically adapted to Allen's rule was a major development in our understanding of human evolution, because it means that a human-like physiology with dependence on sweating as the major (if not the only) cooling mechanism had already been established by 1.5 million years ago.

The size of the individual also led us to reevaluate the estimates of body sizes of the other early *H. erectus* that we knew from East Africa. Our hypothesis that we had found a single individual very far from the mean of stature was falsified (Chapter 11). We have yet to find an early *H. erectus* with a height less than 158 cm and with a body weight less than 51 kg according to our estimations. The biggest estimates we have are 185 cm and 68 kg, which are the estimates of the adult height and weight of the Nariokotome youth. The tall stature of these early *H. erectus* fits in with a concept of these early hominids as being well nourished and healthy. In contrast, as has been pointed out by many authors (see, e.g., Cohen and Armelagos, 1984), there is often evidence to show a decline in health and a reduction in stature associated with the onset of agriculture.

In addition, the specimen has added new information to help resolve the debate about the tempo and mode of human evolution. It now seems clear to us that a major event entailing a period of rapid evolution occurred in our lineage about 1.8 million years ago.

And finally, it is striking to us to see how human this early member of our own genus was over 1.5 million years ago. In limb proportions and body proportions, in tooth features, in the spinal column with its curvatures, in the thorax with the characteristic hominid rib torsion, this skeleton looks human. It is possible to find several primitive features, such as the relatively small brain, but one does have to search for them. A few of these serve as a clear demonstration of the mosaic nature of evolution. But the overwhelmingly human pattern of the skeleton is a strong reminder of how much of our own morphology is deeply rooted in the past.

PART I

THE SITE

C H A P T E R 2

GEOLOGIC SETTING AND AGE

FRANCIS H. BROWN AND IAN MCDOUGALL

A regional stratigraphic framework for the whole Turkana Basin has been developed during the last twenty years. The compositional characterization of numerous volcanic ash layers preserved within the basin was the key to establishing this framework. The composition of glass in the ash layers has been used to establish correlations between stratigraphic sections in different parts of the basin, as well as correlations with sections far from the basin. Some of the ashes contain pumice clasts with alkali feldspar that has been dated isotopically. The geological age of the Nariokotome hominid—about 1.5 million years—has been determined securely by its stratigraphic relationship to these dated ashes.

Pliocene and Pleistocene strata of the Turkana Basin are now reasonably well understood because of the evidence provided by many expeditions over the past two decades. Isotopic ages have been measured on many tuffs by several laboratories, and the temporal placement of most strata is closely controlled. Through chemical analysis of the glass fraction of tuffs, it is now possible to correlate at many levels between strata exposed in the lower Omo Valley, the Koobi Fora region, and along the western edge of Lake Turkana. It is also possible to reconstruct the regional paleogeography with far more precision than was imagined by earlier workers.

Our understanding of the Nariokotome site, and of the evolution of the basin as a whole, has been aided considerably by utilizing information gathered from the Nachukui, Koobi Fora, and Shungura Formations. Strata exposed at the Nariokotome site were laid down by a very large depositional system, and they cannot be properly understood without being placed in the larger context. In this chapter we set out the local physical setting of the site, briefly review the stratigraphy of the major formations in the basin, review isotopic ages of units relevant to the age of the hominid level, and give a short description of geological changes leading up to deposition at the hominid site.

The local setting

Nariokotome is a large ephemeral stream that issues from the Murua Rith range west of Lake Turkana and extends to the lake. Subsurface water has always been available in the channel of Nariokotome, even during the driest years, and small troops of vervet monkeys live along its middle reaches. Downstream from the point where Nariokotome leaves the mountains, the stream has incised its channel through deposits of Pliocene and Pleistocene age that are themselves covered by a coarse gravel lag of late Pleistocene and Holocene age. Small tributaries provide exposures of the older strata on both the north and south banks of the stream. The most prominent outcrops along the lower course of Nariokotome are two cliffs held up by the Nariokotome Tuffs, one north and one south of the river (Figure 2.1). These cliffs are the most likely features to exist far into the future, when other workers may wish to relocate the hominid site. Where the stream flows between these cliffs its course is constricted, but east and west of these its channel is much broader. The Nariokotome hominid site is located on the east bank of a small southern tributary to Nariokotome that enters the latter about 400 m east of the prominent cliff on the south bank of the stream. The west side of this gully has poor exposures of the Lokapetamoi Tuff (see Feibel and Brown, Chapter 3, this volume, for definition).

For purposes of documentation, a photomosaic of the region near the site is given in Figure 2.2 and a local geological map in Figure 2.3. The small extent of exposures of Pliocene and Pleistocene strata shown here are typical for the lower parts of streams in the region. Because the exposures are so limited and separated by such large stretches of cover, tephrostratigraphy is very important for reconstruction of the geology of the region.

Regional stratigraphy

Pliocene and older Pleistocene strata of the Turkana Basin have been termed the Omo Group by de Heinzelin (1983). This group consists of the Shungura, Usno, and Mursi Formations exposed in the lower Omo Valley in southern Ethiopia, the Koobi Fora Formation exposed east of Lake Turkana, and the Nachukui Formation exposed west of Lake Turkana (Figure 2.4). The Usno and Mursi Formations are not further discussed because they substantially antedate the interval of interest here.

Shungura formation. The stratigraphy of the Shungura Formation has been well described by de Heinzelin (1983), and only a few general notes are necessary here. The formation is about 765 m thick and is divided into 12 members on the basis of major tuffs lettered A through L (excepting I) from the base upward (Figure 2.5). Each member comprises all of the strata from the base of the defining tuff up to the base of the next defining tuff and is named for the tuff at its base. A Basal Member is defined as those strata that underlie the lowest major tuff, Tuff A. Each member is subdivided into submembers on the basis of natural stratigraphic breaks, and these are numbered from the base to the top within each member. Less extensive tuffs between the major tuffs are named for the submember within which they lie.

Members A through Submember G-13 are composed principally of fluvial deposits, laid down by the ancestral Omo River. Strata from Submember G-14 through G-26 are of lacustrine origin. From Submember G-27 to the top of the formation (Submember L-6), both fluvial and lacustrine strata are represented. This latter interval is of particular interest in this account, as strata at the Nariokotome site can be demonstrated to correlate with part of Member J of the Shungura Formation.

Koobi Fora Formation. The lithostratigraphy of the Koobi Fora Formation was revised by Brown and Feibel (1986). Regional stratigraphy and paleoenvironments have been described in detail by Feibel (1988). Briefly, the formation is about 550 m thick and consists of eight members, each with a tuff at the base for which the member is named (Figure 2.5). From the base to the top these are the Lonyumun, Moiti, Lokochot, Tulu Bor, Burgi, KBS, Okote, and Chari Members.

The Lonyumun Member is composed primarily of lacustrine strata. The Moiti, Lokochot,

FIGURE 2.1.
The view upstream from the Nariokotome hominid site (NK3), toward the cliffs held up by the Nariokotome Tuffs.

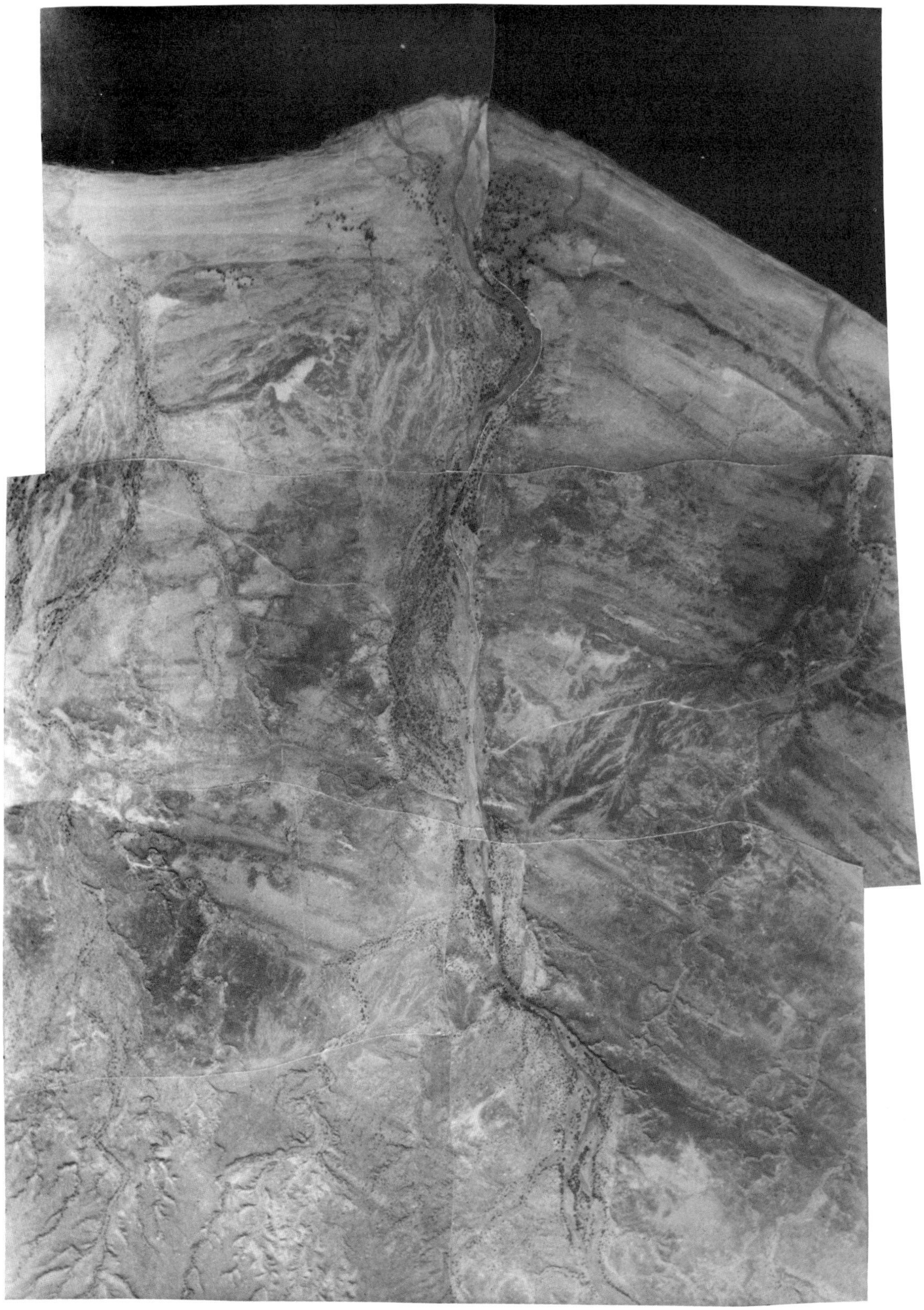

FIGURE 2.2.
Photomosaic of the Nariokotome region. Lake Turkana is the dark region to the right.

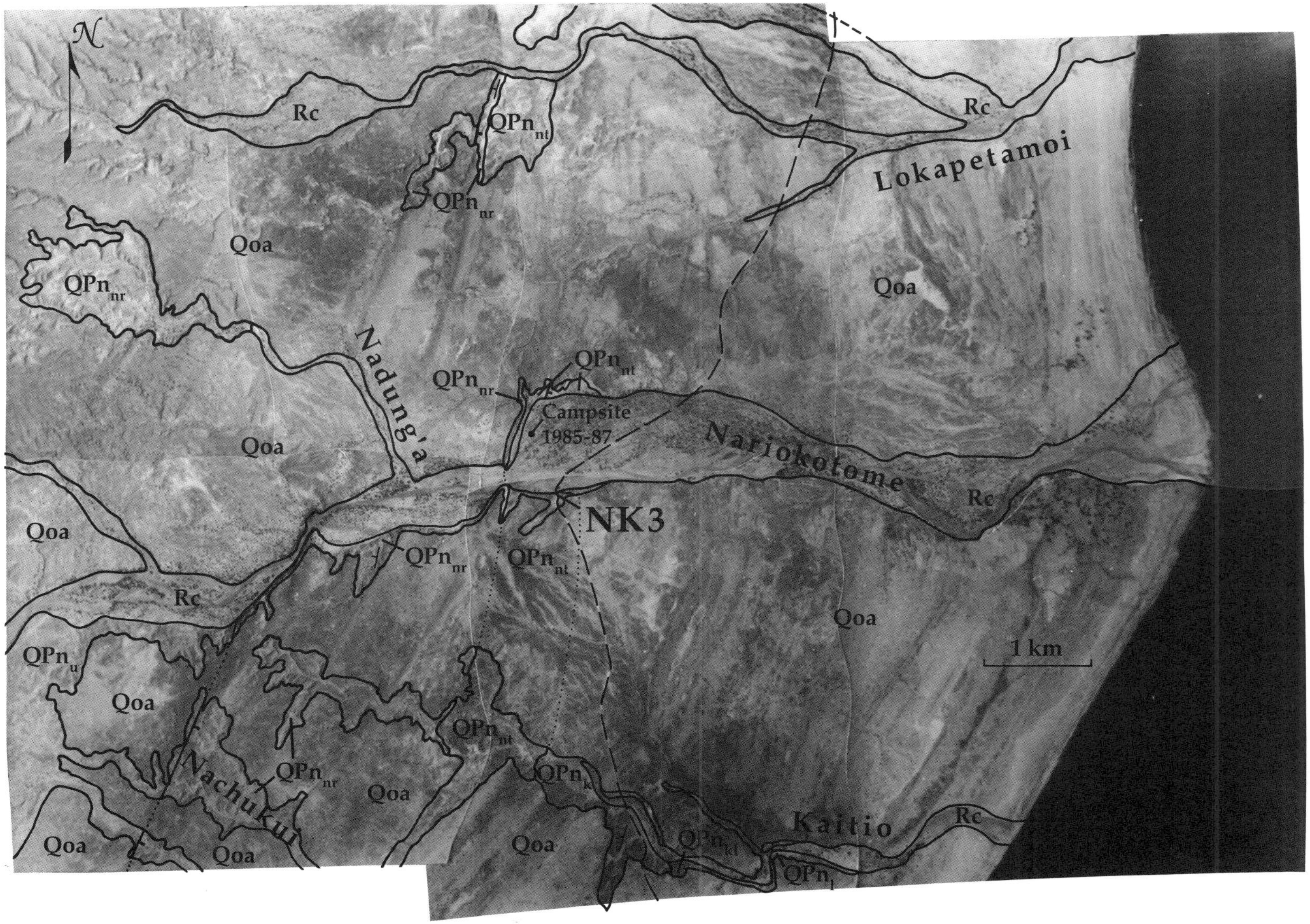

FIGURE 2.3.

Geological map of the region pictured in Figure 2.2. The units of the Nachukui Formation, from oldest to youngest, are the Lokalalei Member (QPn_{l}), the Kalochoro Member (QPn_{kl}), the Kaitio Member (QPn_{k}), the Natoo Member (QPn_{nt}), and the Nariokotome Member (QPn_{nr}). In the southwestern corner of the map, designated QPn_{u}, lie the undifferentiated strata of the Nachukui Formation. Strata equivalent to the Galana Boi Formation, surficial gravels and finer sediments that cover exposures of the Nachukui Formation, are labeled Qoa (Quaternary older alluvium). Alluvium in recent channels is labeled Rc. A few minor exposures, not mappable at this scale, are not shown. The fault lines (dotted where concealed) are marked with ticks on the downthrown side; inferred faults are not shown. Contacts are shown as lines without ticks (dotted where concealed). The heavy dashed line crossing the mosaic from north to south is the road from Kalakol to Lowarengak.

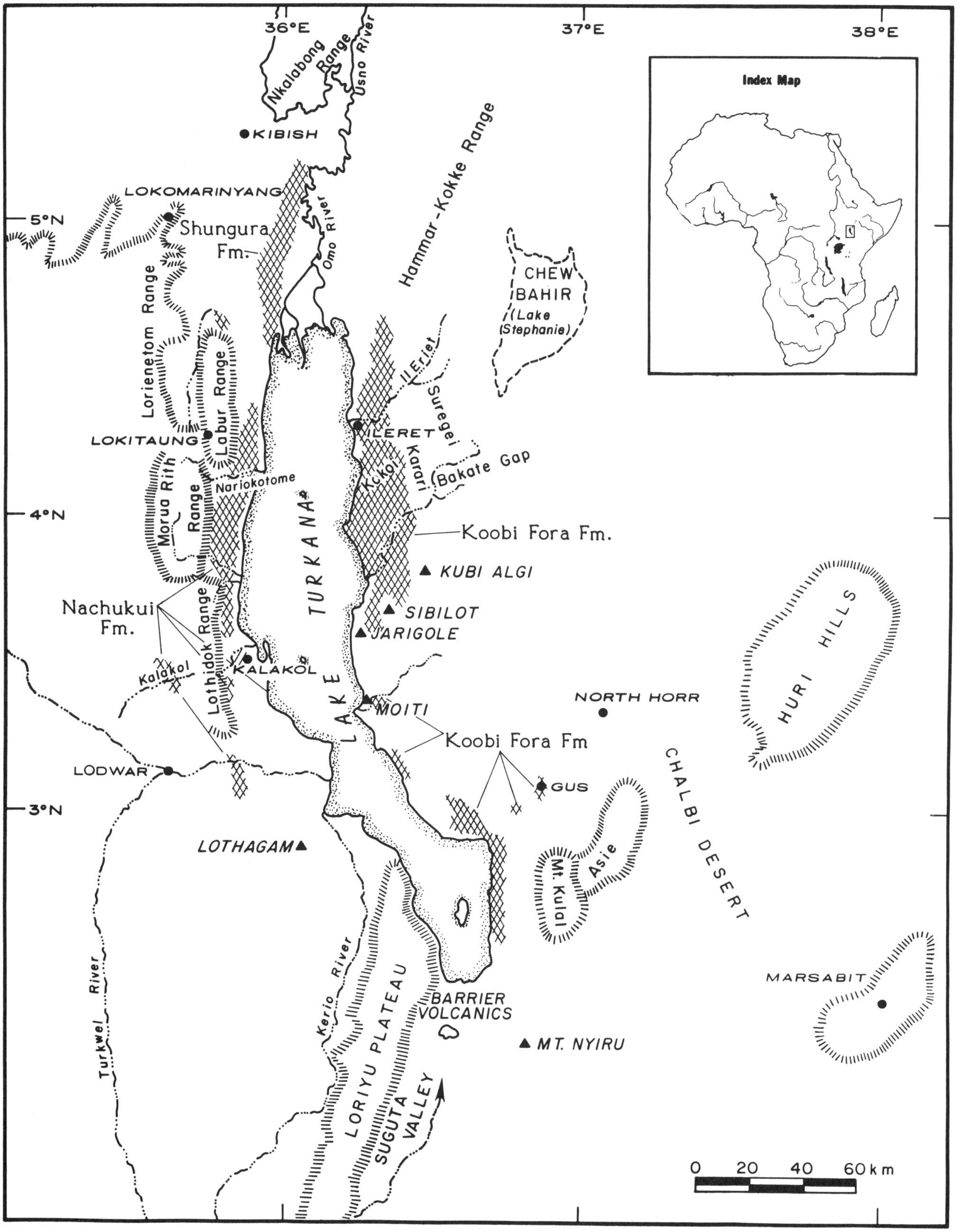
36°E
37°E
38°E
Index Map
Nkalabong Range
Usno River
KIBISH
LOKOMARINYANG
5°N
Shungura Fm.
Omo River
Hammar-Kokke Range
CHEW BAHIR (Lake Stephanie)
Lorienetom Range
Labur Range
Il Eriet
Suregei
ILERET
LOKITAUNG
Moruo Rith Range
Nariokotome
Kokoi
Karari
Bakate Gap
4°N
TURKANA
Koobi Fora Fm.
KUBI ALGI
Nachukui Fm.
Lothidok Range
SIBILOT
JARIGOLE
HURI HILLS
Kalakol
KALAKOL
LAKE
MOITI
NORTH HORR
Koobi Fora Fm
LODWAR
GUS
CHALBI DESERT
3°N
LOTHAGAM
Asie
Mt. Kulal
Turkwel River
Kerio River
LORIYU PLATEAU
BARRIER VOLCANICS
MARSABIT
MT. NYIRU
SUGUTA VALLEY
0 20 40 60 km

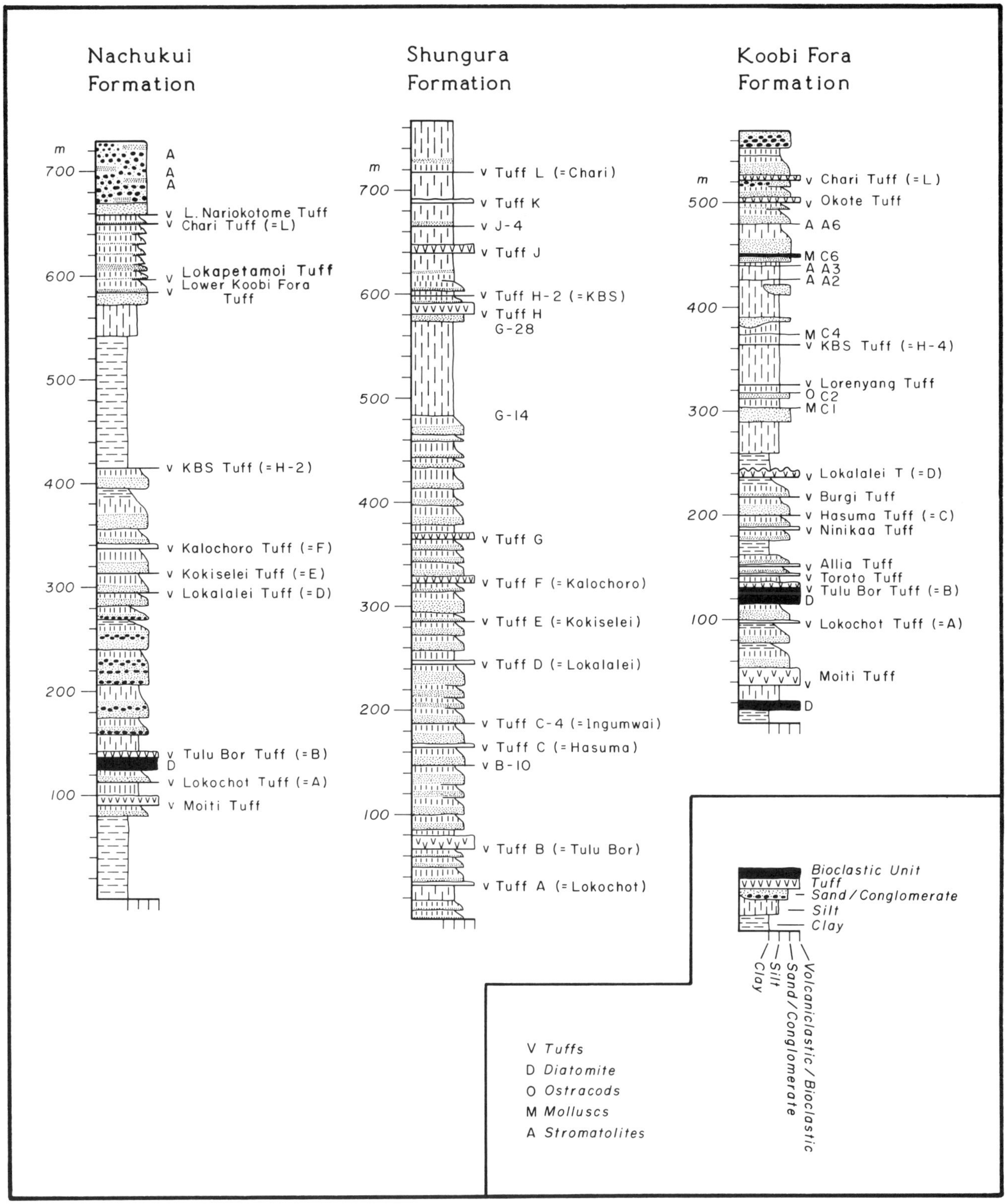

FIGURE 2.4.
Map of the Turkana Basin showing the distribution of Omo Group deposits and place names used in the text. (Adapted from Feibel, 1988.)

FIGURE 2.5.
Composite stratigraphic sections of the Nachukui, Shungura, and Koobi Fora Formations. (From Feibel et al., 1989.)

Tulu Bor, and lower Burgi Members consist dominantly of fluvial strata, but there is also a record of two brief lacustrine intervals, one in the Lokochot Member and one in the Tulu Bor Member. A disconformity separates the lower Burgi Member from the upper Burgi Member; it likely represents an interval of about 0.5 Ma during which no sediments were deposited in the Koobi Fora region. Lacustrine deposition dominates the upper Burgi Member, but the highest part of the member is made of fluvial deposits over most of the area, although lacustrine deposition continues in the southern part of the region. This lacustrine interval corresponds broadly to that in the middle part of the Shungura Formation (G-14 to G-26), although strict temporal equivalence does not obtain. The KBS, Okote, and Chari Members, like the upper part of the Shungura Formation, record alternating fluvial and lacustrine deposition. Of particular importance is the Okote Member, because strata at the Nariokotome site correlate with part of this member.

NACHUKUI FORMATION. Harris et al. (1988) described the Nachukui Formation and divided it into eight members. From the base upward, these are the Lonyumun, Kataboi, Lomekwi, Lokalalei, Kalochoro, Kaitio, Natoo, and Nariokotome Members (Figure 2.5). Again, these members are defined on the basis of tuffaceous marker beds. Lateral facies changes are far more pronounced in the Nachukui Formation than in either the Koobi Fora or Shungura Formations, because the Nachukui Formation was deposited much nearer mountainous terrain on the basin margin. As in the Koobi Fora Formation, the Lonyumun Member is made up primarily of lacustrine deposits. The Kataboi, Lomekwi, Lokalalei, and lower Kalochoro Members are dominantly the product of fluvial deposition, with two brief lacustrine intervals, one in the Kataboi Member and one in the Lomekwi Member, that correspond to those in the Lokochot and Tulu Bor Members of the Koobi Fora Formation. In exposures of these members near the mountains, the sections contain a substantial gravel component where the basinal sediments interfinger with ancient alluvial fans. Unlike the Koobi Fora Formation, the Nachukui Formation does not appear to contain any breaks in deposition. Near the Nariokotome hominid site, the upper part of the Kalochoro Member and all but the uppermost part of the Kaitio Member reflect deposition in a stable lake. Westward, these members are represented by lake margin and alluvial fan deposits. The Natoo and Nariokotome Members again record alternating fluvial and lacustrine deposition in their eastern exposures but document deposition on alluvial fans in the western exposures. The *Homo erectus* skeleton derives from the lowest part of the Natoo Member.

TABLE 2.1.

Tuffs that correlate between the Nachukui, Shungura, and Koobi Fora Formations.

	FORMATIONS		
TUFF	NACHUKUI	SHUNGURA	KOOBI FORA
Chari (L)	X	X	X
Upper Okote (K-α)	X	X	X
Black Pumice (J-7-1)	X	X	X
Lower Koobi Fora (J-6-2)	X	X	X
Morutot (J-4)	X	X	X
Malbe (H-4)	X	X	X
KBS (H-2)	X	X	X
Lokalalei (D)	X	X	X
β-Tulu Bor (B-β)	X	X	X
α-Tulu Bor (B-α)	X	X	X
Lokochot (A)	X	X	X

CORRELATIONS BETWEEN FORMATIONS. Correlations between the Shungura, Koobi Fora, and Nachukui Formations based on the geochemistry of volcanic ashes have been reviewed by Harris et al. (1988) and Feibel et al. (1989). Compositional data on which these correlations are based have been published by Cerling and Brown (1982), Brown and Cerling (1982), Harris and Brown (1985), Brown and Feibel (1985, 1986), and Harris et al. (1988) and will not be repeated here. Several correlations have been made since the last review, and the entire sequence of tuffs in Pliocene and Pleistocene strata of the Turkana Basin is reviewed by Haileab and Brown (1992). There are about 130 compositionally distinctive tuffs within these strata, but only eleven correlate between the three formations (Table 2.1). Each formation, however, contains 30–35 tuffs that correlate to another formation—there are 9 to 14 tuffs that correlate between pairs of formations but not to all three. For example, in addition to the ubiquitous tuffs, ten other tuffs correlate between the Nachukui and Koobi Fora Formation but are not known from the Shungura Formation, and thirteen additional tuffs correlate between the Nachukui and Shungura Formation but are not known from the Koobi Fora Formation. Most of these latter thirteen tuffs lie within the interval represented by the disconformity between the lower Burgi Member and upper Burgi Member during which there was no deposition in the Koobi Fora Formation. The importance of these correlations is that they allow attention to be focused on the interval between the KBS Tuff and the Chari Tuff, in which the Nariokotome site lies. It is possible to restrict this interval further on the basis of geochemical correlations between the formations, and even isolated or geographically restricted ashes can be of considerable importance in local studies (see Feibel and Brown, Chapter 3, this volume).

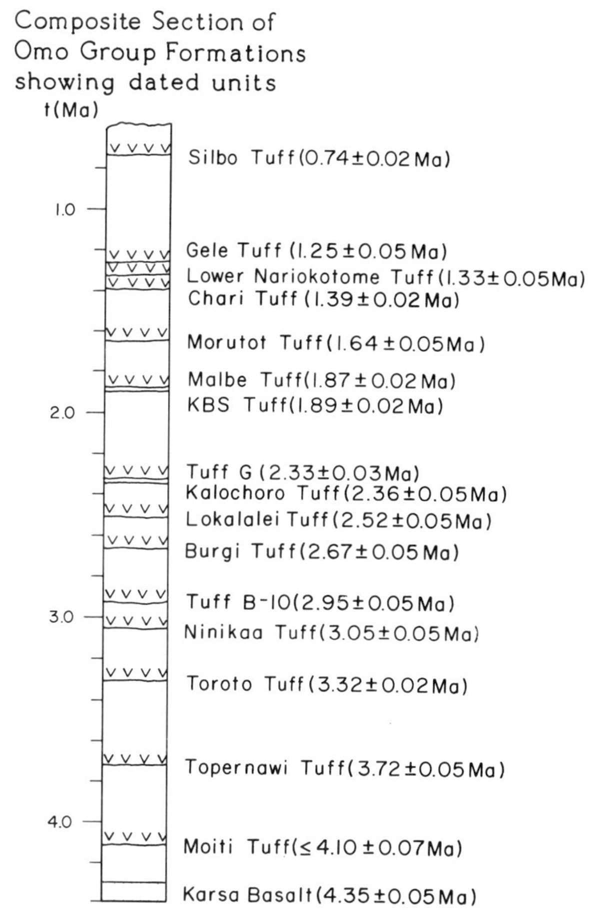

FIGURE 2.6.
Isotopic dates on Pliocene and Pleistocene volcanic materials from the Shungura, Koobi Fora, and Nachukui Formations. Ma = million years before present.

An estimation of the age of the hominid level

The Omo Group deposits of the Turkana Basin are the best-dated hominid-bearing sequence in the world. In spite of early difficulties, the abundant volcanogenic materials available and extensive exposures have made it possible to combine isotopic dates on some 20 levels with a paleomagnetic stratigraphy for the Shungura and Koobi Fora Formations and tephra correlations to establish a tightly knit chronometric framework. The current status of this framework has been reviewed by Feibel et al. (1989) and is repeated here only in graphic form (Figure 2.6). In this section we review the control on the age of strata associated with the Nariokotome hominid site, following a brief summary of the approach used in developing the numerical time framework.

Isotopic dating has been done mainly by the conventional K-Ar method, with some samples measured also by the $^{40}Ar/^{39}Ar$ variant of this method. Measurements have been made exclusively on volcanic materials in the tuffaceous beds—in particular, on alkali feldspar crystals separated from pumice clasts occurring within the tuffs (see McDougall, 1985, for details). Separation of alkali feldspar phenocrysts, usually anorthoclase, from pumice minimizes the problem of contamination by detrital feldspars, the incorporation of which leads to overestimation of ages. Volcanic alkali feldspars are virtually ideal materials for K-Ar dating, and they normally yield highly precise and accurate ages. In general, measurements have been made on at least three, and commonly many more, pumices from the same tuff bed at different localities where possible, as a check on the consistency of the results. The K-Ar clock effectively measures the time since cooling of the alkali feldspars, immediately after eruption. There are good geological reasons for presuming that deposition of a tuffaceous bed normally occurs within a very short interval of the explosive volcanic eruption. Thus the age measured on juvenile feldspar from a tuffaceous bed is likely to closely approximate the depositional age. A more detailed discussion of these questions can be found in Feibel et al. (1989).

In regard to dating of the stratigraphic level from which the *Homo erectus* skeleton was recovered, there are four tuffs that are especially relevant. These are the KBS Tuff, the Morutot Tuff, the Chari Tuff, and the Lower Nariokotome Tuff, in ascending order (Figure 2.6). In addition, the Lokapetamoi Tuff is important in terms of effecting correlations between sections; however, this tuff has not been dated directly.

The KBS Tuff is very securely dated, yielding a mean K-Ar age of 1.88 ± 0.02 Ma. This figure is based upon measurements on feldspars from 13 different pumices in the KBS Tuff collected from diverse localities over a wide area in the Koobi Fora Formation (McDougall et al., 1980; McDougall, 1985). The quoted uncertainty is the standard deviation of the population ($n = 13$), not the standard deviation of the mean. Age spectra by the $^{40}Ar/^{39}Ar$ technique are essentially flat and yield ages indistinguishable from the conventional K-Ar results (McDougall, 1981).

Pumices of the same composition as the Morutot Tuff, but included in the Okote Tuff in Area 131 of the Koobi Fora Formation, were dated at 1.64 ± 0.03 Ma by McDougall et al. (1985) on the basis of measurements on feldspars from 8 pumices. We believe that this age provides a good estimate for the eruptive age of the Morutot Tuff.

The Chari Tuff has been well dated at 1.39 ± 0.02 Ma on the basis of concordant results of 11 of 12 feldspars from separate pumices measured from two well-separated localities in the Koobi Fora Formation (McDougall, 1985). Age spectra on four of these feldspars give ages indistinguishable from the conventional K-Ar ages.

Feldspars separated from two pumices from the Lower Nariokotome Tuff in the Nachukui Formation, near Nariokotome, yielded concordant ages of 1.33 ± 0.02 Ma (Brown et al., 1985).

Overall, these data provide a well-constrained and precise numerical time framework for the stratigraphic interval of interest. As the *Homo erectus* skeleton from site NK3 is stratigraphically above the KBS Tuff and below the Chari Tuff, we are confident that its age lies between 1.88 ± 0.02 and 1.39 ± 0.02 Ma. If we use only these dates and the stratigraphic distances between the KBS Tuff, the level of NK3, and the Chari Tuff at Kaitio, an age of 1.51 Ma for the hominid level is derived by interpolation assuming constant sedimentation rate.

When the ages on the Morutot Tuff were obtained, the tuff itself had not been identified in the section on the Karari Ridge at Koobi Fora from which the pumices came. Resampling of that section demonstrated that the Morutot Tuff is indeed present as a discrete layer that has been removed locally by a channel filled with the Okote Tuff, so we can now apply that date with greater confidence.

A difficulty remains in that nowhere in the Nachukui Formation have we located a section in which the Morutot Tuff and the Lower Nariokotome (or Chari) Tuff are exposed where the hominid level can be unambiguously identified. Nevertheless, Feibel and Brown (Chapter 3, this volume) have shown that to the north of the hominid site, the Morutot Tuff (= J-4) lies 10 to 20 m stratigraphically below the hominid level. Thus the age of the hominid site clearly is less than 1.64 ± 0.03 Ma, the measured age of the Morutot Tuff. A further approach is to make use of yet another tephra layer, the Lokapetamoi Tuff, which overlies the hominid level by 8 m at Nariokotome and extends far to the north. This tuff is defined by Feibel and Brown (Chapter 3, this volume). The Lokapetamoi Tuff lies 37 m above the Morutot Tuff and 28 m below the Lower Nariokotome Tuff; this location yields an approximate age of 1.47 Ma by interpolation, again assuming constant sedimentation rate. This age must postdate the time of burial of the hominid, and extrapolation to the hominid level yields an age of 1.51 Ma.

We can also estimate the age of the hominid at Nariokotome by extrapolation downward using a sedimentation rate computed for the interval (12 m) between the Chari (1.39 Ma) and Lower Nariokotome (1.33 Ma) Tuffs in the section at Nariokotome. This method yields an estimate of 1.56 Ma for the hominid level.

The conclusion to be drawn is that whatever pair of dated tuffs we use for estimating the age of the hominid, and whether we interpolate or extrapolate, the estimates fall within a narrow range from 1.51 to 1.56 Ma. These estimates, based on much greater knowledge of the stratigraphy of the surrounding area, are slightly younger than the range from 1.55 to 1.65 Ma suggested earlier (Brown and Feibel, 1985). Dates on additional units may refine this estimate further but are unlikely to change the age significantly. Should the reader prefer a single number for an age estimate, 1.53 ± 0.05 Ma should cover the range nicely.

Regional paleogeography

From previous work in the basin, the general elements of the paleogeography of the northern part of the basin can be stated with some confidence. Even today the topography is quite subdued within the basin, but it is likely that in the past the topographic gradients from the margins to the center were smaller than today. It is also clear that several highland elements were not in existence until about 2 Ma ago.

Faulting of the Shungura Formation postdates deposition of its youngest strata—that is, the upper part of Member L—and thus took place less than about 1.1 Ma ago. This faulting is quite important for the paleogeography, because the ridge separating Sanderson's Gulf from the floodplain of the Omo River in its lower reaches was formed as a result. Prior to the faulting, one must envision a very broad plain stretching from the Hammar-Kokke Range on the east to the Lorienetom Range on the west. Morphologically, the Korath Range is also very young, and it probably did not exist prior to 1 Ma ago, possibly having begun construction at the same time that the Shungura Formation was faulted. Without the low divide between the Omo River floodplain and Sanderson's Gulf, the Omo River would have been free to wander across this vast plain, and it is also likely that the Kibish River and other small rivers were tributary to it.

In the Koobi Fora region, most of the faulting also postdates deposition of the highest levels of the formation. In particular, the Silbo Tuff (0.74 Ma) is involved with uplift of the Kokoi, the only highland element within the northern part of the region. The Suregei highland was also uplifted (at least in part) rather late in time, because the KBS Tuff is involved in faulting of that block. Drainages from the eastern margin were surely affected by these movements, and it is probable that the present course of Il Eriet postdates uplift of the Kokoi. It is likely that in earlier times it flowed in a southwesterly direction, but it is clear that small streams flowed into the basin from the northeast and east from at least 2 Ma ago to the present. The evidence for these streams is the presence of basaltic pebble conglomerates confined to the northeastern part of the region. The only other highland within the basin, Jarigole, was uplifted sometime after deposition of the Tulu Bor Tuff about 3.3 Ma ago, but it is not known when.

Along the west side of Lake Turkana, a mountainous highland existed, at least from Nariokotome to Topernawi, since earliest times. Volcanic pebble conglomerates derived from the west are prominent parts of the section from 4 Ma to 1.3 Ma. The Labur Range, though, appears to have been partly uplifted in more recent times, sometime between 3 and 1.7 Ma (see Harris et al., 1988). From the steepness of the eastern scarp, it is evident that this uplift still continues.

The elevation difference between the present lake level and the highest preserved Pliocene and early Pleistocene strata is 160 m on the divide between Nariokotome and Kalochoro. Similarly, the topographically highest strata east of the lake on the Karari ridge lie 165 m above the present lake level. In the northern basin of Lake Turkana, the deepest part is about 30 m, so the elevation between the basin margin and the basin center is about 190 m. The structural setting of the strata in either case, however, is quite different. Strata of the Nachukui Formation are in fault contact with older volcanic rocks of the Murua Rith Range, whereas strata of the Koobi Fora Formation rest disconformably on older rocks. It is unlikely, therefore, that the close similarity of elevation of the highest strata results from uplift of the eastern and western highlands. Instead, the present topographic difference most likely results from downfaulting or downwarping of the central part of the basin. If the gradient on the relatively flat surface cut across the older deposits is extended to the basin center from either margin, one finds that the elevation difference from the basin margin to the center was likely on the order of 140 m, some 50 m less than is the case today. This has considerable importance for estimating the width of ancient floodplains of the Omo River when it occupied the region—it is likely that they would have been wider in the past. In addition, the reduced gradient would have made it more difficult for small marginal streams to move coarse sediment very far into the basin.

It is well to keep in mind the order of magnitude of features associated with the Omo River. Today, one rises perceptibly as one approaches the river, and the river does not come into view until one is on its bank. This is because the levees are several meters high. The channel of the river is on the order of 15 m deep, and the width is about 150 m. The channel depth is then about one tenth the vertical elevation difference between the basin margins and the basin center. This alone ensures that when the river was present in the basin, it would dominate the paleogeography to a large extent.

Considerable evidence has been amassed to show that for most of the Pliocene epoch the Turkana Basin was an open basin with a river flowing through it that emptied into the Indian Ocean. Toward the end of the Pliocene, though, conditions began to change with the construction of Mt. Kulal at the southern end of the lake. Lavas of Mt. Kulal overlie older strata, and the KBS Tuff is found intercalated in alluvial fans shed from that mountain toward the northwest. The old exit had been plugged by the growth of a volcano, and a large lake formed within the basin as a result about 2 Ma ago. The abundance of fossil molluscs in this interval at Koobi Fora is proof enough that the lake remained essentially fresh and therefore likely maintained an outlet to the east. Feibel (1988) has carefully documented the progradation of the Omo River southward in the Koobi Fora region at the time of deposition of the KBS Tuff (1.88 Ma). In the Nachukui Formation, however, a rather deep lake persisted at the time of deposition of that tuff, and for a considerable length of time thereafter. In the type area of the Shungura Formation, fluvial deposits lie some 20 m below the level of the KBS Tuff, but in the Kalam region, situated farther southwest, there is little evidence of axial parts of the fluvial depositional system; rather, the deposits consist of siltstones and claystones, with thin intercalated shell beds, and likely represent a northward extension of the lake so well recorded in the Nachukui Formation at this time. This lake, the Lorenyang Lake, resulted in the near complete infilling of the basin by sediments, and the infilling itself influenced the subsequent history of the region. That history is dealt with in much greater detail by Feibel and Brown (Chapter 3, this volume).

Acknowledgments

We thank R. E. Leakey and the National Museums of Kenya for their continued support. This research was funded by the National Science Foundation (grants BNS 84-06737, 86-05687, 88-05371, and 90-07662). We also extend our gratitude to the Australian National University, which provided laboratory and travel support to Ian McDougall and twice hosted F. H. Brown during research trips to Canberra. Technical staff and students at both the University of Utah and the Australian National University have been very supportive over the years, for which we are grateful. Particular thanks go to Sandra Bromley, who drafted the figures without complaint despite the unfamiliar and lengthy Turkana names.

CHAPTER 3

MICROSTRATIGRAPHY AND PALEOENVIRONMENTS

CRAIG S. FEIBEL AND FRANCIS H. BROWN

The geologic setting of the Nariokotome hominid site provides important clues to the environment in which KNM-WT 15000 lived. An evaluation of the microstratigraphy of the site itself, supplemented with data from related exposures throughout the Turkana Basin, allows a reconstruction of the ancient geography of the burial site. The site, now known as NK3, is located on the floodplain of the ancestral Omo River; it was a temporary marsh formed following a period of heavy rains in the Ethiopian Highlands. The channels of the Omo were bordered by narrow belts of trees and would have been an important focus of the local habitats.

The hominid skeleton KNM-WT 15000 derives from an interval of some 15 cm near the base of the Natoo Member of the Nachukui Formation (Harris et al., 1988). At the excavation site itself (NK3), approximately 4 m of section is exposed, including the hominid level. Associated strata exposed nearby along the Nariokotome sand river include a total sequence of some 9 m. Correlative strata exposed in the drainages to the north and south reveal a composite sequence that can be extended to over 25 m in the Nariokotome region.

These sections can be correlated using the abundant tephrostratigraphic markers preserved in this interval. Using further tephrostratigraphic correlations, we can relate strata of the other Omo Group deposits, the Shungura and Koobi Fora Formations. These formations are coeval with this interval of the Nachukui Formation and are important to the reconstruction of the geographic and environmental context in which the hominid lived, died, and was buried.

The approach we take in this study is to detail first the microstratigraphy of the hominid site itself, then to expand the stratigraphic, sedimentologic, and paleontologic scope to the local, regional, and basinal scales, as a framework for discussion of the hominid site. This discussion will expand upon the paleoenvironmental context previously developed for the Omo Group deposits of the Turkana Basin (Feibel et al., 1991).

FIGURE 3.1.
Plan of the NK3 excavation showing areas worked in the 1984–1989 excavation seasons. Locations of the profiles are indicated by the dashed lines within the excavation area.

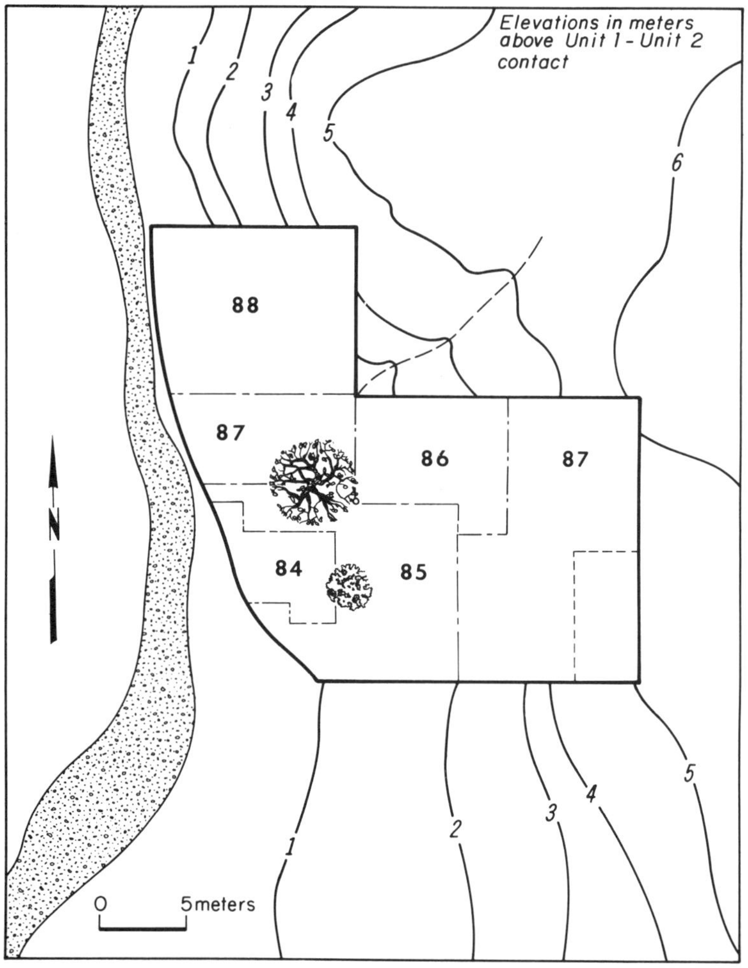

MICROSTRATIGRAPHY OF NK3

The excavations conducted to recover the hominid KNM-WT 15000 offered an opportunity to examine in detail the stratigraphic geometry of some 4 m of section in the lower Natoo Member of the Nachukui Formation. A plan of the NK3 excavation is presented in Figure 3.1, along with the locations of representative profiles of the excavation walls. The excavation provided a detailed view of the sediments occurring from 20 cm below the hominid level to some 3 m above it. Exposures in the excavation provided an extremely good view of the sedimentary sequence (Figure 3.2) and allowed detailed mapping of the local microstratigraphy (Figure 3.3). A geological trench was excavated in 1986 to the north of the site so that additional details of the subjacent strata could be examined.

Although there is a high degree of lateral variation, the sequence exposed at the hominid site can be discussed in terms of ten stratigraphic units exposed in the excavation and an additional unit underlying the floor of the excavation. A representative stratigraphic column showing the described units (based on the north wall of the 1986 excavation) is presented in Figure 3.4, and the units of the sequence are described below.

Unit 0. Immediately beneath the excavation is a massive silty sandstone with footprints and desiccation cracks. The sand is a pale yellowish brown (10YR 6/2) with small limonite mottles. (See Munsell, 1975, for an explanation of the notation used to describe colors.) This unit exceeds 30 cm in thickness in the geological trench to the north of the site. The upper contact is sharp, and the cracks create an irregular upper surface, with about 20 cm relief. Desiccation cracks are infilled with tuff from the overlying unit.

Unit 1. At the base of the excavation is the footprint tuff. This tuff is 5–10 cm thick in most places, but it thickens locally where it fills footprints in the underlying sand. It also occurs as crack fillings extending down into the underlying unit. The tuff is pale yellowish brown (10YR 6/2) toward the base and becomes somewhat darker toward the top. It is highly impure in most places, commonly occurring as a tuffaceous mudstone rather than a true tuff. Pumiceous and stretched shards dominate the ash component of this unit. There is abundant detrital biotite, a characteristic contaminant in the Okote Complex tuffs. Diatoms and sponge spicules occur in the fine-grained fraction of this unit.

Unit 2. This unit is composed of brown mudstone (10YR 6/2) which locally contains hominid fossils. Its thickness is highly variable, as the base of the mudstone infills footprints. The unit averages 80 cm thick in most places and reaches a maximum of 1.6 m in the excavated area. Hominid

FIGURE 3.2.
Sedimentary sequence exposed in the NK3 excavation, September 1988. Scale may be judged by the height of the panga (machete), 0.5 m.

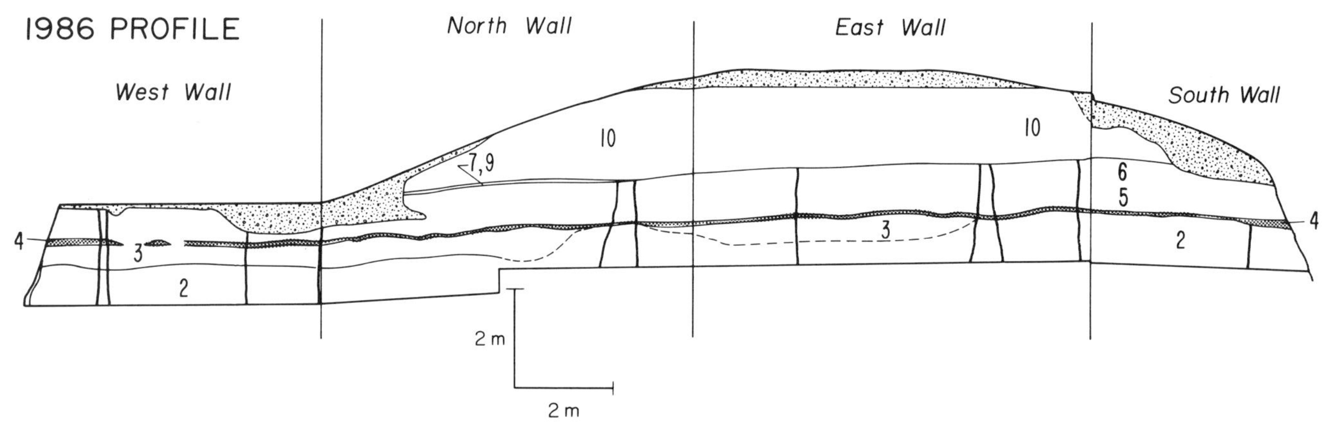

FIGURE 3.3.
Profiles of the excavation walls at the end of the 1986 season. Numbers label the ten stratigraphic units.

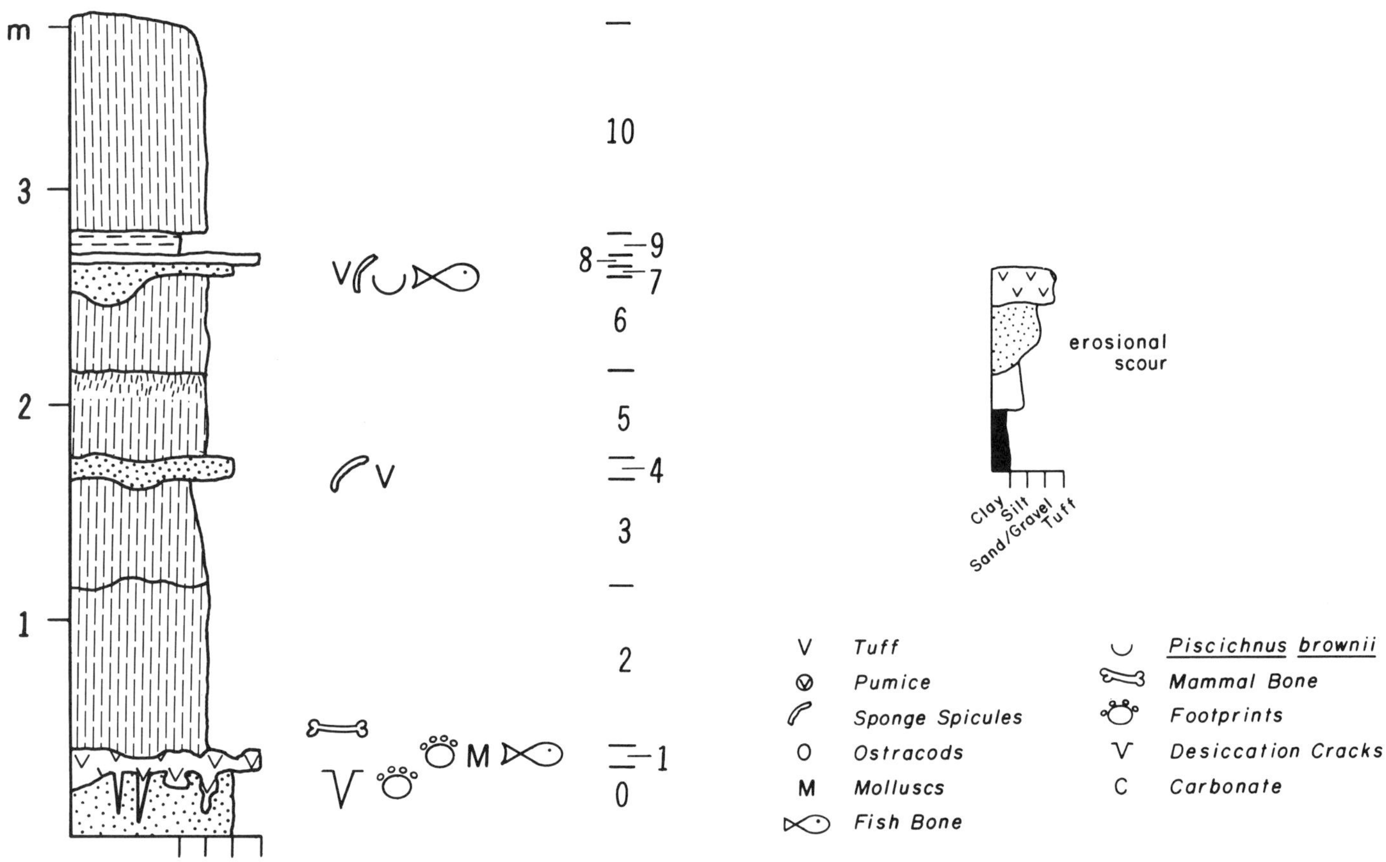

FIGURE 3.4.
Stratigraphic section showing the sequence of units described from the NK3 excavation.

fossils were recovered from the base of this unit. They occur immediately above the contact with the underlying unit, and in some cases (i.e., the teeth) they accumulated within the footprints themselves. Fish bones and gastropod opercula were also recovered from this level. The lower part of Unit 2 is locally tuffaceous, where material from Unit 1 has been reworked upward into the base. Small concretions, 0.5 to 1 cm in diameter and composed of calcite and iron oxides, occur just below the level of the hominid. MnO_2 mottling occurs locally and is particularly associated with the abundant microtubules that penetrate much of the sediment. These tubules, 0.1–0.2 mm in diameter, are a pervasive feature of the finer-grained lithologies in the excavation. Modern root hairs in some of these tubules suggests that they are likely related to recent soil activity, rather than to Early Pleistocene processes. The upper contact of Unit 2 is locally very irregular, and features of the north wall of the 1986 excavation (Figure 3.3) suggest that it is erosional in nature.

Unit 3. Unit 3 is a light gray (10YR 7/2) silt, fining up to clay, and has an undulating upper contact. The unit is massive, and it is laced with the microtubules described from Unit 2.

Unit 4. This unit ranges from a pale brown (10YR 6/2) sandy silt to a very fine sand, locally marked with green mottles. It averages 5–10 cm in thickness but locally thickens to 20 cm. The base of this unit has a relief of nearly 40 cm over the extent of the excavation, indicating minor scouring of the underlying stratum. In the southeast corner of the excavation, where the unit is finer-grained and tuffaceous, it attains a thickness of 10 cm. The tuffaceous facies of this unit contains abundant sponge spicules. Some of the large sediment-filled cracks which cross the excavation originate from the base of this unit.

Unit 5. Unit 5 is composed of a light gray (10YR 7/2) silt, 30 cm thick. The transition from Unit 5 to Unit 6 is marked by an accumulation of clay, recognizable by a slight darkening of color, and a weakly to moderately developed prismatic structure. This clay accumulation is probably the subsurface Bt horizon of a fossil soil. If this is indeed the case, then the Unit 5–Unit 6 transition is a postdepositional feature, related to the depositional break recorded in the top of Unit 6.

Unit 6. This unit, a very pale brown (10YR 7/3) silt, 40 cm thick, has an erosionally scoured upper contact. It is massive and essentially featureless throughout the extent of the excavation.

Unit 7. Known informally as the fish sand, this unit typically ranges in thickness from 0 to 5 cm but occasionally reaches 20 cm, where it rests on an erosionally scoured surface with up to 15 cm relief. The sand, like most units in this section, varies laterally in composition. It is typically very pale brown (10YR 7/3), poorly sorted, and composed largely of carbonate-coated grains, small carbonate concretions, and occasional clay pebbles. Fish fossils are abundant throughout the unit. They consist primarily of small black or brown bone fragments, scales, and vertebrae. Microfossil separates from this sand contain abundant fish teeth, including those of the freshwater stingray *Dasyatis africana* (Feibel, in review), as well as cichlids and pharyngeals of juvenile cyprinids. Small ostracods are very abundant in the sand, and the rare larger forms are also carbonate-coated. Most of the major crack networks obvious in the excavation floor and walls extend downward from the basal contact of Unit 7.

Unit 8. A 0–5-cm-thick tuffaceous sand, this unit contains fossil sponge spicules, ostracods, and fish. It merges laterally with Unit 7 but stands out distinctly where the tuffaceous component and minor clay give it a slightly lighter grayish (10YR 7/1) chroma. Carbonate coatings are not common in this unit. The unit contains abundant fossil fish, and microfossil separates again revealed abundant cichlid teeth and juvenile cyprinid pharyngeals. The ostracods in this unit are abundant and are dominated by occluded valves of *Cyprideis torosa.*

Unit 9. Composed of a very pale olive (10Y 7/2) claystone 10 cm thick, this unit is massive, and recent fracture surfaces are covered with MnO_2 coatings.

Unit 10. Unit 10, a light olive gray (5Y 6/2) silt, forms the local top of the exposures and attains a thickness of at least 1 m in the walls of the excavation. In other exposures it can be seen to reach 5.5 m in total thickness. It is locally overprinted by the recent soil and is channeled into by the gravels capping the exposures.

The low, rounded exposures that make up the site and the surrounding terrain are mantled with a basalt cobble lag of probable late Pleistocene age. This lag is relatively impervious to the highly erosive torrential rains which characterize the local rainy seasons, and thus they preserve the soft Plio-Pleistocene sequence. This gravel cap tends to limit exposure of the soft sediments to erosional gullies and cut-banks of the ephemeral streams draining to the lake.

Sedimentology and Micropaleontology

The sedimentary sequence exposed in the excavation at NK3 is dominated by massive, fine-grained deposits, with three thin intercalations of coarser material. The finer-grained sediments lack the characteristic lamination and horizontal structure typical of lacustrine deposition, and they also lack indications of aquatic fauna. These features, coupled with the occasional prismatic structure suggestive of incipient soil development, imply that the mudstones are of distal floodplain origin. The coarser lithologies show abundant evidence of exposure (footprints, desiccation cracks) and erosional features, but they also preserve a dominantly aquatic fossil assemblage—implying that these intervals were formed by short-term high-energy flooding events followed by periods of immersion under standing water.

The contribution of sand-sized material to the sedimentary accumulation at NK3 is minor. The sands which do occur are dominated by volcanic ash, carbonate grains, and clay granules. The remaining mineral grains are nearly all the size of fine to very fine sand and include two distinct mineral suites. These suites relate to two drainage systems operating in the basin at the time of deposition: an axial drainage system of the ancestral Omo River with headwaters to the north and a marginal system of ephemeral streams coming from the volcanic hills to the west. The sands are arkoses and subarkoses with a variable component of volcanic ash. The feldspars are dominated by fresh sanidines, which likely derive from the pumices carried by the axial (Omo) drainage system. The quartz includes some bipyramidal crystals of like origin. The abundant accessory biotite, being derived from metamorphic source areas to the north, is also an indicator of the axial drainage component. The volcanic ash component could have been derived from the Omo River's axial drainage system, or it could represent locally reworked airfall ash. In the absence of pumiceous clasts, the latter interpretation is possible, but the bulk of the tephra in the Turkana Basin were derived from transport by the Omo system. Its association with detrital biotite suggests that the ash was also a product of the axial transport system. A second source area for the sand components is indicated by the presence of augite and olivine, indicating derivation from the marginal volcanics to the west. This evidence of a dual sediment supply, including the axial and marginal drainages, will be an important factor in our discussion of the paleogeographic reconstruction and depositional systems acting at the time of accumulation.

The sequence at NK3 is not particularly rich in macrofossils. Exceptions include one relatively complete hominid specimen, several varanids, fish, and opercula of the gastropod *Pila* in Unit 2. There are moderately abundant fish remains in the sands of Units 7 and 8, both of which also contain ostracods. Sponge spicules occur in Units 1, 4, and 8.

Brown et al. (1985) first noted the presence of both amphioxea and amphistrongyla of sponges in Unit 4. Subsequent work has shown that these spicules are abundant in Units 1 and 8 as well. All of the spicules identified thus far are megascleres, and they include straight, slightly arcuate, and spinose forms. A study by Williams (1989) tentatively identified two taxa of sponges, the potamolepid *Potamophloios* and an unidentified spongillid. The rather abundant sponge remains encountered in the site are of interest for two reasons. The recognition of potamolepid sponges in East Africa has important implications for reconstructions of Neogene hydrography in Africa, as these sponges are today known only from the Zaire River system and equatorial Africa. Second, the substrate preferences of sponges are significant. Sponges generally do not colonize muddy substrates but rather require a solid, stable object upon which to grow. The sediments associated with the abundant spicules at NK3 would have provided no such support, and thus the presence of aquatic vegetation is strongly indicated. The base provided by branches, reeds, or even larger grasses is sufficient to support sponge growth, and the abundance of spicules suggests that such substrates were readily available.

Fossil ostracods are abundant (greater than 2,000 valves per gram of sediment) in Units 7 and 8. The imprecision of our current understanding of the ecological tolerances of modern African ostracods limits the usefulness of individual species in reconstructing conditions at the site, but collectively their presence is indicative of some general features. Five taxa are represented, according to the identifications made by Feibel (1988). The relatively low diversity of the assemblage is suggestive of waters with concentrated or fluctuating salinities. The overall abundance of ostracods, however, suggests that the waters were well oxygenated. The eurytopic *Cyprideis torosa,* about 80 percent of the fauna, dominates the assemblages. This species has a wide range of salinity tolerance (0.5–150 parts per thousand of total disolved salts, or ppt TDS; Neale, 1988) but it is not commonly a dominant in Plio-Pleistocene ostracod faunas of the Turkana Basin (Feibel, 1988). Its abundance here may indicate that this wide range of chemical tolerance was significant and that water chemistry was highly variable. This species is represented almost entirely by forms lacking any carapace noding, an indication of salinities in the high oligohaline range (ca. 5 ppt TDS). The dominance of *Cyprideis torosa* may indicate bicarbonate-poor waters as well (Carbonel and Peypouquet, 1983).

Four additional taxa are recorded in these sands: *Ilyocypris gibba* (7%); *Plesiocypridopsis newtoni* (6%); a strongly noded and thin-valved form of *Sclerocypris clavularis* (4%), and *Darwinula stevensoni* (3%). *Ilyocypris gibba* ranges from fresh to oligohaline waters (0–3 ppt TDS). Carbonel and Peypouquet (1983) have suggested that the development of noded forms of *Sclerocypris clavularis* is an indication of oligohaline waters (0.5–5 ppt TDS) enriched in organic matter. *Darwinula stevensoni* is an indicator of low-turbidity waters with a salinity range from fresh to oligohaline conditions. The absence of any individuals attributable to *Hemicypris* is noteworthy, as species of this genus are the most common and abundant members of most ostracod faunas throughout the Omo Group (Feibel, 1988); their absence suggests that the concentration of waters likely exceeded that of the modern lake (2.5 ppt TDS; Yuretich and Cerling, 1983). Also lacking from the ostracod fauna is the genus *Limnocythere,* which has been used by Carbonel and Peypouquet (1983) as an indicator of increasing alkalinity.

Overall, the ostracod faunas suggest that a moderate concentration of waters occurred, resulting in salinities in the oligohaline range (0.5–5 ppt) but not increased alkalinity. In relation to the composition of modern Omo River waters (Ferguson and Harbott, 1982), this represents a concentration of 10–100×. Such moderate concentration factors are much more likely in the seasonal floodplain ponds associated with a major river, whereas the last remnants of a major lake would be expected to have much higher concentration effects, particularly with regard to alkalinity.

The presence of opercula of the large gastropod *Pila* in Unit 2 has implications for the chemistry of waters on the site. Since these gastropods secrete shells of aragonite and opercula of calcite, the preservation of one and not the other suggests that water chemistry was such that it dissolved the more soluble aragonite and left the calcite relatively intact. This is in accord with indications from the ostracod faunas that these waters were not markedly alkaline. Selective preservation of opercula without associated shell material has been noted for the riverine marshes of the KBS and Okote Members at Koobi Fora (Feibel, 1988).

Small fish fossils are abundant in Units 7 and 8. These black to brownish-black fossils are most commonly water-worn fragments of fish bones and scales. The microfossil separates from these units, however, yielded large numbers of extremely small fish teeth. This assemblage presents a strong contrast to the more common fish associations reported from the Plio-Pleistocene of the basin, which are dominated by collections of larger fish bones and teeth (Feibel, 1988; Schwartz, 1983). The small size of the material available for fossilization suggests that the fishes present were predominantly juveniles, as one might expect for a very shallow, heavily vegetated habitat. Although the material shows signs of breakage and abrasion coincident with reworking, the presence of very small and delicate elements suggests that this was of relatively short duration. The identifiable taxa recorded from these units, *Dasyatis africana,* Cichlidae, and Cyprinidae, are all suggestive of a shallow littoral habitat. The absence of any large, open-water fishes in the assemblage is noteworthy.

An Interpretation of the Hominid Site

The sedimentary sequence observed in the NK3 excavation is a predominantly fine-grained, conformable interval broken by two erosional contacts, each overlain by thin, coarse-grained strata. The dominance of fine-grained material places the depositional site in a low-energy regime, and the lack of characteristic lacustrine features suggests that the area was more likely a floodplain. Evidence of exposure with little soil development supports other indicators, such as the burial of a relatively complete hominid skeleton, that sedimentation was discontinuous but relatively rapid. Evidence of immersion without indicators of significant water depth implies that a small, shallow body of water of short duration was the primary agent involved. This area was most likely a floodplain marsh, seasonally replenished by floodwaters of the Omo River, which carried most of the sediment to the site. Local runoff carried a minor component of sediment from marginal sources.

One of the more obvious features of the excavation itself was the extensive giant crack network it exposed (Figure 3.5). These cracks could be seen cutting the strata exposed in the walls of the excavation and criss-crossing the excavation floor. As shown in Figure 3.3, the cracks do not cut the entire stratigraphic sequence, however, but rather extend downward from the contacts between Units 3 and 4 and Units 6 and 7. The cracks resulted from two significant events, subsequent to the deposition of Units 3 and 6. The most likely cause of such a major network is a significant lowering of the local water table. This could have resulted from a local hydrographic change, such as lowering of lake level or the switching of Omo River channels away from the site, or it could have been caused by a more regional event, such as a prolonged dry period. During this lowering of the water table, drying of the clay-rich sediment resulted in crack formation, and the cracks were subsequently filled by material from the ground surface, probably concurrently with deposition of Units 4 and 7. Similar crack networks can be seen forming today along Il Naibar at Koobi Fora, where the lowering of the level of the modern lake has resulted in extensive desiccation. Associated with this drying seen today has been a major kill-off of the large *Acacia* trees along the sand river.

Correlative Strata around NK3

In order to discuss the paleoenvironment of the Nariokotome hominid site, we must place the strata associated with the site into a regional and basinal context and view the environmental indicators recorded from the site in association with similar measures from elsewhere in the basin. Our ability to relate strata from scattered exposures around the basin rests on the use of stratigraphic markers. In the local sequence we can occasionally rely on distinctive lithologic markers for correlation, such as the footprint tuff or the fish sand, but these are often somewhat ambiguous and laterally restricted. The best tool for placing the site in a regional and basinal context is the geochemical correlation of the volcanic ashes preserved throughout the sequence. At the level of resolution required for the discussion of environments rele-

FIGURE 3.5.
The NK3 excavation exposed an extensive network of cracks. The crack shown here is filled with sediment. Panga is 0.5 m.

vant to a single fossil, complexities associated with sedimentation and correlation complicate matters. We will discuss first the problems associated with high-precision correlation within the Okote Tuff Complex and associated tuffs, and then consider the related strata and their contribution to the paleoenvironmental context.

The Okote Tuff Complex

Expanding the context of the hominid site outward from NK3 requires that we be able to identify or at least restrict the interval of interest and relate strata from different parts of the basin. As a first approximation, it is relatively easy to restrict the interval of focus to that between the KBS Tuff and the Chari Tuff. Both of these are relatively widespread and continuous markers. Between these two tephra, however, there are at least 42 additional ashes, most but not all of which we can place in a relative stratigraphic order. The high level of volcanic activity which produced these numerous tephrostratigraphic markers so important to dividing this interval also resulted in the production of a series of ashes very similar in composition, some of which are difficult to recognize or discriminate.

The footprint tuff at the base of the NK3 excavation is a correlative of the Lower Koobi Fora Tuff (LKF), an important tephrostratigraphic marker in the Koobi Fora region. In the local exposures west of Lake Turkana, however, the LKF is not particularly common (it is currently known from only two sections). It is also relatively thin compared with exposures elsewhere in the basin. The LKF, whose depositional history is one of the most complex of any tephra in the basin, results from a combination of factors relating to its eruption and accumulation.

Haileab (1988) has shown that during the approximate time of interest (ca. 1.5 Ma) there were two major centers of volcanic activity in the Ethiopian Highlands contributing volcanic tephra to the depositional systems of the Turkana Basin. Each of these centers produced a number of eruptive cycles of distinctive composition, but the characteristics of the resultant tephra link them to a common source area. The difficulty introduced by the high level of volcanic activity at this time comes from several eruptive cycles that produced characteristic tephra over an interval more prolonged than usual. This results in a tephrostratigraphic marker that more closely approximates a zone than an isochronous horizon. The duration of the interval represented by this zone is still quite short, on the order of 50,000 years or less, but it restricts our stratigraphic resolution to a certain degree.

In addition, the majority of these tephra are localized in geographic occurrence and thus make rather unreliable markers (that is, we cannot expect them to occur where we need them). We can make up for the lack of reliability in occurrence with the large number of possible markers, and thus for most local sections there are at least one or two tephra on which to hang correlations.

Compounding these practical difficulties of tephra correlation is the depositional history of one of the key markers within the interval of interest, the Lower Koobi Fora Tuff (LKF), which is one of approximately thirty tuffs which make up the Okote Tuff Complex. The LKF was defined by Brown and Feibel (1985) from Area 101 of the Koobi Fora region, where it occurs as a single depositional unit immediately beneath the Koobi Fora Tuff. Elsewhere, the depositional record of tephra that are geochemical correlatives of the LKF is more complex. In Area 103 at Koobi Fora, just 5 km to the southeast of the type locality, correlatives of the LKF occur through some 15 m of section, which includes more than a dozen depositional units. The lowermost five of these units are remarkably similar to the footprint tuff at the base of the NK3 excavation, with the addition of a thin mollusc concentration at the base of each (Feibel, 1988). Near the middle of the 15 m sequence, however, we find a very different ash, known as the Black Pumice Tuff (BPT), interstratified.

The LKF belongs to Haileab's tephra Group XIIE (Haileab, 1988), which is considered to represent the eruptive products of a single volcanic source that was active over the period approximately from 1.9 Ma to 1.1 Ma. Within this group, analyses of presumed correlatives of the type LKF form a continuum in their geochemical signature. Because of the large number of analyses we have from this continuum, we can see the progressive variation in chemistry of the glass. If we had fewer analyses, however, we might split the group into several distinct compositional entities. Part of the difficulty with this group may lie in varying mixtures of a polymodal glass-shard population, but preliminary microprobe analyses do not indicate that this is a significant factor. What is clear from the stratigraphic evidence is that a group of tephra with very similar composition (and therefore presumably from the same source) were erupted over a period of up to 50,000 years. During this interval other sources were also active, so tephra of different chemical groups (such as the BPT, from Haileab's Group XIID) are interspersed within the primary group. Although we cannot discriminate among LKF correlatives as to early and late products, we can use these secondary tephra to split the LKF sequence and thus to improve our stratigraphic resolution. What is exceedingly interesting for our study of environmental factors is that the ability to resolve the relative order within the LKF interval is not really necessary, as environmental conditions were essentially similar throughout. This may imply that we are dealing with an interval of time considerably less than 50,000 years.

The characteristic features of the LKF in the Nariokotome region suggest that the locus of channel activity at LKF times was on the east side of the basin, hence the thick repetitive sequence there, and that the western exposures of the tuff are the distal floodplain products of a single major flooding event that brought LKF tephra to this

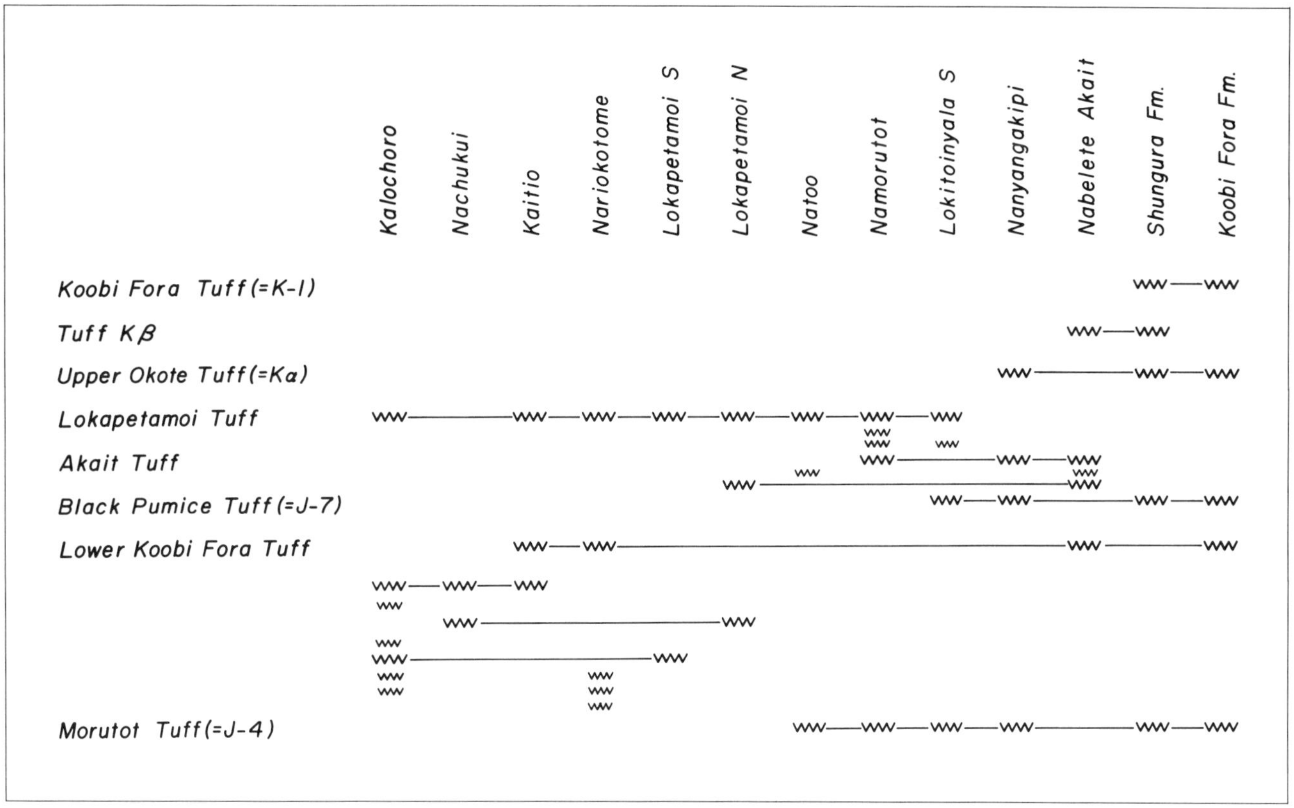

FIGURE 3.6.

Tephrosequence of the study interval. The major correlative ashes are indicated in the bold **V** *pattern, with correlations indicated. Minor, noncorrelative tephra are indicated by the smaller* **v***'s. Note that the absolute position of the minor tephra relative to the markers is not always demonstrable.*

portion of the basin. Where the LKF is not present in the West Turkana sections, we can base correlations on the stratigraphically proximate BPT or on a series of local marker tuffs that lie slightly higher or lower in the section.

Associated tephra

The exposures directly associated with the sequence at the hominid site include a large number of additional tephra, and these allow us to relate disparate exposures from around the region and the basin. The tephrosequence for the interval of study is illustrated in Figure 3.6.

To the north of NK3 and stratigraphically some 10–20 m below the hominid level lies the Morutot Tuff (= J-4). The Morutot was defined by Harris et al. (1988) at Nanyangakipi. It is an important stratigraphic marker known from both the Shungura and Koobi Fora Formations, and it is the unit to which the date of 1.64 ± 0.03 Ma by McDougall et al. (1985) can be applied.

Among the tephra which lie stratigraphically above the hominid interval in the Nariokotome region are two important markers which we define here. The Akait Tuff is a 20 cm, fine vitric tuff in its type locality at Nabelete Akait (K88-3262, see Table 3.1). This tuff shows a high degree of channeling into the associated strata and is pumiceous in the Namorutot section. At Namorutot, the Akait Tuff is stratigraphically overlaid by a second important marker tuff, which we name here the Lokapetamoi Tuff. This tuff was discussed by Harris et al. (1988) as the distinctive tephra 7 m above the Lower Koobi Fora Tuff at Nariokotome. It is characterized by white shards and typically has a thin accumulation of molluscs at the base in the exposures around Nariokotome. Its type locality is on the northernmost of two southern tributaries of Lokapetamoi, where the tuff is 1 m thick (K88-3242, see Table 3.1).

Three additional tephra are discussed here to facilitate understanding of the uppermost parts of the sequence and their basinal relations. The chemical and stratigraphic significance of these units is based in part on recent work by Haileab (1988). A tuff which lies stratigraphically above the Akait Tuff at Nanyangakipi is compositionally identical to a tuff known from the Koobi Fora region, where it has been informally termed the Upper Okote Tuff, and from the Shungura Formation, where it is Tuff Kα. We here define this tephra as the Upper Okote Tuff (= Kα). Its type locality is in Area 131 of the Koobi Fora region (sample K86-2375), where it lies immediately above the tephra K83-1645 discussed by Brown and Feibel (1985). Chemical data on these three correlatives are presented in Table 3.1. At Nabelete Akait, a tephra which overlies the Akait Tuff is compositionally similar to Tuff Kβ of the Shungura Formation, with which it is here correlated (Table 3.1). Stratigraphic relationships in the latter formation allow us to securely place this tephra above the Upper Okote Tuff (= Kα). Finally, the Koobi Fora Tuff in Area 101 of the Koobi Fora region is now known to be chemically identical to Tuff K-1 of the Shungura Formation (Haileab, 1988; Haileab and Brown, 1992), with which it is correlated here. It is not possible at this time to determine from documented sequences the relative stratigraphic relations of these three tephra and the Lokapetamoi Tuff; the higher stratigraphic position of the three tephra indicated in Figure 3.6 is conjectural.

Regional stratigraphic relationships

From the positions of the LKF and other tephra within the associated strata, we can reconstruct

TABLE 3.1.

Compositional data for type and reference analyses of tephra discussed in the text. Fe_2O_3, CaO, and K_2O reported as wt %; all others in ppm.

SAMPLE	Fe_2O_3	CaO	K_2O	Ba	Mn	Nb	Rb	Sr	Ti	Y	Zn	Zr
Akait Tuff—type—Nabelete Akait 88-3263	3.27	0.33	3.80	30	674	110	122	21	1471	81	159	1000
Akait Tuff—reference—Namorutot 89-3386	3.29	0.24	3.57	21	781	119	130	12	1291	88	160	1070
Lokapetamoi Tuff—type—Lokapetamoi N 88-3242	1.47	0.43	3.94	369	516	82	104	50	1434	62	65	408
Lokapetamoi Tuff—reference—Nariokotome 84-2972	1.31	0.29	3.90	1059	488	79	123	203	1440	70	96	302
Upper Okote Tuff(= Kα)—type—Area 131, Koobi Fora 86-2375	5.18	0.22	3.81	23	1869	235	162	10	1894	128	253	1558
Upper Okote Tuff(= Kα)—reference—Nanyangakipi 89-3384	5.05	0.15	3.30	56	1985	216	131	8	2072	117	241	1322
Upper Okote Tuff(= Kα)—reference—Errum TUFF K	5.45	0.33	3.00	53	906	153	118	35	1778	119	262	1430
Tuff Kβ—type—Errum ETH324	5.29	0.23	2.90	19	1587	130	102	5	1977	101	213	1050
Tuff Kβ—reference—Nabelete Akait 88-3265	5.32	0.22	3.22	147	1518	133	95	3	1954	106	223	1041
Koobi Fora Tuff(= K-1)—type—Area 101, Koobi Fora 80-113	5.35	0.30	1.60	10	1862	252	120	6	2176	130	306	1658
Koobi Fora Tuff(= K-1)—reference—Errum 228-7	5.64	0.27	2.80	40	2112	241	143	3	2682	127	290	1576

relationships between isolated sections around NK3 and those throughout the Turkana Basin. This reconstruction will be the basis for the ensuing discussion of stratigraphic relationships and the paleoenvironmental implications that may be drawn from them.

Stratigraphic relationships among the strata exposed some 20 km north and 10 km south of the NK3 site are illustrated in Figures 3.7 and 3.8. The character of the sedimentation pattern reflected in the regional picture is complex. At this level of resolution, the high degree of erosional scouring through this interval becomes quite apparent. Most of the scouring has a relief of only 1–2 m over a lateral distance of several kilometers, and it becomes apparent only when numerous isochronous markers, such as the tuffs, are tracked across the region. A few of the erosional surfaces have relief up to 5–10 m. The channeling is indicative of relatively small streams traversing the area. The presence of pumice clasts in most of the major channels clearly indicates that these are part of the Omo drainage system originating in the Ethiopian Highlands, not local marginal drainages. The erosional scouring illustrated in Figure 3.8, which represents a composite of relative incision, is not intended to depict a north-south profile. Indeed, though the channels were likely wandering to a certain degree, their general trend would have been north-south rather than east-west.

Throughout this interval, interbed contacts are commonly erosional, but where no erosion is evident, they often show land-surface features such as footprints and desiccation cracks. Features of long-term land-surface stasis, such as soils, are not well developed in the interval, however. In addition, both coarse- and fine-grained facies commonly include biotic indicators of aquatic conditions. The juxtaposition of indicators of both standing water and channeled flow, often in the same unit, indicates a system in which the topographic gradient is extremely low and environments are constantly shifting. A similar conclusion can be drawn from the intimate interbedding of subaqueous deposits and subaerial exposure surfaces.

A useful marker of shallow, standing-water conditions, which is well-developed in parts of this sequence, is the trace fossil *Piscichnus brownii* (Feibel, 1987). This shallow, circular depression is identical to the fish nests constructed by cichlid fishes today, and it has been interpreted to represent the record of such activity in the past. Cichlids typically excavate their nests in water of about 1 m depth in relatively clean sands.

The site in the context of the Turkana Basin

The tephrostratigraphy developed above allows for correlations to be made of exposures of the Shungura and Koobi Fora Formations with the strata exposed around the Nariokotome hominid site, and from these correlations a broader-based interpretation of regional paleoenvironments becomes possible. We present first a brief outline of the characteristics of the correlative strata and then discuss their implications.

The Shungura Formation

In the lower Omo Valley, the equivalent strata belong to upper Member J and lower Member K of the Shungura Formation. In those exposures, strata show many features similar to those recorded around Nariokotome, including the development of numerous, small, tuff-filled channels and the interspersal of shallow-water aquatic indicators with evidence of subaerial exposure. In Submember J-7, an exhumed paleolandscape mantled by tuff is closely analogous to correlative deposits and features at Koobi Fora.

The Koobi Fora Formation

The correlative levels of the Koobi Fora Formation east of modern Lake Turkana are the upper KBS Member and lower Okote Member. These strata are highly variable over the Koobi Fora region and reflect a series of different environments.

The type sections for the KBS Member are in Areas 102 and 103 (Brown and Feibel, 1986; Feibel, 1983, 1988). These sections reflect a depo-

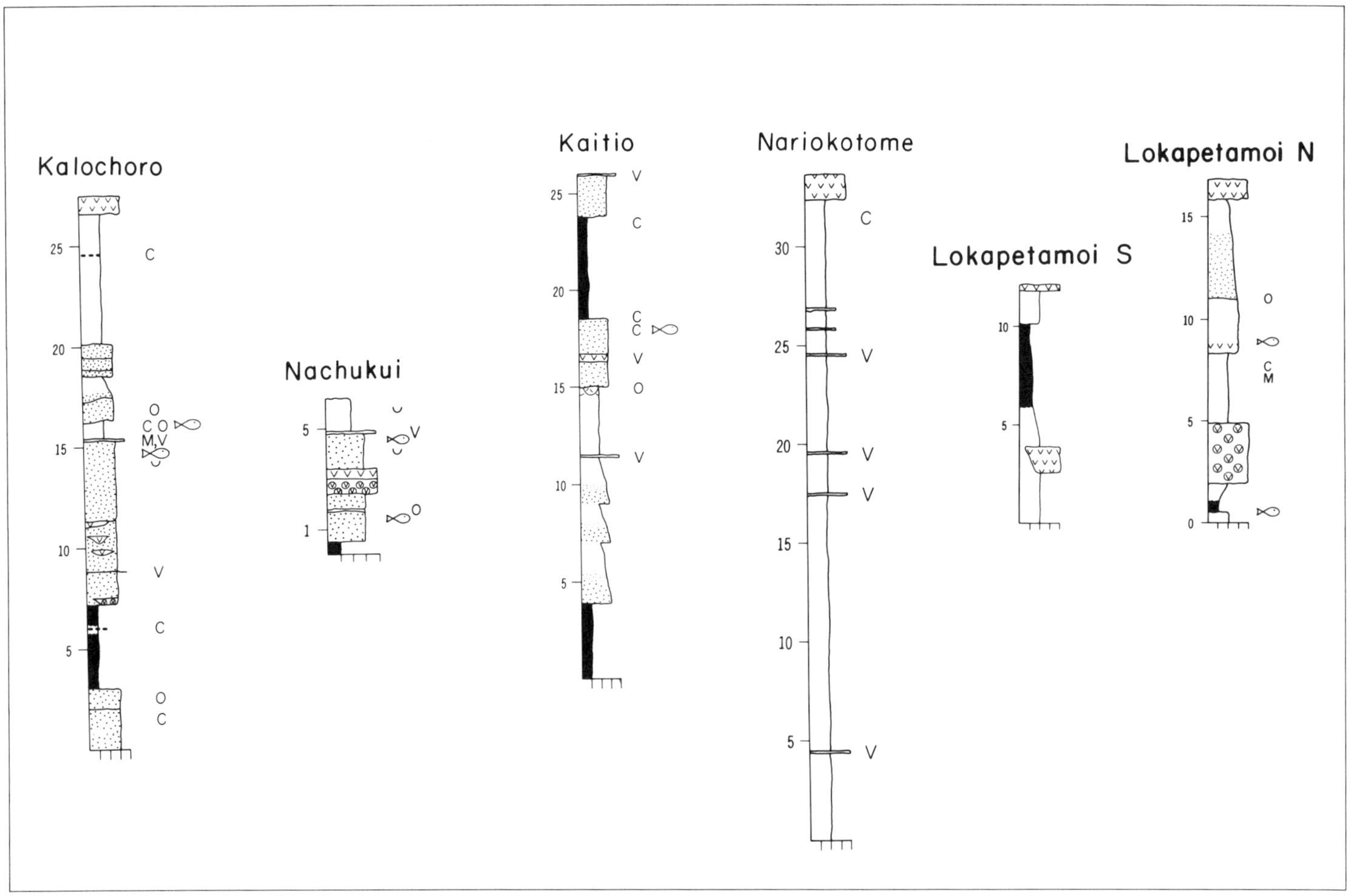

sitional system very similar to that seen in the Nariokotome region, with a mixture of fluvial and shallow lacustrine deposits. One significant contrast to the deposits of both the Nachukui and Shungura Formations at this time is the presence of meandering river deposits both above and below the Lower Koobi Fora Tuff. These intervals contrast sharply with the other fluvial facies not only in depositional style but also in that they are entirely epiclastic, that is nontuffaceous, in nature. These deposits may imply that the unusual nature of the tuffaceous fluvial facies described from the Nachukui and Shungura Formations is due in part to the influence of the abundant volcaniclastic material on deposition. This reasoning is further supported by the character of the tuffaceous fluvial facies at Koobi Fora, which marks the base of the Okote Member. In Areas 101–102–103 near Koobi Fora, as well as in the Karari and Ileret regions, the tuffaceous fluvial facies are identical in character to their correlatives elsewhere in the basin. In Area 103 at Koobi Fora, the Lower Koobi Fora Tuff has at least five levels marked by footprint surfaces. The most significant of these is a surface uncovered in paleontological excavations by Behrensmeyer and Laporte (1981) that included hominid footprints. These footprint levels are in the lower part of the depositional complex of the Lower Koobi Fora Tuff, virtually the same level as the footprint tuff at NK3. The major channeling seen in the Nariokotome region with the Akait and Lokapetamoi Tuffs is also seen at Koobi Fora, where the Koobi Fora Tuff (= K-1) is channeled down to rest immediately on the Lower Koobi Fora Tuff in Area 101.

The broader geographic representation of the Koobi Fora deposits from the study interval reveals a more diverse sampling of local environ-

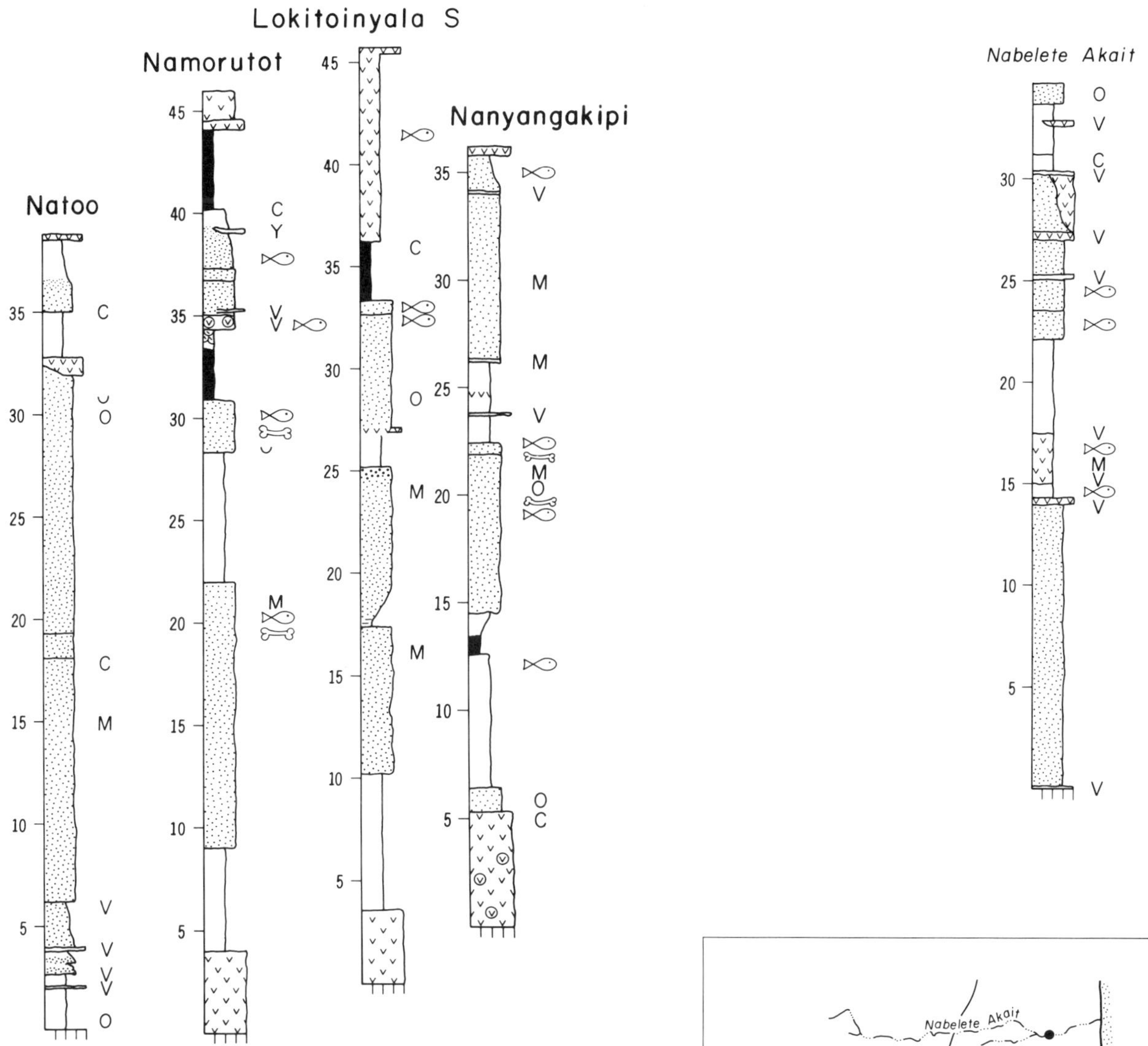

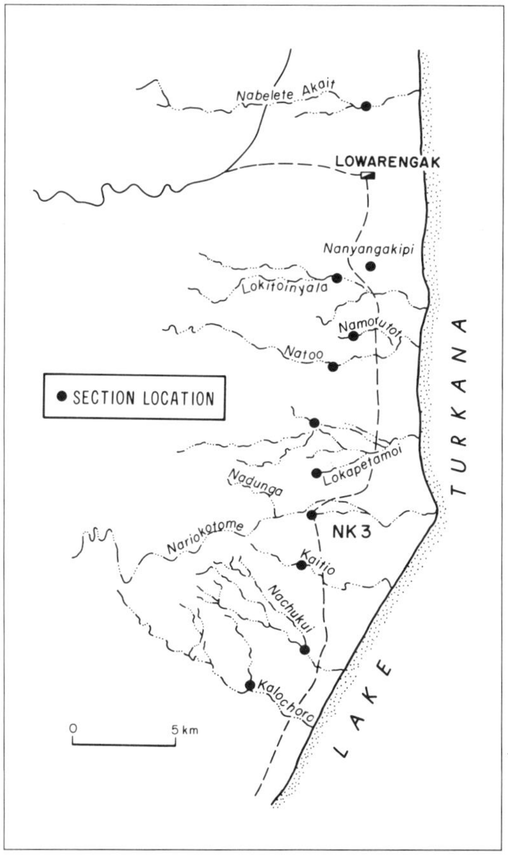

FIGURE 3.7.

Stratigraphic relationships among sections to the north and south of the NK3 excavation. The Lower Koobi Fora Tuff or its approximate stratigraphic level is used as a datum for this interval. Location map shows position of stratigraphic sections.

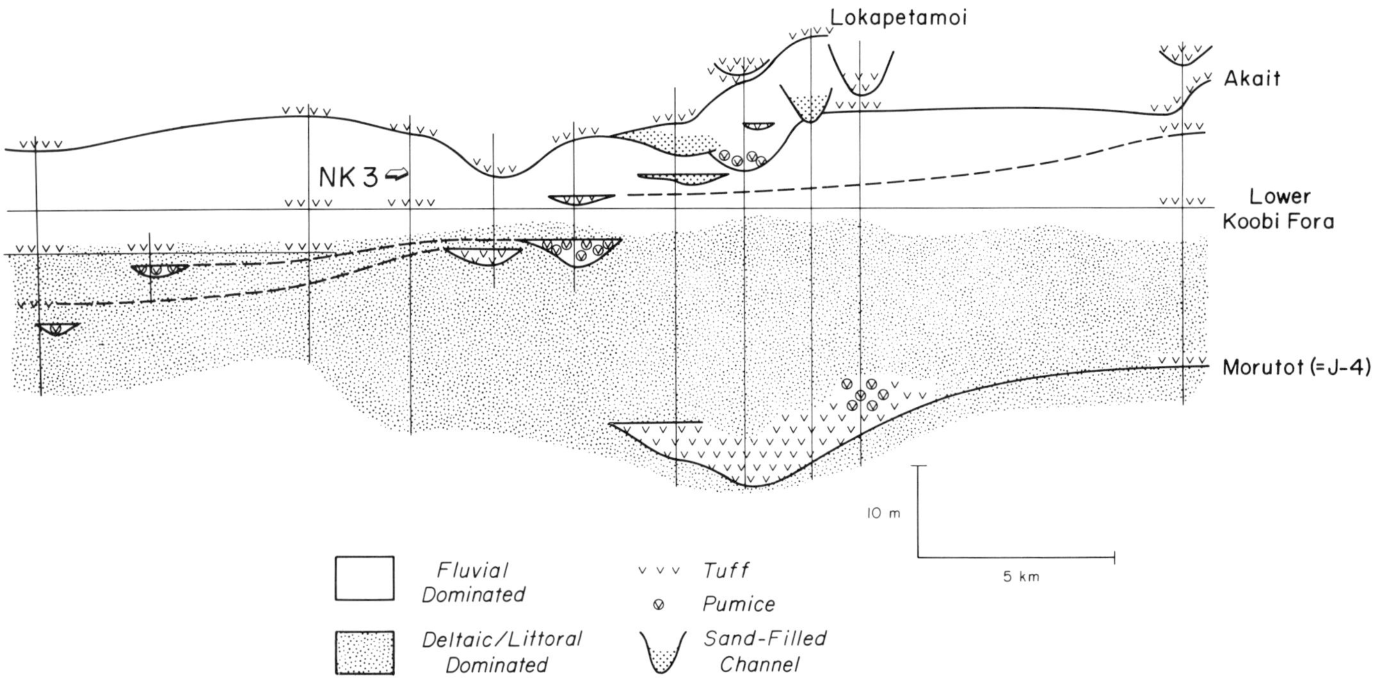

FIGURE 3.8.
Generalized facies diagram based on sections of Figure 3.7 showing channel downcutting and distribution of the deltaic/littoral-dominated and fluvial-dominated deposits.

ments. In the Karari region, sedimentation rates are relatively low, except where the tuffaceous fluvial facies dominate the section. In the nontuffaceous intervals, the low accumulation rates and arid conditions allowed numerous caliche horizons to develop. This evidence from paleosols on the eastern margin of the floodbasin is important in that it demonstrates the aridity of the prevailing climate away from the influence of the perennial rivers.

The Karari region is also witness to the effects of marginal drainages from the northeast. These drainages were ephemeral systems arising in the region around present-day Chew Bahir (Lake Stephanie), which was uplifted slightly between 2.5 and 2 Ma. Although the local relief in this source region was probably less than that in the Murua Rith hills west of NK3, the drainage area was likely greater, as these streams carried a significant component of basaltic gravels into the Karari region. These gravels became the raw material for the cultural activities of early hominids in the region, as is well documented in sites such as FxJj 20 (interstratified within the Okote Tuff, just prior to NK3 times) and FxJj 50 (interstratified within the Koobi Fora Tuff, near the end of the study interval).

Paleogeographic and Paleoenvironmental Reconstructions

Paleogeographic reconstructions for the Turkana Basin at about 1.65 Ma (Feibel, 1988; Brown and Feibel, 1988; Feibel et al., 1991) have had to take into account the intimate relationship between shallow lacustrine and terrestrial facies seen throughout the basin at this time. This particular reconstruction (Figure 3.9) has always been a difficult one, and it is unsatisfactory in many aspects.

The problem reflects in part the difficulties of trying to reconstruct a dynamic landscape at a temporal resolution of about 100,000 years. Landscapes that are changing dramatically at a rate faster than that effectively become blurred in the reconstruction. In addition, it is difficult to determine whether the basin was closed at this time, whether it was open to the east, or whether it was open to the west.

Both of the major features of the landscape at approximately 1.5 Ma were in a state of flux. The lake or lakes in the basin had been steadily filling with sediment since the formation of the relatively deep lake basin (ca. 30 m depth) at about 2 Ma (termed the Lorenyang Lake by Feibel, in preparation). The eastern side of this lake filled within 100,000 years (Feibel, 1988), but a remnant arm persisted on the west side for another 250,000 years or so. The events which occurred at the NK3 site took place just after the end of this particular lake phase.

The fluvial facies of the ancestral Omo River at this time are also indicative of a system in transition. In earlier shifts from lacustrine to fluvial conditions (e.g., in the Lonyumun Member and in the transition from the Lokochot to the Tulu Bor Member at Koobi Fora), sedimentation rates were relatively low, and the replacement of lacustrine facies by fluvial facies is abrupt. In contrast, accumulation at 1.5 Ma was quite rapid and the transition itself is well recorded. Because the depositional front was so broad (some 60 km), the ancestral Omo River broke into a series of smaller channels, and these channels shifted their positions rapidly. As a result, fluvial deposition is characterized by many small, incised, channel-form sands. These channels might well be termed distributary channels, and they correspond in form to a distributary network, but we hesitate to use this term as there is no conclusive evidence that a major lake actually existed at this stage. Previously, we have restricted usage of deltaic terminology to the subaqueous deltaic facies (Feibel, 1988) in order to avoid this problem. It is equally probable that these channels coalesced downstream and the fluvial system continued beyond this "inland delta" much in the manner that the Okavango and Niger Rivers behave today. An additional complication introduced at about this time is the intermittent character of sedimentation. We have suggested previously (Feibel et al., 1991) that the interruptions may in part have been due to the beginnings of piracy of Omo River waters by the Nile drainage system. The presence of tuffaceous deposits, and in particular of pumice clasts, in the deposits discussed here indicates that the Omo River was the major depositional agent for the study interval.

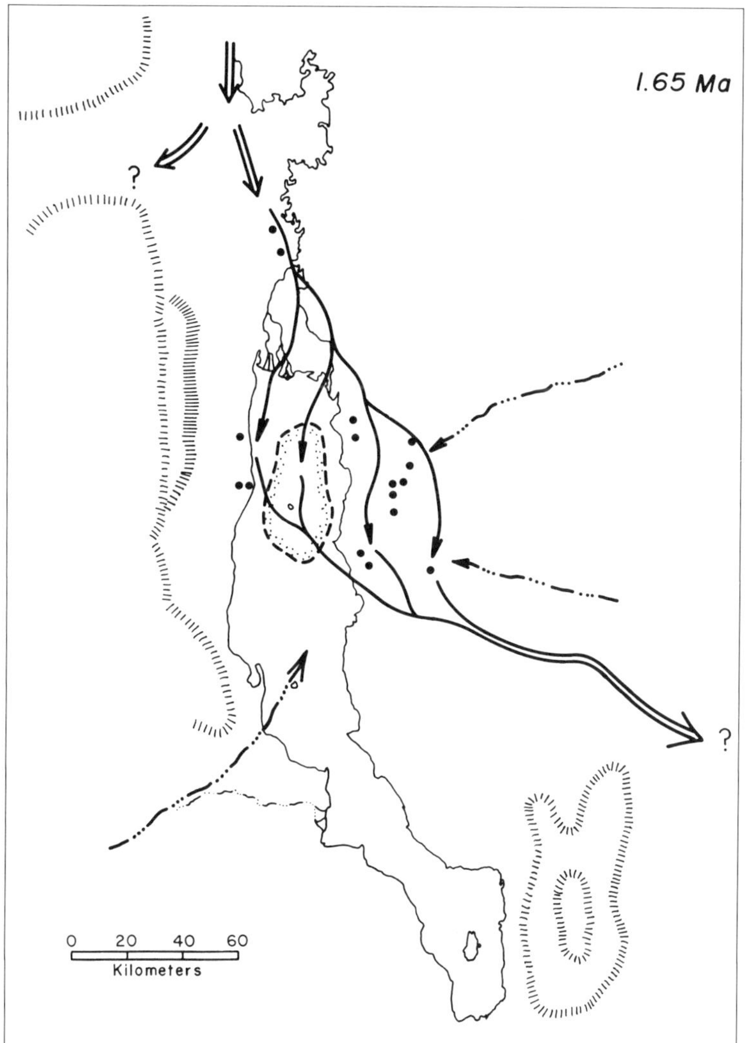

FIGURE 3.9.
Paleogeographic reconstruction of the Turkana Basin at 1.65 Ma (modified after Feibel, 1988). An outline of modern Lake Turkana is shown for reference. Three potential hydrographic configurations are indicated: a distributary network in a closed basin, integration of this network to flow out of the basin to the east, and upstream channel switching into the Nile drainage to the northwest. Dots indicate major exposures of the Okote Tuff Complex tephra.

Because of the frequent volcanic activity at the time, many of these small channels are tuff- or pumice-filled. With the sequential information preserved in these tephra, we can see the rapid lateral shifts of these small channels on a very fine scale. These shifts are the reason that many important markers are often lacking in a given local tephrosequence. In this respect it is instructive to compare the long, linear axial exposures of the Shungura Formation with the broader geographic perspective provided by the Koobi Fora and Nachukui Formations. In the Shungura Formation, some 20 tephra are recognized between Tuff H-2 (= KBS Tuff) and Tuff L (= Chari Tuff) (de Heinzelin, 1983; Haileab, 1988). At Koobi Fora, this same stratigraphic interval has close to 40 tephra, and in the Nachukui Formation we find 23 ashes. Within the basin, only three tephra are found in all three formations, one additional tuff is common to the Koobi Fora and Shungura Formations, one additional tuff correlates between the Nachukui and Koobi Fora Formations only, and a single tuff correlates between the Shungura and Nachukui Formations only (Figure 3.6). The relatively low number of shared tephra is a very good indicator of how transient the small fluvial channels, the distributing agents for these tephra, can be.

FIGURE 3.10.

Paleogeographic reconstruction of the Nariokotome region at 1.5 Ma. The modern lake shore and Nariokotome drainage routes are included for reference, and the hominid site is indicated by a star.

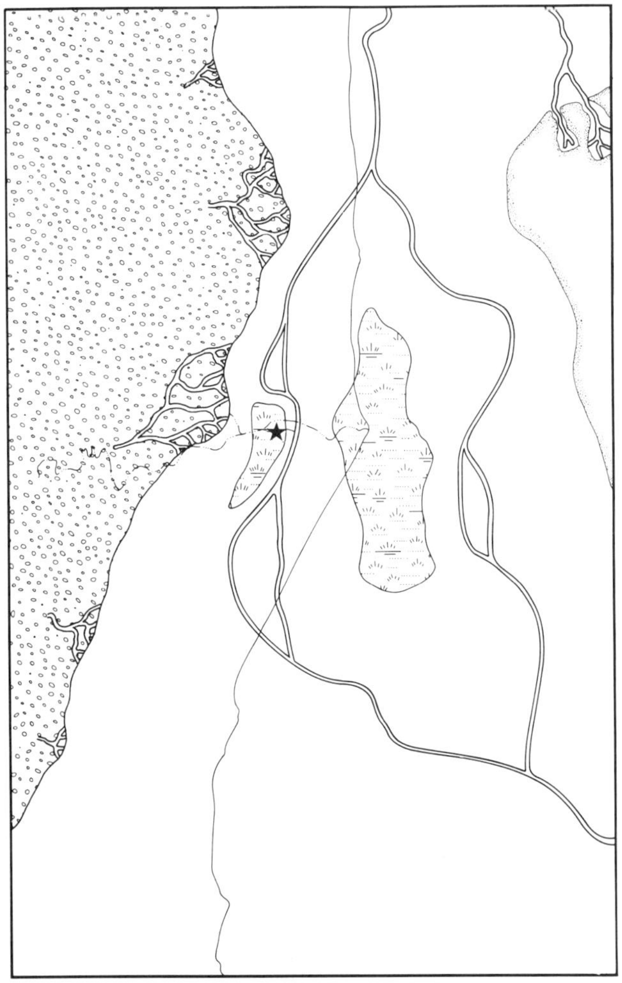

Once we understand the stratigraphic complexity of this particular interval, we can begin to approach the problem of reconstructing the paleogeography and associated paleoenvironments in a slightly different light. In order to re-create a valid model for a particular point in time, such as at the death of the Nariokotome hominid, the sampled interval has to be narrowed to about 0.5 m. Even with the abundant tephra on which to base correlations and refine our stratigraphic resolution, we cannot approach this level of precision. Thus, our reconstruction is a generalization, based on some 2 to 5 m of section, which may incorporate a number of quite significant changes in both the geographic and environmental situation. We do, however, have enough information in this time-averaged and areally averaged interval to posit a reasonable interpretation of the geographic elements present at a given time, and to reconstruct the environmental characteristics which would likely have been associated.

A reconstruction of the paleogeography at the time of the death of the Nariokotome hominid is presented in Figure 3.10. As discussed above, this is a generalization of the geography over an interval of some 50,000 years around this particular event. We cannot be certain that a lake existed at the time the hominid died, but we do know that within the 50,000-year interval, one or more shallow lakes was present, at least locally. Likewise, the course of the Omo River is but an example of the configuration of channels that existed through this time interval.

The paleogeographic reconstruction of Figure 3.10 includes three major geographic elements: the fluvial system, the floodbasin ponds or lakes, and the alluvial fan systems verging to the west. Channels of the Omo River thread their way down from the north. The characteristics of the Omo River fluvial facies seen throughout the Nariokotome region suggest that these channels were relatively straight, perhaps anastomosing to a certain extent but not meandering. Even given the transient nature of these small channels, they would have been the dominant features of the landscape, as they would have been permanent sources of fresh water. The shallow standing water to the east of the site would have changed in character seasonally. The annual Omo River floods would have expanded the size of this body (or refilled a desiccated pan) and freshened the chemistry of the waters. At this time, as throughout much of the Plio-Pleistocene history of the basin, the Omo River floodplain can be considered a seasonally fluctuating floodbasin or lake. With the advent of heavy rains in the Ethiopian Highland source region, the river would swell, overtop its banks, and flood this basin. The temporal and volumetric extent of the floods would control the size and duration of the floodbasin lake. Short, low-intensity rains would produce flooding, which would not completely fill the basin, and last only a matter of weeks. In contrast, extensive rains in the highlands would have flooded the majority of the basin, producing a floodplain lake that would persist for many months. Indeed, minor "lake phases" in the basin can be considered nothing more than the persistence of this flooded state through a number of consecutive years.

What is particularly important about this seasonal flooding for our reconstruction of the local habitats is that the availability of fresh water across the floodplain, particularly during the recessional stage of the floods, would have supported an extensive and lush floodplain grassland. Floodplain grasslands have been described in detail by Vesey-Fitzgerald (1963, 1970) from Central Africa. They are not a characteristic feature of the Turkana region today, and the floodplain grassland is the most important element missed in most attempts at reconstructing the paleoenvironments of the Turkana Basin.

The third geographic element of the basin at Nariokotome times is the alluvial fans of the western hills. These alluvial aprons have become the dominant element in modern-day geography of the region, but in the early Pleistocene they were laterally restricted by the active floodplain of the Omo River and by lower topographic gradients. The Nariokotome River is an important alluvial system extending to the modern lake today, but in the Plio-Pleistocene its influence seldom, if ever, extended as far basinward as it does today.

The paleoenvironmental picture thus reconstructed is a mosaic of features: the perennial channels of the Omo River, the adjacent, seasonally inundated floodbasin, the shallow lakes or ponds that would have persisted in topographic lows, and the better-drained alluvial systems bordering the region to the west (Figure 3.10). Soils and associated vegetation would have varied considerably across this catena. Without the large meandering channel system of the Omo River at this time, the development of forest was likely to have been limited. The fluvial channels would have been bordered by a narrow belt of trees, which would give way laterally to open grassland. This narrow riparian belt, however, would have been an important focus of the local habitats, as it would have offered permanent sources of water. It is difficult to reconstruct the proportion of thornbush, but this vegetation would likely have been concentrated on the local topographic highs within the floodbasin and on the better-drained margins to the west. Much of the floodbasin would have been floored by seasonally flooded grasslands; luxuriant growth would follow flood seasons, and gradually dry off through the remainder of the year. The marsh in which the hominid bones were deposited was likely a flood pan which would have supported thick grass or reeds when water was present. The lack of significant deposition during the paludal intervals suggests that the marshes were, at best, temporary, and perhaps seasonal.

Acknowledgments

We thank R. E. Leakey and the National Museums of Kenya for their continued support. Our research was funded by the National Science Foundation (grants BNS 84-06737, 86-05687, 88-05371, and 90-07662). Thanks also to B. Haileab, S. K. Williams, and M. M. Smith for assistance with laboratory analyses.

C H A P T E R 4

Taphonomy

Alan Walker

An interpretive picture can be pieced together of the circumstances surrounding the death and burial of the Nariokotome hominid on the basis of taphonomic evidence: the distribution, breakage, and other properties of the excavated fossils in relation to the microstratigraphy of the site and of the basin as a whole. Although some of the details are speculative, it is possible to give a reasonably complete account of events during the short interval between the hominid's demise and his sedimentary burial. The clues this specimen has left add vivid color to what we know about the life-styles of early hominids.

Excavations at the site of Nariokotome III (NK3 for short) began in August 1984 and continued for four more seasons. Kamoya Kimeu found a small piece of hominid left frontal among lava boulders and pebble lag on the sloping bank of a small tributary of the Nariokotome sand river. We cleaned the erosion slope by first removing all large boulders and pebbles and then by loosening the thin soil and weathered sediment. Several pieces of hominid braincase were found in this way. Because the hominid fossil is, for the most part, dark brown in color, we decided to screen and water wash all loosened soil. We recovered several more pieces of the cranium as well as some rib fragments by this method. The first pieces of bone that were discovered *in situ* were the left and right maxillas, which were embedded in the underground part of the trunk of a small *Acacia* tree. It seems likely that the tree took root in the damp sediment in the upside-down braincase.

A popular account of Kamoya Kimeu's discovery and the first year's excavation was published in 1985. A site plan of the area around the hominid bones found in 1984 was also published then. Following the discovery we were able to proceed in excavating along the strike of the bone-bearing horizon. The excavation was carried out by hand using Olduvai (or Kitibi) picks—sharpened six-inch nails fastened in an L-shaped piece of wood—a traditional excavating tool in East Africa. Excavation was carried out in 1 m squares. All the excavated sediment from the bone horizon was water washed through window-screen sieves. At first water was brought to the site from Lake Turkana; later, as the volume of excavated sediment increased, the sediment was taken to the lake for screening. The downward limits to excavation proved to be easily found. Below the surface weathering zone, we found an indurated layer beneath the bones that the picks could hardly penetrate. This layer, the top of Unit 1, was seemingly of the same composition as the silt layer with the bones (Unit 2); it was marked by the presence of short, orange root casts about 5 mm in diameter. Although most of the bones in Unit 2 were found toward its base, many of them were positioned horizontally on top of this harder underlying layer. Some of the bones were found in depressions in the harder layer. These depressions are identified as large mammal footprints.

Collecting the Fossils

The bones in Unit 2 are, for the most part, extremely well fossilized. They are dark brown in color with some patches of black manganese dioxide stain. Teeth have dark grey or brown enamel with a bluish tinge. Practically all bones and teeth were encrusted with a hard calcareous layer. In places where modern roots had reached the fossils, there were patches of pale bone and tooth as well as erosion of the bone surface. For example, the posterior surface of the left tibia is eroded by roots. When fossils were discovered, we exposed the upper surfaces of bones or teeth using dental picks and applied plastic hardener (vinyl acetate solution). When dry, bones were strong enough to be lifted without any support. The one exception was the left femur, which had cracked across midshaft and which was taken out of the sediment in a plaster-of-paris jacket.

The positions of bones were measured with reference to fixed steel markers hammered into the deposits. New reference markers were made from time to time to make measurement over large distances easier. Fossils were given temporary field numbers until they were cleaned and identified properly in Nairobi. It was sometimes not possible to make a secure identification in the field because of the encrustation of hard calcareous matrix. Sometimes objects that at first were thought to be bones proved, when cleaned in the laboratory, to be natural casts of the insides of parts of gastropods or calcite-encrusted stones. The hominid bones were cleaned of matrix using Airscribes (Chicago Pneumatic Corp.) and given their accession suffixes in order of return to Nairobi and not in order of body part.

Table 4.1 lists the body parts found of the *Homo erectus* skeleton KNM-WT 15000 together with their identification suffixes. Other fossils, both vertebrate and invertebrate, have not been accessioned but are stored separately as faunal collections from site NK3.

TABLE 4.1.

Skeletal parts of KNM-WT 15000 and their identification suffixes.

A	cranium	AP	left rib 7
B	mandible	AQ	left rib 2
C	left clavicle	AR	lumbar vertebra 1 body
D	right clavicle	AS	right rib 9
E	right scapula	AT	left rib 3
F	right humerus	AU	left rib 8
G	left femur	AV	lumbar vertebra 2 (in part)
H	right femur	AW	right pubis fragment
I	left tibia	AX	left pubis fragment
J	right tibia	AY	distal right rib 1
K	right fibula	AZ	proximal right rib 1
L	left fibula	BA	lumbar vertebra 1 piece
M	distal epiphysis of right femur	BB	sacral vertebra 2 piece
N	left ilium	BC	sacral vertebra 3 piece
O	right ilium	BD	right rib 5 fragment
P	right ischium	BE	right rib 5 fragment
Q	left ischium	BF	left ilium fragment
R	cervical vertebra 7	BG	left ilium fragment
S	thoracic vertebra 1	BH	left tibial epiphysis
T	thoracic vertebra 2	BI	thoracic vertebra 9 piece
U	thoracic vertebra 3	BJ	right rib 5 fragment
V	thoracic vertebra 8	BK	left scapula spine
W	thoracic vertebra 7 piece	BL	left scapula axillary margin
X	thoracic vertebra 10 piece	BM	lumbar vertebra 5
Y	thoracic vertebra 11	BN	right rib 2
Z	lumbar vertebra 3 piece	BO	hand phalanx
AA	lumbar vertebra 2 (in part)	BP	right ulna
AB	lumbar vertebra 4	BQ	hand phalanx
AC	lumbar vertebra 6	BR	? left rib 4
AD	sacral vertebra 1 body	BS	left rib 10
AE	sacral vertebra 4	BT	? hominid rib fragment
AF	sacral vertebra 5	BU	? right metacarpal 1
AG	left rib 1	BV	? left metacarpal 1
AH	right rib 2	BW	lumbar vertebra 3 body
AI	right rib 4	BX	? right metatarsal 1
AJ	right rib 10	BY	sacral vertebra 1 piece
AK	right rib 6	BZ	left ulna
AL	right rib 7	CA	thoracic vertebra 5
AM	right rib 8	CB	left rib 5
AN	left rib 11		
AO	left rib 9		

As well as recording the fossils' position on the site, we tried to record the orientation by measuring the azimuth of the long axis. For most of the bones this was not possible, since they were small fragments with irregular outlines or were strongly curved, like most of the hominid ribs. We were able, however, to measure the long-axis orientation of a total of 94 bones. Dip, when it could be determined, was often close to horizontal.

Figure 4.1 shows a sketch map of the site with the areas of excavation carried out over the 5 years. Much of the skeleton was recovered in 1984. A list of the parts found then is given in Brown et al. (1985). A more complete picture of the site setting is given in Feibel and Brown (Chapter 3, this volume). The irregular shape of the first area of excavation reflects our attempt to follow the initial concentration of hominid bones into the hillside. When we resumed excavation in 1985, we thought that the rest of the skeleton might be found by following the same line of bone concentration, but (apart from a rib and an ulna, which were on the line) the discovery of other parts that year showed that there was more dispersal than we had imagined. Greater volumes of overburden were removed in subsequent years to expose greater areas where bones might be found. By the end of the 1988 season we decided to stop the search. At that point, all certainly identified hominid bones were over 7 m from the excavation walls. The overall site plan is given in Map 1, and an enlargement of a parallelogram-shaped area containing all the certainly identified hominid bones is given in Map 2. (Maps 1 and 2 are inserted in the back of this book.)

Bone distribution

Figures 4.2 and 4.3 indicate the distribution of bone in the excavation. The numbers of bones that fall into each excavated square are recorded in Figure 4.2; bone density is represented graphically in Figure 4.3 by using heavier shading for squares with more bones and lighter shading for squares with fewer bones. If a bone crossed the boundary between adjacent squares it was recorded as being in both.

Of 584 squares, only 190 (32%) contained any bones. It is clear that the bones are not evenly distributed over the site. If the bones had been evenly distributed each square would have contained less than one bone on average; as it is, bone density per square ranges from 0 (the most common) to 20. In the squares containing bones, the average density is 2.76 bones per square. Bone density generally falls off to the south and east, where there were large excavated areas with few or no bones. In contrast, the hominid bones formed a patch to the southwest of the excavated area in a set of nearly contiguous squares that trends from NNW to SSE. Hominid-bone-bearing squares average 5.43 bones per square, and few bones in these squares were nonhominid. Other high-density squares are, without exception, those containing many fragments of chelonian carapace or plastron. These bones often represent either a single bone, broken into several smaller pieces, or adjacent bones separated at the sutures.

Table 4.2 gives a breakdown of distribution of densities over the area. The hominid bones are found in squares of all densities but predominate in those with high densities. Hominid bones account for two-thirds of the bones in squares with 12 to 16 bones and 100 percent of the bones in the highest-density squares of 16 to 20 bones.

The hominid skeleton stands out on the site plan against the more or less random, background scatter of broken bird and mammal bone and isolated bits of fish and aquatic reptile. This concentration of hominid bones in one relatively small area of the site could have been produced in one of two ways. Either the skeleton has not been very well dispersed from the place where the whole body was initially deposited, or the bones have been secondarily concentrated in the place after maceration of the soft tissues.

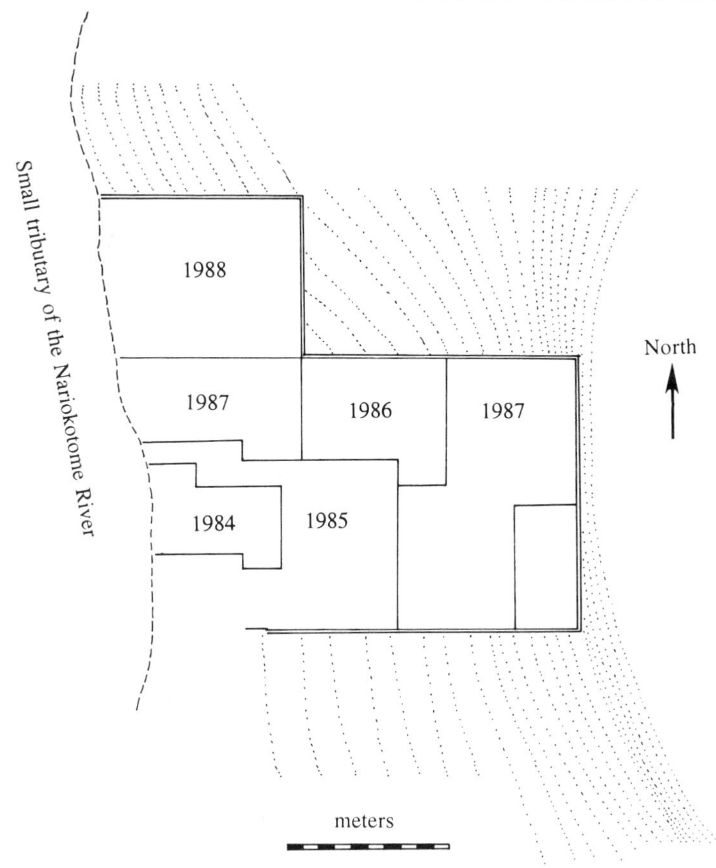

FIGURE 4.1.
Sketch map of the site of Nariokotome III showing the areas excavated in 1984–1988. Form lines are given to show the hill slope.

TABLE 4.2.

Summary of bone densities by squares.

DENSITY (BONES/SQUARE)	ALL BONES	HOMINID BONES
0–1	91	11 (12%)
2–3	57	6 (10.5%)
4–5	24	1 (4.3%)
6–7	4	2 (50%)
8–9	6	2 (25%)
10–11	1	0
12–13	3	2 (66%)
14–15	3	2 (66%)
16–17	1	1 (100%)
18–19	0	—
20–21	1	1 (100%)

Inspection of the site plan suggests that both of these processes took place. One piece of distributional information suggesting that the intact body rested nearby, with subsequent minor dispersal, is that a cluster of teeth were found isolated from the skull. Eleven straight-rooted teeth were found in a small depressed area along with a vertebra, about 3 m to the northeast of the main concentration of hominid bones. This

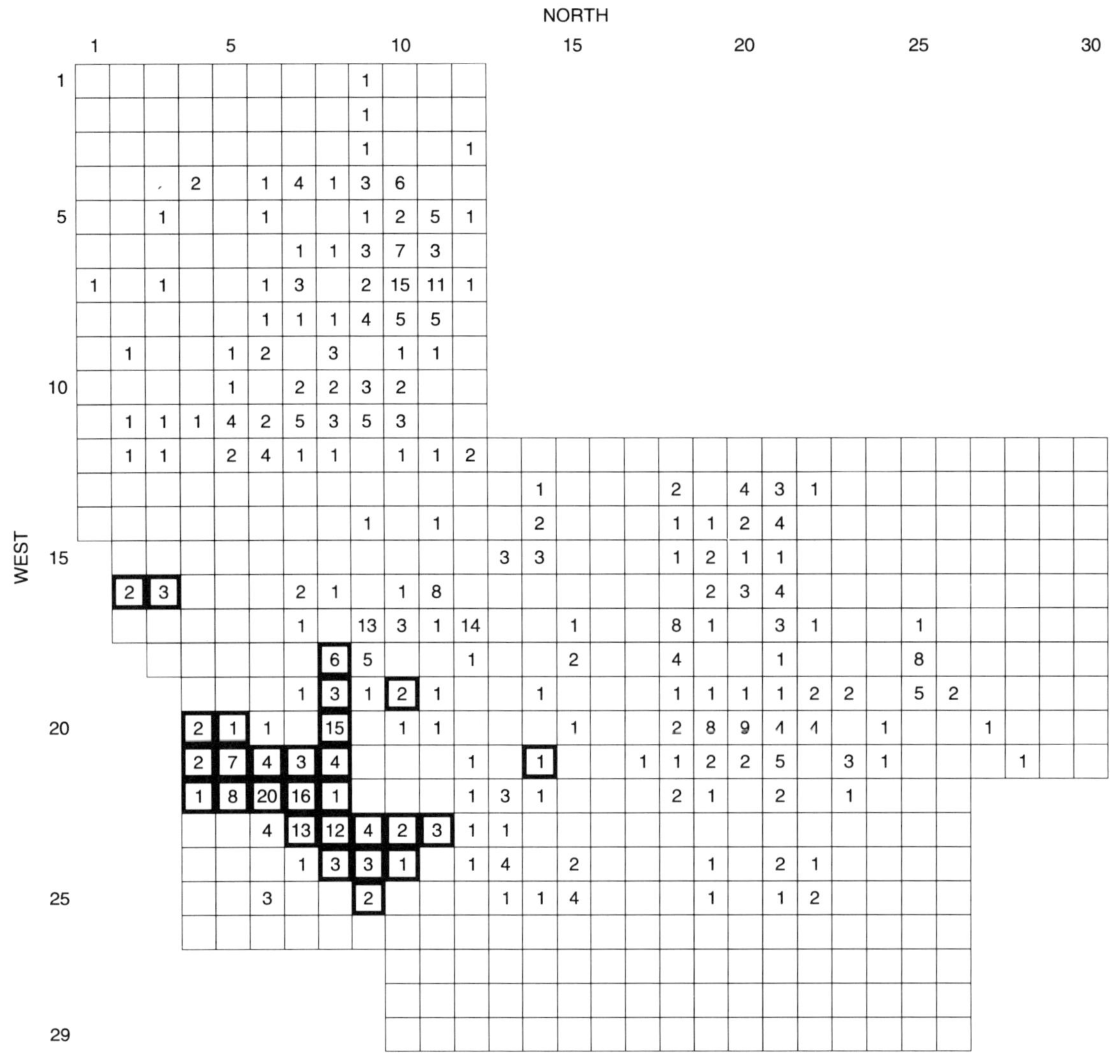

FIGURE 4.2.
Schematic plan of the site showing bone densities. Squares that contained hominid bones are outlined. Each square represents 1 square yard (0.84 square meter).

depression is thought to be a large mammal footprint. On the density map this tooth concentration is marked by a black square to the northeast of the two adjacent black squares that mark the main jumble of skeletal parts. The maxillas were found about 3 m to the west of these teeth and the mandible about 3 m to the south of them. Thus the cranium ended up about 4 m from the mandible. Despite the eventual separation of the maxillas and mandible, the cluster of isolated teeth included both uppers and lowers. Therefore, it seems most likely that the body rotted, lying face down in the water, and that most of the straight-rooted anterior teeth from the front of the jaws dropped out in one spot while the body was largely intact. Later, the body rotted further and the parts were pushed by a gentle current so that the main concentration came to lie in a rough line to the southwest of the initial position. Alternatively, the body could have rotted at the main concentration of bones and

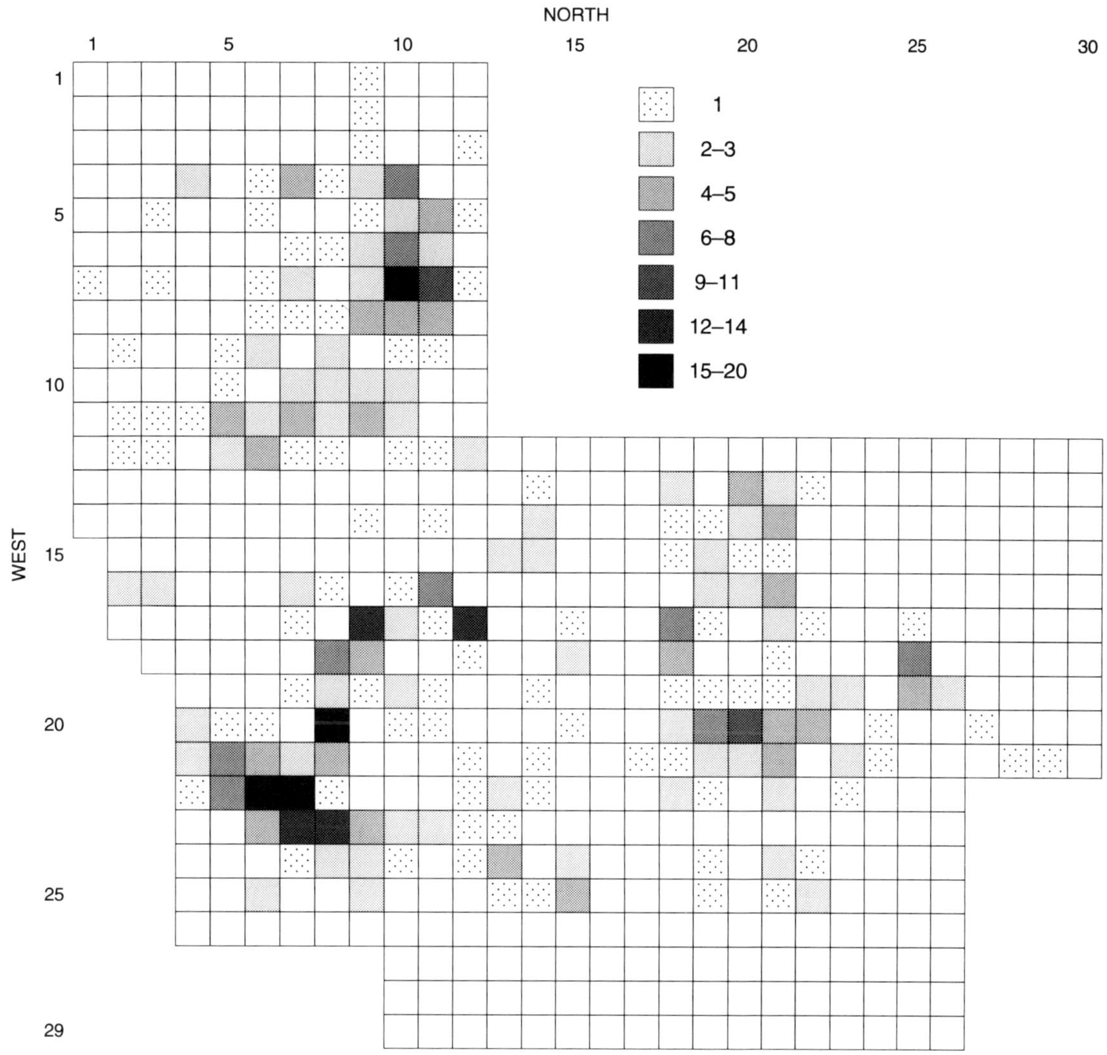

FIGURE 4.3.
Schematic plan of the site as in Figure 4.2, but with shading to represent bone densities.

these teeth initially dropped there. Subsequently the teeth could have been winnowed out by currents and swept to their new position. The first alternative seems more likely because it is more probable that lower and upper teeth of the same hydraulic equivalence would have stayed together after deposition than be brought together from different directions afterwards. It is true that the vertebra and most of the teeth were found in a depression, which was most probably a large mammal footprint, but only minor and local water movements would have been needed to concentrate them in it if they had already been shed nearby.

There are clear cases of close association of adjacent bones, suggesting prior articulation, such as the association of the left scapula and humerus or that of the left femur and left ilium. But, for the most part, elements are separated from their adjacent bones in the skeleton with relatively minor dispersal. Figure 4.4 illustrates the distribution of the hominid bones; the gray lines connect the parts on the site map which once were adjacent in life. This is equivalent to an archeological refitting diagram of stone tool flakes and cores. With the minor exceptions of one or two parts that have become widely separated from the rest of the skeleton, the pattern of lines of association is strongly in the form of a triangle. The base of the triangle is the major concentration of bones. The apex is the cluster of anterior teeth. Lines of association fan out from the tooth cluster across the triangular area. Another set of lines connects the bones in the base of the triangle. This pattern supports the hypothesis that the body was initially located where the tooth cluster was found and that dispersal and disassociation then carried the bones to the southwest; these bones were subsequently spread out along the base of the triangle. (Figures 4.5–4.7 illustrate the distribution of the bones of the skull, the axial skeleton, and the limbs, respectively.)

A word must be said here about three bones that may belong to the skeleton but were situated a long way from the main hominid concentration. These bones are definitely metapodials of a juvenile mammal. If they are hominid, then two are first metacarpals and the other is a first metatarsal. They were found in 1986 about 15 m northeast of the main hominid concentration. Nearby was an immature vertebra that was thought to be hominid but, when cleaned, proved not to be. All three metapodial bones are from immature individuals and lack their epiphyses and articular ends. What remains of the details of morphology is sufficient to cast doubt on their hominid status, but they cannot be securely placed in another taxon either. That they were widely separated from the main concentration of hominid bones is yet another hint that they might not belong to the hominid skeleton.

Bone breakage

Hardly any of the bones at the Nariokotome site are complete. The hominid bones are less broken, on the whole, than the others. Of all the hominid bones recovered, only one epiphysis—the distal humeral epiphysis—is partly fused. All major cranial sutures were unfused also. Those epiphyses that we did find were separated from their long bones. Some of the individual skull bones may have been isolated from each other. Some breakage has occurred following or during this disarticulation.

I believe, on the basis of geological and taphonomic evidence, that most of the breakage was caused by trampling by large mammals. First, hippopotamus and large bovid or giraffid footprints are visible in the tuff below (Unit 1), showing that trampling occurred in the area. Further, the tuffaceous silts that contained the hominid show an irregular upper contact that has all the appearance of footprint disturbance. No obvious trampling marks are visible on the bones, but this may be due to the fine, silty nature of the matrix. Second, several of the bones seem to have been broken where they lay. For example, the left fibula was found with one piece horizontally placed on the hard lower contact and the other piece nearly vertically next to it, as though the intact fibula had been lying in the mud and was stepped on from above. Similarly, the right scapula was found lying horizontally with its humerus lying across it, but the scapula had been crushed until the blade was a mosaic of small pieces held together by matrix.

None of the bones of the skeleton shows any sign of weathering. The fossil hominid does not match Behrensmeyer's (1978) Stage 0 weathering state as defined for recent bones—"Usually bone is still greasy, marrow cavities contain tissue, skin and muscle/ligament may cover all or part of the bone surface"—but it is at an equivalent stage with no cracking. Behrensmeyer reported that her lowest weathering stages (0–1) were more common in bones in the swamps she studied than in other local habitats. The habitat to which weathering evidence points—a marsh or swamp—agrees with the paleoenvironmental reconstruction of the Nariokotome III site.

There is only one bone of the skeleton that shows what might be tooth marks. This is the superior tibial epiphysis (I), which was found with its articular surface facing downward to the northeast of the main bone concentration. This bone is

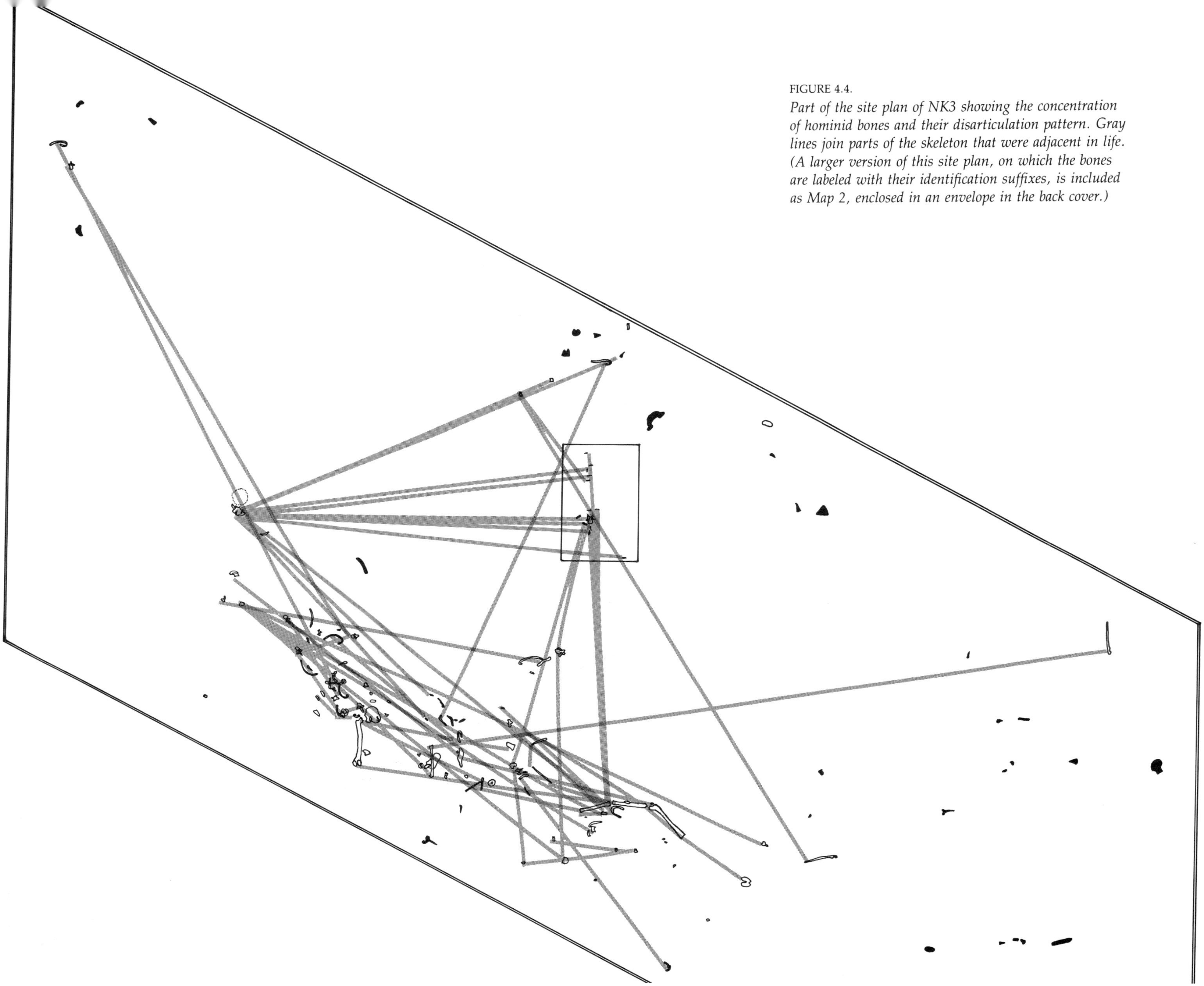

FIGURE 4.4.
Part of the site plan of NK3 showing the concentration of hominid bones and their disarticulation pattern. Gray lines join parts of the skeleton that were adjacent in life. (A larger version of this site plan, on which the bones are labeled with their identification suffixes, is included as Map 2, enclosed in an envelope in the back cover.)

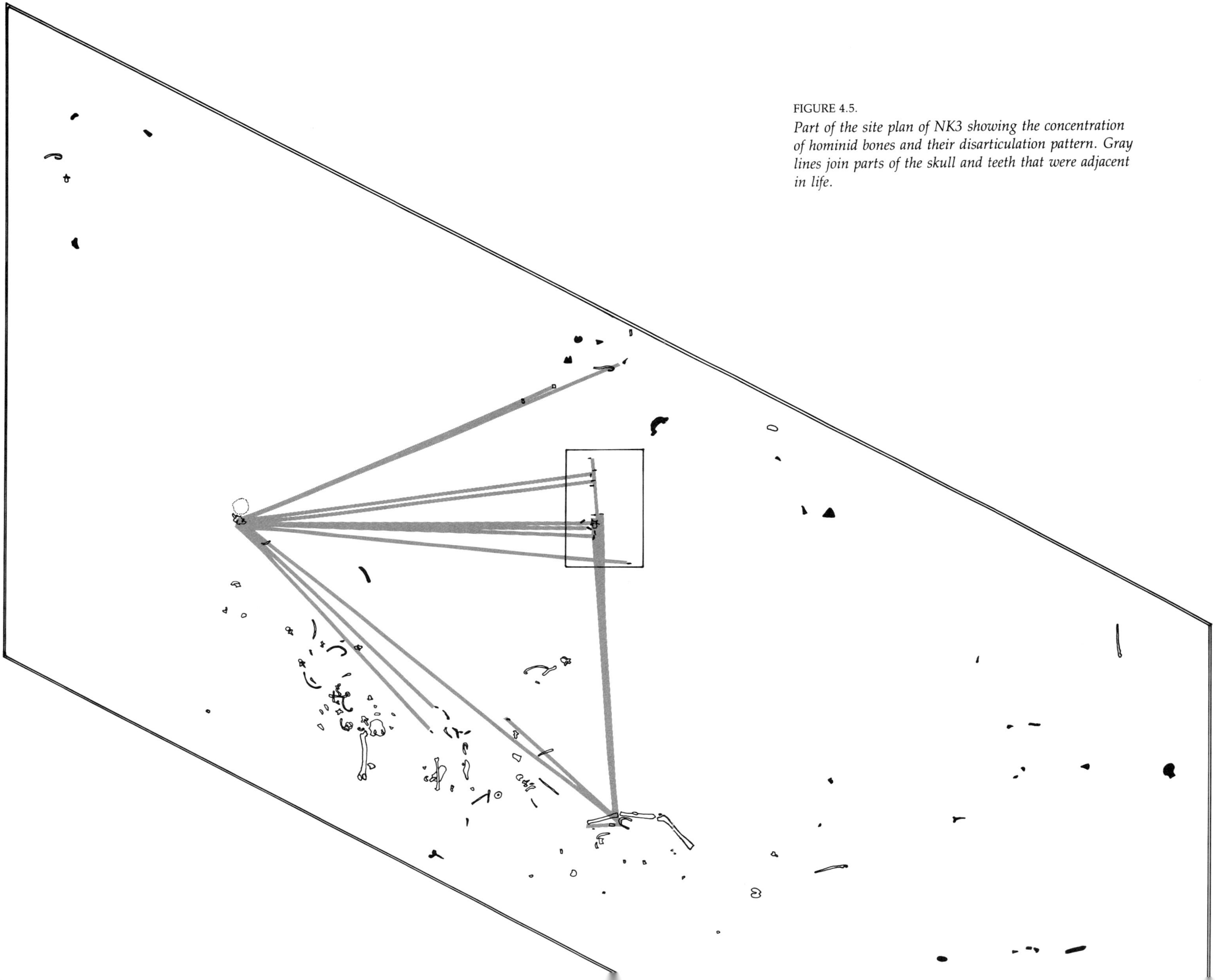

FIGURE 4.5.
Part of the site plan of NK3 showing the concentration of hominid bones and their disarticulation pattern. Gray lines join parts of the skull and teeth that were adjacent in life.

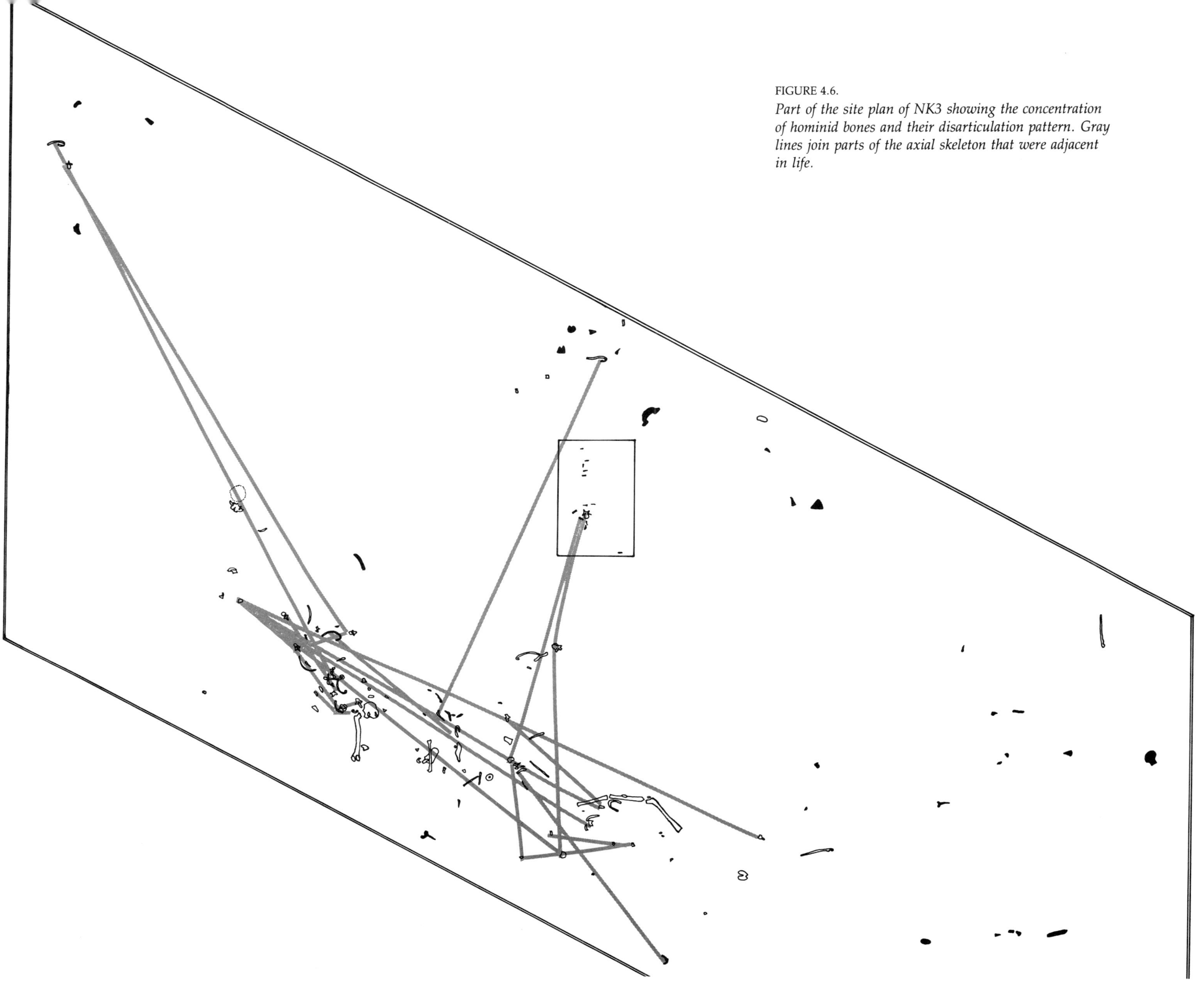

FIGURE 4.6.
Part of the site plan of NK3 showing the concentration of hominid bones and their disarticulation pattern. Gray lines join parts of the axial skeleton that were adjacent in life.

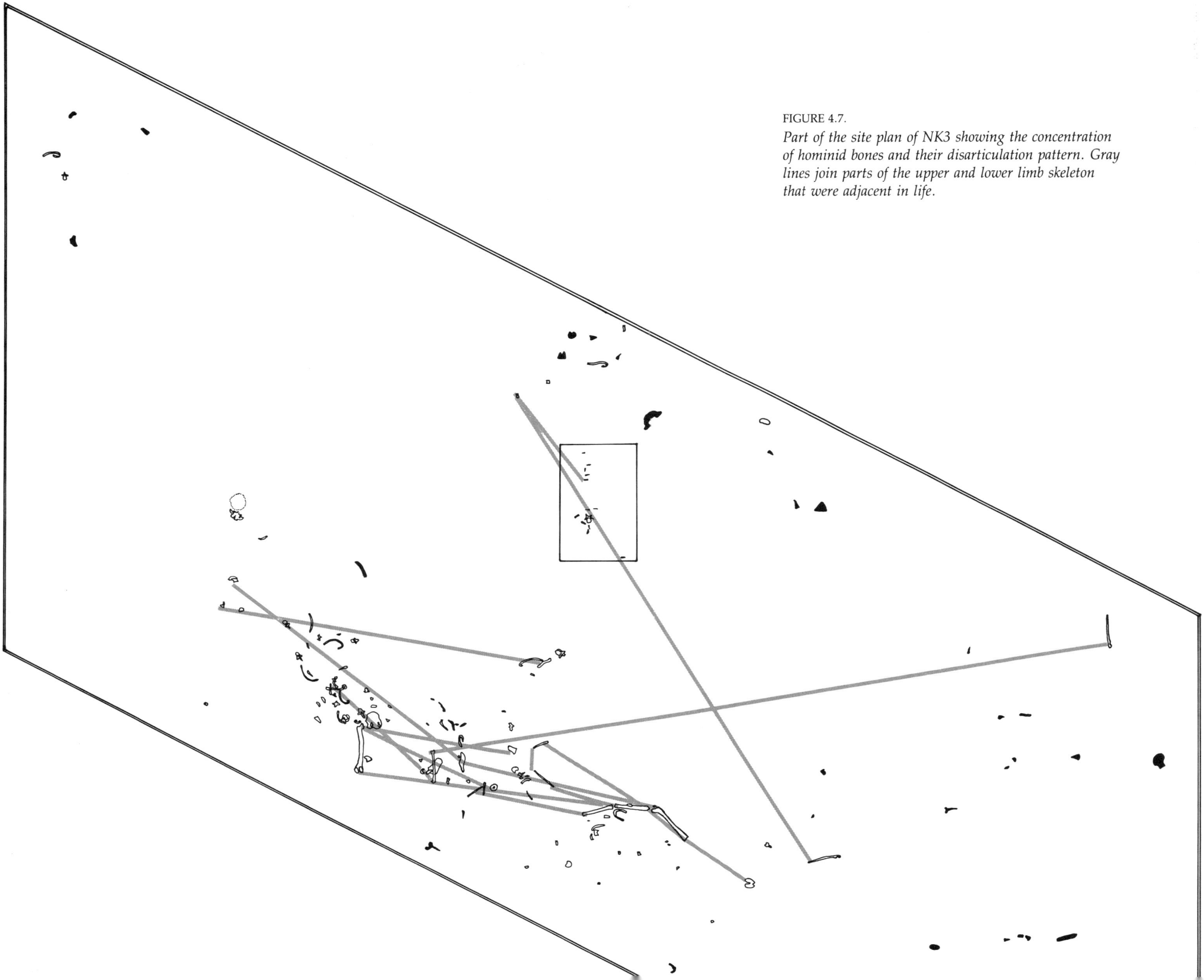

FIGURE 4.7.
Part of the site plan of NK3 showing the concentration of hominid bones and their disarticulation pattern. Gray lines join parts of the upper and lower limb skeleton that were adjacent in life.

eroded around its periphery and shows grooves about 1.0 mm wide on the medial condylar area as well as several smaller ones around the lateral condyle. In view of the fact that there are no tooth marks on any other bones of the skeleton, it is unlikely that these tooth marks were made by a mammalian carnivore. Since there is good evidence that the skeleton was dispersed in a swamp, it seems highly improbable that this epiphysis and only this epiphysis was on dry land. The marks may be the effects of abrasion or root marks, but they are too poorly preserved to say for certain.

Bone orientations

Ninety-four bones were of an elongated shape, and for these we recorded orientations. Other bones or bone pieces were either so curved or so irregular that their long axes could not be determined. Others were oriented nearly vertically. The orientations of these 94 bones are shown in Figure 4.8 as a mirror-image rose diagram. Most of the bones with clear orientations had horizontal or nearly horizontal dips. Two major preferred orientations stand out against the random background orientation. These are roughly north-south and east-west. This cross pattern has been reported upon before and has been interpreted in the light of experimental evidence by Voorhies (1969).

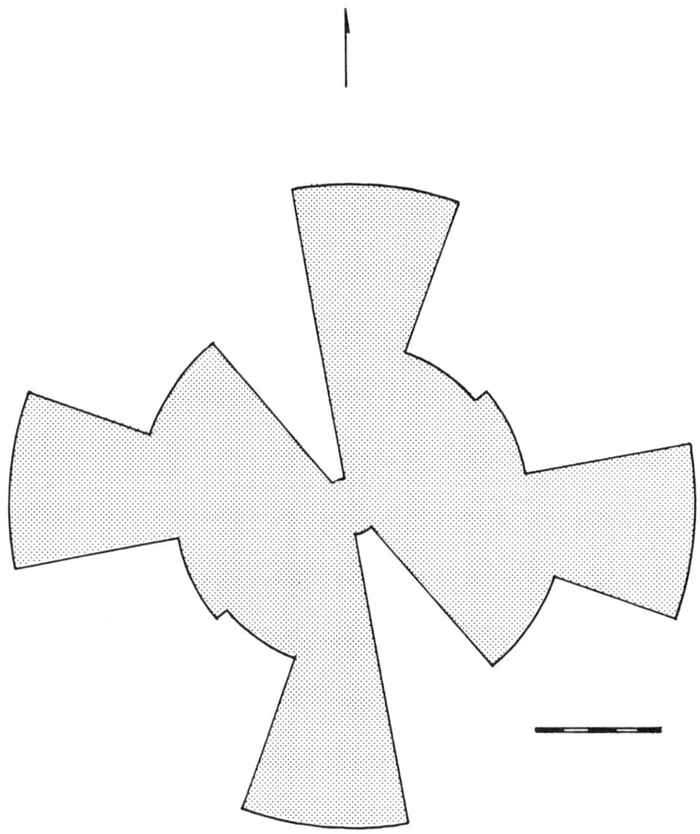

FIGURE 4.8.
Mirror-image rose diagram of the orientations of 94 bones at NK3. Arrow points to north. Scale represents 5 bones.

Voorhies placed bones in a flume (an artificial stream that can be made to contain sediment as well as water) and recorded their orientations at various current velocities. He found that, in shallow water, partly emergent bones showed a strong tendency to take up one of two perpendicular directions—either transverse or parallel to the current. The recorded bone orientations at NK3 are consistent with the conclusion that minor movements in shallow water followed disarticulation. The directional data themselves do not indicate which was the major and which the minor current direction, since with an almost symmetrical cross pattern the main current could have been from any of the four cardinal points of the compass. But the paleogeographic reconstruction provided in this volume by Feibel and Brown gives us a clear indication that the primary current direction is most likely to have come from the north with the main Omo River drainage.

The faunal remains from NK3

Table 4.3 lists the species found in Unit 2—the hominid-bearing level—at the site. The remains of the swamp snail *Pila ovata* caused a minor puzzle at first when we found opercula but no shells. We later learned that the opercula are made of calcite and the shells themselves made of aragonite. Different solubilities of the minerals lead to the preservation of only the opercula (Feibel and Brown, Chapter 3, this volume). After we knew this we were able to detect occasionally the partial natural molds of the inside walls of the shells. This gastropod, which is not common in the deposits of the region, is an air-breathing swamp snail present in the area today (P. G. Williamson, personal communication). Because it is air-breathing, *Pila ovata* inhabits only swamps or vegetated shores of very shallow, permanent or temporary water bodies (Brown, 1980). This fits in with the evidence of the orange root casts at the base of Unit 2 and the deduction by Feibel and Brown (Chapter 3, this volume) that standing vegetation must have been present in the swamp to provide a substrate for the many sponges that are a characteristic of the sediment.

Forty-one of the many opercula were complete enough to measure. Figure 4.9 is a scatter plot of their breadth against length. The relationship between length and breadth is remarkably strong ($r = .982$, $p < .001$). The regression shows that breadth increases at about two-thirds of the length (breadth = .627 × length − .441). Opercula grow by the addition of peripheral rings, and the data suggest that many growth stages are represented. As an annual breeder, *Pila ovata* would leave a disproportiate number of large adult opercula if the swamp in which they lived had lasted for very long. The sample is similar to that which one would get by dredging live snails from a present-day swamp (P. G. Williamson, personal communication). The size distribution of the NK3 *Pila* opercula suggests, then, that the swamp in which the hominid bones lay was of relatively short duration and may only have lasted a year or so.

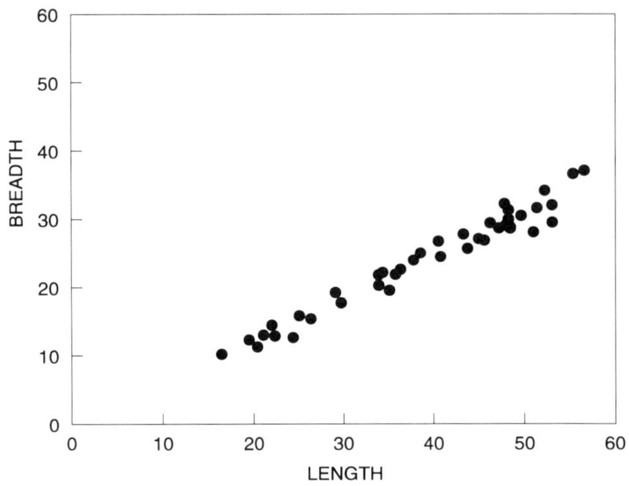

FIGURE 4.9.
Scatter plot of length and breadth (mm) of opercula of PILA OVATA *excavated from site NK3.*

TABLE 4.3.

Species found in Unit 2 at Nariokotome III.

Pila ovata	Air-breathing, shallow-water swamp snail
Clarias sp.	Shallow-water catfish
Clarotes sp.	Catfish
Hydrocynus sp.	Shallow- to deep-water fish predator
Synodontis sp.	Shallow-water spiny catfish
Varanus niloticus	Scavenging and often aquatic lizard
Trionyx sp.	Soft-shelled freshwater turtle
Pelomedusidae spp.	Smooth-shelled water tortoise
Homo erectus	Hominid (specific habitat unknown)
Metridiochoerus sp.	Grazing pig
Hippopotamus aethiopicus	Aquatic herbivore
Hippopotamus gorgops	Aquatic herbivore
Bovidae spp. (duiker- to buffalo-sized)	Grazing and browsing herbivores
Lepus capensis	Grass and herb feeder

The best-preserved of the rest of the fauna are all aquatic animals—either primarily aquatic, as in the case of the fish and turtles, or aquatic for part of the time, such as the monitors, wading birds, and hippopotamuses.

The Death and Burial of KNM-WT 15000

Synthesizing all this taphonomic evidence with the general and local geomorphological setting adduced by Feibel and Brown (Chapter 3, this volume) and the general faunal evidence, we can create a scenario of the circumstances surrounding this particular *Homo erectus* individual's sedimentary burial.

There are few indications of the cause of death of this individual. The only abnormal feature of the KNM-WT 15000 skeleton is a periodontal lesion on the right side of the mandible at the level of the P_4 and M_1. It is apparent that this pocket of inflammatory gum disease was related to the loss of the deciduous second molar, which had occurred not long before death. The deciduous tooth crown was lost in life, but two small indentations mark the site where bits of its root were still in place. These root fragments affected the eruption of the underlying premolar, causing it to rotate counterclockwise relative to the toothrow. It is possible that an inflammatory condition such as this could have led to the death of the boy. Before the advent of antibiotics about half a century ago, death from septicemia from tooth and gum abscesses was common.

The youth was living just before he died in the rich floodplain grasslands on the western side of the Omo River. The river's seasonal flood had left several large swamps that would take most of the year to dry out fully. The grasslands were home to many species of herbivore together with their attendant predators and scavengers. The youth died in or very near one of these swamps. We do not know the cause of death, but it may possibly have been septicemia following gum infection after the shedding of a milk tooth. The state of preservation of the bones indicates that he was not killed by a predator in all probability.

At death, the body either fell in the swamp or was later washed into it by a minor flood. For some time it floated face down in one spot while decomposing. Some of both upper and lower front teeth with simple, straight roots slipped from their sockets here. They were later washed into and around a hippo footprint. The body drifted, decomposing, a few meters to the southwest through the reeds. Bits and pieces of the rest of the body were washed back and forth, mostly in an east-west direction, and perhaps at the swamp's edge. These remains were kicked by passing large herbivores, sucked on by catfish, and chewed on by turtles. In time, only single bones or a few pairs of bones tied together by ligaments were left. Nearly all of these ultimately came to rest in one relatively small patch in a shallow part of the swamp, but one or two bones were washed further away. The bones became embedded in the swamp mud and remained there for over a million and a half years until they began to erode out of the sediments at the side of a small tributary of the Nariokotome.

Acknowledgments

Thanks are due to all the people, too many to name here, who helped with the excavation of Nariokotome III. Special thanks go to John Harris, Pieter Kat, Meave Leakey, Richard Leakey, Emma Mbua, Pat Shipman, and Peter Williamson for their help with the fauna and the taphonomy of the site. Rosa DaCosta and Hannah Grausz kindly helped with the graphics.

CHAPTER 5

THE FAUNAL CONTEXT

JOHN M. HARRIS AND MEAVE LEAKEY

Fossils of other animals found with the Nariokotome hominid skeleton may be used to infer the environmental conditions of the area immediately adjacent to the site 1.5 million years ago. Ungulate species are particularly useful indicators because they often have restricted ecological requirements. Sequential changes in the Pleistocene assemblages of the Turkana Basin reflect changes in both the local habitats and in the communities to which early Homo erectus *belonged. A survey of these changes gives a broader, dynamic view of the area's environment through time.*

Fossil assemblages from the Pliocene and Pleistocene strata exposed to the west of Lake Turkana have been investigated as an outgrowth of the Koobi Fora Research Project—an international and multidisciplinary investigation of the hominid-bearing sediments located near the northeastern shoreline of Lake Turkana in northern Kenya and pursued under the auspices of the National Museums of Kenya. The fossiliferous nature of the rocks exposed in the northwestern portion of the Lake Turkana Basin was confirmed late in 1980, preliminary investigations of the stratigraphy and faunal assemblages were initiated in 1981, and collecting began in earnest in 1984 (Harris et al., 1988).

The West Turkana sequence, like that of Koobi Fora to the east of the lake and of the lower Omo Valley to the north, constitutes a succession of fluviatile and lacustrine sediments intercalated with volcanic ashes. Some of the tuffs serve to define and correlate stratigraphic units from each of these three important fossiliferous areas and act as marker horizons for investigating changes in the faunal assemblages. Many of the fossils recovered from Omo, Koobi Fora, and West Turkana were encountered during the course of surface prospecting, but in most instances the stratigraphic provenance of such surface finds can be determined with respect to subjacent or superjacent tuffaceous horizons. An account of the distribution of fossil mammals within the West Turkana succession was summarized in Harris et al. (1988) and restated in somewhat greater detail in Harris, Brown, and Leakey (1988). For the site abbreviations given here the reader is referred to either of these publications. Nine different Plio-Pleistocene mammalian assemblages have been recognized from West Turkana, one from each member except for the Lomekwi Member (which has yielded three) and Lonyumun Member (which has yet to yield mammalian fossils). The purpose of this chapter is to document the fossil material recovered from the *Homo erectus* excavation site at Nariokotome III (NK3), along with surface finds from the same locality, and to compare the local assemblage from NK3 with others of similar age from elsewhere in the Lake Turkana Basin.

THE NK3 ASSEMBLAGE

Fossil material recovered from the *Homo erectus* (KNM-WT 15000) excavation at NK3 includes opercula of the gastropod *Pila ovata;* bones and teeth of several different kinds of fish (including *Clarotes* and *Hydrocynus*); fragments of trionychid and other chelonian carapaces; jaws, teeth, and vertebrae of the Nile monitor lizard *Varanus niloticus;* and some avian limb bones and phalanges; all were so identified by A. Walker. Mammalian fossil material includes fragments of an immature *Metridiochoerus* mandible, postcranial fragments of *Hippopotamus gorgops* and *H. aethiopicus,* and limb bones and vertebrae of at least two medium- to large-sized bovid species.

NK3 is not a richly fossiliferous locality, which makes the discovery of the *Homo erectus* skeleton all the more remarkable. Among surface materials recovered from elsewhere in this locality are KNM-WT 16001, *Homo erectus* cranial fragments (an individual from a geological horizon later than the level of the excavation); KNM-WT 14865, an upper molar fragment of *Equus* sp.; KNM-WT 17484, a *Kolpochoerus limnetes* M^3 fragment; and KNM-WT 19607, *Megalotragus* sp. horn core fragments.

Most of these mammalian genera have been documented from elsewhere in the Natoo Member, but NK3 is currently the only site in this member from which hominids have been recorded. Two *Metridiochoerus* species, *M. compactus* and *M. hopwoodi,* are documented elsewhere in the Natoo Member by surface specimens while a third—*M. modestus*—has been recovered elsewhere in the basin from strata of equivalent age. The specimen from the excavation is too immature and too incomplete for more positive identification. Likewise, two species of *Equus* occur in the Natoo Member, one of similar size to *E. burchelli* and the other to *E. grevyi,* and others are documented from the correlative unit (Okote Member) in the Koobi Fora Formation; the NK3 specimen is too incomplete for more precise identification. The *Megalotragus* horn cores are the only ones known from the Nachukui Formation. Elsewhere in this unit

TABLE 5.1.

Fossil mammals from the Natoo Member.

Hystrix sp.	*Tragelaphus strepsiceros*
Theropithecus oswaldi	*Pelorovis turkanensis*
Cercocebus/Colobus sp.	*Pelorovis* sp. nov.
Homo erectus	*Hippotragus gigas*
Canis cf. *C. mesomelas*	*Kobus sigmoidalis*
Carnivora indet.	*Kobus kob*
Hipparion cf. *H. ethiopicum*	*Menelikia lyrocera*
Equus cf. *E. grevyi*	Reduncini (medium)
Equus cf. *E. burchelli*	Reduncini (small)
Equidae gen. indet.	*Megalotragus* sp.
Rhinocerotidae gen. indet.	*Connochaetes* sp.
Kolpochoerus limnetes	*Damaliscus* sp.
Metridiochoerus compactus	Alcelaphini (medium)
Met. hopwoodi (?)	Alcelaphini (small)
Hexaprotodon karumensis	*Aepyceros melampus*
Hippopotamus gorgops	*Gazella janenschi*
H. aethiopicus	*Gazella praethomsoni*
Sivatherium maurusium	*Gazella* sp.
Giraffa sp.	*Caprini* gen. indet.

the taxon is represented by teeth. The *Megalotragus* horn core fragments appear to belong to a new species recorded from the Koobi Fora Formation (Harris, 1985) rather than to the species *(M. kattwinkelli)* known from sites farther south, such as Olduvai Gorge (Gentry and Gentry, 1978).

Assemblages from the Natoo Member

Twelve sites have so far yielded fossil remains from the Natoo Member—KL1, KL2, KL5, LK3, LP1, NA1, NN1, NK2, NK3, NK4, and NT1—but only three of these have yielded more than a dozen specimens each (KL1 = 87, NA1 = 25, LK3 = 14). A list of the species recovered to date from the Natoo Member is given in Table 5.1. Fewer mammalian species have been recovered from the Natoo Member than from the subjacent Kaitio Member. Species present in the Kaitio Member but not recovered from later horizons west of Lake Turkana include *Tatera* sp., *Theropithecus* sp., *Australopithecus boisei, Hyaena hyaena, Homotherium problematicus, Deinotherium bozasi, Elephas recki ileretensis, Hipparion cornelianum, Notochoerus scotti, Metridiochoerus andrewsi, Antidorcas recki,* and Caprini sp. B. The absence of *D. bozasi, Loxodonta* sp., *E. recki ileretensis, N. scotti,* and *M. andrewsi* from the Natoo Member appears chronologically significant because these species are not found at correlative or later horizons in the Lake Turkana Basin. However, *Australopithecus, Hyaena, Homotherium,* and *Antidorcas* all occur at horizons equivalent to the Natoo Member in the Koobi Fora region.

Mammal species that are absent from the Natoo Member but present both earlier and later in the succession west of Lake Turkana include *Diceros bicornis, Metridiochoerus modestus,* and *Tragelaphus scriptus.* All three are infrequent constituents of the Turkana Basin Plio-Pleistocene biota. The rhino and bushbuck are closed-habitat species, as *M. modestus* may have been. Their absence from the Natoo Member might be environmentally significant, but the samples from this unit are too small to be conclusive.

Canis cf. *C. mesomelas, Sivatherium maurusium, Hippotragus gigas,* and *Kobus sigmoidalis* are not known later in the West Turkana sequence than the Natoo Member. *Kobus kob* and a small cercopithecoid, currently of uncertain identification (*Cercocebus* sp.), are presently restricted to this member. These distributions may also be an artifact of preservation and/or collecting. The sivathere, jackal, and kob persisted elsewhere in Africa, and the last two are extant species. *K. kob* has been recovered from earlier horizons east of Lake Turkana and is the most common bovid in the temporally equivalent Okote Member. Both *Cercocebus* and *Colobus* are represented in the Okote Member.

Mammalian species absent from the Natoo Member and occurring for the first time at West Turkana in the superjacent Nariokotome Member include *Elephas recki recki, Hippopotamus amphibius, Giraffa* cf. *G. camelopardalis, Syncerus* cf. *S. caffer, Kobus ellipsiprymnus, Kobus leche,* and Caprini sp. A. Of these species, *E. recki recki* and *Kobus leche* are both recorded from the Okote Member in the eastern part of the Lake Turkana Basin and *K.*

ellipsiprymnus has been documented from earlier horizons at Koobi Fora and Omo. Their absence from the Natoo Member may be environmentally significant but could be due to the small sample size of the biota from this unit. Caprines are not presently known from the Koobi Fora sequence; their occurrence in the West Turkana assemblages may be environmentally significant but they are rare. The presence of extant species of hippo, giraffe, and buffalo in the Nariokotome Member, but not at older horizons in the Lake Turkana Basin, appears chronologically significant.

Assemblages from the Okote Member

The Natoo Member is stratigraphically and temporally equivalent to the Okote Member of the Koobi Fora Formation but has a less diverse fauna. A list of the species recovered to date from the Okote Member is given in Table 5.2. Present in the Okote Member assemblages but unrecognized so far from the Natoo Member are: the rodents *Tatera* sp., *Aethomys* sp., *Jaculus orientalis,* and *Thryonomys* sp.; the hominid *Australopithecus boisei;* the elephant *Elephas recki recki;* the suid *Metridiochoerus modestus;* and the bovids *Kobus leche, Beatragus antiquus, Antidorcas recki,* and *Gazella* aff. *G. granti.* Conversely, present in the Natoo Member but missing from the Okote Member are: the jackal *Canis* cf. *C. mesomelas;* and the bovids *Hippotragus gigas, Gazella janenschi,* and Caprini gen. indet. It is possible that some of the fragmentary carnivore, equid, rhino, *Giraffa,* and *Damaliscus* material from the Natoo Member may belong to species that are represented in the Okote Formation by more complete and identifiable remains.

Some of the apparent differences in distribution may be linked to the larger size of the sample from the Okote Member (the Okote sample is larger because the areas of exposure of strata from this time interval on the eastern side of Lake Turkana are larger and the history of collecting from the Koobi Fora Formation is much longer. However, some differences may be due to ecological dissimilarity between the two areas. It is interesting in this respect that both *Hippotragus gigas* and *Gazella janenschi* do not persist above the KBS Member at Koobi Fora and that no fossil caprine has yet been recognized in the strata to the east of the lake. Although present at Koobi Fora from the Upper Burgi Member through the Okote Member, *K. leche* has not been found at West Turkana below the Nariokotome Member. Two small species of *Equus*—*E. tabeti* and *E. burchelli*—have been recognized in the Okote Member; west of the lake, small *Equus* teeth (*E.* cf. *E. burchelli*) have been recovered only from localities along the Laga Kalachoro (KL1, KL6).

TABLE 5.2.

Okote Member faunal list.

Tatera sp.	*M. modestus*
Aethomys sp.	*Hexaprotodon karumensis*
Thallomys quadrilobatus	*Hippopotamus gorgops*
Jaculus orientalis	*Hip. aethiopicus*
Thryonomys sp.	*Sivatherium maurusium*
Hystrix sp.	*Giraffa jumae*
Theropithecus oswaldi	*G. stillei*
cf. *Parapapio* sp.	*G. pygmaea*
Cercocebus sp.	*Giraffa* sp. indet.
Colobus sp.	*Tragelaphus strepsiceros*
Australopithecus boisei	*Pelorovis oldowayensis*
Homo erectus	*Pelorovis turkanensis*
Genetta genetta	*Kobus sigmoidalis*
Dinofelis piveteaui	*Kobus* aff. *K. leche*
Panthera sp. A	*Kobus kob*
Panthera sp. B	*Menelikia lyrocera*
cf. *Homotherium problematicus*	Reduncini gen. & sp. indet.
cf. *Megantereon eurynodon*	*Megalotragus isaaci*
Crocuta crocuta	*Connochaetes gentryi*
Hyaena hyaena	*Damaliscus eppsi*
Elephas recki recki	*Damaliscus* sp. nov.
Ceratotherium simum	Alcelaphini (medium)
Hipparion cf. *H. ethiopicum*	*Beatragus antiquus*
Equus cf. *E. tabeti*	Alcelaphini (small)
Equus cf. *E. grevyi*	*Aepyceros melampus*
Equus cf. *E. burchelli*	*Antidorcas recki*
Equidae gen. & sp. indet.	*Gazella praethomsoni*
Kolpochoerus limnetes	*Gazella* aff. *G. granti*
Metridiochoerus compactus	*Gazella* sp. indet.
M. hopwoodi	

PALEOENVIRONMENTAL INTERPRETATION

Fossil material from the *Homo erectus* excavation comprises a mixture of aquatic and terrestrial species. The only macroscopic invertebrate found is *Pila ovata,* a freshwater snail that is represented by its (calcitic) opercula but not by its (aragonitic) shells. This species is today widely distributed in tropical Africa, where it occupies a variety of niches ranging from permanent swamp to ephemeral aquatic habitats (Brown, 1980). The fish include catfish *(Clarias, Clarotes)* and tigerfish *(Hydrocynus),* which are present today in Lake Turkana. Soft-shelled, side-necked, and hinge-backed turtles are all found today in the vicinity of Lake Turkana (Roger C. Wood, personal communication), but no diagnostic material has been recovered from the excavation locality. The Nile monitor *(Varanus niloticus)* is normally found close to water (Fitzsimons, 1943) and lives in the immediate area today. The exaggeratedly elevated orbits of *Hippopotamus gorgops* are indicative of aquatic adaptation (Coryndon, 1976). It seems probable that *H. aethiopicus* was similarly adapted, though the one known skeleton of this species, recovered from NA1 in 1988, has yet to be prepared and studied.

Metridiochoerus was a hypsodont distant relative of the extant warthog. It seems likely that the two large species of this genus present in the Natoo Member were open-country grazers, but it is possible that the smaller *M. modestus* required denser vegetation for shelter and protection against predators; which species is represented by the very immature individual found at the excavation site has not been determined. The bovid postcranial remains have not been identified to tribe.

The composition of this small assemblage is consistent with a paludal environment, as suggested by the sedimentary context. Orientation of and damage to some of the hominid skeletal elements are consistent with the proposition that the bones were trampled by terrestrial mammals that passed through the swamp (see Chapter 4, this volume). Tracks and prints of a variety of mammals also occur in a tuffaceous horizon that underlies the unit from which the skeleton was recovered.

Stratigraphic evidence suggests that during the Plio-Pleistocene, as today, the Labur and Murua Rith Ranges formed the western margin of the Lake Turkana Basin and were drained by eastward-flowing rivers that fed into a forerunner of the present lake or a major river system (Harris et al., 1988). Paleogeographic reconstruction of the northern part of the Lake Turkana Basin during the time of accumulation of the Natoo and Okote Members (1.65–1.39 Ma: Feibel, 1988; Chapter 3, this volume) indicates the presence of a lake less than one-fourth the size of the present-day Lake Turkana at the onset of this depositional episode. The small size of the body of water might at least in part have been due to partial divergence of its source (the forerunner of the Omo River) via Sanderson's Gulf to contribute to the Nile drainage. The lake evidently expanded during the interval 1.65–1.49 Ma but thereafter contracted and had temporarily disappeared by the time of accumulation of the Chari Tuff (1.39 Ma), and the main drainage of this forerunner of the Omo extended nearly as far south as Nariokotome before diverting eastward to exit the basin.

Those extinct species that have close living relatives with known habitat preferences offer some clue to the environment that prevailed when the fossil species were part of the biota. The Pliocene and Pleistocene bovids from African fossiliferous localities have been particularly helpful in this respect, at least at the tribal level. The bovid samples retrieved from West Turkana localities are of similar composition to those that might be present in the area today were the climate sufficiently humid to support perennial rivers. Thus, regardless of geologic horizon, closed-habitat browsers such as tragelaphine bovids and impalas are abundant at those West Turkana localities adjacent to the hills that form the western margin of the basin. Water-frequenting reduncines predominate at localities that represent the courses of former tributary rivers. Open-habitat grazers, such as alcelaphines and antilopines, are concentrated in those areas representing the lake shore or the floodplain of the ancestral Omo River. Of the four members

TABLE 5.3.

Distribution of bovid tribes in the later parts of the Nachukui and Koobi Fora Formations (%).

TRIBE	KAL	UBU	KAI	KBS	NAT	OK	NAR
Tragelaphini	9	15	32	13	5	21	20
Aepycerotini	11	15	12	9	8	3	12
Bovini	6	3	12	7	9	11	12
Reduncini	38	46	14	37	37	41	28
Alcelaphini	20	18	24	24	23	19	20
Antilopini	17	11	7	9	16	5	8
Total sample size	71	449	89	839	75	155	25

Note: KAL = Kalochoro Member; UBU = Burgi Member; KAI = Kaitio Member; KBS = KBS Member; NAT = Natoo Member; OK = Okote Member; NAR = Nariokotome Member.

in the upper part of the Nachukui Formation (Kalochoro, Kaitio, Natoo, and Nariokotome), the Natoo Member has the smallest proportion of tragelaphines and impalas and the largest percentage of reduncines (Table 5.3). This distribution can be attributed to the fact that the fossiliferous localities in the Natoo Member are sited farthest from the western basin margin and close to the courses of three major rivers (Kalochoro, Nariokotome, and Lowarengak).

The eastern margin of the basin is less clearly defined and the effect of the paleogeography on the distribution of fossil bovid assemblages has yet to be thoroughly investigated. Interpretations by Feibel (1988) and earlier workers suggest that during the Upper Burgi Member much of the present Koobi Fora region was inundated by a large lake and that many of the fossiliferous localities are associated with deltaic accumulations formed by the Omo and other rivers. Thereafter the lake level fluctuated, but the lake shrank during the time represented by the KBS Member and had virtually disappeared before the Chari Tuff was deposited. Most of the fossiliferous sites in the KBS and Chari Members are in fluviatile (channel and overbank) deposits. Tribal representations at individual sites vary greatly. The regional distribution is more uniform although sites along what is now the Karari Ridge were perhaps clothed with more dry grassland and less woodland than other regions during Upper Burgi and KBS Member times. Okote Member assemblages are known mainly from the Ileret and Koobi Fora regions. The Ileret assemblages suggest a greater proportion of wet grassland, whereas those from Koobi Fora suggest a greater proportion of dry grassland and woodland.

Within this context, the large bovid samples from the Koobi Fora Formation (Table 5.3) provide some indication of the environmental changes that took place in the basin during the early part of the Pleistocene. For example, on the east side of Lake Turkana reduncines are the most common tribe throughout the upper part of the Koobi Fora Formation. But whereas the waterbuck-like *Kobus sigmoidalis* and the extinct *Menelikia lyrocera* predominate in the Upper Burgi Member, the proportion of other reduncine species increased in the KBS Member and *Kobus kob* became by far the most common reduncine in the Okote Member, signifying a progressive change in the overall ecology through time. The proportion of bovines represented in the Koobi Fora assemblages increased during the same interval while that of the antilopines decreased, suggesting that the samples were derived from progressively wetter habitats. The proportion of alcelaphines in the biota is relatively constant throughout the upper part of the succession, but an increase in representation of the smaller genera *(Damaliscus, Beatragus, Parmularius)* in the KBS and Okote Members implies a change in the nature of the available grassland, thereby corroborating inferences drawn from other tribes. The Upper Burgi Member assemblage appears to have been associated with a major delta produced by the Omo River at its point of entry into Lake Turkana; the later assemblages are also derived from floodplain and deltaic deposits but come from a time when at least part of the Omo's waters were diverted elsewhere. Perhaps the resultant decrease in extent of the Omo gallery forest explains the smaller percentage of impalas encountered in the KBS and Okote Members.

The Natoo Member bovid assemblage differs from both the temporally equivalent Okote Member assemblage and that of the superjacent Nariokotome Member by having a comparatively larger percentage of open and dry grassland bovids (alcelaphines and antilopines) and fewer woodland or closed-habitat indicators (tragelaphines and impalas). The main difference between the Okote and Nariokotome Members is that samples from the latter (later) member have yielded more impalas but fewer reduncines, suggesting a greater proportion of woodland than in either the Natoo or Okote Members.

The predominance of the kob in assemblages from the Natoo and Okote Members appears significant. This short-grass grazer has a proven physiological need for water (Shoen, 1971) and today requires low-lying flats or gently rolling country without severe seasonal extremes and close to permanent water (Kingdon, 1982). All three prerequisites would have been present in the Lake Turkana Basin adjacent to the river or lake during the accumulation of the Natoo and Okote Members. Under conditions of drought, very large concentrations of kob may gather in low-lying areas where the water table is higher and where green grass is available (Kingdon, 1982); such conditions may also have prevailed on occasion during the later part of the sequences in the Lake Turkana Basin. Kobs are today associated with several larger mammals over the greater part of their range—hippopotamus, topi, hartebeest, buffalo, and cattle—on which they depend for grazing maintenance (Kingdon, 1982); close relatives of these large mammals occur in association with the kob in the Natoo and Okote Members.

Most of the larger herbivorous mammals recovered from both the Natoo and Okote Members were grazers and indicate the presence of both wet and dry grassland on either side of the basin. That woodland was also present is denoted by the occurrence of tragelaphines and impalas among the bovids and is further supported by the primate and giraffid components of the samples, and perhaps by *Kolpochoerus*. It is interesting, though perhaps not significant given the relative size of the samples, that a greater variety of these closed-habitat indicators has been recovered from the east side of the basin than from the west. If a greater proportion of the east side of the basin was given over to woodland or thicket vegetation than in the west, however, it is possible that the absence of *Elephas, Diceros bicornis, Tragelaphus scriptus, Kobus ellipsiprymnus,* and *Australopithecus boisei* from the Natoo Member may not merely be an artifact of sampling. It might also explain why *Hippotragus gigas* and *Gazella janenschi* occur in the (less wooded) KBS and Natoo Members but not the Okote Member.

Summary

Site NK3, from which the *Homo erectus* skeleton was excavated, patently represents a palustrine environment; this conclusion is supported on the bases of the sedimentary facies and the nature of the recovered fossils and on the evidence of trampling both of the hominid skeleton itself and in the underlying sedimentary horizons. The swamp was located in the floodplain of the Omo River system in which, the mammalian fossils suggest, grassland was interspersed with patches of woodland and/or thickets and probably with gallery forest along the main and tributary rivers. The composition and spatial distribution of the West Turkana bovid samples suggest the presence of woodland or thickets on and near the hills that formed the western basin margin. However, comparison of the biota recovered from the temporally equivalent Okote Member indicates that the western part of the basin was, in general, less densely vegetated than the eastern part.

PART II

THE SPECIMEN

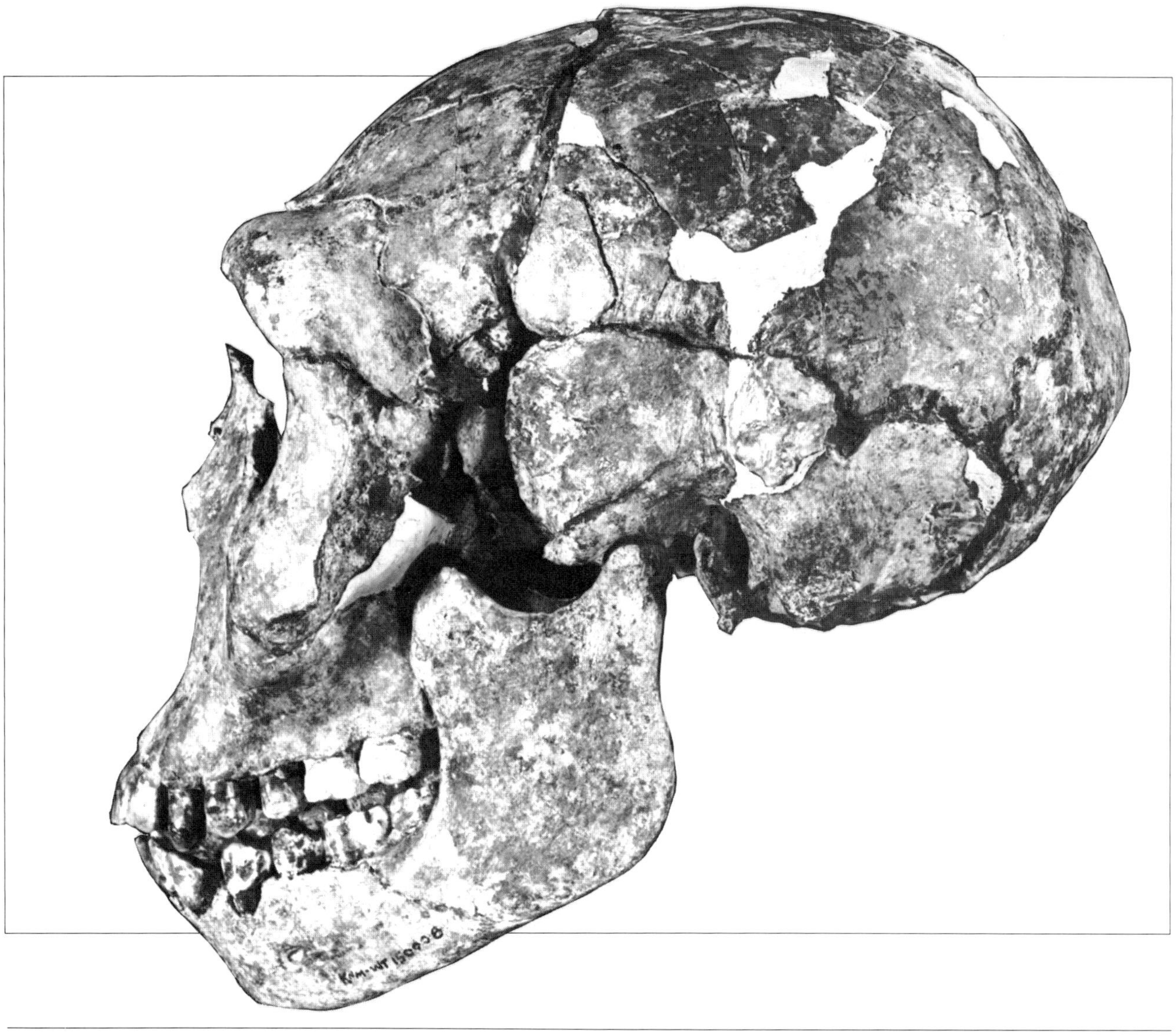

CHAPTER 6

THE SKULL

ALAN WALKER AND RICHARD LEAKEY

The skull of the Nariokotome early Homo erectus *is nearly complete. In this chapter we present descriptions and photographs that document each cranial bone as well as the skull as a whole. This specimen allows us to interpret more fully the other, less complete, hominid skulls found in the Turkana Basin by previous expeditions. Early* Homo erectus *from Africa show considerable cranial variability.*

The skull of KNM-WT 15000 was recovered in about 70 pieces, not counting isolated teeth. The first parts of the skull to be discovered were small fragments of vault. These were found in the soil and pebble lag on the slope below the main fossiliferous horizon. Because they had been exposed on the surface for some time, they did not have much calcitic sediment covering them, but they were chipped and some had corners missing. The broken edges were relatively clean and had no sediment adhering to them. The clean breaks meant that sticking together the small pieces found by screening and washing was relatively easy. The individual vault bones were assembled in this way.

Two large pieces of the skull were never found and, presumably, were washed downstream. The first is a triangular piece of the frontal including the coronal suture on the right side from the midline to the temporal line. The second is a large triangular piece of the right parietal which includes sagittal and lambdoid sutures. It is quite easy to allow for these missing pieces in the endocranial reconstruction because nearly all of the same parts are present on the other side of the cranium.

The mandible, temporals, zygomatics, parts of the sphenoid, nasals, and maxillas were all found with a covering of hard matrix on them. This was removed with pneumatic Airscribes with sharpened carbide tips, and during this procedure particular care was taken to make sure that no matrix was left on the surfaces of the sutures. Several teeth had become separated from the skull (see Chapter 4, this volume). The alveoli for these teeth were filled with matrix. This was removed with Airscribes and a turbo drill with small, pointed diamond dental bits attached to it. It was possible to align the upper and lower teeth in centric occlusion once all the teeth were back in position. This set the angle at which the maxillas faced each other, and the alignment was confirmed by setting the nasals in between them. It was clear, at this stage, that some allowance had to be made for soft tissue that had been in the sutures. The tissue was represented by hard sculptor's wax on the reconstruction. The next step was the placement of the cranial vault part on the mandibular condyles after allowing for the thickness of the articular disk and cartilage in the temporomandibular joint. The set of the face on the cranium was then established by matching the zygomatic sutures with their counterparts on the frontal and maxillas. Some pieces of the cranium cannot be attached. These include bits of the supraorbital tori and parts of the sphenoid.

There is still some plastic deformation to be seen in the skull. In the facial skeleton, the internasal suture is set off from the plane of the intermaxillary suture by about 4.0 mm. The tips of the frontal processes of the maxillas are at different heights. The left is higher than the right by 4.5 mm. There is distortion of the vault, with the top of the cranium moved to the left, which makes the lateral side of the left parietal steeper than it should be and the lateral side of the right parietal less steep. The frontal shows the same distortion, so that the zygomaticofrontal suture on the left is higher than on the right. There is also some rotation of the back of the skull so that a midline passing through the occipital passes through the lateral wall of the right greater palatine foramen.

Unless otherwise noted, all measurements in this chapter are in millimeters. All illustrations of the bones and fragments of KNM-WT 15000 are reproduced at natural size except where noted.

THE CRANIUM IN GENERAL

(FIGURES 6.1–6.6)

The estimated total length of the cranium as it stands is 175 from glabella to inion. Note that glabella can be estimated by examining the curvature of the nasals and the lateral parts of the supraorbital tori which are preserved. Nasion to inion measured directly is 170. The maximum biparietal width of 137 is found low on the bones near asterion. The minimum frontal width is 90.5. Other dimensions, along with comparative measurements from the other early African *H. erectus* specimens, are given in Table 6.1.

The calvaria is sphenoidal in Sergi's terminology (quoted in Martin, 1928) and the maximum cranial breadth is situated far posteriorly. The vertex of the skull is coincident with bregma, but this

TABLE 6.1.

Cranial measurements (mm) of KNM-WT 15000 and other *Homo erectus* specimens.

MEASUREMENT	KNM-WT 15000	KNM-ER 3773	KNM-ER 3883
Length (glabella–inion)	175	183	181.5
Maximum cranial width	141	144	136
Minimum frontal width	90.5	92	88.1
Frontomalare temporale–frontomalare temporale	111.8	(118)	—
Auriculare–auriculare	126.3	132	(133)
Bimastoid breadth	136.8	142	(135)
Ectoglenoid process–ectoglenoid process	122.2	126	—
Entoglenoid process–entoglenoid process	78.5	65	70
Foramen ovale–foramen ovale	54.3	—	53
Asterion–asterion	106	123	115
Orbitale–orbitale	83.5	(85)	—
Zygomatic foramen–zygomatic foramen	111.0	107	—
Bregma–opisthion	129.0	(125)	—

coincidence is caused by a bregmatic swelling on the parietals. Ectocranial superstructures and muscle scars are poorly developed because the skull is that of a juvenile (see Smith, Chapter 9, this volume). Despite this immaturity, the frontal and maxillary sinuses are well developed: the former have invaginated into the brow ridges and the latter have penetrated into the palatine processes. It must be pointed out that, as so often is the case in modern human crania, the left frontal sinus is well developed and the right hardly at all. The sphenoid air sinus had penetrated nearly to the foramina ovale and even the mastoids had developed a pyramidal shape. It is surprising that this amount of pneumatization is present in an individual as young as this. This probably means that the skull would have been very pneumatized had the individual lived to maturity. The cranial outline is a rounded pentagon in posterior view.

The face is large and relatively long. Alveolare to nasion is estimated at 78. This distance was estimated because of the gap between the maxillas where the soft tissue of the intermaxillary suture would be. The bi-frontomalare temporale distance is 111.8. The bi-zygomatic width is 121.6 with no adjusting for bone loss on the left. This must be regarded as a minimum.

The facial skeleton has an interesting profile. Although only the upper halves of the nasal bones survive, their sutural margins on the maxillas are complete. It is certain that the nasals were concave along their length and must have formed a peak in the middle of the profile. The lateral margins of the pyriform aperture are concave in profile and continue almost in a direct line to the anterior margins of the lateral incisors, while there is a very small amount of subnasal prognathism seen in the convexity of the bone over the central incisor roots. The projection of the lower parts of the nasals and the concomitant projection of the frontal processes mean that the frontal processes themselves face anterolaterally. There is a concave break in angle at the position of the infraorbital foramen marked by a groove inferior to it, so that the surface of the maxilla lateral to the break faces more nearly anteriorly. There is then another angulation at the inferolateral corners of the orbits. This is, however, a convex angle, and here the zygomatic bones fold backwards so that their lateral surfaces face almost completely laterally. This second angulation is continuous with the lateral orbital margin.

The pyriform aperture has fairly evenly curved lateral margins that fade out onto the subnasal alveolar region. This makes the inferior border of the aperture blunt. The anterior nasal spine (still bifid at this young age) marks the inferior margin below. The aperture is estimated to be 27 (prosthion to nasospinale) and has a greatest width of 34.7.

The orbits are parallelograms in outline, with both superior and inferior margins sloping downwards and laterally, the medial margin concave laterally, and the lateral margin straight in anterior view. The orbital height is estimated at 38 and the breadth at 35. In side view the orbital margins are almost evenly concave, with the greatest depth of the profile just below the frontozygomatic suture. The take-off of the zygomatic root on the maxilla is blunt, and the inferior margin of the zygomatic itself, seen only on the left side,

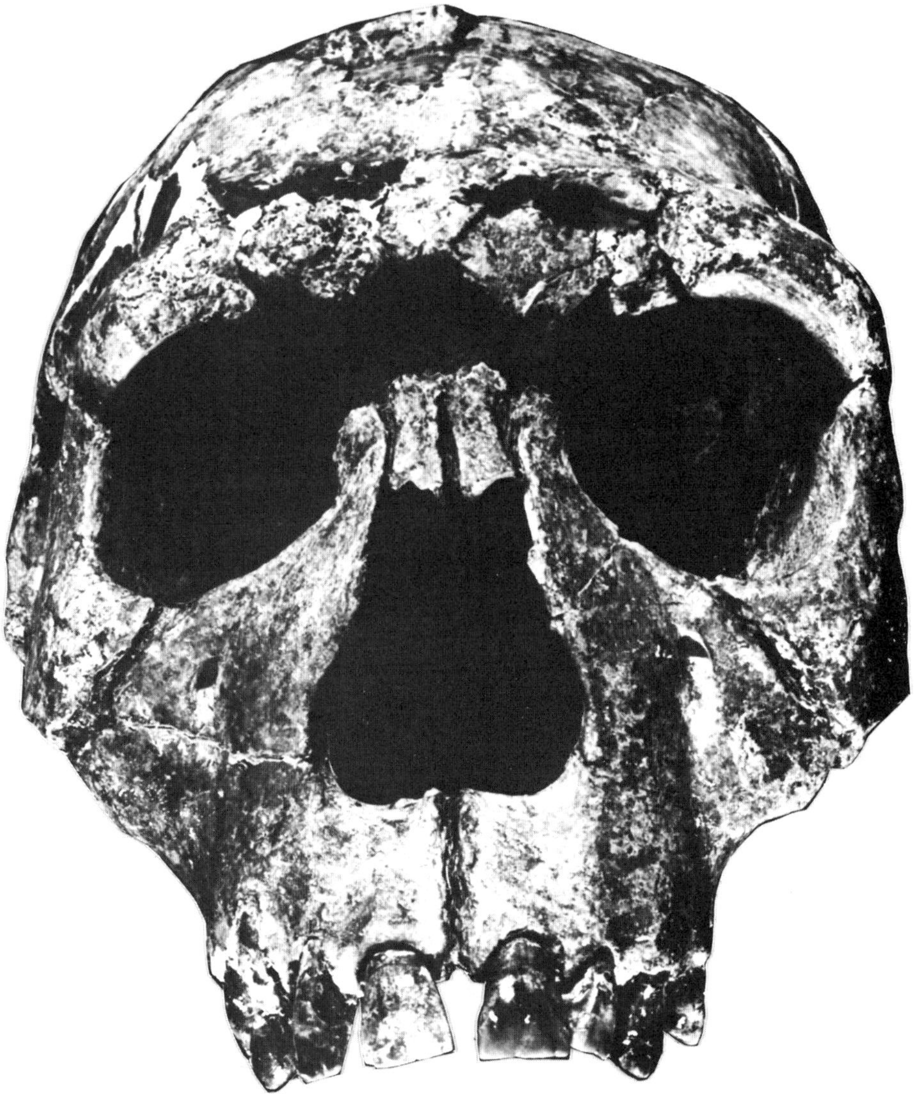

FIGURE 6.1.
Anterior view of cranium KNM-WT 15000 A.

FIGURE 6.2.
Posterior view of cranium KNM-WT 15000 A.

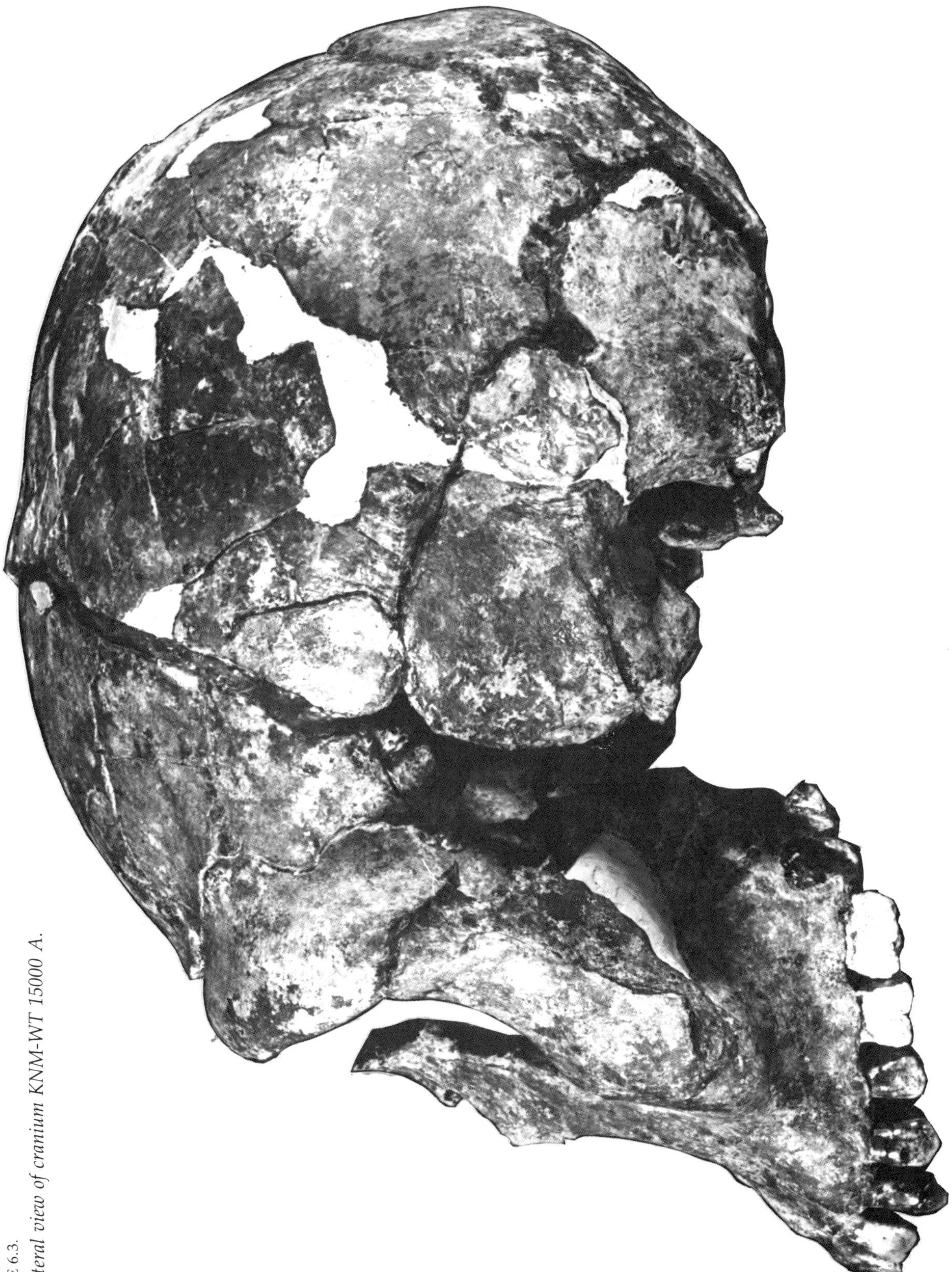

FIGURE 6.3.
Left lateral view of cranium KNM-WT 15000 A.

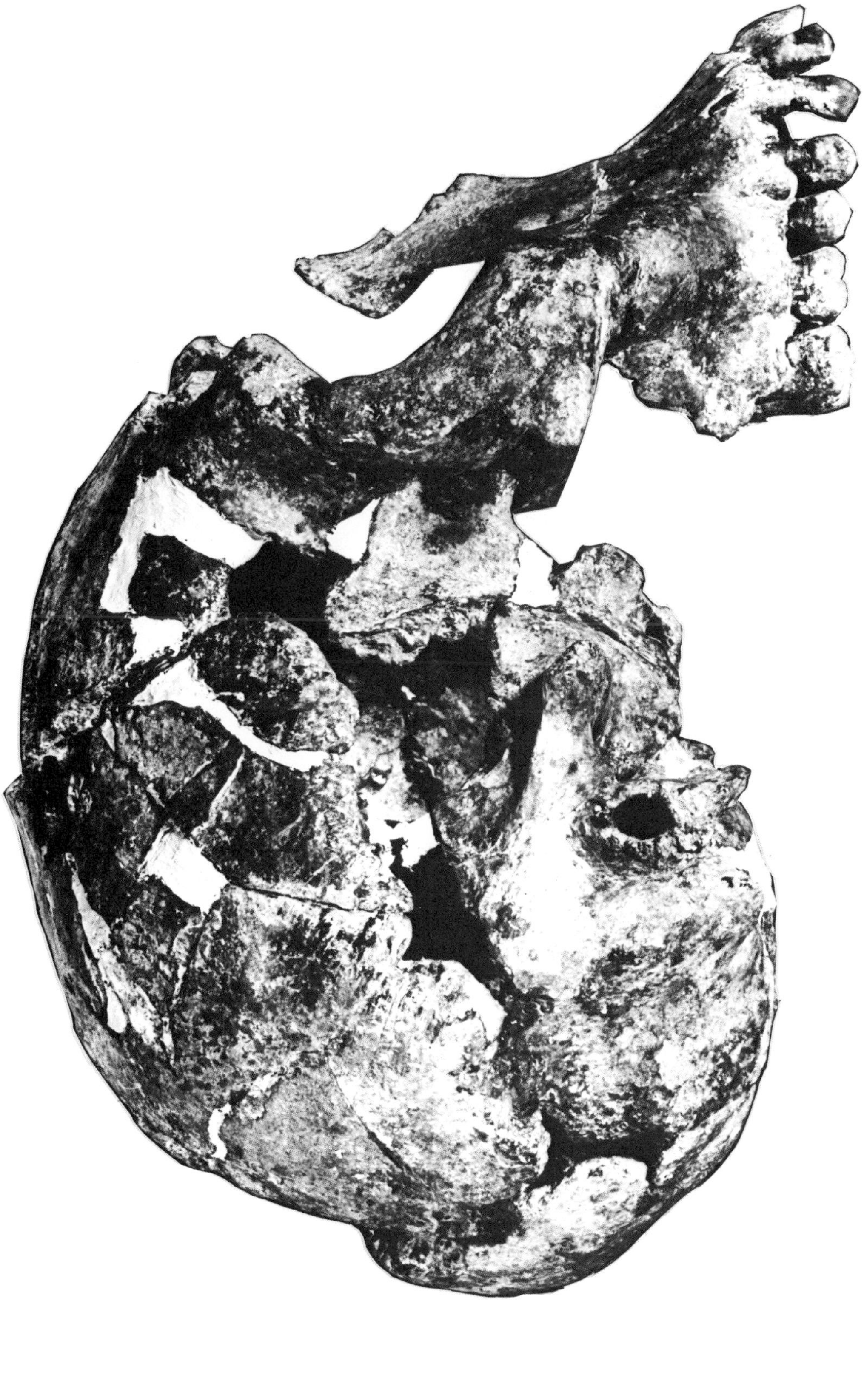

FIGURE 6.4.
Right lateral view of cranium KNM-WT 15000 A.

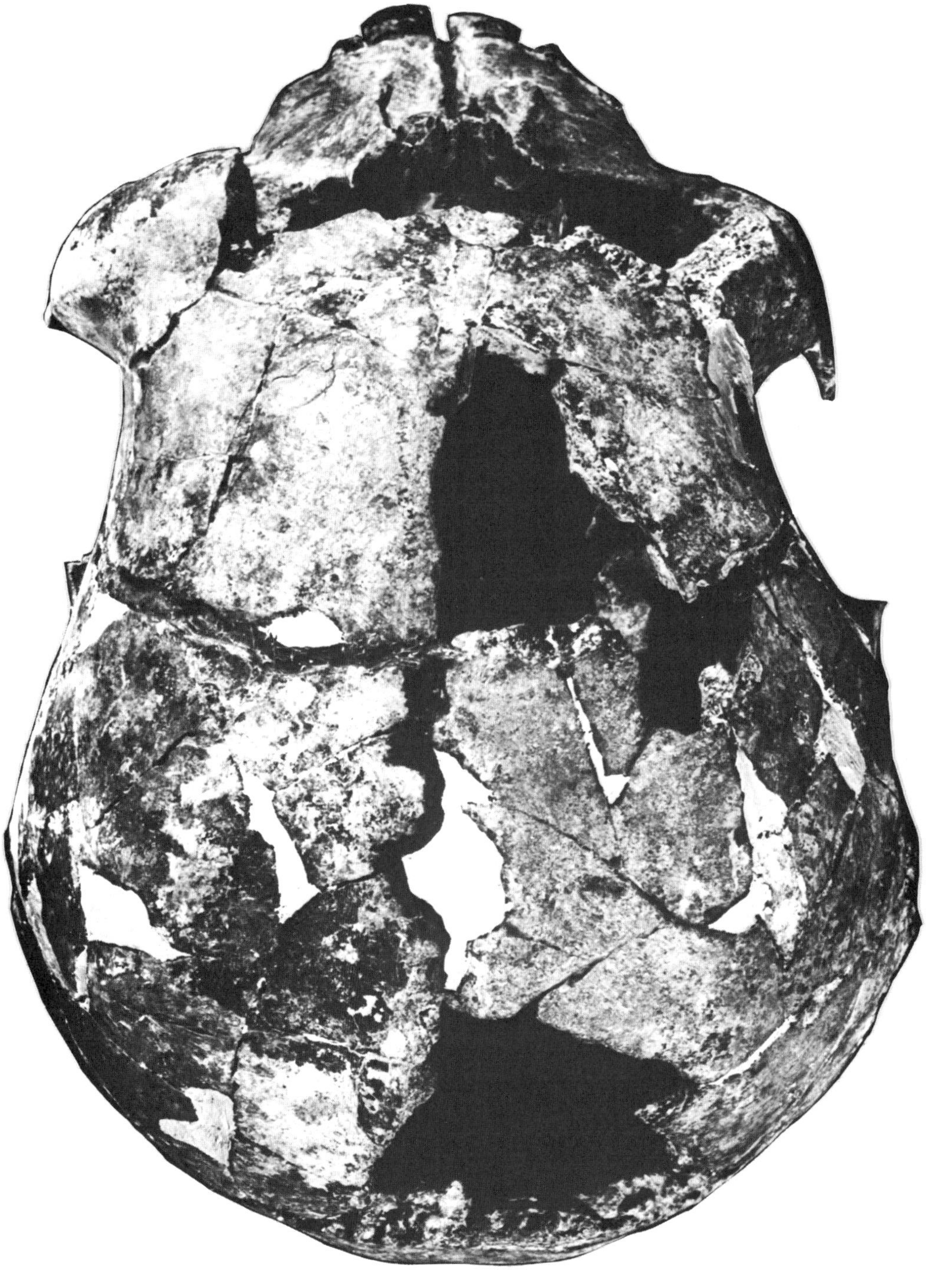

FIGURE 6.5.
Superior view of cranium KNM-WT 15000 A.

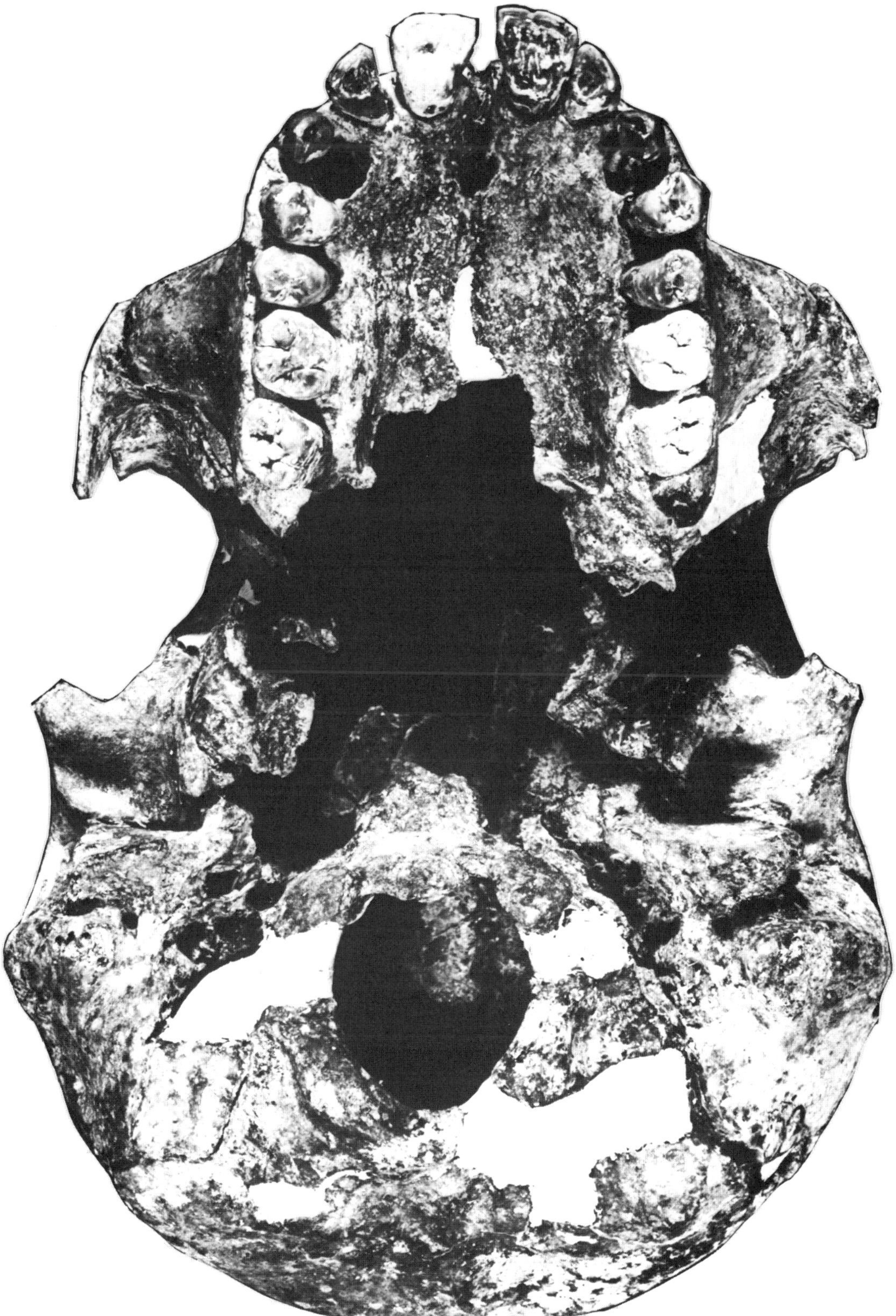

FIGURE 6.6.
Inferior view of cranium KNM-WT 15000 A.

sweeps evenly backwards and upwards to the zygomatic process of the temporal.

The palate is long anteroposteriorly, estimated at 59. It has the form of an even arch anteriorly with slightly diverging cheek-tooth rows. From the level of the premolars posteriorly it is of fairly even depth, but it shallows anterior to them. The incisive foramen is about 4.2 wide and its roof continues anteriorly as a groove between the central incisor roots. Widths of the palate at various levels are given in Table 6.2, together with the same measurements on KNM-ER 3733, the only other early *H. erectus* cranium on which they can be taken.

The frontals show, despite most of the brow ridge being broken, that a moderate supratoral sulcus was present. It had greater concavity behind the middle of the superior orbital margins and less in the midline. The temporal lines are more strongly marked on the frontals than on the parietals. They can be followed from the frontozygomatic suture as they swing towards the coronal suture. They cause a very slight angulation of the squame. They make a small dip inferiorly as they cross the coronal suture and continue across the very center of the parietals. They then curve suddenly back to reach a position just anterior to asterion and continue forwards again across the supramastoid crest. This means that the distance between the temporal lines and the superior nuchal line is about 38 on each side. Further details of the cranium can be found in the descriptions of each bone.

TABLE 6.2.

Palatal measurements (mm) of KNM-WT 15000 and those of KNM-ER 3733.

LOCATION OF MEASUREMENT	EXTERNAL ALVEOLAR WIDTH		INTERNAL ALVEOLAR WIDTH	
	KNM-WT 15000	KNM-ER 3733	KNM-WT 15000	KNM-ER 3733
I^1	27.4	(25)	22.7	(20)
I^2	43.2	(41)	33.5	(30)
dc or C	57.5		—	(32)
P^3	64.8		38.9	35.0
P^4	65.1	(61)	38.3	(33)
M^1	66.8	(63)	37.5	(32)
M^2	66.8	(65)	(40)	(34)

Other cranial specimens of early African *Homo erectus*

Two crania, one more complete than the other, have been found at Lake Turkana. Several other specimens, ranging from hardly identifiable fragments to useful and sometimes associated pieces, are also known. These are compared briefly with the cranium of KNM-WT 15000. A cranium from Olduvai (Rightmire, 1979) is about 300,000 years younger than KNM-WT 15000. It is also compared briefly here with the Nariokotome skeleton. Tooth specimens are compared in Chapter 8, this volume.

KNM-ER 164

(Day and Leakey, 1974)

This is a small piece of parietal which still has matrix stuck on the inside. It has a fragment of lambdoid suture and varies in thickness from 5.0 to 6.9. Its importance is that it is associated with some vertebrae and some hand bones.

KNM-ER 730

(Day and Leakey, 1973; Leakey and Walker, 1985)

This is a partial skull of an old adult. The mandible will be compared in a later section. A fragment of maxilla with roots and an upper canine with the crown worn to the gum appear to be similar to those that would have been present in KNM-ER 3733. A piece of left frontal is preserved. It has a 13.0 thick supraorbital torus. In the same location, the torus in KNM-WT 15000 is 25 thick. The frontal squame is 8.5 thick near bregma. The coronal suture crosses the fragment posteriorly, and it seems that the frontal length was just a little greater than in the Nariokotome specimen. The temporal line takes a dip as it crosses the coronal suture. A piece

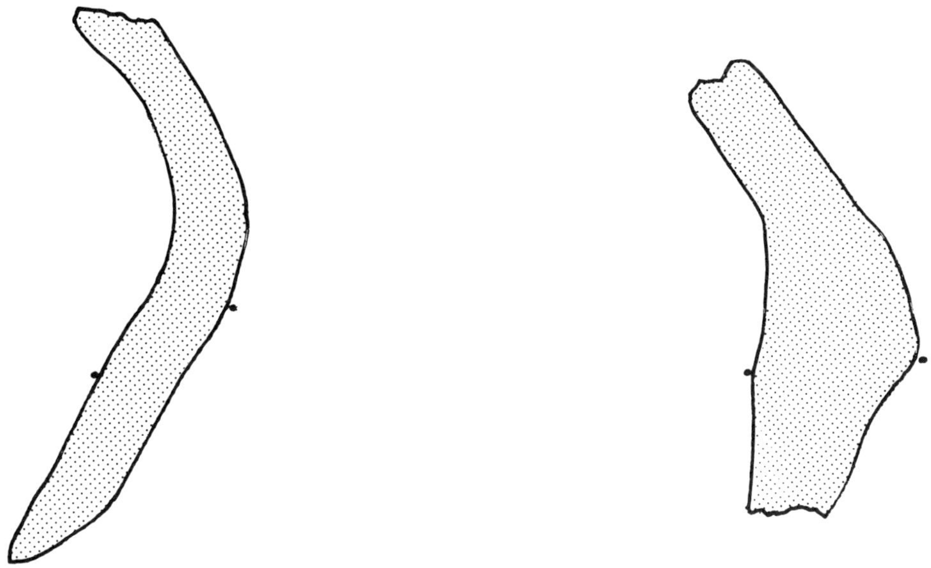

FIGURE 6.7.
Sagittal section of the occipital of KNM-WT 15000 A (left) *and KNM-ER 730* (right).

of parietal is about 9 thick at its center. Although weathered, the temporal lines can be seen and it can be estimated that they came to within 35 of the midline. They reach only to about 40 from the midline in the West Turkana specimen. Because the weathering has exposed bone structure, it is possible only to estimate roughly the length of the parietal as about the same as in the Nariokotome specimen.

The parietal piece articulates with most of an occipital. Its bi-asterionic width is 110, a little larger than the Nariokotome width. The bone is also thicker, being 11.4 at lambda and 13.5 at asterion. The occipital torus is well developed, but softly so, and the superior nuchal line is not undercut by the nuchal area. Sagittal sections are given in Figure 6.7. Inion in KNM-ER 730 is not much higher than endinion. The bone is over 20 thick at inion. This skull would have been about the same size as that of KNM-WT 15000, but with thicker vault bones and heavier tori.

KNM-ER 1466

(DAY ET AL., 1976)

This is a frontoparietal fragment of a thick skull. It is 11.5 at stephanion and 10.9 near bregma. The temporal lines mark, especially so anteriorly, the position of an angulation of the external table. This piece comes from a much thicker cranium than KNM-WT 15000.

KNM-ER 1808

(WALKER, LEAKEY, AND ZIMMERMAN, 1982; LEAKEY AND WALKER, 1985)

This partial skeleton has several pieces of an adult skull. The biggest piece is much of the occipital, which is similar in form to that of KNM-WT 15000, but with a bi-asterionic width of 120 it is a bit bigger. As in that skull, endinion is lower than inion by some distance and the right transverse sinus groove is dominant over the left. The bone is only 13.8 thick at inion, suggesting that the skull was relatively thin. A few parietal fragments are preserved, the thickest being 7.5 thick, so again the skull appears to have been relatively thin. A piece of frontal gives an idea of the shape of the supraorbital tori. They were rounded in section and only 10 thick. In this respect the specimen is very like KNM-ER 3733 and less like KNM-WT 15000. Some teeth from KNM-ER 1808 are mentioned in Chapter 8, this volume, and two pieces of mandible will be mentioned later here. A piece of the right temporal looks very like that in KNM-WT 15000.

KNM-ER 1821

(WOLPOFF, 1984)

This is a small fragment, about 60 square, of left parietal with the junction of the lambdoid and sagittal sutures. It shows a temporal line about 44 from the midline and is about 10 thick at the center and 8.5 thick near lambda. We do not see the angulation at the temporal lines that Wolpoff recorded.

KNM-ER 2592

(UNPUBLISHED)

This is a right parietal piece about 85 long and 53 wide showing parts of the lambdoid and sagittal sutures. It is badly weathered, but is about 10 thick near its center.

KNM-ER 3733

(LEAKEY AND WALKER, 1976, 1985)

This is an adult cranium that has enough of the facial skeleton preserved to provide a good comparison with that of KNM-WT 15000. It was entombed in sands that had extremely hard root casts in them. Several of these were stuck to the cranium and the base was encased in very hard matrix. Surface detail is poor for the most part because of this accumulation, and X-rays show a large number of matrix-filled cracks running through the whole of the vault.

These cracks have left many pieces of vault displaced. The worst damage is in the area of bregma, where a pebble in the matrix has crushed the vault inwards.

The cranium is thicker, the brow ridges thinner, and the face and palate smaller than in the Nariokotome skull. Although the external dimensions are about the same in both, the thicker vault bones make the endocranial capacity of 848 cc a little smaller than the 880 cc of KNM-WT 15000. The ectocranial superstructures are a little more developed than in KNM-WT 15000, so the cranium appears to be longer and lower because of occipital torus development. The temporomandibular joint is also bigger. The facial skeleton is smaller, but it has the same configuration as in KNM-WT 15000. The palate is deeper and narrower, but the molar teeth are a little larger than in KNM-WT 15000. Tooth comparisons are given in Chapter 8, this volume. Because of the juvenile status of the West Turkana skull, the differences between these two crania lead us to consider KNM-ER 3733 as an adult female.

KNM-ER 3883

(LEAKEY AND WALKER, 1985)

This is a calvaria and part of the facial skeleton of an adult individual. It was preserved in soft sediment and has suffered some plastic deformation and much cracking and loss of surface detail. In comparison with KNM-WT 15000, this specimen seems to us to be the closest to what we imagine its adult state might have been. The brow ridges are 13.5 thick and form a continuous torus rather than being interrupted by a strong supraglabellar groove as in KNM-ER 3733. The remaining facial bones are larger and more projecting than in either of the other two crania. Ectocranial superstructures are present in the form of tori and crests. All three crania have mild frontal keeling, bregmatic swellings, and depressions above the external occipital protuberance. The mastoid on KNM-ER 3883 is more prominent than in the others and the temporomandibular joint, with over half of the articular eminence overhung by the zygomatic process, is bigger. The whole base of this skull is crushed upwards, and the cranial capacity of 804 cc reported by Holloway (1983) is a minimum estimate.

KNM-ER 3892

(LEAKEY AND WALKER, 1985)

This is a left frontal piece from an adult. It extends from nearly the midline to pterion and from the left orbital region to the coronal suture. Near bregma the piece is 9.0 thick. There was a supratoral sulcus and the vault curved smoothly up from it. This is like the condition in KNM-WT 15000. The temporal line is not strongly developed and the minimum frontal breadth was probably larger than in the West Turkana skull.

KNM-WT 16001

(UNPUBLISHED)

These are fragments from a second *H. erectus* cranium from Nariokotome but from a higher stratigraphic level some distance away from the excavated site. Worn pieces of parietal and occipital show a thick vault. A parietal piece shows a temporal line and is up to 10 thick. The occipital is about 12 thick at asterion and 16.5 thick on the occipital torus.

OLDUVAI HOMINID 9

(RIGHTMIRE, 1979)

This is a thick and strongly built skull, presumably of a male. It comes from site LLK, Olduvai, and is probably about 1.2 million years old. It is a long and low cranium with massive, forwardly-projecting supraorbital tori that are 19.0 thick in the middle of the orbit and with strongly formed temporal lines and occipital torus. The thickness of the vault bones has been measured in a few places (P. G. Rightmire, personal communication). Near asterion the parietal is 13.5; the center of the parietal is 10.0 and on the temporal line 9.0. The cranial capacity is 1,067 (Holloway, 1975), which is about the same as the mean from Zhoukoudian (Weidenreich, 1943).

Compared with KNM-WT 15000, this cranium looks much more massive, but it does not take too much imagination to see that it could be possible for it to have had a juvenile form like that of the West Turkana fossil. In its general shape, with the greatest cranial breadth far posterior and low on the supramastoid crest and with constricted frontals, the resemblances are clear. It differs only in its robustness, greater pneumatization, and massive brow ridges. The closest African cranium is KNM-ER 3883, which has a cranial base that is extremely similar in all features.

Sagittal craniograms of KNM-WT 15000, KNM-ER 3733, OH 9 (after Maier and Nkini, 1984), and several crania from Zhoukoudian (after Weidenreich, 1943) are given in Figure 6.8 to show some of the range of variation in cranial sagittal section.

THE INDIVIDUAL SKULL BONES

Descriptions of the individual skull bones are given here. Many fossils that are available for comparison from sites around the world are isolated cranial bones or mandibles. These descriptions and photographs should help make comparisons between them and KNM-WT 15000 easier.

THE FRONTAL BONE

(FIGURES 6.9–6.11)

The frontal is represented by all parts except the central portions of the supraorbital torus and glabella, a triangular piece of the squamous part from the midline to about halfway along the coronal suture and the orbital processes. Three small brow-ridge fragments were recovered which cannot be glued onto the main piece. The supraorbital tori are broken away from about the middle of the superior orbital margin on the left to the superolateral corner of the orbital margin on the right. Enough remains of the junction between the tori and the squamous part to see that there was a complete sulcus behind the tori. The frontal sinus development was very asymmetrical. It is well developed on the left, hollowing the torus up to the break, and not at all on the right, where only spongy bone can be seen. The supraorbital tori are 10.4 thick at midorbit and have many vascular foramina across their surface, especially anterosuperiorly.

Because of the relative massiveness of the

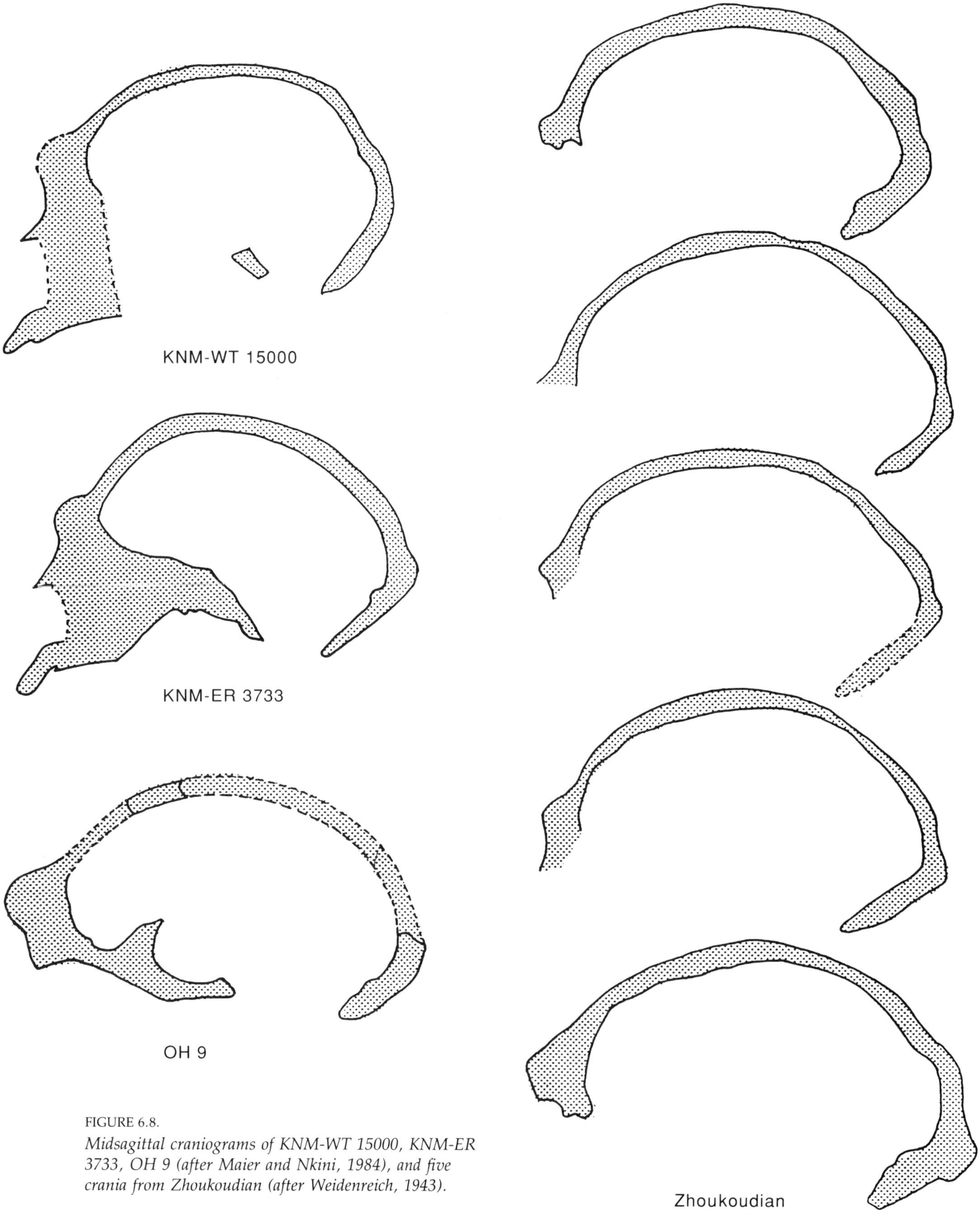

FIGURE 6.8.
Midsagittal craniograms of KNM-WT 15000, KNM-ER 3733, OH 9 (after Maier and Nkini, 1984), and five crania from Zhoukoudian (after Weidenreich, 1943).

FIGURE 6.9.
Anterior view of the frontal of KNM-WT 15000 A.

FIGURE 6.10.
Superior view of the frontal of KNM-WT 15000 A.

FIGURE 6.11.
Inferior view of the frontal of KNM-WT 15000 A.

FIGURE 6.12.
External view of left parietal of KNM-WT 15000 A.

FIGURE 6.13.
Internal view of left parietal of KNM-WT 15000 A.

tori, the zygomatic processes seem short, and the frontozygomatic suture truncates them obliquely. At the suture the tori are 11.4 anteroposteriorly on the left and 11.6 on the right. The temporal lines run upwards for about 25 from the suture before curving sharply through almost 90° and running backwards across the squamous frontal. The inferior temporal line is more prominent than the superior. They are about 7 apart on much of the frontal. The minimum frontal breadth is 90.3 and the breadth across the frontozygomatic sutures (the maximum frontal breadth) is 111.7.

Because the triangular piece of bone from the right side and another small piece from the right pterionic region are missing, the parietal margin is not complete. It is possible by mirror imaging to determine the shape and course of the whole margin. From the pterionic region the coronal sutural surface runs almost straight diagonally backwards to bregma. Inferiorly the sutural surface faces laterally where the parietal overlaps the frontal for a distance of up to 13.0. This overlapping condition is reversed about halfway to bregma, and from then on the frontal overlaps the parietal, though over a lesser distance of up to 7.5. At bregma the bone is 7.9 thick. In the center of the left squamous frontal the bone is 6.1 thick. The internal surface is fairly well preserved and reveals gyral detail. About 30 of the frontal crest is seen before it fades into a feeble sulcus for the superior sagittal sinus. It is asymmetrical and swings sharply to the left anteriorly, perhaps mirroring the asymmetry of the frontal sinus. The orbital plates have broken away, but enough of their origins remains to see that they came off at a sharp angle to the internal surface of the squamous portion.

THE PARIETAL BONES

(FIGURES 6.12–6.15)

The more completely preserved left parietal is missing only a small triangular piece near lambda and another half a dozen individual bone pieces whose edges had been broken before recovery. The right is missing an even larger triangular piece next to lambda and has four other defects anteriorly. This description is based, therefore, on the left.

The bone is convex fairly evenly anteroposteriorly at all levels, but shows a moderate change in convexity in coronal sections about halfway down. This angulation is more apparent anteriorly. The bone shows a mild and centrally placed parietal eminence because of the way in which these curvatures interact.

The frontal border, matching the parietal border of the frontal, is almost straight and the coronal sutural surfaces match those of the frontal reciprocally. The frontal angle is about 90°. The incomplete sagittal border is irregular because of several deviations of the sagittal suture from the midline. The occipital border is irregular for the same reason—that is, the lambdoidal suture is not very straight. The squamous border is not deeply notched. It has many ridges and grooves on its sutural surface and very many fine striations, which run from between 5 to 20 diagonally backwards across the external surface.

The bone is 8.2 thick at bregma, 6.1 near lambda, 10.1 at asterion, and 8.6 at the sphenoid angle. It is 5.0 thick at the parietal eminence. The internal surface of both bones preserves only moderate detail.

THE OCCIPITAL BONE

(FIGURES 6.16–6.20)

The occipital is complete except for small pieces of bone from the left and right mastoid borders, the left posterior margin of the foramen magnum, and the lateral parts around the condyles. The posterior parts of the condyles themselves have been lost, and the edge of the basilar part is abraded. Both the lambdoid and mastoid margins are irregular because of the interdigitations in their sutures. Left and right lambdoid borders do not meet at a sharp angle at lambda and, in fact, it would be difficult to say from the suture alone where lambda was, were it not for a relatively sharp spine projecting on the left lateral side of the superior sagittal sinus groove.

There is no well-developed torus, but rather a low swollen region centered about 27 below lambda and limited above by a supratoral hollow and below by the superior nuchal line. The external occipital crest clearly runs from centrally downturned parts of the superior nuchal line to finish

FIGURE 6.14.
External view of right parietal of KNM-WT 15000 A.

FIGURE 6.15.
Internal view of right parietal of KNM-WT 15000 A.

FIGURE 6.16.
Anterior view of occipital of KNM-WT 15000 A.

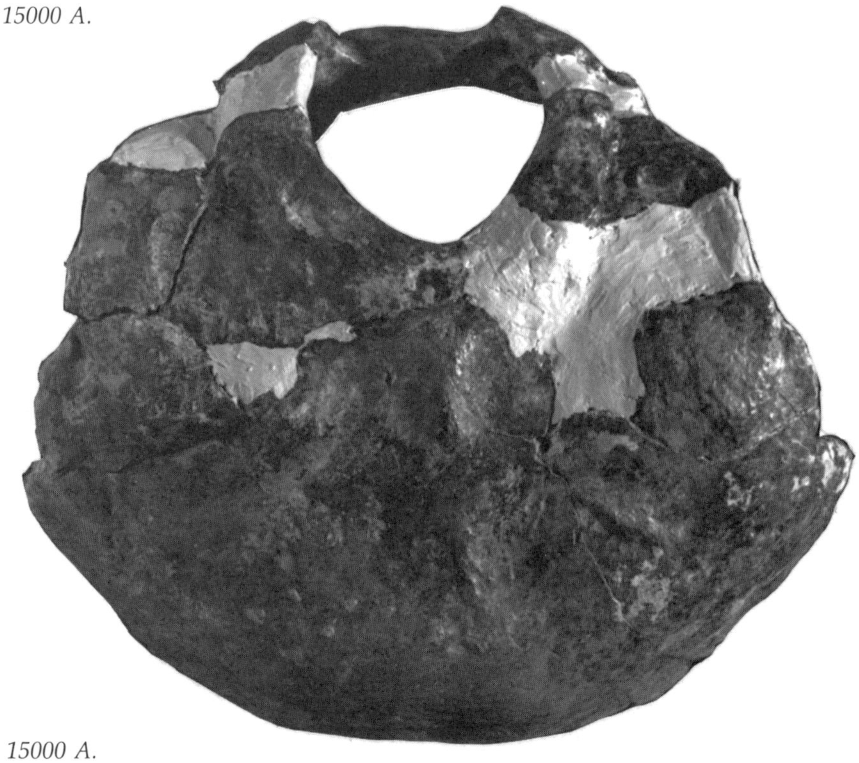

FIGURE 6.17.
External view of occipital of KNM-WT 15000 A.

FIGURE 6.18.
Internal view of occipital of KNM-WT 15000 A.

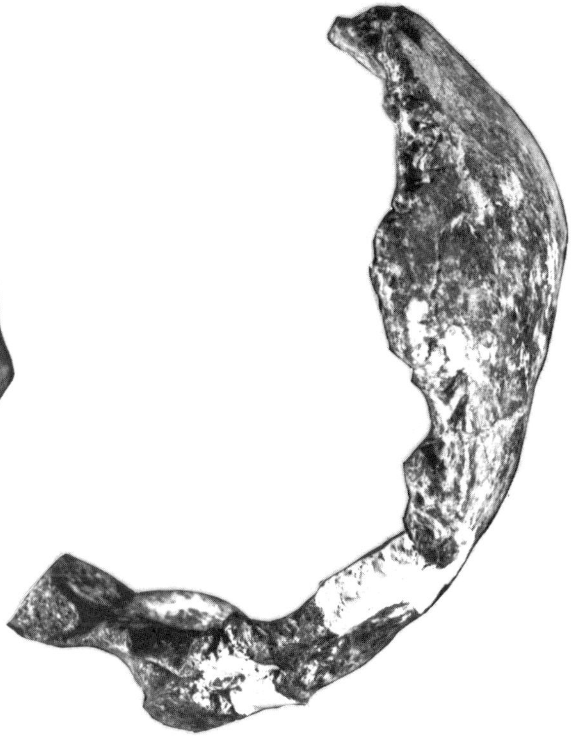

FIGURE 6.19.
Left lateral view of occipital of KNM-WT 15000 A.

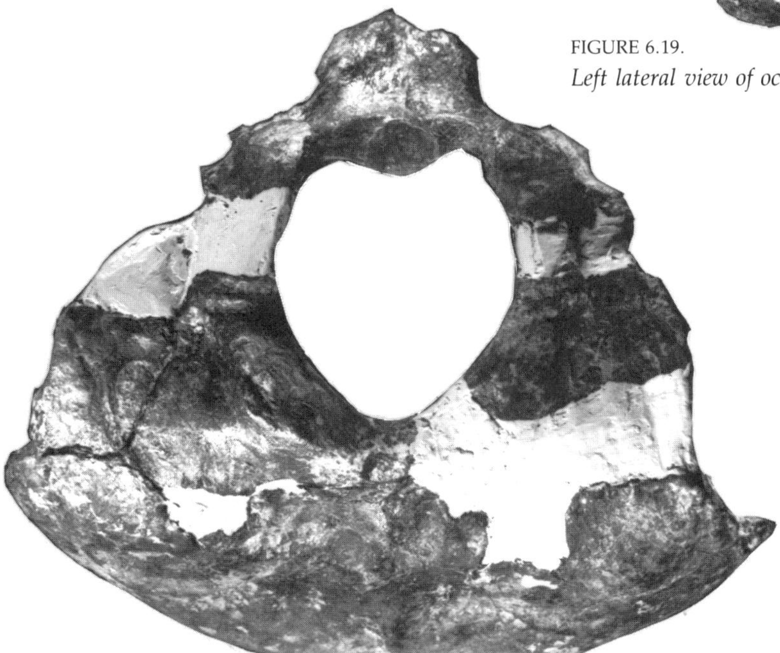

FIGURE 6.20.
Inferior view of occipital of KNM-WT 15000 A.

in a blunt tubercle about 13 from the edge of the foramen magnum. The superior nuchal lines can be followed swinging first upwards and laterally then downwards (where they become grooves) to merge with the inverted edge of the occipitomastoid suture. The inferior nuchal lines are irregular and are marked only by hollow digitations formed by scars of the rectus capitis posterior muscles. The areas on either side of the external crest beneath the superior and inferior lines are depressed somewhat. There are crests on either side of the foramen magnum, roughly paralleling its posterolateral edges about 10 from them. These are especially clear immediately medial to the condylar fossae.

Details of the internal surface are reasonably well preserved. A strong internal occipital protuberance is centered some 33 below the center of the torus. The groove for the superior sagittal sinus is about 5 wide and runs to the right side of the protuberance and turns sharply as the groove for the right transverse sinus. This is about 5 wide too. The left transverse sinus groove is narrower, about 4, and comes off the superior sagittal sinus groove above the protuberance. There are deep hollows for the occipital cerebral poles above the transverse sinus grooves and similar ones for the cerebellum beneath them. The internal occipital crest is wide (over 13) and low, dividing to pass on either side of the foramen magnum. A very small part of the sigmoid sinus groove can be seen on the right. On the left it can be seen running along its whole course to the jugular foramen.

THE TEMPORAL BONES

(FIGURES 6.21–6.31)

Although generally well preserved, neither temporal bone is complete. The left is missing a triangular piece of the squamous part from above porion to the squamous suture and the end of the zygomatic process, and the tip of the petrous part is abraded. The right is missing much of the superior part of the squamous part, the medial end of the petrous part, and the end of the zygomatic process.

The squamous part is thicker anteriorly than posteriorly, being 8.2 in the middle of the sphenoid border, 7.2 at its highest point, and thinning to 6.4 before swelling again towards the parietal notch. The temporal surface is unmarked except for a faint groove for the middle temporal artery. The sphenoid border is relatively short. The zygomatic surface is represented only by its most posterior part. The gutter for the posterior fibers of temporalis is clearly seen, but not wide, being less than 10 over the articular eminence. The thickness here, including the eminence, is 6.1 on the left and 7.4 on the right.

The ectoglenoid process is small and blunt and is actually only the end of the horizontally placed eminence, but the entoglenoid process is large—it makes up the medial side of the mandibular fossa and rises as a plate set at 70° to the plane of the eminence. The postglenoid process is small and blunt. The articular fossa itself is narrow anteroposteriorly (7.7) and wider transversely (20.0).

The tympanic plate is high and set almost at a right angle against the fossa. It increases in height from the edge of the external auditory meatus to a spine just lateral to the styloid sheath. Here it is 21.0 high. Medial to this it is notched to a lesser height, then it turns anteriorly to lie medial to the entoglenoid process. The petrotympanic fissure is seen in the posterior floor of the fossa and between the medial part of the plate and the entoglenoid process. The external auditory meatus is best seen on the right. This is because the left temporal was broken through the petrotympanic fissure and the edges were cemented by matrix before its recovery in two parts.

The meatus itself is oval in outline, with the long axis vertical. It is 10.5 high by 7.3 wide. Its bony margin is thinnest anteriorly (2.7), thickest inferiorly (6.4), and fairly thick posteriorly (4.9). The margin is roughened. The stylomastoid foramen is a small (2.9 × 2.3) opening just anteromedial to the tip of the mastoid process.

The carotid canal, seen just posterior to the medial part of the tympanic plate, is 5.3 in diameter. The mastoid notch, for the posterior belly of the digastric, is narrower on the right than the left (3.3 and 4.6) and the groove for the occipital artery is barely discernible. The mastoid process is marked by a mastoid crest level with the top of the external auditory meatus. The crest runs downwards and forwards to join the relatively sharp anterior margin of the process. The lateral surface of the process is flattened and is inflected towards the stylomastoid foramen. The mastoid crest is more prominent and projects more laterally than the supramastoid crest, which is just the swollen continuation of the lateral border of the gutter for the posterior temporalis fibers.

The groove for the sigmoid sinus is clear and is 9.4 wide on the right and 8.5 wide on the left. Its most superior part undercuts the most lateral part of the crest of the petrous pyramid before the groove swings downwards and medially to the jugular foramen. The crest of the petrous pyramid is otherwise blunt and has only the faintest trace of the groove for the superior petrosal sinus. It becomes more blunt medially. The posterior surface of the petrous is marked by a 10.0 long, slit-like external aperture for the vestibular aqueduct, which is centered 17 medial to the middle of the sigmoid sinus groove at the same level.

The internal auditory meatus is an oval foramen nearly 5 in diameter, centered 35.0 medially from the middle of the sigmoid sinus groove at the same level. The anterior surface of the petrous is fairly flat, with a very low arcuate eminence and a small slit-like opening of the facial canal for the greater petrosal nerve running towards the apex. A meningeal vessel groove is seen running posteriorly just above the angle between the anterior surface of the petrous and the squamous parts.

THE MAXILLAS

(FIGURES 6.32–6.39)

Both maxillas are relatively well preserved. The left is less prepared than the right because during fossilization the wall of the nasal cavity collapsed as a series of bone fragments into matrix filling the maxillary sinus. It was thought best to leave this matrix and bone mixture as it is. Much of the lateral and superior parts have been lost on the right and so the matrix filling the sinus could be removed easily. Much of the left palatine bone and some of the pterygoid processes of the left sphenoid are preserved attached to their maxilla.

Much of the left and less of the right palatine processes are preserved. They are over 6 thick at their sutural contact, which is finely interdigitated. The palatine surfaces are relatively smooth posteriorly, but they become rougher and have many small vascular foramina anteriorly. The grooves for the greater palatine vessels and nerves are not stongly expressed. They run anteriorly from the greater palatine foramen, where the bony palate is angulated at about 90° with the alveolar process, but turn medially as this angulation decreases so that by the time they reach the level of the incisive foramen they are only 7.7 from the median palatine suture.

The incisive canal can be seen running downwards and forwards at an angle of about 40° to the horizontal on each medial surface. Each side of the canal is about 2 in diameter. The superior entrance to the canal is at the level between P^3 and P^4 and the incisive foramen is centered about 12 posterior to alveolare. The foramen opens into a marked median groove that is formed by the alveolar bone here and that closely follows the contours of the inferomedial parts of the central incisor roots.

The nasoalveolar clivus is straight and at an angle of 45° to the alveolar plane. The intermaxillary suture here is thick, up to 9.1 superiorly, and composed of many fine interdigitating bony spines and grooves that run anteroinferiorly. The length of this section from the anterior nasal spine to alveolare is about 27. The spine is made of the two everted edges of the clival bone here and is situated just anterior to a line drawn across the lower margin of the pyriform aperture. It projects only moderately upwards (about 3) into the aperture. The relatively flat clival surface is interrupted by soft grooves, which demarcate the incisor and canine alveolar juga and which are more marked at the alveolar margin.

The pyriform aperture is wider than it is high. It is 33.0 wide at its widest (about one-third the distance from the floor to the top) and is 29.4 high on the left and 27.1 high on the right as measured from the lowest part of its inferior border directly superior to the angle formed by its superolateral margin and the frontomaxillary sutural edge. There is a shallow incisive fossa, more

FIGURE 6.21.
Anterior view of left temporal of KNM-WT 15000 A.

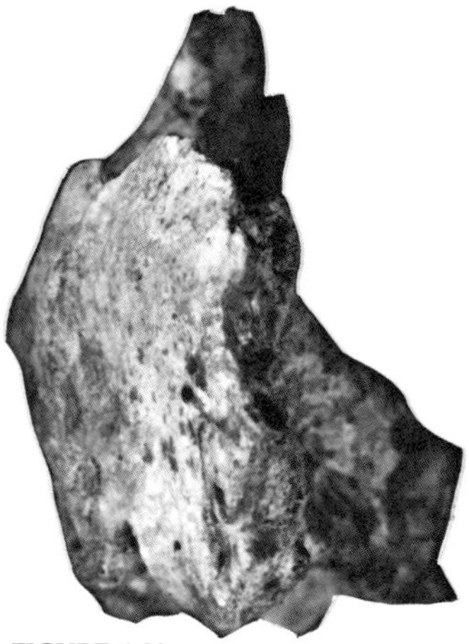

FIGURE 6.22.
Posterior view of left temporal of KNM-WT 15000 A.

FIGURE 6.23.
Medial view of left temporal of KNM-WT 15000 A.

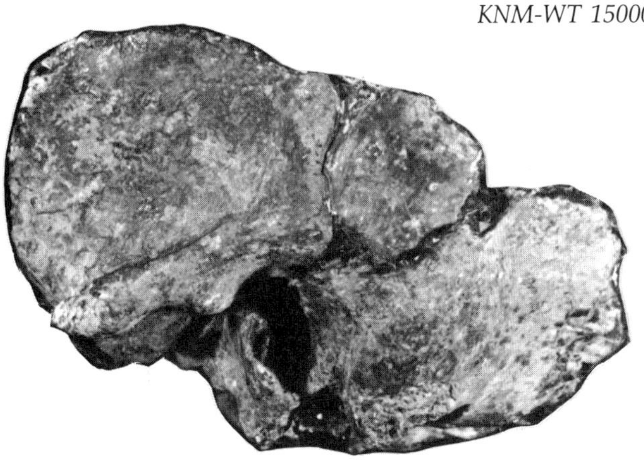

FIGURE 6.24.
Lateral view of left temporal of KNM-WT 15000 A.

FIGURE 6.25.
Superior view of left temporal of KNM-WT 15000 A.

FIGURE 6.26.
Inferior view of left temporal of KNM-WT 15000 A.

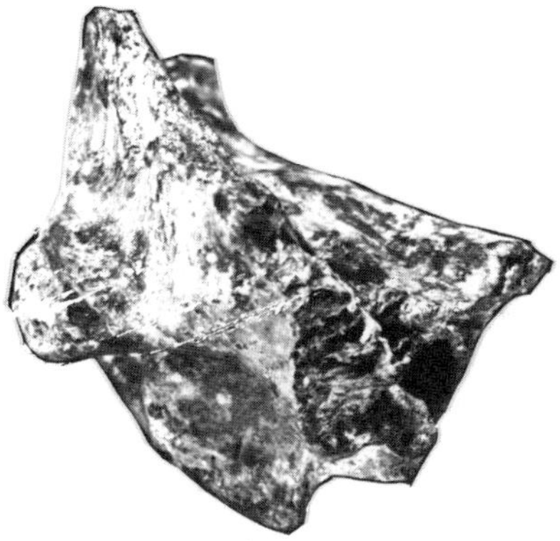

FIGURE 6.27.
Anterior view of right temporal of KNM-WT 15000 A.

FIGURE 6.28.
Posterior view of right temporal of KNM-WT 15000 A.

FIGURE 6.29.
Medial view of right temporal of KNM-WT 15000 A.

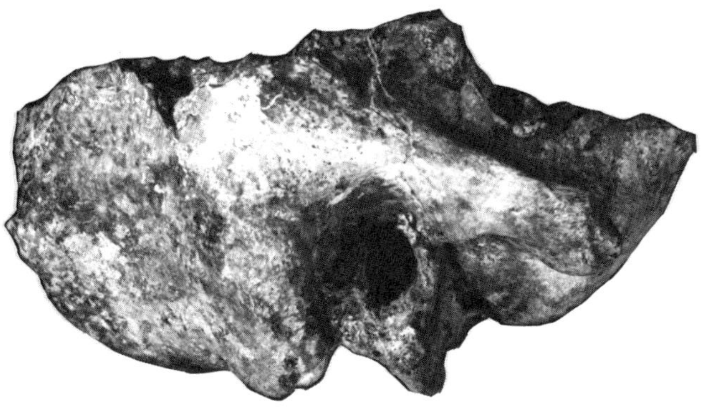

FIGURE 6.30.
Lateral view of right temporal of KNM-WT 15000 A.

FIGURE 6.31.
Inferior view of right temporal of KNM-WT 15000 A.

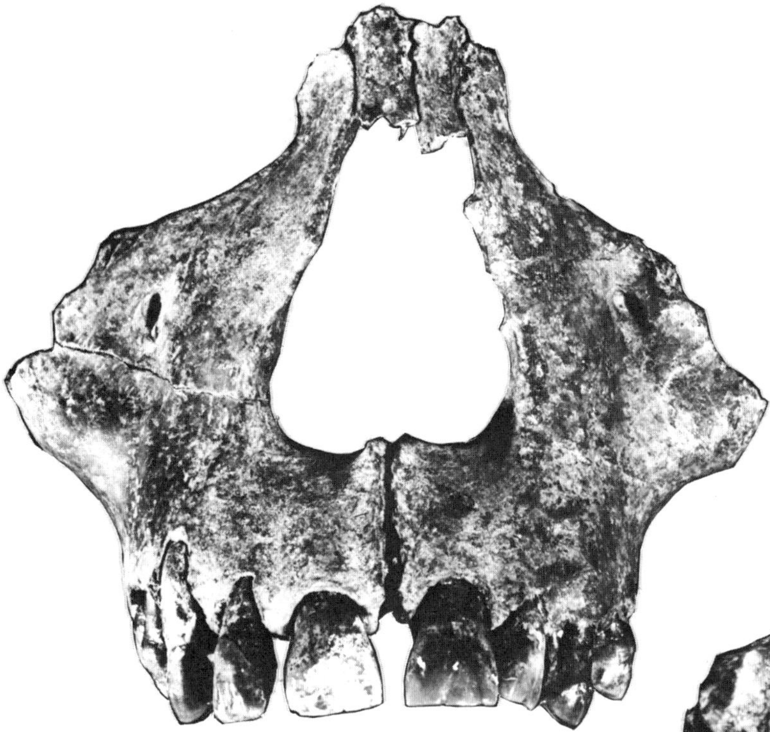

FIGURE 6.32.
Anterior view of left and right maxillas of KNM-WT 15000 A.

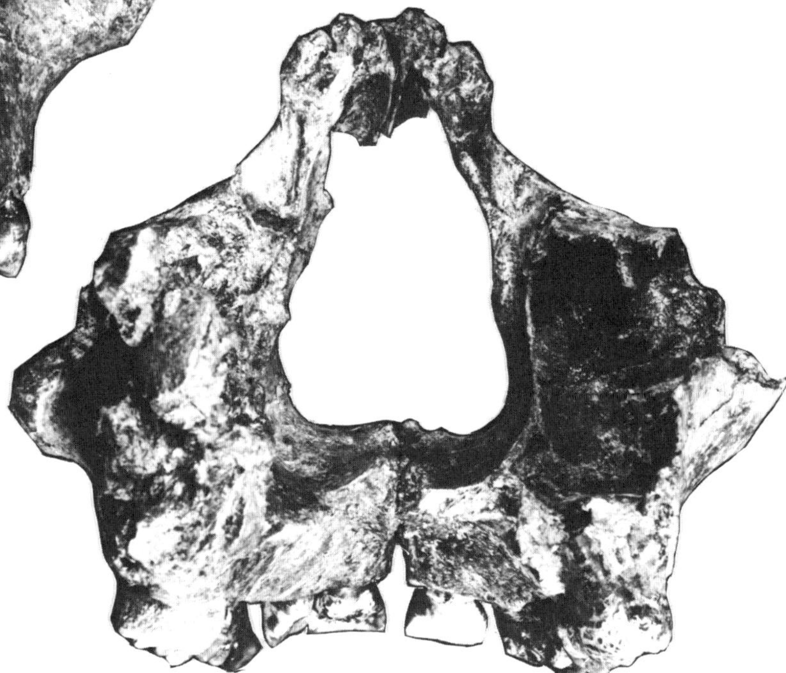

FIGURE 6.33.
Posterior view of left and right maxillas of KNM-WT 15000 A.

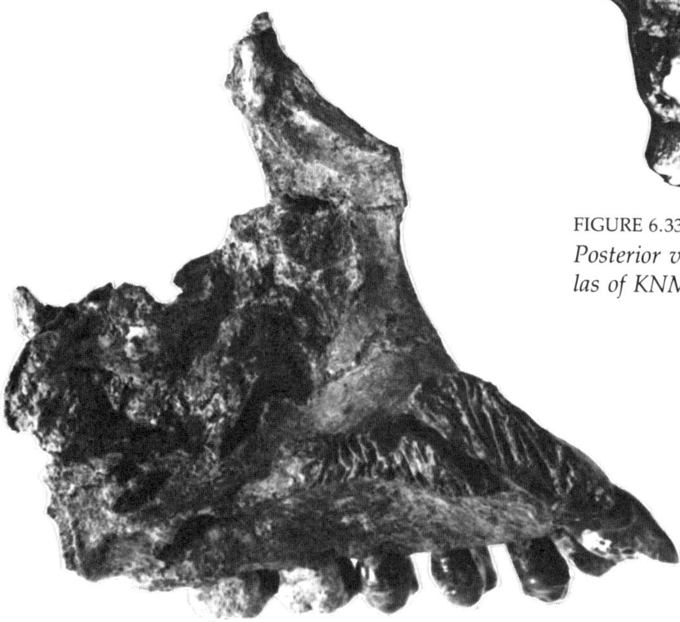

FIGURE 6.34.
Medial view of left maxilla of KNM-WT 15000 A.

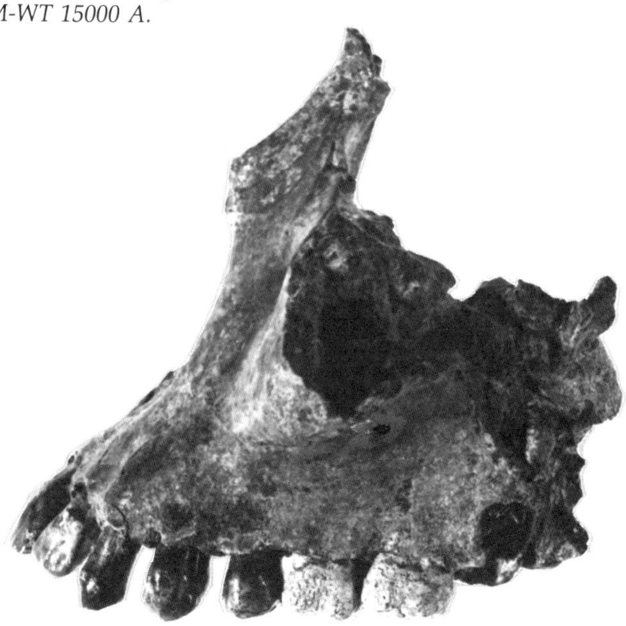

FIGURE 6.35.
Lateral view of left maxilla of KNM-WT 15000 A.

FIGURE 6.36.
Medial view of right maxilla of KNM-WT 15000 A.

FIGURE 6.37.
Lateral view of right maxilla of KNM-WT 15000 A.

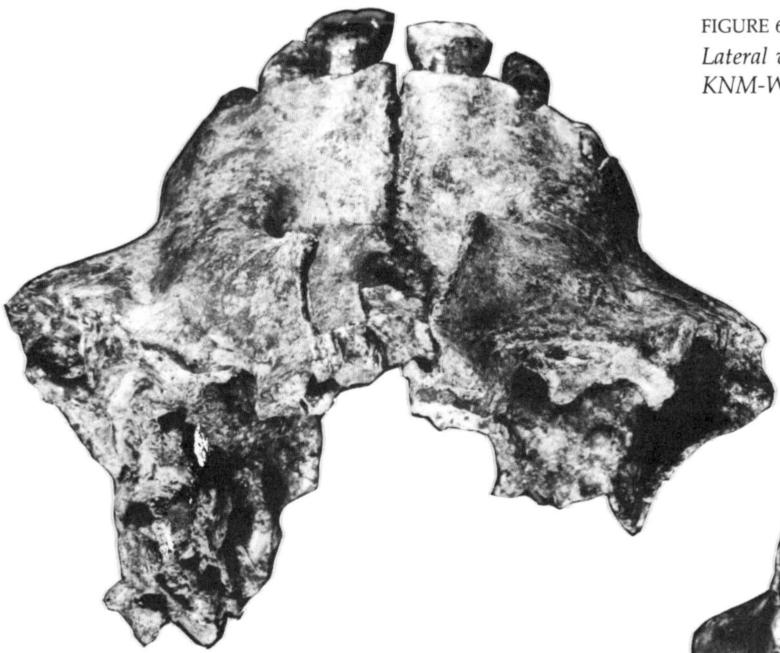

FIGURE 6.38.
Superior view of left and right maxillas of KNM-WT 15000 A.

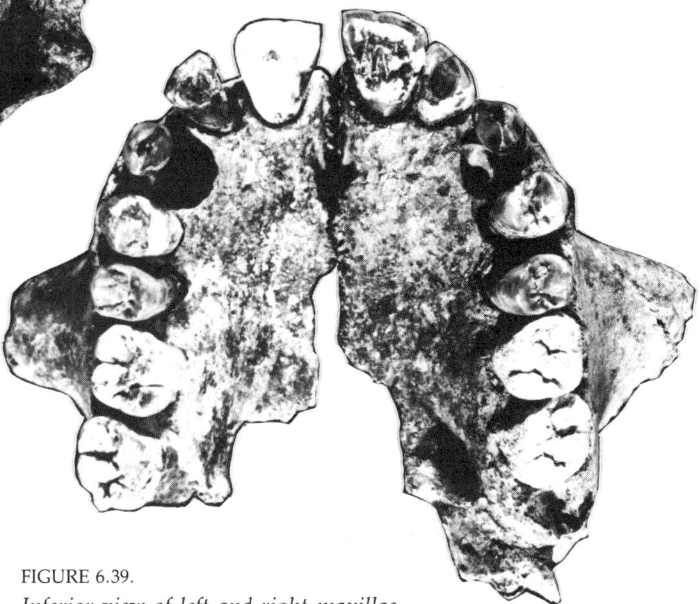

FIGURE 6.39.
Inferior view of left and right maxillas of KNM-WT 15000 A.

marked on the right than the left, just behind the entrance to the aperture.

The zygomatic process is centered about the level of P^4 and M^1, is gently rounded inferiorly, and has only the faintest hint of a malar tubercle. The take off is gradual and there is no clear angulation but rather a very smooth curvature right from the alveolar margin. The anterior surface of the maxilla is also devoid of major features, except for the infraorbital foramen and its inferior groove. The foramen is about 2.5 in diameter, centered 10.0 on the left and 12.0 on the right immediately below the orbital margin, and its groove quickly fades out into the anterior surface itself. This same smooth surface is continued up the frontal process, where the only major changes are the eversion at the inferior part of the nasomaxillary suture and the concavity of the highest portion to match the anterior concavity of the nasal bones. The infraorbital margin is sharp and almost straight throughout and in anterior view is angled sharply to a vertical anterior lacrimal crest. The angle is only about 10 directly from the nearest part of the nasomaxillary suture.

On the right side the orbital surface and the maxillary tuberosity are broken away. The lacrimal groove is exposed and runs almost vertically downwards. It is 3.5 in diameter at its narrowest. The conchal crest immediately anterior to the inferior opening of the nasolacrimal canal is very sharply defined, whereas the ethmoidal crest more superiorly is less so, being just a barely perceptible roughening. All the matrix filling the right maxillary sinus has been removed. The sinus is extensive, evaginating into the frontal process anterior to the nasolacrimal canal, deeply into the palatine process at the level of P^4-M^1, and into the zygomatic process. Excavation between the molar or premolar roots is slight. The bony tube for the infraorbital canal is seen on the anterior wall of the sinus.

Much of the palatine bone and some of the pterygoid process of the sphenoid bone are still attached to the maxillary. They are crushed and distorted, however. The posterior margin of the horizontal palatine plate is preserved medially and shows a flat posteroinferiorly facing surface bordered anteriorly by a sharp crest.

THE SPHENOID BONE

(SEE FIGURE 6.4)

Only fragments of the sphenoid were recovered (which we show not as individual figures but in their places on the basal portion of the cranium). The most important and informative are parts of the greater wing. Left and right posterior parts preserve foramina spinosum and ovale. These both can be attached to their respective squamosal parts. On the left the foramen ovale is 5.5 × 3.6 and on the right it is 4.2 × 4.0. The left fragment has the spine and on the inferior surface the small foramen spinosum can be seen posterolateral to the foramen ovale. Both foramina lie medial to the sphenosquamosal suture. On the superior surface, however, the foramen lies on the suture between the two bones.

A piece of the greater wing is glued to the left frontal. It preserves the lateral wall of the superior orbital fissure and the sharp edge of bone which runs up to the sphenoid angle of the parietal to divide the cranial cavity into anterior and middle fossae. This piece is important in that it gives information about the conformation of the inferior surface of the frontal lobe of the cerebrum. A large piece from the right side has the frontal border, the edge of the superior orbital fissure, the lateral wall of the foramen rotundum, and large parts of the squamosal and zygomatic margins.

THE ZYGOMATIC BONES

(FIGURES 6.40–6.42)

Both bones are preserved, the right being a little more complete in that it has more of the zygomatic process, but the left has more of the orbital process. The surface for articulation with the maxilla has part of the maxilla attached to it on both bones: that of the right includes the uppermost extension of the maxillary sinus, and that of the left is still filled with matrix and some crushed bone plates.

The orbital surfaces are smoothly concave and triangular in outline. The large single zygomaticoorbital foramen is very close to the inferolateral orbital margin. A single zygomaticotemporal foramen lies close to the frontozygomatic suture about midway along its extent. Its course

FIGURE 6.40.
External view of left zygomatic of KNM-WT 15000 A.

FIGURE 6.41.
External view of right zygomatic of KNM-WT 15000 A.

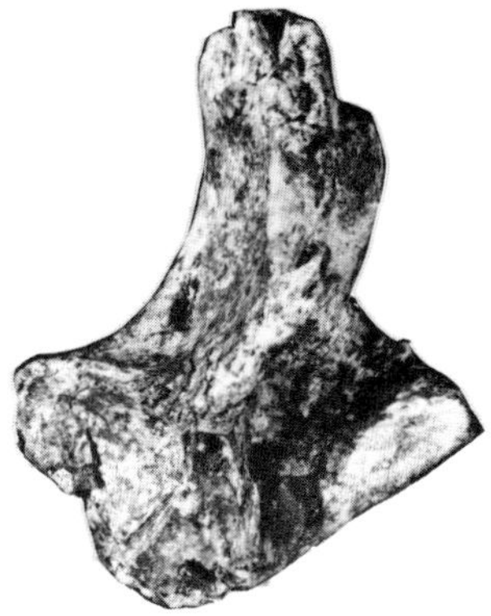

FIGURE 6.42.
Internal view of right zygomatic of KNM-WT 15000 A.

through the frontal process is very oblique, for its opening on the temporal surface is again close to the suture but above the level of the marginal tubercle. Another zygomaticotemporal foramen can be seen in the deepest concavity of the temporal surface. The marginal tubercle is well developed. It is sharply notched superiorly and less so inferiorly. The tubercle has the form of a flange about 3 thick. The zygomatic process is flat and triangular and comes off the maxillary surface in the same plane as it with no angulation. Its inferior margin bears a flat and distinct attachment area for the masseter which runs as far anteriorly as the zygomaticomaxillary suture. On the right there is a single and on the left double zygomaticofacial foramina. These are situated on a rounded swelling below the inferolateral orbital margin.

On the internal surface of the maxillary sinus, which can be seen on the right zygomatic bone, a well-defined groove for the sinus vessels and nerves can be seen running upwards and backwards.

THE NASAL BONES

(FIGURES 6.43–6.45)

The upper parts of both nasal bones are preserved. They had been fossilized separately from the cranium and were covered with a thick calcite crust. They extend from the frontonasal suture for about 18 on the left side and 16 on the right. They are a little asymmetrical, the left being a little wider on its anterior surface than the right. Their anterior surfaces are smoothly concave from superior to inferior and they meet to make an angle of 25°. The binasal width about 8 inferior to the frontonasal suture is 12.0. There is a single vascular foramen on the right, centrally placed about 13 down from the suture. The internasal suture is quite extensive, being up to 7.5 wide. It is relatively flat but begins with a few small interdigitations at the frontonasal suture. There is a flat surface laterally for the frontomaxillary suture. These lateral surfaces fit into distinct notches on the frontal processes of the maxillae. Plastic deformation of the

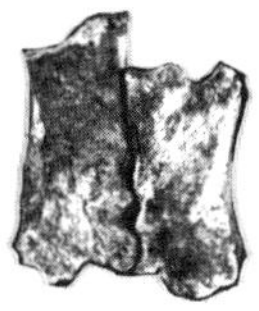

FIGURE 6.43.
Anterior view of nasals of KNM-WT 15000 A.

FIGURE 6.44.
Posterior view of nasals of KNM-WT 15000 A.

FIGURE 6.45.
Left lateral view of nasals of KNM-WT 15000 A.

maxillae makes it difficult to estimate the lower widths of the nasals accurately, but they must each have been about 10 wide at their most inferior. The length of the nasomaxillary suture on the left is 27.0 and on the right 28.0. In articulation with the maxillae the nasals follow the surface contours of the frontal processes without discontinuity, and since the lower parts of the maxillary edges of the suture are everted, it follows that the angle between the two nasals probably became more acute inferiorly.

THE MANDIBLE

(FIGURES 6.46–6.51)

The mandible is complete. Postfossilization cracks run through the body beneath M_2 on both sides and through the symphysis posteriorly to the right canine alveolus, but these hardly distort the specimen. There is sign of a small amount of periodontal disease around the right P_4. A large resorption pocket running forwards from the M_1 mesial root to the distal P_3 root and extending the width of the alveolar region here had affected the eruption of the P_4 so that it is rotated from its proper position 45° counterclockwise. The pocket is 6 deep just in front of the mesiobuccal corner of the M_1 root. Small conical impressions in the sides of the pocket suggest that pieces of the second deciduous molar roots were still in their alveolae.

The symphysis is marked anteriorly by a weak crest, which fades out at a very faint symphyseal tubercle. The anterior surface is gently hollowed on either side of the crest. Digastric fossae are so faint that they are better felt than seen. They reach far forwards on the base and are separated by a distinct mental spine. The spine is about 6 long and points posteriorly. The planum alveolare is convex throughout and blends imperceptibly into a moderate superior torus. The inferior torus, which projects only slightly more posteriorly than the superior, is separated from the latter by a poorly defined genial pit that has a pair of vascular foramina, one on either side. The body is low in height, but quite robust. There is a single mental foramen beneath the mesial root of P_4 on the left and double foramina on either side of the P_4 root on the right. Measurements are given in Table 6.3.

The rami are complete on both sides, but a small flake of bone is detached from the medial side of the right condyle and is cemented by matrix in its displaced position. The ramus is relatively low and joins the body rather obliquely, so that a line drawn through the center of the ramus meets the alveolar plane at an angle of 110°. The oblique line is first seen under the distal part of M_2 about 8 below the alveolar margin. It rises smoothly into the takeoff of the anterior border of the ramus but continues its curvature so that the lower part of the anterior border is concave in lateral view. This curvature is then reversed about 28 above the alveolar plane and the border thickens to make the anterior border of the coronoid process. The process is about 8 higher than the summit of the condyle and is blunt. In lateral profile it resembles a shark's dorsal fin because of the sigmoid curvature of the anterior border.

COMPARISON WITH MANDIBLES FROM EARLY AFRICAN *HOMO ERECTUS*

Several other mandibles have been discovered in the Turkana Basin that can be compared with that of the Nariokotome skeleton. There is not one of equivalent physiological age, however. When making these comparisons, then, like those of the crania, one needs to have a feeling for how much remaining growth the NK3 specimen might have undergone.

KNM-ER 730 (DAY AND LEAKEY, 1973)

This mandible is that of an old adult. There is a lot of alveolar resorption, and the three remaining teeth (the molars on the left) were worn to the gum. Only a trace of occlusal detail remains on the M_3. Comparisons with KNM-WT 15000 B are favorable. The two are similar in general form. KNM-ER 730 is a little more open lingually and was a little deeper in the body. The digastric and genial impressions are also much more marked.

TABLE 6.3.

Measurements (mm) of the mandibular body of KNM-WT 15000.

SIDE	LOCATION OF MEASUREMENT P_3	P_4	M_1	M_2	ORIENTATION OF MEASUREMENT
Left	18.2	18.1	19.5	21.0	transverse
	27.5	27.2	24.4	24.8	superoinferior
Right	18.5	19.0	20.5	21.3	transverse
	25.6	25.3*	23.2	23.5	superoinferior

* Resorbed area.

KNM-ER 731 (DAY AND LEAKEY, 1973)

There is so little left of this bone to examine that not much was said by the original describers. It looks, in all parts preserved, like the mandible of KNM-ER 730. It also came from an individual who suffered from severe periodontal disease.

KNM-ER 820 (LEAKEY AND WOOD, 1973)

This is a juvenile mandible with a mixed dentition. The first molars were in occlusion but had not closed their roots. This specimen is treated by Smith (Chapter 9, this volume). It makes a perfect juvenile stage in a cross-sectional growth series with KNM-WT 15000, KNM-ER 992, and KNM-ER 730.

KNM-ER 992 (LEAKEY AND WOOD, 1973)

This is a nearly complete adult mandible. It is the type of *Homo ergaster* (Groves and Mazak, 1975). It is complete except for 3 incisors and some damage to the rami. We consider that this mandible is a perfect adult model for KNM-WT 15000. The dental differences are discussed in Chapter 8 (this volume). The differences in the body are mainly in height. The ramus has a more developed sigmoidal profile in lateral view and a very slightly steeper posterior border. Masseteric and medial pterygoid muscle scars are strongly developed. The buttress from the interior of the body to the coronoid is stronger. All in all, the differences from the West Turkana specimen are those expected because of the differences in individual ages.

KNM-ER 1808 (LEAKEY AND WALKER, 1985)

Fragment KNM-ER 1808 S is a mandibular condyle on a small piece of neck. It is nearly identical in size and shape to that of KNM-WT 15000. Some mandibular teeth and two eroded fragments of the mandible are preserved also. These are like the equivalent parts of KNM-ER 730.

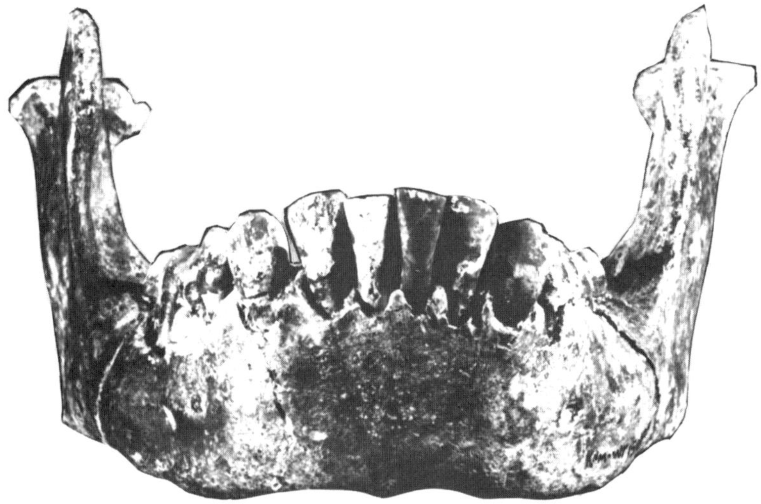

FIGURE 6.46.
Anterior view of mandible KNM-WT 15000 B.

FIGURE 6.47.
Posterior view of mandible KNM-WT 15000 B.

FIGURE 6.48.
Left lateral view of mandible KNM-WT 15000 B.

FIGURE 6.49.
Right lateral view of mandible KNM-WT 15000 B.

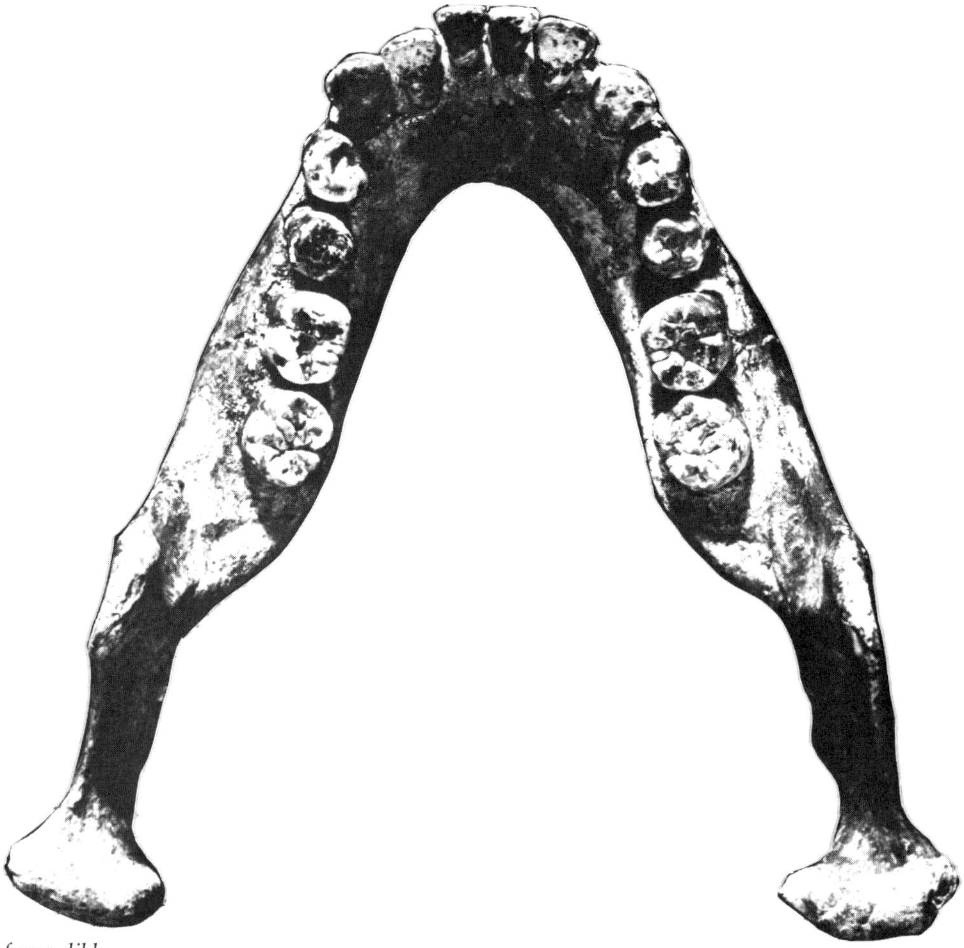

FIGURE 6.50.
Superior view of mandible KNM-WT 15000 B.

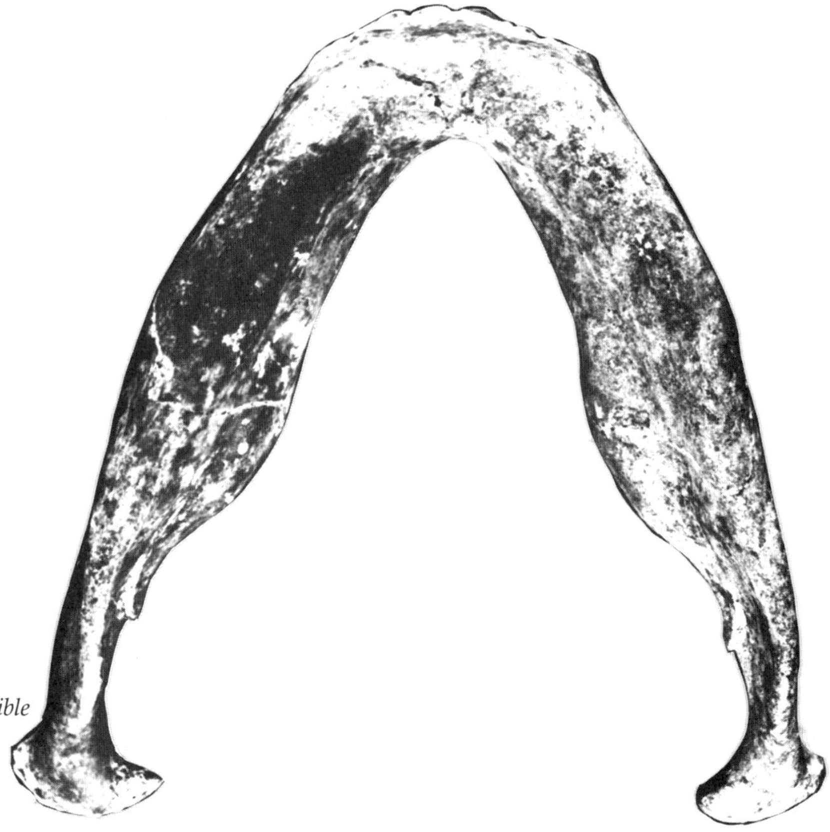

FIGURE 6.51.
Inferior view of mandible KNM-WT 15000 B.

The NK3 excavation site in September 1984, soon after the initial discovery was made. Excavation is being carried out in the shade of a tarpaulin. In the middle ground is the Nariokotome sand river and in the background the Morua Rith hills. (View to the west-northwest.)

The facial skeleton of KNM-WT 15000 was found in the roots of this tree on August 30, 1984. Kamoya Kimeo and Alan Walker are shown here loosening surface soil and unconsolidated sediment. National Geographic photographer David Brill is almost hidden by a small thorn tree. (View to the northwest.)

The right humerus and scapula of KNM-WT 15000 as seen in situ *on September 11, 1984. The humerus lies across the scapula and the scapula itself is crushed down into a large mammal footprint in the base of Unit 2.*

CHAPTER 7

THE POSTCRANIAL BONES

ALAN WALKER AND RICHARD LEAKEY

Nearly eighty pieces of the postcranial skeleton were found in the excavation at Nariokotome. They are documented here by descriptions, measurements, and photographs. These bones, associated as they are with the skull, enable us to identify positively the more fragmentary specimens that have been found at other localities in the Turkana Basin. For many parts of the skeleton of Homo erectus, *the Nariokotome specimens are the first examples to be found.*

Sixteen complete or partial vertebrae and much of the sacrum of KNM-WT 15000 have been recovered. The bones of the axial skeleton were strung out in roughly the order in the body, with upper thoracic bones mainly to the west and sacral parts mainly to the east (see Figure 4.6). Because the erosion slope cut into the deposit from the west we found only one cervical vertebra. The rest of the preserved axial skeleton consists of 18 ribs and some fragments of ribs. Of the upper-limb skeleton, both clavicles, one complete and one fragmentary scapula, and both ulnas were found together with a humerus and some hand bones. Of the lower-limb skeleton, much of the hip bones, both femurs, both tibias, and both fibulas were recovered. In addition, a possible foot bone has been found.

The individual bones are described and illustrated in this chapter and comparative notes are made about those other early African *H. erectus* postcranial bones which can be identified certainly. In the original note of discovery (Brown et al., 1985), we tentatively identified the postcranial bones that we had found in 1984. We did not know then that the skeleton had 6 lumbar vertebrae and we had to make some guesses about the positions of vertebrae whose superjacent or subjacent bones were missing. After the last vertebra was found in 1987, we had a much more complete idea of the segments of the column and so we have revised some of our earlier attributions.

All measurements are in millimeters. A list of all the parts found is given in Table 4.1. Except where noted, all illustrations of the bones and fragments of KNM-WT 15000 are reproduced at natural size.

TABLE 7.1.

Vertebral measurements (mm) of KNM-WT 15000.

ID SUFFIX	VERTEBRA	CANAL HEIGHT	CANAL WIDTH	DORSAL BODY LENGTH	VENTRAL BODY LENGTH	CRANIAL BODY HEIGHT	CRANIAL BODY WIDTH	CAUDAL BODY HEIGHT	CAUDAL BODY WIDTH
R	C7	10.5	21.6	8.7	8.2	13.1	24.7	14.1	23.9
S	T1	10.5	18.7	10.5	10.1	14.2	23.7	14.7	26.2
T	T2	11.9	14.9	11.4	11.2	14.6	22.7	16.2	28.1
U	T3	12.2	14.0	11.6	10.9	16.2	22.5	17.5	26.3
CA	T5	13.1	13.1	11.8	9.8	18.0	23.5	19.2	24.7
W	T7	13.3	13.2	12.7	11.1	20.0	25.1	20.1	28.5
V	T8	13.2	12.2	12.8	11.1	20.5	27.4	21.1	29.3
X	T10	13.2	11.9						
Y	T11	13.6	12.9	15.5	13.7	22.7	30.0	23.3	31.8
AR & BA	L1			17.3	15.6	23.7	30.1	23.5	34.1
AV & AA	L2	12.8	17.2	19.7	17.6		32.9		34.2
Z & BW	L3	12.2	17.2		18.4				
AB	L4	12.6	18.5	17.1	19.0*	27.3	38.6		
BM	L5	14.2	20.0	15.8	19.2	27.5	39.8	28.5	41.2
AC	L6	15.8	21.6	14.9	20.7	27.6	41.9	27.0	41.8
AD & BY	S1		28.0*						
BB	S2		>30.0						
BC	S3	6.4	30.0*						
AE	S4	4.9	16.9	17.5	14.5	8.1	25.0	7.9	17.6
AF	S5		12.8	6.1	17.1	7.3	12.1	7.2	12.0

* Estimated value.

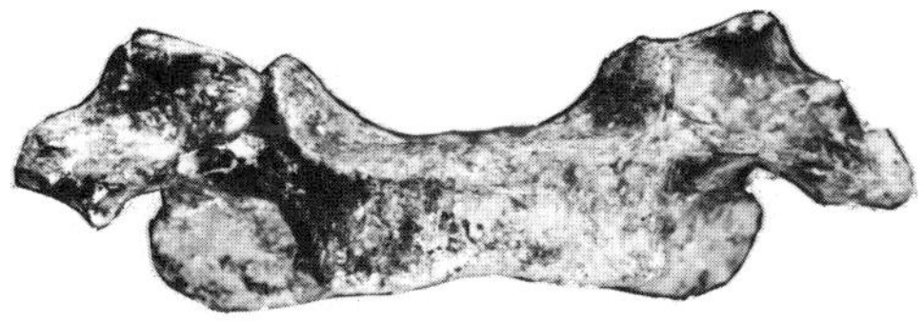

FIGURE 7.1.
Anterior view of cervical vertebra 7 KNM-WT 15000 R.

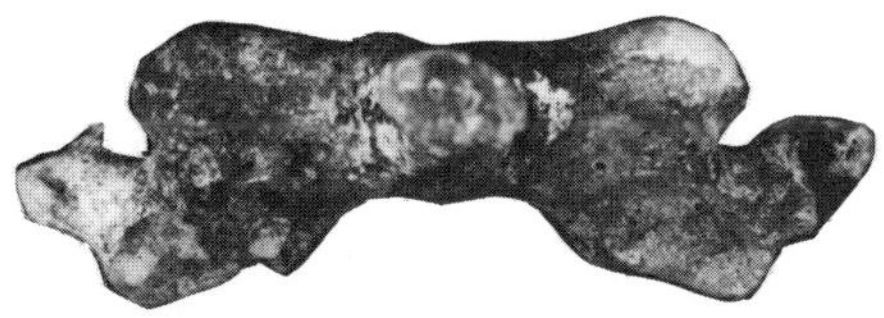

FIGURE 7.2.
Posterior view of cervical vertebra 7 KNM-WT 15000 R.

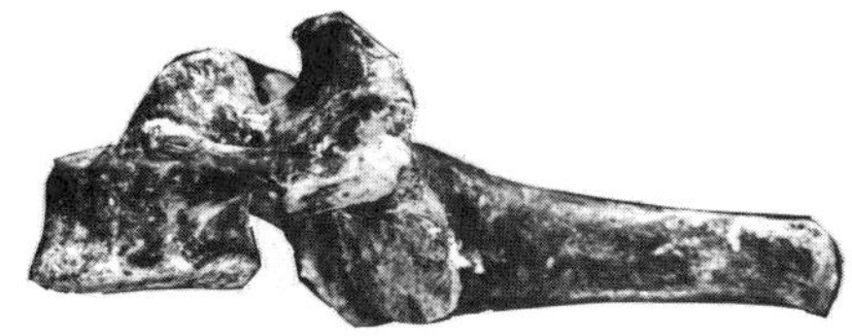

FIGURE 7.3.
Left lateral view of cervical vertebra 7 KNM-WT 15000 R.

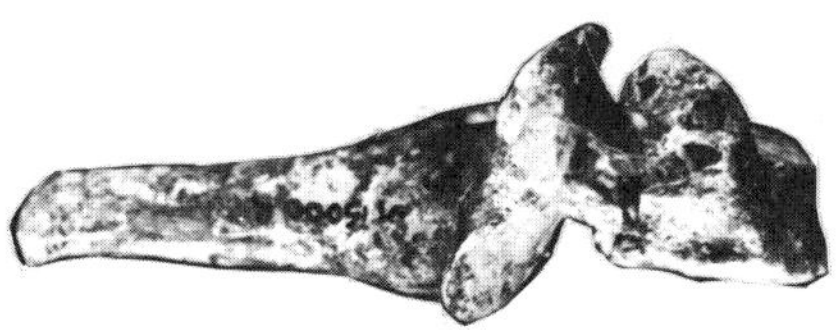

FIGURE 7.4.
Right lateral view of cervical vertebra 7 KNM-WT 15000 R.

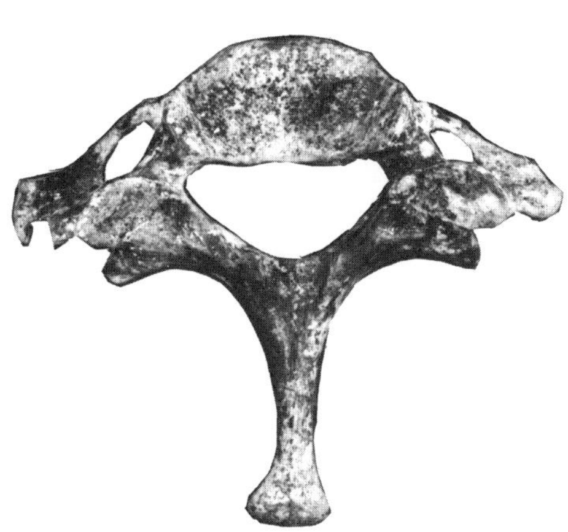

FIGURE 7.5.
Superior view of cervical vertebra 7 KNM-WT 15000 R.

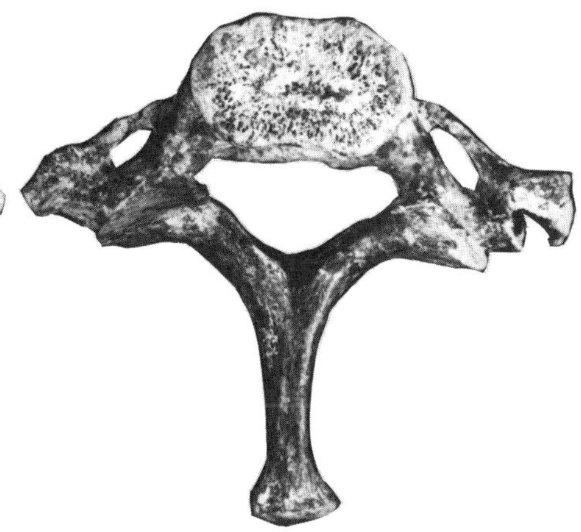

FIGURE 7.6.
Inferior view of cervical vertebra 7 KNM-WT 15000 R.

THE AXIAL SKELETON: VERTEBRAE AND RIBS

The identification of the vertebrae was carried out by comparison with those of modern hominoids. A list of measurements of the bodies and canals of the vertebrae of KNM-WT 15000 is given in Table 7.1.

KNM-WT 15000 R: CERVICAL VERTEBRA 7

(FIGURES 7.1–7.6)

This vertebra is complete except for a tiny piece of bone on the dorsal side of the left transverse process. The body of the vertebra is oval in superior view and more kidney-shaped in inferior view. The superior surface has extremely well-developed lateral articular surfaces, which face medially and a little anteriorly. The inferior surface shows just a trace of being reciprocally-shaped, by being a little beveled laterally. The anterior surface of the body is lipped slightly above and below and has faint hollowings about 7.5 from the midline on each side. The dorsal surface is also lipped above and below and there are bilateral vascular foramina piercing it.

The foramina transversaria are subequal in size and elongated, their long axes running anteromedially–posterolaterally. The left is 8.1 by 2.6 and the right is 5.8 by 2.4. The bony bridge that makes their anterior margins is extremely thin (left

1.4 and right 1.5 superoinferiorly), and the anterior tubercle is well lateral and very close to the posterior tubercle. The pedicles are cylindrical and join to the body with their long axes situated at the level of the body's upper surface. The transverse processes, which are short and not very stout, support the superior articular facets above and the bridges for the foramina in front. The left extends 30.7 from the midline and the right extends 29.8.

The superior facets are pear-shaped in outline, apices medially. They have a slight dorsal concavity and are set on a plane about 140° to the coronal, so that they face upwards and backwards. They measure 11.0 mediolaterally by 8 superoinferiorly (left) and 11.5 by 8.6 (right). The inferior facets are similarly directed, but in the opposite sense. Their shape is, however, more oval in outline and the lateral convexity is more pronounced. Their major dimensions are 10.5 mediolaterally by 8.5 superoinferiorly (left) and 11.9 by 9.1 (right).

The 4.0 thick laminae meet at an angle of about 110° inferiorly and leave a 3.0 wide groove between them where they meet. The dorsal arch of the canal is an even curve superiorly and the canal itself is 21.6 mediolaterally by 10.5 anteroposteriorly. The spinous process begins by taking off smoothly from the laminae, so that it is difficult to determine where it begins. Its length from the dorsal edge of the canal on the superior surface is 29.9, the last centimeter of this being expanded to make the tubercle.

The tubercle itself has a roughened oval surface in dorsal view whose long axis is twisted so that it runs downwards by some 10° or so from left to right. The tubercle is 9.1 mediolaterally by 6.0 superoinferiorly.

A seventh cervical vertebra from an adult early African *H. erectus,* KNM-ER 164, is known from Koobi Fora (Day and Leakey, 1974). It is associated with a first thoracic vertebra, a piece of parietal, and some hand bones. The two vertebrae are still stuck together with matrix, but a canal dimension has been given by the describers as 20 in maximum breadth by 13.5 in maximum height. This is close to the size in KNM-WT 15000 R, but a little less wide and higher. Apart from being adult, these bones are similar to KNM-WT 15000 R and S and, for that matter, to *H. sapiens* bones in their parts preserved.

KNM-WT 15000 S: THORACIC VERTEBRA 1

(FIGURES 7.7–7.12)

This vertebra is complete except for very minor abrasion around the left transverse process. The body is kidney-shaped in superior view and has lateral articular surfaces which are placed posterolaterally, notching the pedicles. The anterior surface is flat and turns an angle to reach the lateral (articular) side. It is lipped slightly by both upper and lower surfaces. The articular facet for the rib seems to be divided into small upper and lower parts by a pit which extends anteriorly. The lower surface is also kidney-shaped and has the middle anterior part of the surface notched upwards. There is a single large vascular foramen centrally placed.

The pedicles are short and stout and take off backwards and slightly upwards. They are 7.4 superoinferiorly and 6.5 mediolaterally (left) and 6.9 superoinferiorly and 7.5 mediolaterally (right). The transverse processes are stout and extend 31.1 from the midline (right) and 27.7 (left). On the unabraded right side it is 10.2 anteroposteriorly by 10.2 superoinferiorly. The costal facet of the transverse process is indistinct and small. The superior articular facets are concave trasversely, roughly oval in outline, and lipped below. The right is 13.5 transversely by 10.4 superoinferiorly, the left is 13.7 by 10.3. They face backwards, slightly upwards, and slightly medially. The inferior articular facets are also concave transversely. They face anteriorly, slightly downwards, and very slightly inwards. They almost lie on a coronal plane. They are 12.5 transversely and 9.4 superoinferiorly (right) and 11.8 by 9.4 (left).

The laminae are thick (6.2 left and 6.3 right) and slightly scarred on the superior edge, and they have strong depressions inferiorly for ligamentum flavum. The superoinferior height is 12.3 right and 12.5 left. The spinous process is strong and projects only a fraction downwards from the horizontal. The long axis of the superior margin makes an angle of 12° with the horizontal. It is 29.6 long from the dorsal edge of the vertebral foramen. It

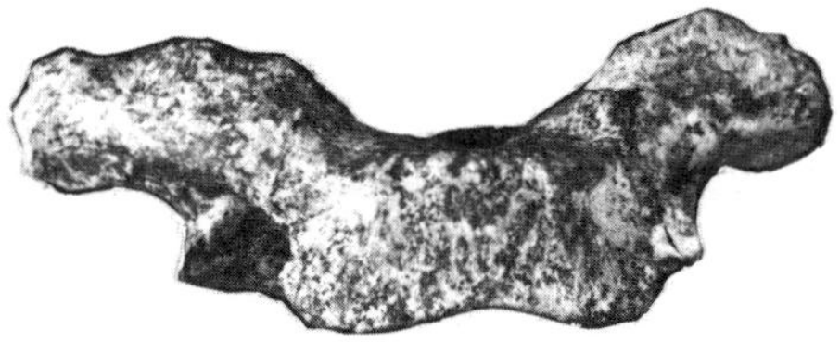

FIGURE 7.7.
Anterior view of thoracic vertebra 1 KNM-WT 15000 S.

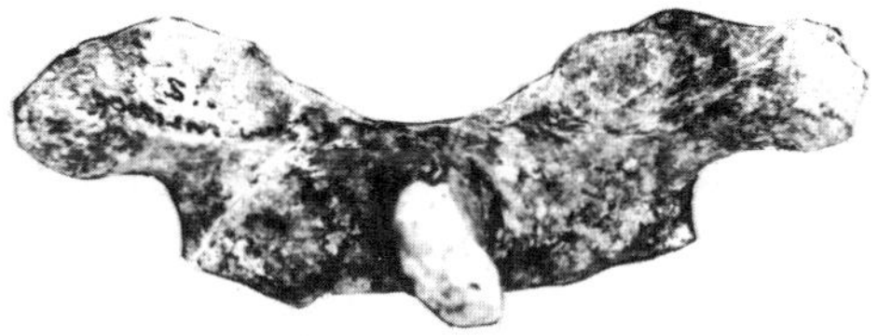

FIGURE 7.8.
Posterior view of thoracic vertebra 1 KNM-WT 15000 S.

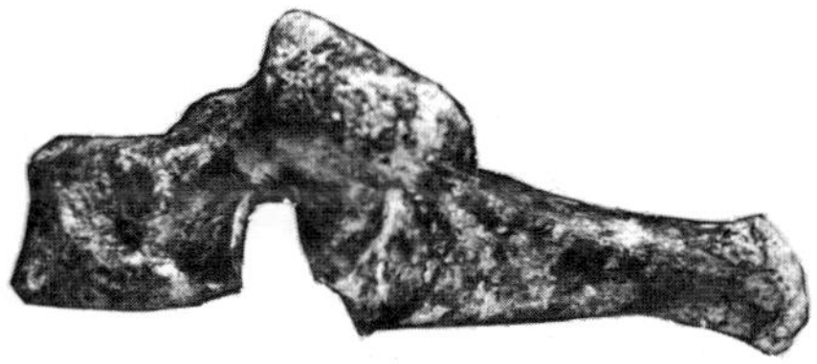

FIGURE 7.9.
Left lateral view of thoracic vertebra 1 KNM-WT 15000 S.

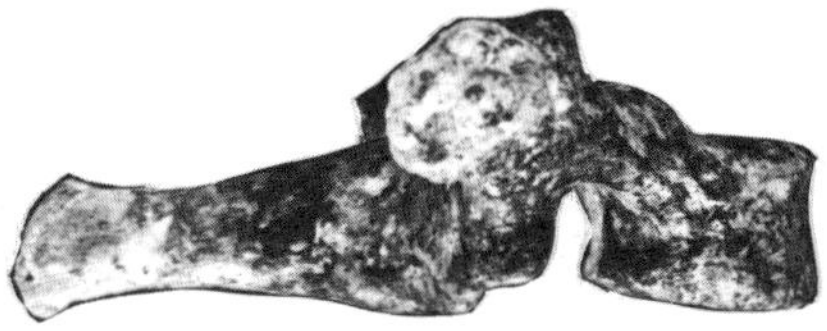

FIGURE 7.10.
Right lateral view of thoracic vertebra 1 KNM-WT 15000 S.

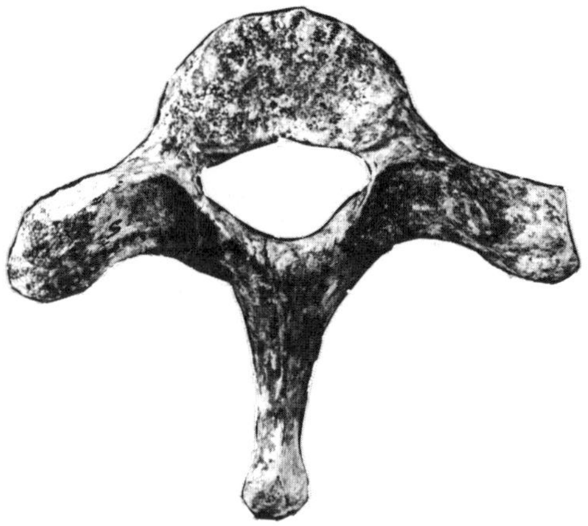

FIGURE 7.11.
Superior view of thoracic vertebra 1 KNM-WT 15000 S.

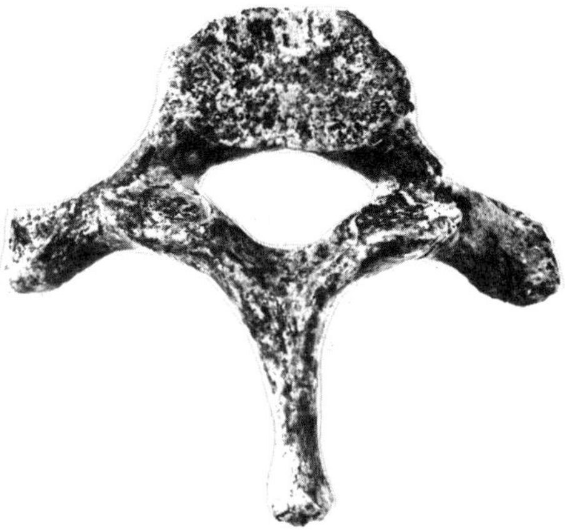

FIGURE 7.12.
Inferior view of thoracic vertebra 1 KNM-WT 15000 S.

ends in a tubercle 4.8 wide and 10.6 high which is rotated counterclockwise in posterior view by 20° from the midline. The vertebral foramen is a rounded triangle in outline, apices laterally and dorsally. Its height is 10.7 and its width 18.8.

The first thoracic vertebra of KNM-ER 164 (Day and Leakey, 1974) is very badly abraded, but what can be seen is similar to KNM-WT 15000.

KNM-WT 15000 T: THORACIC VERTEBRA 2

(FIGURES 7.13–7.18)

This vertebra is complete. The upper surface of the body is quadrangular in outline, with the dorsal border longer than the anterior and the two lateral sides angling anteromedially. The lateral articular facets are posterolateral, as are the pedicles. The anterior surface is hollowed but relatively straight. The lateral surfaces have demifacets top and bottom and are otherwise excavated, especially behind the anterolateral crests. The superior demifacet is 4.1 anteroposteriorly and 3.5 superoinferiorly (left) and 4.7 by 3.4 (right). The inferior facets are set much more obliquely and are 6.8 by 3.8 (left) and 6.9 by 3.6 (right). The inferior surface of the body is kidney-shaped in inferior view. The extended demifacets give a pointed appearance. The superior surface is slightly convex and the inferior one concave anteriorly. The pedicles are stout and are hollowed laterally, being 8.3 superoinferiorly by 5.8 mediolaterally (left) and 8.3 by 6.3 (right). The demifacets are set well dorsally,

FIGURE 7.13.
Anterior view of thoracic vertebra 2 KNM-WT 15000 T.

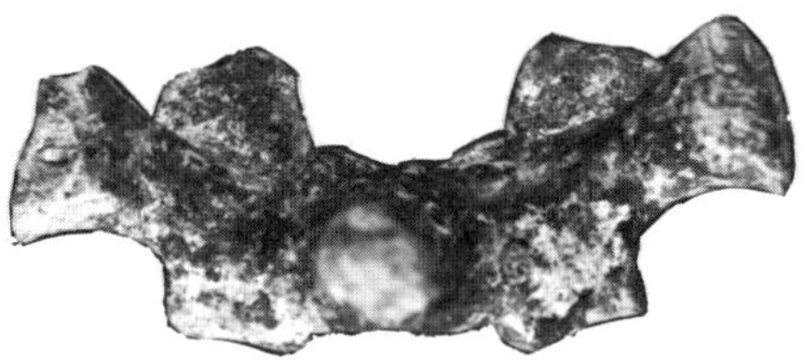

FIGURE 7.14.
Posterior view of thoracic vertebra 2 KNM-WT 15000 T.

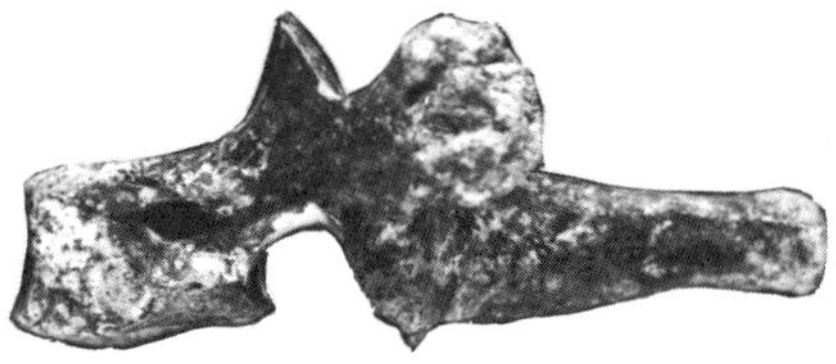

FIGURE 7.15.
Left lateral view of thoracic vertebra 2 KNM-WT 15000 T.

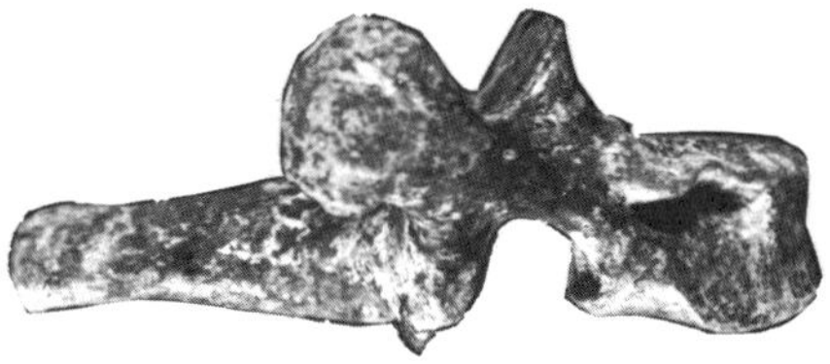

FIGURE 7.16.
Right lateral view of thoracic vertebra 2 KNM-WT 15000 T.

FIGURE 7.17.
Superior view of thoracic vertebra 2 KNM-WT 15000 T.

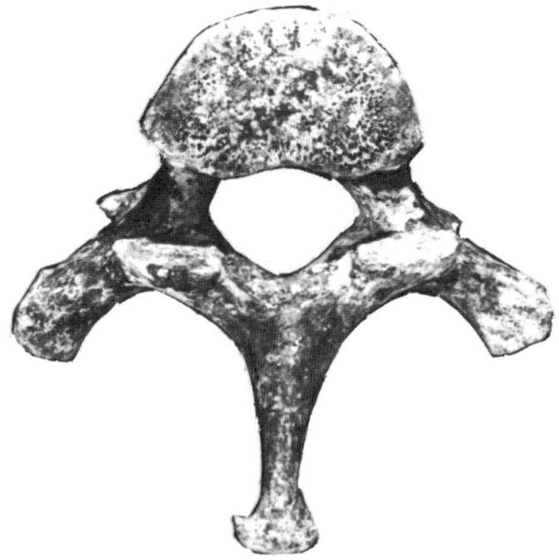

FIGURE 7.18.
Inferior view of thoracic vertebra 2 KNM-WT 15000 T.

near the base of the pedicle. The superior articular facets are plane and face dorsally and slightly upwards and laterally. The left is bigger than the right. It is 9.7 superoinferiorly by 13.5 mediolaterally and the right is 8.8 by 12.5. It excavates the lamina more than the right.

The inferior articular facets are virtually plane and face anteriorly and slightly downwards. They are rounded equilateral triangles in outline, with apices anterior, posterior, and lateral. They are 11.0 mediolaterally by 11.2 superoinferiorly (left) and 10.8 by 10.0 (right). The transverse processes are strong and semicylindrical, and they end in a blunt tubercle that is 9.5 mediolaterally by 13.7 superoinferiorly (left) and 10.1 by 13.6 (right). The costal facets on them are oval and are 8.6 superoinferiorly by 6.9 (left) and 9.2 by 7.4 (right). The transverse processes make an angle of 55° to the sagittal.

The laminae are thick and convex dorsally. They have narrow superior and broader inferior ligamentum flavum pits. They are 7.3 dorsoventrally by 11.8 superoinferiorly (left) and 6.9 by 11.9 (right). The spine is 28.7 long from the dorsal midline of the vertebral foramen. Its rounded superior margin slopes very gently downwards, but the straight first two-thirds of its inferior margin is horizontal. The last third is angled downwards a little. The underside of the spine has only a faint central groove. It ends in a roughened tubercle that is rotated 45° counterclockwise in dorsal view. The tubercle is 6.4 by 8.9. The vertebral foramen

FIGURE 7.19.
Anterior view of thoracic vertebra 3 KNM-WT 15000 U.

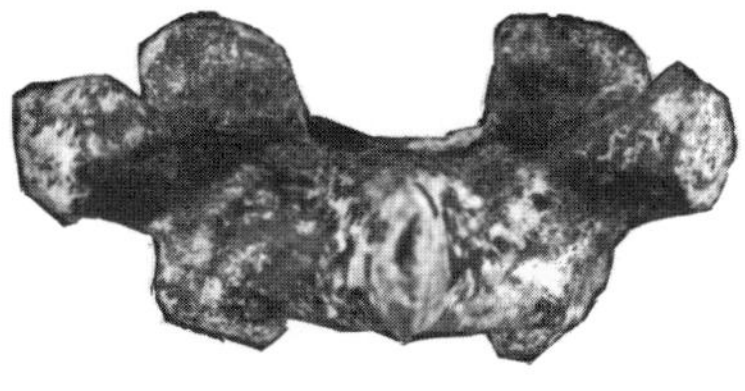

FIGURE 7.20.
Posterior view of thoracic vertebra 3 KNM-WT 15000 U.

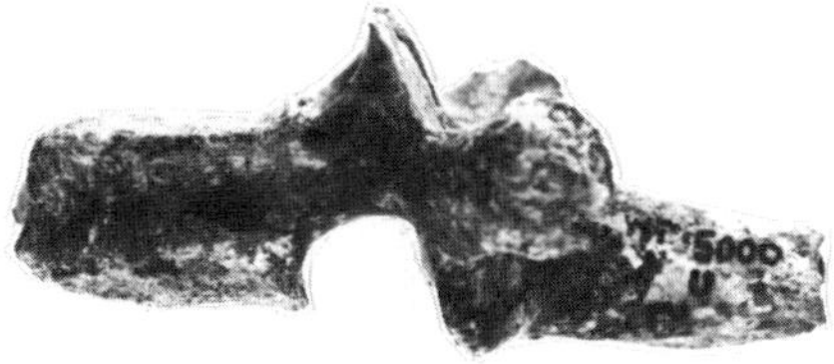

FIGURE 7.21.
Left lateral view of thoracic vertebra 3 KNM-WT 15000 U.

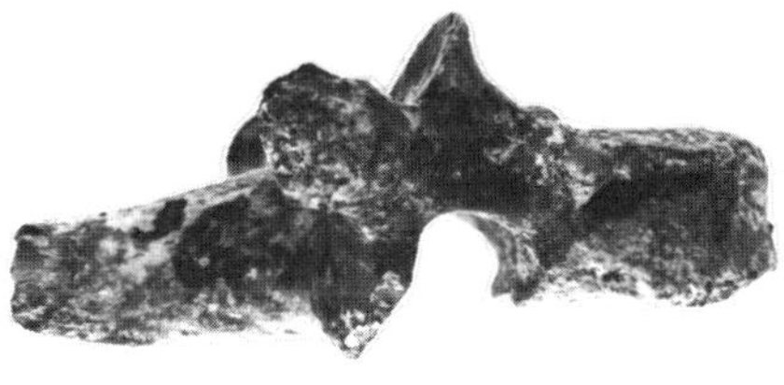

FIGURE 7.22.
Right lateral view of thoracic vertebra 3 KNM-WT 15000 U.

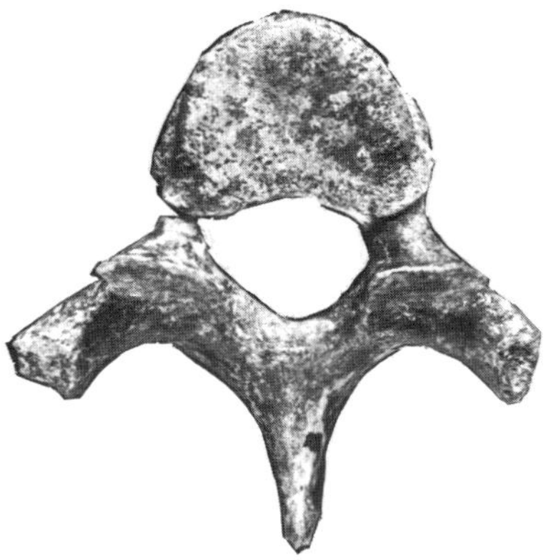

FIGURE 7.23.
Superior view of thoracic vertebra 3 KNM-WT 15000 U.

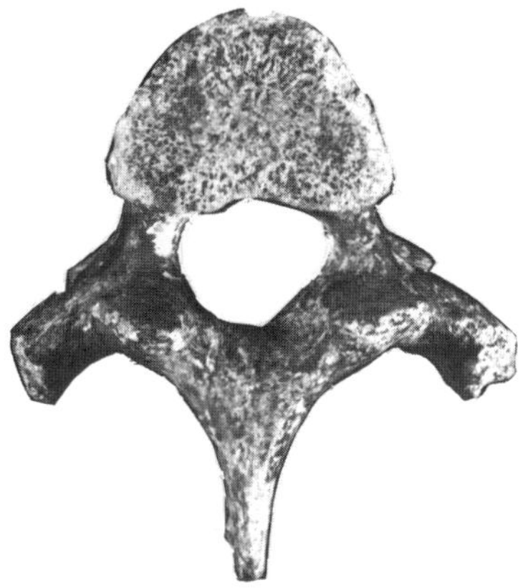

FIGURE 7.24.
Inferior view of thoracic vertebra 3 KNM-WT 15000 U.

is pentagonal in outline, with the laminae, pedicles, and anterior margin making sides.

KNM-WT 15000 U: THORACIC VERTEBRA 3

(FIGURES 7.19–7.24)

This is complete except for the tip of the spinous process and minor abrasion around the margin of the inferior surface and the margin of the right inferior articular facet. The superior surface is heart-shaped in outline, and the lateral articular facets are situated posterolaterally at the pedicles. It is gently notched dorsally. There is a faint anterior keel, and the lateral sides are hollowed deeply between the demifacets. The superior demifacets are nearly vertical and the horizontal ones merely flanges on the inferior surface. The superior demifacets are 5.7 superoinferiorly by 4.6 (left) and 3.1 by 5.3 (right). The left inferior demifacet is abraded but the right is 3.5 superoinferiorly by 9.6. The body is 11.9 thick. The inferior surface is heart-shaped in outline. The pedicles are short and have a ridge where they meet the body. This makes their superior surface notched. They are hollowed on their lateral surfaces. They are 7.4 superoinferiorly by 5.1 (left) and 8.1 by 5.8 (right).

The superior articular facets are in the same plane, are flat, and face backwards and slightly upwards. They are roughly oval in outline and they are 8.9 superoinferiorly by 11.9 (left) and 10.5 by 11.6 (right). Because of their orientation they excavate the superior part of the lamina. The inferior articular facets are flat and lie in the same plane. They face anteriorly and slightly downwards. The unabraded left facet is 10.5 superoinferiorly by 9.5. The transverse processes are stout and short and swell out to their ends. They are 6.9 by 7.6 superoinferiorly (left) and 7.0 by 7.9 (right). Their ends are roughened and the costal facets indistinct.

The laminae are thick and convex dorsally. Two strong ligamentum flavum scars notch their anterior margins, and less distinct inferior ones fill in the under surface between the posterior articular facets. They are 11.8 superoinferiorly by 7.5 (left) and 12.4 by 6.7 (right). The superior border of the spinous process is quite sharp and runs backwards and slightly downwards from the horizontal. The inferior border is even closer to the horizontal and slightly upturned at its distal end. It has a faint crest running along its underside. The vertebral foramen is pentagonal in outline, with the laminae and pedicles and anterior border making the sides.

KNM-WT 15000 CA: THORACIC VERTEBRA 5

(FIGURES 7.25–7.30)

This vertebra, which was found in 1987, is complete except for minor abrasion. The body is heart-shaped in both superior and inferior views. The spinous process is long—from the superior edge of the canal to the tip is 29.4. But it is thin near the tip, being only 2.1 thick near the poorly developed tubercle. The latter is more nearly a crest, rather than a tubercle. The process swings out from transverse processes across the laminae so as to overhang the inferior facets. The canal is pentagonal in outline and is 13.1 both dorsoventrally and transversely. The pedicles are placed well superiorly on the body and the lateral profiles of the superior margins swing superiorly and posteriorly up to the superior articular facets.

The demifacets are best seen on the right side, where the superior is 5.4 superoinferiorly by 7.1 anteroposteriorly. The posterior one is smaller and eroded. The laminae are convex dorsally and measure 14.0 superoinferiorly (left). The superior articular facets are oval in outline and plane, being set at a very slight angle to each other. Ligamentum flavum impressions are weak, about 4.0 wide and separated by a small midline keel. The inferior articular facets are also oval in outline, the right being slightly damaged inferiorly. The left is pointed in its most inferior part. A single, large (5.0 by 3.0) vascular foramen is seen in the body dorsally. The transverse processes are strong, club-shaped, and 51.2 between the tips. The lengths from the most inferior incutting of the superior articular facets are about 18.0. The processes are 6–8 thick by 14.8 superoinferiorly (left) and 7.5 by 12.7 (right), with the left being better preserved. A nearly circular rib facet, about 7.0 in diameter on the right process, faces almost directly laterally.

KNM-WT 15000 W: THORACIC VERTEBRA 7

(FIGURES 7.31–7.36)

This is complete except for cracking and crushing of the left transverse process and minor abrasion exposing spongy bone at the edges of the body. It was recorded as the ninth thoracic vertebra in Brown et al. (1985), but the discovery of more vertebrae since then has made us revise this attribution. The upper surface of the body is a little more than a semicircle in outline, with "lateral" articular facets abutting the pedicles. There are a few faint radial ridges around the periphery in the front. The anterior and lateral surfaces of the body are hollowed, the former slightly and the latter greatly, especially between the demifacets. The superior demifacets are continuous with the upper surface and set well back against the pedicles. Those on the left are abraded, but those on the right are 5.9 dorsoventrally by 4.9 (superior) and 9.2 dorsoventrally by 4.4 (inferior). A few small vascular foramina are visible around the lateral surface. The inferior surface of the body is almost a semicircle in outline. The greatest width is posteriorly. There are faint radial ridges peripherally. The pedicles are attached so that their superior borders are almost in line with the plane of the superior surface. They are hollowed on their lateral sides. They are 7.6 superoinferiorly and 4.5 mediolaterally (left) and 8.3 and 4.9 (right).

The superior articular facets are flat and face almost directly dorsally, with just a little tilt to face laterally. They are roughly oval in outline and 10.9

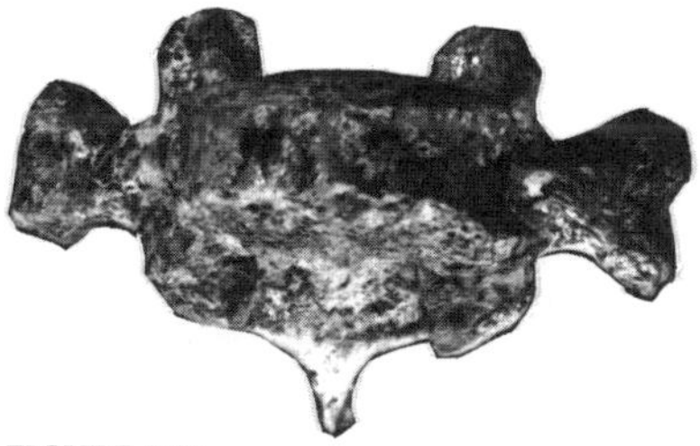

FIGURE 7.25.
Anterior view of thoracic vertebra 5 KNM-WT 15000 CA.

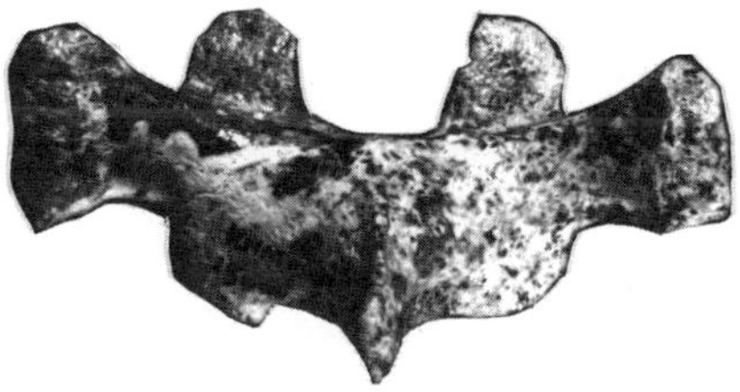

FIGURE 7.26.
Posterior view of thoracic vertebra 5 KNM-WT 15000 CA.

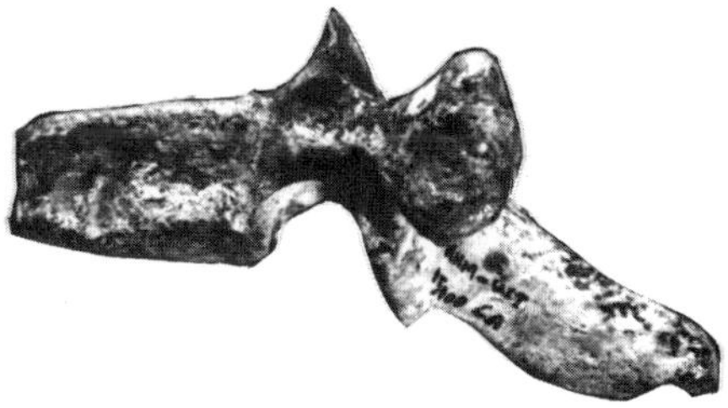

FIGURE 7.27.
Left lateral view of thoracic vertebra 5 KNM-WT 15000 CA.

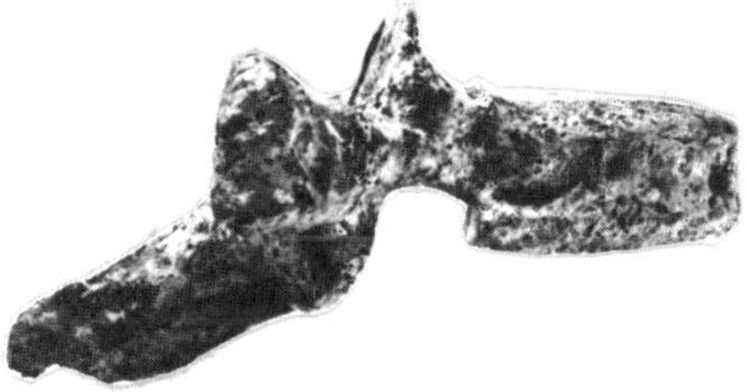

FIGURE 7.28.
Right lateral view of thoracic vertebra 5 KNM-WT 15000 CA.

FIGURE 7.29.
Superior view of thoracic vertebra 5 KNM-WT 15000 CA.

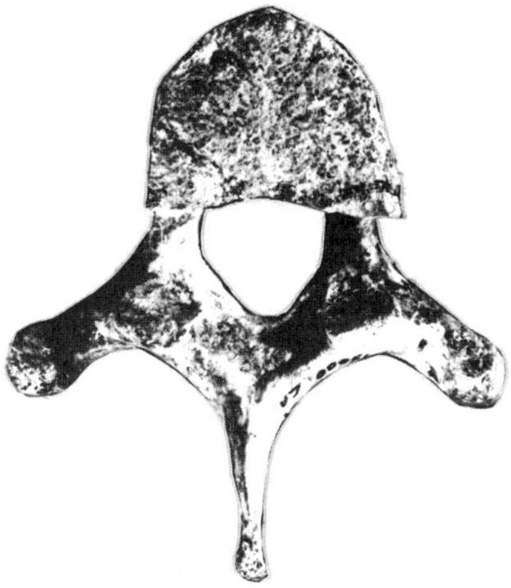

FIGURE 7.30.
Inferior view of thoracic vertebra 5 KNM-WT 15000 CA.

dorsoventrally by 9.3 (left) and 11.1 by 8.8 (right). The laminae are convex superiorly and the ligamentum flavum scar is a 3.6 wide strip notching the anterior border. The left is 6.6 dorsoventrally by 14.8 superinferiorly and the right is 6.7 by 14.6. The inferior ligament scar is a wide V between the inferior articular facets. The inferior facets are concave in the transverse plane and flat in the other. They are set to face anteriorly and little medially. They are roughly triangular in outline, with sides medially, inferiorly, and laterally; the medial side curves laterally to meet the superior apex. They are 10.0 superoinferiorly by 8.4 (left) and 10.5 by 9.8 (right).

The transverse processes are stout and end in a club-like tubercle. The right projects about 19 from the base of the anterior articular facet, horizontally and posterolaterally. The tubercle is 12.4 superoinferiorly by 9.2 dorsoventrally. The small, indistinct costal facet can be seen on the underside of the tubercle. The spinous process is thin and long. It is 35.2 from the laminar notch. Its anterior border is very sharp and notched near the terminal tubercle. The tubercle itself is 12.0 superoinferiorly by 3.7 transversely. The upper margin slopes downwards and backwards at an angle of 38° and the inferior one parallels it for the first half. At this point the margin is interrupted by a blunt tubercle. The rest of the inferior margin is then concave in profile as it hooks towards the terminal tubercle. The vertebral foramen, slightly distorted in outline, is hexagonal, with pedicles, laminae, and two

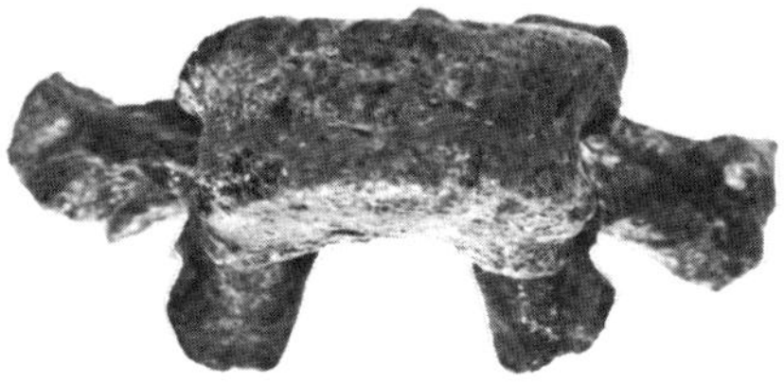

FIGURE 7.31.
Anterior view of thoracic vertebra 7 KNM-WT 15000 W.

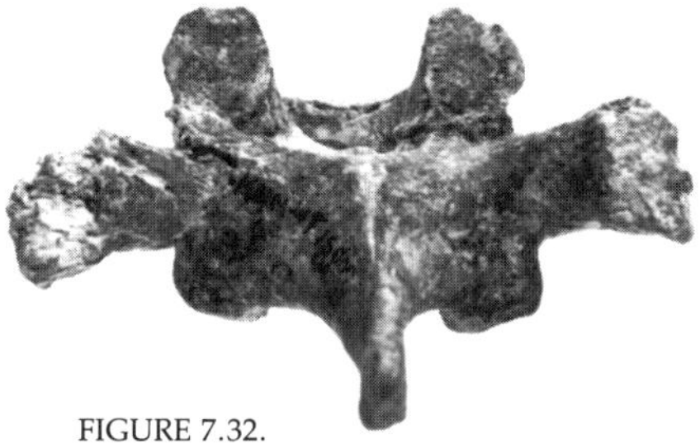

FIGURE 7.32.
Posterior view of thoracic vertebra 7 KNM-WT 15000 W.

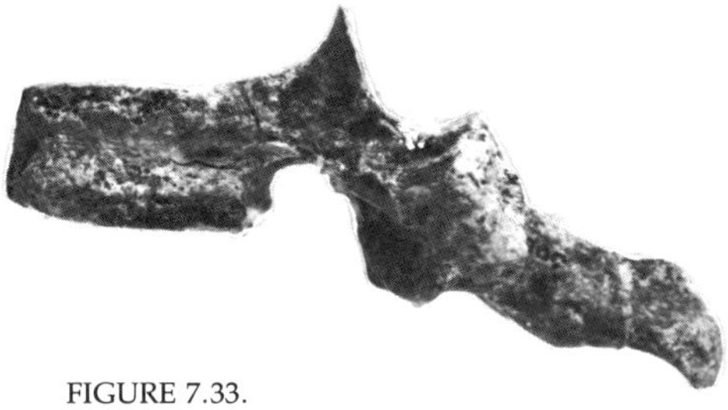

FIGURE 7.33.
Left lateral view of thoracic vertebra 7 KNM-WT 15000 W.

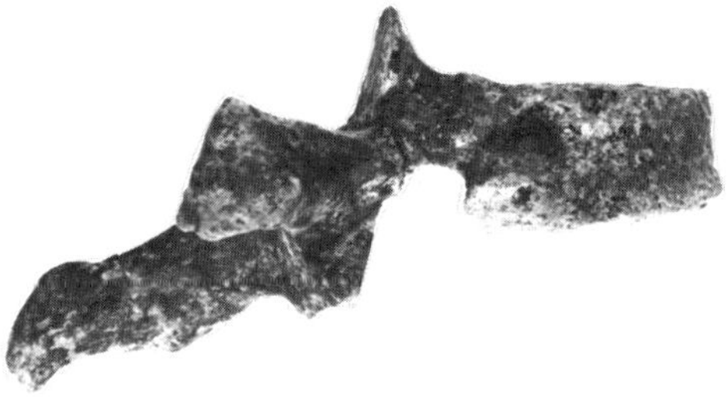

FIGURE 7.34.
Right lateral view of thoracic vertebra 7 KNM-WT 15000 W.

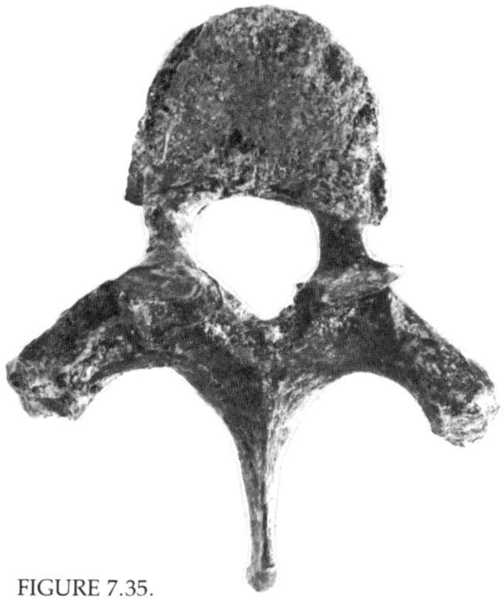

FIGURE 7.35.
Superior view of thoracic vertebra 7 KNM-WT 15000 W.

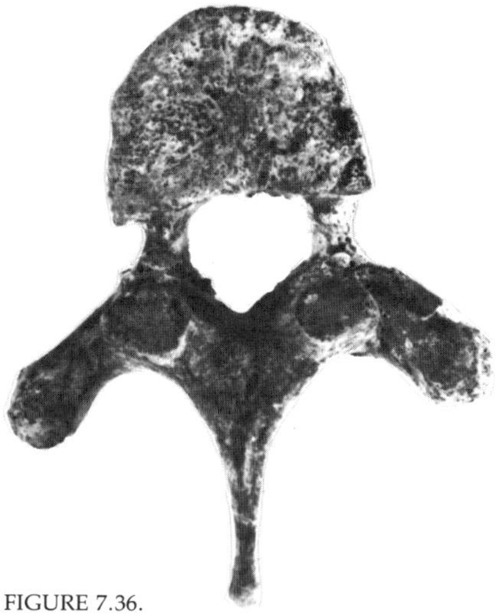

FIGURE 7.36.
Inferior view of thoracic vertebra 7 KNM-WT 15000 W.

inclined sides of the posterior margin of the body making sides.

KNM-WT 15000 V: THORACIC VERTEBRA 8

(FIGURES 7.37–7.42)

This is complete except for the left transverse process, which has lost its end. The superior surface of the body is heart-shaped and is saddled near the pedicles by the upward extension of the "lateral" articular facets and the merging of the demifacets with the surface itself. It has faint radial grooves around the anterior circumference and is widest across its middle. The anterior surface is slightly grooved, but the lateral surfaces are distinctly grooved and the grooves continue into the space between the demifacets. The demifacets have different orientations, the superior nearly vertical and the inferior set at a low angle. They are 5.0 superoinferiorly by 6.6 dorsoventrally (superior left) and 4.8 by 10.3 (inferior left). The superior right is displaced by a small crack and cannot be measured accurately. The inferior right is 4.3 by 9.5. The body is 13.2 thick. The inferior surface is a little more than a semicircle in outline. It too has a few radial grooves and ridges on its anterior circumference, but it is otherwise smooth. The pedicles are directed straight backwards. They have a ridge where they join the articular superior surface, thus making their superior margin

FIGURE 7.37.
Anterior view of thoracic vertebra 8 KNM-WT 15000 V.

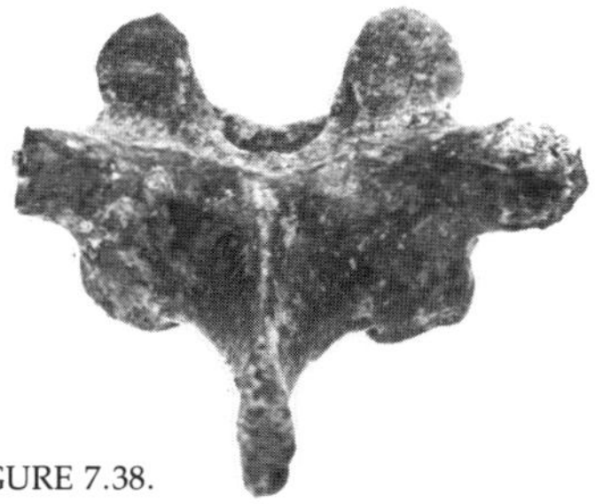

FIGURE 7.38.
Posterior view of thoracic vertebra 8 KNM-WT 15000 V.

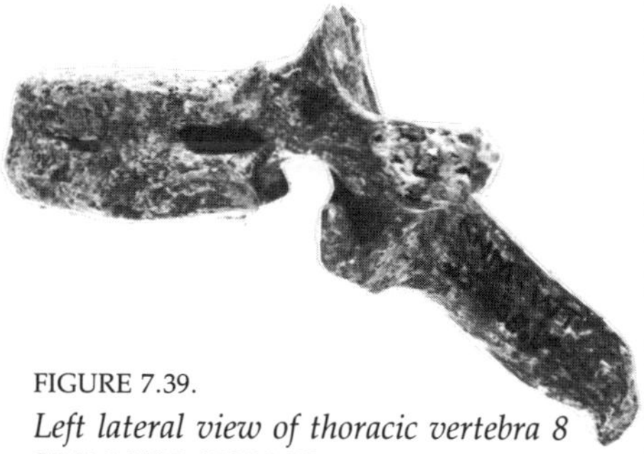

FIGURE 7.39.
Left lateral view of thoracic vertebra 8 KNM-WT 15000 V.

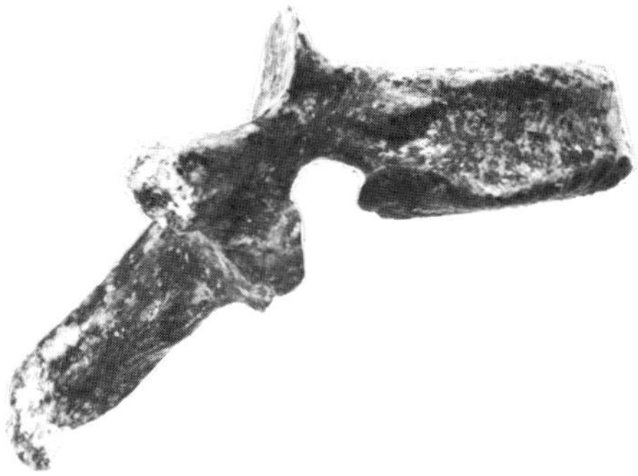

FIGURE 7.40.
Right lateral view of thoracic vertebra 8 KNM-WT 15000 V.

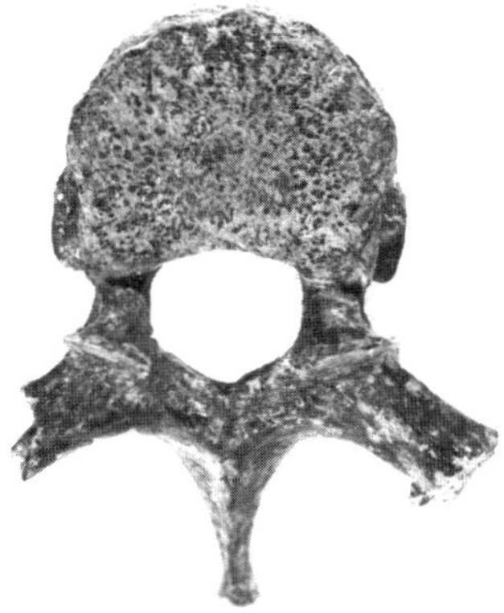

FIGURE 7.41.
Superior view of thoracic vertebra 8 KNM-WT 15000 V.

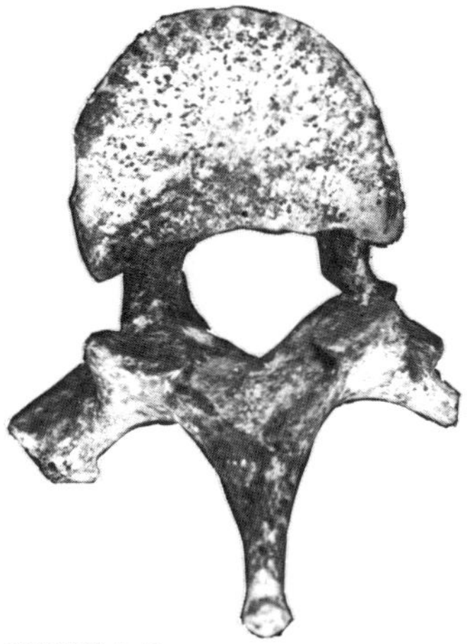

FIGURE 7.42.
Inferior view of thoracic vertebra 8 KNM-WT 15000 V.

notched. They are distinctly hollowed on their lateral sides. They are 4.7 mediolaterally by 8.2 superoinferiorly (left) and 5.2 by 8.5 (right).

The superior articular facets are flat and face backwards and slightly upwards and laterally. Roughly oval in shape, they are 9.7 superoinferiorly by 8.0 mediolaterally (left) and 11.2 by 9.1 (right). The inferior articular facets are more like rounded triangles in outline, with sides dorsally, medially, and anterolaterally. They are slightly concave and face forwards and slightly downwards and medially. They lie so that their curvatures are on part of a circle. They are 9.1 superoinferiorly by 8.2 mediolaterally (left) and 10.4 by 9.3 (right). The right transverse process is stout and ends in a blunt tubercle, which is slightly crushed. It is keeled a little superolaterally. It is 8.3 superoinferiorly by 7.3 dorsoventrally.

The laminae are fairly thin, are convex superiorly, and bear deep bilateral scars for ligamentum flavum. There is another ligamentum flavum scar, triangular in shape, filling the laminar surface between the inferior facets. The spinous process is 33.7 from the laminar notch, is very sharp on its superior and inferior margins, and ends in a thin bilaterally compressed tubercle. This latter is 4.2 mediolaterally by 10.1 superoinferiorly. There are slight excavations on either side of the spine's inferior crest in the middle of its length. The spine is angled downwards at 45° to the horizontal. The

FIGURE 7.43.
Anterior view of thoracic vertebra 9 KNM-WT 15000 BI.

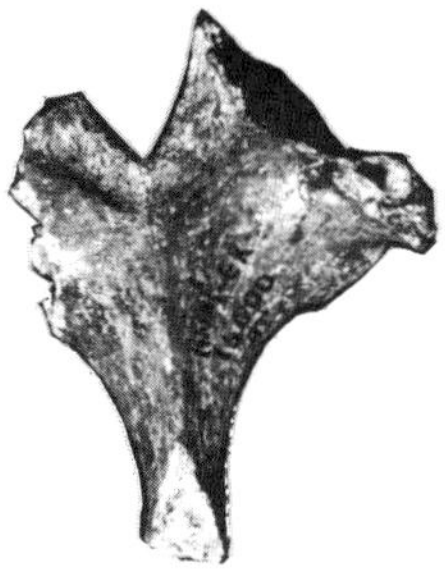

FIGURE 7.44.
Posterior view of thoracic vertebra 9 KNM-WT 15000 BI.

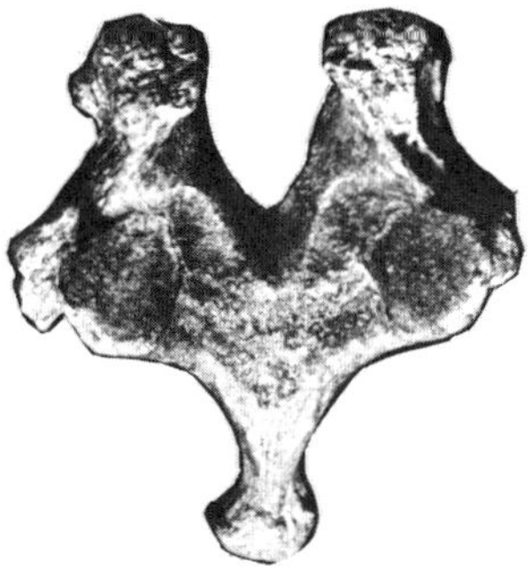

FIGURE 7.45.
Anterior view of thoracic vertebra 10 KNM-WT 15000 X.

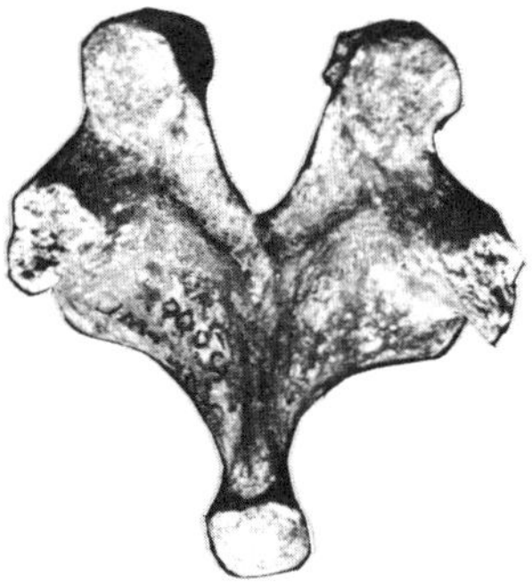

FIGURE 7.46.
Posterior view of thoracic vertebra 10 KNM-WT 15000 X.

vertebral foramen is roughly circular in outline. There are two vascular foramina about 3 apart in the middle of the dorsal surface of the body.

KNM-WT 15000 BI: THORACIC VERTEBRA 9

(FIGURES 7.43–7.44)

This piece of lamina, spinous process, and base of the right transverse process is from a lower thoracic vertebra and can be placed in our series as from thoracic vertebra 9. The laminae are thin and superiorly convex. The left is 5.4 dorsoventrally and 14.9 superoinferiorly. The ligamentum flavum scar is well marked and is a 4.5 wide depression bordering the superior notch. The right inferior articular process is preserved and is roughly rectangular in outline; it measures 11.0 superoinferiorly by 11.6 across the diagonal of the rectangle. The inferior ligament scars are not very pronounced. The facet is slightly concave in both major directions and faces anteriorly and a little downwards and inwards.

KNM-WT 15000 X: THORACIC VERTEBRA 10

(FIGURES 7.45–7.46)

This is the laminae, pedicles, and spinous process of a thoracic vertebra and we can place it in our series as the tenth. The pedicles are short, are grooved superiorly, and have sharp inferior borders. They are 5.9 mediolaterally (left) and 6.0 mediolaterally (right) and 10.9 superoinferiorly (right only; the left is broken inferiorly).

The flat superior articular facets face backwards and slightly laterally. They are roughly oval in outline and are 10.6 superoinferiorly by 8.4 mediolaterally (left) and 11.0 superoinferiorly by 9.6 mediolaterally (right). The inferior articular facets are roughly oval in outline, are slightly concave, and face anteriorly and slightly medially. They are 9.8 superoinferiorly and 9.3 mediolaterally (left) and 11.6 by 10.0 (right). The transverse processes have been broken off near their base, but can be seen to be angling off at 30° to the coronal plane. The right at its base is 7.7 superoinferiorly by 6.4 dorsoventrally.

The laminae are thin and have strong triangular scars separated by a spine for the superior

FIGURE 7.47.
Anterior view of thoracic vertebra 11 KNM-WT 15000 Y.

FIGURE 7.48.
Posterior view of thoracic vertebra 11 KNM-WT 15000 Y.

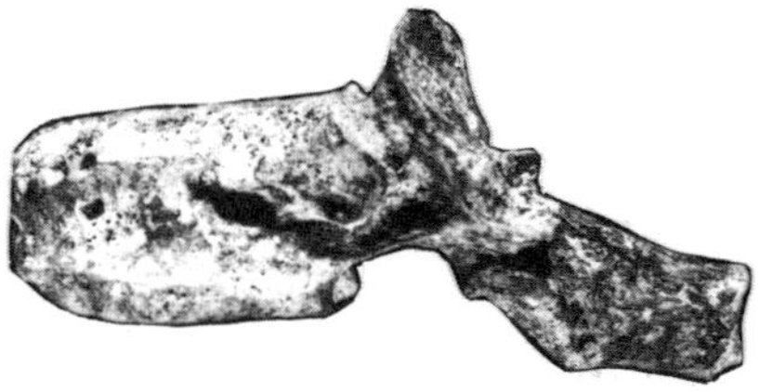

FIGURE 7.49.
Left lateral view of thoracic vertebra 11 KNM-WT 15000 Y.

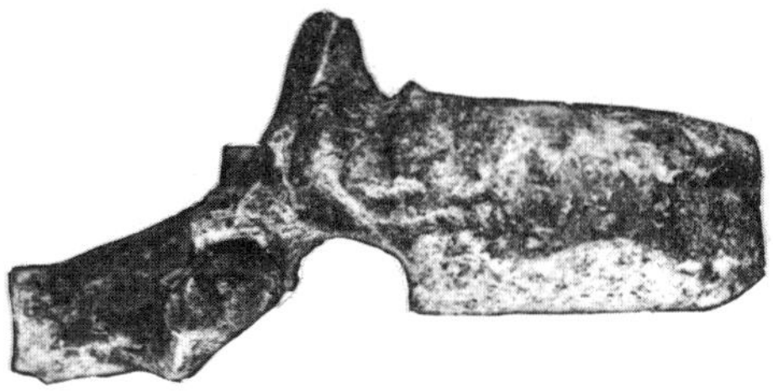

FIGURE 7.50.
Right lateral view of thoracic vertebra 11 KNM-WT 15000 Y.

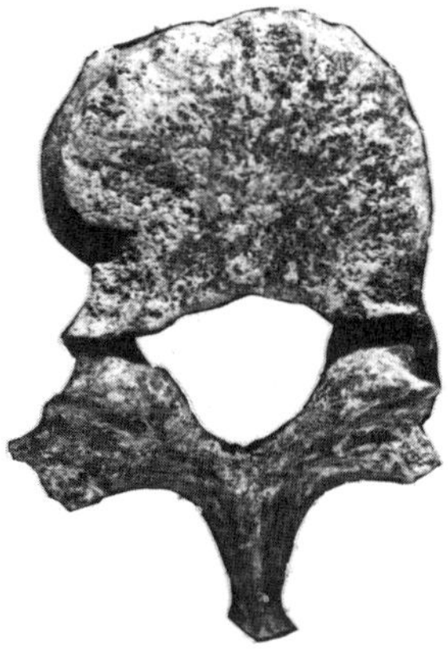

FIGURE 7.51.
Superior view of thoracic vertebra 11 KNM-WT 15000 Y.

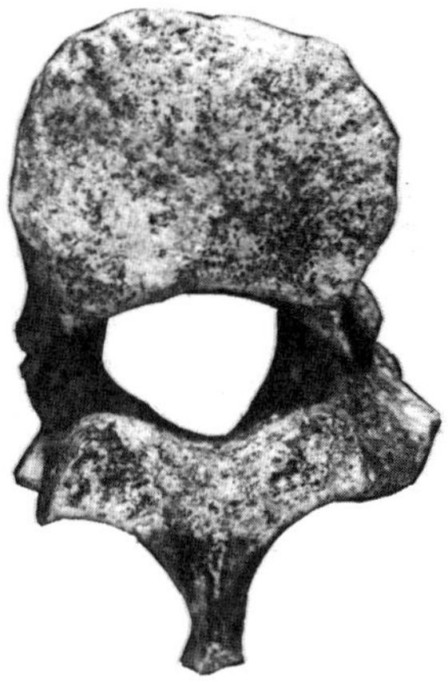

FIGURE 7.52.
Inferior view of thoracic vertebra 11 KNM-WT 15000 Y.

ligamentum flavum. The inferior scars are irregular and seem to occupy the whole of the underside except for the facets. The laminae are 5.1 thick by 15.0 (left) and 5.1 by 14.7 (right). The spinous process is short, only 27.1 from the anterior laminar notch. The inferior margins of the laminae swing round and smoothly join the lateral margins of the triangular spine. The spine is strongly ridged above and flat below so that it is triangular in section. It ends in a roughened, knob-like tubercle 7.8 mediolaterally by 7.1 superoinferiorly. This is rotated slightly counterclockwise in dorsal view. The spinous process angles downwards at about 45° (as estimated from the pedicle attachment).

KNM-WT 15000 Y: THORACIC VERTEBRA 11

(FIGURES 7.47–7.52)

This vertebra is missing the transverse processes, parts of the inferior articular processes, and the tip of the spine, and it is abraded around the margins of the body. The superior surface of the body is heart-shaped in outline. Abraded on the left lateral side, it still shows coarse radial ridges and grooves at the circumference anteriorly. The sides of the body are hollowed, gently anteriorly and deeply laterally, below and in front of the costal facet. The costal facet is large and is built partly on the body and partly on the pedicle. The left is 8.7 dorsoventrally by 8.4 superoinferiorly, the right is 11.5

FIGURE 7.53.
Anterior view of lumbar vertebra 1 KNM-WT 15000 AR.

FIGURE 7.54.
Superior view of lumbar vertebra 1 KNM-WT 15000 AR and BA.

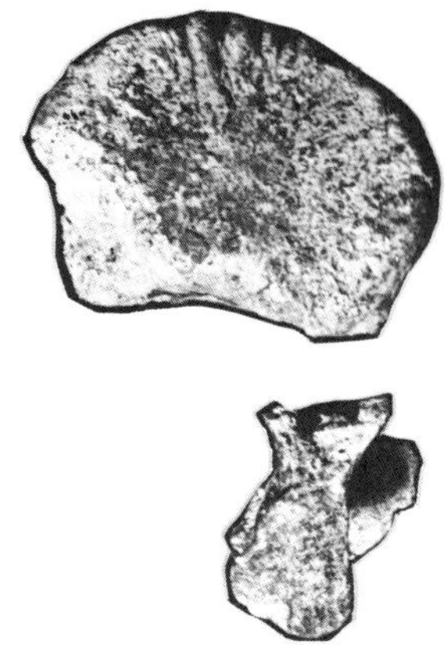

FIGURE 7.55.
Inferior view of lumbar vertebra 1 KNM-WT 15000 AR and BA.

dorsoventrally by 8.4 superoinferiorly. They are raised above the surfaces they are built on. The inferior surface of the body is almost circular in outline, except for the dorsal notch. It is 25.4 dorsoventrally by 31.8 wide. It is abraded on the right lateral and left anterior parts, but coarse radial ridges and grooves can be seen circumferentially.

The pedicles are short and stout. They are concave laterally where the costal facet is built out from them. They are 11.2 dorsoventrally by 7.5 mediolaterally (left) and 10.4 dorsoventrally by 7.2 mediolaterally (right). The vertebral canal is a rounded triangle in section, with apices anteriorly, laterally, and dorsally. The superior articular facets are flat and face dorsally and a little laterally. The left is 10.3 dorsoventrally by 9.6 mediolaterally and the right is 10.8 dorsoventrally by 9.1 mediolaterally. The laminae are almost in the same plane as these facets. They are 13.7 superoinferiorly and 5.7 dorsoventrally (left) and 15.4 by 5.6 (right). The ligamentum flavum scars extend from the medial side of the superior articular facet and meet at the midline. They are 5.0 wide superiorly and 3.7 at the midline. The preserved part of the spine is 22.5 long, and the entire spine would not have been much longer, because it is swelling out for its tubercle at the break. It is 8.9 dorsoventrally just in front of the break. The inferior facets are abraded, but enough remains of the right to see that they faced laterally and a little anteriorly. The broken bases of the transverse processes show that they came off directly inferior to the superior facets and that they were separated from the superior margin of the inferior facets by a 2.7 wide groove.

KNM-WT 15000 AR & BA: LUMBAR VERTEBRA 1

(FIGURES 7.53–7.55)

This vertebra is made of two parts, a vertebral body (AR) and a fragment of lamina with an inferior articular facet (BA).

AR is a broken lumbar vertebral body. The pedicle is broken at its base on the left, but a portion of the body itself is broken off near the pedicle on the right. The body was originally kidney-shaped in both superior and inferior views. The superior surface has radial ridges and grooves around the circumference extending up to 9 towards the center. The anterior surface is very slightly hollowed, but the hollow is deeper on the

FIGURE 7.56.
Anterior view of lumbar vertebra 2 KNM-WT 15000 AV.

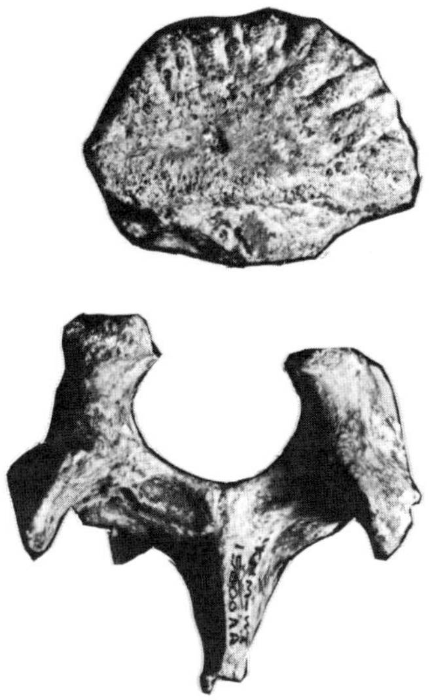

FIGURE 7.57.
Superior view of lumbar vertebra 2 KNM-WT 15000 AA and AV.

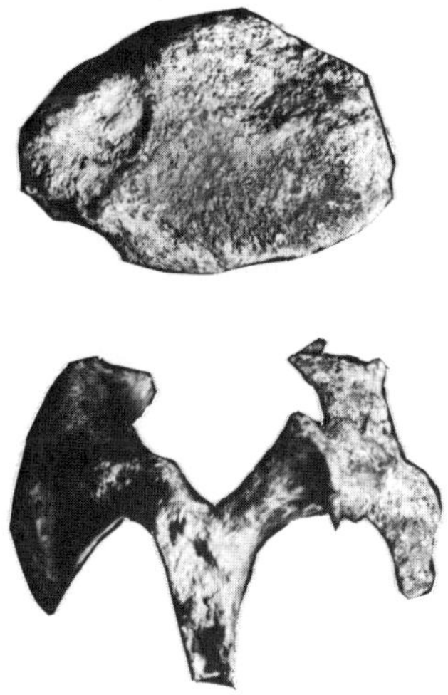

FIGURE 7.58.
Inferior view of lumbar vertebra 2 KNM-WT 15000 AA and AV.

sides. There are single vascular foramina at the deepest part of the hollow on each side. The inferior surface has ridge-and-groove patterns similar to the markings on the superior surface. These are, however, less pronounced on the inferior. There is a single large, oval vascular foramen on the anterior surface of the neural canal.

BA is a fragment that includes parts of left and right laminae and the right inferior articular facet. Not much can be said except that the right lamina is 6.3 thick dorsoventrally and that the articular facet is flat along its length and convex across it, so that its surface forms part of a cylinder. The facet is 13.7 by 10.6.

There is no other first lumbar vertebra known from early African *H. erectus.*

KNM-WT 15000 AA & AV: LUMBAR VERTEBRA 2

(FIGURES 7.56–7.58)

AV is an incomplete body of a lumbar vertebra. It has lost some bone next to the pedicles on both sides and fairly large pieces from the anterior part of the undersurface. The superior surface has strong radial ridges and grooves around its circumference extending up to 10 towards the center. The anterior surface is not much hollowed, but hollowing increases as one proceeds laterally. There are multiple vascular foramina in the depths of the hollow. The inferior surface is smooth over the part preserved.

AA is the pedicles, superior articular facets, left lamina and inferior articular facet, much of the right lamina, and the base of the spinous process. The preserved margins of the vertebral foramen are nearly circular in section, 17.3 wide by an estimated 14 dorsoventrally. The pedicles are stout, 11.3 dorsoventrally by 7.0 for the undamaged right, and slightly hollowed laterally. The anterior articular facets are large, rounded triangles in cross-section and concave transversely, plane superoinferiorly. They are part of inside cylinders. They face medially and a little dorsally. The left is 13.3 superoinferiorly by 11.9 dorsoventrally and the right 12.5 superoinferiorly by 12.3 dorsoventrally. The ligamentum flavum impressions are deep ovoid pits separated by the sharp anterior crest of the spinous process. The laminae are wide and thick. The complete left one is 17.9 superoinferiorly and 6.6 thick. The one inferior articular

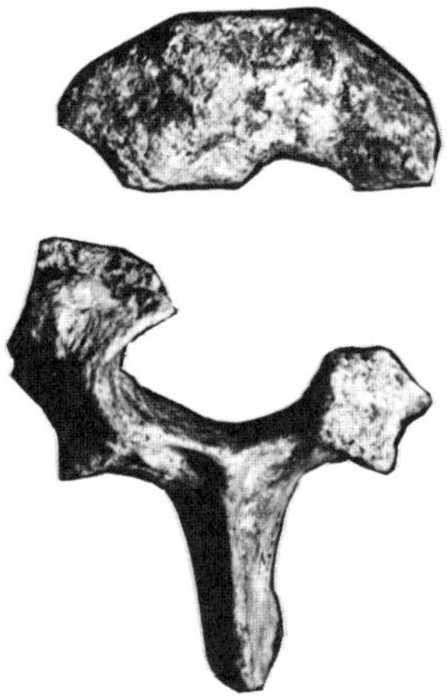

FIGURE 7.59.
Superior view of lumbar vertebra 3 KNM-WT 15000 Z and BW.

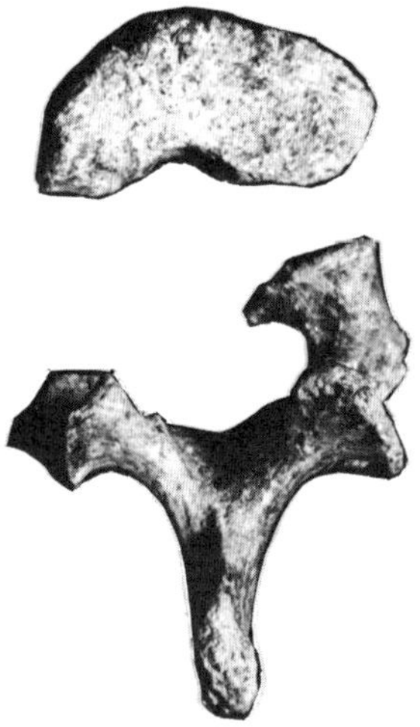

FIGURE 7.60.
Inferior view of lumbar vertebra 3 KNM-WT 15000 Z and BW.

facet is 13.4 superoinferiorly by 9.8 dorsoventrally. It is part of a cylinder facing almost directly laterally, its long axis superoinferior.

There is no other second lumbar vertebra known for early African *H. erectus.*

KNM-WT 15000 Z & BW: LUMBAR VERTEBRA 3

(FIGURES 7.59–7.60)

Fragment Z is the laminae, spinous process, inferior articular facets, and part of the left pedicle. Note that on the casts available there appears to be a small tag of bone on the superior edge of the left lamina, but this was matrix and has been removed since the molding. Only the inner surface of the left pedicle remains. It is enough to give an estimate of the vertebral canal width. The left lateral wall of the canal shows the contour of the roughly oval canal, whose greatest depth is more anterior than halfway the roughly estimated height of 13.0.

The laminae are broad (17.9 left and 17.7 right, superoinferiorly) and thick (7.3 left and 6.9 right, dorsoventrally). The spinous process takes off superiorly at the laminar notch and so completely separates the two strong ligamentum flavum impressions. These are 6.7 by 2.8 (left) and 7.4 by 3.3 (right). They parallel the superior laminar border. The superolateral edges of the laminae are marked by downward- and inward-running ridges from the lateral sides of where the superior facets would be to the medial sides dorsal to the inferior facets. The inferior facets are convex transversely and plane superoinferiorly. They are thus parts of cylinders. They face laterally and a little anteriorly. They are 10.9 dorsoventrally by 13.4 superoinferiorly (left) and 9.3 by 13.3 (right). The ligament impressions between them are strong. The spinous process springs from the complete superoinferior length of the laminae and begins broadly and attenuates slowly. The laminae actually rise to meet it posteriorly, making a tented roof over the canal. The spine is 27.9 long from the laminar notch. The superior border is very sharp and the inferior has a sharp crest for the free part of the spine after the laminae have been fused. In lateral view the terminal tubercle has two edges, a superior, longer one (13.5) and a shorter inferior one (8.8), angled to each other at about 90°.

BW is an abraded fragment of vertebral body. Not much can be said about it, but the part of the surface for the epiphysis that remains is probably the superior. It has several strongly marked radial grooves around the periphery of the surface. This specimen was found in 1984 (fragment A on the original site plan) and was given the suffix BW to replace an abraded vertebral body found in 1986 that upon further examination was determined not to be hominid.

FIGURE 7.61.
Anterior view of lumbar vertebra 4 KNM-WT 15000 AB.

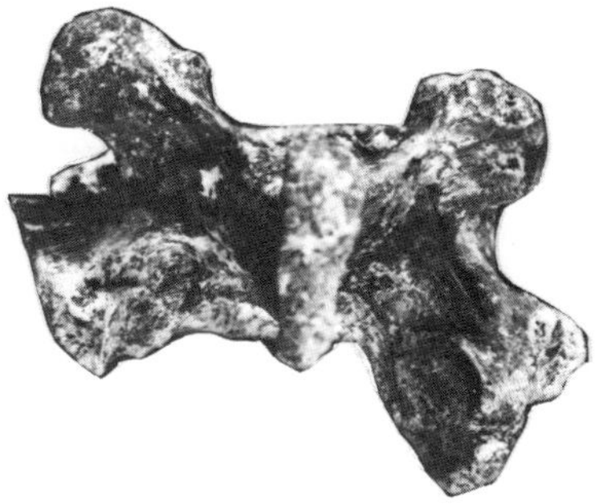

FIGURE 7.62.
Posterior view of lumbar vertebra 4 KNM-WT 15000 AB.

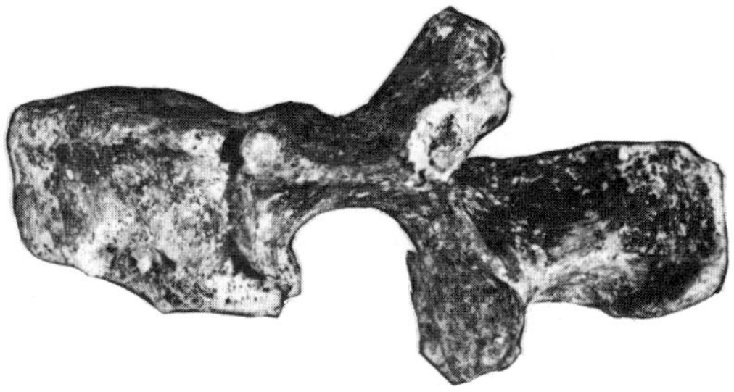

FIGURE 7.63.
Left lateral view of lumbar vertebra 4 KNM-WT 15000 AB.

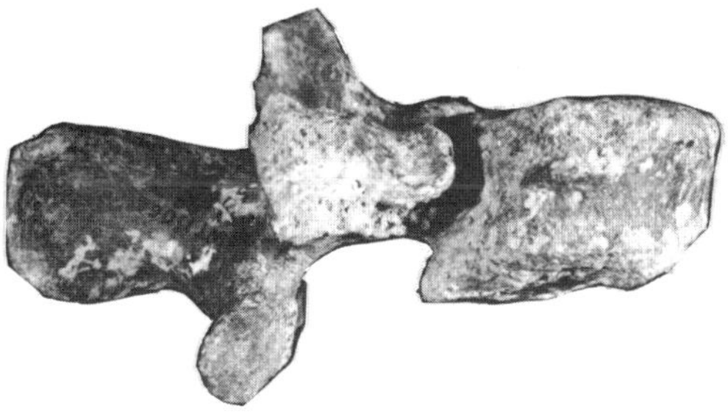

FIGURE 7.64.
Right lateral view of lumbar vertebra 4 KNM-WT 15000 AB.

FIGURE 7.65.
Superior view of lumbar vertebra 4 KNM-WT 15000 AB.

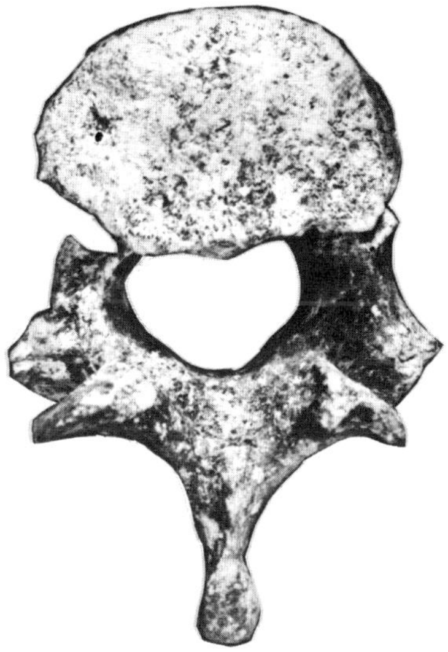

FIGURE 7.66.
Inferior view of lumbar vertebra 4 KNM-WT 15000 AB.

There is no other third lumbar vertebra known for early African *H. erectus.*

KNM-WT 15000 AB: LUMBAR VERTEBRA 4

(FIGURES 7.61–7.66)

This is complete except for abrasion of the inferior surface of the body and some bone missing at the junction of the body and pedicles, from the transverse processes, and from the right superior articular process. The superior surface of the body is semicircular in outline, with marked radial grooves all round the periphery. The sides of the body are more hollowed laterally. There are a few small vascular foramina on the sides. The inferior surface has had all anterior and lateral detail removed by abrasion. The vertebral canal is kidney-shaped in cross-section. The pedicles are very stout, the complete left being 8.9 superoinferiorly by 10.0 mediolaterally. A lateral crest runs up to the base of the transverse process.

The superior articular facet is oval in outline and concave medially. It faces medially and a little dorsally. It is 13.9 superoinferiorly by (11.0) transversely. The last measurement is estimated because of abrasion. The laminae are thick and deep, 6.0 dorsoventrally by 17.5 superoinferiorly (left) and 6.0 by 17.5 (right). They are excavated anteriorly by long, tear-drop-shaped ligamentum flavum pits running from medial to the superior articular processes to the midline (3.2 left and 5.0 right wide there) and separated by the front of the spine and by less deep pits lateral to them and inferior to the superior articular facet. The inferior articular facets are set on flaring processes and are slightly convex transversely. They face dorsally and laterally at an angle of 45° to the midline. They are rounded triangles in outline, being 11.9 mediolaterally by 12.8 superoinferiorly (left) and 12.8 by 9.9 (right).

The spinous process is rectangular in lateral view and has a sharp horizontal superior border. The laminae fuse at a very dorsal level to make the spine inferiorly. Here the crest begins as a small bump between the two laminae and runs dorsally for 10.5 before blending with the thick base of the terminal tubercle. The spine is 28.3 long and the terminal tubercle is 16.6 superoinferiorly by 6.3 mediolaterally. The base is the thickest part and turns under to meet the inferior crest. The underhang is about 10 long, but it is difficult to measure because of the rounded posteroinferior angle. There is a single midline nutrient foramen.

There is no other fourth lumbar vertebra known for early African *H. erectus*.

KNM-WT 15000 BM: LUMBAR VERTEBRA 5

(FIGURES 7.67–7.72)

This vertebra is complete except for the left transverse process and slight abrasion on the left lateral side of the superior surface of the body. It was found in 1985. In the original note (Brown et al., 1985) we listed KNM-WT 15000 AC as the fifth lumbar. We knew that it was the presacral vertebra but, because of damage to the inferior part of the body of KNM-WT 15000 AB, which we correctly identified was the fourth lumbar, we could not tell that there was an intervening vertebra that was yet to be found. It is worth pointing out here that the only other published early hominid for which a lumbar vertebral count is possible, Sts 14 from Sterkfontein, also has six lumbar vertebrae. As Robinson (1972) said, it is not possible with a sample of one to know if six was the modal number of lumbar vertebrae in *Australopithecus africanus*, but the finding of the same feature in the first known *H. erectus* skeleton makes it more likely. In a very large survey, Schultz and Straus (1945) recorded only 4.8 percent of people with six lumbars and almost 92 percent with five.

The superior body surface is kidney-shaped in outline. It has the radial ridges and grooves peripherally and is raised a little just anterior to the pedicles. The sides of the body are overhung all round by the superior edge but slope away at an angle of about 10° below the overhang. There are several large vascular foramina around the sides. The lower surface is oval in outline and slightly hollowed centrally with marked radial grooves anteriorly. It is 28.3 dorsoventrally by 42.0. The pedicles are thick and compressed superoinferiorly because of the low takeoff of the transverse process, which starts to swing out at the very base. They are 9.0 superoinferiorly by 12.4 mediolaterally (left) and 8.8 by 11.0 (right).

The superior articular processes overhang the superior surface of the body by half the length of their facets. These are concave transversely and plane superoinferiorly. Rectangular in outline, they are 12.0 superoinferiorly by 15.2 mediolaterally (left) and 11.3 by 14.0 (right). Inferior to each is a small depression about 6 in diameter and medial to those are 4.0 wide strong ligamentum flavum pits. The laminae are 6.2 dorsoventrally by 18.1 superoinferioly (left) and 5.7 dorsoventrally by 17.7 (right). The spinous process is rectangular in lateral view and projects horizontally just below the level of the superior body surface. It is sharp anteriorly, measures 27.2 long, and ends in a tubercle 19.2 superoinferiorly by 5.4 transversely. This is extended on to the superior and inferior margins of the spinous process by about 10. The inferior margin is short because of the dorsal joining of the laminae. It is 17.2 long and has a thin crest. The inferior articular processes flare out from the posterior parts of the laminae and so a notch is formed between them and the transverse pro-

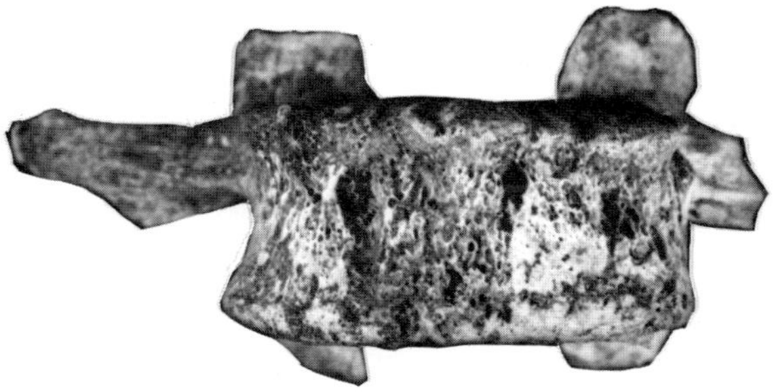

FIGURE 7.67.
Anterior view of lumbar vertebra 5 KNM-WT 15000 BM.

FIGURE 7.68.
Posterior view of lumbar vertebra 5 KNM-WT 15000 BM.

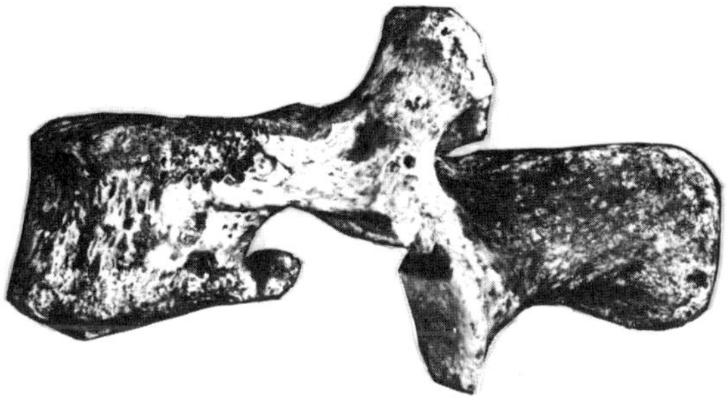

FIGURE 7.69.
Left lateral view of lumbar vertebra 5 KNM-WT 15000 BM.

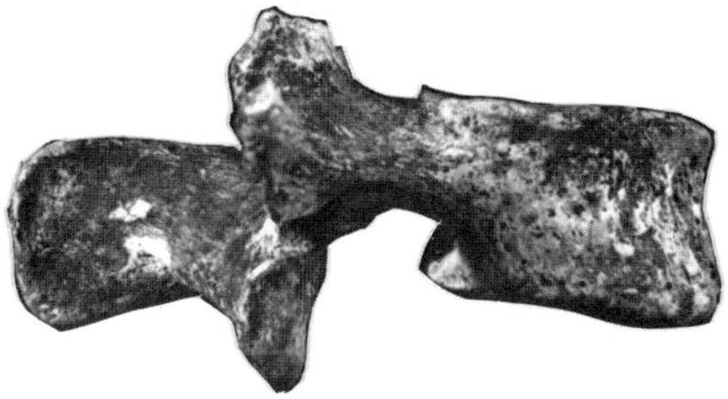

FIGURE 7.70.
Right lateral view of lumbar vertebra 5 KNM-WT 15000 BM.

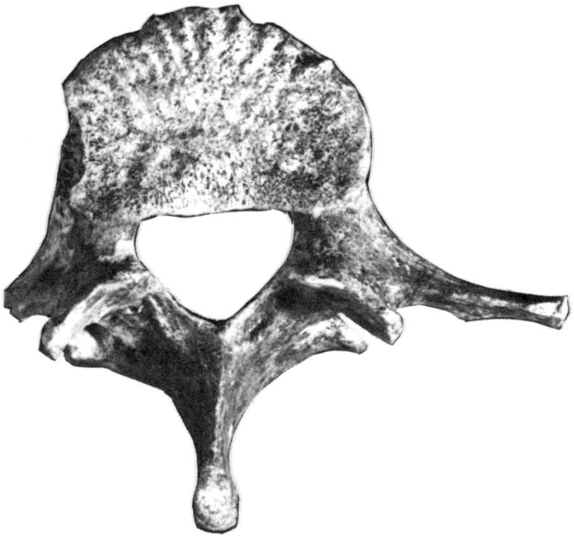

FIGURE 7.71.
Superior view of lumbar vertebra 5 KNM-WT 15000 BM.

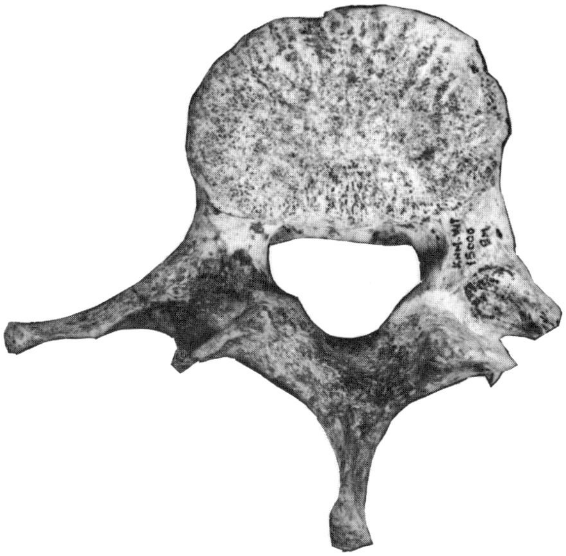

FIGURE 7.72.
Inferior view of lumbar vertebra 5 KNM-WT 15000 BM.

cesses. This is nearly enclosed to form a foramen on the left. It is 5.2 superoinferiorly and the right notch is 7.3 superoinferiorly.

The inferior facets are flat, quadrangular in outline, and asymmetrical because of the left notch development. They are 15.0 superoinferiorly by 12.2 mediolaterally (left) and 13.8 by 11.0 (right). They face anteriorly, slightly laterally, and very slightly downwards at angles of 50° to the midline. The transverse process, seen completely on the right, is 21.1 long from the lateral side of the superior facet. It is thin and compressed anteroposteriorly. It has a flat dorsal surface and a keeled ventral one. The anterior border is sharp and straight, the dorsal one angled, the medial half running slightly dorsally, and the lateral half running slightly ventrally. It ends in a small, blunt tubercle 6.0 superoinferiorly by 4.2 dorsoventrally. What remains of the left side has an extra groove on the inferior side of its base. This may be related to the notch formation. There is a single nutrient foramen.

KNM-WT 15000 AC: LUMBAR VERTEBRA 6

(FIGURES 7.73–7.78)

This vertebra is complete except for crushing damage of the right pedicle, the left superior articular process, and the base of the spinous process. The superior surface of this wedged body is semicircular in outline. It has well-defined radial grooves and ridges extending about 8 towards the center around the periphery anteriorly. The sides of the body are concave, but even the lateral sides are not deeply excavated. There are a few scattered vascular foramina over them. The inferior surface of the body is elliptical in outline. Its greatest width is 41.8 and its midline depth is 27.2. The center of the surface is slightly hollowed out.

The pedicles are massive and compressed superoinferiorly: the blunt transverse processes arise from the pedicles in such a way that it is difficult to say which is pedicle and which is process except distally. The complete left is 15.6 mediolaterally by 9.2 superoinferiorly. The transverse process, complete on the left, is little more than a blunt tubercle. It projects only 14.0 lateral to the side of the body, that is to say, 34.9 from the midline. It is separated from the superior articular process by a groove 8.8 wide. It has a flat inferior face, oval in outline, 10.7 dorsoventrally by 6.4 mediolaterally, which is set with a slight counterclockwise twist on the process. The superior articular processes completely overhang the superior surface of the body. The facet is roughly oval in outline, with the longest diameter being mediolateral. The complete right facet is 14.3 superoinferiorly by 13.6 mediolaterally. It is very slightly concave transversely and faces dorsally and medially at an angle 50° to the midline. Immediately inferior to the facet there is a deep, well-delineated pit. This is 10.0 wide on the left and 6.0 wide on the right.

The vertebral canal is triangular in outline, widest near the body. The spine is short, 23.3 long. Its superior border is thin and angled downwards and backwards at 25°. Because the laminae are so large they fuse well dorsally and leave the spine with only about 7.0 of inferior border. The terminal tubercle is 14.2 superoinferiorly by 7.0 transversely and is oval in outline. The laminae are 5.2 dorsoventrally and 14.0 superoinferiorly (left) and 4.2 and 15.1 (right). In inferior view they meet at an angle of 110°. The posterior articular facets face anteriorly, slightly laterally, and very slightly downwards. They are a little concave in both diameters. They are roughly triangular in outline, with apices superiorly, inferiorly, and laterally. They are 15.0 superoinferiorly by 14.0 mediolaterally (left) and 15.2 by 11.9 (right). There is a double nutrient foramen in the center of the dorsal surface body.

There are no other fifth or sixth lumbar vertebrae known for early African *H. erectus.*

KNM-WT 15000 AD & BY: SACRAL VERTEBRA 1

(FIGURES 7.79–7.80)

This vertebra has the right superior articular process preserved separately as fragment BY. The right lateral part is missing, and the body is broken and crushed to the right of the midline in its inferior part. The size of the base can be estimated by doubling that of the left half. In order to estimate the size of the superior surface, we can use the inferior surface of vertebra AC. Although the

FIGURE 7.73.
Anterior view of lumbar vertebra 6
KNM-WT 15000 AC.

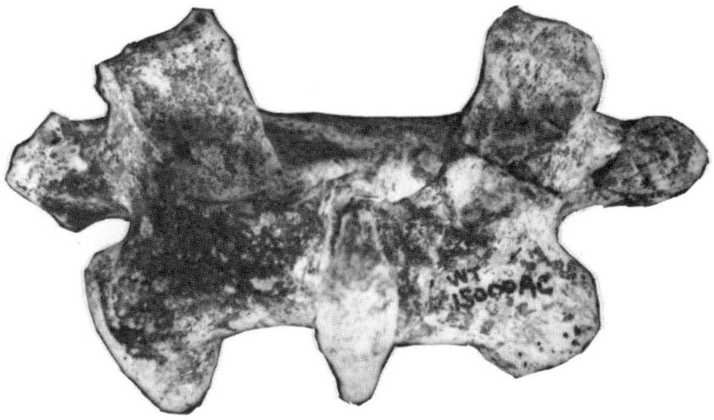

FIGURE 7.74.
Posterior view of lumbar vertebra 6
KNM-WT 15000 AC.

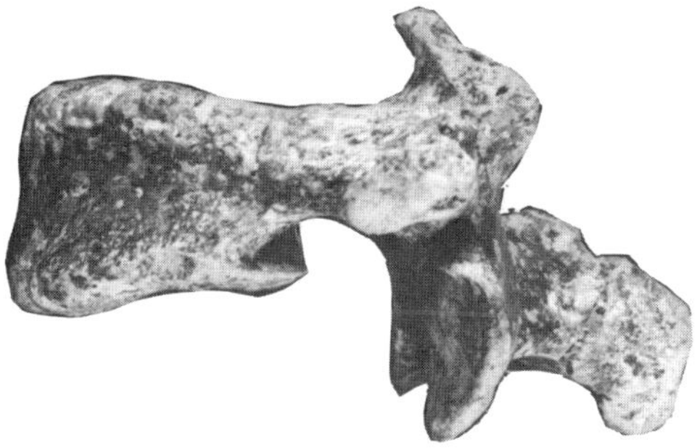

FIGURE 7.75.
Left lateral view of lumbar vertebra 6
KNM-WT 15000 AC.

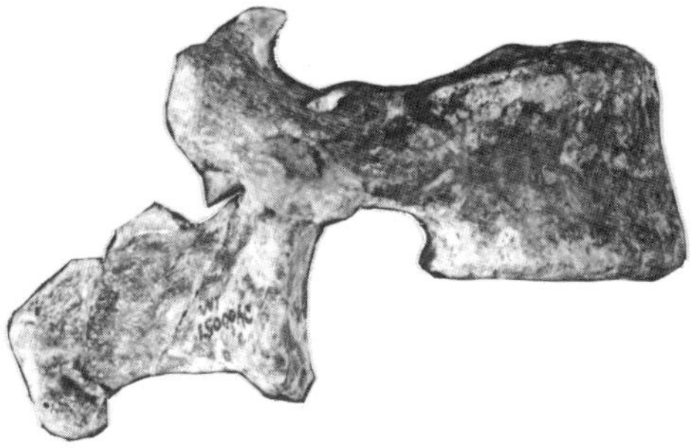

FIGURE 7.76.
Right lateral view of lumbar vertebra 6
KNM-WT 15000 AC.

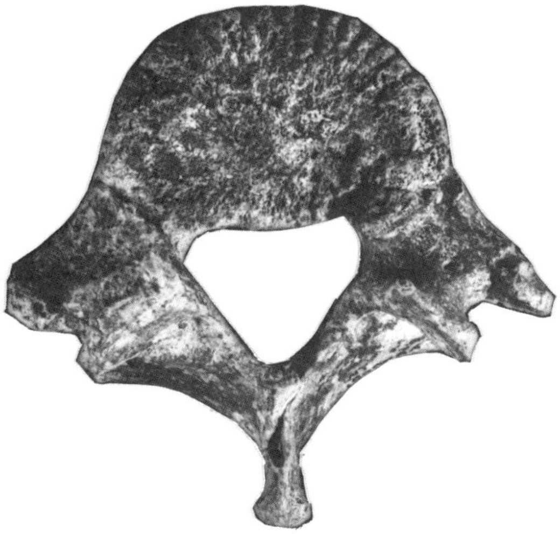

FIGURE 7.77.
Superior view of lumbar vertebra 6
KNM-WT 15000 AC.

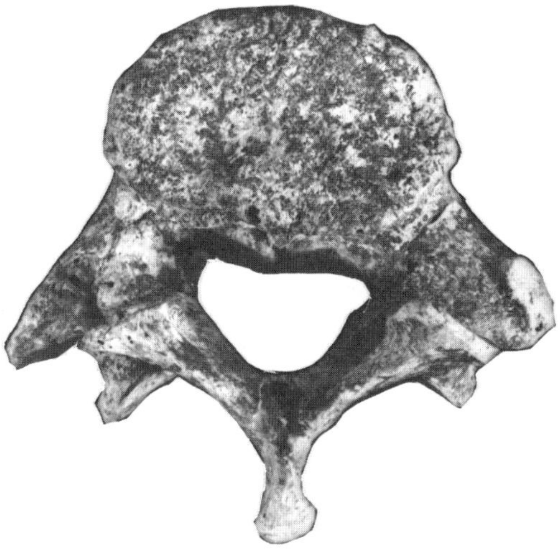

FIGURE 7.78.
Inferior view of lumbar vertebra 6
KNM-WT 15000 AC.

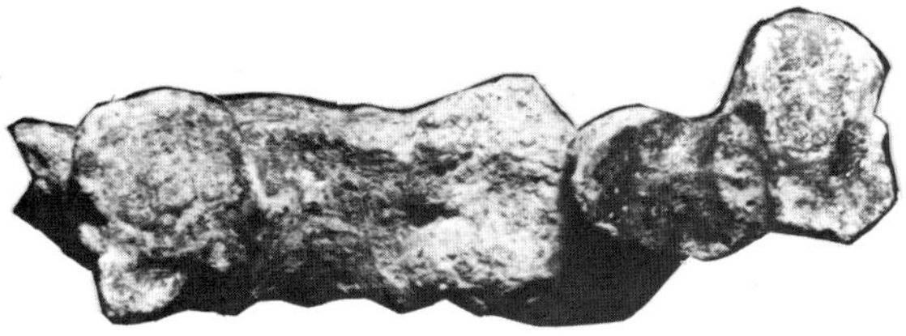

FIGURE 7.79.
Posterior view of sacral vertebra 1 KNM-WT 15000 AD and BY.

FIGURE 7.80.
Superior view of sacral vertebra 1 KNM-WT 15000 AD and BY.

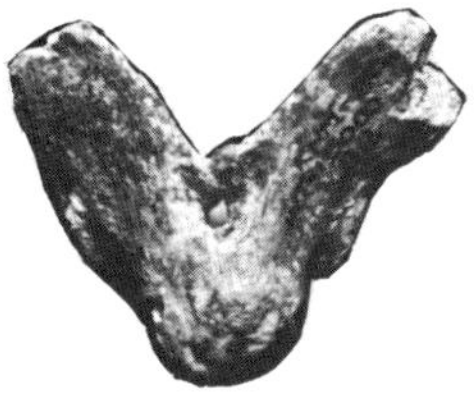

FIGURE 7.81.
Posterior view of sacral vertebra 2 KNM-WT 15000 BB.

base of the body is crushed, it it possible to see that there was a strong sacral prominence. The superior vertebral notch, cutting into the base of the superior articular process, is 4.6 wide. As it passes laterally, a second narrower groove undercuts the process on the lateral side. It is difficult to see how much of the lateral part has been abraded, but it extends laterally 34.0 from the midline. The main part of the lateral part is flat and faces anteriorly and laterally. There is a slight crest swinging round from the anterior margin of the body about 12 in front of the articular process. There is no discernible mammillary process.

The left articular facet is flat and faces almost directly dorsally. That this orientation is not due to deformation is confirmed by the position of the inferior articular facet on the last lumbar vertebra. It is roughly circular in outline and is 13.0 superoinferiorly and 13.7 mediolaterally. The small piece of bone with the right superior articular process (BY) also confirms that the left piece is not distorted. That facet is 13.4 superoinferiorly by 13.6 mediolaterally.

No other sacral vertebrae are known for early African *H. erectus,* but an abraded partial sacrum is associated with other cranial and postcranial fragments in KNM-ER 3735 from Koobi Fora that are between 1.88 and 1.91 Ma in age. In the parts preserved they are similar, but the body of the first sacral vertebra in KNM-ER 3735 is smaller in both transverse and anteroposterior diameters.

KNM-WT 15000 BB: SACRAL VERTEBRA 2

(FIGURE 7.81)

This is the laminae and crest of the second sacral vertebral segment, which was assigned to the third vertebra in our original note (Brown et al., 1985). The laminae are thin and are joined in the midline to form an angle of 110° in superior view. Their anterior margin is very sharp and the posterior ones roughened. The left side is a little abraded, but the right has a distinct tubercle about 7 wide centered 8.5 from the midline. They are 2.7 dorsoventrally by 14.1 superoinferiorly (left) and 2.0 by 12.9 (right). The crest is marked by a very strong rounded tubercle, 8.9 superoinferiorly by

11.5 wide. Between its superior surface and the laminar notch there is a very deep pit, triangular in outline, that is formed where the laminae divide so that the anterior limbs of the division join at the notch proper and the posterior limbs become everted and join at the crest tubercle. The distance from the notch to the tip of the crest tubercle is 15.1.

KNM-WT 15000 BC: SACRAL VERTEBRA 3

(FIGURES 7.82–7.83)

This is the right half of the third sacral vertebra. It includes the abraded crest, the lamina, part of the body, and the lateral process. The body part that is preserved reaches just to the midline. The lateral process extends 34.5 laterally from the midline, where its edge is abraded. The body narrows rapidly laterally as it runs to join the lamina to make the lateral process. The lamina is 4.7 thick dorsoventrally and 14.3 superoinferiorly. Its medial and superior crest is notched by a ligamentum flavum scar. This is 7.5 long and begins in a pit just to the side of a midline crest, which in turn lies superior to the crest. The crest here is really a thick midline tubercle. Inferolaterally to the crest is an abraded 6.8 wide second tubercle which grows out from the inferior border of the lamina. The inferior border of the lamina swings upwards again as it runs to the lateral process and makes the superior border of the next dorsal sacral foramen down. The lateral process is 20.0 superoinferiorly. The complete sacral canal was a flattened oval in section.

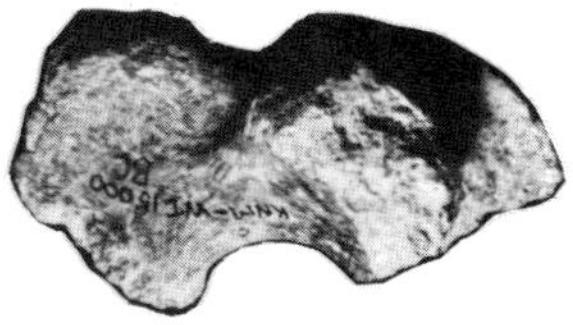

FIGURE 7.82.
Anterior view of sacral vertebra 3 KNM-WT 15000 BC.

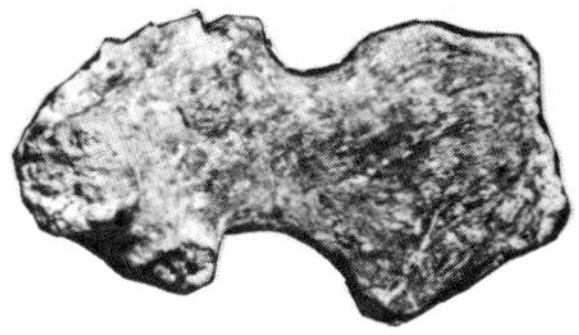

FIGURE 7.83.
Posterior view of sacral vertebra 3 KNM-WT 15000 BC.

KNM-WT 15000 AE: SACRAL VERTEBRA 4

(FIGURES 7.84–7.85)

We called this vertebra the last sacral in our original note (Brown et al., 1985) but have since determined that it is the penultimate one. This bone is complete except for the left lateral process and the right sacral cornu. The body is cracked across by an oblique break and recemented, but this does not greatly affect the morphology. The body is thin, being an elongated oval in cross-section. It tapers rapidly inferiorly. The lateral processes are wing-like and extend to 27.0 from the midline (note that the right side only is preserved to the

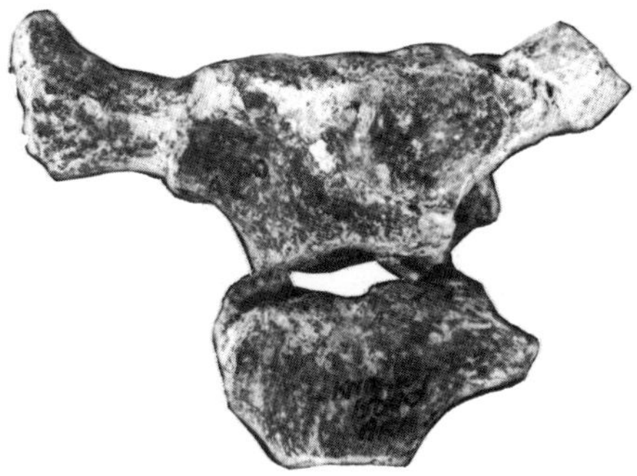

FIGURE 7.84.
Anterior view of sacral vertebrae 4 and 5 KNM-WT 15000 AE and AF.

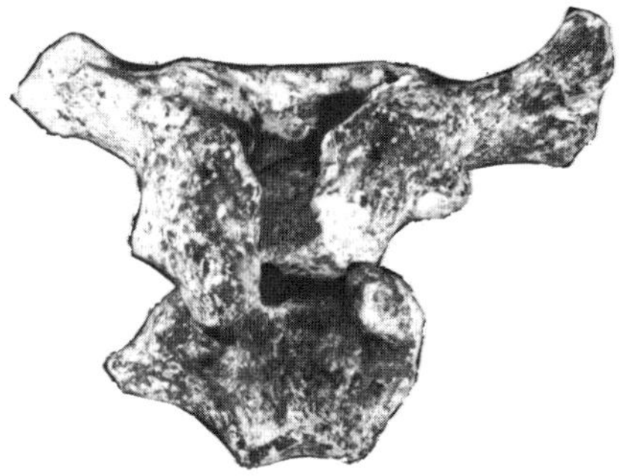

FIGURE 7.85.
Posterior view of sacral vertebrae 4 and 5 KNM-WT 15000 AE and AF.

end). The superior border of the process is concave in anterior view, making the inferior margin of the sacral foramen. Its very lateral border is 14.1 long superoinferiorly. The two laminae are relatively thick (5.9 right and 5.8 left) and end in two rugose sacral cornua with an accessory tubercle centered 10.2 lateral to it (measured on the left side only). The upper surface of the junction of the lamina to the body is grooved to form the exit for the sacral nerves. The laminae are 8.6 superoinferiorly (left) and 10.4 (right). The sacral canal is 16.9 wide and, although a distinct sacral hiatus is present, if an imaginary line is drawn over the hiatus to complete the canal posteriorly, it is 4.9 dorsoventrally.

KNM-WT 15000 AF: SACRAL VERTEBRA 5

(FIGURES 7.84–7.85)

We called this vertebra the first coccygeal in our original note (Brown et al., 1985), but it has now been determined that it is the last sacral. It is complete except for the tip of the left horn and the right lateral process. The body is flat and has an irregular upper surface. Its anterior surface is flat mediolaterally but concave superoinferiorly. It tapers to the inferior surface. The horns are thick hooks arising at the lateral corners of the body and turning upwards and backwards to be congruent with the sacral horns. The lateral process, broken off on the right side, is a small, flat triangular plate of bone which extends 16.5 laterally from the midline.

THE RIBS

(FIGURES 7.86–7.87)

Table 7.2 lists the ribs we have identified. Some of the pieces we had listed before are possibly not hominid, and other smaller pieces, mainly found in the sieves, are not included in the photographs or text. The ribs are dealt with more fully in Chapter 13 (this volume) and only brief descriptions are given here.

Rib 1. The left first rib (AG) is complete. The right first rib consists of proximal (AZ) and distal (AY) fragments with the intervening middle piece missing. This description is of the left. The head is small, 3.2 dorsoventrally by 2.7 superoinferiorly, and is oval in outline. The neck is long and thin. From the head to the tubercle is 17.0. In midneck it is 6.2 dorsoventrally by 4.6 superoinferiorly. The tubercle is marked and is 9.9 dorsoventrally by 7.1 superoinferiorly. It is set at an angle of 115° to the long axis of the neck. The body is very flat and the scalene impressions can hardly be felt. In midshaft it is 3.3 thick superoinferiorly by 17.7 wide. The distal end is V-shaped with a 16.0 long dorsal limb and an 8.5 long anterior limb. Here the bone is 5.9 thick superoinferiorly and 17.7 wide. The angle of the distal end is 90°.

Rib 2. The left (AQ) is practically complete. The right is in two parts. There is a larger proximal piece (AH) and a smaller distal one (BN). Piece AH is missing part of the head, and the end which joins BN is crushed, which means that the two cannot be joined properly. The left has a crack in the middle which has been filled with wax. It is tightly curved proximally and has a narrow neck and a large, cylindrically convex articular facet on the tubercle. It becomes markedly flatter about halfway along its length and ends in a slightly expanded distal end.

TABLE 7.2.

The ribs of KNM-WT 15000.

	ID SUFFIX	
RIB NUMBER	Left	Right
1	AG	AY & AZ
2	AQ	AH & BN
3	AT	—
4	BR*	AI
5	CB	BD, BE, & BJ
6	—	AK
7	AP	AL
8	AU	AM
9	AO	AS
10	BS	AJ
11	AN	—
12	—	—

* Probably nonhominid.

Rib 3. Only the left (AT) is preserved. It is broken just distal to the tubercle and has lost its distal end.

Rib 4. The rib which we had identified as rib 4 on the left (BR) is probably not from a hominid, although it was found near the main skeletal cluster of bones. Only proximal part of the right rib 4 (AI) has been found. The head is reduced to a thin point of bone with the two facets. The tubercle is pronounced and the body, in contrast to the two superjacent ribs, is flattened so that it has a distinct lateral surface.

Rib 5. On the left, rib 5 (CB) is complete except for the head and tubercle and two small portions of the shaft that were spalled away. On the right, rib 5 is made of three fragments (BD, BE, and BJ).

Rib 6. There is only the right rib 6 (AK). It runs from the tubercle to nearly the sternal end but is damaged at both.

Rib 7. Both seventh ribs are preserved. AP is the left and AL the right. Both are complete except for the heads and necks. The right is more flared distally.

Rib 8. Both eighth ribs were found. The left (AU) is missing about 50 mm distally, but the right (AM) is complete except for abrasion of the distal extremity. The heads are flattened and the articular facet on the tubercle is plane with a circumferential lip.

Rib 9. Both ninth ribs were found. The left (AO) has the distal end missing and the right (AS) has even more of the distal shaft missing. The necks are compressed dorsoventrally.

Rib 10. Both tenth ribs are complete along their lengths, but the left (BS) is damaged in the region of the posterior angle and the right (AJ) has a crushed head and proximal neck.

Rib 11. AN is the left eleventh rib. It is missing some of the body distally.

There are no ribs from other *Homo erectus* specimens with which to compare those of KNM-WT 15000.

The Shoulder Girdle and Upper Limb Skeleton

KNM-WT 15000 E, BK, & BL: The Scapulae

(Figures 7.88–7.92)

Most of the right (E) and a fragment of the left scapula (BK joined to BL) are present. The right was fossilized in a fragmented and trampled state, but it has been reconstructed fairly satisfactorily. The left fragment, which consists of the base of the spine and the axillary border, was found in two pieces, the spine *in situ* and the border washing out on the surface a little distance away. This piece is undistorted and is important for that reason.

The left fragment consists of an 80 long piece of the axillary border and associated part of the blade (BL) and a 45 long piece of the spine and associated blade (BK). The spine is inclined strongly dorsally so that its axis makes an angle of about 40° with the blade at the axillary border. It is 8.7 thick at its origin from the blade and has a rounded lateral margin. One large and one very small vascular foramen are found in the angle between the blade and the spine. The axillary border is quite sharp and in lateral view curved, concave costally. The dorsal surface becomes gently convex into the infraspinous fossa almost directly medial to the crest. On the costal surface there is a soft groove, no doubt for the insertion of teres minor. Immediately medial to this is a stout, rounded ridge which swings laterally to join the axillary border at the inferior break. Although there is root-acid damage to the dorsal part of the border, the remnant of a small, rough infraglenoid tubercle can be seen.

The right scapula is complete except for the end of the acromion. When found *in situ* it was in a broken and distorted condition lying under the right humerus, with most of the blade being

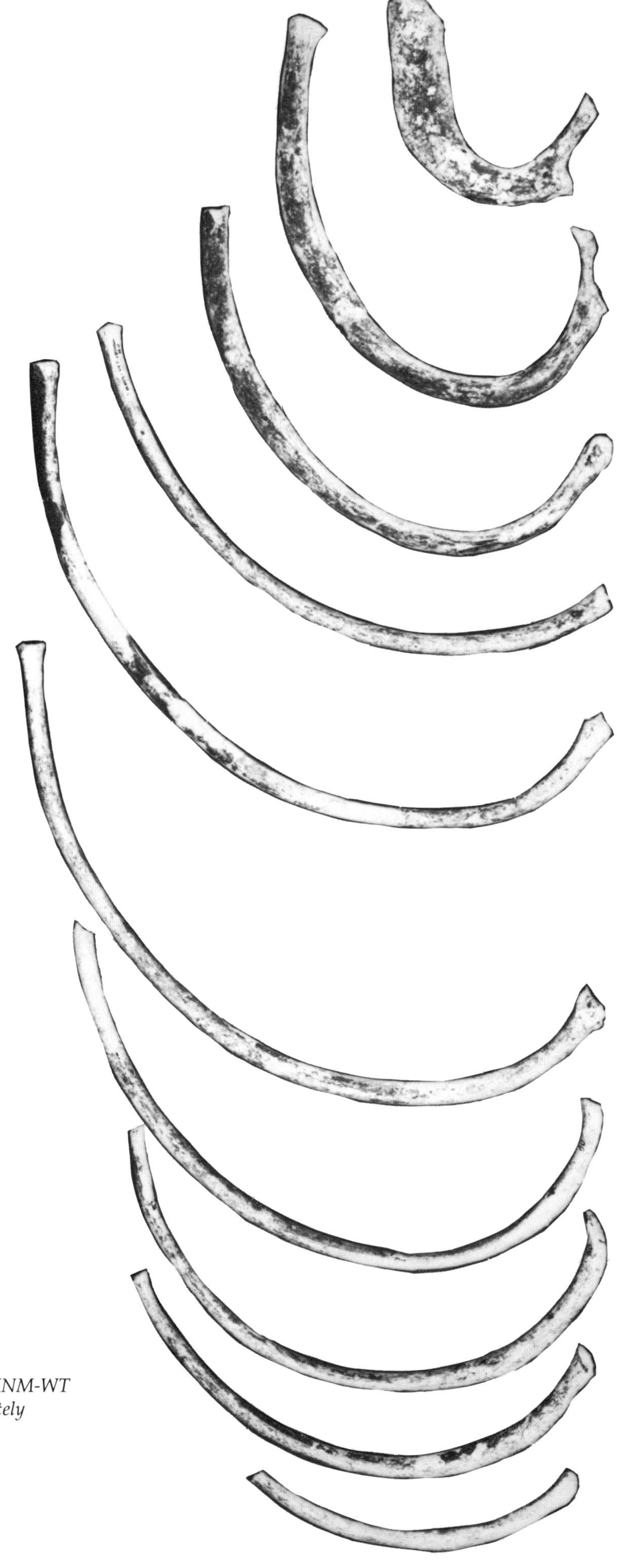

FIGURE 7.86.
Superior view of left ribs of KNM-WT 15000. (Shown at approximately 50 percent natural size.)

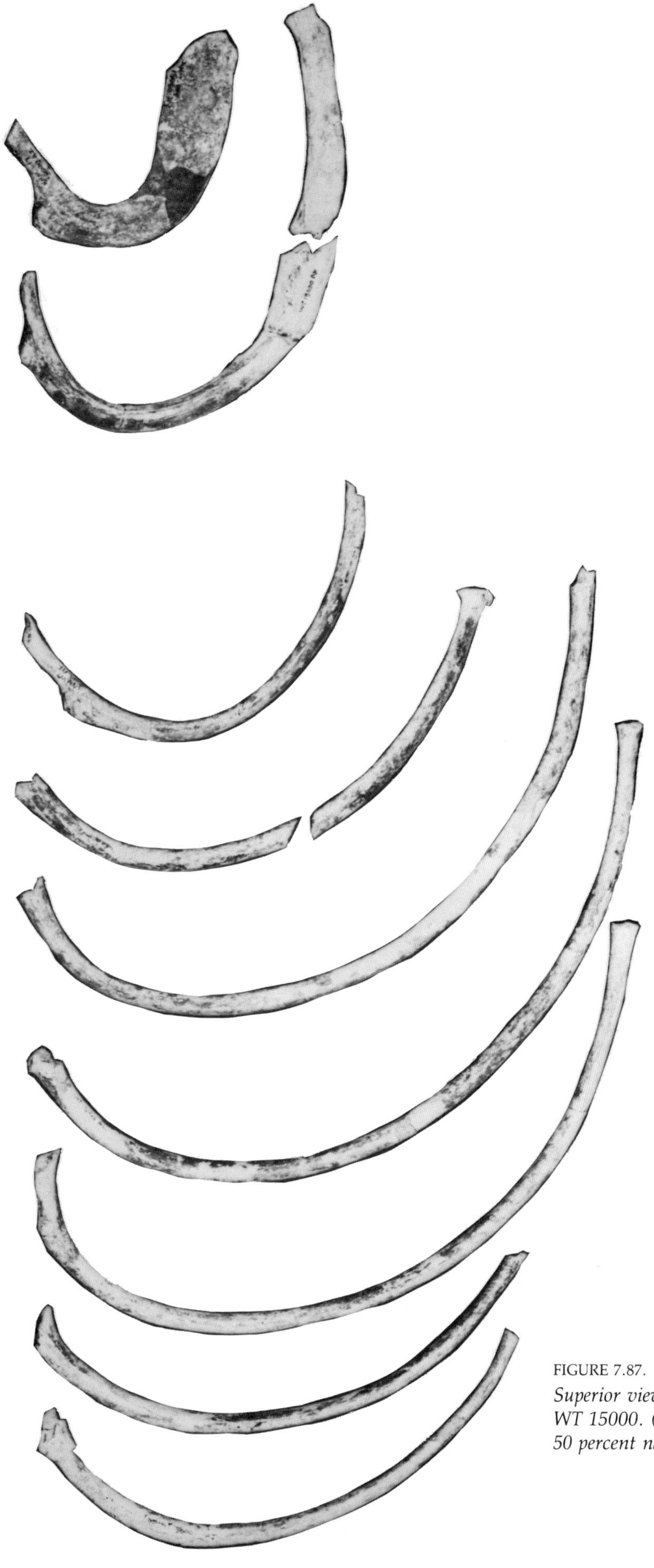

FIGURE 7.87.
Superior view of right ribs of KNM-WT 15000. (Shown at approximately 50 percent natural size.)

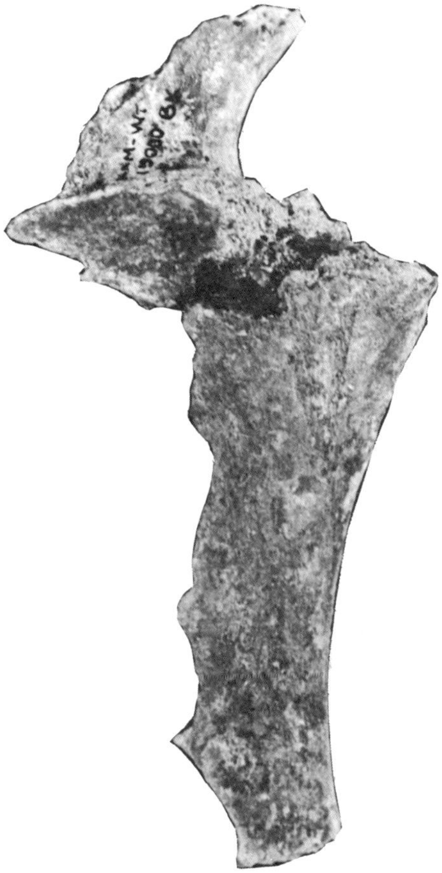

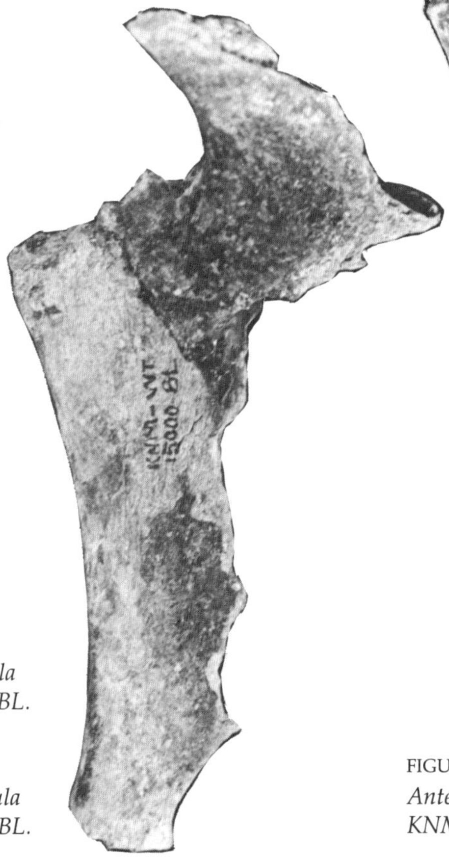

FIGURE 7.88.
Anterior view of left scapula KNM-WT 15000 BK and BL.

FIGURE 7.89.
Posterior view of left scapula KNM-WT 15000 BK and BL.

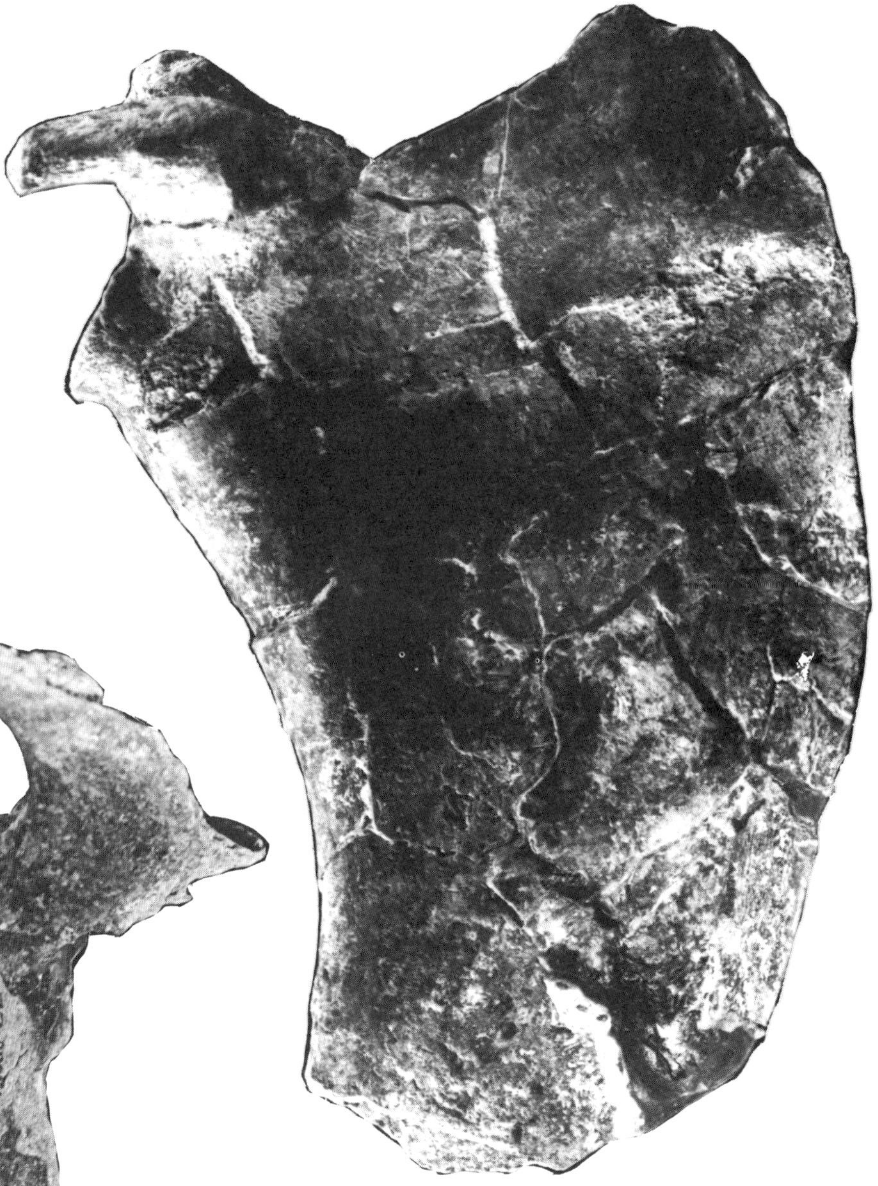

FIGURE 7.90.
Anterior view of right scapula KNM-WT 15000 E.

FIGURE 7.91.
Posterior view of right scapula
KNM-WT 15000 E.

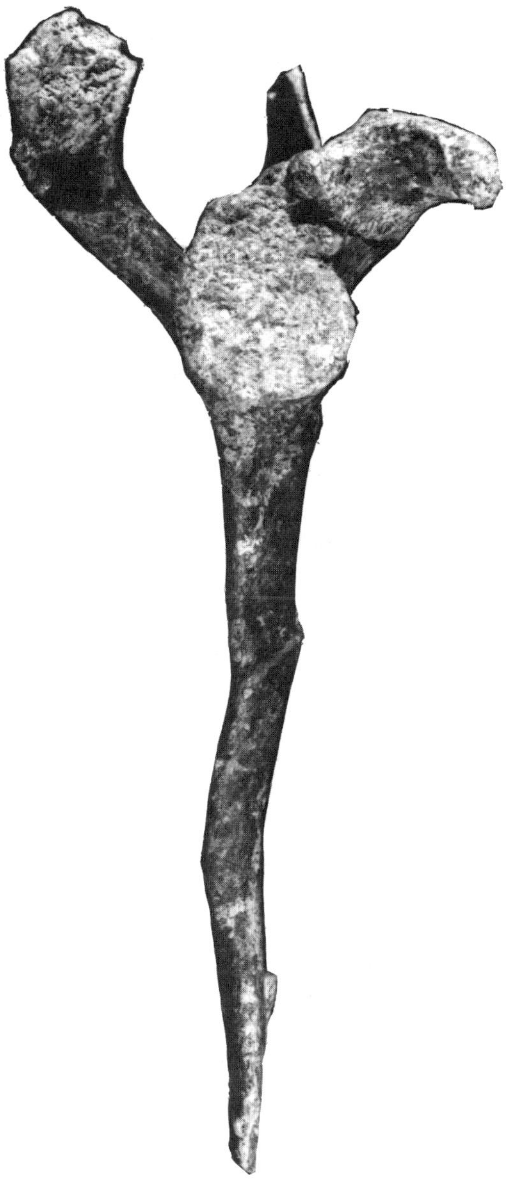

FIGURE 7.92.
Lateral view of right scapula
KNM-WT 15000 E.

cracked and the individual plates of bone separated or folded on one another. It has been possible to realign most of these, but in cases where hard matrix fills cracks running obliquely through the bone, such as the two on the axillary border, realignment is not possible on the original, although it would be on a cast. Being from a juvenile, the vertebral border and inferior and superior angles were still cartilaginous. The bone here is therefore ill-defined. The coracoid process was found fossilized separately just next to the glenoid and has been replaced, but it had not fused to the main part of the bone at the time of death. The glenoid surface was still mainly cartilaginous, and no evidence of bony centers other than for the body and the coracoid were found. The coracoid center appears close to fusion with the body, but there was no sign of a subcoracoid center. The coracoid fuses in modern people sometime between the sixteenth and eighteenth year.

In posterior view the scapula is elongated superoinferiorly. From the superior to the inferior angle is 154.0 and there are small parts missing from each. From the most inferior point on the glenoid surface to the vertebral border just medial to the base of the spine is 101.6. The length of the spine, from its notch near the glenoid to the vertebral border, is 77.9. The chord length across the concavity of the axillary border is 102.5. The vertebral border in dorsal view is convex throughout, the greatest change in convexity being found where it turns to meet the inferior angle.

The supraspinous fossa is small relative to the infraspinous fossa and is overhung by the spine, which is directed upwards and outwards. The inferior surface of the spine is, in fact, smoothly continuous with the surface of the infraspinous fossa. There is hardly a discernible point of angulation where the spine takes off from the blade. The base of the spine at the vertebral border is interesting in that it shows two ridges running to the border. The first and more superior is the continuation of the superior edge of the spine. The second is a well-marked roughening which starts about half the way along the spine and runs downwards from the superior border of it. This first parallels the border of the spine then turns inferiorly, so that by the time it reaches the vertebral border it is 35.0 from the superior ridge. The rest of the infraspinatus fossa is relatively featureless, except where there is a small ridge separating the teres major insertion from the infraspinatus one.

The superior border is slightly concave in dorsal view and bears a distinct (7.5 wide, 7.0 deep) scapular notch. The costal surface is gently hollowed behind the infraspinous fossa, more so deep to the area where the spine attaches on the dorsal surface, and is convex behind the suprascapular fossa. Again, as in the left scapula, there is a pronounced ridge running from the posteroinferior part of the edge of the glenoid, behind the teres minor insertion area and joining the axillary border about two-thirds down its length. The axillary border is very concave throughout in dorsal view.

The glenoid surface itself is overhung slightly by the articular portion of the base of the coracoid process. It is teardrop-shaped in outline, being 33.6 high by 22.7 wide. There is only a feeble infraglenoid tubercle. The coracoid process rises immediately lateral to the scapular notch. Its posterior surface is damaged at the epiphyseal cartilage region and the base is quite swollen there. It tapers rapidly so at the lateral end it is 6.8 thick by 9.0 high. From the notch to the end is 37.5. Most of the acromion is missing. It has been broken off medial to the clavicular articular surface, but in any case its separate ossification centers would not have been fused.

Parts of a scapula of another early African *H. erectus* are known. These are KNM-ER 1808 AC and AD (Walker et al., 1982; Leakey and Walker, 1985). This partial skeleton shows evidence of widespread periostitic disease and is thus of reduced analytical value. However, there is relatively little periostitic bone deposition on the larger piece from the axillary border and none on the the base of the scapular spine. The concave nature of the axillary border can be confirmed, as is the curvature and angulation of the well-built base of the spine. The spine of KNM-ER 1808 is more strongly built than that of KNM-WT 15000 and seems to have curved over the supraspinous fossa more, but this is probably due to the former being adult at death. Even the strong folding of the spine over the supraspinous fossa and the large nutrient foramen at the base of the spine can be found in the two specimens.

KNM-WT 15000 C & D: THE CLAVICLES

(FIGURES 7.93–7.96)

Both clavicles are present. The left is lettered C and the right D. The total lengths are 130.5 (right) and 130.4 (left). Both are essentially complete, although there is minor crushing and bone loss on the sternal end of the right and the acromial end of the left. The acromial end has a flattened superior surface and a concave inferior one. In superior view the acromial end finishes in a triangular point with an apex of about 110°. The acromial articular surface is slightly hollowed and roughened and in lateral view has a teardrop-shaped outline, with the apex anterior. It measures 14.5 anteroposteriorly by 9.3 superoinferiorly (right side only because of damage to the left). The superior clavicular surface is lipped slightly at its junction with the acromial articular surface. The anterior border is sharp laterally, starting at the apex of the acromial articular facet. In superior view it is smoothly concave in its lateral part, then begins to be convex at an angulation 42 (chord length) from its lateral start. After this angle the convexity has a radius of curvature greater than the radius of the previous concavity and the crest loses its sharpness. It continues for another 49.0 (chord length) until another angulation, this time marking the change from convexity to a medial concavity in the border. This last concavity is more marked on the left clavicle, possibly because of damage to the right. The posterior border is simpler. This border is blunt all along its length and starts laterally with a relatively straight section of about 30 on the right and 35 on the left before becoming smoothly concave for the rest of its length.

The superior surface is flat laterally, except for the acromial lipping, and it remains flat almost to the lateral anterior angulation. Between this flat surface and the anterior border is a shallow groove, much better developed on the right, which is 22.0 long by 3.8 wide (right) and 17.0 long by 3.0 wide (left). This groove is for the attachment of the anterior fibers of the deltoid. The deeper groove on the right may be an indication that the individual was right-handed. After the deltoid groove the whole body of the bone is twisted by about 30° so that the superior surface faces anterosuperiorly. The surface now is no longer flat, but convex anteroposteriorly. It is a little flattened in the area where pectoralis major would attach. At its sternal end the surface rises to reach the sternal articular facet.

The inferior surface is convex throughout its length, except for where the trapezoid line should be. Here it is concave. A low, blunt conoid tubercle is situated about 32 from the lateral extremity. The costoclavicular ligament impression is better developed on the left than the right. A faint linear impression along the middle of the inferior surface may mark the attachment of subclavius. Posterior to this mark are the nutrient foramina. There is a single large one on the left about 54 from the lateral end and three small ones on the right, the largest centered 64.7 from the lateral apex.

The sternal articular surface shows no signs of the pseudoepiphysis and is quite rugose. The left, which is complete, is oval in outline, being 14.6 anteroposteriorly by 17.8 superoinferiorly. In midshaft the anteroposterior diameters are 12.8 (right) and 13.6 (left). The superoinferior diameters are 7.9 (right) and 7.6 (left).

Another early African *H. erectus* clavicle is known. The specimen includes most of a right clavicle of the diseased skeleton KNM-ER 1808 (Leakey and Walker, 1985). In general it is very similar to the clavicles of KNM-WT 15000 and looks like a modern *H. sapiens* clavicle. The ligament markings are much more strongly developed than in KNM-WT 15000, and it has a strongly lipped oval costoclavicular ligament depression and a ridged conoid ligament crest that is developed enough to be seen in superior view. The cross-section of the sternal end is also more rounded and less flattened than in the West Turkana specimen. The KNM-ER 1808 clavicle would

FIGURE 7.93.
Superior view of left clavicle KNM-WT 15000 C.

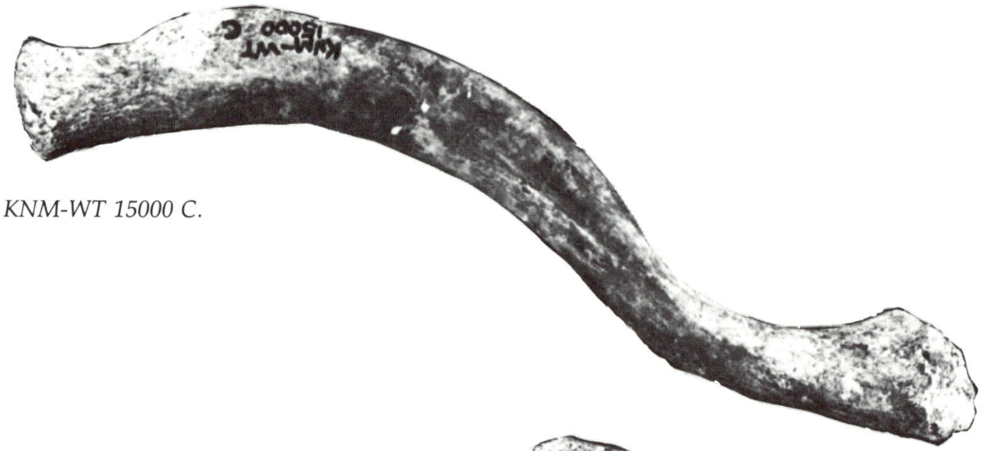

FIGURE 7.94.
Inferior view of left clavicle KNM-WT 15000 C.

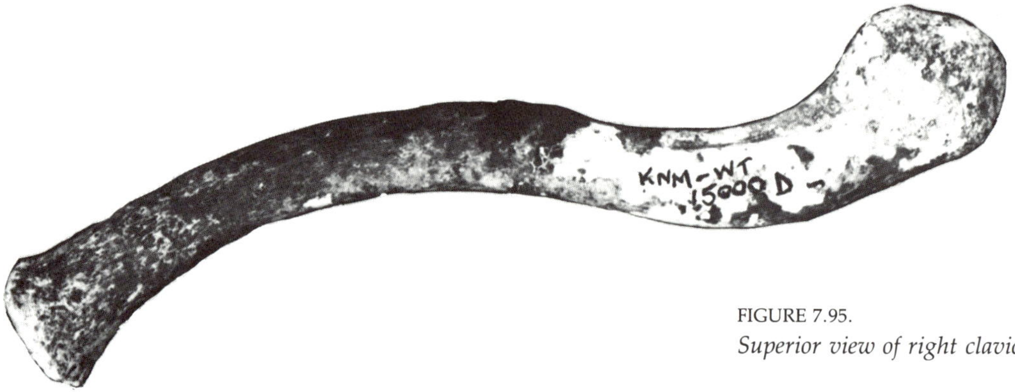

FIGURE 7.95.
Superior view of right clavicle KNM-WT 15000 D.

FIGURE 7.96.
Inferior view of right clavicle KNM-WT 15000 D.

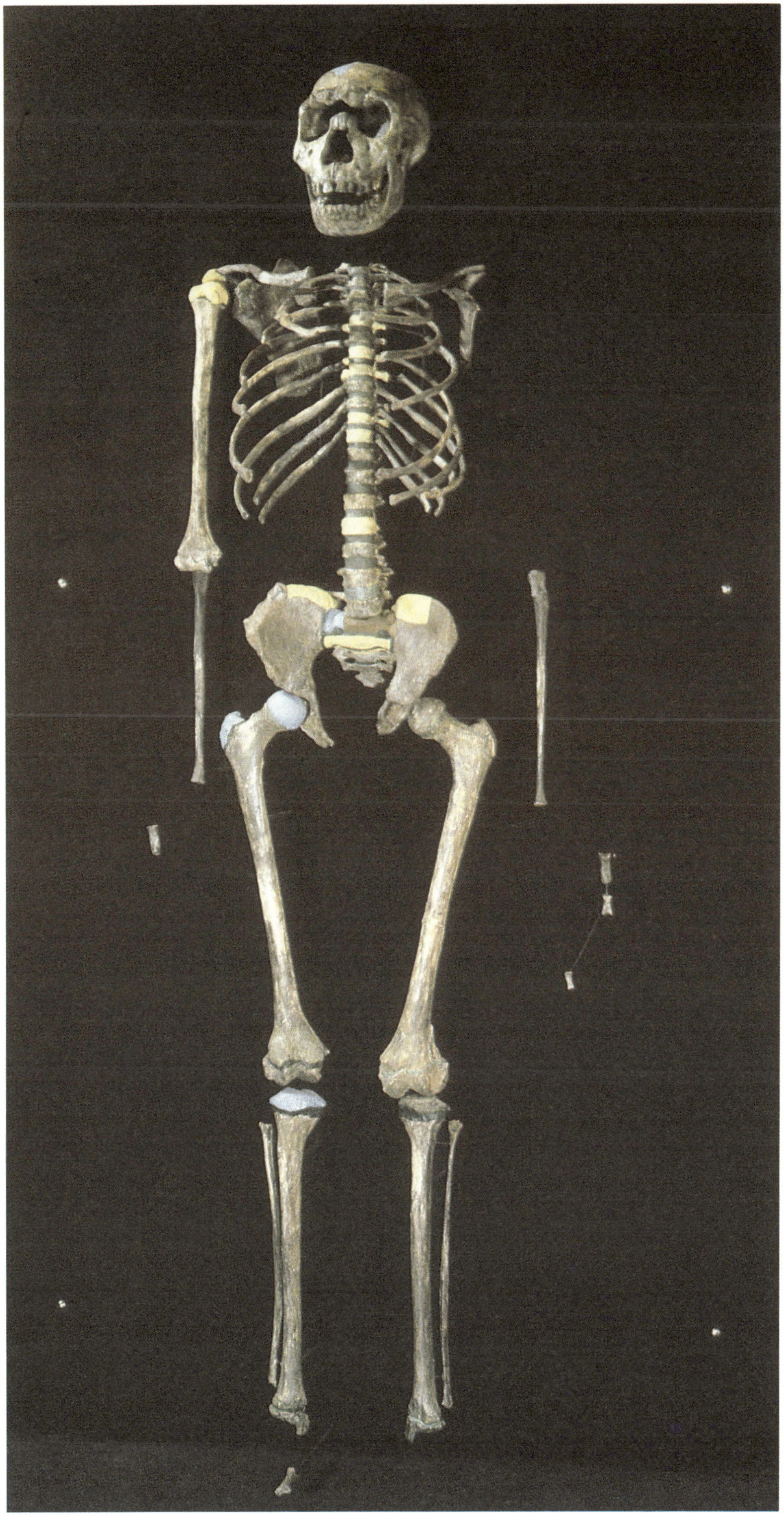

The Nariokotome Homo erectus *skeleton.*

Excavators working in the northeast corner of the NK3 site at the end of the 1986 season. The "floor" shown here is the excavated base of Unit 2. The massive and pale Unit 6 can be seen running across the section. Cracks filled with deposits continuous with those of Unit 7 run down through Unit 6 and lower and can be seen in the walls as well as criss-crossing the floor. These attest to a prolonged period of lowered water table after the deposition of Unit 6. (View from the southwest.)

By the end of the 1986 season, large piles of overburden and matrix had been removed and were awaiting water screening. (View to the east.)

have been only a little larger than those of the juvenile. The differences can be accounted for on the basis of their different individual ages.

KNM-WT 15000 F: THE HUMERUS

(FIGURES 7.97–7.103)

Only the right humerus is preserved. It is missing its superior epiphysis and the epiphysis of the medial epicondyle. The lateral half of the distal epiphysis was found fossilized separately and has been replaced in its original position. The trochlear half was still attached to the metaphysis and it appears that fusion with the shaft had started. The epiphyseal line can still clearly be seen posteriorly, but close examination of the anterolateral junction shows that it was fused there (see Figures 7.102 and 7.103).

As it stands the piece is 301.2 long. Allowing for the missing superior epiphysis, we estimate the humerus would have been 319.0 long at the time of death. The bone is undistorted and because of the individual's youth is only lightly marked by the major features usually seen on adult humeri. The intertubercular groove is wide and shallow. At the epiphyseal surface it is 8.0 wide, and the groove disappears distally because the crest of the lesser tubercle fades rapidly into the shaft surface. The crest of the greater tubercle persists to join the medial margin of the deltoid tuberosity. The low deltoid tuberosity ends at about 148 below the top of the bone. Its lateral margin is a low crest which spirals upwards to meet the lateral border about 70 below the top of the bone. There is a single large nutrient foramen on the medial border at the level of the distal end of the deltoid tuberosity. Below the deltoid tuberosity the shaft becomes more triangular in cross-section because the posterior surface becomes more flat towards the olecranon fossa and supraepicondylar crests are developed.

At midshaft the anteroposterior diameter is 29.9 and the mediolateral one is 16.7. The groove for the radial nerve, which is barely perceptible, is almost at midshaft. The midshaft dimensions are 29.9 anteroposteriorly and 16.7 mediolaterally. Because the shaft is so straight, the crest separating the trochlea from the capitulum is set almost exactly on the long axis of the bone. This is not true of the anterior ridge, which makes up the front of the shaft below the deltoid tuberosity. The anterior ridge swings medially away from the long axis some 70.0 above the distal end of the bone. It does not reach the epiphysis, however, because it is excavated by the coronoid fossa. The fossa is 17.4 wide by 7.3 proximodistally and is well defined. The radial fossa is less deep and its margins poorly defined.

The lateral epicondyle is small. It projects barely 2.0 lateral to the lateral surface of the capitulum. It is centered about 16 posterior to the anterior surface of the capitulum and is surmounted by a rounded extensor crest. The medial epicondyle, despite not having its epiphysis attached, is large, projecting 14.0 medial to the medial border of the trochlea. In medial view its epiphyseal surface is oval in outline, being 18.4 proximodistally and 13.5 anteroposteriorly. The capitulum is well rounded and is 16.6 in diameter. It is set well anterior to the anterior surface of the shaft. The trochlea is not deeply grooved; its narrowest diameter of 16.2 is set 13.8 from its medial border. The crest separating the capitulum from the trochlea is clearly seen. The groove between the crest and the capitulum is 1.7 deep and the lateral side of the crest is 4.3 long. The olecranon fossa is shallow, about 8 deep and centered so that its lateral margin is almost in line with the axis of the shaft. It is 19.9 wide by 14.5 superoinferiorly. The total bi-epicondylar width (without estimating for a medial epiphysis) is 52.0 and the total condylar width is 40.8. A total bi-epicondylar width is estimated to have been about 55.0.

There are other humeri from early African *H. erectus*. KNM-ER 1808 (Leakey and Walker, 1985) has a piece of the distal left humerus and most of the shaft of the right, but both are heavily obscured by new periostitic bone. The least obscured piece, the distal shaft, KNM-ER 1808 T, is very similar in the parts preserved to KNM-WT 15000 F, but the supinator crest is stronger and its margin is convex when viewed from the front, not straight. KNM-ER 1591, listed by Howell (1978) with a query, is now known not to be hominid.

FIGURE 7.97.
Anterior view of right humerus KNM-WT 15000 F. (Shown at approximately 80 percent natural size.)

FIGURE 7.98.
Posterior view of right humerus KNM-WT 15000 F. (Shown at approximately 80 percent natural size.)

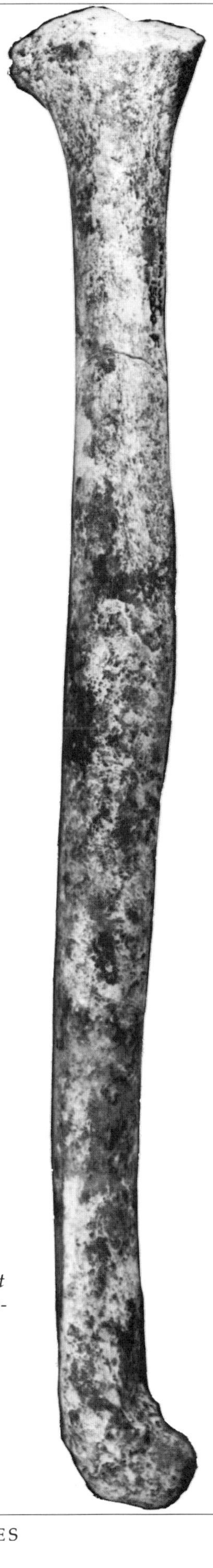

FIGURE 7.99.
Medial view of right humerus KNM-WT 15000 F. (Shown at approximately 80 percent natural size.)

FIGURE 7.100.
Lateral view of right humerus KNM-WT 15000 F. (Shown at approximately 80 percent natural size.)

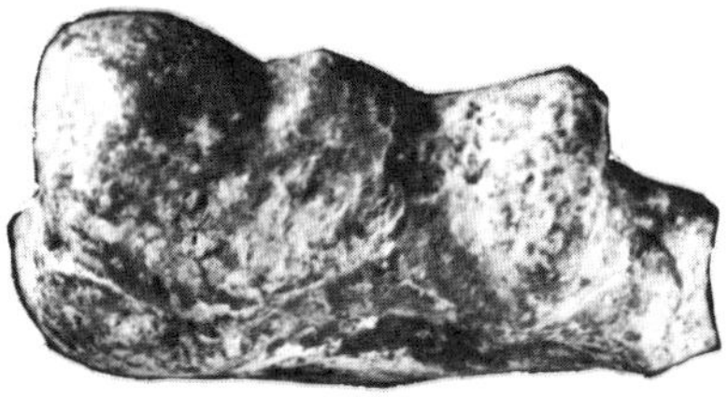

FIGURE 7.101.
Distal view of right humerus KNM-WT 15000 F.

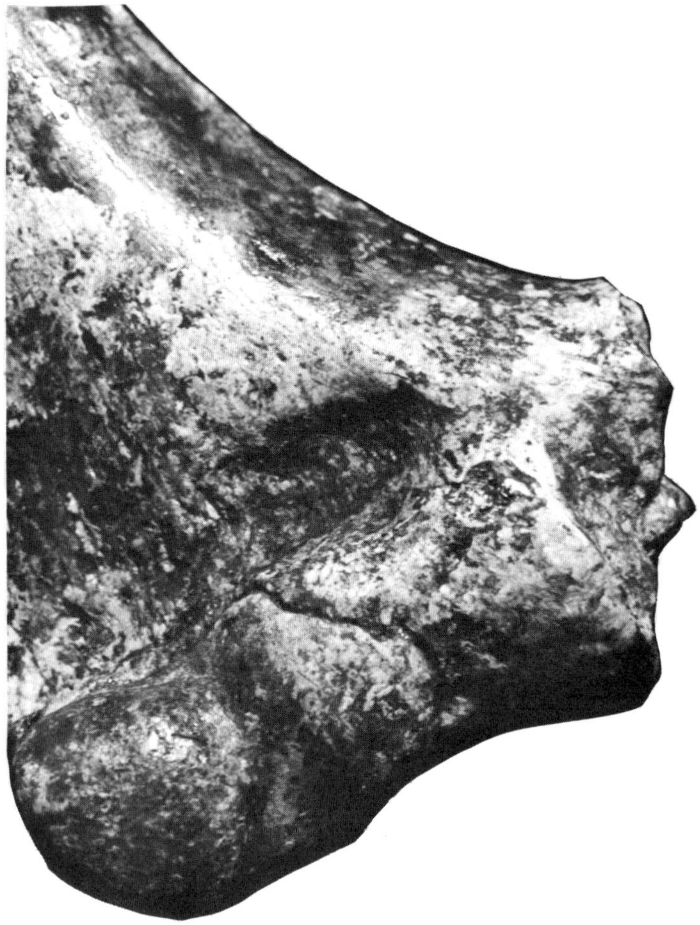

FIGURE 7.102.
Anterior view of distal epiphysis of right humerus KNM-WT 15000 F. (Enlarged view.)

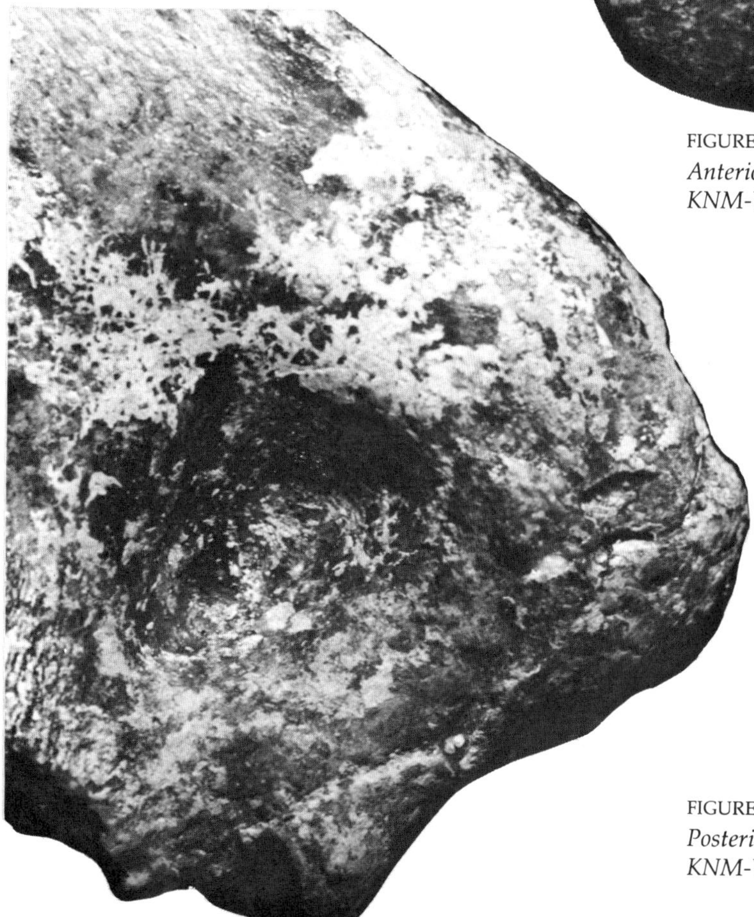

FIGURE 7.103.
Posterior view of distal epiphysis of right humerus KNM-WT 15000 F. (Enlarged view.)

KNM-WT 15000 BP & BZ: THE ULNAS

(FIGURES 7.104–7.111)

Both ulnas are complete except for their epiphyses. They were found in 1985 and have the suffixes BP (right) and BZ (left). There is a tiny patch of root damage around the proximal end of the right olecranon and tip of the coronoid process. The total lengths of the bones as they stand are 257.2 (left) and 260.3 (right). Initial estimates of the total length of both bones were made in order to calculate limb proportions, but they have been superceded by better estimates. The method of estimating complete radial and ulnar lengths from epiphyseal lengths is given in Chapter 11. The estimated total length of 270.0 is based on the mean diaphyseal length of the two sides.

The epiphyseal surface of the olecranon process is a roughened oval, 20.5 mediolaterally and 16.2 anteroposteriorly (right side). It faces posterosuperiorly at an angle of about 45°. The trochlear notch is shallow (8.0 deep), 25.0 long and 15.7 wide at its narrowest point, which is about in the middle of its length. The same values for the right ulna are 7.5, 24.7, and 16.0. The notch is saddle-shaped, both olecranon and coronoid parts are convex, and the latter is slightly keeled. The radial notch has the form of a slightly depressed rounded triangle with one side on the supinator crest, the other the junction between the notch and the trochlear notch, and the last distal and running up to the coronoid process. The distance from the anterior side to the opposite apex is 11.6 and from the junction of the two articular surfaces to the opposite apex 11.7 (both left side).

The ulnar tuberosity for the insertion of brachialis is a feeble swelling on the crest running from the coronoid process down the anterior part of the shaft. It does little more than flatten out the concavity here. The crest is hollowed more medially than laterally. The anterior border of the shaft is concave throughout its length, with greater curvatures at either end where the coronoid process rises and the shaft swells out to meet the head. The posterior border is convex throughout its length except for the swelling to the head. The lower two-thirds are straighter than the upper third. In anterior or posterior view the shaft is mostly straight, but the whole of the proximal end, from a point roughly 35 distal to the coronoid process, is angulated from the rest of the shaft to make an angle of about 10° to the long axis of the shaft.

The crest running from the coronoid process fades out on the anterior surface about one-third of the way down the shaft. Thereafter the anterior surface is smoothly congruent with the convex medial surface from the interosseous border to the posterior border. The interosseous border is nothing more than the angulation between the flat lateral surface and the anterior one, and there is hardly a trace of roughening along it. It begins about 30 distal to the coronoid process and ends about the same distance from the distal end.

The distal metaphyses are fairly symmetrical and swell out evenly from the distal shaft, except for the region of the groove for extensor carpi ulnaris. In distal view the epiphyseal surfaces would be almost circular in outline but for these grooves. Faint radial ridges mark the periphery of the epiphyseal surfaces and there is a single raised bump, situated not quite centrally, in the slightly hollowed surface. The dimensions of the distal surfaces are 5.8 by 15.1 (anteroposteriorly by mediolaterally) for the left and 5.8 by 15.5 for the right.

Fragments of ulna of other early African *H. erectus* are known. KNM-ER 803 (Day and Leakey, 1974) includes a piece of left ulna shaft from the level of the coronoid process to two-thirds of the way down the shaft. This adult specimen shows much stronger crests and muscle insertion areas than do the juvenile bones described here, but it appears that it was about the same total length. KNM-ER 1808 (Leakey and Walker, 1985) also has pieces of ulna preserved, but these are so heavily encrusted with periostitic bone that useful comparisons cannot be made. The ulna from Bed II Olduvai Gorge, OH 36, assigned by Day (1986) to *H. erectus*, belongs instead to *Australopithecus boisei*. The recently published ulna from Baringo (Solan and Day, 1992) is very similar to the bones described here, but adult. The Kapthurin site from which it comes is probably more than half as young as the site of Nariokotome III.

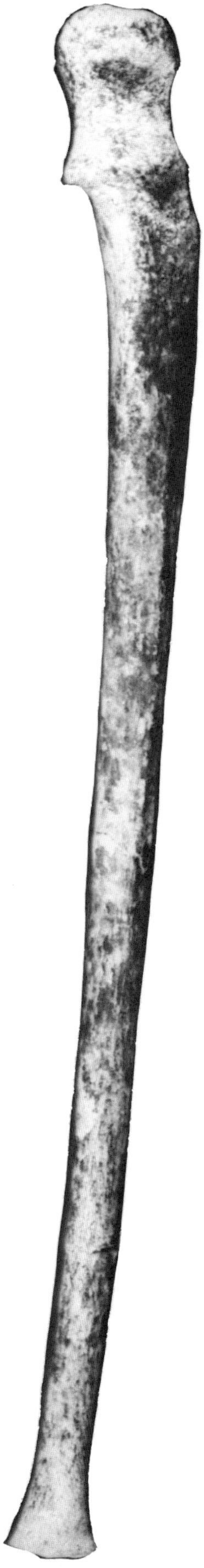

FIGURE 7.104.
Anterior view of left ulna KNM-WT 15000 BZ. (Shown at approximately 90 percent natural size.)

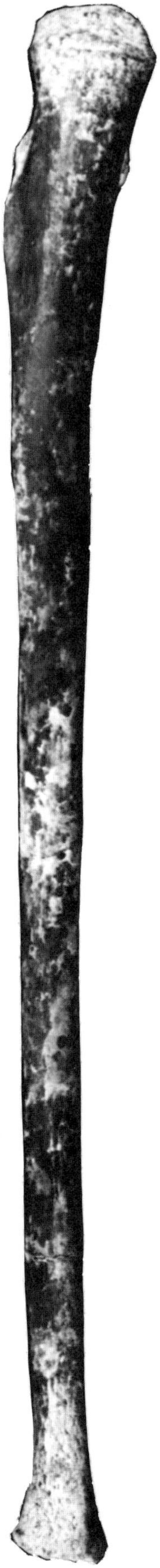

FIGURE 7.105.
Posterior view of left ulna KNM-WT 15000 BZ. (Shown at approximately 90 percent natural size.)

FIGURE 7.106.
Medial view of left ulna KNM-WT 15000 BZ. (Shown at approximately 90 percent natural size.)

FIGURE 7.107.
Lateral view of left ulna KNM-WT 15000 BZ. (Shown at approximately 90 percent natural size.)

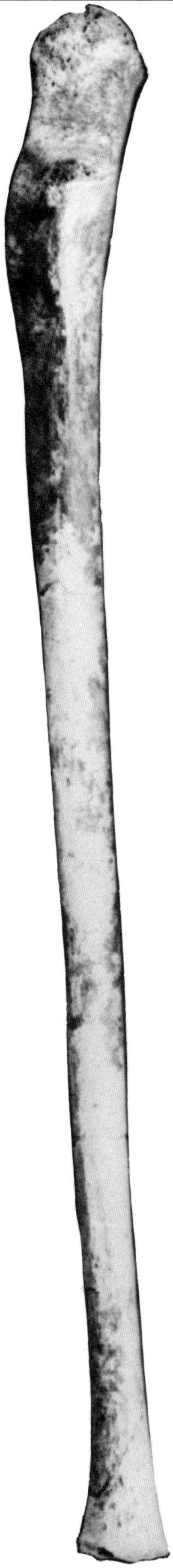

FIGURE 7.108.
Anterior view of right ulna KNM-WT 15000 BP. (Shown at approximately 90 percent natural size.)

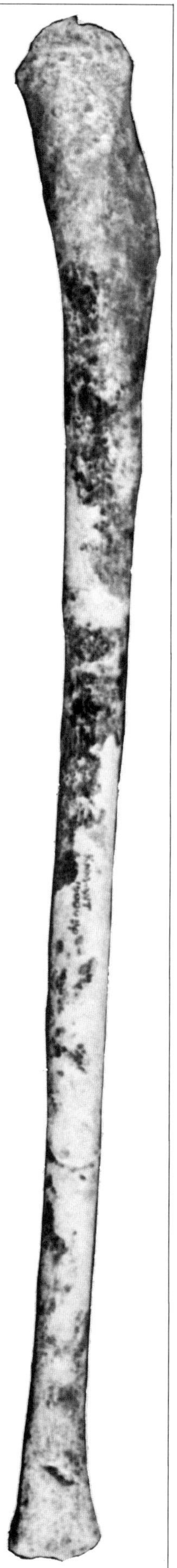

FIGURE 7.109.
Posterior view of right ulna KNM-WT 15000 BP. (Shown at approximately 90 percent natural size.)

FIGURE 7.110.
Medial view of right ulna KNM-WT 15000 BP. (Shown at approximately 90 percent natural size.)

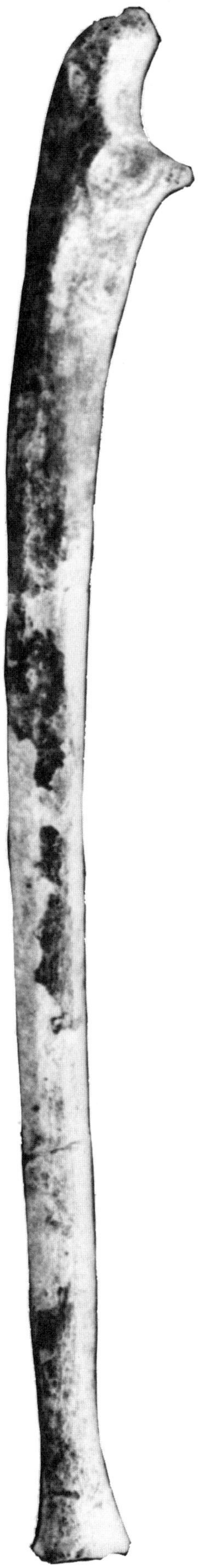

FIGURE 7.111.
Lateral view of right ulna KNM-WT 15000 BP. (Shown at approximately 90 percent natural size.)

FIGURE 7.112.
Palmar view of ? metacarpals 1 KNM-WT 15000 BU and BV.

FIGURE 7.113.
Dorsal view of ? metacarpals 1 KNM-WT 15000 BU and BV.

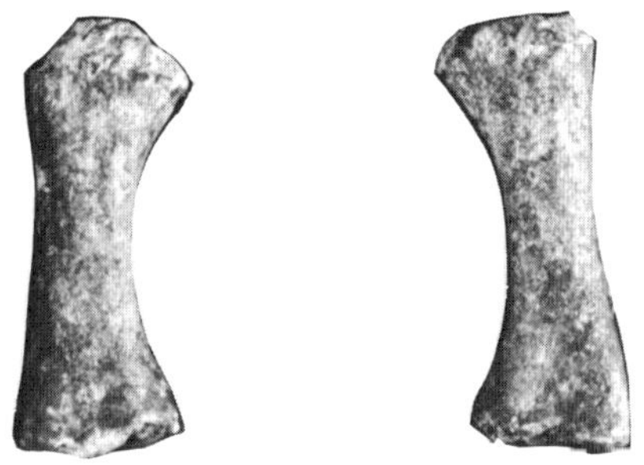

FIGURE 7.114.
Medial view of ? metacarpals 1 KNM-WT 15000 BU and BV.

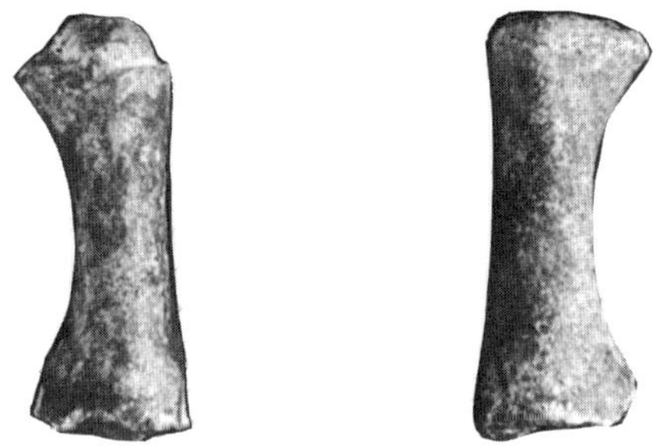

FIGURE 7.115.
Lateral view of ? metacarpals 1 KNM-WT 15000 BU and BV.

KNM-WT 15000 BU & BV: FIRST METACARPALS

(FIGURES 7.112–7.115)

What might be the shafts of both first metacarpals are preserved. BU is the right and BV the left. As they stand they are 35.2 long (left) and 34.7 long (right). The epiphyses are missing and the heads are abraded so that nothing useful remains distally. The proximal epiphyseal surface is roughly quadrangular in outline, and the dimensions are: anteroposterior, 14.6 (left), 13.7 (right); and mediolateral, 12.4 (left), 12.6 (right). The shafts are roughly triangular in section; they have a flat to slightly concave lateral surface, with the sharpest angle anteriorly where the opponens muscle attaches. The medial surface is concave proximodistally and convex in the other direction. The posterior surface is slightly concave proximally and convex distally. The distal end is peculiar in that there is a very well-developed process on the medial side. This beak-like feature has been seen only once before in hominid first metacarpals, and that is in SK 84 from Swartkrans. That is a smaller bone and is from an adult. Its total length is about two-thirds what that of the KNM-WT 15000 bones would have been when complete. Nevertheless, the inferred presence of a beak-like process in these first metacarpals has led Susman (1988) to conclude that it is a special and derived characteristic of *H. erectus*.

There are, however, some problems with the attribution of these two bones to KNM-WT 15000. They are mammal metapodials and the nearest ones to them that we have found in our searches of juvenile comparative material are those of subadult human first metacarpals. However, there are many extinct mammal species from the Pleistocene of the Lake Turkana region that have

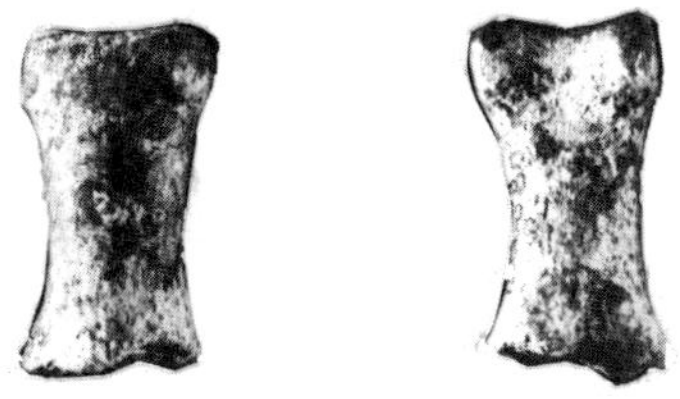

FIGURE 7.116.
Dorsal view (left) *and palmar view* (right) *of proximal phalanx of thumb KNM-WT 15000 BQ.*

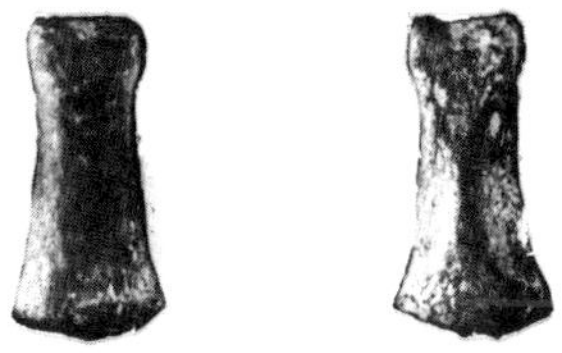

FIGURE 7.117.
Dorsal view (left) *and palmar view* (right) *of middle digital phalanx of hand KNM-WT 15000 BO.*

unknown or unrecognized postcranial skeletons. It is possible, then, despite the fact that they are of the right epiphyseal state to belong to our skeleton, that they belong to a juvenile of another mammal. They also come from a part of the site where the only hominid bones are not certainly attributed. The presumed first metatarsal comes from this part of the site, as does a vertebral body that we first thought belonged to the hominid individual but that proved not to be upon later examination.

KNM-WT 15000 BO & BQ: HAND PHALANGES

(FIGURES 7.116–7.117)

There are two phalanges, each without the epiphysis. The larger (BQ) is taken to be the left proximal phalanx for the thumb. This conclusion is based on size and faint asymmetry. It is obviously a proximal phalanx from the first ray. It seemed to us to be clearly from the hand, but Day (1986, Fig. 82, p. 234), using a photograph by David Brill of a mount made by one of us, challenged our classification by blacking out the bone on the photograph. We measured a small series of hand and foot first proximal phalanges of modern *H. sapiens*. Because the epiphysis was unfused and is missing from the fossil, we chose to measure the maximum distal transverse breadth. To get some normalization for size and to keep errors of scale low, we also measured maximum femoral head diameters on the same individual skeletons. Fig 7.118 shows a bivariate plot of the phalanx measurements against femoral head values. The fossil is marked by a star and falls clearly within the cluster of thumb phalanges.

The proximal epiphyseal surface is elliptical in outline, with the long axis running anteromedially to posterolaterally. The major and minor axes are 12.1 and 10.6. The bone as it stands is 23.8 long. The shaft changes in cross-section from being nearly circular proximally to a very flat el-

FIGURE 7.118.
Bivariate plot of distal widths (mm) of proximal hand and foot phalanges against the diameters of femoral heads for a small sample of modern humans.

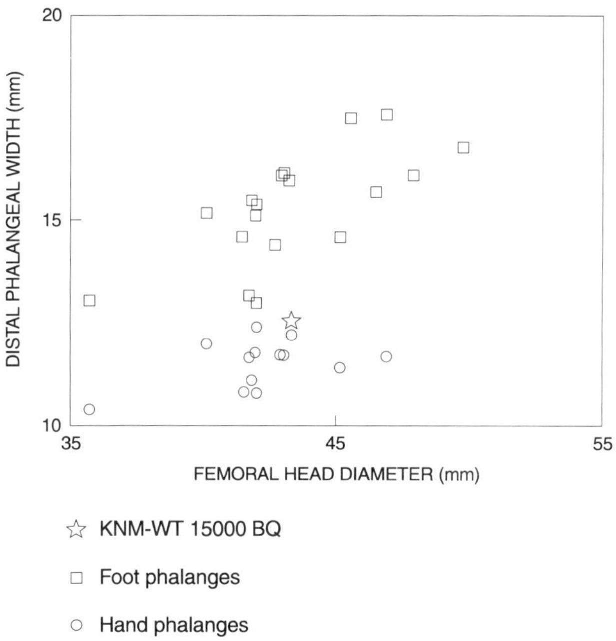

lipse distally. The distal articular surface is 12.7 wide and 6.9 in thickness. It is almost symmetrical and its medial and distal articular faces are smoothly concave, without any obvious central notch.

The second phalanx (BO) is a middle one from the hand. As it stands it is 20.5 long without its proximal epiphysis. The epiphyseal surface is 7.6 anteroposteriorly by 9.8 mediolaterally. The shaft is only very slightly curved. Most of that curvature is distal and really is formed by the slightly palmar tilt of the articular surface. The dorsal surface is smooth and unmarked. The palmar surface has very strong bilateral excavations for the annular part of the digital tendon sheath. These extend for about half the length of the shaft and their excavation leaves a proximal midline keel over the proximal half. This keel is about 6 wide proximally and narrows to 2.1 in midshaft before fading out on the distal palmar surface. The distal articular surface is a simple semicylindrical one, 7.5 wide by 3.7 dorsoventrally.

There are no comparable hand phalanges known of early African *H. erectus*.

The Pelvic Girdle and Lower Limb Skeleton

KNM-WT 15000 N, O, P, Q, AW, AX, BF, & BG: Hip Bones

(Figures 7.119–7.124)

Most of both ilia and ischia are preserved and two fragments of pubis. In the original note (Brown et al., 1985) the pubic pieces were both listed as from the right side, but our new reconstruction (Chapter 10) makes it more likely that they are left and right pieces. The bones have suffered from trampling damage and have also lost their pseudoepiphyses, if they had them. The triradiate suture was still unfused, of course, so the meeting of the ilia and ischia cannot be determined precisely. Because some parts are preserved on only one side and not the other, measurements of the whole pelvis were made on a reconstruction. As can be seen in Chapter 10, most of the pelvis can be reconstructed satisfactorily with mirror imaging of some parts of the pelvic bones and sacral elements.

The right ilium (O) has the following parts missing: fragments around the suture, part of the ala and crest posteriorly, the anterior superior spine. It also has the posterior superior spine eroded and a 65 by 18 section of the ala and crest crushed and folded down on it. The left ilium (N) has the following parts missing: fragments around the suture, the posterior superior spine, some parts of the ala and crest. It also has part of the tuberosity abraded. Two pieces of the crest (BG and BF) were found separately. The right ischium (P) has small parts near the suture missing, the tuberosity is abraded, and the ramus is lost. The left (Q) similarly is missing small parts near the suture and has lost the ramus and part of the tuberosity. Two major cracks on the inner surface have led to some slight displacement. Two pieces of the pubis are preserved (AW, right; AX, left). They cover part of the internal surface only and preserve the anterior margin of the obturator foramen from the obturator groove downwards.

The sacral articular surface is preserved best on the right. It consists of two subequal leaf-shaped parts meeting anteroinferiorly at an angle of about 90°. These surfaces are separated by a deep groove running towards the angle. The upper part is 34.9 from the angle to the most superior part and is a maxiumum of 10.3 wide. The lower part reaches from the angle to the crest between the two posterior spines and is 37.9 long and up to 20.5 wide. The area of the medial side of the tuberosity between and behind them is smooth and has rather large undulations running towards them. The distance from the angle to the (abraded) tip of the posterior superior spine is 55.9.

The greater sciatic notch covers more than a right angle. The left sciatic notch is more angled than the right. The distances from the sutural surface to the posterior inferior spine are 28 (right) and 29 (left) and the depths from these chords into the notches are 11.3 (right) and 12.6 (left). The thickness of the iliac body here is 20.0 (right) and 20.5 (left). This would really be the thickness at the arcuate line if there were one. The height of the tuberosity from the posterior inferior spine superiorly is 55.6.

The alae have two thickenings, one backwards from the acetabular area to the articular surface (the arcuate line area) and another where the acetabulocristal buttress was developing. The beginning of the buttress can be seen above the posterior part of the iliac lunate surface. Anterior to it and posterior to the anterior inferior spine is a shallow groove running anterosuperiorly. Between these two thickenings the ala is very thin indeed. Near the crest in the middle it can be measured as only 1.2 thick. This thinned area is concave on both outer and inner surfaces. The pseudoepiphyseal surface of the anterior superior spine is preserved on the left. The base of this spine surface is 19.7 above the top of the epiphyseal surface for the anterior inferior spine. This latter is 20.2 by 10.8, elongated along the axis of the anterior border. The notches between the two spines and between the inferior spine and the acetabular margin are about the same depth, but note that part of the left superior acetabular margin is crushed.

The lunate surface is not preserved well on either piece, but it is 19.4 wide near the suture posteriorly on the right. The two pieces of crest, which were preserved separately, show the irregular grooved surface for the pseudoepiphysis. The larger fragment (BF) has over 50 preserved and it is gently sinusoid in outline. It is taken to be towards the posterior part and is thickening towards the posterior superior spine. The smaller fragment (BA) is very close to the anterior superior spine and the crest here is 13.3 thick.

The pieces of pubis (AW and AX) are not very enlightening. They are both just parts of the inner surface of the pubic ramus and preserve that thin bit of bone which forms the anterior and inferior edge of the obturator foramen. The fragments can be oriented by position of the obturator groove. The obturator foramen is oval in outline and was about 50 anteroposteriorly and 30 superoinferiorly if these pieces have been correctly aligned in the reconstruction.

The two ischial pieces are much more informative. The acetabular notch was at least 18 wide, for that is the distance from the lower part of the lunate surface to the broken edge of the acetabular fossa at the level of the notch. The lunate surface is best preserved on the left fragment. At its narrowest (posterior) point it is 16.4 wide and at its widest it is 19.0. The surface ends in a very clear set of finger-like processes and ridges which formed the bony part of this bit of the triradiate suture. The distance between the edge of the acetabulum and the everted lip of the nearest part of the tuberosity is 9.6 (left) and 8.8 (right). On the right, although abraded a little, the tuberosity is 26.2 wide transversely by 45.6 superoinferiorly. It is pear-shaped in outline, apex inferiorly, and has an irregular, roughened surface.

There are other early African *H. erectus* hip bones available for comparison. The best is Olduvai Hominid 28 (Day, 1971), an associated left hip bone and femoral shaft from Bed IV. This specimen is quite a lot younger than the Nariokotome specimen (Hay, 1976)—it is probably less than 1 million years old—but the femur is very like other *H. erectus*. The hip bone, from a large female, is massive and has a great lateral flare, an immense acetabulocristal buttress, and a strongly developed horizontal buttress from the acetabulum to the posterior iliac spine. The bone of the iliac blade between these two buttresses is extremely thin. KNM-ER 1808 (Leakey and Walker, 1985) also has pelvic parts preserved. Two iliac fragments show, as far as one can tell, the same general morphology as that seen in KNM-WT 15000, but the external surfaces are covered with abnormal bone deposits. KNM-ER 3228, a right hip bone from the upper Burgi Member of Koobi Fora that is about 1.95 Ma (Rose, 1984; Feibel et al., 1989), is complete except for the pubis and ischiopubic ramus. From an adult male, it is very similar indeed to OH 28 and, from what one can see, to KNM-ER 1808. Allowing for the juvenile state of KNM-WT 15000, it is remarkably like that specimen too. Had the Nariokotome juvenile lived to maturity, it would have developed much stronger muscle scars and buttresses. These were only incipiently developed at the time of death.

FIGURE 7.119.
Medial view of left pelvic bone KNM-WT 15000 N and Q.

FIGURE 7.120.
Lateral view of left pelvic bone KNM-WT 15000 N and Q.

FIGURE 7.121.
Medial view of right pelvic bone KNM-WT 15000 O and P.

FIGURE 7.122.
Lateral view of right pelvic bone KNM-WT 15000 O and P.

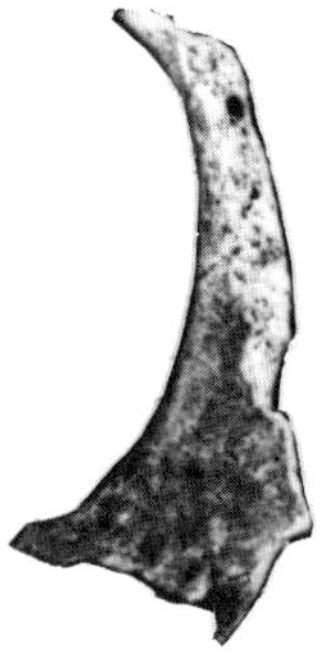

FIGURE 7.123.
Internal view of left pubic fragment KNM-WT 15000 AX.

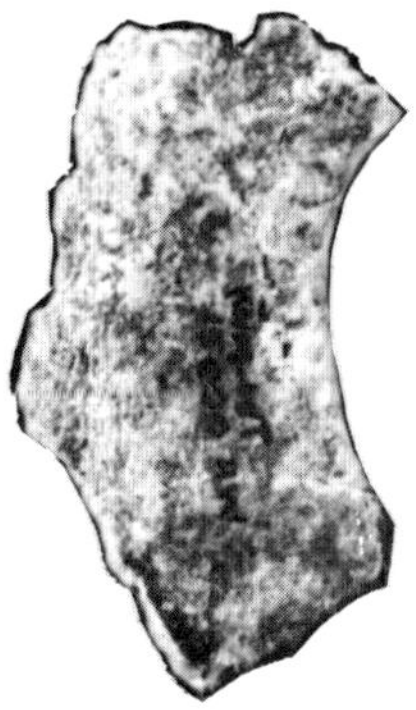

FIGURE 7.124.
Internal view of right pubic fragment KNM-WT 15000 AW.

KNM-WT 15000 G, H, & M: FEMURS

(FIGURES 7.125–7.134)

Both femurs are represented. The left (G) is complete, but the right (H) is missing the epiphyses for the head and greater trochanter and its distal epiphysis (M) cannot be attached because of distortion. Both femurs have suffered crushing of the cancellous bone tissue of the distal metaphysis and epiphysis. Even the distal epiphysis of the left femur was fossilized next to its metaphysis and crushed into the shaft slightly out of place. It has not yet been repositioned. A near midshaft break in the left femur was damaged by roots, and some bone fragments around the crack are not in the correct alignment. The head and greater trochanter epiphyses were found separately and have been replaced. The description given here is based on the left femur.

As it stands, with the distal epiphysis still pushed proximally and crushed, especially on the lateral side, the total length is 430.3. The length from the tip of the greater trochanter to the most distal part (medial condyle) is 426.2. A total length of 432 is estimated after allowing for crushing and the thicknesses of the epiphyseal cartilages.

The hemispherical head is 46.0 in diameter. In medial view the fovea is situated with its center 8.6 behind the center of the head. It is oval in outline, 17.8 long by about 11 wide, with the major axis running posterosuperiorly at an angle of 25° to the horizontal. It has quite a strong lip on its superior and posterior border, but inferiorly its border is rounded and opens into a shallow groove anteriorly. The neck is long. The distance from the lateral border of the shaft to the head along the neck axis is 110.2. That from the long axis of the shaft to the head along the neck axis is 85.0. The same measurements to the epiphyseal line are 84.2 and 59.3. In the middle of the neck the anteroposterior diameter is 23.8 and the superoinferior diameter is 36.2. In cross-section the neck is an ellipse with the posterior surface flattened and the contour interrupted by a mild point where the posterior and inferior surfaces meet. Vascular striae are found all round the neck, but particularly strong ones are seen running from the superomedial surface to the lesser trochanter. There is no obvious groove for the tendon of the obturator externus. The neck-shaft angle is 110°. The greater trochanter has its apex, which is posteriorly placed, 10.0 below the top of the head. There is a strong, irregular pit in the region of the insertion of the gluteus minimus laterally. The lesser trochanter is missing its epiphysis and faces almost directly posteriorly. The epiphyseal surface is an oval one about 20 long by 13 wide, with the long axis set at about 30° to the long axis of the shaft.

The shaft is strongly platymeric superiorly with subtrochanteric dimensions of 31.0 (mediolateral) by 29.5 (anteroposterior). The medial border of the shaft is seen in anterior view to be almost straight. A 33 long lateral gluteal hollowing is bordered posterolaterally by a roughened and linear gluteal tuberosity, which begins at the level of the inferior margin of the lesser trochanter. The hollowing is also bordered anterolaterally by a soft crest, which forms a bulge on the lateral margin of the shaft in anterior view. A clearly marked third trochanter marks the proximal end of the gluteal tuberosity and is situated almost level with the lesser trochanter. It is about 17 long (proximodistally) and about 9 wide (mediolaterally). By midshaft the shaft has become much less platymeric, with dimensions of 24.3 (mediolateral) by 24.5 (anteroposterior). The narrowest part of the shaft is the midshaft region. The linea aspera is a clearly defined crest only for about 40 in the midshaft region. Above and below this crest the lateral and medial limbs of the linea fade into the bone surface and are more difficult to trace, particularly medially. The region of the metaphysis just superior to the patellar groove is marked by several large vascular foramina. The popliteal surface is gently convex mediolaterally but is disrupted by crushing laterally. Both epiphyses are badly crushed, though the left one shows its shape more reliably. The patellar groove is about 8 deep and 34 wide (to the crests of the marginal ridges) superiorly, and it is difficult to determine the extent of the buildup of the lateral margin. Both condyles are damaged, but the intercondylar groove can be estimated to have been about 21 wide posteriorly. Most epicondylar features are obscured by cracking and crushing, but a maximum bi-epicondylar width of about 80 is indicated.

The femur is quite well represented in the early hominid fossil record. There are several early African *H. erectus* ones that can be compared with those of KNM-WT 15000, but none of them is complete. From Koobi Fora, KNM-ER 737 (Day and Leakey, 1973), KNM-ER 803 (Day and Leakey, 1974), and KNM-ER 1808 (Leakey and Walker, 1985) seem to be about the same geological age and certainly seem to be from the same taxon as KNM-WT 15000. From Olduvai, OH 28 seems to be of the same sort too. Day (1971), following Weidenreich (1941), gave a list of features that, taken together in combination, were typical of *H. erectus*. These include: lack of much anteroposterior bowing of the shaft, narrowest part of the shaft placed distally, convexity of the medial border of the shaft, pronounced platymeria subtrochanterically, thickness of shaft cortex (medullary stenosis). The other early African *H. erectus* femurs, as far as one can tell, have all these features. Even the femur of KNM-ER 1808, which is thickly covered by abnormal bone, has, underneath it, a femoral conformation that closely resembles that of KNM-ER 737 (Leakey and Walker, 1985). In the case of the juvenile KNM-WT 15000, some of these features are present but others are not. These latter may have appeared had the youth lived to maturity. The Nariokotome femurs do show, however, the following: lack of anteroposterior bowing, pronounced subtrochanteric platymeria, a very thick shaft, and a narrow medullary cavity (as judged by plane-film X-ray).

Femurs from early in the East African fossil record also have some of these same features, as Kennedy (1983) has pointed out. KNM-ER 1481 (Day et al., 1985) has them. Although this *H. erectus*-like feature appears in the fossil record as early as about 1.89 Ma (KNM-ER 1481 date given by Feibel et al., 1989), this does not mean that this species itself appeared that early, as Kennedy thought. Trinkaus (1984) pointed out that KNM-ER 1481 was much more probably from another early species of the genus *Homo* and that this pattern of femoral morphology has been typical of our lineage from about 2 Ma until the appearance of anatomically modern humans. If Kennedy's scheme were followed there would be *H. habilis* skulls but not femurs and *H. erectus* femurs but not skulls from deposits at Lake Turkana about 1.9 million years of age.

KNM-WT 15000 I, J, & BH: TIBIAS

(FIGURES 7.135–7.145)

These are represented by both shafts, both distal epiphyses, and the left superior epiphysis. The left tibia has the suffix I and the right J. The left distal epiphysis is BH. The epiphyseal surfaces of the shafts have been eroded somewhat. Those of the right tibia are more complete. Both malleoli were displaced from their epiphyses and there is missing bone on their medial surfaces. One or two

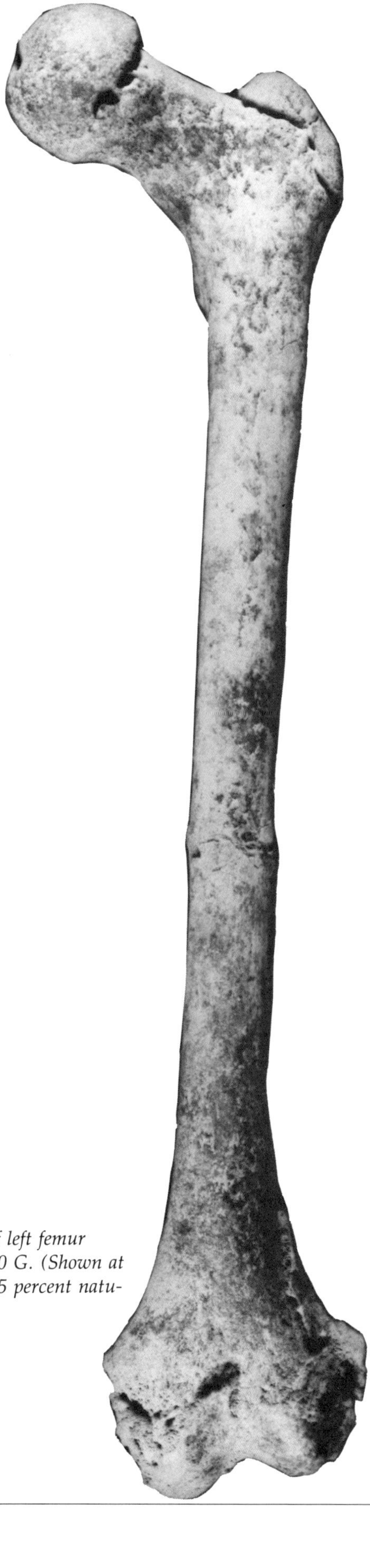

FIGURE 7.125.
Anterior view of left femur KNM-WT 15000 G. (Shown at approximately 55 percent natural size.)

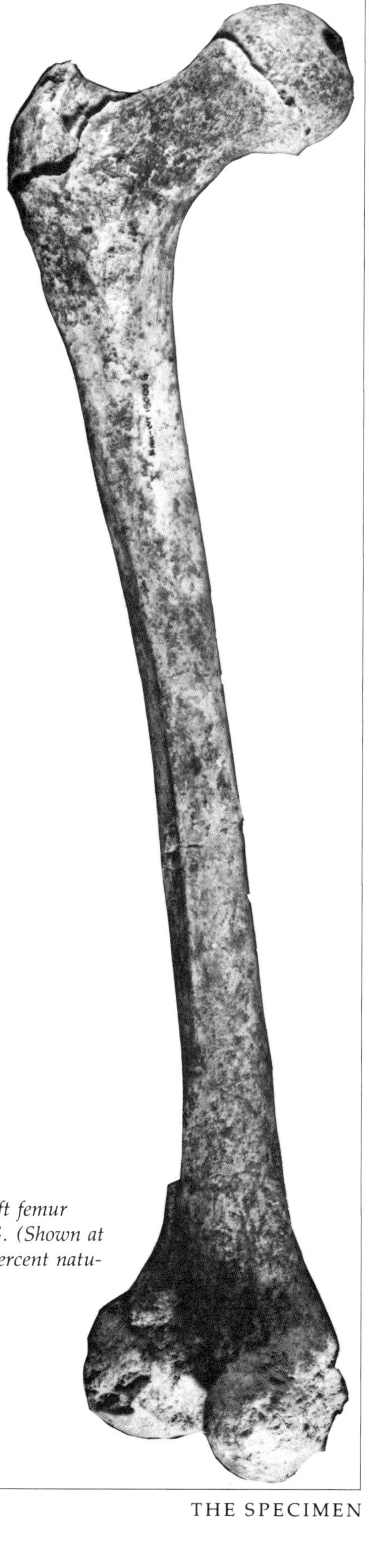

FIGURE 7.126.
Posterior view of left femur KNM-WT 15000 G. (Shown at approximately 55 percent natural size.)

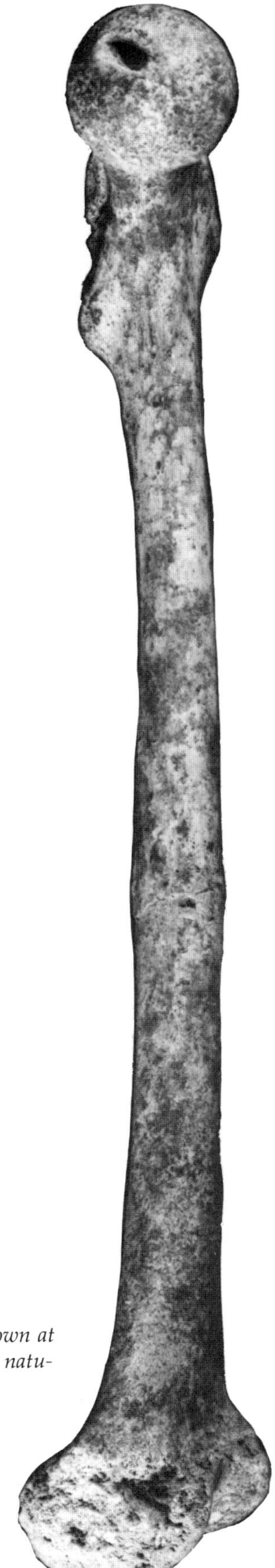

FIGURE 7.127.
Medial view of left femur KNM-WT 15000 G. (Shown at approximately 55 percent natural size.)

FIGURE 7.128.
Lateral view of left femur KNM-WT 15000 G. (Shown at approximately 55 percent natural size.)

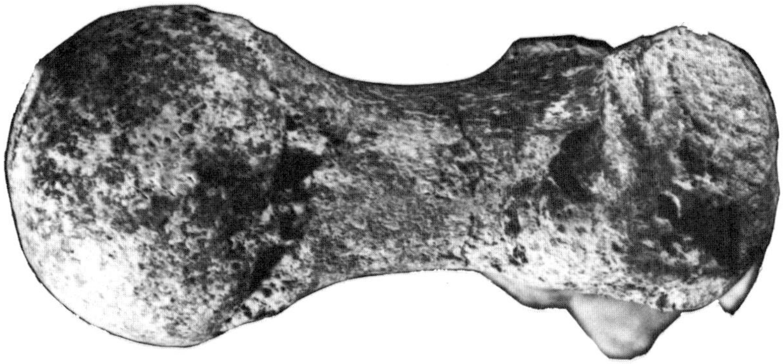

FIGURE 7.129.
Proximal view of left femur KNM-WT 15000 G.

FIGURE 7.130.
Distal view of left femur KNM-WT 15000 G.

FIGURE 7.131.
Anterior view of right femur KNM-WT 15000 H. (Shown at approximately 55 percent natural size.)

FIGURE 7.132.
Posterior view of right femur KNM-WT 15000 H. (Shown at approximately 55 percent natural size.)

FIGURE 7.133.
Medial view of right femur KNM-WT 15000 H. (Shown at approximately 55 percent natural size.)

FIGURE 7.134.
Lateral view of right femur KNM-WT 15000 H. (Shown at approximately 55 percent natural size.)

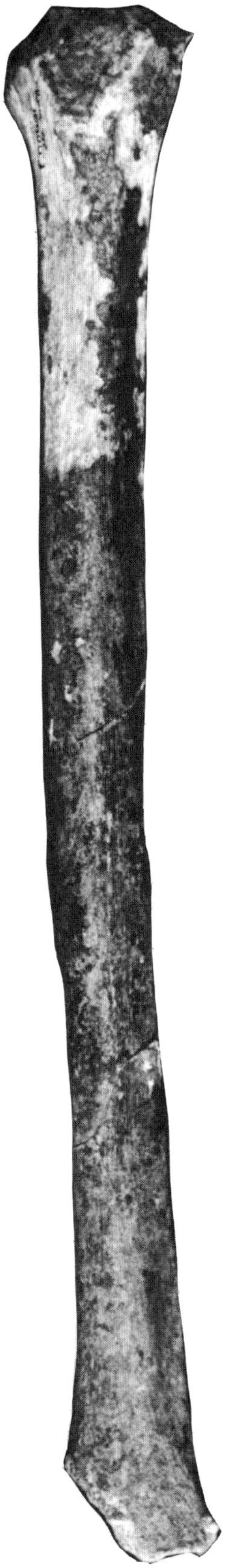

FIGURE 7.135.
Anterior view of left tibia KNM-WT 15000 I. (Shown at approximately 68 percent natural size.)

FIGURE 7.136.
Posterior view of left tibia KNM-WT 15000 I. (Shown at approximately 68 percent natural size.)

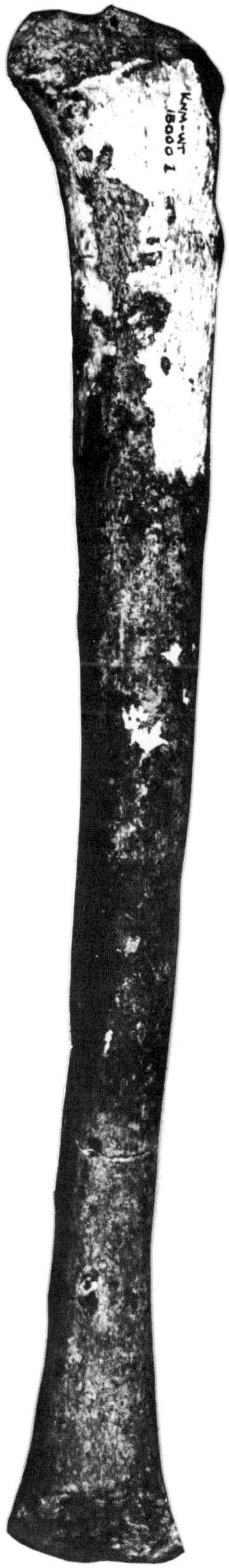

FIGURE 7.137.
Medial view of left tibia KNM-WT 15000 I. (Shown at approximately 68 percent natural size.)

FIGURE 7.138.
Lateral view of left tibia KNM-WT 15000 I. (Shown at approximately 68 percent natural size.)

FIGURE 7.139.
Proximal view of proximal epiphysis of left tibia KNM-WT 15000 I.

FIGURE 7.140.
Distal view of distal epiphysis of left tibia KNM-WT 15000 BH.

FIGURE 7.141.
Anterior view of right tibia KNM-WT 15000 J. (Shown at approximately 68 percent natural size.)

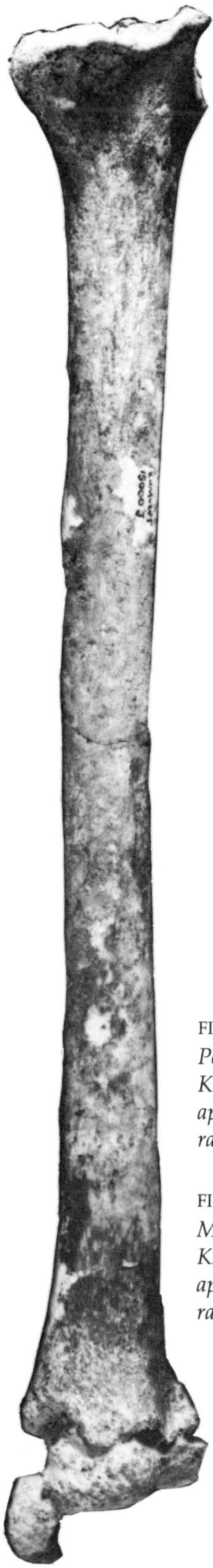

FIGURE 7.142.
Posterior view of right tibia KNM-WT 15000 J. (Shown at approximately 68 percent natural size.)

FIGURE 7.143.
Medial view of right tibia KNM-WT 15000 J. (Shown at approximately 68 percent natural size.)

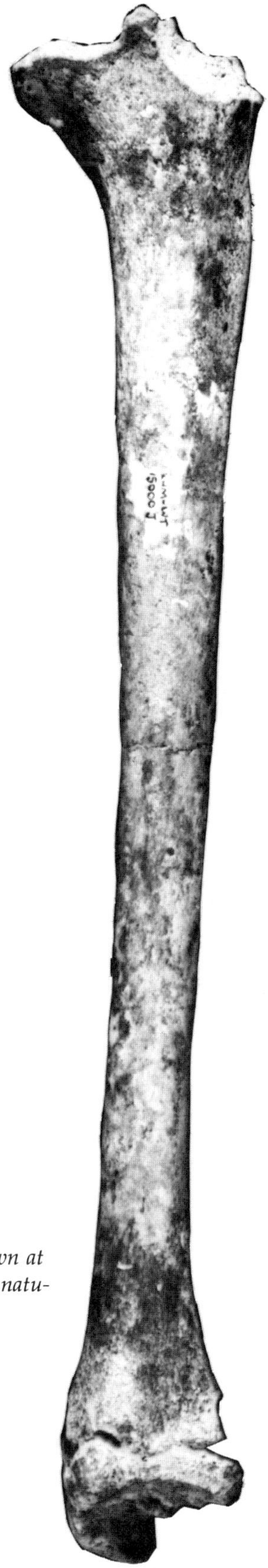

FIGURE 7.144.
Lateral view of right tibia KNM-WT 15000 J. (Shown at approximately 68 percent natural size.)

FIGURE 7.145.
Distal view of distal epiphysis of right tibia KNM-WT 15000 J.

patches of root damage have obscured detail on the shafts. This is particularly so on the upper quarter of the left tibia. Here root erosion has exposed underlying bone structure and caused the formation of a peculiar series of grooves that cut across the surface of the softened bone. The root-damaged bone is cream in color. Allowing for missing bone and estimating the thicknesses of the epiphyseal cartilages, we estimate the total tibial length to be 380 and the bicondylar length 375.

The superior epiphysis is pear-shaped in outline, the apex medially and the posterior border slightly concave. It is abraded, especially around the medial border. The anteroposterior diameter is 42.2 from the anterior rugosity to the posterior notch and the transverse diameter is 61.5. The epiphysis is thick anteriorly and thinner posteriorly. A 15.5 thick anterior rugosity runs all across the anterior part of the epiphysis. It is separated from the articular surface by a vertical step. The anterior intercondylar area is not very well delimited, but there is a 6.0 broad pit immediately anterior to the lateral intercondylar eminence. The posterior intercondylar area is more clearly defined and triangular in outline, apex anteriorly.

Superiorly, the shaft shows a depressed, elongated tibial tuberosity area about 30 long and 9.0 wide where the anterior part of the proximal epiphysis had not fused to the shaft. The lateral margin of the area is continous with the sharp upper part of the anterior border of the shaft. The upper part of the metaphysis, which supports the epiphysis, flares out fairly evenly from the shaft, but it is strikingly twisted on the shaft so that the long axis of the top of the metaphysis is turned by about 60° from the coronal. The shafts become more triangular in cross-section by midshaft but widen again at the distal metaphysis to become more circular in section. The midshaft dimensions are 24.4 anteroposteriorly and 20.4 mediolaterally. The medial border is only faintly developed, being a distinct ridge for only its middle third. The interosseous border is not itself very strongly marked but is a straight, obtuse crest. The soleal lines are relatively feebly marked. Large nutrient foramina are found just inferior to their lower limits about 10 distal from the epiphyseal surface. The midshaft diameters are: anteroposterior, 24.5 (left) and 24.0 (right); mediolateral, 20.4 (left) and 20.4 (right).

The left distal epiphysis is better preserved on the whole, but the right malleolus is complete whereas the left is missing bone on its medial surface. The talar surfaces are distinctly wedged, with the fibular side being about 40 anteroposteriorly and the malleolar side about 26. The surface is concave anteroposteriorly, but the ridge for the groove in the talar trochlea is hardly seen. The malleolus is set on the distal articular surface at an angle so that its lateral surface faces slightly anteriorly as well as laterally. This angulation, together with this surface having a slight convexity, adds to the impression of a wedged trochlear surface. The malleolus is stout, 12.5 mediolaterally by 26.5 anteroposteriorly. In medial view its outline is quadrangular, with the stronger apex anteriorly and a lesser one towards the rear. The anterior margin is concave, leading to the impression that the stronger apex is hook-shaped. The posteromedial border is raised to accommodate the tibialis posterior tendon laterally.

There are several tibias that could belong to early African *H. erectus.* They are all from the Pleistocene of Lake Turkana. KNM-ER 803 (Day and Leakey, 1974) has most of a left tibial shaft. KNM-ER 741 (Leakey et al., 1972), although described as *Australopithecus,* is almost certainly *H. erectus.* KNM-WT 19700, an unpublished fossil, is heavily rolled but probably is one. The tibias from KNM-ER 1808 (Leakey and Walker, 1985) are definitely of this taxon, but they are very badly affected by external abnormal bone deposition. There are general resemblances, of course, between these adult bones and the juvenile tibias of KNM-WT 15000, but the latter are considerably more gracile and their features are less well defined. The tibial parts of KNM-ER 1481 (Day et al., 1975) have all the features to be expected in slightly smaller and geologically earlier tibias, as might be anticipated following the discussion about the femurs.

KNM-WT 15000 K & L: FIBULAS

(FIGURES 7.146–7.153)

Both fibula shafts are preserved but no epiphyses. The left (L) is more complete since it preserves its superior epiphyseal surface and just possibly a very small part of the inferior surface. The right (K) is broken at both ends. The bones are long and thin and have a striking curvature, so that in lateral view the anterior border is very concave and the posterior border convex. If a line is drawn through the centers of the ends of the bone, the center of the shaft at the midshaft level is some 15.0 posterior to the line and the anterior border itself has even curved posterior to the line.

The total lengths, as the fibulas stand, are 321.5 for the left and 297.6 for the right. The superior epiphyseal surface is oval in outline and is 23.5 anteroposteriorly by 15.6 mediolaterally. It tapers rapidly to the neck, about 33 from the upper surface. The diameters at the neck are 9.9 (anteroposterior) by 5.6 (mediolateral) for the left and 10.1 (anteroposterior) by 6.1 (mediolateral) for the right.

The anterior border is blunt for its upper half and sharply crested for its lower until just above the swelling for the malleolus. The medial surface is flat, nearly completely so on the right and interrupted more by the interosseous crest on the left. The posterior border is sharply crested throughout on the right and only in the upper half on the left. In posterior view the shafts are also seen to be curved, this time so that they swing slightly towards the tibias at midshaft. The lateral surface is convex throughout, giving the bone a rounded triangular appearance in cross-section. The midshaft diameters are 10.5 (anteroposterior) by 9.1 (mediolateral) for the left and 12.5 by 8.9 (mediolateral) for the right.

Single nutrient foramina are found 114 below the upper epiphyseal surface on the posterolateral edge of the left fibula. On the right a foramen is in nearly the same position but about 10 lower, as judged by the position of the fibular neck.

Fragments of other early African *H. erectus* fibulas are known. That from KNM-ER 1808 (Leakey and Walker, 1985) is not too covered with diseased bone to rule out comparison here. The lower part of the left fibula KNM-ER 1808 U has the same general configuration as the fibulas of KNM-WT 15000 but shows none of their characteristic curvatures. Anterior and posterior borders are very straight. The profile of the lateral surface is straight in anterior view except for the inferior swelling for the malleolus. KNM-ER 803 O, listed as a fragment of left fibula by Day and Leakey (1984), is very close in general form to those of KNM-WT 15000, but we believe that this small fragment is from the right side. Again no strong curvatures are seen.

KNM-WT 15000 BX: ? FIRST METATARSAL

(FIGURE 7.154)

This might be a right first metatarsal from the skeleton. It is a metapodial and has the right texture and developmental age, but it has some peculiar features and is from a part of the site that has yielded only questionable pieces. We have not been able to match it in our comparative searches, but, as with the first metacarpals, we recognize

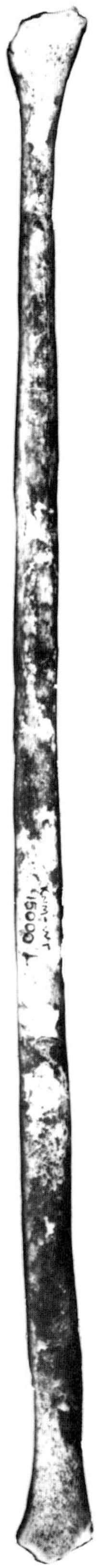

FIGURE 7.146.
Anterior view of left fibula KNM-WT 15000 L. (Shown at approximately 68 percent natural size.)

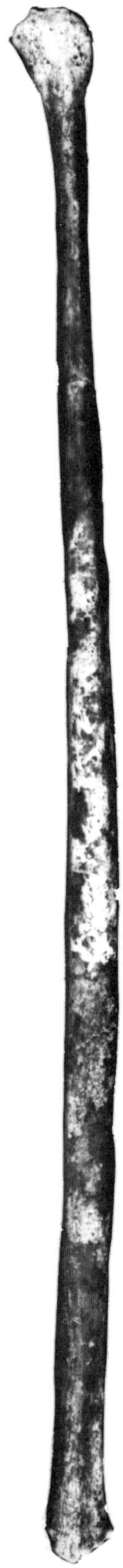

FIGURE 7.147.
Posterior view of left fibula KNM-WT 15000 L. (Shown at approximately 68 percent natural size.)

FIGURE 7.148.
Medial view of left fibula KNM-WT 15000 L. (Shown at approximately 68 percent natural size.)

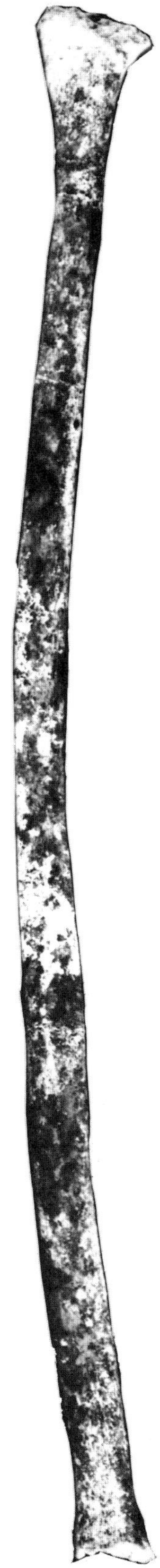

FIGURE 7.149.
Lateral view of left fibula KNM-WT 15000 L. (Shown at approximately 68 percent natural size.)

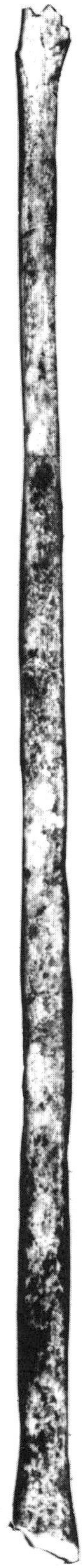

FIGURE 7.150.
Anterior view of right fibula KNM-WT 15000 K. (Shown at approximately 68 percent natural size.)

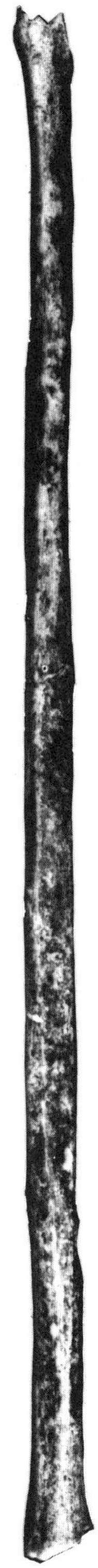

FIGURE 7.151.
Posterior view of right fibula KNM-WT 15000 K. (Shown at approximately 68 percent natural size.)

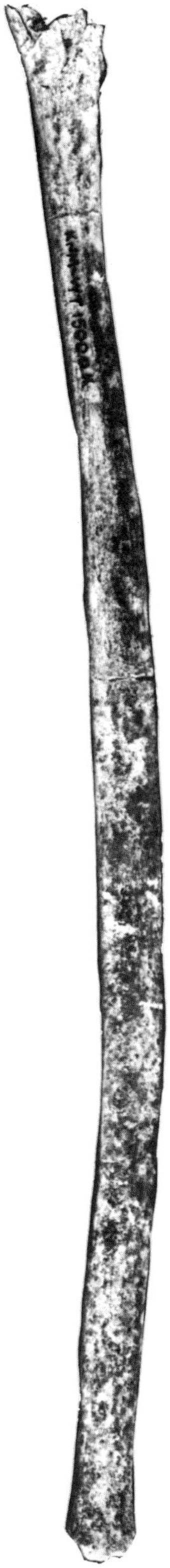

FIGURE 7.152.
Medial view of right fibula KNM-WT 15000 K. (Shown at approximately 68 percent natural size.)

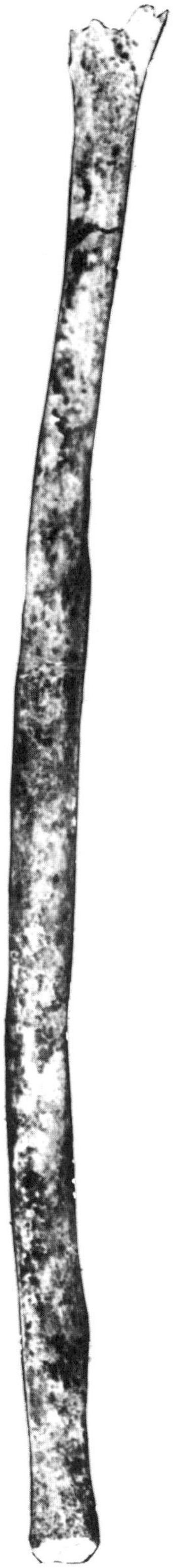

FIGURE 7.153.
Lateral view of right fibula KNM-WT 15000 K. (Shown at approximately 68 percent natural size.)

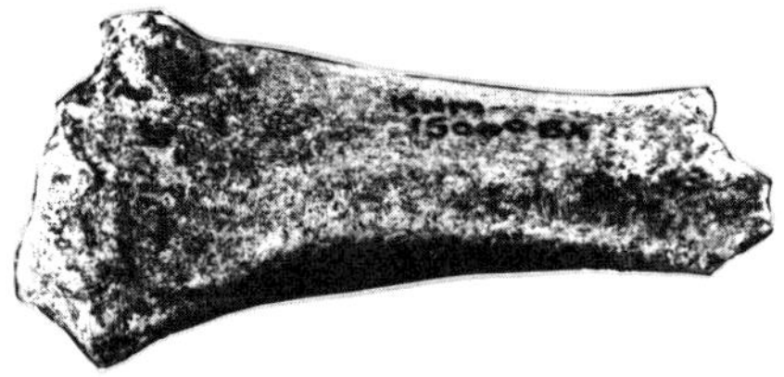

FIGURE 7.154.
Lateral view of ? right metatarsal 1 KNM-WT 15000 BX.

that it may belong to another extinct mammal whose postcranial skeleton is unknown or unrecognized.

The proximal epiphysis is missing and the head broken off. As the piece stands it is 52.0 long. The epiphyseal surface is rough and abraded some. It is 24.5 dorsoplantarly and 14.9 mediolaterally. The proximal third of the shaft is flat medially and concave laterally to give a semicircular cross-section. At midshaft the bone twists so that at the distal break it is 14.7 dorsoplantarly by 16.5 mediolaterally. The distal half has a prominent rounded ridge running down the middle of the medial surface. About 4.0 plantarwards to this is a thin crest, running parallel to the ridge. The ridge and crest are not seen in hominid first metatarsals and this alone should encourage us to be skeptical that this bone belongs to the skeleton. We have placed this specimen on record here because it might just belong and because someone with more experience than we have in the identification of juvenile mammal metapodia might be able to identify it.

CHAPTER 8

THE DENTITION

BARBARA BROWN AND ALAN WALKER

Because they are very durable, teeth make up a large part of the sample of hominid fossils that have survived over the ages. Furthermore, because they develop early in life and change later only by wear and breakage, not remodeling, they express genetic information more directly than do other parts of the skeleton. Analyses of fossil dentition have therefore been of great importance in studies of human origins. The teeth of the Nariokotome hominid are documented in this chapter by descriptions, measurements, and photographs. They are compared with other early Homo erectus *teeth from East Africa and with the later sample from Zhoukoudian in China.*

Specimen KNM-WT 15000 is extremely unusual among fossil hominids in that it has almost all its teeth. The upper right M_3 germ is missing because the piece of maxilla containing it was never found. M^3 germs within the mandible are not visible by X-ray. Either they had not yet developed, or they had broken into indistinguishable pieces within their crypts (see Chapter 9, this volume). Our purpose here is to provide a photographic, metric, and descriptive record of the teeth and to determine to what extent they aid in our taxonomic and phylogenetic assessment of the specimen. There have been a number of recent attempts to define *Homo erectus* in cladistic terms (Andrews, 1984; Bilsborough and Wood, 1986; Rightmire, 1984, 1990; Stringer, 1984; Wood, 1984), but almost without exception these have not taken into consideration dental traits.

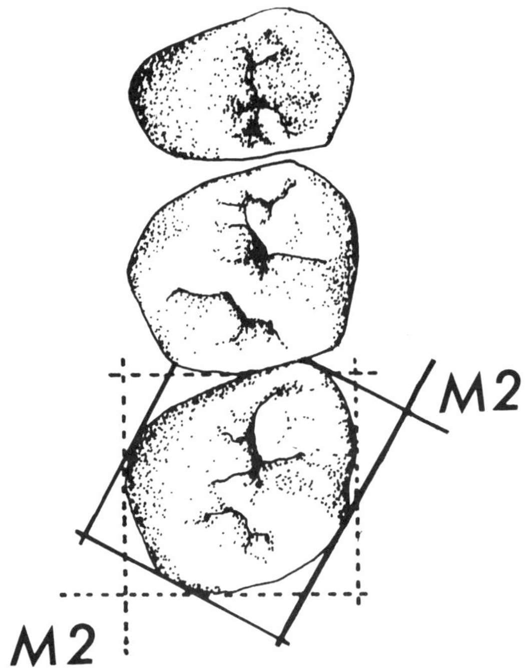

FIGURE 8.1.
Two measurements of the upper second molars.

Walker (Chapter 4, this volume) gives details of the positions of the teeth when found. Map 2 shows their exact locations. The maxillae were discovered embedded in the roots of a small *Acacia* tree at the edge of the excavation dug in the first season. The only maxillary teeth still in place were the molars and the left I^2 and the left $\underline{C}$ germ. The right maxillary tuberosity and its enclosed M^3 were never recovered. They were probably lost to surface deflation before the site was found. The mandible was recovered in an inverted position 5 meters east of the maxillae and it had the molars, the canines, and three premolars still in place. The right P_4 was not found until the following season. All the teeth still missing at the end of the 1984 season were discovered in one small area to the north of the first excavation. The teeth vary in surface quality. Many of them have poor surface detail because of root etching. Even those that retain good surface detail were so little worn in life that they are virtually unusable for microwear studies.

Table 8.1 gives the measurements of the teeth. Figure 8.1 shows the measurement technique for M^2. Figures 8.2 and 8.3 are occlusal views of the lower and the upper sets of teeth, respectively. Figures 8.4 and 8.5 show four views of the roots of those teeth that were initially separated from the skull. All measurements are given in millimeters, unless noted otherwise.

Mandibular teeth

The mandible has a complete set of teeth, excluding the unerupted M_3s. From plane-film X-ray examination, we conclude that either the M_3s are not yet developed or they are fragmentary pieces within poorly defined crypts (Figure 8.6). Computerized tomography (CT) examination is required to aid in distinguishing the state of the M_3 germs. The retromolar triangle is quite large, however, and there are gubernacular canals for the M_3 at its apex. There is one small pathological lesion on the mandible. Alveolar resorption has occurred on the right side of the body, from the mesial aspect of the P_4 buccally extending to about halfway down the completed M_1 mesiobuccal root. On the lingual side the resorption is less deep and is primarily adjacent to the P_4. Buccally, about half of the undeveloped P_4 root is exposed. Approximately 3 of the distobuccal root of the RM_1 is also exposed. Teeth which had erupted shortly before death in both mandible and maxilla have not undergone mesial drift to set up definitive occlusal relationships. In the mandible there are gaps between all teeth except for the right I_1 and I_2 and the right M_1 and M_2. The right P_4 is rotated counterclockwise from its proper position by about 45 degrees. The long axes of the M_2s are set at an angle to the long axes of the M_1s so that the mesiobuccal corners of the M_2s are touching the distal surfaces of the M_1s. This is most likely because the M_2s had only just erupted.

Lower first incisors

The LI_1 is very well preserved. Approximately one-third of its root is exposed labially. Some of the incisor alveolar matrix had to be removed before the teeth were repositioned, however. It may be that they would have been seated lower in life. There is a small chip on the labial occlusal edge. Its edge is worn slightly at an angle, the most medial aspect is the highest, and it drops laterally. Along the edge, dentine is exposed as a thin line bordered mesially and distally by slightly thicker dentine islands. The RI_1 is weathered on its labial surface so that it is pitted and discolored. This damage extends along the exposed portion of the labial root. A small aspect of the lingual surface is also weathered on this tooth. Both I_1s are symmetrically narrow teeth at the cervix. The lingual

TABLE 8.1.

Tooth measurements (mm) of KNM-WT 15000.

		MAXILLARY				MANDIBULAR			
TOOTH	SIDE	Mesiodistal	Buccolingual	Crown height	Root length	Mesiodistal	Buccolingual	Crown height	Root length
I1	R	11.5	9.4	13.0	21.4	6.5	6.9	9.6	20.4
	L	12.0	9.3	12.9	20.9	6.6	6.8	10.1	19.9
I2	R	8.4	8.5	11.2	18.2*	7.9	8.2	11.3	20.5
	L	8.6	8.4	13.3		7.5	8.3	11.1	20.5
d$\underline{c}$	R	8.0	6.9	6.5	17.9				
	L	7.8	6.7	6.5	17.7				
C	R	9.6	10.2	11.9	14.8*	8.9	9.2	11.3	
	L					8.9	9.6	11.3	
P3	R	8.7	11.7	9.4		9.2	10.5	9.8	
	L	8.5	11.9	9.4	14.5*	8.9	10.1	10.5	
P4	R	8.1	11.7	7.9		9.0	10.2	9.1	10.1*
	L	8.5	11.8	8.2	13.5*	9.0	10.1	9.4	
M1	R	12.2	12.2	6.9		12.7	11.1	7.7	
	L	11.3	12.1	6.7		12.4	11.0	7.6	
M2†	R	12.2	12.4	8.0		12.9	11.8	7.8	
	L	11.8	11.3	7.3		12.5	12.0	7.8	
M2′†	R	13.4	12.5						
	L	13.3	11.8						

Note: Tooth measurements have been amended since original publication of data in Brown et al., 1985.

* Roots still incomplete.

† M2 represents tooth measurements parallel and perpendicular to the tooth row. M2′ represents maximum length and breadth measurements recorded along the rhomboid shape (see Figure 8.1.). These correspond to an angle approximately 40 degrees from the tooth row.

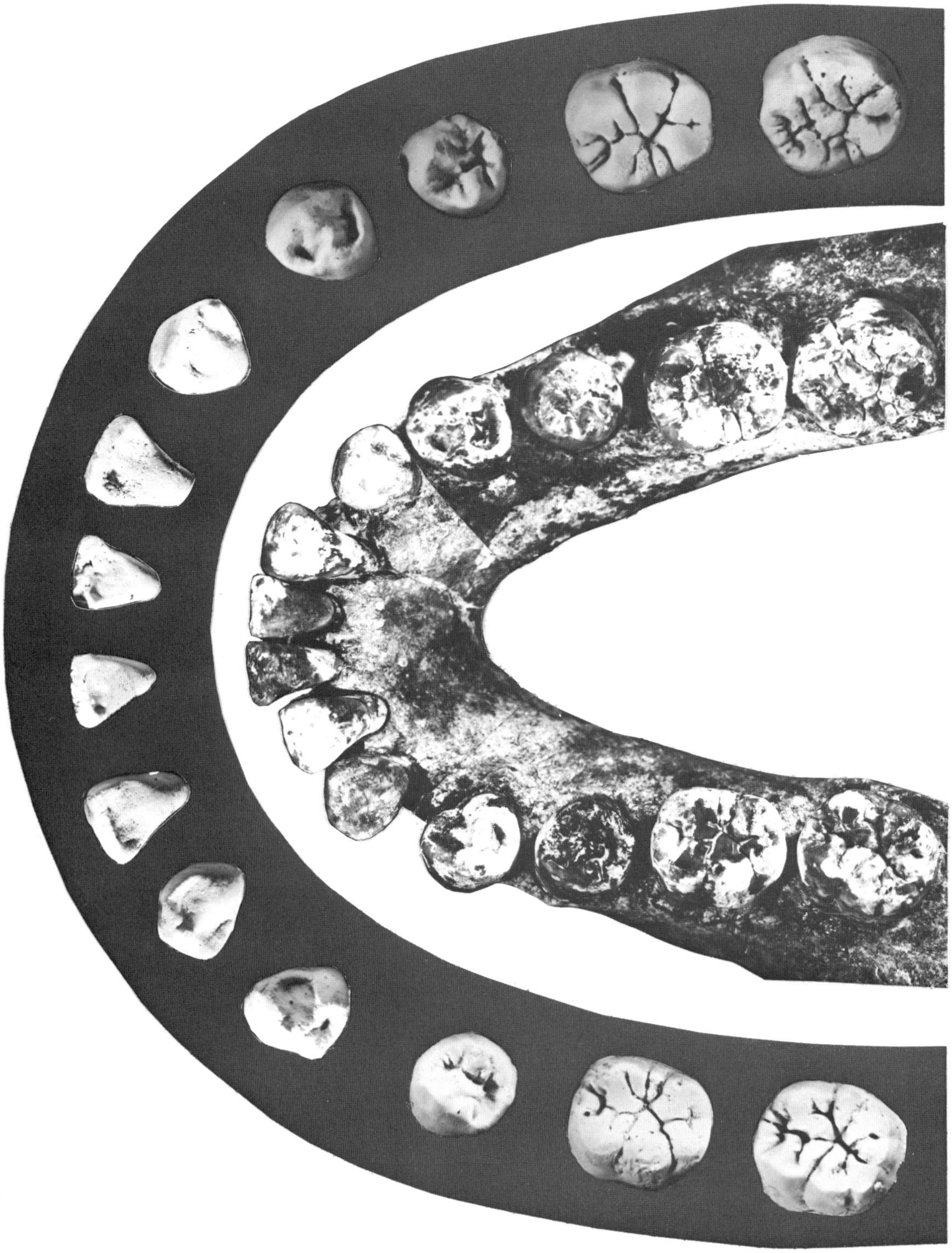

FIGURE 8.2.
Occlusal view of mandibular teeth of KNM-WT 15000 B. Single photographs of coated teeth are displayed on a black outline. (Shown at approximately twice natural size.)

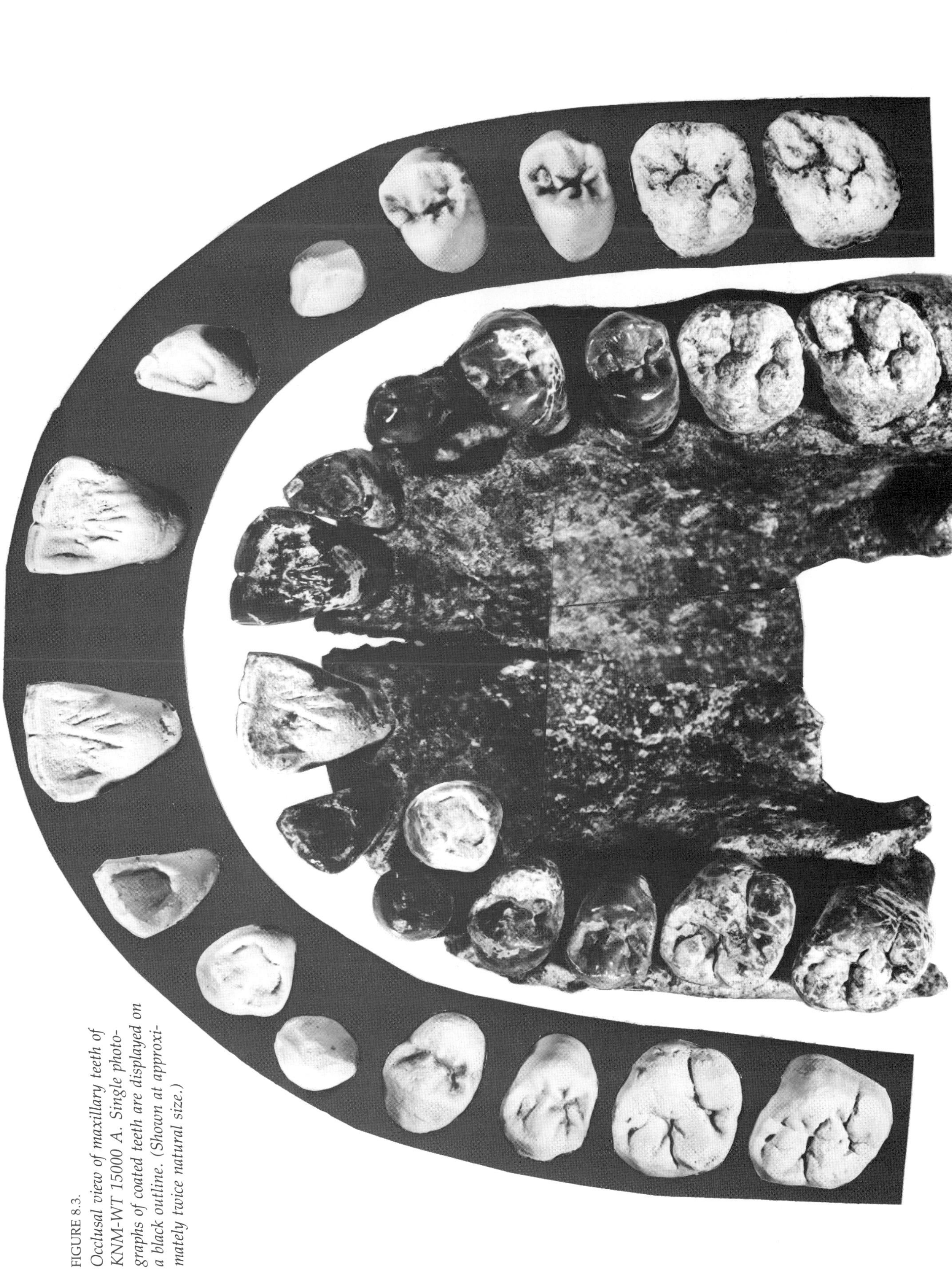

FIGURE 8.3.
Occlusal view of maxillary teeth of KNM-WT 15000 A. Single photographs of coated teeth are displayed on a black outline. (Shown at approximately twice natural size.)

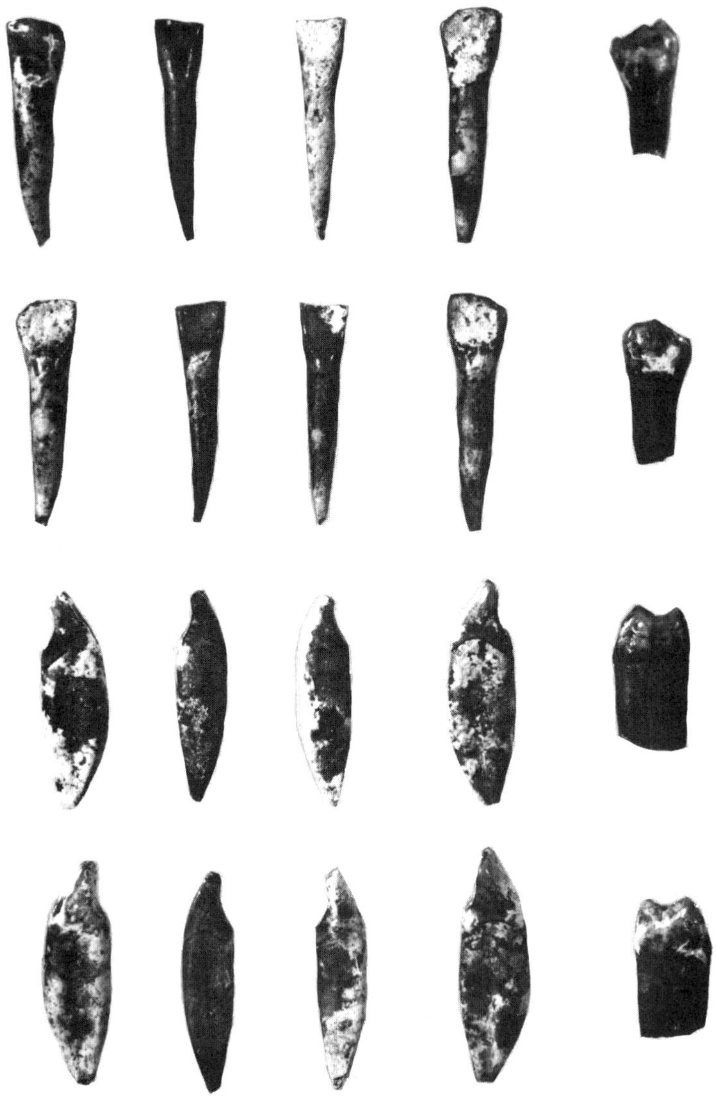

FIGURE 8.4.
Isolated mandibular teeth of KNM-WT 15000 B. Top row: *labial/buccal view;* second row: *lingual;* third row: *mesial;* bottom row: *distal.* From left to right: LI_2, LI_1, RI_1, RI_2, RP_4.

cingulum is weakly developed, the mesial and distal marginal ridges extend along the sides of the teeth to approximately half the height of the tooth enamel. The lingual fossa is very shallow. Perikymata are visible along the labial and lingual surfaces of the LI_1 and labial surface of the RI_1. The crown is set so that the mesial border is continuous with the mesial edge of the root, but the distal border meets the distal root edge at an angle. Both labial and lingual cervical bases appear to be at the same level. There are small interstitial facets for the adjacent central incisor but none for the lateral incisor on the left. Damage prevents one from seeing whether a facet was on the right tooth, but there is a clear facet on the undamaged LI_2. Root development is complete. The roots are strongly compressed mesiodistally and faintly grooved on either side, the distal grooves being deeper. Each I_1 root measures 4.5 by 6.7 at the cervix.

Lower second incisors

The I_2s exhibit minimal wear. The RI_2 is pitted and weathered along its labial surface. The LI_2 is only partly weathered along approximately three crack lines. A small line of dentine is exposed along the occlusal edge. The wear facet has just affected the corner of the incisive edge. There is no facet for either adjacent tooth on the left incisor. There is a small facet for the central incisor on the RI_2 but there is no facet for the canine. The lateral incisors are larger and more asymmetrical than the central incisors such that the crown is higher mesially than distally. The labial surface is more mesiodistally convex than that of the central incisor. The lingual cingulum is also about as developed as in the central incisors. The mesial and distal marginal ridges fade into lingual fossae before appearing again in the occlusal margin. The ridges are nearly equal in development and occur only on the upper half of the crown. The distal marginal ridge blends in with the distolabial enamel surface just at the level where the cingulum forms the inferior border of the lingual fossa. Labially the I_2s are relatively smooth. Perikymata are visible on the LI_2. The roots do not appear to have quite finished their development because the apical canals are not as closed as those of the central incisors. The roots are mesiodistally compressed and reasonably strongly grooved on both faces. At the cervix they measure 5.2 by 8.3 (left) and 5.3 by 8.7 (right).

Lower canines

The canines had just erupted and do not appear to have come into full occlusion. Small facets of

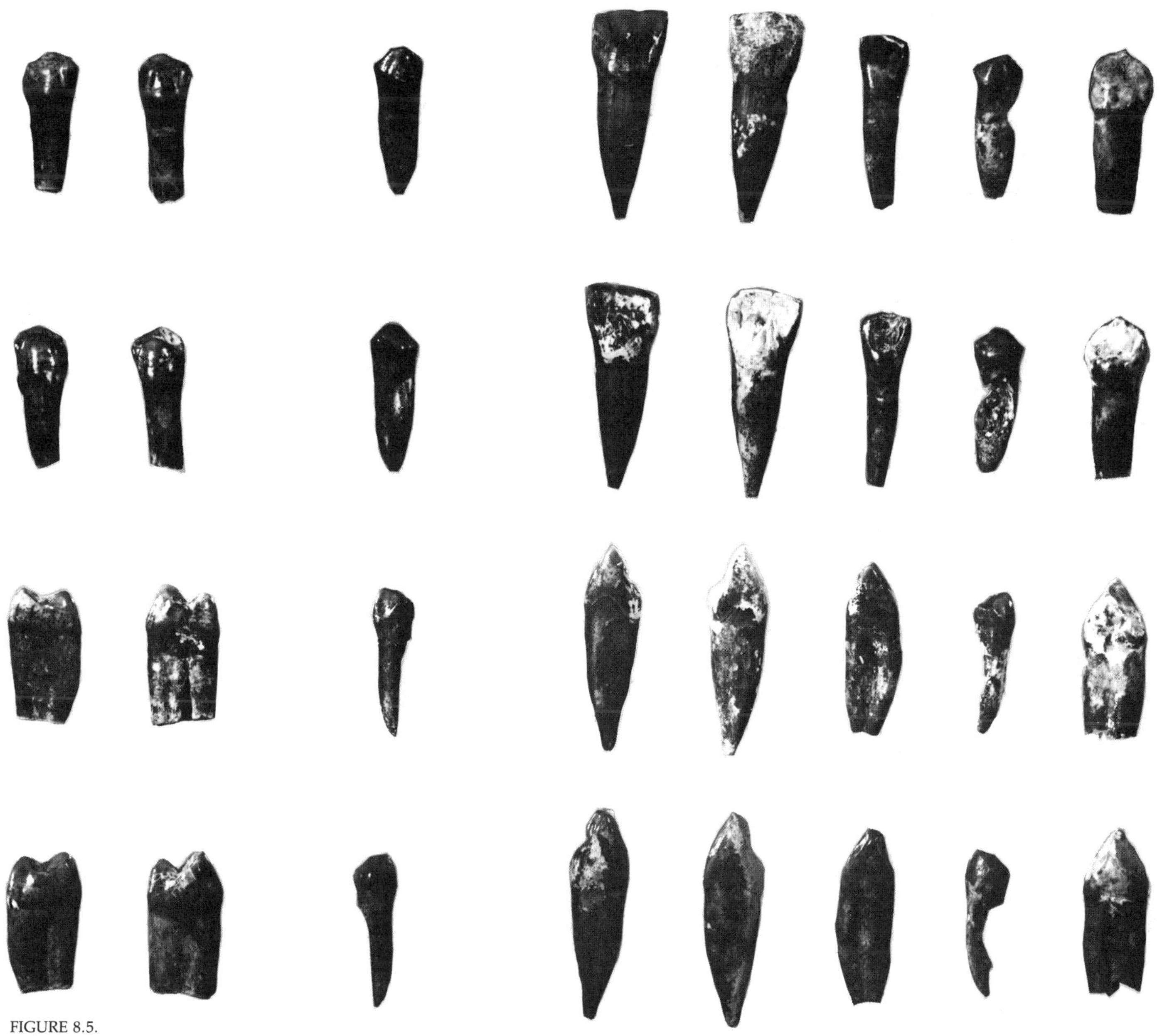

FIGURE 8.5.
Isolated maxillary teeth of KNM-WT 15000 A. Top row: *labial/buccal view;* second row: *lingual;* third row: *mesial;* bottom row: *distal.* From left to right: LP^4, LP^3, $Ld\underline{c}$, LI^1, RI^1, RI^2, $Rd\underline{c}$, $R\underline{C}$.

wear caused by occlusion with the lateral wear facet on the I_2s can be seen on the apex extending onto the mesial marginal ridge. Small indentations occur halfway along the distal cusp ridges at about the place where the upper $d\underline{c}$s would have contacted them, but it is likely that these are developmental features rather than artifacts of occlusion. The canines have a prominent lingual ridge separating mesiolingual and distolingual fossae, which, in turn, are bordered by marginal ridges. The canine apex is at the superior aspect of the lingual ridge. It blends with the mesial and distal cusp ridges. The mesial marginal ridges meet the mesial cusp ridges at about 90 degrees. This gives the crown its asymmetry because the distal marginal ridge meets its cusp ridge at a much wider

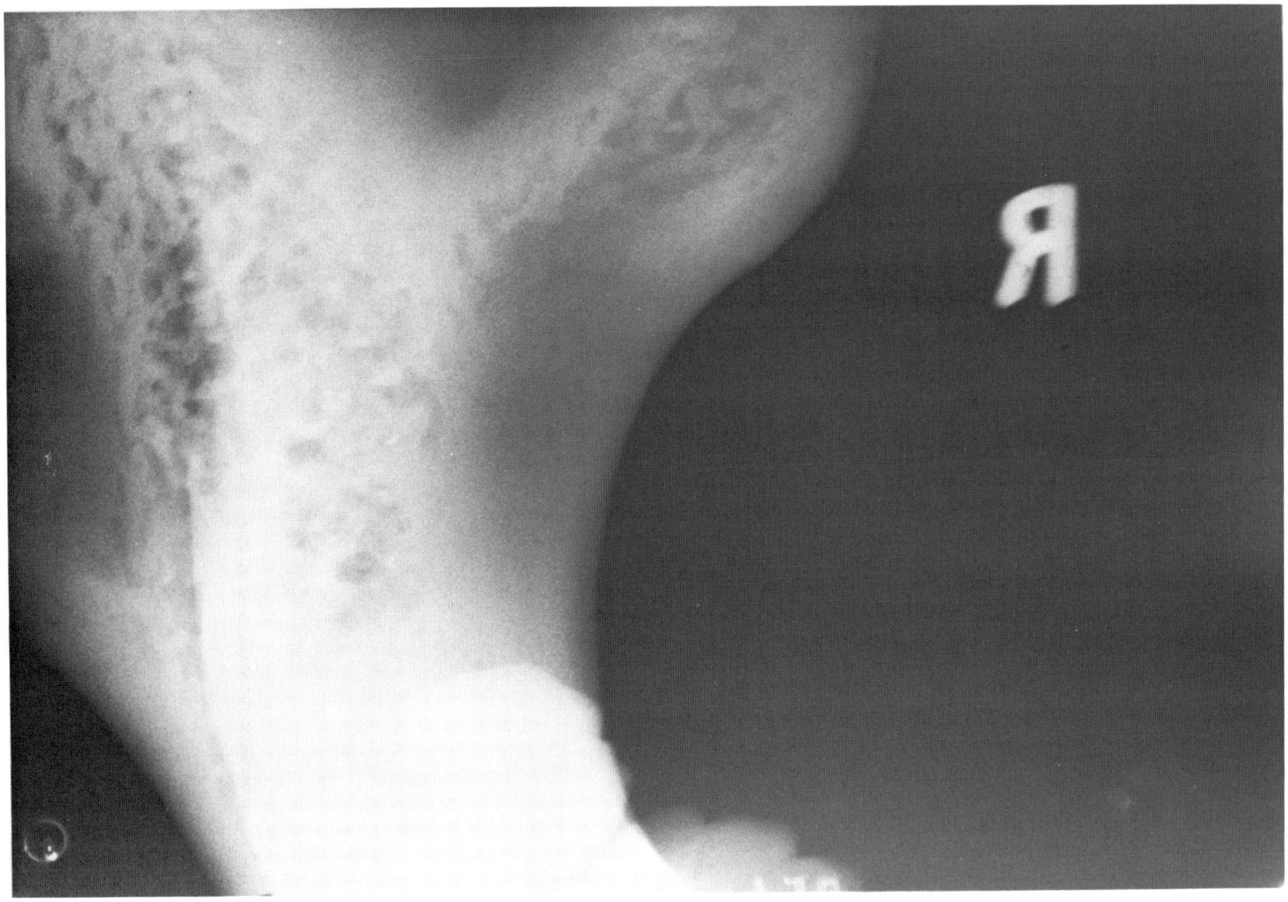

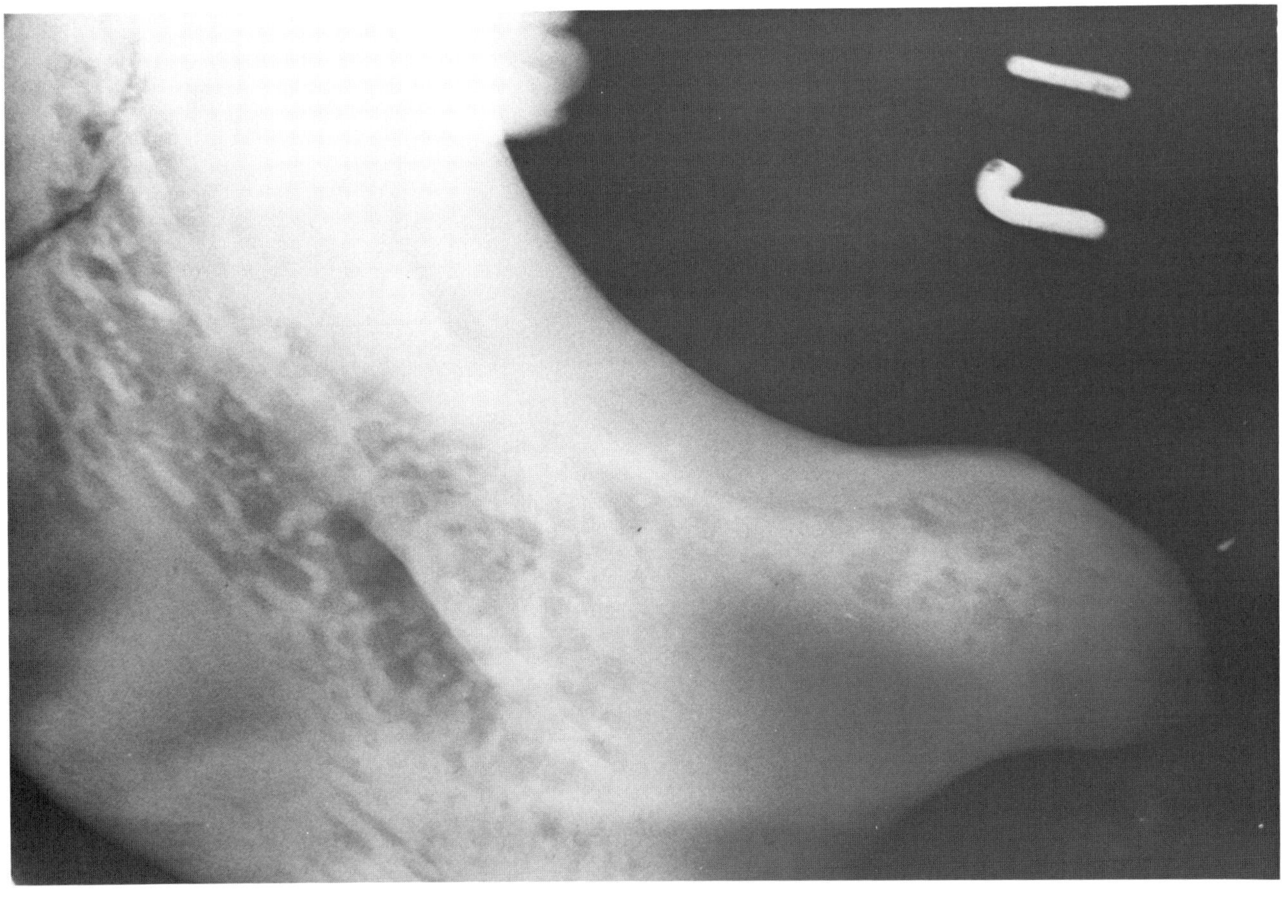

angle. The distal border of the cervix is relatively horizontal until it dips slightly to form a lower enamel extension along the labial enamel surface. The distolingual fossae are deeper than the mesial fossae. Perikymata are visible. A cross-section of the crown is almost square in shape. On the X-ray image, the canine roots can be seen to have grown only to the length of those of the lateral incisors. The apices are still at least 5.5 wide.

Lower third premolars

Both P_3s are in reasonable condition. They form a rounded triangle in outline and are narrowest mesially. Wear facets caused by the upper d$\underline{c}$s are on the mesial cusp ridges of the buccal cusps. There is slight wear on the distal cusp ridge of the buccal cusp. Wear is not apparent on the lingual cusps. Parts of each P_3 are pitted and weathered but none of the occlusal anatomy is obscured. The P_3s have relatively deep mesial and distal foveae. The distal fovea of the RP_3 as well as the mesial fovea of the LP_3 are especially pronounced. The foveae are separated by a prominent triangular ridge which blends with the triangular apices of the buccal and lingual cusps. The mesial and distal marginal ridges of the buccal cusp are thick and somewhat bulbous compared with the occlusal ridge of the canines. They each have an angular curve which blends with the lingual cusp apex. The buccal cusps are higher and more prominent than the lingual cusps. The mesial marginal ridge on the LP_3 is notched at the base of the fovea and that on the right is not. The alveolar margin is quite low along the buccal aspect of the P_3 roots. The roots are about as much exposed as those of the canines, but approximately 4 of the buccal LP_3 root is exposed in a crescentic margin, which suggests resorption rather than post-fossilization breakage. The alveolar bone is quite thin along the buccal margins of the anterior teeth. Root development is evident from the X-rays. Root formation is very incomplete and there are no signs of apical constriction.

FIGURE 8.6.
X-ray of KNM-WT 15000 B; note the absence of clear M_3 tooth germs.

Lower fourth premolars

The P_4s are relatively large and bulbous, as are the P_3s. They exhibit little to moderate weathering. Slight wear along the occlusal surface of the buccal cusp apex is visible, especially at the buccal surface of the LP_4. Because of alveolar disease around the dM_2, eruption of the RP_4 was affected such that the tooth is rotated counterclockwise from its normal position by about 45 degrees. The lingual cusps are quite prominent, they are almost as tall as the buccal cusps. The buccal and lingual cusps are separated by a central groove, which joins the mesial and distal foveae. The distal foveae are quite deep and broad. The distal marginal ridge is therefore much lower on the crown than the mesial marginal ridge. Beading, as small accessory ridges, decorates the edges of the foveae, especially the distal marginal ridge. The occlusal ridge along the buccal aspect of the buccal cusp is irregular. The distobuccal angle of the lingual cusp forms a style along the buccal surface. There are no styles evident on the P_3s or along the mesial marginal ridge of the P_4s. The occlusal surface of the lingual cusp is also irregular. The cusp is quite pointed. Its surface slopes to two grooves separating a fold. This is evident on the RP_4. The LP_4 has indistinct grooves with a similar small enamel fold, but this looks more like an outgrowth of the mesiolingual cusp. The roots are incomplete. Only 10.1 of the right root has developed. It is mesiodistally compressed, being 7.2 by 9.8 at the cervix. There is a faint mesial and a slightly stronger distal groove running down the two main faces. The root is roughly rectangular in cross-section.

Lower first molars

Both lower M_1s have discolored, weathered, and pitted areas along the distobuccal angles and the lingual surfaces of their enamel. The mesiolingual cusp of the RM_1 is slightly worn along its central aspect. The mesiolingual cusp of the LM_1 is slightly worn near the base of the central fovea along its buccal aspect. Both teeth have wear facets along the buccal and lingual aspects of the mesiobuccal and distobuccal cusps, the lingual aspect of the distal cusp, and the buccal aspect of the distolingual cusp. The cusps on the M_1s are all relatively equal in height, except for the larger mesiolingual cusp. The mesiobuccal cusp of the RM_1 is slightly larger than that of the LM_1. Both mesial and distal foveae are quite small. They are either grooves (as in the mesial foveae) or small pits (as in the distal foveae). All five cusps are separated from each other by clearly marked grooves. These are the buccal and lingual grooves running approximately through the middle of the crown, a central groove connecting the mesial and distal foveae, and a distobuccal groove delineating the distal cusp from the distobuccal cusp. The mesial marginal ridge is divided by small grooves radiating from the mesial fovea. The mesiolingual cusp is also subdivided along its distal portion by two additional grooves causing the formation of two additional enamel folds, the most distal of which is a tuberculum accessorius mediale (TAM) (Weidenreich, 1937). A deflecting wrinkle is present. The X-rays show the roots to be complete with the mesial and distal plates coming towards each other. They appear long and narrow and their pulp chambers, as expected for a young individual, are taurodont.

Lower second molars

Both lower M_2s are well preserved. The occlusal surfaces are discolored, and there is weathering on the mesiolingual border of both teeth. The RM_2 is less weathered than the LM_2 and wear facets are more prominent on the LM_2. The wear facets occur on the buccal aspect of the mesiolingual cusps and along the deflecting wrinkle. They also occur on the distal portion of each M_2 on both the distolingual and distal cusps. Like the M_1s, the M_2s have deflecting wrinkles and TAMs. However, they also have more folds of enamel than the M_1s. The mesial marginal fovea is distinct and fairly wide buccolingually (ca. 4 compared with ca. 2 in the M_1s), but it is constricted mesiodistally. Five small grooves radiate from the foveae to subdivide the mesial marginal ridge into four enamel folds. The central groove connects the mesial fovea with the distal fovea along the approximate midline. It is shifted slightly from the midline because of the presence of the large deflecting wrinkle. On the RM_2 the central groove has almost been obliterated by wear. Lingual and buccal grooves join the central groove just mesial to the distal fovea. Another small groove runs buccally from the mesial part of the central groove to split the lingual portion of the mesiobuccal cusp. This groove does not extend as far as the buccal border but ends just below the cusp apex. Two additional grooves radiate from the central fovea to bifurcate the distobuccal cusp or separate the distal cusp from the distolingual cusp. This splitting either causes a buccal shift in the distal cusp, with an extra cusp at the distal marginal border, or results in a low distal cusp shifted distally and a bifurcated distobuccal cusp. This is referred to by Weidenreich (1937) as a tuberculum sextum. The M_2 roots can be seen on the X-rays. As expected, they are very incomplete and their apices are wide open.

Maxillary teeth

Upper first incisors

Both crowns are extremely large and show only a trace of wear. The left central incisor is a little better preserved than the right. The crown of the right incisor is damaged somewhat by root acid so that its surface detail is imperfect. Both marginal ridges are prominent. There is a wear facet along the incisal edge of the distal ridge. A mesial interproximal wear facet was formed directly adjacent to the occlusal surface of the mesial ridge. This is because the axis of the tooth and root runs obliquely upwards and laterally, bringing the crowns together at their mesial tips. The crown is much wider at its occlusal surface than at the cervix. Wear facets are less distinct on the more weathered RI^1, however. The intermaxillary sutures were not synostosed. Each tooth exhibits minimal wear along the incisal edge. Dentine is exposed as a thin line along the edge, interrupted at the mesial one-third by a mesiolabial (intermamelon) groove. A distolabial groove is evident only as a small indentation along the incisal edge. Before wear these grooves separated three major mamelons, the two larger medially and the smaller distally. The labial surface is rugose with shallow asymmetrical surface grooves. The lingual surface is distinctly shoveled. It exhibits a prominent lingual cingulum with ridges on its surface and finger-like prolongations extending into the lingual fossa. The prolongations become less distinct as they blend into the lingual fossa. The root is conical, subtriangular in cross-section, and tapers gradually along its length. There are equally developed mesial and distal grooves. The development of the root is complete. At the cervical margin both left and right I^1s are 8.6 by 8.6.

Upper second incisors

Like the central incisors, the lateral incisors are shoveled lingually and they exhibit similar wear. There is a midline indentation along the occlusal edge that suggests a labial groove which separated two main mamelons. The occlusal surface is concave on the LI^2 possibly because of the indentation. The occlusal surface of the right I^2 is flatter. It appears that wear has exposed dentine only on the mesial half of it. Both mesial and distal marginal ridges are thick and prominent, especially along the lingual surface, causing a relatively deep lingual fossa. The margins blend at the base, forming a lingual cingulum which has a midline groove or V. As on the I^1s, the labial surfaces have soft longitudinal grooves on them. Root closure was incomplete. On the right tooth, found separately, the root apex has an oval apical foramen 3.8 by 1.9 wide. At this stage of development the (right) root is 18.2 long as measured from the highest enamel on the lingual surface. It is compressed mesiodistally (5.4 by 8.7 at the cervical margin) and is distinctly grooved down the center of both its mesial and lingual surfaces.

Upper deciduous canines

The deciduous $\underline{c}$s are still present although their roots were being resorbed in advance of permanent canine eruption. Both deciduous canines are well preserved, the left slightly better. Resorption of the root had progressed further on the right and a chip of root is missing from the edge of the right side's resorption zone after fossilization. The extent of the resorbed area on the left is a patch about 8 by 4 centered roughly one-third up the lingual side of the root. On the right it is a 9 by 5 excavation and the upper and lower parts of the roots are joined only by a thin isthmus. Both d$\underline{c}$s are very worn along their occlusal surfaces. Their crowns are nearly symmetrical, with the mesial crest running a little less steeply to the single cusp than the distal one. Dentine is exposed as a circle blending into a 1.5–2 line, which extends to the mesial and distal borders. The teeth look triangular from the labial aspect. Slight mesial and distal marginal grooves are present along the lingual surface near the cingular edge. They are almost worn away, however, and the cingulum is poorly defined. The cingulum is low, about 3 high. The lingual fovea is only a very shallow concavity from the crown to the basal border. Crown heights are 6.1 for both left and right d$\underline{c}$s. The lengths of the roots as measured from the highest labial enamel to the apex are 17.7 (left) and 17.9 (right). The roots are oval in section at the cervix but are labiolingually flattened towards the apex. This flattening is, at least in part, due to resorption.

Upper canines

The L$\underline{C}$ still remains deep in its crypt, and only the tip of the crown can be seen. The R$\underline{C}$ is isolated from the skull. It had fallen from its socket and was found separately. The R$\underline{C}$ root socket appears larger than the left, suggesting that the R$\underline{C}$ was further along in the eruption sequence. This conclusion is also supported by greater root resorption of the Rd$\underline{c}$. The R$\underline{C}$ is well preserved, and it exhibits little to moderate weathering. The crown is slightly asymmetrical with a shorter mesial than distal crest running up to the single main cusp, and the length of the mesial marginal ridge is greater than the distal. The crown has a pointed apex, which angles slightly mesially and is more pronounced distally. Just distal to the apex is a small groove or indentation along the occlusal edge. At this point, the occlusal margin slopes superiorly towards the angle between the junction of the occlusal edge with the distal marginal ridge. Both mesial and distal marginal ridges are pronounced. They border clear distal and mesial marginal grooves and a shallow canine fossa. The cervical line is relatively straight around the whole circumference, but it dips slightly on the mesial aspect. The labial surface exhibits prominent perikymata. There is a mesiobuccal groove, and a slightly less perfect distobuccal groove is represented only by a faint distal concavity, which is enhanced by weathering. The occlusal surface is unworn. The two rounded marginal ridges join to form a complete 5.0 high cingulum around the central part of the lingual surface. The basal cingulum is present, therefore, but not pronounced. There is also a small notch distally in the same plane as the indentation along the occlusal edge. The root had not finished developing. Since there was no apical closure, the root was only 14.8 long as measured from the highest labial enamel. It is compressed mesiodistally (7.1 by 9.9 at the cervical margin) and grooved down the middle of both mesial and distal faces. The mesial groove is more pronounced towards the apex. The apical canal is figure-8 shaped in outline and is 6.8 by 4.3.

Upper third premolars

The P^3s are bulbous, nearly symmetrical, bicuspid teeth. Some wear exists along the mesial edge of the buccal cusp, giving each tooth an angular appearance and blunting the mesiobuccal style. The mesial and distal foveae are relatively deep and prominent. They are connected by a central groove. The lingual cusp is also bulbous, but not quite as tall as the buccal cusp. Three small grooves cutting into the lingual cusp form the central groove. Both teeth are minimally weathered. The mesial cervical line and root area is only shallowly concave. The buccal and lingual surfaces are not parallel, the cusp apices being set much closer together than the cervical buccolingual diameter. The roots are incompletely developed, about 14 long, mesiodistally flattened, and figure-8 shaped in cross-section. The superior half of the mesial surface is very deeply grooved so that two separate root tips are nearly formed.

Upper fourth premolars

The P^4s are bulbous, rounded, symmetrical bicuspids. They show little to no wear. The buccal cusp and lingual cusp are similar in height and overall size, but the buccal cusp is slightly larger. The mesial fovea is relatively deep and has an extra fold of enamel within the buccal portion of the fovea. The distal fovea is also subdivided by two strong folds of enamel radiating from the distobuccal and distolingual aspects of each cusp. The RP^4 has a transverse ridge, which is kinked at the midline, rather than exhibiting separate triangular ridges. The RP^4 is also more rounded and less conical than the LP^4, which appears more pinched and has deep yet less broad foveae. Root development is still incomplete. The oval cross-section is broken by shallow mesial and deeper distal longitudinal grooves.

Upper first molars

The LM^1 is more weathered, pitted, and discolored than the right. They are roughly square in occlusal outline. Generally, all cusps appear to be similar in height. The mesiobuccal and distobuccal cusps are relatively equal in size. The mesiolingual cusp is slightly larger than they are, and the distolingual cusp is slightly smaller. The mesiolingual and distobuccal cusps are connected by a concave ridge. The distal fovea joins a marked lingual groove which separates the mesiolingual from the distolingual cusp. The short buccal groove extends at a mesial angle after it separates the mesiobuccal from the distobuccal cusp. There are two depressions separating a small enamel fold in the mesial fovea. The mesiobuccal cusp has an additional groove in it radiating from the central or mesial groove. The distobuccal cusp on the RM^1 has either a small developmental defect or chip missing on its distobuccal corner. This is not evident on the LM^1. Complete root development can be seen on X-ray. The apices of the roots extend to a similar depth as those of the premolars. The mesial M^1 roots are approximately 13.5 as measured from the X-ray. We expect to see much greater detail when CT scans are made of this fossil.

Upper second molars

These teeth are rhomboidal in outline and, like the M^1s, the left is more weathered than the right. The mesial face is even more inclined and the distal part of the tooth is much narrower than the mesial so that it appears to taper distally. Cusps of both

M^2s are similar in height to the other cusps. The mesiobuccal cusps appear enlarged because they are the mesiobuccally projecting corners of the rhomboid. The rhomboid shape is enhanced by a relatively large mesial marginal ridge. The mesiolingual cusp is quite large and expanded. The other three cusps are similar in size, though the distolingual cusp is slightly smaller than both mesiobuccal cusp and distobuccal cusp. The mesial marginal groove is relatively pronounced and makes a right-angled groove to connect with a more central fovea. The distal marginal groove is distinct as a fairly deep small pit. The mesiolingual cusp is subdivided slightly distally with a small groove radiating from the distal marginal fovea. The exposed RM^2 lingual root is still wide open with only ca. 10 root development. An estimated measurement of the mesial root of the LM^2 from X-ray is about 12.

Upper third molars

The left M^3 can be seen deep within its crypt, because alveolar bone has been broken away over the maxillary tuberosity. The incomplete crown shows its two buccal cusps through the bony defect. These apices of the cusps are about 6.5 apart. Although it is difficult to be certain, there seems to be a well-developed distal marginal ridge delineating a posterior fovea. An estimate of the length of the germ from the X-ray is 9.0.

Examination of the perikymata

Perikymata of KNM-WT 15000 are best represented on the lower left second incisor and canine. Detailed epoxy resin replicas were made of these teeth and examined under a scanning electron microscope (see Teaford and Oyen, 1989, for methods). Figure 8.7 shows montages of scanning electron micrographs of the labial surfaces of these two teeth. A count of perikymata from incisal edge towards the cervix was made for each tooth. There are more than 96 perikymata on the LI_2 and more than 90 perikymata on the LC. We do not feel that these minimum perikymata numbers add substantial new information about the age of the individual to that presented by Smith (this volume). The count for the LI_2 is close to that of the better-preserved LI_2 of KNM-ER 820, another African early *H. erectus,* and both canine and incisor showed a condensation of perikymata towards the cervical border similar to that described for modern *Homo* and other East African fossil hominids (Bromage and Dean, 1985; Dean, 1987a,b).

When found, the left lower central incisor had a small chip broken from the upper part of the crown. This small triangular piece was unglued from the tooth and an accurate epoxy replica was made in the same way as for the perikymata. Figure 8.8 shows some enamel detail as seen in the scanning electron microscope. Because this purely fortuitous break runs diagonally across the crown, the features are of limited use, but they serve to show that at this point about halfway down the crown there are no missing perikymata. Regular Retzius' striae can be seen matching perikymata on the surface.

Systematic and phylogenetic implications

It has already been claimed (Leakey and Walker, 1976; Walker and Leakey, 1978) that certain specimens from Koobi Fora dating between about 1.8 and 1.5 Ma are representatives of early *Homo erectus.* This assessment has not gone unchallenged (Andrews, 1984; Wood, 1984; Stringer, 1984) and there seems to be an emerging consensus that these fossils should not be called *H. erectus* (Andrews, 1984; Groves, 1989; Stringer, 1984; Wood, 1984, 1991). Whatever the final outcome, it is clear that KNM-WT 15000 is another member of the population from which specimens such as KNM-ER 992, 3733, and 3883 were derived. KNM-ER 992 is the type specimen of *Homo ergaster* (Groves and Mazak, 1975). Groves (1989) has given an account of why he considered this specimen to be allied with those like KNM-ER 1805 and 1813 rather than with those like KNM-ER 3733 and 3883. He refers the latter to an unnamed species of *Homo.* He relied heavily on the evidence from KNM-ER 730, a

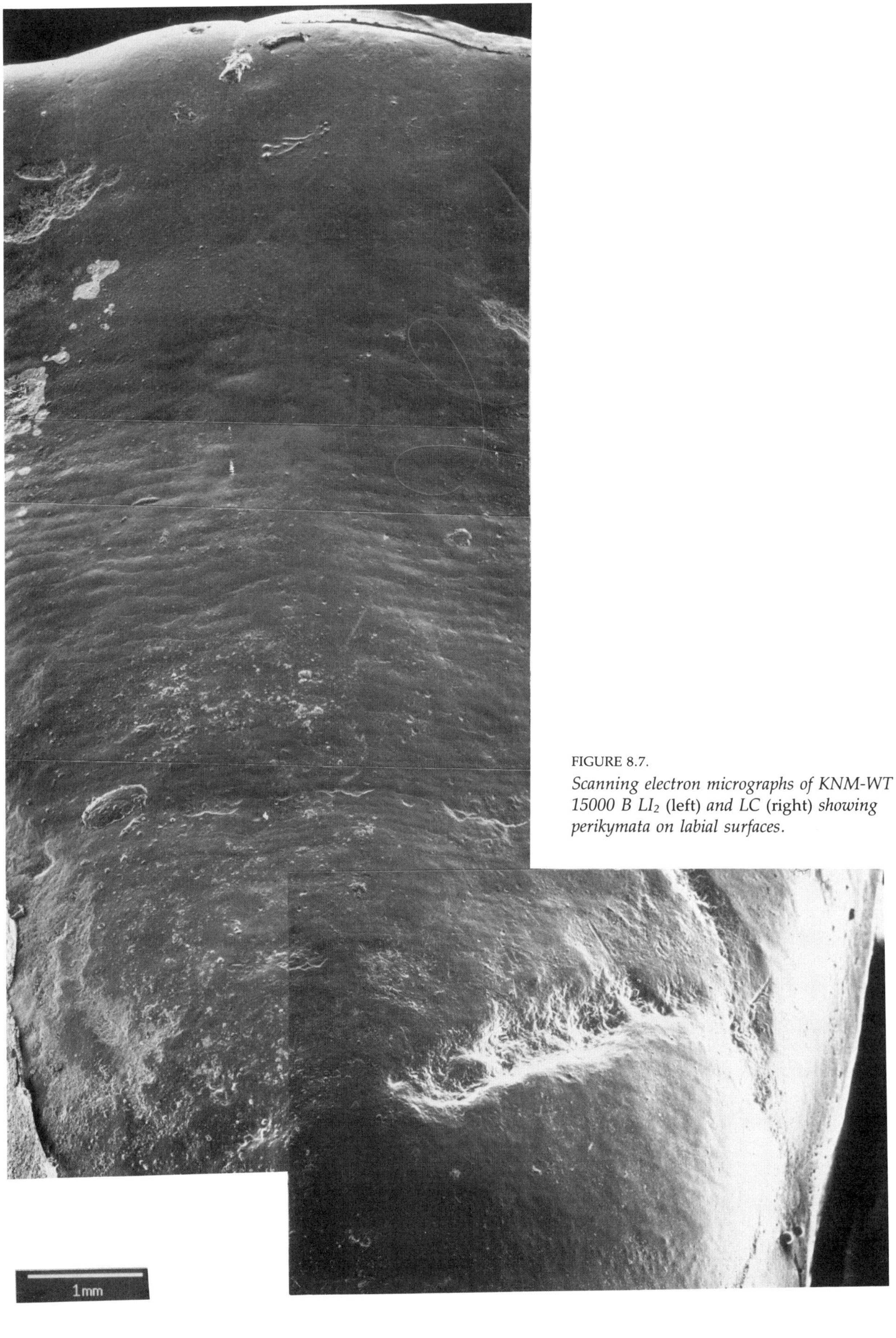

FIGURE 8.7.
Scanning electron micrographs of KNM-WT 15000 B LI_2 (left) *and LC* (right) *showing perikymata on labial surfaces.*

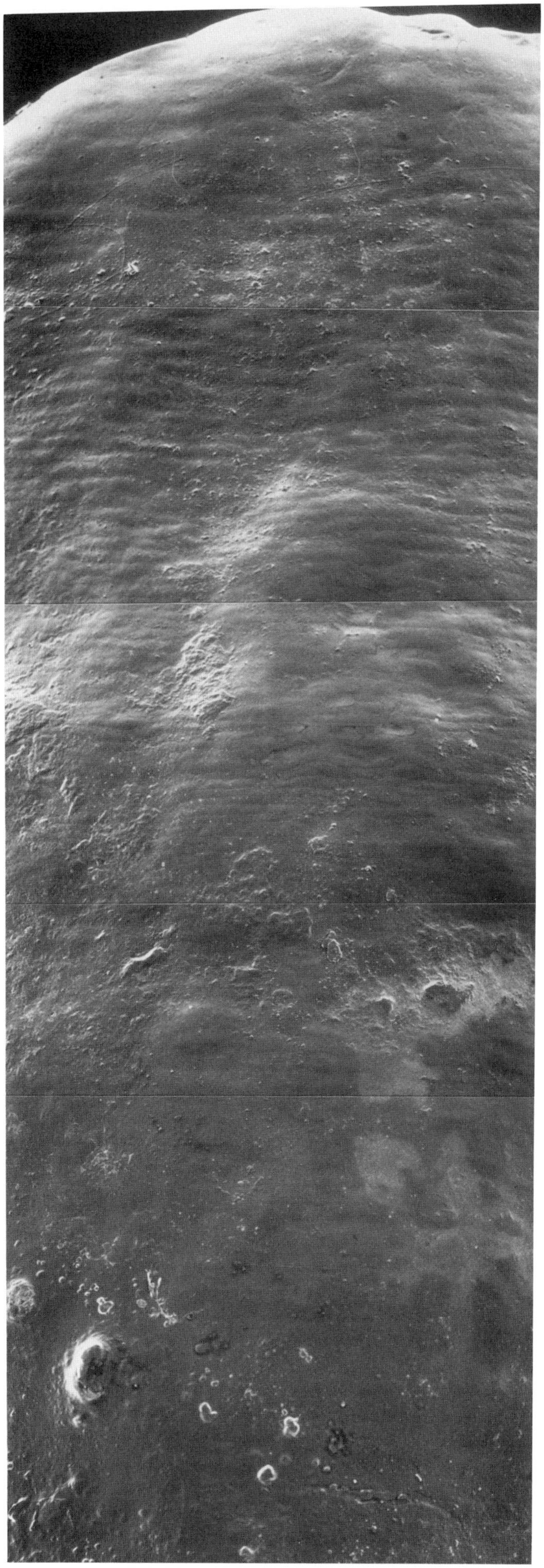

fragmentary skull with heavily worn lower teeth but with clear affinities to KNM-ER 3733 and KNM-ER 3883. We will show that KNM-WT 15000, in those teeth that can be compared, is extremely like the type of *H. ergaster* and, further, that KNM-ER 730 is almost certainly a very old individual of the same type.

There is confusion at present because some workers consider that there is only one species of *Homo* about 1.5–2.0 Ma (Wolpoff, 1980; Skelton et al., 1986; Johanson et al., 1987; Tobias, 1991), while others suggest that there are at least two (Walker and Leakey, 1978; Rightmire, 1984, 1990; Wood, 1984, 1991; Stringer, 1984, 1986; Groves, 1989).

Other Turkana Basin Early *Homo erectus*

In this section we compare the teeth of KNM-WT 15000 with those of other early *Homo erectus* specimens from the Lake Turkana Basin. This is to document the range of variation and to establish the position of KNM-WT 15000 within that range. We have also simulated wear on casts of the unworn teeth of this specimen so that we can compare it with other worn teeth. Isolated teeth are not considered here because of the likelihood of confusion with those of *Homo habilis* (in the broadest sense). See Tables 8.2 and 8.3 for comparative measurements taken from Day and Leakey, 1973, 1974; Day et al., 1976; Leakey and Walker, 1985; Leakey and Wood, 1974.

Mandibular teeth

The only decently preserved lower incisors are those of KNM-ER 820. This is a mandible of a juvenile with all four shoveled incisors erupted. The individual's age has been estimated to be 7.1 years by Bromage and Dean (1985), using perikymata increments, and 6.5 years by Smith (1986), using an independent method. Both estimates are younger than the original estimate of 8–9 years by Leakey and Wood (1973). The wear on KNM-ER 820 is a little less than on KNM-WT 15000, and the teeth are a little smaller in overall dimensions, but otherwise they are extremely similar.

TABLE 8.2.

Mandibular tooth measurements (mm) of various Turkana Basin *Homo erectus* specimens.

ID NUMBER KNM-ER	SIDE	I1 m/d	I1 b/l	I2 m/d	I2 b/l	C m/d	C b/l	P3 m/d	P3 b/l	P4 m/d	P4 b/l	M1 m/d	M1 b/l	M2 m/d	M2 b/l	M3 m/d	M3 b/l
730	L											>11.7	>11.7	>12.0	>11.6	>13.0	>11.5
820	R	6.1	6.4	7.5	6.7							12.2	10.8				
	L	6.0	6.2	7.3	6.9							12.3	10.7				
992	R					9.3	9.3	9.2	11.2	8.6	11.1	11.9	10.7	13.2	12.5	13.0	12.1
	L			7.0	6.9	9.0	8.1	9.5	11.1	8.4	11.1	12.0	10.9	13.0	12.3	12.8	12.3
1502	R											13.4	11.4				
1507	L											13.3	11.1				
1808	L							*9.0	9.5							13.6	12.0

Note: m/d = mesiodistal; b/l = buccolingual.

* Estimated value.

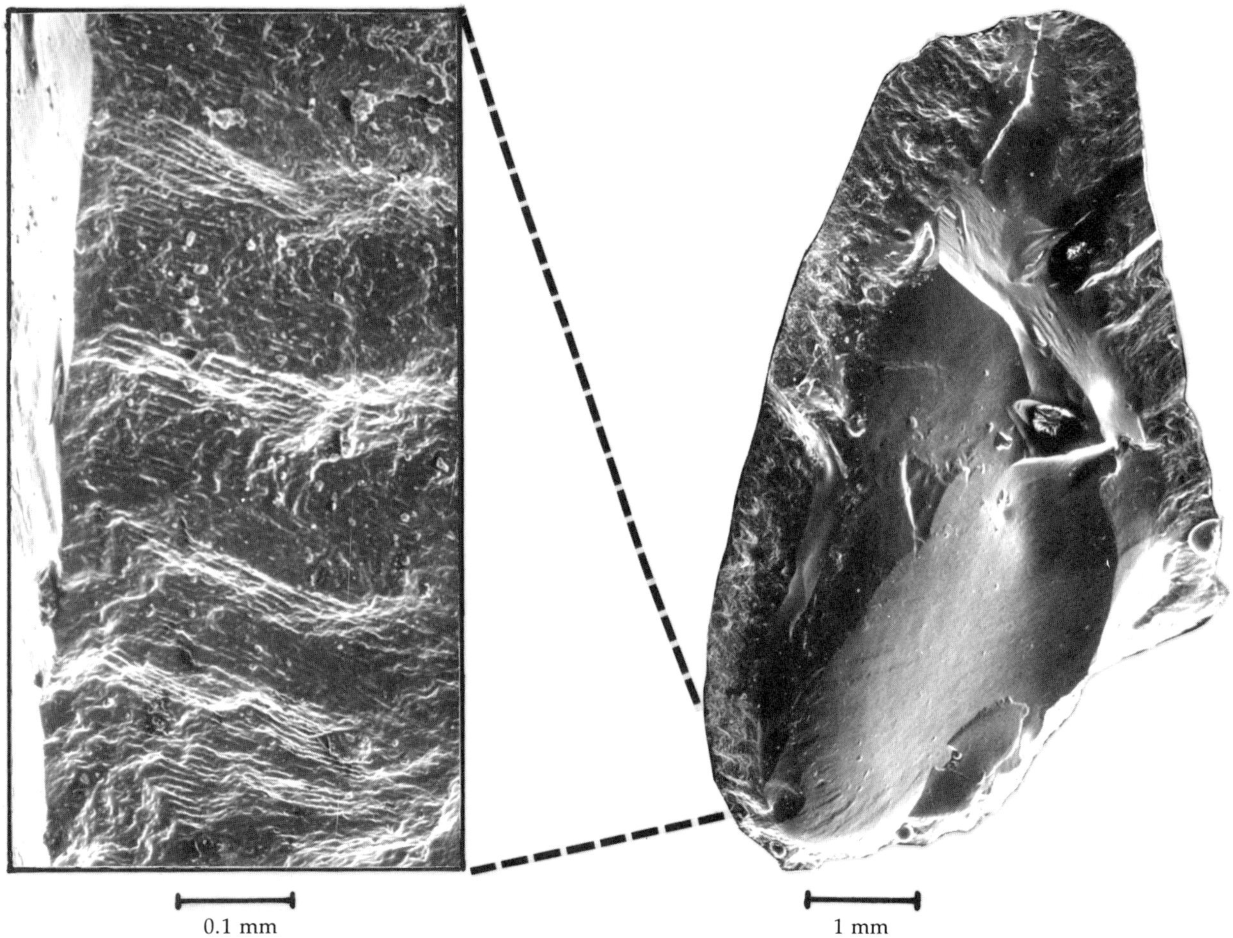

FIGURE 8.8.

Scanning electron micrographs of fragment of LI_1 of KNM-WT 15000 B showing Retzius' striae.

Mandible KNM-ER 992 has both canines, but the left has a large fragment missing from the lingual side. The right canine is slightly larger than in KNM-WT 15000 in both major diameters. Although extremely similar in overall morphology, the RC of KNM-ER 992 appears to have deeper mesial and distal foveae when compared with our simulated wear cast. An eroded and much worn upper central incisor was reported for KNM-ER 1808 (Leakey and Walker, 1985), but upon reexamination we now think that this tooth crown is more likely to be a left lower canine. This very worn tooth is exactly the same size as our cast with simulated wear and, again, it has a stronger mesial marginal ridge and fovea (Figure 8.9).

Third premolars are known for KNM-ER 992 and 1808. Both P_3s are well preserved on KNM-ER 992. Flat wear has blunted the buccal cusps to expose dentine. Comparison with our simulated wear casts shows these teeth to be a little larger (Figure 8.9). They have almost exactly the same morphology as the right P_3 of KNM-WT 15000 in that the mesial marginal ridge is not notched. The very badly eroded left P_3 of KNM-ER 1808 is a little smaller buccolingually than that of KNM-WT 15000 and KNM-ER 992, but it has, as far as one can tell, the same configuration as the left P_3 of KNM-WT 15000.

Only KNM-ER 992 has P_4s available for comparison (Figure 8.9). These are the same size mesiodistally but about 1 broader. If the strong interstitial wear is taken into account, it is clear that the KNM-ER 992 premolars were bigger than those of KNM-WT 15000. These four teeth have the same general configuration, but it appears that the mesial foveae in KNM-WT 15000 were much less deep. The KNM-ER 1808 tooth is so eroded that it is difficult to place in proper orientation. Details are heavily obscured.

M_1s are available for comparison in KNM-ER 820, 992, 1502, and 1507. The M_1 of KMN-ER 730 was worn in life to the gum and has also been broken subsequently. With interstitial wear it was nearly 12.0 long, but no crown details remain. All the remaining teeth are very similar in size and proportions to those of KNM-WT 15000. KNM-ER 1502 and 1507 are slightly larger mesiodistally even though they have little interstitital wear mesially and none distally. With the exception of KNM-ER 730, all these teeth show considerable similarity in cusp size and arrangement. They all have medial tubercles and deflecting wrinkles. The tubercles are less distinct on the more worn teeth of KNM-ER 992.

The M_2s of KNM-ER 992 are bigger in both major diameters and their morphology is generally similar to that of the M_2s of KNM-WT 15000. Mesiodistally, the M_2s of KNM-ER 992 are more elongate. The M_2 in KNM-ER 730 cannot be compared morphologically, but it is almost the same size as the M_2s of KNM-WT 15000.

TABLE 8.3.

Maxillary tooth measurements (mm) of various Turkana Basin *Homo erectus* specimens.

ID NUMBER KNM-ER	SIDE	I1		I2		C		P3		P4		M1		M2		M3	
		m/d	b/l	m/d	b/l	m/d	b/l	m/d	b/l	m/d	b/l	m/d	b/l	m/d	b/l	m/d	b/l
803	R	10.2	7.5														
	L					9.3	8.5										
807	R											*13	*14		*14	11.0	13.0
1808	L							8.5	13.8					11.5	13.0		
3733	R							8.7	12.0	8.1		12.5		12.4	13.9		
	L									8.1	12.1	*12.0					

Note: m/d = mesiodistal; b/l = buccolingual.

*Estimated value.

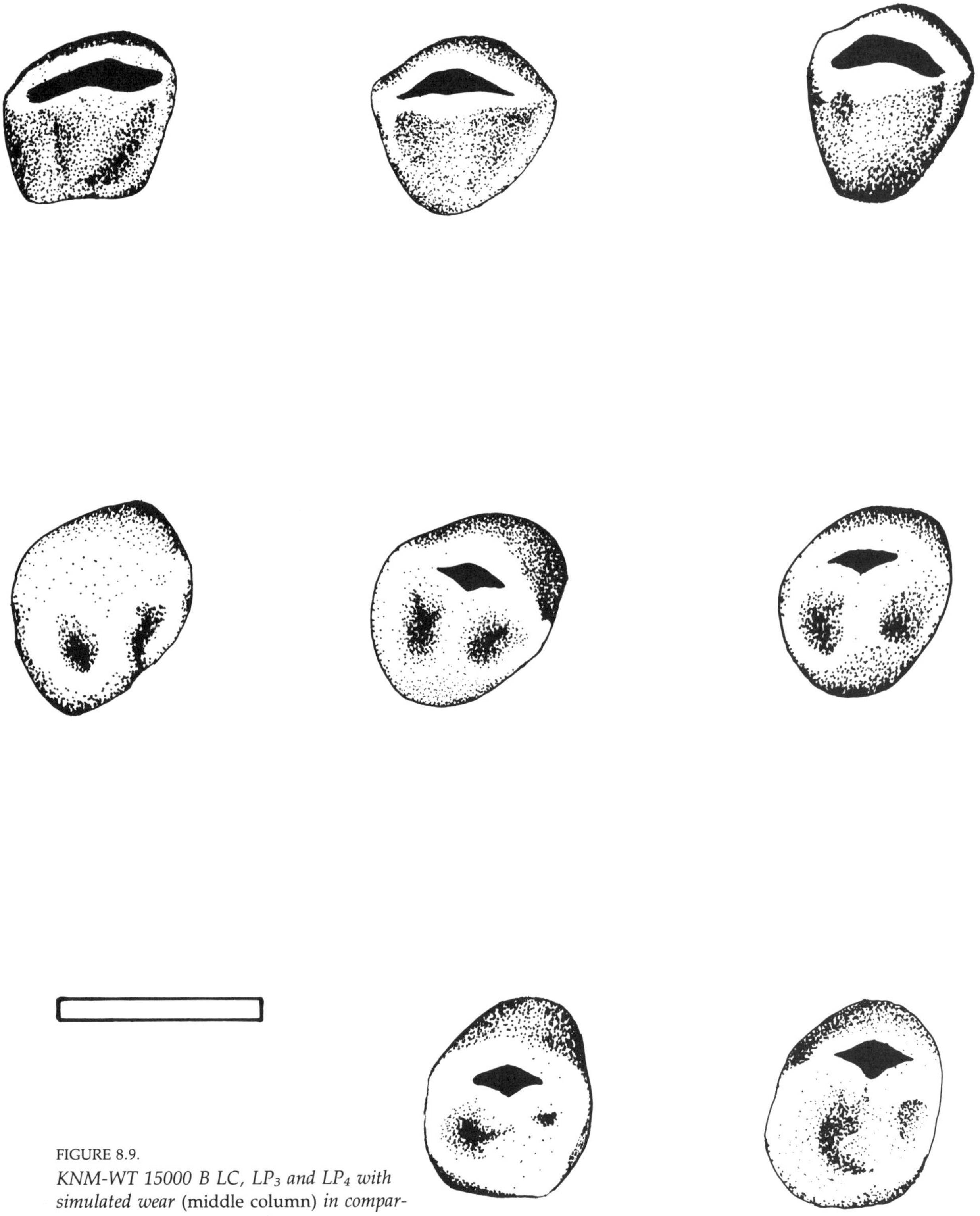

FIGURE 8.9.
KNM-WT 15000 B LC, LP_3 and LP_4 with simulated wear (middle column) *in comparison with the same teeth of KNM-ER 1808* (left) *and KNM-ER 992* (right). *Teeth of right column reversed. Bar = 1 cm.*

MAXILLARY TEETH

There is a very worn I^1 associated with the partial skeleton KNM-ER 803. It had the same overall morphology but was smaller than the central incisors of KNM-WT 15000. The mesiodistal diameters are 10.2 for KNM-ER 803 (Day and Leakey, 1974) and 11.5 (right) and 12.0 (left) for KNM-ER 15000. Practically all that remains of the tooth crown in KNM-ER 803 is the basal tubercle, but the edges of the mesial and distal marginal ridges show that this tooth was shoveled. Our simulated wear cast has the same occlusal outline (Figure 8.10). Only the canine of KNM-ER 803 can be compared with KNM-WT 15000. It is much smaller than the KNM-WT 15000 canine but shows the same deep marginal ridges joined by a notched basal cingulum. Our simulated wear cast has the same occlusal outline (Figure 8.10).

KNM-ER 3733 has a right P^3 which is relatively complete. A mesiodistal break through the midline has resulted in a missing enamel flake from the crack distally and another from the mesial marginal ridge near the labial border. Portions of the roots are present. They have been broken away along the cervical border of the distolingual quarter. The P^3 is about the same size as those of KNM-WT 15000. It is moderately worn, exhibiting interproximal wear facets along the distal marginal ridge of enamel near the midline and at the mesial marginal ridge. It also has wear facets that are more developed buccally. Dentine is exposed only on the buccal cusp. This tooth has prominent mesial and distal labial styles. The mesial style is accompanied by a more prominent V-shaped groove than is the distal style. These styles are absent on KNM-WT 15000. There are both mesial and distal marginal foveae. The distal marginal fovea has a central groove that extends to but does not cross the midline towards the mesial marginal ridge which separates the foveae and connects both cusps. In summary, the third premolars in these two early hominid specimens are rather similar in their simple overall plan, but that of KNM-ER 3733 has more strongly marked features. The left P^3 of KNM-ER 1808 is larger buccolingually than that of KNM-ER 3733. Despite its heavier wear, it is clear that the two foveae were connected by a strong groove and that there was no ridge connecting the two cusps.

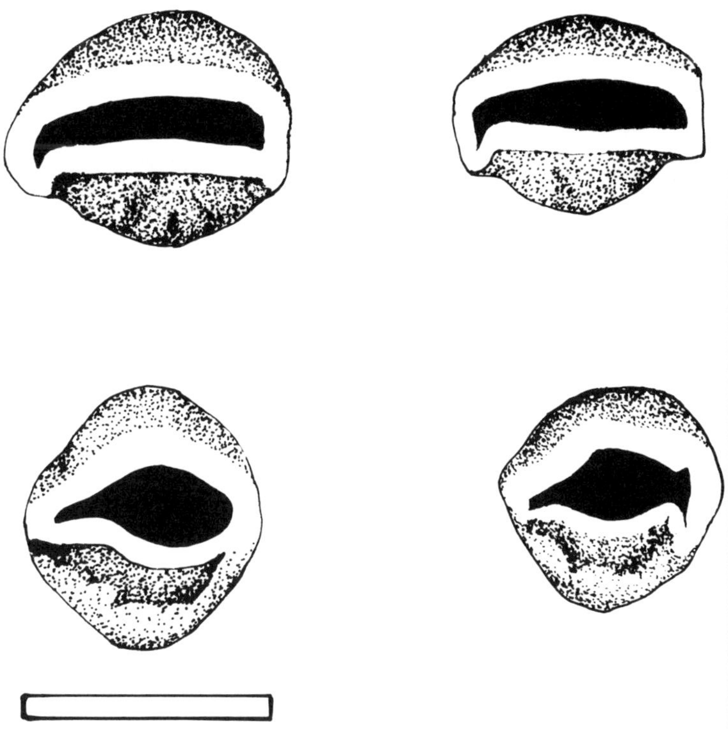

FIGURE 8.10.
KNM-WT 15000 A RI^1 *and R*$\underline{C}$ *with simulated wear* (left) *in comparison with teeth of KNM-ER 803* (right, *canine reversed*). *Bar = 1 cm.*

KNM-ER 3733 has a left P^4 and the buccal half of the RP^4. The LP^4 will be used primarily for comparison. The cervical margin is broken mesially. Like the P^3, the P^4s are both worn. The buccal cusp has tiny dentine exposures and is relatively flat, worn to nearly the same height as the mesiolingual cusp, which is worn but has no dentine exposure. There is an interproximal wear facet along the mesial marginal ridge near the buccal aspect of the crown opposite the buccal limit of the mesial marginal groove. The other interproximal facet extends over much of the length of the distal marginal ridge. The distal marginal ridge is relatively smooth. It borders a deep and broad fovea, which has two small grooves running towards each cusp that result in two extra folds in the medial aspect of the fovea. The mesial fovea is narrow and less deep than the distal. The two

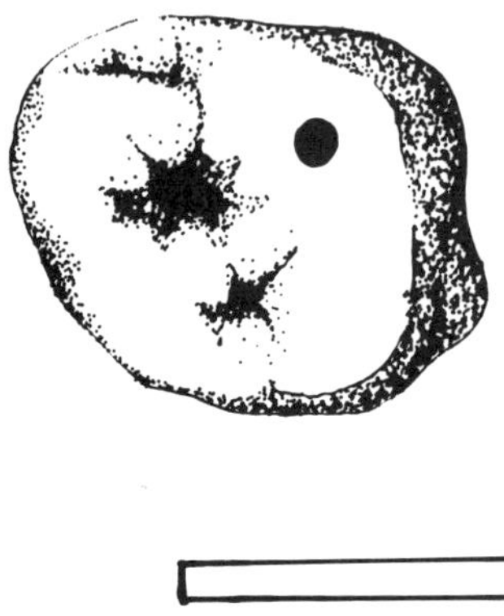

FIGURE 8.11.
RM^2 of KNM-WT 15000 A (left, *with simulated wear*) *and KNM-ER 3733* (right). *Bar = 1 cm.*

cusps are connected by a transverse ridge, which has a faint central groove in it. The tooth is larger than the LP^4 of KNM-WT 15000, especially in its buccolingual dimension. The styles are not as prominent as those of the P^3. The placement of the styles adds to the rectangular shape of the tooth. There are no grooves along the buccal enamel surface as in the P^3 of KNM-ER 3733. In this respect, the P^{4}s of KNM-ER 3733 are more like the premolars of KNM-WT 15000.

Specimens for M^1 comparison are KNM-ER 807, which has a cracked M^1, and KNM-ER 3733, which has incomplete M^1 crowns. The M^{1}s of KNM-ER 3733 are heavily worn but were probably square in occlusal outline. Dentine is exposed on all four main cusps. The dentine islands of the lingual cusps are just coalescing on the RM^1 but not quite on the LM^1. Both M^{1}s are missing their buccal portions of enamel. Primary dentine is exposed on the buccal cusps of the RM^1. There is an interstitial wear facet on the mesial aspect of the left tooth, but that on the right has been broken away. There appear to be faint wear facets on the preserved portions of the distal border. The heavy wear makes comparison of morphological details between this specimen and KNM-WT 15000 impossible, but the teeth of the latter are smaller, at least in the mesiodistal dimension.

There is an eroded M^2 crown of KNM-ER 1808, one complete M^2 and another partial crown for KNM-ER 3733, and the distal half of a crown for KNM-ER 807. The M^{2}s of KNM-ER 3733 are also heavily worn. Only the mesiolingual cusp has dentine exposed. Although the M^{2}s of KNM-ER 3733 are larger than those of KNM-WT 15000, they also have a rhomboidal occlusal outline with the distal buccolingual diameter smaller than the mesial. Our simulated wear cast resembles the RM^2 of KNM-ER 3733 down to the tiniest wrinkle in occlusal surface detail (Figure 8.11). The buccal half of the M^2 of KNM-ER 807 is like that of KNM-ER 3733 and has a similar wear state. The eroded crown of KNM-ER 1808 also has an overall shape similar to that of the M^{2}s of KNM-WT 15000.

In our population of early *Homo erectus* there is clear evidence of maxillary M^3 reduction. Although the M^3 germ of KNM-WT 15000 had not yet finished its growth, it is not likely to have grown to the size of the M^{2}s. The M^3 in KNM-ER 807 is reduced relative to the M^2 and has only two main, mesially placed cusps and a beaded distal marginal ridge behind them. In the case of KNM-ER 3733, there are no M^{3}s preserved and no roots of M^{3}s showing in the spongy bone of the maxillary tuberosity. Small, distinct, distal interstitial facets can be seen on both M^{2}s, however. Either both M^{3}s and their roots were lost in life without leaving traces of disturbance in the cancellous bone, or else these teeth were very reduced and had proportionately tiny roots. Leakey and Walker (1985) claimed the latter to be more likely. Thus all three specimens show M^3 reduction.

Subocclusal morphology

Crown structure and root development within the maxilla and mandible of KNM-WT 15000 can be seen radiographically (Figure 8.12). The roots of most of the single-rooted teeth of KNM-WT 15000 have been described with the teeth themselves, but some can be seen only in X-ray. The penetra-

FIGURE 8.12.
X-rays of KNM-WT 15000 A left and right maxillas (top), *KNM-WT 15000 B mandibular symphysis* (center), *and KNM-WT 15000 B left and right mandibular corpora* (bottom). *The matrix in both lower central incisor sockets was drilled out in order to replace each tooth, and the radiolucent patch below each apex is an artifact from the drilling.*

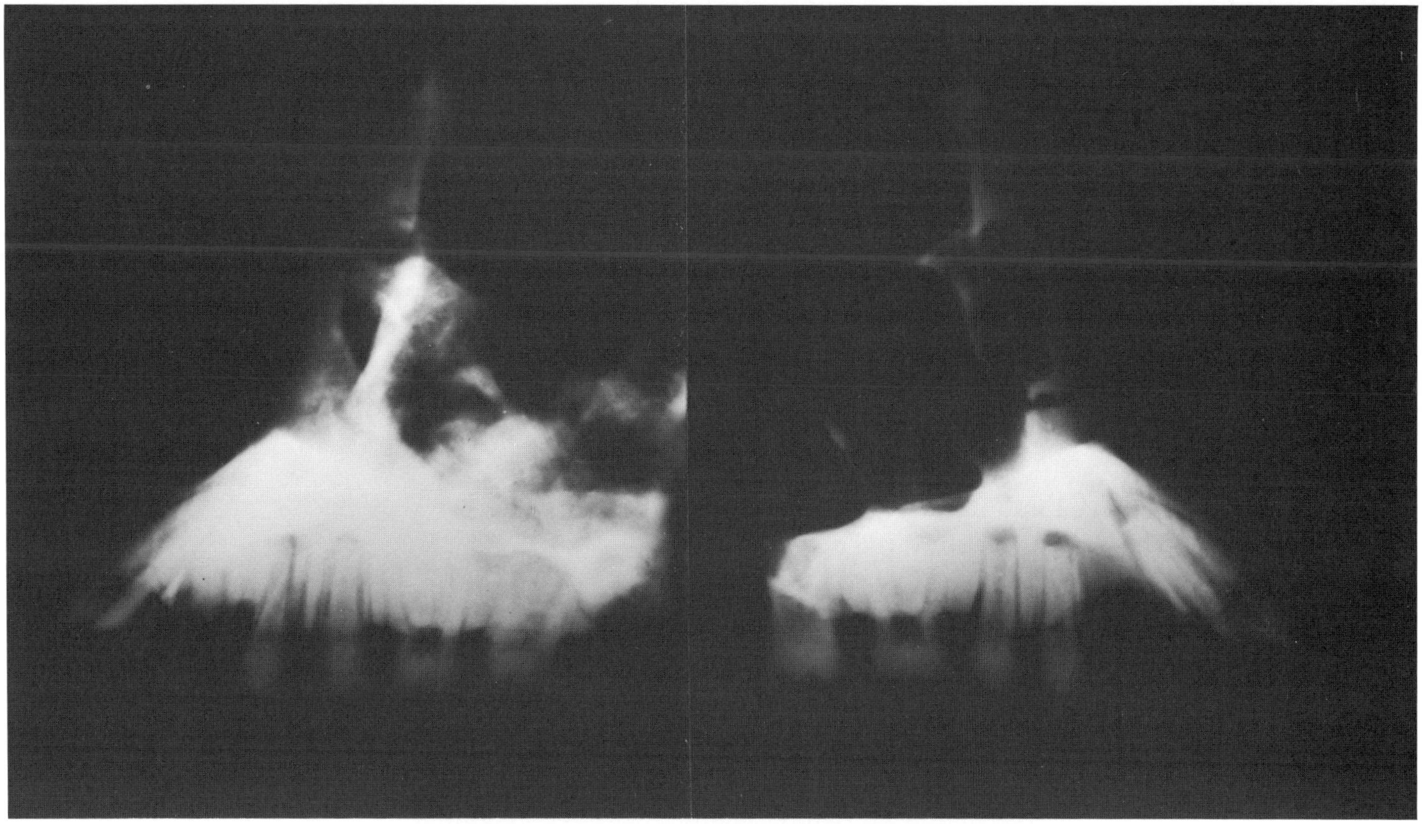

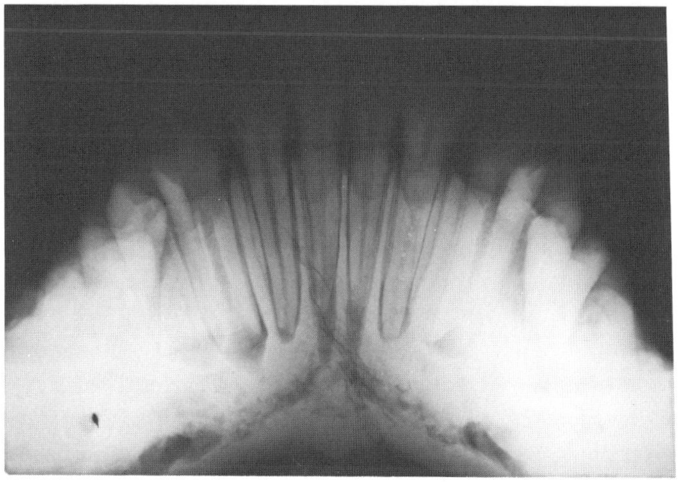

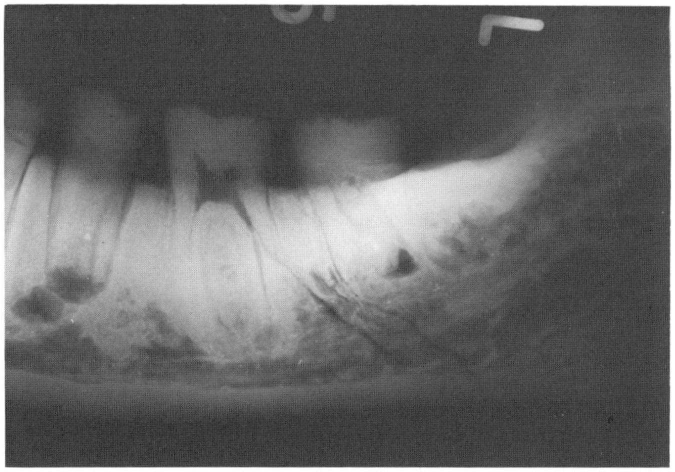

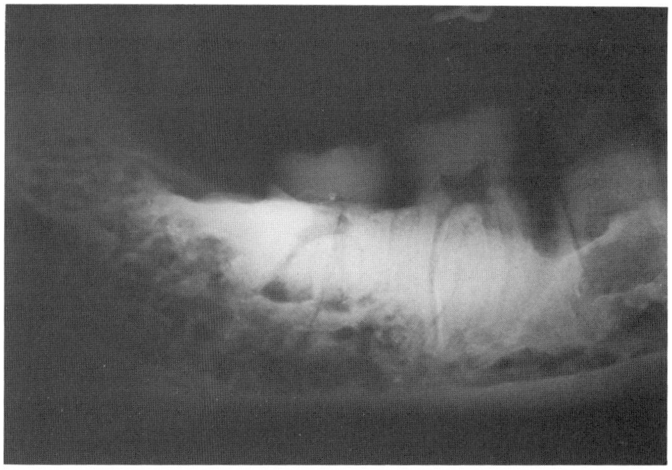

tive phase of root development (Kovaks, 1971) is evident as slight constrictions on the apical part of the incisor roots themselves. The penetrative phase can also be seen on the developing roots of the postcanine teeth on X-ray. A slight indentation on the lingual aspect of the LP^4 root surface may also represent the penetrative phase. The mandibular symphyseal X-ray shows that the completed root length for the lateral incisors is the same as that of the undeveloped canines. The central incisor roots are slightly shorter. The apices of each incisor are situated approximately halfway down the body. The internal structure and orientation of these tooth roots are like those of the Zhoukoudian *Homo erectus* mandible B IV seen in X-ray (Weidenreich, 1937, figure 302, plate XXXI). Both upper and lower premolars of KNM-WT 15000 appear to be single-rooted and have taurodont pulp chambers. The premolars are mesiodistally compressed and grooved, however, and two pulp canals are visible at each base. The completed right P^3 might have developed as a Tomes' root (Wood et al., 1988). This is a kind of double root where the bifurcation occurs along the apical portion of the root rather than near the cervix. It can appear as one root in X-ray because double pulp canals may be superimposed on one another and the bifurcation may be quite near the apex. A bifurcation does appear to have been developing along the right P^3 root and may have formed later in the roots of the other premolars. Premolar root form is quite variable in *Homo erectus* (Weidenreich, 1937) and other fossil hominids (Wood et al., 1988), and a Tomes' root is present in KNM-ER 992A and a number of the Zhoukoudian premolars. Molar roots of KNM-WT 15000 appear long and tubular. They retain a similar width from the cervix nearly to the apex rather than tapering gradually to a point. The mandibular first molar roots of KNM-WT 15000 are like those of the first molars of KNM-ER 820 and KNM-ER 730 and the first and second molars of KNM-ER 992 (Figure 8.13). Root length to crown height indices (root length/crown height × 100) for mesial and distal M_1 roots of KNM-WT 15000, KNM-ER 992, and two Zhoukoudian mandibles (KI and GI) were calculated from X-ray measurements and are given in Table 8.4.

TABLE 8.4.

Root length to crown height indices for M_1.

SPECIMEN	SIDE	MESIAL ROOT	DISTAL ROOT
KNM-WT 15000	R	29	27
	L	30	31
KNM-ER 992	R	32	32
	L	33	30
Zhoukoudian			
BI	R	30	27
GI*	L	12	19

*Mandible GI has very worn teeth.

EARLIER TURKANA BASIN *HOMO* SPECIMENS

The teeth of KNM-WT 15000 were also compared with those of three earlier specimens of *Homo* from the Turkana Basin (Tables 8.5 and 8.6). These specimens are KNM-ER 1590, KNM-ER 1802, and KNM-ER 1813 (Day et al., 1976). Most teeth of both

TABLE 8.5.

Mandibular tooth measurements (mm) of an early Turkana Basin *Homo* specimen.

ID NUMBER KNM-ER	SIDE	P3		P4		M1		M2	
		m/d	b/l	m/d	b/l	m/d	b/l	m/d	b/l
1802	R	10.4	12.1	11.3	12.1	14.6	13.2	16.5	14.3
	L			11.4	12.0	14.7	13.3	16.6	14.1

Note: m/d = mesiodistal; b/l = buccolingual.

KNM-ER 1590 and KNM-ER 1802 are much larger than those of KNM-WT 15000, but the anterior teeth of KNM-ER 1813 are much smaller and the cheek teeth about the same size as those of KNM-WT 15000. All the preserved permanent teeth of KNM-ER 1590 except the central incisor have mesiodistal and buccolingual dimensions which far exceed those for KNM-WT 15000. KNM-ER 1590 consists of a fragmentary juvenile cranium with a number of deciduous and permanent teeth. Comparisons were made with maxillary LI^1, $Ld\underline{c}$, $L\underline{C}$, $R\underline{C}$, LP^3 germ, RP^4 germ, RM^1, and LM^2. The teeth of KNM-ER 1813 consist of maxillary $L\underline{C}$–M^3 and RI^2–$\underline{C}$, and RM^3. Although the central incisor crown is badly damaged, the mesiodistal diameter can be estimated. They are all of a similar shape to those of KNM-WT 15000. KNM-ER 1802 is a mandible with RP_3–M_2 and LP_4–M_2. Like the maxillary teeth of KNM-ER 1590, the mandibular teeth of KNM-ER 1802 are much larger than those of KNM-WT 15000. They have minimal to moderate wear. Some wear facets are evident buccally, but no dentine had been exposed.

Mandibular teeth

The RP_3 of KNM-ER 1802 appears very large and bulbous. Although the buccal cusp is worn slightly, both buccal and lingual cusps are similar in height. A central groove separates each cusp and connects a deep mesial fovea with a deeper and more broad distal fovea. The P_3 of KNM-ER 1802 is more similar in morphology to the P_4s than to the conical P_3s of KNM-WT 15000. Mesial and distal styles are present along the buccal surfaces

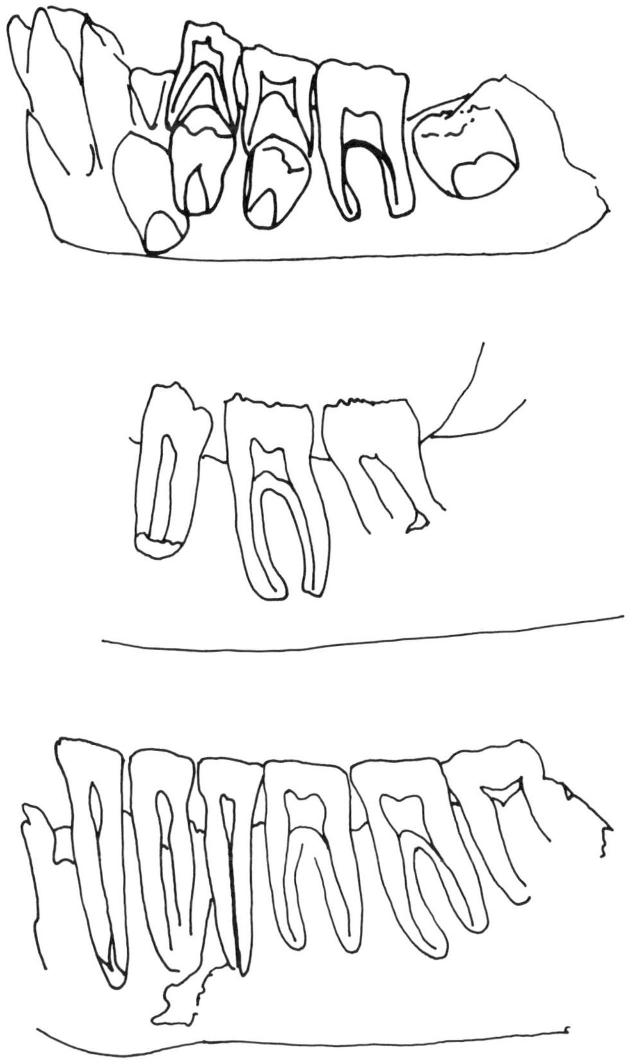

FIGURE 8.13.
Tracings from X-rays of the mandibles of KNM-ER 820 (top), *KNM-WT 15000 B* (middle), *and KNM-ER 992 B* (bottom), *arranged to show longitudinal age series.*

TABLE 8.6.

Maxillary tooth measurements (mm) of early Turkana Basin *Homo* specimens.

ID NUMBER KNM-ER	SIDE	I1		I2		C		P3		P4		M1		M2		M3	
		m/d	b/l	m/d	b/l	m/d	b/l	m/d	b/l	m/d	b/l	m/d	b/l	m/d	b/l	m/d	b/l
1590	R					11.4	12.2	10.1	13.3	10.1	14	13.3	13.8				
	L	11.8	8.1			11.3	12.5	10.2	13.6	10.5	13.6	13.5	14.0	13.9	16.8		
1813	R	*9		6.5	5.8	8.2	8.4									11.7	13.5
	L					8.0	8.4	8.0	11.1	9.0	11.3	12.2	12.5	12.2	13.6	11.5	13.0

Note: m/d = mesiodistal; b/l = buccolingual.
*Estimated value.

of the KNM-ER 1802 P_3, which is also similar to the P_4s of KNM-WT 15000. The P_4s of KNM-ER 1802 are nearly square in outline. Both buccal and lingual cusps are relatively equal in height and overall occlusal area. The mesial marginal groove is broad buccolingually and deep. The distal groove is deep and quite expanded. It has an enlarged distolingual ridge with less prominent distobuccal ridges. Thus, these teeth appear to be more molariform than the premolars of KNM-WT 15000.

The M_1s and M_2s of KNM-ER 1802 and KNM-WT 15000 are similar, although the cusps on KNM-ER 1802 are larger and more bulbous. Each M_1 has a TAM. Deflecting wrinkles are not as distinct on KNM-ER 1802 as on KNM-WT 15000. In addition, the beading along the mesial marginal ridges of the M_1s and M_2s, as described for KNM-WT 15000, is not apparent on the molars of KNM-ER 1802.

MAXILLARY TEETH

The LI^1 of KNM-ER 1590 is similar to that of KNM-WT 15000. Simple simulated wear can produce a pattern in which the main facet turns onto the distal marginal ridge. This simulated wear pattern is like that in KNM-WT 15000, but there is no intermamelon groove. The distal marginal ridge curves from the distal-occlusal angle onto the lingual surface to make the crown shovel-shaped. However, KNM-ER 1590 does not possess an enlarged lingual tubercle with fingerlike prolongations along its base, as is seen in KNM-WT 15000. It has only a lingual swelling or thickening along its basal border. This swelling extends to the approximate midline of the lingual surface at the formation of the lingual fossa. Thus, the incisor of KNM-ER 1590 is smaller buccolingually than the incisor of KNM-WT 15000. Mesiodistal measurements for the central incisor are, however, quite similar between both KNM-ER 1590 and KNM-WT 15000. KNM-ER 1813 has only a broken RI^1 crown, but it preserves the interstitial facet for I^2 and the mesiodistal diameter can be estimated accurately. The same diameter in KNM-WT 15000 is 28 percent bigger. The RI^2 of KNM-ER 1813 also exhibits a shovel shape, and its mesial and distal marginal ridges are as prominent as in KNM-WT 15000. The lateral incisor of KNM-ER 1813 is much smaller, however.

The Ld$\underline{c}$ of KNM-ER 1590 is less worn than the Ld$\underline{c}$ of KNM-WT 15000. Although both have apical wear, the wear extends along the mesial and distal marginal ridges on KNM-WT 15000. The lingual surface of KNM-ER 1590 is ridged and rugose. There is a longitudinal mesial marginal groove and a more prominent, expansive distal fovea with a small ridge bisecting it. The distal fovea is also quite concave compared with that of KNM-WT 15000. There is evidence, despite the greater wear, that a slight mesial marginal groove remains on the Ld$\underline{c}$ of KNM-WT 15000.

The upper canines of KNM-ER 1813 are also smaller than those of KNM-WT 15000. Although the canine of KNM-ER 1813 is worn, the overall shape is similar in both. The canines of each specimen have pronounced mesial and distal marginal ridges. Those of KNM-ER 1813, however, do not have a small central tuberosity along their lingual cingulum like that seen in KNM-WT 15000. The canine germs of KNM-ER 1590, on the other hand, are very large and appear taller relative to their width when compared with those of KNM-WT 15000. The canine germs from each specimen exhibit an indentation along the occlusal edge. Like the incisors, however, the canine lingual fossa of KNM-ER 1590 is relatively shallow. The lingual surface has longitudinal ridges or tubercles which are bordered by deep grooves. The grooves bisect each mesial and distal marginal ridge. As a result, the marginal ridges do not form a cingulum or continuous rim along the lingual surface of the base.

Although the LP^3 was probably still incomplete, it has a prominent central ridge between the buccal and lingual cusps. This ridge divides the occlusal fovea into a smaller, relatively deep mesial fovea and a larger, deeper distal fovea such as can also be seen on the P^4 of KNM-WT 15000. The overall shape of the tooth and the associated prominent wrinkles are also similar to those of the LP^4 of KNM-WT 15000. The cusps are rounded and have an uneven occlusal rim bordering the occlusal fovea. The buccal surface of the LP^3 of KNM-ER 1590 exhibits mesial and distal styles. The mesial style is separated from the buccal cusp by a distinct, relatively wide, triangular depression. The distal style is separated by a narrow, more vertical groove. The worn P^3 of KNM-ER 1813 looks like the P^3 with simulated wear of KNM-WT 15000 but is slightly smaller. The P^4 of KNM-ER 1813 is larger than those of KNM-WT 15000. Morphologically it differs from the P^4 of KNM-WT 15000 because it does not show the buccolingual expansion of the distal fossa. Unfortunately, the RP^4 of KNM-ER 1590 is too immature for adequate comparisons.

Although larger, the M^1s of KNM-ER 1590 are very similar in morphology to those KNM-WT 15000, but KNM-ER 1590 has a buccolingually broad mesial marginal ridge. It has cusps which are faceted within the occlusal fovea, and there are prominent wrinkles and a fairly deep posterior fovea. A lingual extension of the mesial marginal ridge curves around the base of the mesiolingual cusp. The LM^2 is also quite large. Like the M^2s of KNM-WT 15000, it is rhomboidal in outline and has a pronounced mesiobuccal cusp. This tooth also shows the wrinkling commonly associated with a pre-eruption state. The mesial marginal ridge extends along the mesial edge and forms a relatively deep mesial fovea. Both central and distal foveae are quite deep in each specimen. There is evidence of a cingulum along the lingual aspect of the mesiolingual cusp. This appears as a small ridge of fingerlike prolongations extending from the distolingual cusp along the base of the mesiolingual cusp. A cingulum is not evident on either M^2 of KNM-WT 15000. Both the M^1 and M^2 of KNM-ER 1813 also look relatively similar to those of KNM-WT 15000. In contrast to the metrics reported for the anterior teeth, the molars of KNM-ER 1813 are a little larger than the molars of KNM-WT 15000. A shift in overall size is evident among the teeth of KNM-ER 1813 when compared with KNM-WT 15000 such that the anterior teeth of the former are relatively much smaller, the P^3s are approximately the same size, and the other post-canine teeth are relatively larger in KNM-ER 1813 than in KNM-WT 15000.

Zhoukoudian Specimens of *Homo erectus*

Tooth morphology of KNM-WT 15000 is strikingly similar to that of *H. erectus* from Zhoukoudian. Weidenreich (1937) described, measured, and compared the teeth. We have used his descriptions here, but we were aided by our examination of casts of selected, best-preserved teeth kindly provided by Ian Tattersall. Weidenreich numbered the teeth from this locality according to a simple scheme. Each type of tooth (such as an incisor, whether upper or lower or right or left or lateral or central, whether isolated or not) received a sequential number; for example, lateral incisor #66 and central incisor #59 belong to the same mandible and were found *in situ*. See Tables 8.7 and 8.8 for comparative measurements.

Mandibular teeth

The I_1s are very similar in both morphology and size. Both sets of incisors have relatively nondescript labial and lingual surfaces with slight swellings along the basal aspect of the lingual surfaces. In both samples the I_2s are larger than those in the I_1s. Although overall morphology is the same, the I_2s of KNM-WT 15000 are much larger than the Zhoukoudian sample.

The canines of KNM-WT 15000, like the Zhoukoudian sample, are characterized by having an appearance similar to that of an incisor. This similarity is greater in KNM-WT 15000, whose canines do not exhibit extensive lingual grooves or ridges. The mesial and distal marginal ridges blend with a basal tubercle which extends towards the apex as a broadened midline ridge separated by shallow mesial and distal foveae. Some of the Chinese specimens have mesial and distal grooves rather than broad foveae. Their basal tubercle is not pronounced and the lingual fossa is relatively shallow with indistinct ridges. One tooth (#72 illustrated in Weidenreich, 1937) looks less sculptured and therefore more similar to the canine of KNM-WT 15000. These teeth share similar morphological features of the buccal surface and apex, however.

TABLE 8.7.

Mandibular tooth measurements (mm) of Zhoukoudian *Homo erectus* specimens.

ID NUMBER	SIDE	I1 m/d	I1 b/l	I2 m/d	I2 b/l	C m/d	C b/l	P3 m/d	P3 b/l
5	L	6.0	6.4						
54	R	6.8	6.5						
55	L	6.2	6.1						
57	R	6.7	6.2						
58	L	6.3	6.7						
59	L	6.2	6.8						
135′		6.7	5.8						
8	L			7.2	7.1				
10	R			7.2	7.1				
11	L			6.4	6.8				
12	R			6.3	6.4				
59	R			6.8	6.7				
60	L			7.0	7.1				
62	L			7.0	7.0				
63	L			6.8	7.0				
64	L			7.0	7.3				
66	L				7.3				
17	L					9.0	10.4		
18	L					8.5	8.7		
69	L					8.1	8.2		
70	L					8.4			
71	R						9.5		
72	L					9.0	10.1		
73	L					8.1	8.2		
74	L					8.9	9.3		
20	R							9.3	10.0
21	R							8.2	9.2
22								8.4	9.3
23	R							7.9	10.2
24								8.2	9.4
80	R							8.8	9.1
81	R							8.8	10.1
82	R							9.0	10.6
83	L							9.1	10.7
84								9.0	
85	L							9.8	10.2
103′								8.1	10.8
132′								8.2	10.3
Z. p*	L							8.3	8.9

Note: m/d = mesiodistal; b/l = buccolingual.

*Zdansky's premolar (from a collection in Sweden).

TABLE 8.7 *(continued).*

Mandibular tooth measurements (mm) of Zhoukoudian *Homo erectus* specimens *(continued).*

ID NUMBER	SIDE	P4 m/d	P4 b/l	M1 m/d	M1 b/l	M2 m/d	M2 b/l	M3 m/d	M3 b/l
29	R	9.2	11.1						
30	R	8.7	9.6						
89	R	8.2	8.0						
90	R	9.0	9.8						
91	L	8.5	11.0						
92			8.2						
93	L	8.7	10.9						
96	R			11.8	11.2				
97	R			11.9	11.1				
98	R			13.6	12.6				
99	R			13.4	12.2				
100	L			13.2	12.5				
101				11.3					
102	L			13.0	12.6				
34	L			12.2	10.9				
35	R			12.5	12.6				
36	L			14.1	12.8				
37	R				12.0				
38	L			9.9	10.1				
137′	L			13.0	11.7				
147′				12.3	11.7				
106	R					11.9	11.4		
107	R					12.9			
109							12.9		
110	L					12.5	12.7		
111	L					12.6	12.6		
43	L					12.8	11.1		
44	L					13.1	11.5		
45	R					12.1	12.0		
138′						11.3	11.5		
114	R							10.0	10.0
115	R							12.9	12.4
116	L							12.0	12.3
117								10.0	10.2
50	L							13.8	
51								12.2	12.1
52	L							12.2	11.4
131′	R							12.7	10.8
134′								10.6	10.7
136′								10.9	10.8

TABLE 8.8.

Maxillary tooth measurements (mm) of Zhoukoudian *Homo erectus* specimens.

ID		I1		I2		C		P3		P4	
NUMBER	SIDE	m/d	b/l	m/d	b/l	m/d	b/l	m/d	b/l	m/d	b/l
1	L	9.9	7.6								
2	R	9.8	7.9								
3	L	10.8	7.5								
4	L	10.7	8.1								
53			7.5								
6	R			8.3	8.2						
7	R			8.2	8.1						
141′					8.0						
13	L					9.3	9.9				
14	R					9.6	10.6				
15	L					9.1	9.8				
16	R					10.5	10.4				
67						9.6	10.4				
68	L					8.5	9.8				
19	R							9.2	12.8		
77	L							8.7	12.6		
78	R							7.4	10.5		
142′								8.0	11.6		
25	L									7.9	11.3
26										8.3	11.2
27	L									8.3	12.1
28	R									7.2	10.3
86										8.9	12.5
87	R									7.3	10.8
88	L									7.4	11.2
133′	L									8.8	11.7
143′	L									7.3	11.1

Note: m/d = mesiodistal; b/l = buccolingual.
*Zdansky's molar (from a collection in Sweden).

The premolars of the Zhoukoudian sample are distinguished by accessory ridges along the occlusal surface of the two main cusps. In the P_3s the cusps are connected by a meandering ridge. A meandering ridge is present on the KNM-WT 15000 P_3s. As with the other anterior teeth, however, the surfaces of the premolars are smooth relative to the Chinese ones. They have neither accessory ridges occlusally nor longitudinal grooves along a style on their buccal surfaces. These grooves are described by Weidenreich as being associated with a cingulum rising along the mesial aspect of the tooth towards its apex. The P_4s of KNM-WT 15000 and the Zhoukoudian sample are very similar. They have mesially located cusps causing the distal portion of the tooth to be lower in height than the mesial. The distal fovea appears fairly deep and has accessory ridges radiating from its center. Although Weidenreich stressed the oval cross-section of P_3 crowns, the

TABLE 8.8 *(continued)*.

Maxillary tooth measurements (mm) of Zhoukoudian *Homo erectus* specimens *(continued)*.

ID NUMBER	SIDE	M1 m/d	M1 b/l	M2 m/d	M2 b/l	M3 m/d	M3 b/l
31		13.1					
32	R	11.3	11.7				
33	L	12.1	13.4				
94	R	10.0	11.7				
95	L	10.2	12.3				
140′	L	11.1	13.7				
144′		10.6	12.4				
39				10.5	12.3		
40	L			12.2	12.2		
41	L			11.1	13.2		
42	R			11.4	12.4		
104	R			10.3	12.8		
105	L			10.2	12.8		
145′				10.6	13.4		
46	R					9.1	10.9
47	L					9.4	11.3
48	R					9.9	12.0
49	R					8.7	10.4
112	R					10.1	12.5
113	L					10.4	12.1
146′						9.8	12.5
Z.m*	R					9.8	12.0

occlusal outlines of KNM-WT 15000 P_3s resemble triangles with rounded corners. Some Zhoukoudian premolars also have this shape.

Mandibular molars of both groups are very similar morphologically and, in both, M_2s are larger than M_1s. One difference, other than overall size, is that molars of the Zhoukoudian sample appear more square than rectangular. This is likely due to the presence of a basal swelling or cingulum (in Weidenreich's terminology) on the Zhoukoudian teeth. In addition, this observation takes into account both the interstitial wear and tooth germ morphology of the sample. There is no evidence of a cingulum as such on KNM-WT 15000 molars.

Maxillary teeth

The Zhoukoudian central incisors are distinguished by an infolding of enamel along both mesial and distal borders which effects a concavity along the lingual surface, thus forming a shovel shape. The incisors of KNM-WT 15000 are also shovel-shaped and have pronounced basal tubercles and finger-like prolongations. This is especially evident when they are compared with the Zhoukoudian specimen #2. The incisors of KNM-WT 15000 are much larger in both mesiodistal and buccolingual dimensions. The greater buccolingual dimension is caused by a more bulbous basal tubercle.

The lateral incisors also have strong infoldings of enamel. In this case it extends onto the base, forming a thickened basal border or cingulum in KNM-WT 15000 rather than the basal tubercle seen in the Zhoukoudian sample. There is a similar notch in the basal cingula and tubercles in both samples, however. The two lateral incisors from which an accurate measurement could be taken for the Chinese sample are a germ and an unerupted tooth which had not developed completely. A cast of tooth germ #6 shows a greater infolding of mesial and distal marginal ridges than that seen in KNM-WT 15000. A pronounced wedge shape described by Weidenreich is not evident in KNM-WT 15000.

The canines of the Zhoukoudian sample have very distinct morphologies. Like the incisors, they have well-developed lingual prolongations and vertical grooves. One canine in particular has a basal cingulum with finger-like prolongations reminiscent of the central incisors. The canine of KNM-WT 15000 is not as sculptured. Its morphology is more similar to that of the upper lateral incisors. It also has prominent mesial and distal marginal ridges blending basally to form a lingual cingulum. The resulting lingual fossa is not as large or robust as that in the Zhoukoudian sample. Overall dimensions place KNM-WT 15000 slightly

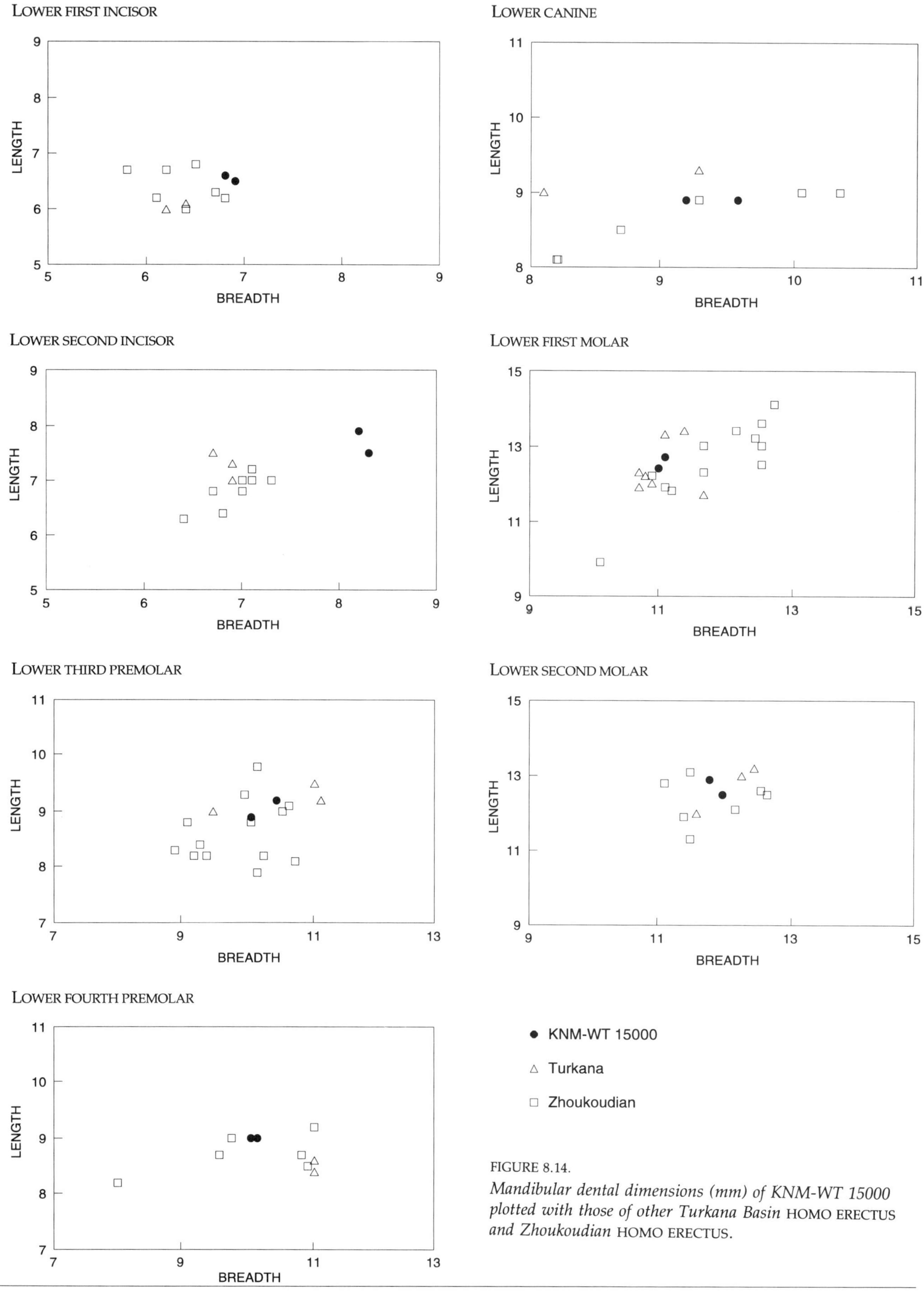

FIGURE 8.14.
Mandibular dental dimensions (mm) of KNM-WT 15000 plotted with those of other Turkana Basin HOMO ERECTUS *and Zhoukoudian* HOMO ERECTUS.

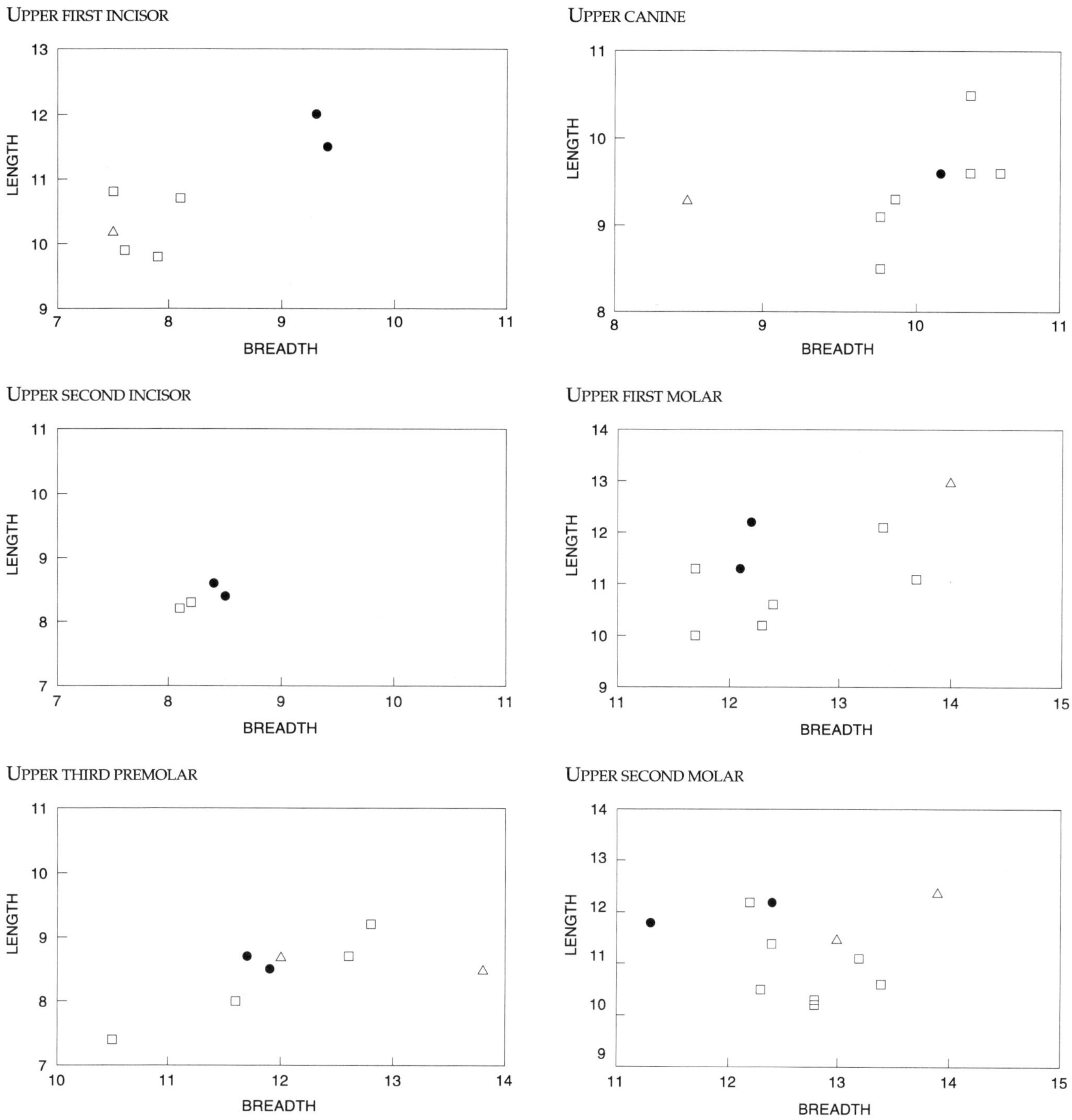

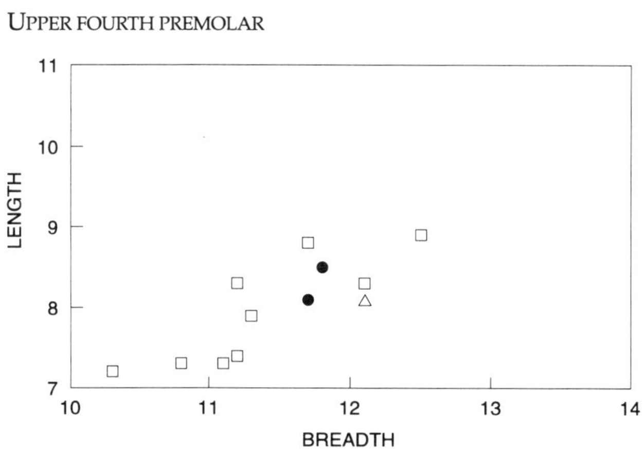

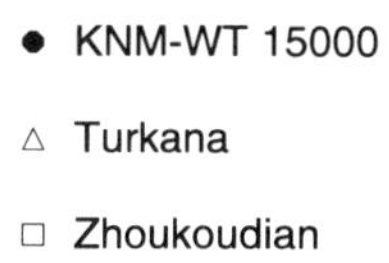

FIGURE 8.15.

Maxillary dental dimensions (mm) of KNM-WT 15000 plotted with those of other Turkana Basin HOMO ERECTUS *and Zhoukoudian* HOMO ERECTUS.

below the mean for the Zhoukoudian sample, which includes large and small specimens, assumed by Weidenreich to be male and female, respectively.

Zhoukoudian premolars are characterized by a robust shape especially in the buccolingual dimension. The P^3s are the most robust among the anterior dentition. One isolated P^3 from Zhoukoudian was particularly well preserved with little wear. Compared with the P^3 of KNM-WT 15000, this premolar has remarkably similar morphology, although it is more robust and bulbous along the basal aspect of the lingual and buccal cusps. This swelling is evident on the lingual aspect of the P^4s. The mesial and distal faces of the premolars therefore appear very rounded in the Chinese sample, whereas they appear more upright or vertical in KNM-WT 15000. Metrically the premolars of KNM-WT 15000 are much smaller than the one male tooth recorded in the Chinese sample, and they are within the range of the smaller female teeth.

Like the canines and premolars, the molars of the Zhoukoudian sample have well-formed basal swellings and accessory ridges along the occlusal surface. Weidenreich (1937) claimed these ridges are extremely well developed in the molars. The M^2s are distinguished by a rhomboid shape, which, as mentioned above, also characterizes the M^2s of KNM-WT 15000. The latter seem to be more rhomboid than the sample from China, which are more square. Unfortunately the casts representing a portion of the Chinese sample are those of tooth germs which may not have developed completely. Also, the Zhoukoudian molars, which are complete enough for measurement, are all considered female except one. Metrically, the molars of KNM-WT 15000 fit within the female range.

Conclusions

We have discussed the similarities in tooth morphology among the teeth of KNM-WT 15000, early *Homo erectus* from the Turkana Basin, and Asian *Homo erectus* from Zhoukoudian. Many of the described morphologies are found in all these specimens, and characters such as shovel-shaped incisors and certain enamel wrinkling patterns are seen in the teeth of some modern populations. These features are likely to be primitive retentions for the post-*habilis Homo* clade. Metrically, KNM-WT 15000 falls within the range of length/breadth dimensions plotted for African and Asian *Homo erectus* specimens (Figures 8.14 and 8.15). The only exceptions are the upper central and lower lateral incisors. Both these teeth are expanded labiolingually by a pronounced lingual basal swelling.

Although tooth morphology and dimensions are often not included in cladistic analyses of *Homo erectus,* Tobias (1991) has identified dental characters that appear to discriminate among *Homo habilis, Homo erectus,* and known australopithecine species. Most of these characters effectively sort *Homo habilis* and *Australopithecus.* However, many features listed by Tobias are shared by *Homo habilis* and *Homo erectus,* and only a few discrete metric traits, as defined by Tobias (1991), can reliably differentiate between the two species. When features of the KNM-WT 15000 dentition are compared with those listed by Tobias, the specimen is generally placed within the *Homo erectus* group. Many of the features do not definitively distinguish the specimen from *Homo habilis,* and some are similar to the australopithecines. Although we believe KNM-WT 15000 is a good representative of early *Homo erectus* dental morphology and metrics, it appears that dental anatomy and metrics do not decisively sort early East African *Homo erectus* from other early species of *Homo.* Because KNM-WT 15000 is such a complete and well-preserved specimen, it will, however, help us determine which characters of the dentition are primitive and which derived in our lineage.

Acknowledgments

We thank Chris Dean, Hannah Grausz, Saleem Malik, Mark Teaford, Ian Tattersall, Steve Ward, and Bernard Wood for their kind and courteous help.

PART III

ANALYTICAL STUDIES

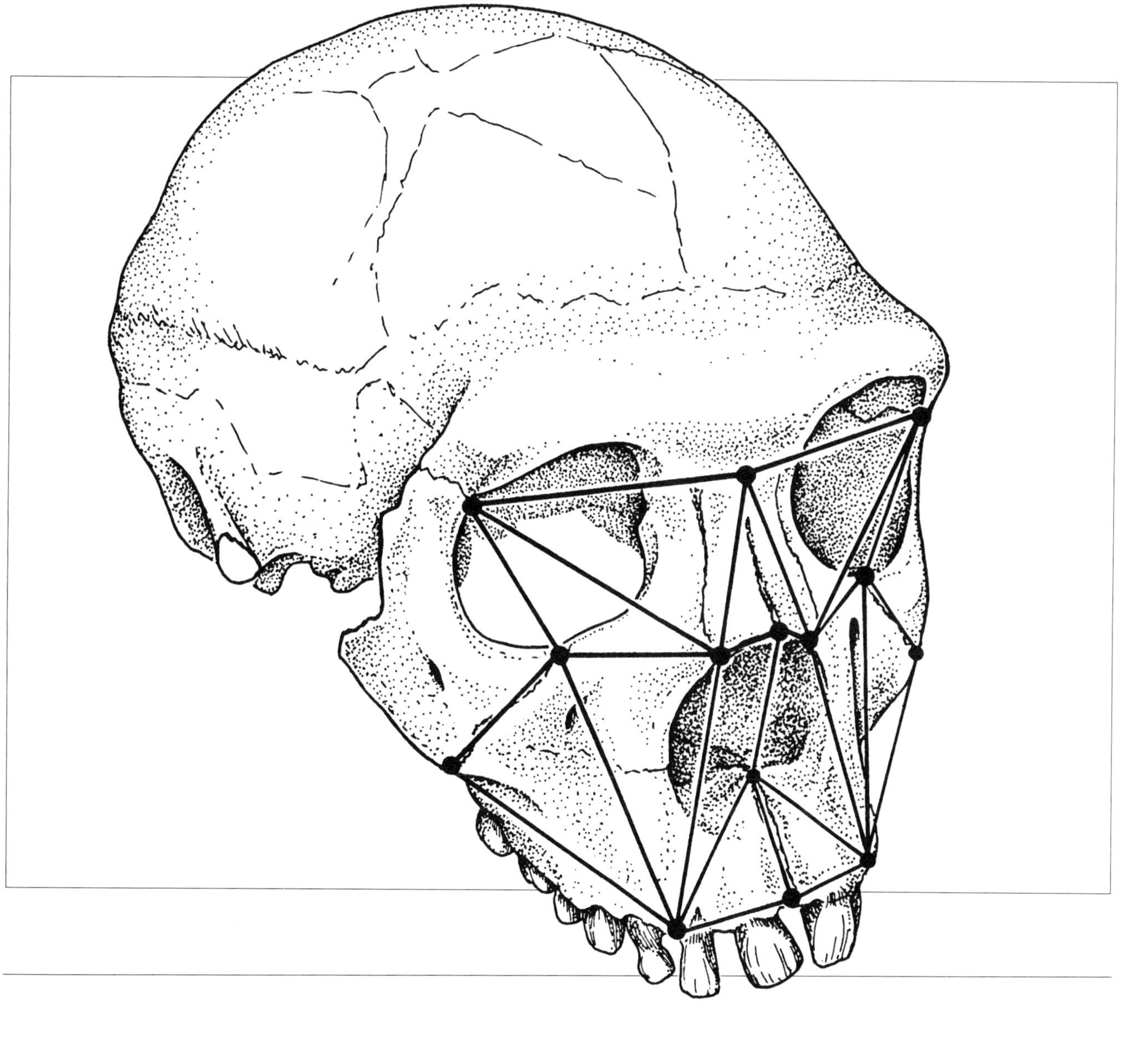

CHAPTER 9

THE PHYSIOLOGICAL AGE OF KNM-WT 15000

B. HOLLY SMITH

Humans are unique in many ways, among them the prolongation of the different stages of growth in an individual's development. The Nariokotome boy has much to teach us about the life history of early hominids because he died before reaching maturity. Of great importance is the size of his brain, for brain size and life history strategies are linked in mammals. Given that this specimen lived just at the beginning of the period of brain expansion in hominids, its study may produce special insights into the evolution of the human condition.

All hominid fossils tell us something about our evolutionary history, but fossils of individuals who died as juveniles are especially valuable. The recovered skull and skeleton of KNM-WT 15000 preserves a nearly complete record of a moment in time in the growth and development of the hard tissues of an individual hominid. This record is in principle equivalent to a cross-sectional record of skeletal maturation of any living child in a modern human growth study, except that in this instance the growth record is 1.5 million years old. Some aspects of the life history of KNM-WT 15000 can be reconstructed by comparing his growth record to standards based on living humans and other living primates. Perhaps most important, KNM-WT 15000 offers information on an earlier evolutionary stage in human growth and development.

Humans have a unique and extreme life history among living mammals. Despite a modest body size, humans have a long gestation period, a long period of infant and child dependency, and the longest life span of any mammal (see Dobzhansky, 1962; Sacher, 1975). Development from conception to adulthood takes approximately twice the length of time for humans as for great apes. Seen in a broad context, humans invest an enormous amount of effort in feeding, protecting, and training offspring for the twenty years or so until adulthood. Human growth curves also have unusual features: a seeming prolongation of a childhood period with slow to moderate growth, followed by a period of extremely high growth velocity called the "adolescent growth spurt" (Tanner, 1962; Laird, 1967; Bogin, 1988). In some higher primates, pubescent males undergo a growth spurt in weight (Tanner, 1962), but the occurrence, in both male and female adolescents, of a defined growth spurt in linear dimensions that is of high magnitude and easily demonstrable appears to distinguish humans from nonhuman primates (Bogin, 1988) and from other mammals (Tanner, 1962). One step in trying to explain the evolution of these characteristics of human life history is to try to establish when they evolved, or in what manner. Such information may be used to support or reject explanatory models that demand a particular sequence of appearance of features of development, morphology, or behavior.

This study assesses the physiological age of KNM-WT 15000 on the basis of comparisons with living primates and addresses the question of the contribution of this specimen to our understanding of growth and development in *Homo erectus*.

Methods of estimating age

Many techniques exist to estimate the age of mammalian subjects. Mammalian skeletons can be placed in a relative framework of age fairly easily because mammals share determinate growth and regularities of skeletal maturation. Thus, a mammalian skeleton usually can be categorized with confidence as infant, juvenile, younger adult, or older adult. In addition, many tissues show predictable changes with age, changes that can be used to assign a more precise physiological age (see Kimura, 1977, a 1,460-item bibliography on age determination in mammals). Dental or skeletal ages, estimates based on the formation of teeth or the extent of ossification, are the best-known physiological ages applied to mammals, but there are many more measures of age: body size, tusk size, dental attrition, bone histology, organ histology, sexual maturation, or eye lens weight.

A growing body of evidence suggests that a variety of animal tissues grow in increments that respond to astronomical cycles (Neville, 1967; Dean, 1987b; Kimura, 1977; Fisher, 1987) and that preserved lines of incremental growth track these external clocks with fair reliability. Thus, counts of increments of growth can provide a physiological age that may be closely or even precisely tied to chronological age. For this kind of age determination, teeth have the advantage that original growth lines are not obscured by later remodeling, as bones are. Unfortunately, most methods of examining incremental lines begin by sectioning teeth, a problem for studies of rare fossil material. Lines of growth visible on the surface enamel of incisors have been used to estimate age for several early hominids (Bromage and Dean, 1985; Dean 1987a), a method appropriate for very young juveniles who died during or soon after formation of incisor crowns (birth to age three or four years in hominoids). At death, KNM-WT 15000 was already too old for this nondestructive technique. Perhaps an estimate based on lines of incremental growth will become available for KNM-WT 15000 in the future, since this is an expanding area of current research (e.g., Beynon and Wood, 1987; Dean, 1987a, 1989), but at present more conventional physiological ages can be provided by the state of the dentition, skeletal epiphyses, and body size. First, it is necessary to describe the state of maturation of the dentition and skeleton.

Indications of the Age of KNM-WT 15000

Several teeth of KNM-WT 15000 were recovered separately from the facial bones. For these teeth, sharp casts were made of the entire tooth by S. Kazinga before they were replaced in the jaws. For the present study, these casts were examined along with casts of the reconstructed mandible and maxilla and radiographs of the actual specimen. Nearly every mandibular and maxillary tooth could be observed on at least one side of the dentition.

The dentition of KNM-WT 15000 clearly indicates a subadult because the permanent dentition was not yet complete. The reconstructed mandible has permanent incisors, canines, premolars, and first and second molars erupted (see Walker and Leakey, Chapter 6, this volume). The maxilla shows permanent incisors, premolars, and first and second molars erupted also, but it retains deciduous canines in place; the right permanent maxillary canine crown is clearly visible at the base of the deciduous canine root, which was apparently undergoing resorption. No third molars were erupted. Even on casts, wear facets are clearly visible on second molars, and anterior teeth also show light wear. Although 26 permanent teeth were erupted, probably only ten of these (first molars, central incisors, and lower lateral incisors) were completely formed (Figure 9.1).

Developmental stages of teeth of KNM-WT 15000 were judged using the ordinal scale of Moorrees et al. (1963). In this system, teeth are graded according to sequential stages based on proportion of crown, root, and root apex completed (Table 9.1). It is standard practice to modify this scale slightly or allow interpolations (Moorrees et al., 1963; Anderson et al., 1976). Here, stages based on division of crowns or roots into thirds are recognized and intermediates are allowed between stages. During the process of assigning stages, teeth were drawn at actual size and projected final length was dotted in following natural contours, as reproduced in Figure 9.1. Measurements of finished and projected segments were used to help judge the proportion of the root completed. Stages were assigned blindly, without referring to standards for any species.

TABLE 9.1.

Stages of development of tooth formation and their abbreviations from Moorrees et al. (1963), with additional stages (*).

STAGE	ABBREVIATION
Initial cusp formation	Ci
Coalescence of cusps	Cco
Cusp outline complete	Coc
Crown ½ complete	Cr½
Crown ⅔ complete*	Cr⅔
Crown ¾ complete	Cr¾
Crown complete	Crc
Initial root formation	Ri
Initial cleft formation	Cli
Root length ¼	R¼
Root length ⅓*	R⅓
Root length ½	R½
Root length ⅔*	R⅔
Root length ¾	R¾
Root length complete	Rc
Apex ½ closed	A½
Apex closure complete	Ac

In the maxilla, central incisor roots are fully formed, and a pinhole-sized opening is still present at the root apex (Figure 9.1). The root is at the radiographically defined stage "apex complete," although the distinct aperture at the apex is consistent with recent closure. The upper lateral incisor lags behind the first in formation, between

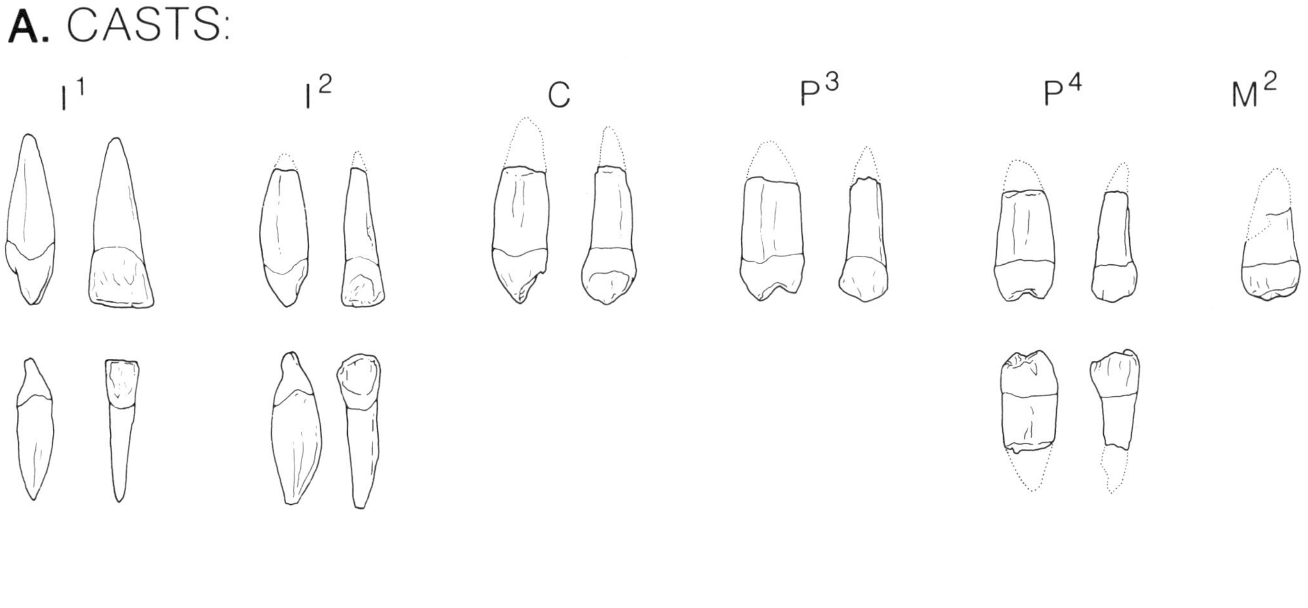

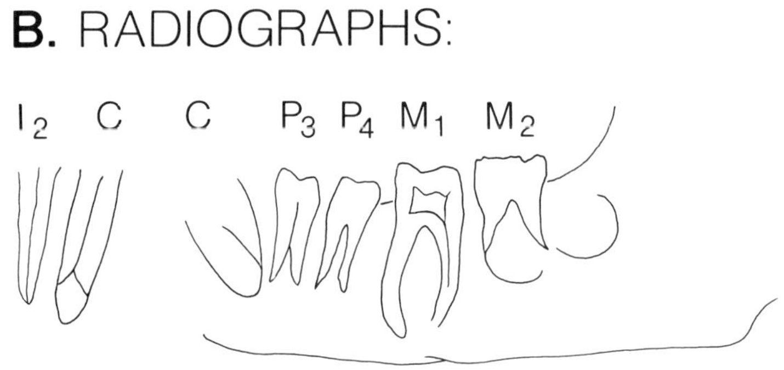

FIGURE 9.1.
Teeth of KNM-WT 15000 drawn from sharp casts (A) *and from radiographic views* (B). *Dotted outlines project remaining growth. The second radiograph was taken from an oblique view, which elongates roots of cheek teeth. Note that most teeth of KNM-WT 15000 were incompletely formed. Some radiographic views of the mandible suggest crypt formation (as here), but others show nothing at this position (see Brown and Walker, Chapter 8, this volume); maxillary third molars were present but are not drawn here.*

the stages "root complete," or "Rc," and "apex half-complete," or "A½." The upper canine is estimated to have completed some 70 percent of its adult length and is recorded as "R⅔." Upper P^3, P^4, and M^2 have substantially immature roots, judged as R⅔, R¾, and R⅔, respectively.

In the mandible, both incisors are complete, although the pinhole-sized aperture on the I_2 root suggests recent closure before death. The lower canine is clearly visible on a radiograph of the symphyseal region and on an oblique view of cheek teeth; its development is scored as R¾. Premolars and the second molar are substantially immature in radiographic view: premolars are in between the stages R½ and R⅔, and M_2 is at R½. First molars are fully complete.

The crown of the right maxillary third molar has been broken out of its crypt and lost, but the left crown remains intact. This crown can be seen well through a window in the alveolar bone and poorly on a radiograph. It appears to be approximately "crown ⅔," or Cr⅔, a stage too premature for even the beginning processes of eruption. In the mandible, gubernacular canals for third molars are clearly present on both sides, suggesting that crowns were developing here also; further, size and location of the canals are perfectly appropriate for human and chimpanzee juveniles of this dental age who are developing third molars. Some radiographs seem to suggest a crypt for lower third molars, but others show no sign of crypt or crown (see Brown and Walker, Chapter 8, this volume).

Advanced crypt formation would rule out typical third molar agenesis, which is not characterized in human children by substantial crypt or crown development which later fails or resorbs (Stanley Garn, pers. comm.). Mandibular third molar crowns may have been lost postmortem or they may have failed in development, although "agenesis" seems an imperfect description when gubernacular canals show that something did generate at the third molar position. Dissection of the third molar area would probably resolve the question of the premortem condition of mandibular third molars, but this information is not essential for a calculation of dental age. For this purpose, the stage of development of the maxillary molar is assigned to the mandibular third molar. This is justified because developmental timing of mandibular and maxillary molar crowns (including third molars) is nearly identical in humans (Anderson et al., 1976), something that also appears to hold for great apes (Dean and Wood, 1981). Omitting a value for the mandibular third molars would slightly lower the overall mean dental age. As will be seen below, results are more conservative by including an estimate for mandibular third molars in the mean dental age.

Three other types of observations are relevant to a determination of the physiological age of KNM-WT 15000: state of epiphyses, stature, and sex. Major ossification centers of long bones had appeared and most remain unfused; of particular importance, the triradiate cartilage still separated primary elements of the innominate. Epiphyses of the distal humerus, however, had fused to some degree (see Walker and Leakey, Chapter 7, this volume). The proximal parts of the elbow joint (trochlea and capitulum of the humerus at least, possibly the lateral epicondyle) were in the process of fusing, but the distal parts (radius head and olecranon of the ulna) remained ununited. Stature at death is presently estimated at circa 160 cm, and minimum estimates have been made of 153 (Ruff and Walker, Chapter 10, this volume) or 155 cm (Feldesman and Lundy, 1988). Sex has been assigned as male on the basis of an overall assessment of the sciatic notch, skull robusticity, and general size of the skeleton (Ruff and Walker, Chapter 10, this volume).

Models of growth and development

Among models we might choose to reconstruct extinct species, Tooby and DeVore (1987) distinguish between those that are "referential," based on a specific living genus or species, and those that are "conceptual," derived from a general principle discerned from actual species but for which there may be no exact known representative. Tooby and DeVore (1987) disparage referential models for early hominid behavior, but these models always play an important role in initiating and shaping analyses that may lead to conceptual models. Use of close living relatives as models for extinct species is certainly appropriate for studying characteristics that are heritable and thought to be conservative in evolution; this aspect of conservatism may increase the importance of referential models for studies of tissues relative to those of behavior. Indeed, it is hardly possible to frame the relevant questions about the maturation of KNM-WT 15000 without beginning with referential models.

Referential models can differ considerably in their breadth of scope: broad models can help determine what features of animals are universal; narrow ones can be used to propose more detailed resemblances. Some characteristics of development in other species were probably shared by KNM-WT 15000 because a basic plan of life history is common across higher primates. Other characteristics require many more assumptions to reconstruct because close living relatives differ in these traits. This evaluation of KNM-WT 15000 considers both broad and narrow referential models, the success of these models, and a conceptual model as well.

Broad referential models

Primate life history falls naturally into periods that can be divided by eruption of the teeth, as shown in Figure 9.2 for three catarrhine primates: humans, chimpanzees, and macaques. Following

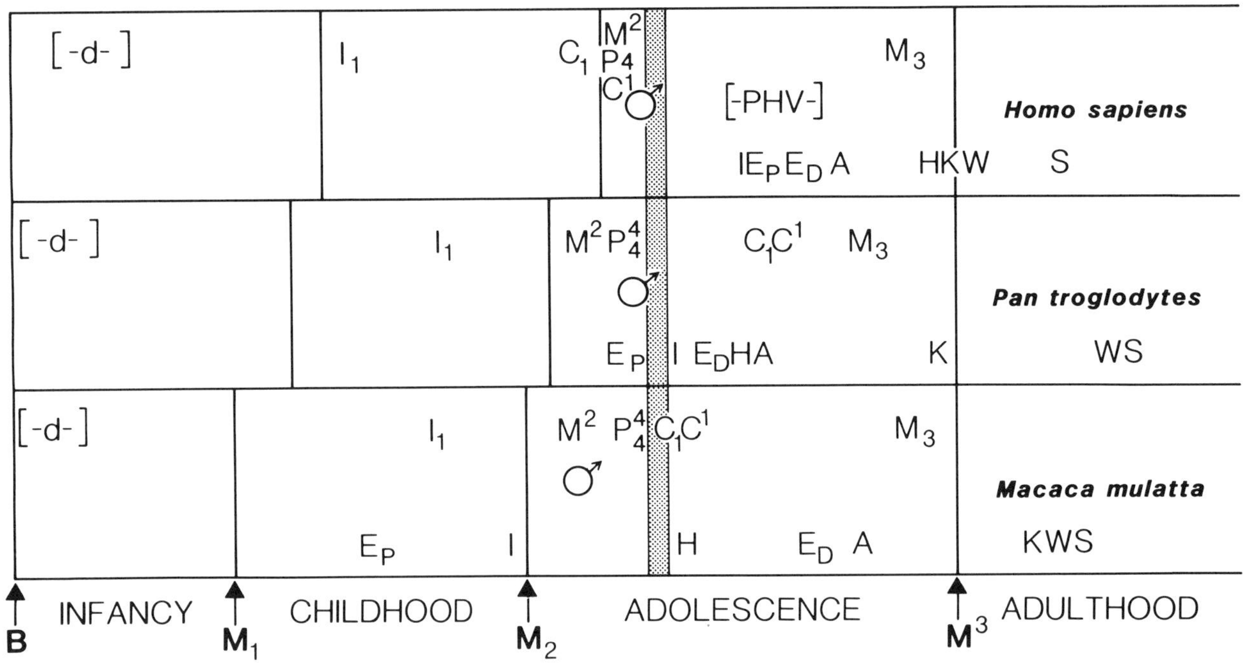

FIGURE 9.2.
Stage of development of KNM-WT 15000 (shaded bar) *seen in the perspective of relative life history in monkeys* (M. MULATTA), *apes* (P. TROGLODYTES), *and humans* (H. SAPIENS). *The juvenile period is divided into infancy* (B = *birth*), *childhood, and adolescence by the emergence of the molar teeth M1, M2, and M3; adulthood is defined to begin after emergence of last permanent tooth. The bar for each genus shows events in tooth emergence at top and events in epiphyseal fusion at bottom; the male symbol marks a rise in serum testosterone level, an early sign of puberty. Abbreviations for teeth: -d-, period of emergence of deciduous teeth; permanent teeth indicated in standard form* (I, C, P, M). *Abbreviations for epiphyses:* I, *triradiate cartilage of innominate;* E, *elbow (subscripts designate proximal and distal elements);* H, *hip (proximal femur);* A, *ankle;* K, *knee;* W, *wrist;* S, *shoulder.* PHV *denotes mean time of peak height velocity (the adolescent growth spurt) across a wide range of human populations. Note the similarity of proportional life history in higher primates. KNM-WT 15000 is most parsimoniously placed by recent emergence of cheek teeth* (shaded bar); *canine emergence and bone age tend to disagree as to relative age within and between models. (Data sources: Hurme and van Wagenen, 1953, 1961; Dahlberg and Menegaz-Bock, 1958; Krogman, 1962; Kerley, 1966; Knorr et al., 1974; Friedlaender and Bailit, 1969; Brown et al., 1971; Martin et al., 1977; Rose et al., 1978; Cheverud, 1981; Little et al., 1983; Watts, 1986a.)*

Schultz (1956a, 1960), emergence of the first permanent molar marks the end of infancy. This is a natural marker because age of first molar eruption approximates age of weaning in nonhuman primates (Smith, 1991a), signifying a major decrease in dependency of the offspring on the mother. Emergence of the second molar performs fairly well as a measure of the beginning of adolescence in these primates, a period usually defined to begin with puberty (e.g., Bogin, 1988, p. 16). In Figure 9.2, the male symbols indicate one early sign of male puberty, a rise in serum testosterone. Most physical, physiological, and behavioral manifestations of sexual maturity of these catarrhines remain in the adolescent period when the period is defined as post-M2 emergence (Kraemer et al., 1982; Harvey and Clutton-Brock, 1985). For these three species, completed eruption of the dentition occurs near the end of the increase in linear body dimensions, although weight continues to increase in adulthood. Completed tooth eruption is often used as a marker of the beginning of adulthood in primates, although it consistently precedes fusion of some skeletal epiphyses in primates and other mammals (see Gavan, 1953; Schultz, 1960; Shigehara, 1980).

In terms of chronological years of age, Figure 9.2 defines the periods of infancy, childhood, and adolescence, respectively, as (all ages in years): birth to 1.4, 1.4–3.2, and 3.2–5.8 (macaques); birth to 3.3, 3.3–6.5, and 6.5–11.4 (chimpanzees); and, birth to 5.9, 5.9–11.3, and 11.3–18.0 (humans). Inevitably, arbitrary decisions are included in such charts, and events could shift relative to each other to some degree if different population values or laboratory studies were substituted for those chosen here. The chimpanzee chart is the only one to fulfill the ideal of basing all tissue systems (tooth eruption, hormones, epiphyses) on a single sample, that of the Yerkes Primate Center; data for macaques and humans are necessarily from more diverse sources.

It is evident from Figure 9.2 that the catarrhine life cycle shares much in common when scaled relative to completion of the dentition. Using the last tooth eruption as an age of 100%, one could say that all three species erupt deciduous teeth at age 5% and erupt cheek teeth P4 and M2 at circa 60–70%. Males of all three species also undergo a rise in serum testosterone at 60–70% relative age. In skeletal maturation, all three species fuse the last of the major epiphyses (shoulder) at about 110–120%; indeed, the chart would not change much if adulthood had been defined by the last epiphyseal fusion. The species share the same approximate sequence of fusion of epiphyses: elbow, hip, ankle, knee, wrist, shoulder (Schultz, 1956b, 1960).

Although the life cycles of these primates are overwhelmingly homologous, some traits differ. Eruption of canines differentiates the three species to some extent; these teeth appear particularly early in humans and particularly late in the chimpanzee. Other traits differ in a stepwise manner. From monkeys to apes to humans, there is a tendency for many events to drift to slightly later relative times. This drift is strongly expressed for two elements of skeletal maturation: the proximal elbow (distal humerus) and the triradiate cartilage of the innominate. These elements unite at a very early relative age in monkeys, a later time in apes, and the latest time in humans. Despite the great difference in the proximal elbow joint, the distal part of the same joint (proximal radius and ulna) unites at much the same relative age for all three. Lastly, humans experience a growth spurt in adolescence (PHV, "peak height velocity," in Figure 9.2), at ca. 75% relative age in males, depending on the population chosen to represent humans; this growth pattern may not be strictly paralleled in other mammals (see below).

In his development, KNM-WT 15000 has a distinctive dental age and a distinctive bone age, but the two do not always agree in the broad model based on three catarrhines. In terms of dental age, KNM-WT 15000 died shortly after erupting central cheek teeth P4 and M2 (see below), which places him at 65–75% relative age for all three taxa—thus, about two-thirds of the way to adulthood. Even this simple model suggests, however, that his canine eruption follows a pattern less extreme than that shown by either humans or chimpanzees, a timing that fits with the intermediate pattern of the macaque. In terms of skeletal age, fusion of at least some elements of the proximal elbow but not of the triradiate cartilage suggests an age of circa 75% as a human, circa 65–70% as a chimpanzee, and circa 40–45% as a macaque. Clearly these disagreements complicate our assessment of his physiological age.

The broad model succeeds in providing background information on catarrhine growth and it also firmly places KNM-WT 15000 as an adolescent. Comparison with the model finds, however, that KNM-WT 15000 is not a perfect match to average conditions for any referent—either canines differ from cheek teeth, the dentition differs from the skeleton, or both. As would be expected, KNM-WT 15000 appears most grossly out of synchrony with the macaque model; a macaque of this dental age should have long before fused all elements of the distal humerus in addition to uniting the primary elements of the pelvis. Even if we limit comparisons to hominoids, the broad model merely brackets a possible physiological age to about 65–75% of age of young adulthood. This may not be precise enough to answer all our questions about KNM-WT 15000, because remaining growth differs extensively for juveniles who have lived 65% versus 75% of the time to adulthood (see below). Coming to a more precise conclusion about physiological age, or any conclusion about chronological age and remaining growth for KNM-WT 15000, requires either a narrower referential model or a conceptual model of life history, both of which make many more assumptions about processes of growth and development in *Homo erectus* than does the broad referential model of Figure 9.2.

Narrow referential models

The broad referential model considered the relative age of KNM-WT 15000, but his development can also be evaluated strictly as if he were a human child. Here, dental age is examined in detail; dental development is then combined with more limited observations on the development of the postcranial skeleton.

The human model

Dental age. Dental age can be assessed from two main aspects of tooth development: eruption/emergence of the teeth (*emergence* refers to cutting the gingiva, whereas *eruption* is the more general term denoting the entire process) or formation of crowns and roots. A large body of information is available on the age of tooth emergence through the gingiva in humans, the marker most frequently used in studies of tooth eruption in the living. The main limitation of emergence data is that emergence is an "on/off" signal rather than a continuous event. Still, a dental age can be based on emergence by taking a midpoint between mean age of last tooth emergence and mean age of next expected tooth emergence.

Given the state of eruption of the teeth (see above) of KNM-WT 15000, it is easy to establish which tooth would be next to emerge: the maxillary canine. This would be the next tooth to erupt in any hominoid primate, and a substantial time gap would remain until emergence of third molars (see Dean and Wood, 1981, p. 117). The last (previous) tooth emergence is more debatable; this may have been any of the lower canines or the upper or lower third or fourth premolars or second molars, as all these teeth emerge fairly closely in time in living humans, particularly fourth premolars and second molars (see review in Jaswal, 1983). The condition of the teeth and jaws of KNM-WT 15000 suggests the fourth premolars were last to erupt. Whether these or the second molars are chosen as the last teeth to have erupted makes only a minor difference to the dental age, but it is preferable to choose second molars to avoid artifacts in modern data for premolar emergence introduced by caries of deciduous molars and by agenesis of P_4 (see Smith and Garn, 1987).

Table 9.2 shows mean age of tooth emergence of canines and second molars in a worldwide sample of males of living populations of *Homo sapiens*. Several points can be observed in these data. First, a number of studies find that world populations break into two groups according to age of emergence: Eurasians versus all others. Eurasians, here represented by Denmark, India, and China, show later ages of molar emergence than other groups around the world (Dahlberg and Menegaz-Bock, 1958; Friedlaender and Bailit, 1969). The phrase *twelve-year molars* is apt for Eurasia, where typical ages for M2 emergence range near 11.8–12.6 years, but a better rendering is *eleven-year molars* for the rest of the world, where mean age of M2 emergence varies around 10.7–11.7 years of age. No single factor accounts completely for patterns in worldwide distributions, and the contribution of genes versus the environment is not understood (Friedlaender and Bailit, 1969; Eveleth, 1986); however, most workers would choose non-Eurasians as a model for prehistoric peoples for two reasons. First, if environmental factors such as attrition, caries, or degree of masticatory forces are important determinants of emergence, then indigenous non-Eurasians may be better models for humans in the past. Second, if emergence is largely genetically determined, it seems more likely, arguing simply by geographic distribution, that the Eurasians are "derived," compared with the rest of the world. Thus, an overall mean of determinations for non-Eurasian groups is used here to evaluate tooth emergence in KNM-WT 15000.

There is one discrepancy that should be pointed out between KNM-WT 15000 and data in Table 9.2. It is typical for the canine to precede the second molar in emergence (the sequence C^1M^2) in extant human groups (Jaswal, 1983; Smith and Garn, 1987). Clearly, KNM-WT 15000 would have had the atypical sequence M^2C^1. This variant sequence is known to occur today in the maxilla of male Australian aborigines, American blacks, and American whites, at frequencies of 13%, 12%, and 11%, respectively (Barrett et al., 1964; Smith and Garn, 1987). In computing a dental age, it is still appropriate to take a midpoint between mean ages of emergence of C^1 and M^2, even though it is not possible for a child to be both ≤10.7 and ≥11.4 years. In unpublished work, Stanley Garn and I confirmed empirically in 1979 the trivial expectation that human subjects showing tooth *X* early in emergence sequence also show emergence of that tooth at an age earlier than typical for the group. Because the same finding holds for a tooth late in sequence, the appropriate age estimate for a subject with a variant sequence is still the midpoint of mean ages of emergence of the teeth in question. It is neither practical nor desirable to separate emergence standards by emergence sequences since some sequence variation is the rule rather than the exception in humans (Smith and Garn, 1987). Taking all the above into consideration, one would conclude that the dental age of KNM-WT 15000 based on emergence is 11.1 years.

It is often preferable to base a dental age on tooth formation rather than emergence, since formation is a continuous process through the juvenile years not limited to a few marker events. The problem is that tooth formation is primarily known for Eurasian-derived children, and little data exist

TABLE 9.2.

Mean age of emergence of canines and second molars in a world-wide sample of recent human males.

CONTINENT and/or GROUP	AGE OF EMERGENCE (years) LOWER CANINE	UPPER CANINE	LOWER SECOND MOLAR	UPPER SECOND MOLAR	SOURCE
Non-Eurasians					
Africa[a]					
Ghana	10.6	10.9	11.3	11.4	Houpt et al. (1967)
Uganda	10.1	10.6	10.7	11.0	Krumholt et al. (1971)
Australia-New Guinea[b]					
Australia	10.0	10.5	11.2	11.5	Brown (1978)
New Guinea	9.3	9.5	11.0	11.3	Friedlander and Bailit (1969)
North America					
Inuit	9.5	11.1	10.8	11.4	Mayhall et al. (1978)
Pima	10.8	11.7	11.3	11.7	Dahlberg and Menegaz-Bock (1958)
Mean:	10.1	10.7	11.1	11.4	
Eurasians					
Denmark	10.5	11.2	11.9	12.4	Helm (1969)
India	10.5	11.5	11.8	11.9	Singh (1980)
China	10.5	11.2	11.9	12.6	Lee et al. (1965)
Mean:	10.5	11.2	11.9	12.3	
Grand mean:	10.2	10.9	11.3	11.7	

a. One-half year is added to means for Ghana and Uganda to correct for downward grouping of ages to age at previous birthday in both studies.
b. Aboriginal inhabitants.

for other human populations. Despite this lack of information, there are reasons to suggest that worldwide variation in tooth formation may be small compared with that known for tooth emergence. In humans, it has been established that nutrition, caries, and tooth loss can affect emergence (Eveleth, 1986; Alvarez and Navia, 1989), whereas tooth formation appears comparatively resistant to nutritional effects (Lewis and Garn, 1960). In addition, experimental data from nonhuman models suggest that tooth emergence is more affected by environmental factors than is tooth formation (Tonge and McCance, 1973). Thus, although it is clear that population differences in tooth formation timing do exist (Owsley and Jantz, 1983; Chertkow, 1980), they should be less than those associated with emergence timing. Indeed, standards of tooth formation for Europeans have successfully estimated chronological age to within reasonable limits for Asian and South American children adopted in Sweden (Crossner and Mansfeld, 1983). Judging from worldwide trends in emergence data, the direction of differences should be for earlier formation of cheek teeth in non-Eurasian peoples (which would give non-Eurasians a younger dental age for a given developmental stage).

Table 9.3 shows human dental ages appro-

TABLE 9.3.

Developmental stages of teeth of KNM-WT 15000 and chronological ages associated with those stages in living humans and great apes.

		CHRONOLOGICAL AGE (years)		
		HUMAN MALES[a]		GREAT APES[b]
TOOTH	STAGE OF DEVELOPMENT[c]	Moorrees et al. (1963)	Anderson et al. (1976)	Dean and Wood (1981)
Maxilla				
I1	Ac	—	(>10.6)	(>6.5)
I2	Rc–A½	—	10.1	6.2
C	R⅔	—	9.5	8.2
P3	R⅔	—	9.9	6.6
P4	R¾	—	10.6	7.0
M1	—	—	—	—
M2	R⅔	—	11.4	6.6
M3	Cr⅔?	—	12.3	6.7
Mandible				
I1	Ac	(>8.1)	(>9.2)	(>6.5)
I2	Ac	(>9.3)	(>9.9)	(>6.7)
C	R¾	9.9	10.2	8.6
P3	R½–⅔	9.7	10.0	6.4
P4	R½–⅔	10.5	10.5	6.6
M1	Ac	(>9.4)	(>10.0)	(>5.7)
M2	R½	10.6	10.5	6.2
M3	[Cr⅔?][d]	11.5	12.3	6.7
Mean dental age		10.4	10.7	6.9

a. Human standards based on North American white males. In both cases, ages for stages are interpolated from original sources as in Smith, 1991c; parentheses indicate minimum ages for fully formed teeth.

b. Ages for stages interpolated as in Smith, 1989b.

c. Stages defined in Table 9.1; the stage of both M3's is uncertain.

d. The stage of the maxillary third molar used to estimate age for missing mandibular third molars.

priate to developmental stages observed in KNM-WT 15000 (the age of KNM-WT 15000 by great-ape standards, also given in the table, will be considered below). More than twenty statistical studies of the age of permanent tooth formation exist for human children (reviewed in Smith, 1991c); the standards of Moorrees et al. (1963) are used here because the study is based on a large sample and because it represents the best statistical method of analyzing the data. Results based on Anderson et al. (1976) are also shown because that study contains data for both maxillary and mandibular teeth (rare in the literature), although its statistical technique is slightly less desirable. Both studies are based on North American children of European descent, Ohio and Massachusetts children in the case of Moorrees et al. (1963) and Canadian children of British descent in Anderson et al. (1976). To date, it appears that there are no comprehensive studies of tooth formation in non-Eurasian-derived populations that are satisfactory in sample size and statistical methodology (Smith, 1991c).

Both Moorrees et al. (1963) and Anderson et al. (1976) present their tooth formation data as "age of attainment" or "transition" schedules; these schedules can and should be modified for use in age prediction or they will consistently underestimate age (Goldstein, 1979, p. 64; Smith, 1991c). Full conversion of Moorrees et al. (1963) data to age prediction tables appear in Smith (1991c); these tables have been used to assign dental ages to each tooth of KNM-WT 15000 (Table 9.3). Results give KNM-WT 15000 a dental age equivalent to that expected for a human male of about 10.4–10.7 years. Anterior teeth tend toward younger ages and posterior teeth toward older ages, indicating some degree of patterned difference of KNM-WT 15000 from these particular standards of tooth formation.

Table 9.4 summarizes assessments of dental age in KNM-WT 15000 based on human standards. Though the precise result of estimating age depends on how one selects or combines data from all aspects of tooth development, some techniques are preferable to others. Methods A–E are listed in order of preference. Preferred methods give younger ages, 10.5–11 years of age. The less strict methods, those based on ignoring all but second molars or those ignoring tooth formation, suggest ages of 11–12 years. Overall, a dental age of circa 11.0 years appears to be most consistent with all the data. In terms of relative age, this places KNM-WT 15000 at 63% the time to young adulthood. The simple model in Figure 9.2, which places him by eruption of cheek teeth, ignoring the canine, placed him at 67% relative age, or 12 years of age. The difference between these two estimates seems minor.

Some alterations of procedure could lower the tooth formation age of KNM-15000: (1) using non-Eurasian standards of tooth formation (because limited data suggest, but do not demonstrate, that these would give slightly younger ages); (2) omitting the tentative third molar as-

TABLE 9.4.

Estimates of dental age of KNM-WT 15000 obtained from five methods of interpreting developmental data using standards for human males, listed in order of preference.

CRITERION	DENTAL AGE
A. Tooth formation	
Average of ages from two studies (10.42, 10.66)	10.5
B. Tooth development	
Six formation ages and two emergence ages averaged for a dental arch (maxilla)	10.7
C. Tooth eruption	
Midpoint between age of last emergence (upper second molar) and next expected emergence (upper canine):	
Africans, (11.2 + 10.8) ÷ 2	11.0
Non-Eurasians, (11.4 + 10.7) ÷ 2	11.1
World-wide, (11.7 + 10.9) ÷ 2	11.3
D. Second molar development	
Combined ages from root formation and emergence	11.1
Age of males with M2 emerging at an early stage (root ½) of development (known only for Europeans; Grøn, 1962)	11.3
E. Second molar eruption	
Age is greater than given mean age:	
Africans	≥11.2
Non-Eurasians	≥11.4
World-wide	≥11.7

sessments, which would lower dental age to 10.1–10.2; or (3) changing sex assignment, which would reduce dental age by approximately one-half year from the male value because females form and erupt teeth earlier than males (Adler, 1959; Moorrees et al., 1963). Even for recent, well-preserved specimens, juvenile skeletons are not sexed with 100% confidence. Although there are legitimate causes for caution, making these changes would not overturn the analyses or conclusions presented below.

There is no way to assign a valid confidence interval to estimated age of death of KNM-WT 15000. The literature disagrees as to the width of confidence intervals for dental age as an estimator of chronological age, even within a single extant human population. In a series of tests based on Asian, South American, and Scandinavian subjects, two studies claimed accuracy to within ±6 months of actual age in all cases (Gustafson and Koch, 1974; Crossner and Mansfeld, 1986) whereas another found that accuracy declined to within ±2 years in older children (Hagg and Matsson, 1985). The specific method used here has been tested in a small preliminary trial on European-derived children, where it proved very accurate (Smith, 1991c). However, no confidence interval can assure us that *Homo erectus* development took place on a human time scale—that requires a separate line of inquiry.

Skeletal age. Epiphyseal fusion can be observed to provide a skeletal age for KNM-WT 15000. The distal humerus has four centers of ossification, and its fusion is a process rather than a single event, although it has been simplified in Figure 9.2. KNM-WT 15000 apparently had united the earlier-fusing elements of the distal humerus (capitulum to trochlea, possibly the lateral epicondyle) but not the latter elements (capitulum to shaft, medial epicondyle); the primary elements of the pelvis remain ununited. Krogman (1962) cites Modi and Flecker for age of fusion of the early parts of the distal humerus at 11–15 years and 12–15 years (see also Ruff and Walker, Chapter 10, this volume). The triradiate cartilage is said to disappear at 15–16 (Modi cited in Krogman, 1962, pp. 64–65). Krogman (1962, p. 33) lists fusion of the distal humerus overall as circa 14–16 for white males and gives union of the primary elements of the pelvis as 13–15 years. Overall, ignoring data on the dentition, these studies of skeletal maturation could be interpreted to place KNM-WT 15000 broadly as between 11 and 15 years of age, or more narrowly as circa 13–13.5 years.

It is also possible to assign KNM-WT 15000 an age based on his body size, just as any human child could be assigned an approximate age based on stature. At 160 cm (63 inches), KNM-WT 15000 is tall for an early adolescent. For comparison, an average stature of 152.5 cm (60 inches) is usually reached at mean values between 12 and 15 years of age in human males, depending on the population (Falkner and Tanner, 1986; Little et al., 1983). Even in tall populations like the Dinka, Turkana, and Masai, it is typical to reach 160 cm at late adolescence, at approximately age 15 or later (Roberts, 1960; Little et al., 1983; see also MacKay and Martin, 1952), and it is surprising how similar different human populations are at juvenile ages (see Roberts, 1960). If KNM-WT 15000 was truly 11 years of age at his death, he was tall by any human population standard. Phrased another way, KNM-WT 15000 has a stature age more appropriate to a 15-year-old.

The great-ape model

Recent work has questioned the previously accepted idea that early hominids share human-like maturation patterns, extending doubts even to early representatives of *Homo* (Bromage and Dean, 1985; Bromage, 1987; Smith, 1986, 1989b, 1991b). Thus, it may be instructive to compare KNM-WT 15000 to a great-ape model.

Dental age. The sequence of tooth emergence, M^2C^1, fulfills expectations based on chimpanzees; Schultz (1940) reports this sequence in 100% of subjects (compared with about 11% of human males). According to schedules of tooth emergence based on data from chimpanzees (Nissen and Riesen, 1964), emergence of the central cheek teeth of KNM-WT 15000 suggests an age of ≥7.5 years, although eruption of the lower canines but not the upper suggests a much older animal, of circa 9 years.

Just as in the case of human dental age (above), estimates based on tooth formation give slightly younger ages than those based on emergence. Dean and Wood (1981) provide a schedule of tooth formation in great apes. The schedule is preliminary because it is based on many fewer data than the human studies and on few observations of animals of known age, but it is a great improvement over previously existing data. According to these data, most teeth of KNM-WT 15000 give dental ages of 6.2–6.7 years (see Table 9.3). The canine, however, is far more advanced than cheek teeth, giving an age >8 years. Overall, a great-ape schedule of tooth formation suggests an age of circa 7 years, or 62% relative age (see above); a schedule for tooth emergence suggests 7.5–8 years (66–70% relative age) or up to 9 years, including the canine (79% relative age).

Skeletal age. It is somewhat simpler to assess skeletal age by chimpanzee standards. Kerley's (1966, p. 75) study of Yerkes chimpanzees places fusion of the distal humerus (excluding the medial epicondyle) at about 7.5 years of age and disappearance of the triradiate cartilage of the pelvis at 8 years. Thus, KNM-WT 15000 has a skeletal age of circa 7.5 years (66% relative age) according to a chimpanzee model, quite close to his dental age from a chimpanzee model. It is not reasonable to assign a stature age to KNM-WT 15000 based on a chimpanzee model because it is obvious that he was not only a biped, but a long-legged one. It remains possible, however, to consider some aspects of body size in terms of a great-ape model if we take a different approach (see below).

Evaluating referential models

Which referential model is most consistent with the data? In evaluating the growth of an extinct species, it is appropriate to compare consistency of events observed with those expected from a living model. In the case of KNM-WT 15000, so much is preserved of skull and skeleton that several growth systems can be examined for consistency with referential models: internal consistency of dental ages; dental age versus skeletal age; dental age versus body size; and skeletal age versus body size. In terms of growth assessment, the only major lack for KNM-WT 15000 is hand bones. A well-preserved hand might allow a highly detailed examination of skeletal age because hand-wrist maturation is the standard tool in child growth assessment (Greulich and Pyle, 1950, and others). Even without any of the postcranial skeleton, however, it would still be possible to examine the internal consistency of dental ages in some detail, as in Mann (1975) and Smith (1986). Recently, numerical interpretation has begun to augment description in assessing development of the early hominid dentition; three such approaches are outlined below.

Internal consistency of dental age

(1) Allocation. One approach to choosing between referential models is to design a simple numerical allocation procedure that sorts specimens of living and fossil taxa into the categories "like humans" or "like great apes" (Smith, 1986, 1989b). This is illustrated in Figures 9.3 and 9.4, in which dental development of KNM-WT 15000 is plotted on summary charts of tooth formation for humans and for African apes. These figures illustrate dental development as in Table 9.2, except that substages of tooth formation stages are plotted on charts using an identical set of simple rules for each genus to standardize comparisons (these rules assume that the development of crowns is linear with time and that root development is linear for the first three-quarters of development, the last one-quarter of time being devoted to apical closure; Smith, 1986, 1989b). In each chart, approach to a straight line indicates degree of consistency with the standards of development. Charted on human standards (the top graphs of both figures), KNM-WT 15000 can be seen to show progressive advancement of more posterior teeth. Charted on African-ape standards, it is strikingly

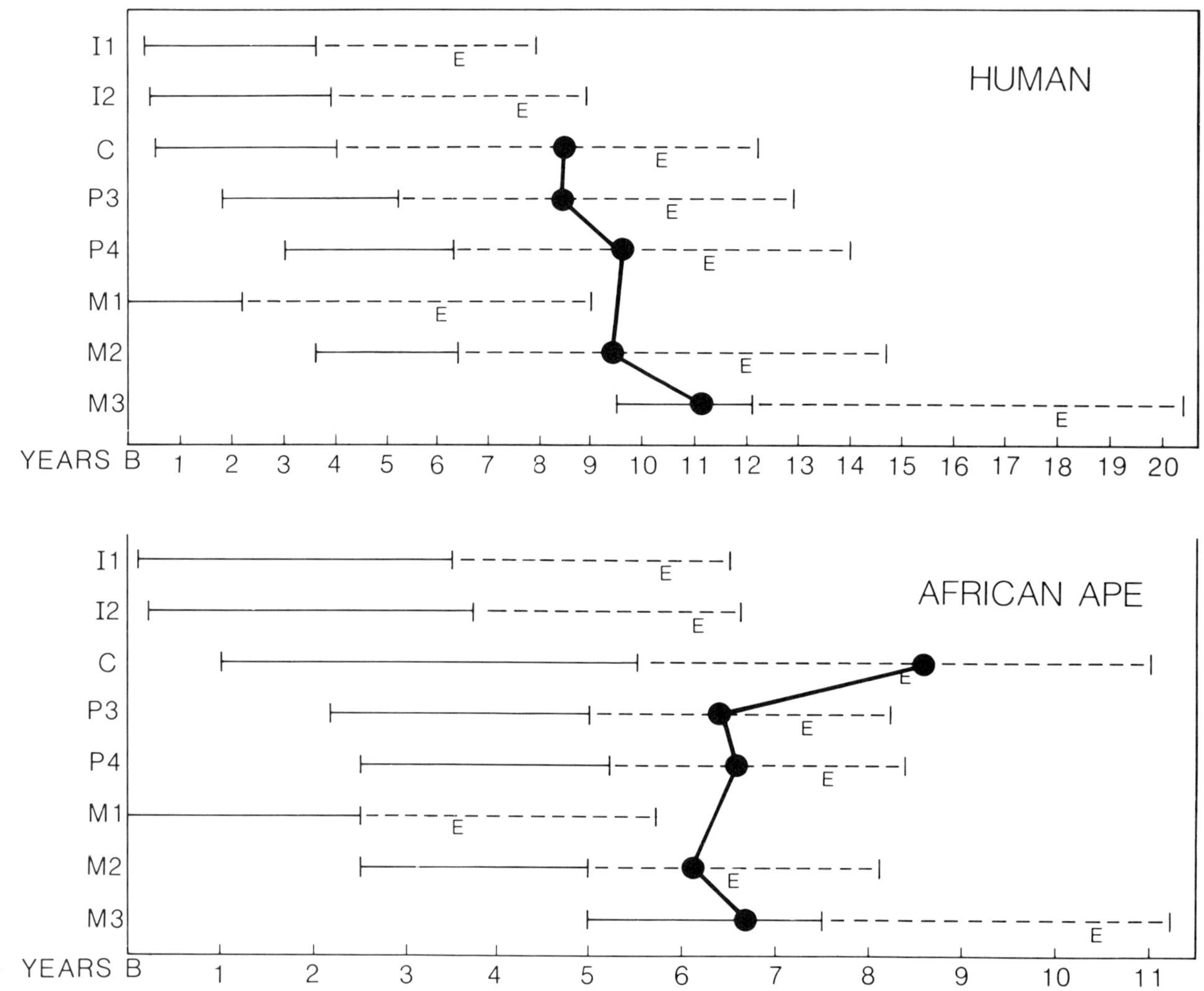

FIGURE 9.3.
Development of mandibular teeth of KNM-WT 15000 charted on standards for humans (top) *and African apes* (bottom); *note that M_3 development is estimated from M^3. Solid lines represent duration of crown formation; dotted lines, root formation; and* E, *emergence of teeth. (Charts modified from Dean and Wood, 1981; data plotted as in Smith, 1989b.) Chronologic scale for human chart denotes first attainment of stage, and the scale should be shifted left for age prediction (Smith, 1991c); this is not critical for analysis of developmental pattern. Points in a vertical line would represent a perfect fit to standards. At top, the mandible of KNM-WT 15000 shows a gradient of advancement in posterior teeth. At bottom, development of cheek teeth appears to fit the pattern for great apes but relative canine development is greatly out of synchrony with other teeth.*

evident that the canine development of KNM-WT 15000 is far too advanced compared with other teeth. In addition, Figure 9.4 compares the maxilla of KNM-WT 15000 to the only available data for *Homo habilis,* the maxillary teeth of the younger juvenile KNM-ER 1590. The *Homo habilis* specimen appears much more out of synchrony with human standards than does early *Homo erectus,* at least suggesting that the two species may differ in developmental pattern.

The degree of fit to the charted standards can be expressed numerically by computing the standard deviation over dental ages given by each tooth (Smith, 1986). If individual teeth tend to agree on dental age (a good fit of subject to standards), the standard deviation of those ages is small. Conversely, widely varying dental ages (poor congruence of subject and standards) result in high standard deviations of dental age. For a single specimen, degree of fit to any developmental standard can be measured by the standard deviation of dental age achieved under that model.

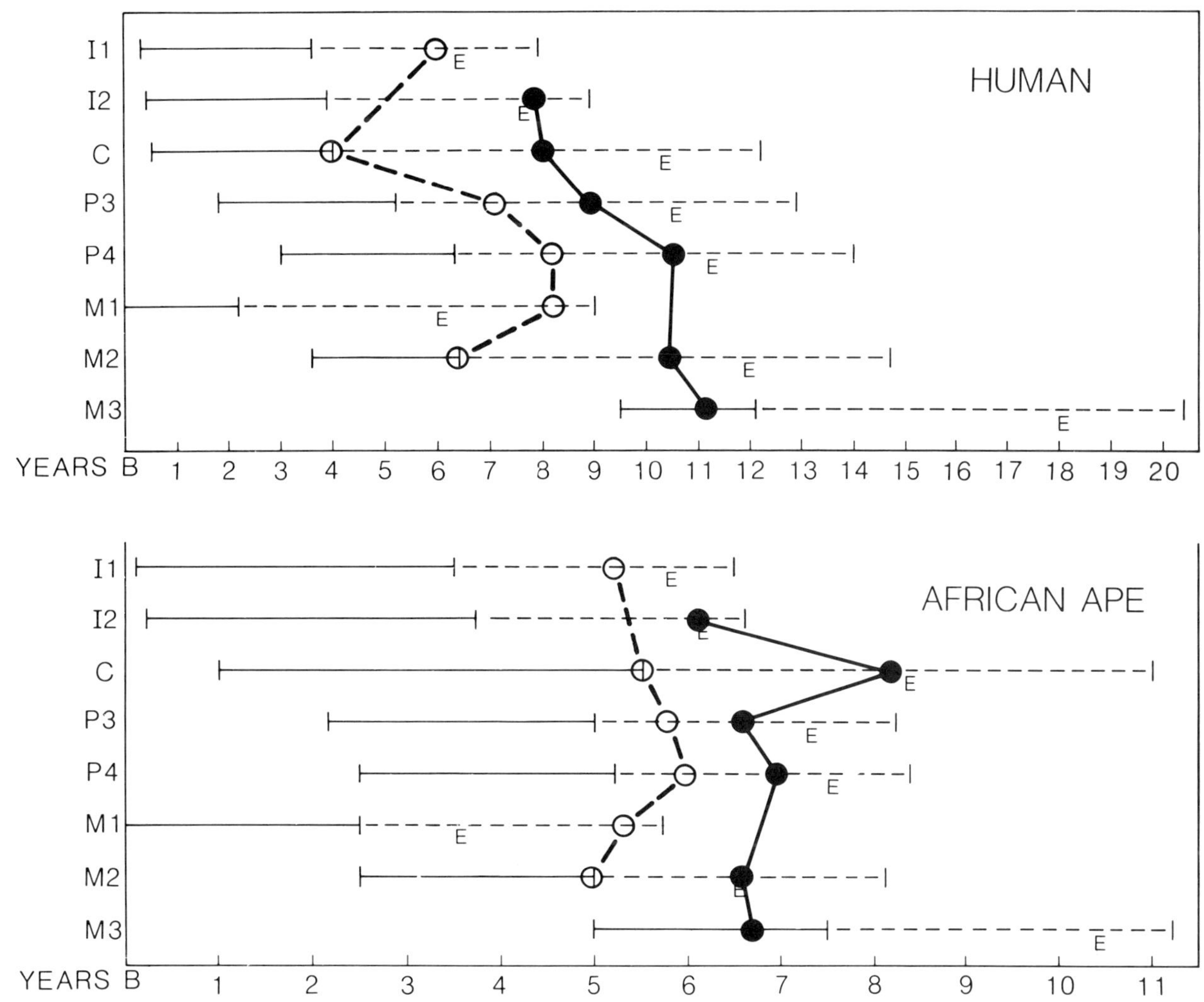

FIGURE 9.4.
Development of maxillary teeth of KNM-WT 15000 (black circles) *charted on standards for humans* (top) *and African apes* (bottom) *and compared with that of KNM-ER 1590* HOMO HABILIS (open circles). *(See Figure 9.3 for definition of symbols.) The development of the maxillary teeth of KNM-WT 15000 echoes the findings for the mandible. The* H. ERECTUS *specimen appears to fit human standards much more easily than does the representative of* H. HABILIS; *KNM-ER 1590 appears significantly more primitive in developmental pattern than KNM-WT 15000.*

This standard deviation is in units of absolute years; it must be adjusted for scale before comparing it to the deviation achieved under another model based on an animal with a different life span. Multiplication of great-ape dental ages by 1.80 equalizes the length of the period of development of the dentition in humans and that in great apes, resulting in an expression of the standard deviation of great-ape dental age in "human years." The best fit to a model is indicated by the lowest standard deviation of scaled dental age.

Small test samples of recent humans and great apes have been shown to be correctly sorted by this procedure (Smith, 1989b). When applied to fossil hominids (Table 9.5), the procedure sorts fossil specimens fairly consistently within each species. Two variants in the calibration of growth of tooth roots have been used previously for great apes (C1 and C2 in Table 9.5). Both variants successfully allocate test samples of *Homo sapiens* and *Pan* to their proper categories (Smith, 1989b), but, for *H. erectus,* small differences in calibration may be more important and both are reported here.

Table 9.5 lists hominid fossils in approximate stratigraphic order. At top, it is evident that more recent hominids usually resemble humans more than apes in developmental pattern. Proceeding down the table to species further removed from modern humans, specimens increasingly tend to resemble apes in developmental pattern. If a more or less continuous series of changes characterized the evolution of hominid dental development, a dichotomous allocation such as this would eventually split a single species sample into two groups. For this dichotomous test, the split point or transition between ape and human developmental patterns appears between *Homo habilis* and *Homo erectus*. It should be noted that the allocation procedure has differing sensitivity to the mandible and maxilla because dental ages are assigned from *mandibular* standards (because maxillary dental development is less well known). Thus, even for an individual human, the maxilla should appear more deviant from this standard than the mandible. The allocation procedure may draw the split point between human and ape patterns differently for mandibles versus maxillae—witness the equivocal result for the maxilla of KNM-WT 15000. The allocation procedure itself is not perfectly consistent within species, although these inconsistencies would greatly lessen if specimens with less than five observed teeth were removed. In any case, overall results in Table 9.5 clearly suggest transition from more ape-like patterns of development to more human-like patterns occurring by the appearance of *Homo erectus*.

(2) Comparison of coefficients of variation in samples. One problem with the allocation procedure above is that the human standards are more accurate for humans than ape standards are for apes (Smith, 1989b). A second analysis can be designed which discards the ape standards and compares all samples against the human standard only. Dental ages are assigned to teeth as in (1) above, based on a human standard (Smith, 1986). The standard deviation of dental ages obtained over teeth from an individual then expresses how much the individual deviates from standards. Division by mean age creates the coefficient of variation *(CV)* of dental age, compensating for the fact that absolute variance in dental age depends on the mean. This allows comparisons to be made between different subjects; note that transformation to *CV* was not necessary in (1) above because the standard deviation was never compared between subjects. The resulting *CV* gives a measure of distance from a human standard, a measure that can be computed for almost any juvenile mammal. This "developmental distance" from humans is analogous to the immunological distance measured by comparing a range of mammals to anti-human sera (e.g., Sarich, 1968). There is one limitation: ideally, *CV* of dental age should be calculated over precisely the same teeth for each individual. In Figure 9.5, *CV* has been calculated for specimens with data for six or seven teeth, with the added requirement that all fields of the dentition are represented; only one incisor or one premolar is allowed to be missing (two older individuals with first molars with completed development were included). This restriction does limit sample size for comparison; for example, both Teshik Tash and the mandible of KNM-WT 15000 are not included here because all incisors have finished formation (the maxilla of KNM-WT 15000, however, is included because one developing incisor was observed). Possible values for *CV*s of human dental age could range from nearly zero (a perfect fit to standards) to 100 or more.

The distribution of attained *CV*s is shown in Figure 9.5 for successive samples of humans, hominid fossils, and extant great apes. Humans shown here are a small but varied lot: four European-derived children, one Inuit, one child of unknown origin (from Mann et al., 1987) and two archeological specimens from the Middle East of several thousand years ago (data from Smith, 1989b). These humans show a range of *CV*s centering about 10, and the single Neandertal is easily subsumed in this distribution. It is evident that the maxilla shows higher deviations than the mandible for each taxon—which is reasonable given that *CV* measures deviation from standards of *mandibular* development. As shown at Figure 9.5 bottom, great apes develop very high *CV*s, between 20 and 40. Gracile australopithecines develop *CV*s that appear as deviant from human standards as those of great apes. The combined distribution for early *Homo* appears to be intermediate between these

TABLE 9.5.

A comparison of fossil hominids with dental age standards of humans and African apes, fit to standards measured by the standard deviation of dental age.

TAXON AND SPECIMEN	JAW	NUMBER OF TEETH OBSERVED	STANDARD DEVIATION OF DENTAL AGES (years)			BEST FIT[b] C1/C2
			HUMAN SCHEDULE	APE SCHEDULE[a]		
				C1	C2	
Homo sapiens neanderthalensis						
Gibraltar	md	7	0.47	1.34	1.39	H/H
Teshik Tash	md	6	1.20	2.02	1.50	H/H
"Archaic" *Homo sapiens*						
Ehringsdorf	md	4[c]	0.83	3.69	2.69	H/H
Homo erectus						
Zhoukoudian B–I[d]	md	7	1.02	1.81	1.36	H/H
KNM-WT 15000	md	5	1.09	2.26	1.75	H/H
KNM-WT 15000	mx	6	1.40	1.63	1.25	H/A
KNM-ER 820	md	6	0.88	1.26	0.87	H/+
KNM-ER 1507	md	4[c]	0.97	0.54	0.54	A/A
Homo habilis						
KNM-ER 1590	mx	6	1.58	0.91	0.68	A/A
Australopithecus africanus						
Sts 24	md	3[c]	1.44	0.10	0.10	A/A
Sts 24	mx	6	1.39	0.99	0.55	A/A
Sts 57	mx	4[c]	0.96	1.45	1.45	H/H
Sts 151	md	6	1.18	0.72	0.46	A/A
Sts 151	mx	6	1.19	0.50	0.26	A/A
Taung	md	7	1.10	1.04	0.86	A/A
Australopithecus afarensis						
LH 2	md	4[c]	0.40	0.76	0.92	A/H
LH 3	md	5	1.05	0.64	0.57	A/A
LH 3	mx	6	1.16	0.93	0.87	A/A
LH 6	mx	4[c]	0.93	0.57	0.94	A/+

Note: md = mandible; mx = maxilla.

a. Multiplying dental ages for great apes by 1.8 equalizes length of the juvenile periods in humans and apes and expresses σ for great apes in "human years." C1 and C2 represent two methods of calibrating root-length growth for apes, as in Smith, 1986 and 1989b, respectively.

b. The lowest standard deviation is judged as evidence of best fit to human (H) or African ape (A) standards of development; "+" indicates a tie. Results vary slightly by choice of root-length calibration for apes (C1/C2).

c. Note that finding is based on a small sample of individual's teeth.

d. Formation stages corrected from Smith, 1989b, after discovery of additional views in Weidenreich, 1936, 1937.

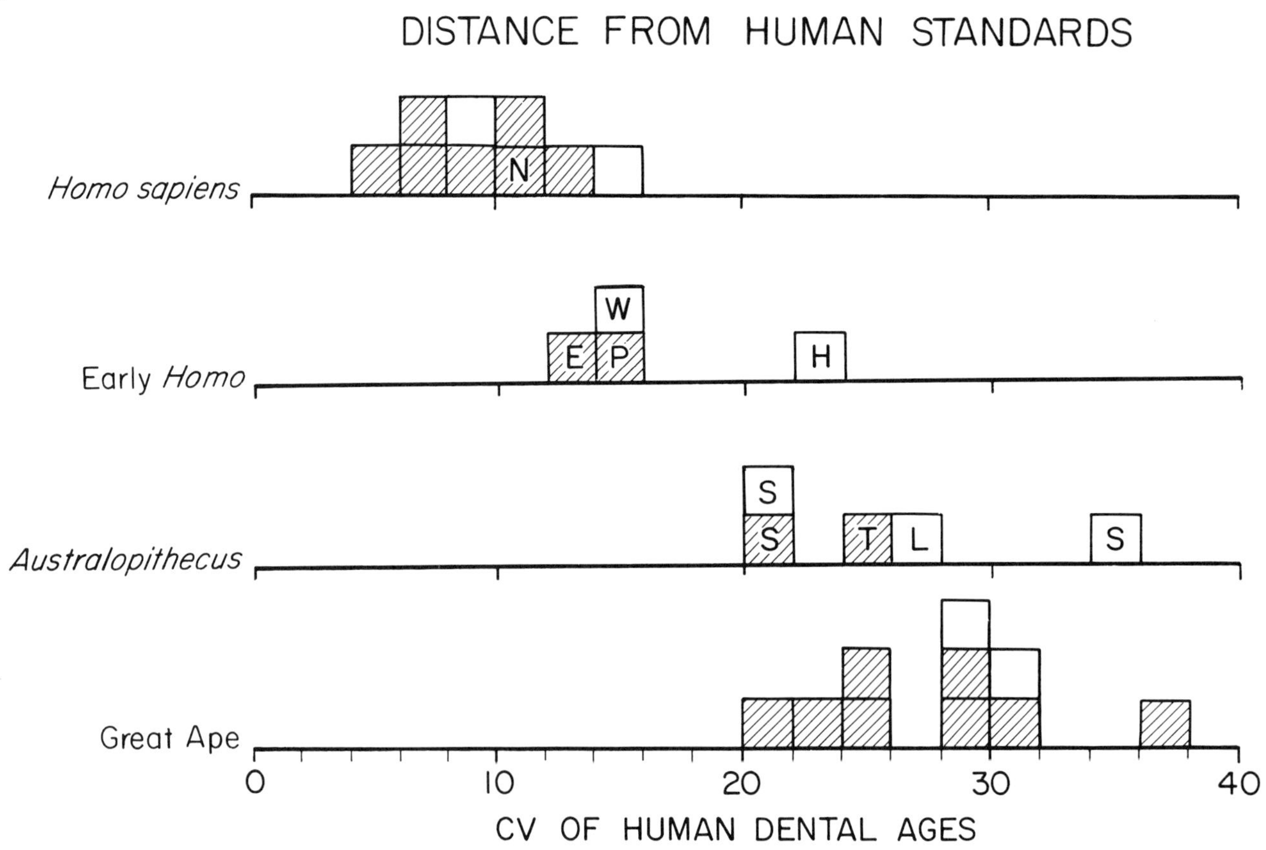

FIGURE 9.5.
Distance from standards of dental development among North American whites for samples of HOMO SAPIENS, *fossil hominids* (lettered boxes), *and extant great apes. Distance measured by the coefficient of variation* (CV) *of dental age over all available teeth in a dental arch. Each box represents a single individual; shaded box = mandible; open box = maxilla.* N *is the Gibraltar 2 Neandertal child; the three* H. ERECTUS *are* W *(KNM-WT 15000),* E *(KNM-ER 1820), and* P *("pekinensis," the B-I mandible from Zhoukoudian); and* H, *KNM-ER 1590, is a* HOMO HABILIS. *Gracile australopithecines* (S, *Sterkfontein;* T, *Taung;* L, *Laetoli) show coefficients in the range of great apes* (PAN *and one specimen of* PONGO). *Fossil specimens of the genus* HOMO *appear to be part of an intermediate distribution showing significant overlap with* HOMO SAPIENS. *(Data from Smith, 1991b.)*

two extremes, overlapping with both preceding and following samples. Taking *Homo erectus* alone, specimens appear to be out of the range of gracile *Australopithecus* and within the range of *Homo sapiens*. Even at these small sample sizes, distributions shown in Figure 9.5 have significantly different medians ($p < .0004$). Pairwise comparisons of medians suggest that *Homo sapiens* differs significantly from all other distributions but that *Pan* and *Australopithecus* do not differ from each other.

(3) Comparison with individuals. Dental development of KNM-WT 15000 can also be compared with individuals rather than with population standards. In Figure 9.6, KNM-WT 15000 is compared with 17 great apes (unpublished data collected from wild-shot specimens in British museums) matched to the same developmental stage of the mandibular second molar. For comparison, a range of values for 17 humans (raw data from Nolla, 1952) is also shown. In several ways KNM-WT 15000 is a puzzle: his posterior teeth show values that are perfectly typical of chimpanzees, but his anterior teeth are out of the chimpanzee range. In relative development of anterior teeth, KNM-WT 15000 touches the human range, although it is

typical for human subjects from both Africa and Europe to show more advanced development of anterior teeth at this stage of M_2 development (Chertkow, 1980).

In Figure 9.6, there is no sign that the Nariokotome specimen reflects any special delay in the development of third molars. In third molar development, KNM-WT 15000 matches the median expected for great apes precisely and shows a more well-developed crown than that typical of humans. This finding appears robust to the choice of human reference population, because substituting in Australian aboriginals as the referent (data from Fanning and Moorrees, 1969) does not effect the range of M_3 development shown in Figure 9.6. It should be remembered, however, that data for third molar development for KNM-WT 15000 are taken from the maxilla only, and this weakens the comparison, although studies of both humans and apes have observed no systematic differences in crown formation of maxillary versus mandibular third molars (Anderson et al., 1976; Dean and Wood, 1981).

Mann (1975) proposed that the development of third molars relative to first and second molars is the key developmental distinction between humans and great apes, and the key to understanding the evolution of growth rates in the Hominidae. Figure 9.6, however, shows that, while humans and apes do differ in relative for-

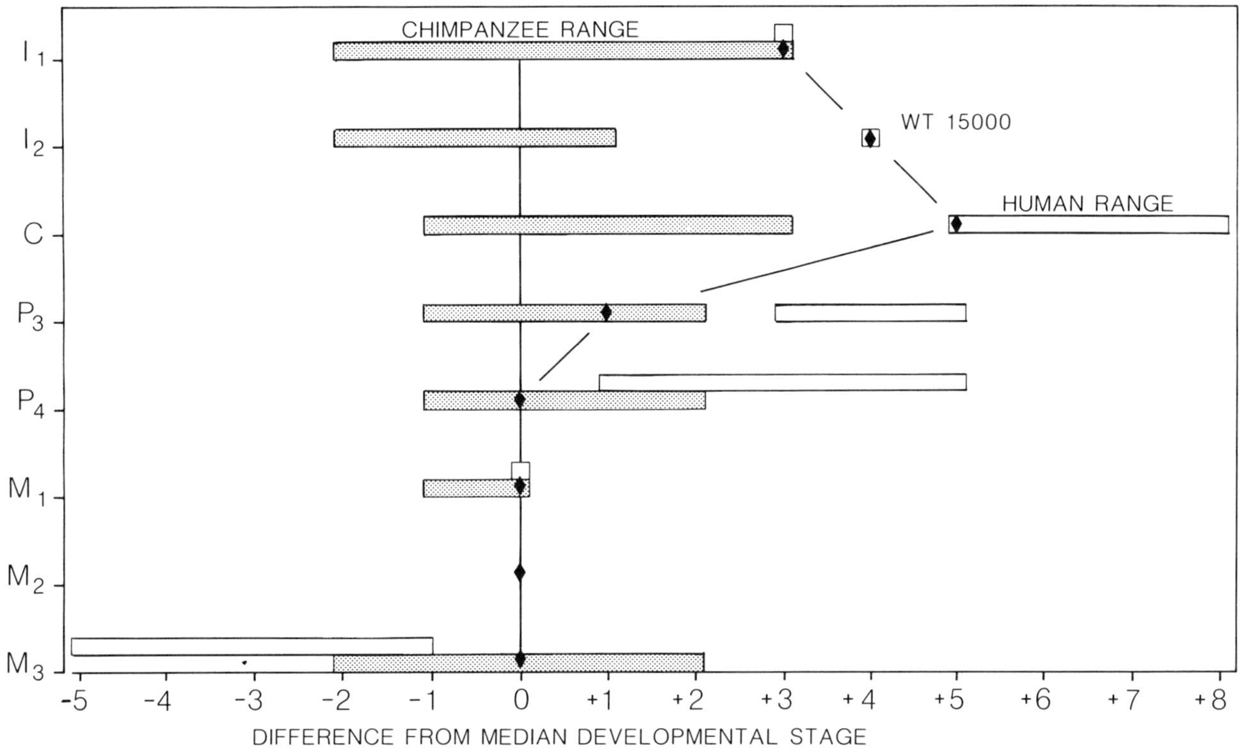

FIGURE 9.6.
Tooth-formation stages of KNM-WT 15000 (black diamond) *compared with those of chimpanzees and humans with equivalent second molar development (stage R½); mandibular teeth shown throughout, except for M3 of KNM-WT 15000. Shaded bars show ranges of stages observed in 17 chimpanzees (Smith, unpublished); white bars give range for 17 humans (raw data from Nolla, 1952, but also in agreement with data in Fanning and Moorrees, 1969). Stages are defined strictly as in Moorrees et al., 1963. Zero indicates the median found in chimpanzees, and all data are plotted by stage of difference from the chimpanzee median. Note that KNM-WT 15000 shows developmental stages for cheek teeth that match expectations for chimpanzees. However, anterior teeth show development more advanced than that found in chimpanzees.*

mation of second and third molars, this is not by any means the largest difference between the two—indeed, there is a more substantial and systematic difference between humans and apes in the formation of anterior relative to posterior teeth. It is in the relationship of these two sets of teeth that KNM-WT 15000 departs from expectations based on great apes by showing more advanced development of anterior teeth, out of the range for chimpanzees and just touching the human range in some cases.

(4) Summary. Given the comparisons made here, a few features of the maturation of KNM-WT 15000 appear unusual relative to a chimpanzee model of development. Specifically, the central incisor, lateral incisor, and canine of KNM-WT 15000 are each substantially advanced in comparison with the development of cheek teeth as observed in chimpanzees. In terms of overall dental age versus skeletal age, however, KNM-WT 15000 appears fairly consistent with a chimpanzee model.

The dental age of KNM-WT 15000 appears slightly more internally consistent with a human developmental pattern than with a chimpanzee pattern, although choice of one referential model over another depends on specifics of the decision rule. On one hand, Mann's (1975) view, that the relative development of cheek teeth reflects actual rate of growth, would see KNM-WT 15000 as very like a chimpanzee. On the other hand, a count of the total stages of dental maturation (Figure 9.6) that separate KNM-WT 15000 from the range of extant species makes KNM-WT 15000 appear closer to humans than to apes (by a hair), and allocation and *CV* analyses—numbers (1) and (2), above—definitely emphasize his resemblance to humans. On a human scale, KNM-WT 15000 has a dental age of about 11 years, a skeletal age of 13–13.5 years, and a stature age of about 15 years. It is intriguing that skeletal maturation and body size tend to agree with each other more than either does with dental age.

All in all, neither referential model is completely satisfactory. Because neither fits really well, choice of one model over another requires that we tabulate similarities, weight characters, or discriminate characters as significant or discardable. Taken together, analyses of the dentition (allocation, coefficients of variation, and comparison with individuals) are more easily interpreted to mean that KNM-WT 15000 occupies an intermediate position with respect to humans and African apes, matching developmental patterns of neither species precisely. It seems clear, for example, that the pattern of dental development seen in KNM-WT 15000 and in other *Homo erectus* differs in a human direction from great apes and gracile *Australopithecus,* although it may not be fully modern in the relationship of anterior to posterior teeth.

Conceptual models

It is possible to break through the limitations imposed by referential models by considering conceptual models of growth. Sacher (1959) first noted a strong relationship between life span and brain weight in living mammals. Subsequent research has found that correlations with brain weight extend to almost every conceivable measure of species life history, suggesting that maturation rate as a whole is tied to brain size (Sacher and Staffeldt, 1974; Sacher, 1978; Harvey and Clutton-Brock, 1985; Smith, 1989a). Primates offer some of the strongest evidence on these points (Sacher, 1975; Harvey and Clutton-Brock, 1985; Smith; 1989a).

It can be shown in a series of primates that age of emergence of permanent teeth is an accurate and stable marker of life history across primate species, and that age of emergence of the first molar correlates at $r = 0.98$ with species brain weight for 21 primate species (Smith, 1989a). Since brain and body weight can be estimated for extinct species in the fossil record, it is possible to predict life history parameters based on equations developed from living primates. Predictions for age of tooth emergence and life span in early *Homo erectus* describe a species with a life history virtually intermediate between that of living great apes and humans (Smith, 1989b, 1991b). Early *Homo erectus,* with a mean brain size for the species estimated at 810 cc, is predicted to erupt M1 at 4.5 years and potentially live some 15 years longer than *Pan.* In almost all parameters, this turns out to be a life history almost exactly in between that of humans and great apes. With such an intermediate life history, it would be expected that M2s would emerge at about 9 years of age, and the age of KNM-WT 15000 using this conceptual model would be in the range of 9–10 years.

For the purposes of argument, Figure 9.7 outlines an explicit conceptual model of development for early *Homo erectus.* This model has been constructed by: (1) incorporating elements from Figure 9.2 that virtually must appear at approximately the relative time shown because timing of these events is broadly shared by catarrhines; (2) averaging the relative timing of events from human and chimpanzee models; and (3) incorporating events whose relative order is determined by KNM-WT 15000 himself (order of emergence of canines and status of epiphyses). The time scale and the rationale for averaging over human and chimpanzee life cycles are both based on studies of brain and body weight in living primates that predict that early *H. erectus* possessed a life history halfway between these two living species (Smith, 1991b). The model itself is similar to one in Smith (1991b), although it is much more detailed; its construction differs from that only in setting early adulthood in modern humans at 18 years rather than 20. The time scale itself may seem the most provocative part of the model, but it actually proposes a pace of growth for *Homo erectus* that differs from modern humans by a factor of only 1.50.

The main purpose of making an explicit model, as in Figure 9.7, is to make predictions that can be tested. Any juvenile specimen of early *Homo erectus* can serve to test or refine the sequence of events shown in the model. The time scale can be tested or adjusted as more data become available from juveniles aged on the basis of incremental lines of growth.

A prediction about the adolescent growth spurt is perhaps the most speculative aspect of the model. Figure 9.7 leaves out "peak height velocity" (PHV in Figure 9.2), implying that the adolescent growth spurt is not present in *Homo erectus.* A true conceptual model of attainment of adult size in early *Homo erectus* requires a rationale about the significance of the adolescent growth spurt: What are its correlates? What is its function? Is there reason to expect it to be present more than a million years ago? At present, the idea that the growth spurt might not exist in fully modern form in early hominids is based on an emerging concept

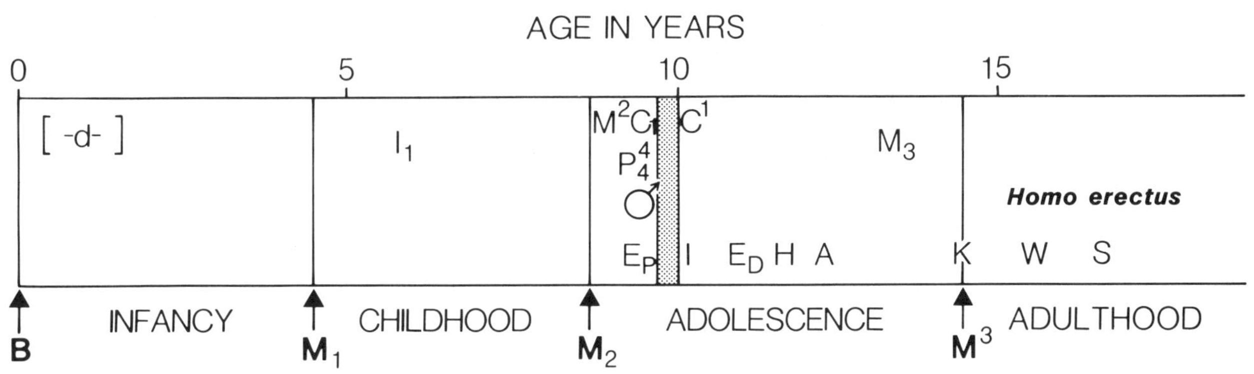

FIGURE 9.7.

A conceptual model of life history proposed for early HOMO ERECTUS; *symbols and abbreviations as in Figure 9.2. Shaded bar shows position of KNM-WT 15000. This proposal involves less speculation than one might think because life history has a substantial inertia across higher primates; many events could hardly be far out of place. Sequences of other events are determined by KNM-WT 15000 (cheek teeth, some epiphyses) or are based on the premises that brain and body size are excellent predictors of life-history parameters and that early* HOMO ERECTUS *had a life history intermediate between that of living* HOMO *and* PAN. *According to this model (compare with Figure 9.2), a* HOMO ERECTUS *youth of 14.5 years was entering young adulthood and was comparable in maturation to a 5.8-year-old macaque, an 11.4-year-old chimpanzee, and an 18-year-old modern human.*

of the adolescent growth spurt, and also on the small amount of information on growth in size provided by KNM-WT 15000 himself.

THE PROBLEM OF THE ADOLESCENT GROWTH SPURT

It is well known to paleontologists that human brain size, bipedal locomotion, and life span are unique among primates, but less well known that the human growth curve is also unique. In humans, adolescence is characterized by a substantial increase in the velocity of growth that occurs characteristically around the age of 12 years in the female and 14 years in the male. The adolescent growth spurt affects the axial and appendicular skeletons as well as bones of the face and skull (Tanner, 1962; Brown et al., 1971). This growth spurt has been observed in virtually all human groups, although prolonged undernutrition may delay the growth spurt or alter its amplitude (Tanner, 1962). In well-nourished U.S. males, peak velocities in adolescent growth are attained after age 12 and typically near age 14 (Tanner, 1962); in nonindustrialized populations the spurt tends to occur later, for example, by some two years in the case of the Turkana (Little and Johnson, 1987).

The evolutionary history of the adolescent growth spurt becomes important in the case of KNM-WT 15000 because he has the dental age of a juvenile who has yet to experience the adolescent growth spurt (Chertkow, 1980) but the stature and bone age of a juvenile who is well into it (Roberts, 1960; Falkner and Tanner, 1986; Little and Johnson, 1987). Comparing KNM-WT 15000 with studies of dental versus skeletal maturation emphasizes the disparity of the two systems. A study of South African black and white children (Chertkow, 1980) found that skeletal signs of the onset of the growth spurt (based on hand-wrist maturation) coincide with later stages of dental maturation than are shown by KNM-WT 15000. Thus, the human growth spurt typically comes at a stage of dental development later than that shown by KNM-WT 15000. A study of Finnish children (Leinonen et al., 1972) found a fairly small range in skeletal ages, from 11–11.5 years, in boys with dental maturation scores equal to that of KNM-WT 15000 (a score of 90 by the authors' technique). Comparing KNM-WT 15000 to non-white standards should only increase the disparity between bone and skeletal age, because, as Chertkow's (1980) data suggest, black children probably have faster dental maturation but slower skeletal maturation than white children. Thus, KNM-WT 15000 doesn't slot easily into an average position on charts of human development, although he would be accommodated at some percentile.

Laird (1967) and Bogin (1988, 1990) show that growth curves of humans cannot be fit by the same general mathematical functions that are used for other mammals. They argue that humans have inserted an extra phase into the developmental period between weaning and sexual maturation, a phase that serves to prolong childhood and to delay achievement of adult size. Bogin (1990) argues that the selective advantage underlying this pattern of growth is to extend the amount of time the child can learn from and be protected by adults. In addition, the pattern by which an adolescent reaches adult size influences the manner in which he enters into adult social roles. Rapid growth in stature might also be a social signal of approaching adulthood comparable to emergence of adult canines in nonhuman primates. The growth spurt, like the extended length of the juvenile period and long life in humans, might be ultimately related to enlarged brain size, increased intelligence, and the long periods of training necessary to make use of such a brain in learning a complex informational and behavioral repertoire (Smith, 1990), and it is by no means clear that the adolescent growth spurt particularly, or the human growth curve generally, should be fully developed at a modern human level in earliest *Homo erectus*.

Nonhuman primates appear to show simple decays in growth velocity throughout adolescence (see Bogin, 1988). There appears to be no distinct increase in adolescent growth velocity in chimpanzees that is comparable to the human spurt (Gavan, 1953; Gavan and Swindler, 1966). Recently Watts and Gavan (1982) reanalyzed Gavan's data on Yerkes chimpanzees, suggesting that there is a very small growth spurt. However, Watts (1986b, p. 157) locates it at 11.0 years of age in male chimpanzees, an age which indicates young adulthood rather than adolescence. Such a late age is not easily reconciled with Kerley's (1966) dissections of Yerkes chimpanzees (including some animals from Gavan's original study), which found growth in major long bones to ". . . level off at about ten years of age and there is little additional growth in length after age twelve" (1966, p. 74). More recently, Tanner et al. (1990) place a suggested growth spurt in tibial length of female rhesus macaques at a more reasonable age (about 2 years in females), but individual growth curves illustrate multiple dips and peaks in growth rate velocity rather than a single, clearly defined one. In agreement with Bogin (1988, pers. comm.), neither case makes a clear and convincing homology with the human adolescent growth spurt.

The point that humans and chimpanzees grow up on different trajectories does not depend on the insistence that only humans have a true adolescent growth spurt. Gavan's (1953) studies of chimpanzees showed that chimpanzees and humans attain their adult size in different ways. Gavan (1953) found that, when both taxa are scaled proportionately, adolescent chimpanzees have attained a higher percentage of their adult body size than humans matched for relative age. Contrary to chimpanzees, *humans reserve a significant proportion of their growth until after the onset of adolescence.*

Table 9.6 illustrates this point by comparing remaining growth for adolescent chimpanzees and humans at homologous dental ages, at age 7 years for chimpanzees and age 11 years for humans. These ages can be regarded as homologous for several reasons: they are 61–62% relative age, they are appropriate markers of earliest adolescence for each genus, and they are the tooth formation ages given to KNM-WT 15000 from chimpanzee and human models. In Table 9.6, the growth achieved by an early adolescent is expressed as the percentage an adolescent can expect to increase in size in adulthood relative to his present size. Young adolescent captive chimpanzees and well-nourished humans from Ohio will almost (but not

TABLE 9.6.

Percent an adolescent male is expected to increase in size relative to his present size for 7-year-old chimpanzees and 11-year-old humans.

PARAMETER	YERKES CHIMPANZEES (Gavan, 1953)	HUMANS: OHIO WHITES (Gavan, 1953)	HUMANS: AUSTRALIAN ABORIGINALS (Abbie, 1967)	HUMANS: TURKANA ABORIGINALS (Little et al., 1983)
Weight				
Body weight[a]	88	79	103	125
Body segment lengths				
Hip breadth[b]	25	28	—	—
Trunk length	14	18	—	—
Thigh length	17	31	—	—
Knee height	18	22	—	—
Leg length	13	20	—	—
Stature or approximation to stature				
Trunk + leg + thigh	14	23	—	—
Stature	—	23[c]	23	31

a. Expressed as percent of adult weight achieved at age 7 *(Pan)* or 11 *(Homo)*, these are (left to right): 53, 56, 51, and 44.

b. Bi-trochanteric width; body segment measurements as defined in Gavan, 1953.

c. Stature growth estimated from Ohio white males from the Fels Institute longitudinal study (Garn, 1966); even tall boys with tall parents show a 22% increase. Body segment measures are from the Brush Ohio studies.

quite) double their body weight by young adulthood. Leaner humans, represented here by Australian aboriginals and Turkana, would more than double their adolescent weight by adulthood. Stature comparisons are slightly complicated because stature is not measured for nonhuman primates. Body segment measures can be used to approximate stature because it can be shown for humans that proportionate growth in trunk + leg + thigh length is comparable to that for stature. Remarkably, both Ohio whites and Australian aboriginals share a predicted 23% increase. Available studies of the Turkana (pastoralists who presently occupy the region of northwest Kenya where KNM-WT 15000 was found) describe slow growth over a long time period; the stature of an 11-year-old is expected to increase by nearly a third by adulthood. However, the extreme figure for the Turkana might be modified when longer-term records are completed on these subjects (Little and Johnson, 1987, p. 699), and the 23% figure is probably more reasonable. Overall, a young chimpanzee adolescent would be expected to increase in "stature" by some 14%, whereas a comparable human adolescent would be expected to increase some 23% or more. Thus, at early adolescence, chimpanzees have completed more of their adult growth than have humans.

Table 9.6 may not offer the best method to use to predict adult stature for KNM-WT 15000 (see Ruff and Walker, Chapter 10, this volume), because it is based on simple averages of stature per age group and because it extrapolates from a strictly estimated dental age, whereas stature is often predicted for human children on the basis of skeletal age rather than even on chronological age (see Preece, 1988). The example is included here to make the more general point that humans and chimpanzees grow up on different trajectories. One explanation for the advanced body size and skeletal age of KNM-WT 15000 relative to his dental age is that he followed a more chimpanzee-like pattern in the attainment of body size. In other words, members of *Homo erectus* may have attained much more of their adult size by early adolescence than modern humans typically do because the species had not yet evolved the distinctive human pattern of suppressing growth in late childhood and subsequently "catching up" through the human adolescent growth spurt.

Summary and Conclusions

Several points about the life history of KNM-WT 15000 can be made with some confidence because higher primates share a basic plan of life history (Table 9.7). Comparing KNM-WT 15000 with a broad model of catarrhine development indicates that it is more parsimonious (i.e., it gives rise to fewer discrepancies) to gauge his relative age by maturation of the cheek teeth than by canine development or epiphyseal fusion. Eruption of the cheek teeth of KNM-WT 15000 describes an early adolescent who lived about two-thirds of the length of the juvenile period of his species and died substantially before the final fusion of epiphyses and cessation of growth. Very fine-scaled assessment of formation stages of all teeth suggest a slightly younger relative age (63%), but the two estimates are certainly quite close. At this approximate relative age, male higher primates (macaques, chimpanzees, and humans) generally have

TABLE 9.7.

Inferences about the life history of KNM-WT 15000 that are robust over a broad range of living referents *(A)* and those that depend on many more assumptions about growth and development of early *Homo erectus (B).*

A. Broad referential model based on higher primates
   1. Died at early adolescence; survived about two-thirds of the length of the total juvenile period in his species
   2. Died substantially before the completion of the growth period, before reaching adult size and proportion of axial and appendicular skeletons, including face and skull
   3. Died after initiating but before completing physical and behavioral sexual maturation

B. Narrow referential or conceptual models required
   4. Absolute chronological age at death: 7, 9, or 11 years?
   5. Projected stature growth: 14% or 23%?

begun puberty, as measured by a rise in serum testosterone; this rise is generally correlated with altered social behavior in all these species (Martin et al., 1977; Rose et al., 1978; Kraemer et al., 1982; Bogin, 1988).

As Table 9.7 points out, answers to the questions "Precisely how old was he in years at his death?" and "How tall would he have been if he had lived to adulthood?" require many more assumptions about growth and development in early *Homo erectus* because living higher primates differ in the time scale on which they grow up and in their manner of attaining adult size. A modern human referent projects many requirements onto the growth of *Homo erectus* because modern human growth standards are enormously affected by the presence of rapid and substantial adolescent growth.

In-depth consideration of tooth maturation of KNM-WT 15000 suggests a pattern of events in *Homo erectus* that is slightly more like humans than like great apes, but one that is probably not fully like extant humans. The dentition of KNM-WT 15000 retains some primitive characteristics in dental development, such as the emergence sequence M^2C^1, advanced development of posterior teeth relative to anterior teeth, and advanced development of third molars relative to second.

In light of Adolph Schultz's work (1960), findings for the teeth of KMN-WT 15000 reflect an important shift towards the human condition. Schultz viewed relative development of anterior and posterior teeth in primates as a response to postnatal growth rate. He noted that higher primates show an increasing disparity in development of anterior and posterior teeth across the *scala naturae,* a disparity that culminates in humans. He explained this as a product of slowed postnatal growth, both overall growth and growth of face and jaws particularly. The dental development of KNM-WT 15000 fits Schultz's (1960, pp. 12–13) explanation remarkably well, fulfilling what he would predict as a developmental pattern typifying a growth rate intermediate to that of great apes and modern humans.

Relative to dental age, the skeletal maturation of KNM-WT 15000 is advanced by human standards, retarded by macaque standards, and about as expected by chimpanzee standards. If we must choose between dental and skeletal systems to judge chronological age (even relative chronological age), the dentition is usually regarded the better estimator because of its greater robusticity to environmental effects (Lewis and Garn, 1960; Garn et al., 1965a,b; Niswander and Sujaku, 1965; Murchison et al., 1988; Tonge and McCance, 1973). However, skeletal age may be regarded as superior to chronological age in assessing maturational status of a living child for some purposes or in predicting his future growth (e.g., Acheson, 1966; Preece, 1988; Helm, 1990).

Results of this investigation can be taken in two ways. If one is convinced that *Homo sapiens* is the only appropriate model for growth and development of *Homo erectus,* then KNM-WT 15000 was circa 11 years of age at death, with a skeletal age of 13–13.5 years and a stature more appropriate to a 15-year-old. If the sex assignment of KNM-WT 15000 were changed to female, the relative value of these ages would not improve in consistency: dental age would reduce to about 10.5 years, skeletal age to 11–12 years, and stature age would remain at about 15 years. KNM-WT 15000 displays several features of maturation that are probably primitive, in the relative timing of events within the dentition and in early fusion of the distal humerus relative to dental maturation. Compared with modern human children, KNM-WT 15000 appears unusual in development of dentition relative to size and maturation of skeleton. He may fit in a distribution of living humans at some percentile, but present data suggest he cannot be made to appear ordinary.

On the other hand, the somewhat uneasy fit of KNM-WT 15000 to human growth standards may reflect the fact that *Homo erectus* did not grow up precisely in the modern human manner, and conceptual models of growth and aging should be explored. A conceptual model that links primate life history to brain size (Sacher, 1975; Smith, 1989a,b, 1991b) predicts that early *Homo erectus* life cycle and life span were intermediate to those of extant apes and humans. A comparable conceptual model of attainment of adult size is not yet fully articulated; however, Bogin (1988, 1990) suggests that the distinctive human growth curve also evolved in response to demands of increasing intellectual capacity and behavioral complexity. The unique skeleton of KNM-WT 15000 stands at a point near the very beginning of the great increases in brain size in the human lineage; this fossil makes a substantial contribution towards the recovery of the form and pattern of the evolution of human life history from the fossil record.

Acknowledgments

I thank A. Walker and R. E. F. Leakey for inviting me to study the dental development of KNM-WT 15000. For permission to study comparative material included here, I thank the governments of Kenya and Tanzania, R. E. F. and M. G. Leakey at the Kenya National Museum; P. V. Tobias and R. J. Clarke at the University of the Witwatersrand; C. K. Brain, E. Vrba, and A. Turner at the Transvaal Museum; P. Andrews and C. Stringer at the British Museum (Natural History); M. C. Dean at University College London; C. F. Merbs at Arizona State University; D. R. Pilbeam at Harvard University; and the Primate Foundation of Arizona. I thank C. Ruff and A. Walker for discussions on KNM-WT 15000, and Tracey Crummett for comments on the manuscript. Bonnie Miljour drew Figures 9.1, 9.2, 9.6, and 9.7; Teryl Lynn drew Figure 9.5. This research is supported by grants BNS-8717738 and BNS-9020974 from the U.S. National Science Foundation.

CHAPTER 10

THE RECONSTRUCTION OF THE PELVIS

ALAN WALKER AND CHRISTOPHER B. RUFF

The bony pelvis is of major importance for body support and locomotion. It also encloses the birth canal. The demands of locomotion and birth on the human pelvis conflict with each other. Though the heads of human neonates are large, much brain growth is postponed until after birth. This means that human infants require more care for longer periods of time than other primate infants. The availability of the pelvis of KNM-WT 15000 allows us to ask whether early Homo erectus *had already developed this uniquely human trait.*

The reconstruction of the pelvis of KNM-WT 15000 involved not only the placing together of the two pelvic bones with the sacrum but also the reconstruction of all three pelvic elements. As the individual was a juvenile, the individual pelvic and sacral components were still separated. Adding to this difficulty, there was damage due to trampling before fossilization. Descriptions of the individual pieces as they were preserved can be found in Walker and Leakey (Chapter 7, this volume).

Neutral-colored casts of the fossils were prepared and joined together with a mixture of hot sculptors' paraffin wax and plaster of paris. This mixture has the advantage of being pourable when hot and relatively strong and stable when cold. Parts known only for one side could be modeled in mirror image for the other. These and missing parts were modeled with the same material, following existing contours and, if necessary, using the early hominid pelvic bone KNM-ER 3228 (Rose, 1984) and a modern human pelvis as a guide. Final details of the reconstruction were carved in the wax.

The bones

Left hip bone

(Figure 10.1)

The left hip bone is made of five pieces. These are the main part of the ilium (N), a piece of the iliac crest near the anterior superior iliac spine (BG) and another near the posterior superior iliac spine (BF), the superior part of the ischium (Q), and a small part of the pubis from near the anterior border of the obturator foramen (AX). The postioning of the most anterior of the two smaller iliac fragments was easier than the posterior. Although bone is missing from the lateral aspect of the piece, there is a contact between the inner surface and the main part of the ilium and there is nearly contact between the adjacent epiphyseal surfaces for the anterior iliac spine. Piece BG shows the junction of the epiphyseal surfaces for the anterior superior iliac spine and the iliac crest. The posterior piece was positioned by following the contour of the anterior iliac crest and by matching the inflection point where the iliac crest turns towards the posterior superior spine with the inferior part of the same inflexion, which can be seen just above the auricular surface on the main piece. The ischial piece has a cracked and displaced spine region, but it can be placed relative to the main iliac part by aligning the (abraded) acetabular surface on the ilium with that on the ischium and by following the contours of the internal surface. The thickness of the Y-shaped epiphyseal cartilage was estimated by positioning the two parts of the acetabulum on the femoral head while keeping the internal surfaces congruent. The placing of the final piece was possible because of the recognition that this small piece of pubis had part of the obturator groove on it. The ischial fragment also showed the superior border of the obturator foramen and the two pieces could be positioned so that the margins of the foramen ran smoothly into one another. The anterior part of the pubic ramus and the ischiopubic ramus was also modeled by referring to these obturator foramen margins.

Right hip bone

(Figure 10.2)

The right hip bone is made up of three pieces. These are the main iliac part (O), the right ischium (P), and a small part of the pubis (AW). A piece of the upper edge of the iliac blade had been broken before fossilization and was crushed against the main piece. This was cut off from the main piece in the reconstruction and repositioned as it would have been in life. The irregular superior edges of the piece reach nearly as far as the iliac crest. There is some crushing of the ischial spine region, but this does not affect the positioning of the ilium and ischium very much. Fitting the two main bones together is relatively easy because the acetabular parts of these bones are reasonably well preserved. The small pubic piece was placed in the same way as the piece on the left, by using the orientation of the obturator groove. This piece has more of the internal pubic surface preserved and this serves to provide a minimum length for the pubic ramus. This inner surface is also concave

FIGURE 10.1.
Lateral view of reconstructed left hip bone of KNM-WT 15000.

mediolaterally and the curvature was used to help determine the internal pubic contour. Both pubic rami and symphyseal regions were left unreconstructed until the sacrum and the two pelvic bones could be articulated.

COMPARISONS

The only well-preserved early *H. erectus* with which a comparison is possible is OH 28 (Day, 1971), but an early *Homo* hip bone from Koobi Fora (Rose, 1984) also makes an instructive comparison. Both these bones have acetabula about the same size as that in the reconstruction. The other fragments of hip bone from early *H. erectus* are listed in Chapter 7, this volume.

OH 28 is a left adult female hip bone. The iliac crest, the pubic ramus, and the ischiopubic ramus are missing. There is some distortion and there are tooth marks and splintering due to chewing by a carnivore. The major differences between this and the reconstructed hip bones of KNM-WT 15000 are these. (1) In the Olduvai fossil the greater sciatic notch is much more open and the superior limb of it is about twice as long. (2) The lateral buttress from the acetabulum to the iliac crest is extremely well developed in OH 28, and only feebly so in the juvenile. (3) The iliac tuberosity is much more well developed in the adult. (4) The ischial tuberosity is more developed and medially rotated in the Olduvai fossil. All these features can be related to two differences between the bones—the sex and the degree of maturity.

KNM-ER 3228 makes a similar comparison with KNM-WT 15000, except in this case it is also a male. Here the differences one can find are only of degree of development of buttresses and ligament attachments.

SACRUM

(FIGURES 10.3–10.5)

The sacrum was reconstructed from the following parts:

(AD) A little more than the left half of the first sacral vertebra and the superior right articular process, which is preserved separately.

(BB) The laminae and crest of the second sacral vertebra.

(BC) The right half of the third sacral vertebra.

(AE) The fourth sacral vertebra.

(AF) The fifth sacral vertebra.

It should be pointed out that these identifications are not quite as they were in the original note on the specimen (Brown et al., 1985). In that first paper we recorded them as the first, third, fourth, and fifth vertebrae and had AF as the first coccygeal vertebra. This was due to unfamiliarity with the details of variation in unfused sacra in apes and humans, since unfused sacra are frequently left in a ligamentous preparation in museum collections.

The reconstruction of the sacrum was carried out by first making mirror images of the parts preserved only on one side. In this way most of the first and third vertebrae could be reconstructed without recourse to imaginative modeling.

The parts of the first vertebra that are missing after this are laminae and spine, the inferior part of the body (this was crushed), and the thin edges of the alae. A good part of the auricular surface is preserved, however. The outline of the upper surface of the body was checked against the inferior surface of the undistorted body of the last lumbar vertebra. No attempt was made to model the laminae and spine of this vertebra.

The second vertebra was missing everything but the laminae and spine and so reconstruction involved the making of a body and lateral elements. The body was based on the widths and heights of the first and third bodies. The lateral elements were built up by examining the auricular surfaces of the pelvic bones and knowing that the lateral elements of the third just reach the inferior margins of the auricular surfaces without articulating with them. The widths were also checked by examining the fit of the laminae and spinous processes with those of the second vertebra above. These parts fit together, with the inferior ones nested inside the V of the superior ones. The width of the third vertebra itself was established by checking this relationship.

The third vertebra had its right lateral element, a small part of the body and lamina and

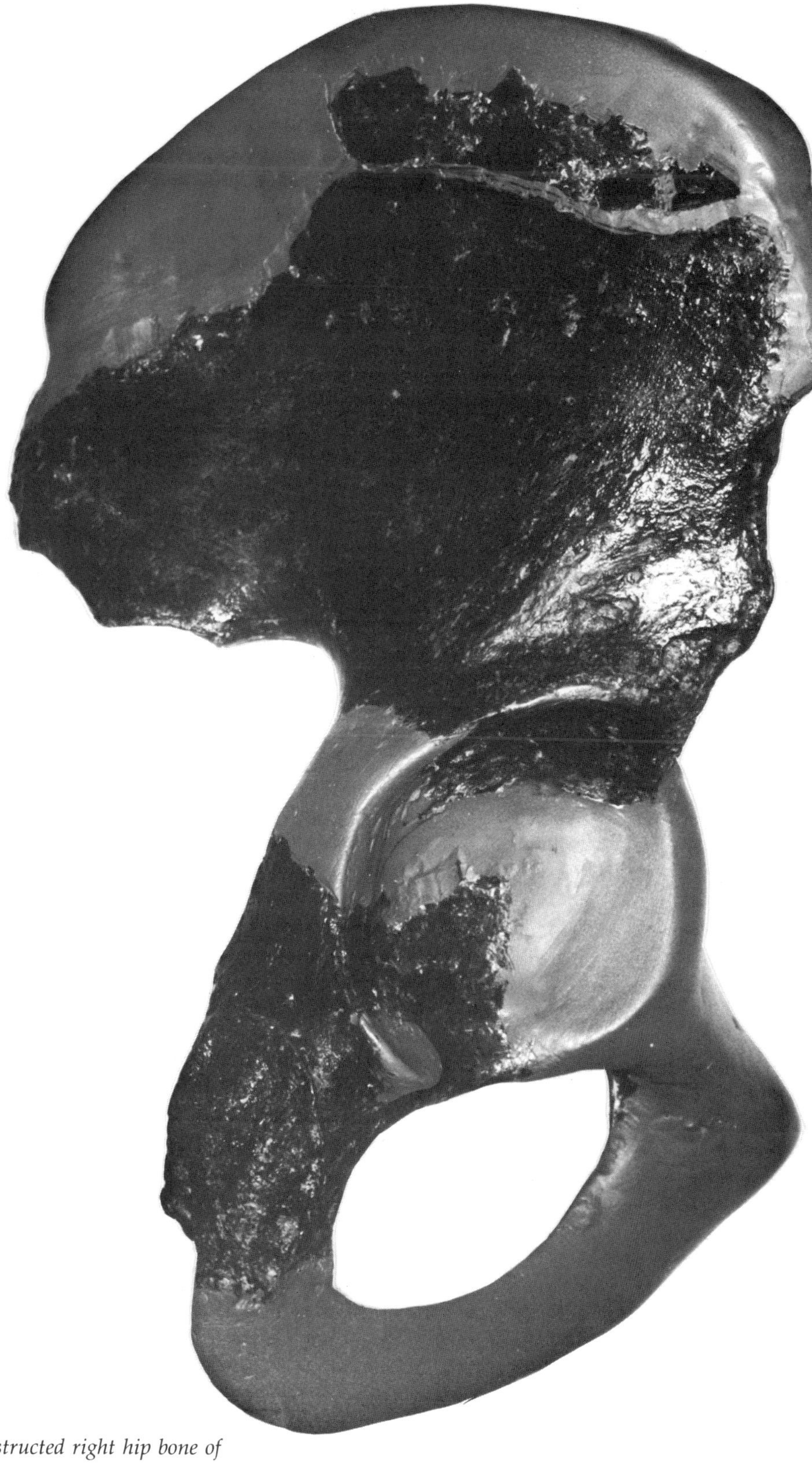

FIGURE 10.2.
Lateral view of reconstructed right hip bone of KNM-WT 15000.

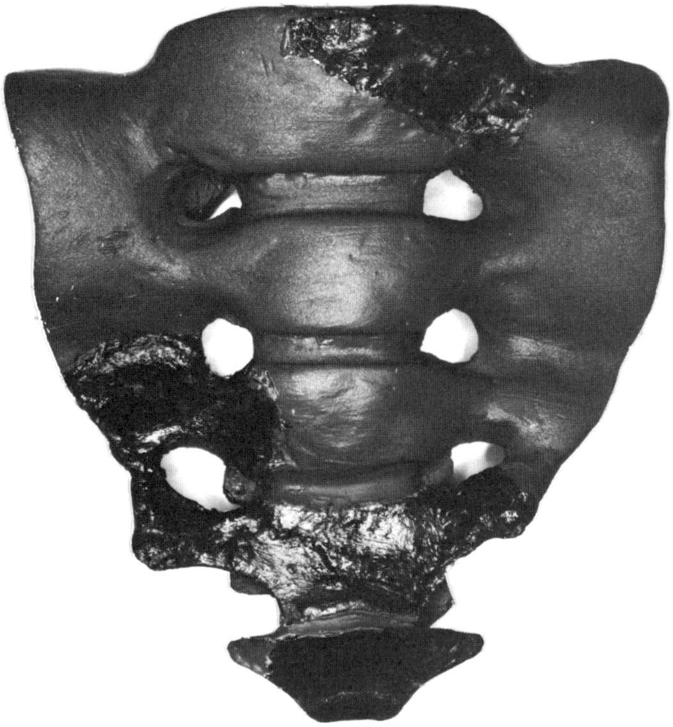

FIGURE 10.3.
Anterior view of reconstructed sacrum of KNM-WT 15000.

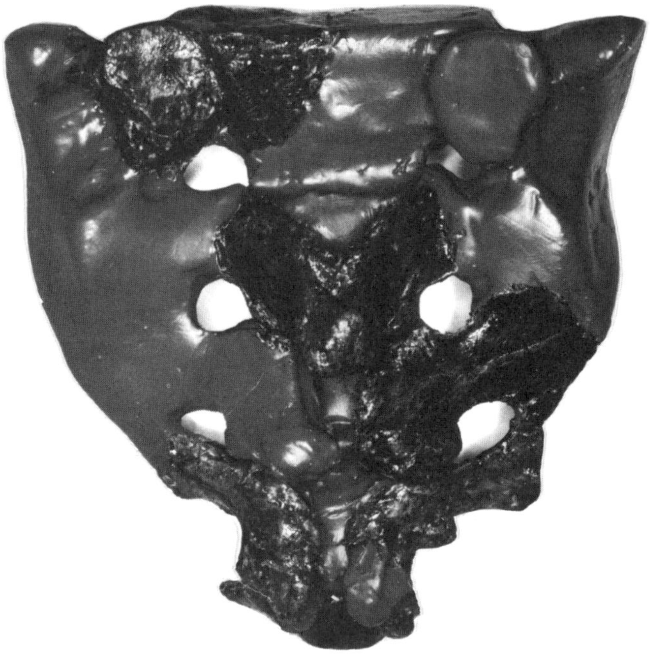

FIGURE 10.4.
Posterior view of reconstructed sacrum of KNM-WT 15000.

spine preserved. The laminae had not fused at the spines, but it was possible to estimate the distance between left and right spines by placing them under the fused laminae of the second vertebra under which they fitted tightly.

What little reconstruction was necessary for the last two vertebrae was done by mirror imaging.

The whole sacrum was put together after the reconstruction of the individual vertebrae by the addition of imaginary intervertebral discs to fill the spaces left by the missing unfused epiphyses of the bodies. The thicknesses of these discs were judged by checking the contacts between the lateral elements of the last three vertebrae, the contact between the laminae and spines of the second and third vertebrae, and the match between the auricular surfaces on the pelvic bones and the partly reconstructed ones on the first and second vertebrae. The internal curvature of the sacrum was controlled by the fit of the spinous processes, the contacts between the pelvic bones and the first vertebra, and the near contact between them and the third vertebra while keeping the anterior surfaces of the bodies on the same curving plane.

With the three major parts of the pelvis reconstructed except for the pubic symphysis region, the sacrum and pelvic bones were put together at the auricular surfaces. Minor asymmetries were seen that were probably due to several factors. These included prefossilization ones due to asymmetrical growth, postfossilization ones due to trampling, compression by sediment load, and modelling artefacts. These factors were adjusted as far as possible and the pelvic bones pivoted

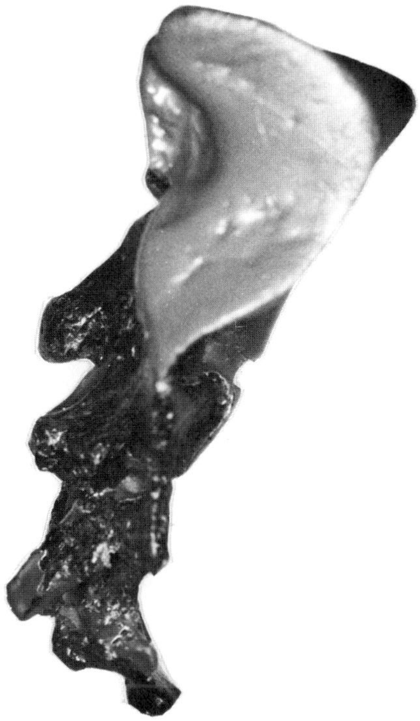

FIGURE 10.5.
Right lateral view of reconstructed sacrum of KNM-WT 15000.

around their articulations with the sacrum until a satisfactory configuration and length of the pubic bones was brought about. Rotation of the pelvic bones was mainly constrained by the auricular surfaces, but a minimum length of the pubis is constrained by the right pubic piece (AW) and a maximum by the two pubic bones meeting. A compromise mid-position was chosen. Figures 10.6–10.8 show the reconstructed pelvis. Table 10.1 gives measurements of its major dimensions.

Discussion

As we noted briefly in the original announcement of the find (Brown et al., 1985), there were more important insights to be gained from the discovery of this pelvis than just confirming the essentially human nature of the pelvis of *Homo erectus.* We recognized that something could be said about the very important development of secondary altriciality in our lineage. Gould (1985) developed a popular essay on this theme following the publication of our note.

We were following the lead of Gould (1977) and Martin (1983), who borrowed the term *secondary altriciality* from Portmann (1941). Humans are unique among primates in having secondarily altricial infants. All other primates change their steep fetal rate of brain growth to a slower one at birth. A nonhuman primate is born, typically, with about a third to half its adult brain weight. The rest of the brain weight is added at a much slower rate. In humans the fast fetal rate of brain growth continues until the infant's first birthday. Although human neonates also are born with about a third of their adult brain weight, this extended postnatal period of fast growth results in one-year-old humans having very large brains. More than another third of the brain weight is added during the first year (see Martin, 1983, for details and especially his figure 11). Martin went on to ask when this unique growth pattern first appeared. He concluded, unlike Portmann (who thought that obstetrical mechanics were not important here), that the pelvic limitation on increased fetal brain development was probably reached just before the emergence of *Homo erectus.* Prior to this, he argued, modifications of the rate and/or extent of

TABLE 10.1.

Major dimensions (mm) of the reconstructed pelvis of KNM-WT 15000.

Dimension	mm
Bicristal breadth	225.0
Bi-acetabular breadth (in notch)	102.0
Maximum inlet (transverse) diameter	100.0
Maximum inlet (anteroposterior) diameter	88.0
Estimate of interspinous transverse diameter (allowing for the fact that the right spine is crushed)	105.0
Transverse diameter between centers of ischial tuberosities	126.0

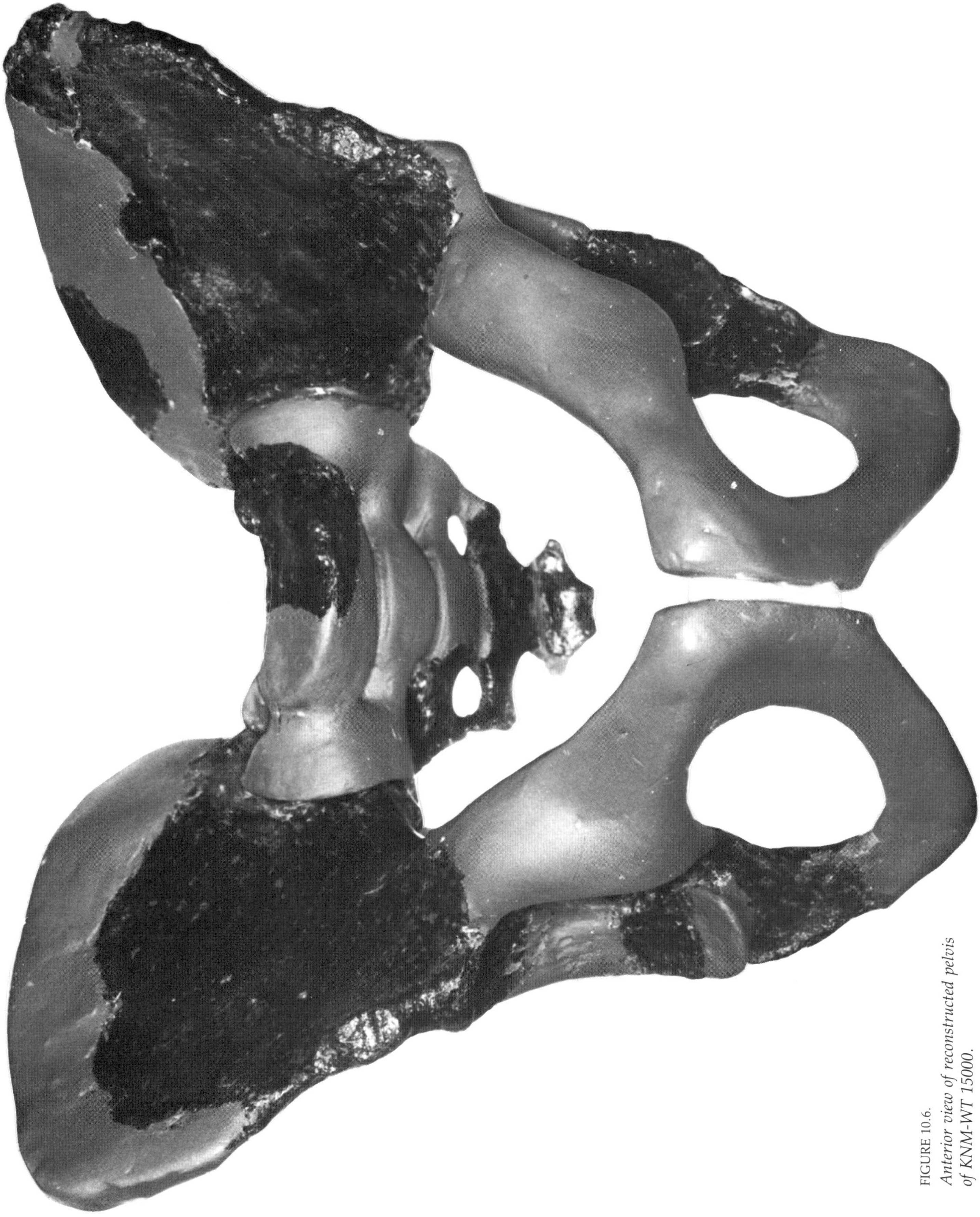

FIGURE 10.6.
Anterior view of reconstructed pelvis of KNM-WT 15000.

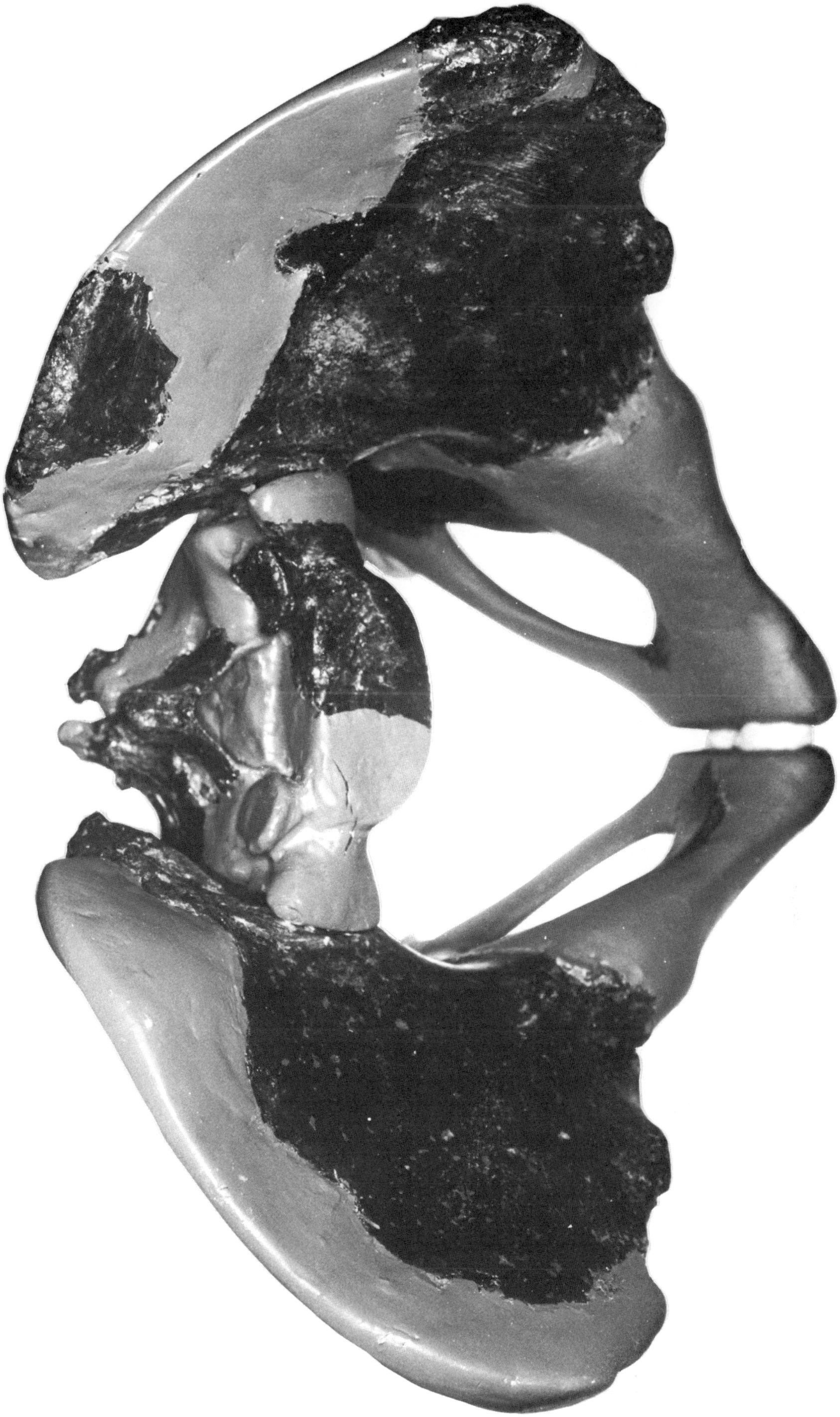

FIGURE 10.7.
Superior view of reconstructed pelvis of KNM-WT 15000.

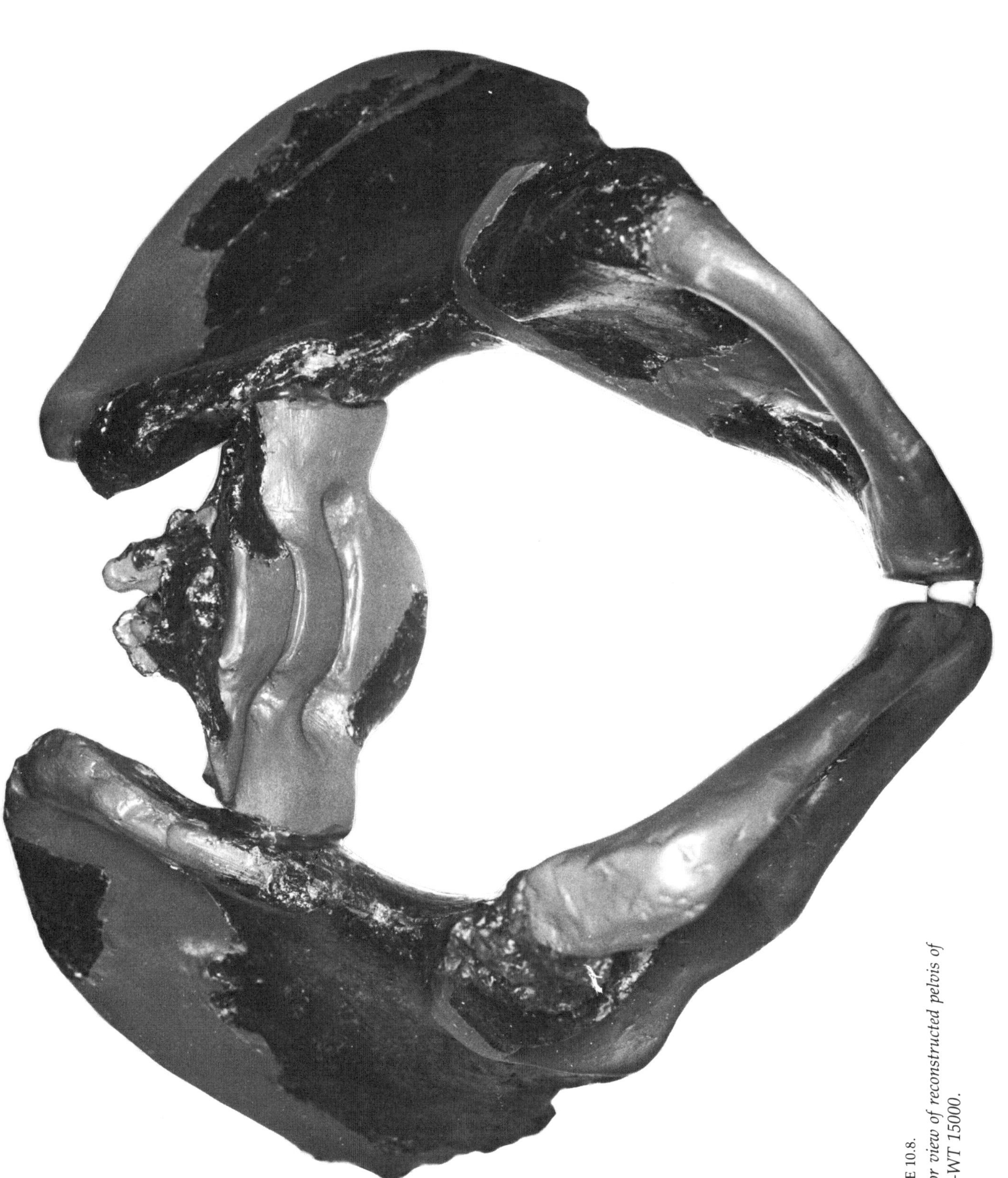

FIGURE 10.8.
Inferior view of reconstructed pelvis of KNM-WT 15000.

fetal growth could have produced the relatively large brain sizes typical of early hominids. He set a limit of about 850 cc (for adult cranial capacity) after which the development of the secondary altricial condition would have become necessary. In thinking this he was, as we shall see, remarkably prescient.

With the finding of the pelvis of KNM-WT 15000, we had a chance, for the first time, of estimating the birth canal dimensions. With this estimate, and knowing the adult cranial capacity, we could ask whether *Homo erectus* infants were secondarily altricial. Some people may say that the final answer comes from making so many assumptions which themselves were based upon others that it is an answer not worth knowing. However, the assumptions made are not frivolous and may well be correct. Besides, the ideal specimen may not be found for a long time. It has taken almost exactly a century since the original finding of *Homo erectus* to find this one!

These are the assumptions:

1. The attribution of male sex to the individual is correct.
2. This species would have grown as modern humans do at least from the age of about 11 years on.
3. The amount of sexual dimorphism in pelvic size and shape was about the same in this species as that seen in modern human populations.

With regard to the first point, we have done further work on the attribution of sex, which has been based since the finding of the specimen on the large size of the facial skeleton and thickness of brow ridges in a youth together with visual inspection of the sciatic notch.

We undertook computerized curve analysis of the sciatic notches of the reconstructed pelvis using the method described in Walker (1984) for crania. Ten male and ten female modern human hip bones were taken from the Hannam-Todd Collection of the Cleveland Museum of Natural History. These individuals were of known sex from the death certificate. Jennie Smith kindly took accurate replicas of the sciatic notches, using cold-cure silastic rubber for the molds and dental plaster for the positives. These were then sectioned along their lengths to give curves which ran from the articular surface to the sciatic spine. These curves were digitized using a Hewlett-Packard microcomputer and software written by one of us (A.W.). The programs standardized the curves for length and computed average, maximum, and minimum curves for each sex. These were then compared and the degree of mismatch of male and female curves calculated. It was found that the minimum curves gave the best discrimination between the sexes, so these were used for our comparisons. We then digitized notch curves for the KNM-WT 15000 reconstruction, Olduvai Hominid 28, and KNM-ER 3228. KNM-ER 1808 was not complete enough to digitize.

Table 10.2 gives the results of the computer comparison. There is slight asymmetry in the reconstruction. Part of this is really present in the fossil and part may be due to the reconstruction or distortion. The left side of the Nariokotome hip bone is very close in profile to that of the purported male *H. habilis*, KNM-ER 3228. It is next closest to the modern human male minimum curve and last to the modern female minimum curve. The same is true for the right side, but the match with the older fossil is not quite as good. The

TABLE 10.2.

Percentage distances between curves of greater sciatic notches (standard deviations in parentheses).

	KNM-WT 15000			KNM-ER
	Left	Right	OH 28	3228
Female minimum	13.1 (8.9)	13.3 (6.0)	9.5 (7.2)	13.7 (9.2)
Male minimum	9.0 (8.9)	9.3 (5.8)	12.2 (5.3)	8.7 (9.8)
OH 28	13.3 (7.3)	16.8 (9.3)	—	13.8 (7.1)
KNM-ER 3228	1.1 (0.1)	6.9 (4.4)	13.8 (7.1)	—

Note: See Walker, 1984, for method.

presumed female, OH 28, is closest to the modern female minimum, next to the modern male, and farthest from KNM-WT 15000 and KNM-ER 3228. The latter is closest to KNM-WT 15000 (left), then to the modern males, then to modern females, and is least close to OH 28. Thus it appears that pelvic differences between the sexes were present in early *Homo* and that KNM-WT 15000 has a male sciatic notch shape.

In Chapter 11 on body size and proportions, we showed (Figure 11.9) that the bi-iliac breadth of the pelvis grows isometrically with stature in modern boys. Because of this work, we feel fairly comfortable in making use of a simple ratio to estimate the adult bi-iliac breadth of KNM-WT 15000 by extrapolation. Given that the specimen has a pelvic inlet breadth of 100 mm at 11 years, we estimate a breadth of 115.6 mm by adulthood. We do not know the exact amount of sexual dimorphism in pelvic inlet size in early *Homo erectus,* since we have only isolated pelvic bones for two female individuals (Olduvai Hominid 28 and KNM-ER 1808) besides the pelvis of KNM-WT 15000. But the fact that a pelvic bone from a large male individual of *Homo habilis* (or the species antecedent to the early *Homo erectus*), KNM-ER 3228, is extremely similar to these other fossils, taken together with the observation that all these early hominid pelvic bones appear similar to those of modern humans, encourages us to think that it was probably not very different. Females of modern human populations usually have larger transverse pelvic inlet diameters than males. The amount of the size difference varies, but Williams et al. (1989) state that the female transverse diameter is just over 4% bigger and Tague (1989) found the same value (4%) for 4 populations. Applying this correction gives an adult female *Homo erectus* transverse diameter of 120 mm. Note that this is likely to be an overestimate, since KNM-WT 15000 would have grown up to have been the tallest early hominid ever found had he lived to maturity and there was probably some size difference between the sexes in early *Homo erectus* (Ruff and Walker, Chapter 11, this volume).

Data gathered by Tague (1989) show that the mean transverse diameter of the pelvic inlet for American whites is 134 mm. A series of healthy pregnant United States women were monitored by ultrasound by Levi and Erbsman (1975), who recorded the major dimensions of their fetuses. These included the maximum fronto-occipital diameter, which is the critical one for the engagement of the full-term fetal head. In order to estimate what full-term fetal head could be engaged in a pelvis with a transverse diameter of 120 mm, some allowance must be made for soft tissue within the birth canal. They report other researchers' mean full-term fronto-occipital diameter of 124 mm, but for other populations. Levi and Erbsman (1975) themselves give 122 mm for 40-week Belgian fetuses. Taking the first measurement, so as to bias the answer in favor of larger, rather than smaller, neonatal endocranial volumes, we arrive at an estimate of 10 mm for soft tissue in the birth canal. Subtracting this from 120 mm gives an estimated fronto-occipital diameter of 110 mm for a *Homo erectus* neonate. We can make sense of this estimate by comparing it with the diameters recorded by Levi and Erbsman. It matches the diameter for a 32–33-week human fetus. Dobbing and Sands (1973) collected data on fetal brain weights at various weeks of gestation. We can use their data to ask what the full-term brain weight of a *Homo erectus* would be. The value predicted from their equation is about 200 grams, but taking into account the scatter of points around the calculated best-fit curve, we believe it could be as high as 240 grams.

Martin (1983) has documented in detail the uniquely human trait of continuing a high brain growth rate after birth. This feature makes the human neonate "secondarily altricial" (Portmann, 1941). The human infant is born with about 384 grams of brain, but the brain grows by more than 2.5 times in the first year of life. A one-year-old child has 1,000 grams of brain, most adults have 1,400 grams (1,345 ml capacity, as noted by Tobias, 1971, multiplied by 1.04, the specific gravity of the central nervous system). If we apply this reasoning to the case of *Homo erectus,* the child is born with about 200 grams, adds a further 500 grams by the end of its first year, and then adds only another 200 grams to adulthood. This makes a total of 900 grams. Begun and Walker (Chapter 14, this volume) estimated the adult endocranial volume of KNM-WT 15000 as about 909 ml, so a human-like growth pattern gives a nearly perfect estimate. If, however, a great-ape growth pattern is used to estimate adult brain size, then the adult value would be between 440 grams (gorilla model) and 600 grams (chimpanzee model), both of which are very low. In conclusion, then, it would appear that *Homo erectus* had, by 1.5 Ma, secondarily altricial neonates and that the pattern of growth of the fetal brain was almost, if not fully, as derived as that in modern humans.

Martin (1983), made the following statement, which we are now in the position to test:

> As a final point relating to the fossil evidence for human evolution, it is worth noting that pelvic limitation of increased foetal brain development was probably reached about one and a half million years ago, just prior to the emergence of *Homo erectus,* as currently recognized in the fossil record. Prior to that time, increase in relative brain size above the great ape level could probably have been achieved by modification of the rate and/or extent of foetal growth without any need for significant postnatal postponement of a foetal brain growth pattern. However, subsequent to the attainment of a cranial capacity of about 850 cm^3, increasing development of the "secondarily altricial condition," was probably involved as postnatal continuation of foetal growth patterns became progressively more necessary.

The evidence we have put forward in this chapter would suggest that this "secondarily altricial condition" was already well established by 1.5 Ma, earlier than anticipated by Martin, and that enlargement of the birth canal and the related reduction in the lengths of the femoral necks was attained later. There are two reasons why a narrow canal might have persisted in *H. erectus* rather than being selected against when larger-brained infants were being selected for. First, it might be that the mechanical advantages given by bipedal locomotion benefiting from narrow pelvises and long femoral necks could not be compromised until later in human evolution. Second, perhaps the narrow body build necessary for a large hominid living in open country on the equator (see Chapter 11, this volume) could not be compromised until later. These two reasons are not, of course, mutually exclusive. We must caution that we have not been able to use neonate and pelvic dimensions from modern populations who are adapted for the hottest, driest climates.

CHAPTER 11

BODY SIZE AND BODY SHAPE

CHRISTOPHER B. RUFF AND ALAN WALKER

Estimating body size and shape for the Nariokotome individual is relatively straightforward because the skeleton is nearly complete. Because body proportions vary significantly with climate in modern human populations, it is also possible to determine whether early Homo erectus *individuals had body shapes adapted for the hot, arid climate in which they lived. Early hominids from this region are found to be unexpectedly tall; this conclusion has had many consequences for our efforts to deduce life history strategies and their evolution in our lineage.*

It has been amply demonstrated that body size has a direct relationship to many other biological and behavioral characteristics of a species (Calder, 1984; Schmidt-Nielsen, 1984). Stature and body weight estimates of early hominids have been used for a variety of purposes, for example, to calculate relative encephalization (Begun and Walker, Chapter 14, this volume), infer social organization (Foley and Lee, 1989), and reconstruct ecology (Shipman and Walker, 1989). Because KNM-WT 15000 has the most completely preserved skeleton of any early hominid, his body size (and shape) can be estimated with relatively greater accuracy than has previously been possible in any hominid other than *Homo sapiens*.

In this chapter his stature and body weight, both as a juvenile (at the time of death) and projected as an adult, are calculated using several different techniques, including some developed specifically for this analysis. In the course of the study we include some general comments regarding methods of height and weight estimation in fossil hominids. Because the reconstruction of body size (height and weight) from skeletal material is partly dependent on assumptions regarding body proportions, and because body proportions themselves are of intrinsic interest and relevance to interpretations of early hominid adaptations, we begin with an analysis of some general body shape characteristics of KNM-WT 15000.

Before doing this, though, we need to mention briefly our rationale for the chronological age assignment of 11 to 12 years for KNM-WT 15000 used in the following analyses. Smith (Chapter 9, this volume) has demonstrated that an age of about 11 years is most consistent with his overall dental development, an estimate based on modern human standards. Union of long bone epiphyses with shafts, or the primary elements of the os coxae, does not begin to occur in modern males until at least 14 years of age (Krogman, 1962). Thus, since none of these events had yet occurred in KNM-WT 15000, his skeletal development can only be used to place an *upper* limit of about 14 years on his age at death. However, a less often recognized skeletal maturational event does generally occur prior to 14 years in modern males—the union of the trochlea and capitulum (and also the lateral epicondyle) of the humerus, prior to their joint union with the humeral shaft. Fusion of these elements in boys has been reported at 12 to 15 years (Flecker, 1932/33; Australian whites) and 11 to 15 years (Modi, 1957, cited in Krogman, 1962, p. 64; East Indians), with an apparent median or mode of about 13–14 years. That these elements *were* fused in KNM-WT 15000 (at least the capitulum and trochlea) suggests a *skeletal* age for him of somewhat more than 11 years. This could mean that the coupling of dental and skeletal maturation in *Homo erectus* was slightly different from the pattern in modern humans, or perhaps simply that KNM-WT 15000 was near the extremes for both dental (retarded) and skeletal (advanced) development for his chronological age. In either event, 11 to 12 years would seem to be the best compromise figure to use for his chronological age at death in, for example, extrapolations of remaining growth to adult stature, and this is what we have done.

Body shape

In reconstructing and interpreting the body morphology of KNM-WT 15000, two major issues must be addressed: the appropriateness of different modern reference samples and the effect of his juvenile status. The first issue is common to any analysis of this kind; the second is more specific to this fossil but is also relevant to many human archeological and paleontological samples. We address the general issue of reference samples here and turn to the effects of his juvenile status in later sections dealing specifically with body shape and size estimation.

We note at the outset that only human reference samples will be considered, since it is evident that the general postcranial proportions of KNM-WT 15000 are well within the modern human range and outside the range of nonhuman hominoids (see below and other chapters of this volume). This decision leaves two basic considerations in evaluating the appropriateness of different modern reference samples. First, do modern humans themselves vary enough in body proportions to require attention to this factor (variation in proportions) at all? Second, given the elapsed time and further evolution of the genus *Homo* since the early Pleistocene, is it valid even to attempt matching a particular modern human reference sample or samples to KNM-WT 15000? For example, with regard to stature estimation, some researchers have contended that "in early hominid fossil material, broad racial affiliation is mostly guesswork and use of a race-specific formula is pointless" (Feldesman et al., 1990, p. 370).

Two general types of body proportions are of particular interest here: the length of the limbs relative to stature and the width of the body relative to stature. Together these two proportions effectively encapsulate the most important components of general body shape in humans. The first is particularly important in estimating stature from long bone lengths, and the second in relating estimated stature to estimated body weight. It will be argued here that both types of proportions vary significantly and systematically in modern humans as well as fossil hominids and that this variation is largely determined, or at least constrained, by climatic adaptation.

Limb length proportions

Variation in relative limb length has been shown to be correlated with ambient temperature in many species, including humans. Populations living in colder climates have relatively shorter extremities, a principle known generally as "Allen's rule" (e.g., see Hesse et al., 1951; Mayr, 1963). The underlying rationale for this observation is that relatively longer limbs lead to a relative increase in body surface area because the limbs, being narrower cylinders than the trunk, constitute a disproportionate share of the total body surface area (also see Figure 11.5, below). Thus, a relative increase in limb length will increase body surface area more than body mass, promoting heat loss, and *vice versa*. The heat-regulating properties of the limbs, particularly the distal elements of limbs, have been well documented by human physiologists (Bazett, 1968; Robinson, 1968). For living humans clinal variation in relative limb length has been demonstrated in various ways: as a ratio between combined upper and lower limb length over body weight (Schreider, 1964), as arm span over stature or conversely, trunk length over stature (relative sitting height) (Roberts, 1978), and as upper or lower limb length relative to trunk length (Eveleth and Tanner, 1976). The relatively longer limbs of living Africans and Australian aborigines compared with Europeans is apparent early in childhood, persists into adulthood, and cannot be explained by nutritional differences (Eveleth and Tanner, 1976; Eveleth, 1978; also see Stinson and Frisancho, 1978; Tanner et al., 1982). The limbs of African populations are also relatively and absolutely thinner than those of Europeans, which is again probably not explained entirely by differences in nutrition (Eveleth and Tanner, 1976).

In skeletal remains, limb length relative to body size is difficult to estimate precisely or without circular reasoning (e.g., without using limb bone lengths themselves to estimate stature). Fortunately, this question can be approached in a different way by examining *intra*limb length proportions, which *can* be determined quite precisely from skeletal remains. This is because the relatively longer limbs characteristic of modern human tropical populations are largely a result of relative lengthening of the distal elements of the limbs (Martin, 1928, pp. 394–395; Schultz, 1937; Roberts, 1978; Trinkaus, 1981). That is, as most commonly expressed, the crural (tibia/femur length) and brachial (radius/humerus length) indices are high in tropical populations and progressively decrease with a decline in mean annual temperature (Trinkaus, 1981, figures 3, 4; also see below). In contrast, the *inter*membral index (i.e., upper/lower total limb length) shows no such climatic correlation, as would be expected (Trinkaus, 1981; Schultz, 1937).

In addition to this correspondence in geographical variation among modern humans, the developmental basis (see Trinkaus, 1981) and thermoregulatory consequences (Roberts, 1978) of relative elongation of the total limb and elongation of the distal relative to the proximal limb segments are similar. Thus, it is reasonable to use variation in intralimb proportions as a guide to variation in relative limb length to total body size in fossil hominids.

In Table 11.1 we have listed brachial and crural indices for KNM-WT 15000, along with comparative data for living sub-Saharan Africans and Europeans and, for contrast, a sample of Neandertals (Trinkaus, 1981). *Maximum* long bone lengths (including epiphyses for KNM-WT 15000) were used in calculating both indices. Maximum length of the radius for KNM-WT 15000 was estimated from the average length of the preserved right and left ulnar diaphyses using ratios between average diaphyseal and total lengths of the ulna and radial/ulnar lengths measured radiographically in modern 11–12-year-old boys (Maresh, 1970, tables F-5, F-6). It should be noted that the maximum radial length obtained by this method —255 mm—is slightly greater and the brachial index correspondingly slightly higher than originally estimated (Leakey and Walker, 1989). The maximum length of the humerus (including the missing proximal epiphysis) of 319 cm was reconstructed by comparisons with similarly sized modern specimens, and this figure is consistent with estimates derived from a juvenile skeletal sample from Pecos Pueblo (Ruff et al., in prep.) as well as modern radiographic samples (Maresh, 1970).

Modern human comparative samples were taken from Schultz (1937) and Trinkaus (1981). In addition, the combined Western European and Near Eastern Neandertal sample described by Trinkaus (1981) is included as an example of premodern hominids from nontropical environments. In all cases the lengths given for the humerus, radius, femur, and tibia were converted to maximum lengths, where necessary, using conversion formulae derived from a sample of modern femora. Schultz (1937) gave absolute lengths for only the humerus and femur. His brachial and crural indices were used to derive average radial and tibial lengths, and then all lengths were converted to maximum lengths and the indices recalculated. Similarly, the femoral bicondylar length used by Trinkaus (1981) was converted to maximum length and the average crural index recalculated. For simplicity, and because the sex of some of the fossil specimens is uncertain, the values for males and females were averaged for all samples, although modern human males show a consistently higher brachial index than females (Trinkaus, 1981).

TABLE 11.1.

Brachial and crural indices.

SPECIMEN OR SAMPLE	BRACHIAL INDEX	CRURAL INDEX
KNM-WT 15000	79.9	88.0
Living African	76.4–78.7	82.8–85.8
Living European	72.9–74.0	78.4–83.1
Neandertals (Trinkaus, 1981)	74.4	78.9

As shown in Table 11.1, the ranges for living African and European sample means are completely distinct for the brachial index and barely overlap for the crural index, with Africans having higher indices in both cases, as expected. The intralimb indices for KNM-WT 15000 are both above the living African ranges. The Neandertal means are well below the African ranges and within (crural) or slightly above (brachial) the European ranges.

Because comparisons between ratios can be misleading if used over a wide size range (Trinkaus, 1981), the data were also analyzed as bivariate plots of distal versus proximal limb segment lengths, with results presented in Figures 11.1 and 11.2. In addition to Europeans and sub-Saharan Africans, a number of other population samples from either tropical/subtropical or temperate/arctic regions were included in the analysis (Schultz, 1937; Trinkaus, 1981; see figure legends). The results confirm those shown in Table 11.1: KNM-WT 15000 has intralimb proportions similar to those of living tropical/subtropical humans, while Neandertals most closely resemble living temperate/arctic populations.

In fact, particularly with regard to the lower limb (Figure 11.2), both KNM-WT 15000 and Neandertals tend towards the extremes for living populations from lower or higher latitudes; in other words, they are "hyper-tropical" and "hyper-arctic" in proportions, respectively. KNM-WT 15000 falls more than 1 standard deviation (S.D.) above the mean (converted) crural index for five of the six living tropical/subtropical male samples where individual distributional data are available, and more than 2 S.D. above the mean for three of the

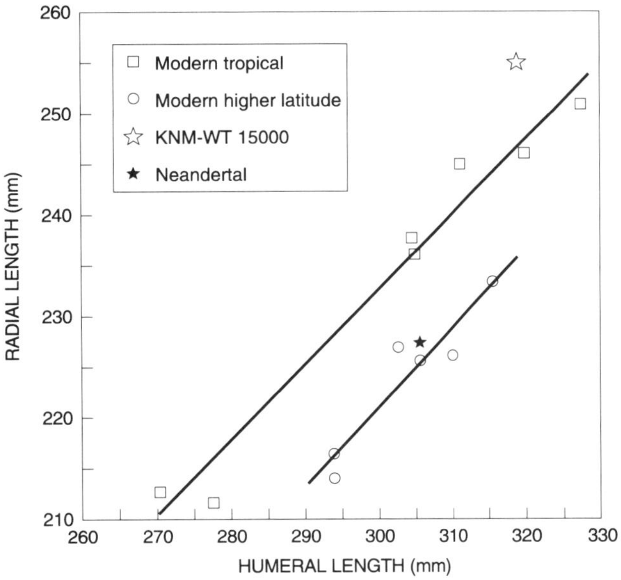

FIGURE 11.1.
The relation of radius length to humerus length in tropical and higher-latitude modern humans, KNM-WT 15000, and Neandertals. (Data from Schultz, 1937; Trinkaus, 1981; and present study.) Tropical populations include two samples of U.S. blacks, Australian aborigines, Bushmen, Pygmies, Egyptians, and Melanesians; higher-latitude populations include two samples of U.S. whites, Lapps, Yugoslavians, and two samples of Eskimos. All data are averages of males and females, maximum bone lengths.

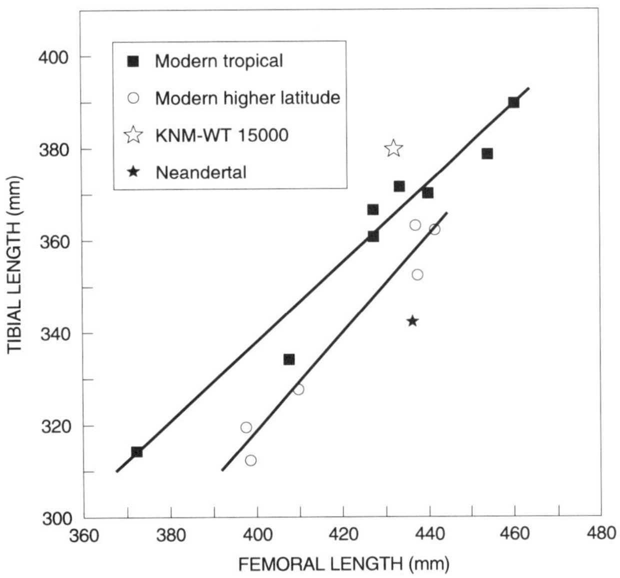

FIGURE 11.2.
The relation of tibia length to femur length in tropical and higher-latitude modern humans, KNM-WT 15000, and Neandertals. (Same samples and sources as in Figure 11.1, with the addition of a Bantu sample to the modern tropical sample.)

samples. The upper limb comparison is slightly complicated by the larger average brachial index for males. However, even if males alone are considered, KNM-WT 15000 falls above the mean brachial index for all of the modern comparative samples and in the top 20% of (converted) values for male American blacks in Schultz's (1937, figure 4) study. Neandertals, most of which were male or probably male (Trinkaus, pers. comm.), fall just below the male temperate/arctic regression line in Figure 11.1.

In addition to the sex difference in the brachial index alluded to above, it is possible that there are age changes in intralimb proportions that could affect these comparisons of the juvenile KNM-WT 15000 with the modern adult samples. We examined this question using radiographically measured bone lengths reported from two modern cross-sectional growth studies (Anderson et al., 1963, table 1; Maresh, 1970, tables F-4 to F-8). Ratios of tibia/femur and radius/humerus lengths (including epiphyses) were calculated from the reported yearly means for each bone for boys in the age range 10 to 18 years. (Sample sizes occasionally differ slightly for the "paired" length measurements, but any discrepancies are quite minor.) No correction for radiographic magnification was made, since all long bones within a limb should have been subjected to about the same amount of distortion in any one individual. Results from the study that included all four bones (Maresh, 1970) are shown in Figures 11.3 and 11.4. These show that the crural and brachial indices do indeed exhibit significant variation over this age range, apparently because of differences in the timing and possibly magnitude of growth velocities of the different bones (specifically, distal limb elements seem to undergo an earlier adolescent growth spurt than proximal limb segments). Despite this age-related variation, however, both of the intralimb proportions for 11–12-year-old boys are very similar to those for 18-year-old boys. Beyond 18 years little change in the ratios would be expected, since all long bone epiphyses except possibly the humeral head are generally fused by about this time (Krogman, 1962). Similar analysis of the other growth sample (Anderson et al., 1963), which included only the femur and tibia, also indicates virtually no difference in a tibia/femur ratio between 11–12-year-old boys (.7895) and 18-year-old boys (.7890). (Note that measurement techniques varied between the two studies, so that absolute values of ratios vary; this also applies to comparisons with the adult values in Table 11.1, which are all based on osteometric measurements of museum specimens.) Both available growth series are measures of middle-class U.S. whites and thus may not be entirely representative of growth in other modern populations, to say nothing of African early *Homo erectus*. As far as they go, however, the data indicate that direct comparisons of KNM-WT 15000 with adults in this particular analysis should not introduce any strong bias.

Thus, variation in intralimb length proportions, and thus very likely limb length/body size proportions, is significant and shows strong clinal variation among both living and earlier hominids. KNM-WT 15000 exhibits definitely tropical or even "hyper-tropical" limb proportions.

RELATIVE BODY BREADTH

Another way to alter surface area or body mass is to change the shape of the whole body, by altering its breadth or height, or both. Variation in body breadth/height among modern and earlier hominids and its significance has been discussed in detail elsewhere (Ruff, 1991). We present here a summary of this analysis and its application to KNM-WT 15000.

The human body can be modeled most simply as a cylinder, with stature taken as the height of the cylinder and some body breadth taken as the diameter (Figure 11.5). If this is done, it can be shown that changes in height have no effect on surface area/body mass (SA/BM) as long as breadth is kept constant (Figure 11.5). Conversely, a change in breadth will always change SA/BM regardless of height, with an increase in breadth leading to a reduction in SA/BM and *vice versa*. (This is actually just a more general treatment of Allen's rule, applied to the whole body rather than extremities.) Two predictions follow from this geometric observation: populations living in colder climates should have absolutely wider bodies than populations in warmer climates, regardless of stature; and populations within the same general climatic (temperature) zone should have similar absolute body breadths, regardless of stature. The second prediction has another implication: taller populations should be *relatively* more linear (i.e., have a smaller breadth/height ratio) than shorter populations living within the same general climatic zone.

The dimension that will be used here for

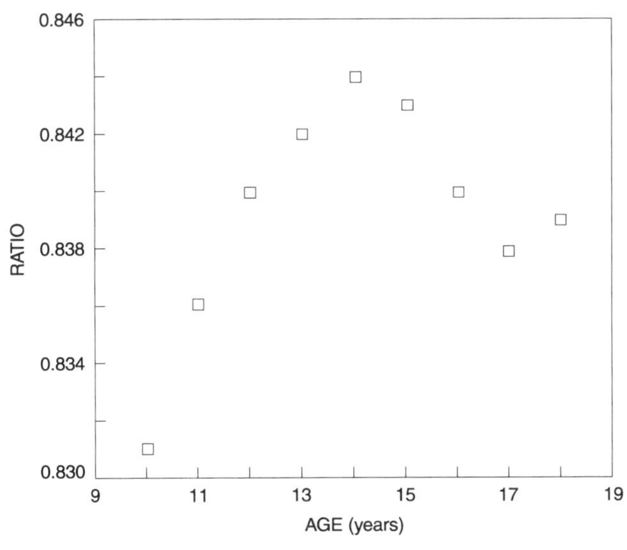

FIGURE 11.3.
Changes with age in the ratio of tibial to femoral length (crural index) in boys from 10 to 18 years of age. (Data from Maresh, 1970.)

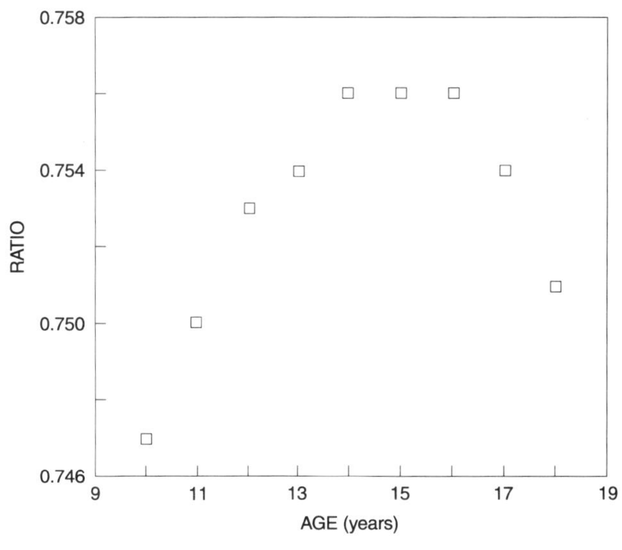

FIGURE 11.4.
Changes with age in the ratio of radial to humeral length (brachial index) in boys from 10 to 18 years of age. (Data from Maresh, 1970.)

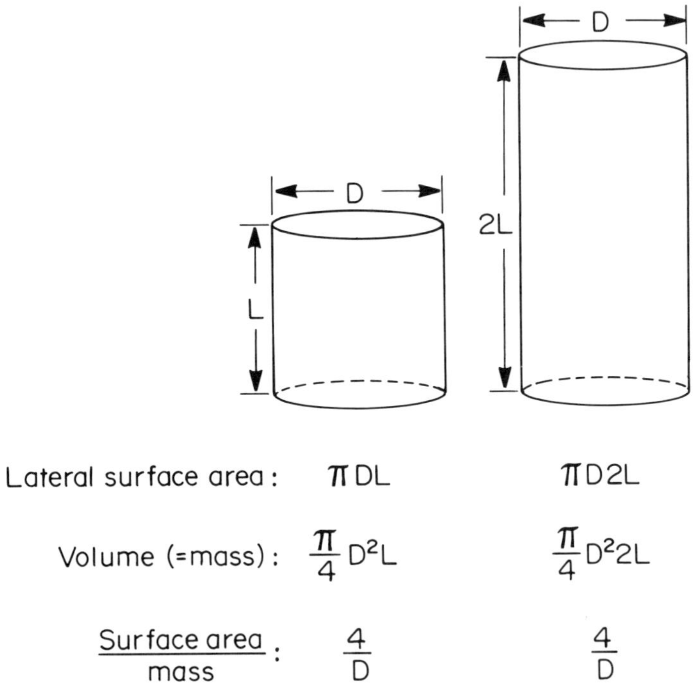

FIGURE 11.5.
The cylindrical model of the human body. Change in height (L) *produces no change in the ratio of surface area to body mass, as long as breadth* (D) *remains constant. Conversely, a change in breadth will alter the ratio regardless of height. (Reprinted from Ruff, 1991.)*

body breadth is bi-iliac breadth, or maximum (mediolateral) pelvic breadth taken across the iliac crests. This measurement has a number of advantages over other possible dimensions (e.g., bi-acromial, bi-trochanteric, or chest breadths), including good available samples for modern living populations, low measurement error, less dependence on limb morphology, and, most important, direct comparability between living and skeletal samples. Bi-iliac breadth, stature, and the cylindrical model (Figure 11.5) accurately represent the major variations in body mass and skin surface area in modern human populations (Ruff, 1991).

To test the predictions of the model, 71 sex/population means for bi-iliac breadth and stature in living adults were obtained from the anthropometric literature (see Ruff, 1991, for details). These can be divided into four broad regional/climatic groups: sub-Saharan African, southeast Asian, European, and northern Asian, which can be considered tropical, subtropical, temperate, and arctic/subartic, respectively. A bivariate plot of the ratio of bi-iliac breadth/stature (body linearity) in these populations is shown in Figure 11.6. Exponential functions through each of the four groups (see Ruff, 1991) and theoretical curves that would result if bi-iliac breadth were actually held constant over different statures are also plotted.

As shown in Figure 11.6, the predictions of the cylindrical model are borne out by the living human data. Populations in progressively colder climates have progressively wider bodies, regardless of stature. Within similar temperature zones, body breadth varies relatively little (the regressions fit through each group and the theoretical curves representing constant body breadths are almost parallel), and thus taller populations become relatively thinner populations. This visual impression is supported by statistical analysis: regression slopes of the four groups show no significant variation, while line elevations are all significantly different (see Ruff, 1991). Other analyses indicate a strong correlation between absolute bi-iliac breadth and latitude, which remains highly significant when stature is controlled for. Other body breadth dimensions also show the same latitudinal clines (Ruff, 1991).

Table 11.2 lists values for bi-iliac breadth, stature, and bi-iliac breadth/stature for KNM-WT 15000, KNM-WT 15000 projected as an adult, an

TABLE 11.2.

Bi-iliac breadth and stature (cm).

SPECIMEN OR SAMPLE	BI-ILIAC BREADTH[a]	STATURE	BI-ILIAC BREADTH/ STATURE
KNM-WT 15000	23	160	.144
KNM-WT 15000 (projected size of adult)	26.6	185	.144
Living African	23.1–26.3	137–175	.148–.174
Living European	27.4–29.8	156–176	.160–.188
Kebara 2 Neandertal	31.8	170	.187

a. 0.5 cm added to measurements of the fossil specimens to account for soft tissue (see Ruff, 1991).

adult Neandertal (Kebara 2; see Rak and Arensburg, 1987), and the ranges of variation in sex/population means for these dimensions in living adult Africans and Europeans (from Figure 11.6). The bi-iliac breadth of KNM-WT 15000 was measured directly from the reconstruction of his pelvis (see Chapter 10, this volume). His statures as a juvenile and as an adult were estimated using techniques described in the next section of this chapter; his bi-iliac breadth as an adult was estimated by assuming that bi-iliac breadth/stature does not change significantly from late childhood to adulthood (see below). And we added 0.5 cm to the bi-iliac breadths of the fossil specimens to account for soft tissue (Ruff, 1991).

In terms of both absolute and relative body breadths, KNM-WT 15000 clearly groups with living African rather than European populations. In fact, although not shown here, *all* early hominids from Africa whose body breadths can be estimated, including some very short australopithecines (AL 288-1 and STS 14), have absolute breadths within the modern African range (Ruff, 1991). In contrast, the Neandertal, from a colder climate, groups much more closely with Europeans (and northern Asians; see Figure 11.6). Thus, the predictions of the cylindrical thermoregulatory model hold true for earlier hominids as well as living humans. (Note also that this conclusion is not dependent upon the precise stature estimates for KNM-WT 15000.)

In terms of living human comparisons, the closest match in general body shape to KNM-WT 15000 can be found among certain Nilotic populations. Hiernaux's (1968) compilation of adult male African anthropometrics gives a mean stature of 185 cm and a mean bi-iliac breadth of 26 cm among the Nuer (unmatched samples, thus not included in Figure 11.6). These are quite close to those projected for KNM-WT 15000 as an adult (Table 11.2). Somatological photographs of adult male Dinka and Shilluk (Roberts and Bainbridge, 1963) give estimates of bi-iliac breadth/stature ratios of about .134–.147, again quite similar to the breadth/height "shape" ratio for KNM-WT 15000 (see Figure 11.7 for an example of one of Roberts' photographs).

KNM-WT 15000's juvenile status could call

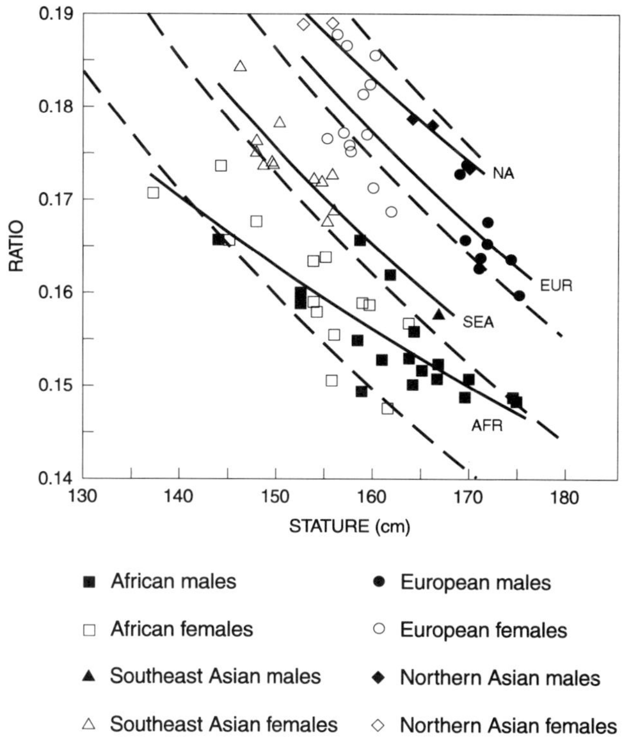

FIGURE 11.6.
Relative body linearity (bi-iliac breadth/stature) in modern human populations, grouped into four regional-climatic subgroups: AFR, *Africans;* SEA, *Southeast Asians;* EUR, *Europeans;* NA, *Northern Asians. Solid lines are exponential functions fitted through each group (pooled sexes); dashed lines are theoretical "isobars" that would result if bi-iliac breadth were actually held constant across different statures. (Reprinted from Ruff, 1991; see text and this reference for further details.)*

into question the use of his bi-iliac breadth/stature ratio in these analyses, if this body shape parameter changed significantly from late childhood to adulthood. In fact, as shown in Figure 11.8, while there appears to be age-related variation in the ratio among juveniles, there is no consistent change in the index from 11–12 years to 17–18 years among African and African-descent boys (data from Eveleth and Tanner, 1976). In two of the samples plotted there, the ratio is somewhat higher at the end of adolescence, in one sample it is much lower, and in one sample (the only East African [Tanzanian] sample with available data) it is about equivalent.

Another way to address this question is to

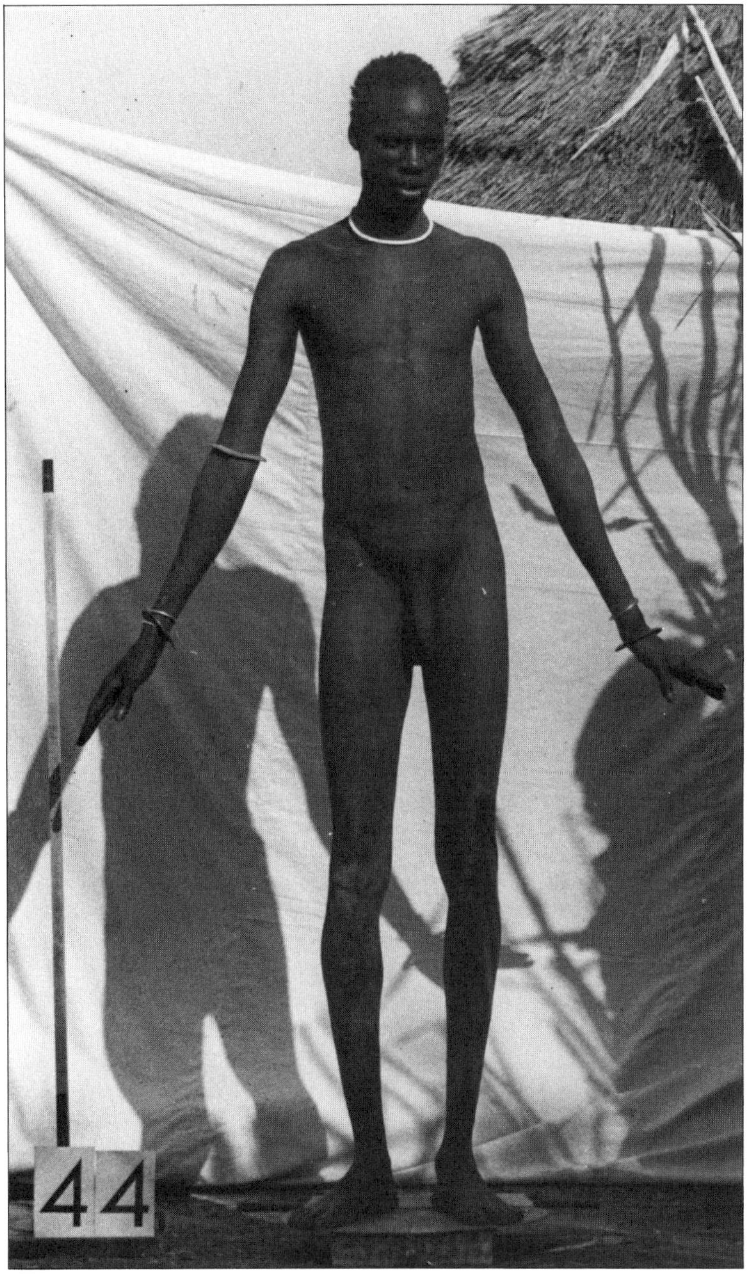

FIGURE 11.7.
An example of Allen's rule in modern HOMO SAPIENS. *This young man is a Dinka from the Southern Sudan. The Dinka live on the same latitude and about 200 km west of Nariokotome. This individual has trunk and limb proportions almost identical to those of KNM-WT 15000. The Dinka have, of course, no closer genetic links to early* HOMO ERECTUS *than any other people on earth; the resemblance is due entirely to physiological convergence. We think this body shape is a good example of Allen's rule in action. (Photo courtesy of D. F. Roberts.)*

carry out log-log regressions of bi-iliac breadth on stature within growth samples and examine the data for correspondence with isometry (i.e., a slope of 1.0). This has been done in Figure 11.9 for the same African and African-derived samples shown in Figure 11.8. The value for KNM-WT 15000 (as a juvenile) has also been plotted, along with a theoretical isometric line through the center of the distribution. The plot shows that although there is some variation in the slopes, most of the data are fit remarkably well by an isometric relationship. The lower slope among American blacks may be due to a pre- or early-adolescent increase in fat in well-nourished boys (e.g., Parizkova, 1973, figure 1), which leads to an increase in bi-iliac values in the younger groups (Figure 11.8). KNM-WT 15000 falls well within the bivariate distribution.

Finally, we can examine variation in bi-iliac breadth/stature among modern juveniles age-matched to KNM-WT 15000. Figure 11.10 plots bi-iliac breadth against stature for five samples of 10–12-year-old boys from Africa and five samples of U.S. white boys of the same age (Eveleth and Tanner, 1976; means at each year), along with KNM-WT 15000. Although the comparison is complicated by the overall smaller size of the available African samples, it is apparent that even at this age the difference in absolute (and relative) bi-iliac breadth between temperate and tropical populations is marked and follows the same pattern as in adults (Figure 11.4). An extension of the African regression line into a larger body size range (which seems justified, given the results shown in the previous figure) exactly intersects the point for KNM-WT 15000, who falls well below the U.S. white regression line.

Therefore, the juvenile status of KNM-WT 15000 should not introduce any significant bias into this analysis. As a boy he falls exactly where predicted by modern tropical populations, and his general body shape in all likelihood would have been maintained into adulthood.

The general correspondence in body shape of KNM-WT 15000 with modern Nilotic populations introduces another point regarding climatic adaptation and body form among humans. Both comparative geographic studies and direct physiological tests indicate that humidity is an important factor in determining the efficiency of heat loss (Ruff, 1991). Specifically, longer, more linear bodies are better adapted for heat loss in dry open environments, where evaporative heat loss from sweating is very effective. All modern-day tall "elongated" Africans (e.g., Nilotics) are restricted to such environments. In contrast, in very humid, closed environments (such as tropical rainforests), sweating is relatively ineffective, and the best way to avoid overheating is simply to restrict body mass (e.g., Pygmies).

This strongly implies that KNM-WT 15000, with his long, linear physique, would *not* have been well adapted for a relatively closed wet environment, but rather would have been limited to open or at least mixed environments. The broader implications of this conclusion for *Homo erectus* and early hominids in general will be discussed at the end of this chapter.

Conclusions

It is clear that the two questions posed at the beginning of this section must both be answered in the affirmative: basic body proportions do vary significantly among modern humans (and fossil hominids), and from all available evidence KNM-WT 15000 groups strongly with modern tropical populations in this regard. It should be emphasized that this association is based *not* on proposed "racial" or "ethnic" affiliation *per se* (*contra* Feldesman et al., 1990) but, rather, on basic thermoregulatory principles that should be universally applicable. As one modern example, both Australian aborigines and sub-Saharan Africans have relatively long legs (Eveleth and Tanner, 1976); however, this is not a result of a close genetic tie but rather a common biological adaptation to a hot climate. Indeed, given his general environment (Chapter 3, this volume), it would be very surprising if KNM-WT 15000 were *not* proportioned in a manner generally similar to modern tropical populations.

This indicates that care must be exercised in choosing and interpreting the results of stature

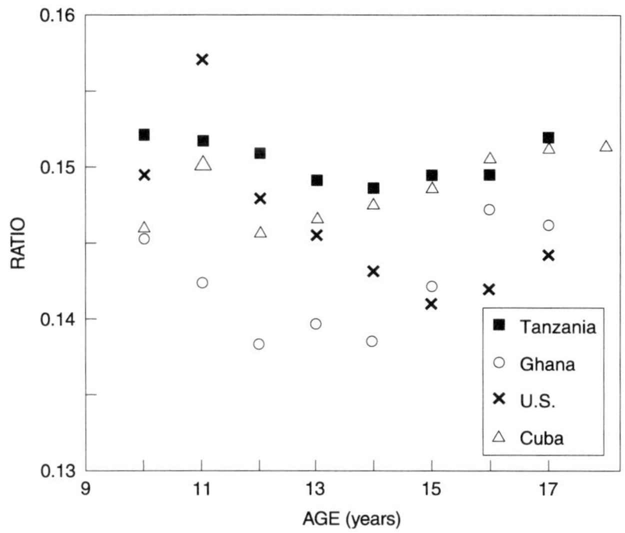

FIGURE 11.8.
Changes with age in relative body breadth (bi-iliac breadth/stature) among males from Africa or of African descent from 10 to 18 years of age. (Data from Eveleth and Tanner, 1976.)

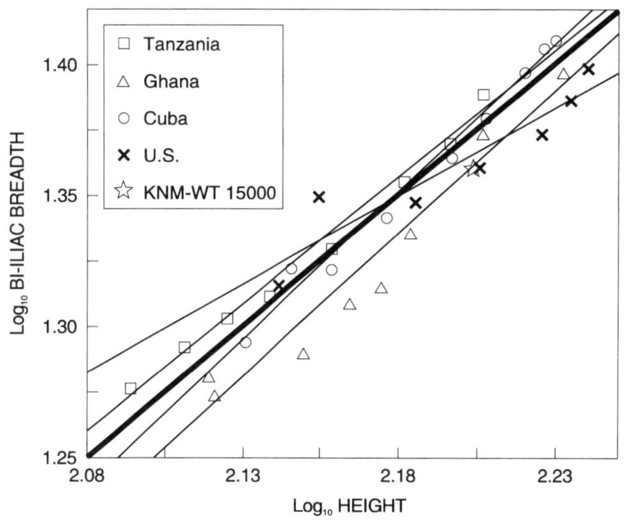

FIGURE 11.9.
Log-log plot of bi-iliac breadth on stature in 10- to 18-year-old males from Africa or of African descent and in KNM-WT 15000. Thin solid lines are linear regressions through each sample; the thick solid line is theoretical isometry, drawn to show the relation when there is no change in breadth/height proportion. (Data from Eveleth and Tanner, 1976, and present study.)

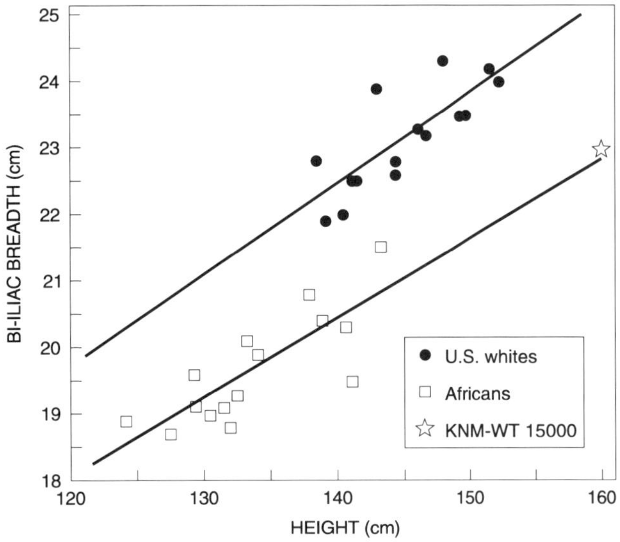

FIGURE 11.10.
Body proportions (bi-iliac breadth versus stature) of 10- to 12-year-old boys: U.S. whites, African blacks, and KNM-WT 15000. Solid lines are linear regressions through each group. (Data from Eveleth and Tanner, 1976, and present study.)

and body weight reconstruction formulae for KNM-WT 15000. Ideally such formulae would be derived from similarly tall, thin modern tropical populations. However, as shown below, information of this sort is rare (although in the one case where it is available, it is very enlightening). More data exist for other, shorter native sub-Saharan African population samples and for American blacks, although in the latter case caution must be exercised since the body proportions of American blacks are not precisely representative of native Africans (see below). Juvenile data from appropriate reference samples are even rarer; for example, we know of no data that could be used to derive stature reconstruction formulae from long bone lengths of African children. In some cases this limitation affects the choice of reconstruction method used, as shown in the next section.

Stature

The body size of KNM-WT 15000 as a juvenile, at the time of his death, and his projected size as an adult will be estimated here. In some ways estimates of his size as an adult are of more interest than his juvenile size, since these allow direct comparisons with the great majority of other available fossil (and modern) hominid specimens. Also, most applications of body size data, such as calculations of relative encephalization, have been based on adult values.

There are two possible approaches to calculating KNM-WT 15000's stature as an adult: reconstruct his stature as a juvenile from his preserved skeletal remains and then project the change in this stature to adulthood, or project the remaining growth to adulthood of his skeletal remains and then estimate his stature from these adult skeletal values. To our knowledge there are no stature estimation formulae *per se* available for juveniles and no data at all on skeletal size in juveniles of known stature other than for Euroamerican samples. Particularly given the observed changes in some body proportions during growth (see Figures 11.3 and 11.4 above), it is obvious that caution must be exercised in using adult stature reconstruction formulae for juveniles. On the other hand, some data are available on changes in skeletal size during growth to adulthood in both modern Euroamerican as well as other population samples, and there are many techniques for stature estimation of adults, including non-Euroamericans. Thus, the second approach—calculating KNM-WT 15000's remaining skeletal growth and then his adult stature from the size of his adult skeleton—has much to recommend itself. Both approaches will be employed here but the results of the second will be favored.

In either event, in estimating the adult body size of KNM-WT 15000 certain provisional assumptions must be made. First, it must be assumed that he had a growth curve generally similar to that of modern children. For example, it will be assumed here that he would have undergone an adolescent growth spurt. Also, it will be assumed, as discussed earlier, that he was chronologically 11 to 12 years of age at death. Finally, because of the general robusticity of the skeleton, particularly the skull (see Chapter 6, this volume), it will be assumed that KNM-WT 15000 was male. Any of these assumptions could be debated (and are considered further in other chapters), but in lieu of better evidence to the contrary they will be accepted here as working hypotheses.

Stature estimates for KNM-WT 15000 will be based here upon the lengths of his long bones, especially those of the lower limbs. These have been shown to give the smallest errors in estimation of stature in modern populations (Krogman, 1962). Potential alternative methods of stature reconstruction using the whole skeleton (e.g., the Fully technique; see Feldesman and Lundy, 1988) are hampered by the juvenile status of KNM-WT 15000 and the absence of vertebral body epiphyses and missing cervical vertebrae.

As emphasized by other investigators (Krogman, 1962), when using any stature reconstruction formula for skeletal remains it is quite important to measure the bones in exactly the same way as measurements were taken in the original derivation of the equation. In all of the following analyses, care was taken to ensure that this was the case, with specific bone length measurements noted in the text and tables; in analyses of modern radiographic data, distortions or differences in technique were corrected for as precisely as possible.

MODERN REFERENCE SAMPLES

From the discussion presented earlier, it is apparent that the best estimates of body size in KNM-WT 15000 will be derived from modern reference samples with the most similar general body proportions—namely, tropical populations. What is not as clear is whether reference populations with a tropical ancestry, such as American blacks, are similar enough in body proportions to use in such reconstructions. This is an important issue, since American blacks have been included in some of the largest and best-documented stature reconstruction studies (Trotter and Gleser, 1952, 1958).

The ratios of (maximum) tibial length to stature in 11 male population samples—3 U.S. white, 3 U.S. black, and 5 African black—are plotted in Figure 11.11. In all cases the tibia/stature ratio was calculated using the means for each dimension given by the original investigators. The values for U.S. whites and blacks were derived from data reported by Trotter and Gleser (1952, table 5; 1958, table 3) and include both general (Terry Collection) and military samples. The general samples range in age from the third through ninth decades, while the military samples are younger, ranging from 18 years through the fourth (blacks) or fifth (whites) decades. The values for African blacks were derived from data reported by Feldesman and Lundy (1988, table 1; South African Bantu), and Allbrook (1961, table 2; all others). Feldesman and Lundy reported only "bicondylar" tibial length (not including the proximal tibial intercondylar spines); this measurement was adjusted to maximum tibial length by multiplying by a factor of 1.026, a formula derived from data given by Trotter and Gleser for both lengths (1952, table 5, for black males).

The Ugandan samples studied by Allbrook (1961), in a rarely cited but important paper, pre-

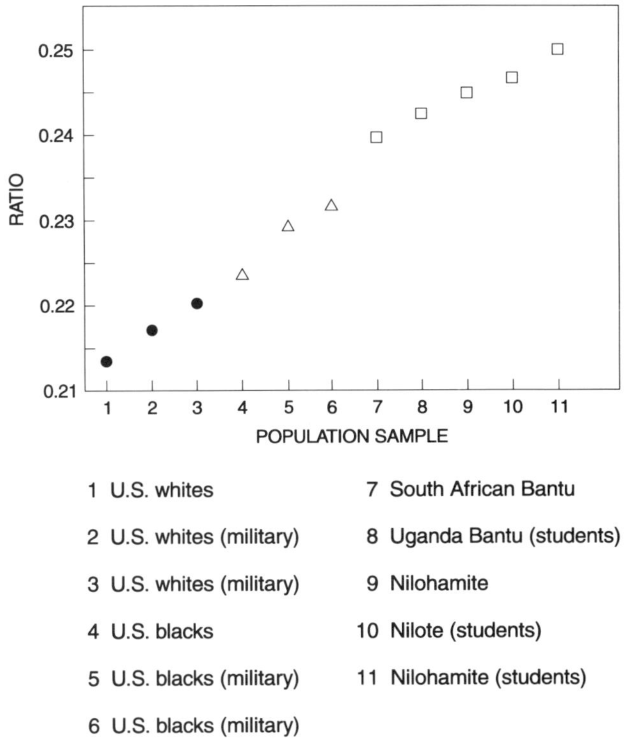

FIGURE 11.11.
Maximum ratios of tibial length to stature in modern adult males: U.S. whites (circles), *U.S. blacks* (triangles), *and African blacks* (squares). *(Data from Trotter and Gleser, 1952, 1958; Allbrook, 1961; and Feldesman and Lundy, 1988.)*

sent a special case since bone lengths were measured in living subjects. He describes his tibia measurement as follows: with the subject seated and the knee semi-flexed, "the left tibial length was measured with spreading calipers from the medial condyle (as it becomes palpable and diverges anteriorly from the articulating femoral condyle) to the tip of the medial malleolus (sphyrion)." This measurement seems to be equivalent to bicondylar length of the tibia. However, he later compares these lengths to *maximum* lengths of ethnically matched skeletal samples (and shows no significant difference), suggesting that he considered the measurement equivalent to maximum tibial length *in vitro*. Also, the figures Allbrook gives for a sample of young British soldiers stationed in Uganda (1961, table 2) seem to indicate that what he measured was closer to maximum tibial length. (The tibial length/stature ratio in the British sample is similar to *maximum* tibial length/stature in a U.S. white male military sample [Trotter and Gleser, 1958], but well above the range for *bicondylar* tibial length/stature in a similar U.S. sample [Trotter and Gleser, 1952].) Therefore, we have considered his reported tibial measurement to be maximum length, both in this analysis and in the application of his stature estimation formulae to KNM-WT 15000 below. Allbrook also measured ulnar length in his subjects; this measurement was clearly maximum length.

Figure 11.11 clearly illustrates the variability in tibial length/stature present among modern human populations, as well as the relative lengthening of the distal element of the lower limb among tropical or tropical-ancestry populations, as discussed earlier in relation to the crural index. However, the figure also clearly shows that in this regard *U.S. blacks are as different from African blacks as they are from U.S. whites*. If the Ugandan data were considered to represent tibial bicondylar rather than maximum length, the difference would be even greater.

A similar but less pronounced pattern is evident for the femur. Feldesman et al. (1990) compiled femur/stature ratios for a large number of population samples and demonstrated a significant difference in the ratio between black (mostly American) and white Euroamerican samples (black ratio .2725, white ratio .2664). Several of the U.S. samples were military. As shown in Figure 11.11 for the tibia, and also in their compilation for the femur, military samples almost always have higher long bone/stature ratios than samples drawn from the general population. This is almost certainly due to the younger average age of military samples, which usually include many 18- and 19-year-olds (e.g., see Trotter and Gleser, 1952, 1958).[1] The mean femur/stature ratio for the one male African black sample listed by Feldesman and Lundy (not including a Pygmy sample), a South African Bantu sampling of the general nonmilitary population, is .2750. The mean ratio for the three nonmilitary U.S. black male samples in their list is .2721, with all samples falling below the African sample (range .2710–.2727). Furthermore, given the results for the tibia shown in Figure 11.11, it is likely that had data for taller African samples been available, the difference between African and American femur/stature ratios would have been greater (also see note 1, at the end of this chapter).

Thus, the long bone/stature proportions of American blacks do *not* appear to be representative of native African blacks (at least those from South and East Africa); rather, they seem intermediate in some respects between Africans and U.S. whites, due possibly to genetic admixture, environmental effects, or other factors. Since KNM-WT 15000 was shown earlier to have tropical or "hyper-tropical" body proportions, we concluded that his stature will therefore be most accurately reconstructed by using only similarly proportioned tropical reference samples and that equations based on data from American blacks are not appropriate.

This leaves only two good sources for stature estimation in KNM-WT that we are aware of: Feldesman and Lundy's (1988) South African Bantu sample, from which they derived Model I (least squares) and Model II (major axis) regression formulae on the basis of femoral and tibial lengths; and Allbrook's (1961) Model I regression formulae based on tibial and ulnar lengths in four Ugandan samples. All samples are adult (18 years or over); Feldesman and Lundy's original sample included both males and females, although only the results for males will be used here; Allbrook's samples are all male. Characteristics of each reference sample

TABLE 11.3.

Modern reference samples (adult males) for estimates of stature.

SOURCE	REFERENCE POPULATION	*n*	AGE RANGE (years)	BONE LENGTH (cm) (MEAN ± S.D.)[a] FEMUR	TIBIA	ULNA
Feldesman and Lundy (1988)	South African Bantu	175	18–65	44.8 ± 2.3	38.1 ± 2.3	—
Allbrook (1961)	Ugandan Bantu	185	18–29	—	39.7 ± 2.2[b]	28.4 ± 1.5[b]
	Ugandan Nilohamite[c]	30	—	—	43.3 ± 2.4	30.6 ± 1.3
	Ugandan Nilohamite[d]	7	19–21	—	42.3 ± 4.2	30.4 ± 2.1
	Ugandan Nilote	7	20–26	—	41.7 ± 2.0	29.1 ± 1.6

a. Femur: bicondylar length; tibia: bicondylar length (Feldesman and Lundy, 1988) or maximum length (Allbrook, 1961); ulna: maximum length.

b. Mean of three groups; S.D. shown is average S.D. of the three groups.

c. "Mature young" tribesmen.

d. Students.

are given in Table 11.3, including sample size, age range (when available), and average lengths of the long bones used to derive stature estimation equations.

Corresponding long bone lengths for KNM-WT 15000 are shown in Table 11.4 (including epiphyses). The ulnar maximum length was derived from its preserved diaphyseal length using ratios between diaphyseal and total length in modern 11–12-year-old boys (Maresh, 1970, table F-6), as noted previously. The tibial and femoral lengths were measured after placing the preserved epiphyses in realistic anatomical position (see Walker and Leakey, Chapter 7, this volume).

Using reference samples of the same general size as the specimen in question avoids problems in extrapolation that can be particularly marked when using Model I (least squares) regressions (see Feldesman and Lundy, 1988; Jungers, 1988a). If similarity in general skeletal size (i.e., bone lengths) is used as a criterion, it is apparent from comparison of Tables 11.3 and 11.4 that the two Bantu samples provide a reasonably good match to KNM-WT 15000 at his time of death (within 1 S.D. for tibia and femur length, 1.5 S.D. for ulna length). The other three Ugandan samples, with longer bone lengths, are better matched with the estimated adult skeletal size of KNM-WT 15000, as shown below, although only one of these (Nilohamite tribesmen) is really large enough in number to derive reliable regression formulae.

TABLE 11.4.

Long bone lengths (cm) of KNM-WT 15000 at time of death.

	FEMUR	TIBIA	ULNA
Maximum length[a]	43.2	38.0	27.0
Bicondylar length	42.9	37.5	—

a. Including epiphyses; ulnar length estimated from preserved diaphyseal length of 25.7 cm (average right and left sides), using diaphyseal/total length ratio of 1.045 for 11–12-year-old boys derived from Maresh, 1970, table F-6.

JUVENILE STATURE OF KNM-WT 15000

If we are to apply these adult stature reconstruction formulae to KNM-WT 15000, it must be assumed that the proportions of long bone lengths to stature are not significantly different in adult males and 11–12-year-old boys. Once again we turned to modern radiographic growth data (Maresh, 1970; Anderson et al., 1963) to examine this question. The mean femur/stature and tibia/stature ratios for 11–12-year-old and 18-year-old males in these two samples are shown in Table 11.5. All bone length measurements were corrected for magnification and/or differences in measurement technique.[2]

The results in Table 11.5 show that both femur/stature and tibia/stature ratios are almost identical among 11–12-year-old and 18-year-old

TABLE 11.5.

Bone length/stature ratios in 11–12- and 18-year-old U.S. white males.

REFERENCE	FEMUR LENGTH/STATURE 11–12 YEARS	FEMUR LENGTH/STATURE 18 YEARS	FEMUR LENGTH/STATURE %Δ	TIBIA LENGTH/STATURE 11–12 YEARS	TIBIA LENGTH/STATURE 18 YEARS	TIBIA LENGTH/STATURE %Δ
Maresh, 1970, table F-13	.2733	.2727	−.2	.2303	.2325	.9
Anderson et al., 1963, table 1	.2730	.2725	−.2	.2240	.2236	−.2

Note: Maresh, 1970, lengths were corrected for magnification; Anderson et al., 1963, lengths were corrected for different measurement technique; see text.

males in these samples. This similarity is a result of the same general phenomenon shown earlier in Figures 11.3 and 11.4: just prior to the main adolescent growth spurt, males (and females) have long bone proportions similar to those of adults. Younger children have relatively shorter limbs, and early to mid-adolescents (13–16-year-old males, 11–14-year-old females) have relatively longer limbs because of the accelerated growth of the long bones at this time (our calculations from sources cited above plus others; also see Eveleth and Tanner, 1976, figure 176, p. 231). Fortunately, this suggests that adult stature estimation equations using these bones may be applied to KNM-WT 15000 without too much error, again assuming (by necessity) that his growth curves were basically similar to those of modern populations. (This issue is discussed again at the end of the section on adult stature, below.)

Table 11.6 lists the stature predictions for KNM-WT 15000 at his time of death using regression equations from the South African and Ugandan Bantu samples (Table 11.3). Both Model I (least squares) and Model II (major axis) formulae were available for the South African sample; results of both methods, and different combinations of bones, are shown in Table 11.6.

All of the estimates are consistent with a stature of about 160 cm (5′3″) for KNM-WT 15000 at his time of death. This is also the best estimate arrived at by Feldesman and Lundy (1988), although they used slightly different figures (i.e., maximum femur length rather than bicondylar length). Estimates derived from the two samples, from different bones, and from different methods of line fitting are very similar, ranging between 156 and 163 cm. Although, as noted previously (Brown et al., 1985; Feldesman and Lundy, 1988), KNM-WT 15000's stature would actually have been slightly less than this, due to his lower cranium, we have not made this adjustment here, for two reasons. First, one of the primary purposes of deriving his stature is to use this figure to help estimate his body mass, and the slight reduction in cranial height would have a minimal effect in this regard (particularly given the general robusticity of his skull). Second, stature estimates for other early hominids do not include this correction, with the possible exception of Schmidt's (cited in Geissmann, 1986) whole body reconstruction of AL 288-1. For comparative purposes of stature *per se* among fossil hominids, the above figure is most useful.

Although specific confidence limits of individual stature estimates are not listed in Table 11.6, in this situation, with large sample sizes and the predictor variable relatively close to the mean of the reference sample (Table 11.3), 95% confidence intervals are almost exactly equal to plus or minus twice the standard errors of estimate (Zar, 1984). Thus, these techniques can be considered to predict stature to within about ±5–9 cm with 95% confidence; better estimates (±5–7 cm) are achieved by using the lower limb bones, as expected. Given the number of uncertainties in predicting KNM-WT 15000's stature, including his juvenile status, we would not want to apply these ranges too literally here; however, they do give some indication of the degree of expected error. The fact that different samples, bones, and techniques yield very similar results lends some additional confidence in the estimate. Also, the 160 cm estimate appears reasonable relative to a whole body reconstruction of his skeleton (including the extrapolated addition to cranial height). This issue is considered again from a different perspective in the section below on estimated adult stature, where some evidence is presented that suggests a possibly lower stature as a juvenile.

Finally, it should be noted that the stature estimate arrived at here is some 4–8 cm lower than that originally estimated (Brown et al., 1985) using Trotter and Gleser's (1952) formulae for American blacks and Whites, and it is about 2 cm lower than Feldesman et al.'s (1990) estimate based on a world-wide "generic" femur/stature ratio. Allbrook (1961) demonstrated a similar 2–12 cm overestimate of stature in his East African samples using Trotter and Gleser's U.S. black and other more "generic" formulae. This is expected given the variation in body proportions in both living and fossil hominids demonstrated earlier. Because of this variation, the reference samples used in the present study should provide more accurate stature estimates for KNM-WT 15000.

ADULT STATURE OF KNM-WT 15000

STATURE ESTIMATED FROM ADULT BONE LENGTHS

The average percentage changes in long bone lengths from 11 or 12 years to 18 years in males of five modern population samples are listed in Table 11.7. The samples include three U.S. white and two Amerindian (recent archeological) samples. Because each study differed somewhat in measurement technique and design, and because of the importance of the resulting figures for the present analysis, we will consider these differences in some detail.[3]

All of the studies of U.S. whites were radiographic. In only one of these (Anderson et al., 1963) was true longitudinal growth of individuals measured; this study also had no problems with radiographic magnification. About half of these measurements were of the normal leg of individuals with polio, although the authors reported finding no significant difference in growth or average size (in males) between these subjects and the rest of the sample. The other two studies reported cross-sectional growth data. In most cases sample sizes were relatively large ($n > 50$), except for some of the data on 18-year-olds (with n as low as 18). Gindhart (1973) reported difficulty in measuring some of the larger tibiae in older males—this plus sampling variability may explain the average reduction in tibial bone length from 17 to 18 years in males in her sample (we took the average of 17 and 18 years for our calculations). Data in this study were corrected for magnification, although the author used a constant magnification

TABLE 11.6.

Stature estimates of KNM-WT 15000 at time of death.

REFERENCE SAMPLE[a]	DIMENSION	REGRESSION METHOD	S.E.E. REG.[b] (cm)	PREDICTED STATURE (cm) OF KNM-WT 15000
South African Bantu	Bicondylar femoral length	Model I	2.8	159
		Model II	—	156
	Bicondylar tibial length	Model I	2.8	163
		Model II	—	161
	Femoral + tibial lengths	Model I	2.4	160
		Model II	—	159
		Mean		160
Ugandan Bantu	Maximum tibial length	Model I	3.7	160
	Maximum ulnar length	Model I	4.6	160
		Mean		160

a. Allbrook, 1961, table 3; Feldesman and Lundy, 1988, tables 2 and 3; also see Table 11.3, above.

b. Standard error of estimate of the regression line. The 95% confidence limits for individual predicted values are about ±2 S.E.

TABLE 11.7.

Percent change in long bone lengths from 11 or 12 years to 18 years in boys.

SAMPLE	FROM 11 TO 18 YEARS				FROM 12 TO 18 YEARS			
	FEMUR	TIBIA	HUMERUS	RADIUS	FEMUR	TIBIA	HUMERUS	RADIUS
U.S. whites[a]	24.2	26.0	29.8	30.2	18.2	19.5	24.6	23.9
U.S. whites[b]	—	26.6	—	28.8	—	21.1	—	22.6
U.S. whites[c]	23.0	23.3	—	—	16.9	16.7	—	—
Pecos Pueblo[d]	25	27	27	31	18	21	22	23
Eskimo[e]	24	26	—	35	—	—	—	—
Best estimate for all boys	23.5	25	29	29	17.5	19	23	23

a. Maresh, 1970, maximum length (including epiphyses), corrected for magnification when possible (see text).

b. Gindhart, 1973, diaphyseal length.

c. Anderson et al., 1963.

d. Ruff, unpublished data, maximum length (including epiphyses).

e. Y'Ednak, 1976, diaphyseal length. Humerus % change not included because data are based on only 1 older individual.

factor for each tube-film distance, which may have slightly inflated the length values for the largest (i.e., 18-year-old) bones, thus slightly increasing the percentage changes shown in Table 11.7. Maresh's (1970) femoral and tibial lengths were corrected for radiographic magnification and adjusted for differences in bone length, as described earlier. From the information available in the literature (Green et al., 1946; Gindhart, 1973), however, it was not possible to make such bone length–adjusted corrections for the upper limb bones. Therefore, it is almost certain that the percentage growth changes for the humerus and radius shown in Table 11.7 for this sample are overestimates. This is also suggested by comparison with the Gindhart (1973) data (and is one reason why the radius rather than ulna was examined here—Gindhart did not include the ulna in her study).

The two recent archeological samples are both Amerindian, but from very different environmental and genetic backgrounds (Pecos Pueblo, New Mexico, and Kodiak Island Eskimo and Aleut). Besides being cross-sectional in nature, they suffer from small sample sizes at each measured age. This is particularly true for the Eskimo-Aleut sample studied by Y'Ednak (1976), where data for only 2–3 males in each of the age groups examined were available. Average bone lengths were presented for "10–11.9"-year-old males, with age based mainly on dental eruption (canines and premolars erupted, second molars not erupted), and "16–17.9"-year-old males, which apparently included individuals in which all long bone epiphyses (except the proximal humerus) were fused (thus justifying their use here as 18-year-olds). The younger group is considered here to have had an average chronological age of 11 years (see discussion of dental age standards in Smith, Chapter 9, this volume). The Pecos Pueblo juvenile sample (Ruff et al., in prep.) includes a larger number of individuals—four 10–12-year-olds and five 17–19-year-old males; furthermore, with the exception of one missing radius, all four of the bones were available for each individual (i.e., they are matched). No attempt was made to sex individuals under 15 years of age in this sample; however, as long as this part of the age range was relatively equally sampled between the sexes (so that chronological/skeletal ages are not biased), this should have little effect, since long bone lengths in boys and girls are very similar (<3% average difference) prior to about 14 years (see references above for the three U.S. white samples). The average chronological age of the younger Pecos group, using dental eruption criteria, is 11 years. Bone length values for 12 years were derived by fitting age curves (DWLS routine in SYSTAT SYGRAPH; Wilkinson, 1989, p. 260) to the Pecos Pueblo juvenile sample (n = 23–24, 5–19 years, males only above 14 years), and graphically determining the appropriate values. Because of the various uncertainties involved in the two archeological samples, they are listed here mainly to corroborate the much more complete, but also genetically and environmentally limited, U.S. white data.

Taking all of the above factors into account, we determined the average "best estimates" of growth remaining in the femur, tibia, humerus, and radius from 11 and 12 years to maturity, in boys, which are shown at the bottom of Table 11.7. (The humerus is not used in our stature estimates but is included here for completeness.) The estimates for the two archeological samples, except for the radius in the Eskimo sample, are quite similar to those obtained for the U.S. white samples, lending some confidence that the figures are generally applicable to modern populations. Also, percentage changes in diaphyseal and total long bone lengths appear to be similar.

Table 11.8 shows the extrapolated adult maximum lengths of the femur, tibia, and ulna of KNM-WT 15000, estimates based on his juvenile dimensions and the best estimates of remaining growth in each bone from Table 11.7. Growth of the ulna was assumed to be similar to that of the radius. Also, percentage change in bicondylar length was assumed to be the same as in maximum length.

Clearly, of the available modern reference samples for stature estimation (Table 11.3), the three Ugandan Nilohamite or Nilote samples studied by Allbrook (1961) have long bone dimensions most similar to those extrapolated for KNM-WT 15000 as an adult. His adult tibia length is within 2 S.D. of the mean tibia length for the larger Nilohamite tribal sample, even if the more extreme 11 year to adult extrapolation is used, and is within 1 S.D. of the mean using the 12 year to adult extrapolation. His adult tibia length also falls well within a +2 S.D. range of the Nilohamite student sample, regardless of his age at death, and just within a +2 S.D. range (12 year age at death only) of the Nilote sample, although here caution must be exercised since these samples were so small (n = 7). His ulna is within 2 S.D. of the mean lengths for all three samples if the 12 year to adult extrapolation is used, and just beyond 2 S.D. of the mean for the Nilohamite tribal sample if the 11 year to adult figure is used. This is also made clear in bivariate plots of individual statures against bone lengths published in Allbrook's report (1961, figures 2a,b): 2 Nilohamite or Nilotic men in his sample had tibia lengths *above* the *larger* figure extrapolated for KNM-WT 15000 as an adult; several more had tibia lengths above the smaller figure. No individuals had ulnae as long as the larger extrapolated figure for KNM-WT 15000, but at least one had an ulna length longer than the smaller adult extrapolation. In contrast, KNM-WT 15000's adult long bone lengths (tibia, ulna, and femur), regardless of which age at death is used, are well above 2 S.D. from the mean lengths of any of the Bantu samples—East or South African—listed in Table 11.3 (also see Allbrook, 1961, table 2), although a few individuals in Allbrook's combined Bantu sample had tibia lengths that reached the smaller adult value extrapolated for him.

In terms of both skeletal size and sample

TABLE 11.8.

Extrapolated adult long bone lengths (cm) for KNM-WT 15000.

BONE (dimension)	LENGTH AT TIME OF DEATH	ADULT LENGTH IF AGE WAS 11 YEARS AT DEATH[a]	ADULT LENGTH IF AGE WAS 12 YEARS AT DEATH[a]
Femur (maximum)	43.2	53.4	50.8
Femur (bicondylar)	42.9	53.0	50.4
Tibia (maximum)	38.0	47.5	45.2
Tibia (bicondylar)	37.5	46.9	44.6
Ulna (maximum)	27.0	34.8	33.2

a. Estimates based on Table 11.7.

size, the Nilohamite tribesmen sample in Table 11.3 appears to be the most appropriate available modern reference sample for estimating KNM-WT 15000's adult stature. Results using the regression equations for this sample are shown in Table 11.9. Results using the two much smaller Nilohamite-Nilote samples are also shown for comparison. Estimates based on samples that do not overlap KNM-WT 15000 in bone length are placed in parentheses.

As expected, the regressions of stature on tibia length always have smaller associated standard errors of estimate than the ulnar estimates, and thus are more reliable. Tibial estimates of adult stature for KNM-WT 15000 using the Nilohamite tribesmen reference sample average 184 cm over the two ages at death. Ulnar estimates using the same sample average 188.5 cm, with the higher estimate for the 11 year age at death based on a length outside the range of the reference sample. Stature estimates using the other two reference samples are smaller, ranging between 173 and 185 cm (average over two ages at death). Variability in stature estimates over all three samples is more than twice as great when the ulna us used as when the tibia is used, again as expected given the less direct relationship of ulnar length with stature.

Taking all of the above into account, the average best estimate of adult stature in KNM-WT 15000 is about 185 cm (6′1″). Based on the Nilohamite tribesmen tibial estimates, the most reliable, the 95% confidence interval of this estimate would be about ±4 cm (also see above). Of course, the true potential error is higher than this, since there is some individual variation around the mean growth figures we used to calculate his adult long bone lengths (Table 11.7), and we have had to make various other assumptions discussed previously. Even if this confidence interval is doubled, however, it still indicates a fairly circumscribed range of possible adult statures for KNM-WT 15000—from 177 to 193 cm, or 5′10″ to 6′4″. As noted earlier, this does not include the effect of his lower cranium, which would result in an actual stature a few centimeters lower relative to living populations.

Finally, it is of interest to compare this stature estimate with estimates that would result from using other, less appropriate modern reference samples. The somewhat shorter Ugandan Bantu sample also studied by Allbrook gives adult stature estimates for KNM-WT 15000 of between 178 and 185 cm, similar to those given by the Nilote sample, which is also not too different in bone lengths (Table 11.3). The regression equations for the South African Bantu sample (Feldesman and

TABLE 11.9.

Estimates of adult stature of KNM-WT 15000, using extrapolated adult long bone lengths.

REFERENCE SAMPLE[a]	BONE[b]	S.E.E. REG. (cm)[c]	PREDICTED ADULT STATURE (cm)[d] IF AGE AT DEATH WAS— 11 YEARS	12 YEARS	MEAN (cm)
Nilohamite tribesmen	Tibia	1.9	186	182	184
	Ulna	3.0	(191)	186	188.5
Nilohamite students	Tibia	3.8	180	175	177.5
	Ulna	5.4	(174)	172	173
Nilote students	Tibia	3.7	182	178	180
	Ulna	5.1	(188)	182	185

a. Allbrook, 1961, table 3.

b. All maximum lengths (see text).

c. Standard error of the regression; all Model I regressions.

d. Bone lengths from Table 11.8. Parentheses indicate estimates based on reference samples whose range of bone lengths does not include those for KNM-WT 15000.

Lundy, 1988), who are still shorter, give generally lower stature estimates, averaging 176 cm (Model II only, 180.5 cm). Feldesman et al.'s (1990) "generic" femur/stature index of .2674 gives a much higher average stature estimate of 195 cm, while his "black" (actually, mostly American black) index of .2725 gives an average stature of 191 cm. Trotter and Gleser's (1958) American black regression formulae give stature estimates based on the femur, tibia, and femur + tibia of between 179 and 189 cm, averaging 185 cm. None of these samples is as large in bone lengths as the Nilohamite-Nilote samples favored here for stature estimation, and some do not have the appropriate body proportions (see above). However, the estimated values all range between about 175 and 195 cm, with a midpoint of about 185 cm, the same value obtained using the best modern reference sample.

STATURE ESTIMATED FROM JUVENILE STATURE

The average percentage changes in stature from 11 or 12 years to 18 years in boys are listed for five types of modern populations in Table 11.10. The figures shown include the means for 19 European samples, 5 U.S. white longitudinal samples, 2 U.S. black samples, 3 African black samples (all of the above from Eveleth and Tanner, 1976), and an Australian aborigine sample (Abbie, 1967).[4] The consistency in results of the different studies is remarkable, all clustering around 23% (11 to 18 years) and 18.5% (12 to 18 years). Since mean stature in males appears to increase slightly from 18 years to the early twenties, because of continued growth of the axial skeleton, even in affluent populations (Trotter and Gleser, 1958) these percentages are probably slight underestimates of the total average change to adult stature. The difference appears to be on the order of about 1 cm (Trotter and Gleser, 1958), or an additional .5–1% in Table 11.10. However, at least one of the present samples (Australian aborigines; Abbie, 1967) does *not* show such an increase, and the exact degree of change may be obscured by positive or negative secular trends in stature (Tobias, 1985; also see note 2, below). Since virtually all growth data have been reported only to 18 years of age, we will use the figures in Table 11.10 without correction, but consider that they may lead to slight underestimates of adult stature.

TABLE 11.10.

Percent change in stature from 11 or 12 years to 18 years in boys.

SAMPLE	FROM 11 TO 18 YEARS	FROM 12 TO 18 YEARS
European[a]	23.1	18.7
U.S. White[b]	22.4	17.9
U.S. Black[c]	24.3	18.4
African[d]	22.1	18.9
Australian aborigine[e]	23.7	17.9
Mean	23	18.5

a. Mean of 19 samples (Eveleth and Tanner, 1976, appendix table 3).
b. Mean of 5 longitudinal samples (ibid., appendix table 25).
c. Mean of 2 samples (ibid., appendix table 40).
d. Mean of 3 samples (ibid., appendix table 40).
e. Abbie, 1967.

The extrapolated adult stature of KNM-WT 15000 using the average figures shown in Table 11.10 is 189.5 cm (11 years at death) and 197 cm (12 years at death). These are substantially higher than the stature estimates based on extrapolated adult *bone* length in KNM-WT 15000 (Table 11.9 and above), and it could be argued that the difference should be even greater given the additional correction suggested in the previous paragraph. It is difficult to reconcile the two types of estimates, especially since some of the same samples (e.g., U.S. whites studied by Maresh, 1970) were included in calculations of both remaining long bone growth and remaining increase in stature (Tables 11.7 and 11.10). The lack of agreement suggests that: (a) the stature of KNM-WT 15000 as an 11–12-year-old was overestimated, and/or (b) the average percentage changes in long bone length shown in Table 11.7 are too small for KNM-WT 15000, and/or (c) the average percentage changes in stature shown in Table 11.10 are too large for KNM-WT 15000.

Deciding among these alternatives is complicated by the use of ratios, both in estimating remaining growth and in evaluating juvenile and adult proportions, which can sometimes give unexpected results if applied over a wide size range. Another major problem, as noted earlier, is in the application of adult stature estimation techniques to KNM-WT 15000 as a juvenile. Although average bone length/stature proportions seem similar in 11–12-year-old and adult males (Table 11.5), there is no guarantee that *regression* equations developed for adult males (e.g, those in Table 11.3) will necessarily be directly applicable to juveniles of this age range. Also, given the slight reduction in long bone length to stature proportions after age 18, discussed above, the femur/stature and tibia/stature ratios for 11–12-year-old boys may actually be slightly above those for true adults (21+ years), rather than essentially equivalent, as shown in Table 11.5. As tests of these possibilities, we applied two adult stature estimation techniques—Feldesman et al.'s (1990) average white femur/stature ratio and Trotter and Gleser's (1958) femur regression equation for whites—to the boys in Maresh's (1970) sample (correcting femur length for magnification as described earlier). The adult femur/stature ratio predicted stature of 10-year-old boys almost perfectly, but overestimated the stature of 11-year-olds by 2.6 cm and 12-year-olds by 4.7 cm. The stature of 18-year-old males was still overestimated by 4.2 cm (a smaller *percentage* error, however, than that for the younger males). Trotter and Gleser's stature estimates were uniformly much too high in the age range 10–12 years (11–12 cm), but they became progressively better with age and were almost perfect at age 18 years. This latter result might suggest, indirectly, that the adult regression equations used here (Table 11.3) are not appropriate for KNM-WT 15000 as a juvenile. However, it must also be noted that Trotter and Gleser's adults were much taller than Maresh's younger adolescents, while the samples we used for stature estimation were within the same size range as KNM-WT 15000 as a juvenile. If the average femur/stature ratio of .2750 for Feldesman and Lundy's South African black males (cited in Feldesman et al., 1990) is applied to KNM-WT 15000's juvenile femur, a stature of 157 cm is obtained, 3 cm shorter than our estimate based on the regression equations (Table 11.6). Reducing this by another 3.5 cm, a correction based on the above analysis of Maresh's data for 11–12-year-old boys (average of 2.6 and 4.7 cm overestimate), yields a stature of 153.5 cm. This is precisely the juvenile stature that would produce an adult stature of 185 cm if the statural growth figures given in Table 11.10 were applied (average over two ages at death).

Thus, it is possible that our stature estimate of 160 cm for KNM-WT 15000 at his time of death, though a reduction from our earlier estimate, is still too high and should be reduced further by as much as 6–7 cm (to about 5′). However, there are still the other possibilities that either his remaining growth in long bone length or stature (Tables 11.7, 11.10) were misestimated. Here the problem is not an age difference but rather a body size difference: KNM-WT 15000 was larger than the great majority of 11–12-year-olds used to calculate these percentages. In terms of skeletal size, the U.S. white samples are closest, with 13–14-year-old males equivalent in femoral and tibial lengths to KNM-WT 15000 (when corrected for magnification and differences in measurement technique); the individuals from the archeological samples were even smaller. It is difficult to say whether being taller for his age than these reference samples would have had any marked effect on the percentage change in KNM-WT 15000's size to adulthood or on his body proportions as an adult, even assuming that he followed modern growth curves.

Conclusions

Because of the greater uncertainty in KNM-WT 15000's juvenile stature, we consider the adult stature estimates based on extrapolated bone lengths (which we *can* measure with certainty) to be more reliable than those based on extrapolated juvenile stature. These estimates all cluster around 185 cm, or 6′1″, not including the slight reduction in stature due to reduced cranial height relative to modern populations. His stature as a juvenile is tentatively estimated as about 160 cm (again without cranial height reduction), although this should be viewed as a maximum estimate, since there is some reason to believe that he was actually several centimeters shorter at death. The problem here is that we do not really have appropriate modern juvenile reference samples, matched with KNM-WT 15000 in both age and body size, and we have no stature estimation techniques for juveniles that might allow us to factor in the age/size variation. The adult stature estimate should be more certain, if we are correct in assuming a modern human-like adolescent growth pattern in skeletal size.

BODY WEIGHT

Body weights (masses) of earlier hominids have been estimated by a number of investigators on the basis of preserved postcranial skeletal size, mainly articular or external diaphyseal breadths (McHenry, 1975, 1976, 1988, 1991a; Oleksiak, 1986, Rightmire, 1986; Jungers, 1988b). We have extended this general approach to include true diaphyseal cross-sectional geometric properties, which exhibit a close relationship to body mass in living primates (Ruff, 1987), and have applied both diaphyseal and articular regression equations to the estimation of body mass in the Miocene hominoid *Proconsul* (Ruff et al., 1989).

These kinds of approaches have an advantage over other traditional methods (e.g., tooth size or bone length regressions, impressionistic comparisons to living populations) in that they employ skeletal structural features that bear a direct functional relationship to weight bearing, either through joints or long bone shafts. As long as there is an appropriate modern reference sample (or samples) of known body mass with which to compare such properties, they should provide the most accurate means of estimating body weight from skeletal remains. "Appropriate" here refers to "mechanically appropriate," referring to samples that are subjected to generally the same kinds of mechanical stresses in their postcranial skeletons.

Unfortunately, in the case of early hominids, there is no "mechanically appropriate" modern reference sample. It can be demonstrated that all post-Pleistocene humans have undergone a significant decline in skeletal robusticity, defined as diaphyseal cross-sectional size relative to body mass, compared with earlier hominids, including early anatomically modern *Homo sapiens*, Neandertals, and *Homo erectus* (Ruff et al., 1993). This structural change probably reflects a general and marked decline in muscular strength and/or activity level in recent human populations—in other words, a basic change in the mechanical environment. Nonhuman primates have both muscular strength and activity levels which are probably more similar to early hominids, but none is bipedal. Articular size shows a more constant relationship to (supported) body mass, regardless of activity level (Ruff et al., 1993), although it may be influenced by other more localized factors, such as degree of joint excursion (Ruff, 1988) or specific joint mechanics (Ruff, 1991). There are also some as yet unexplained peculiarities in the scaling of joint size within humans, for example, a highly positively allometric relationship of femoral head size to body mass that can have a major effect on weight predictions (Ruff, 1988; McHenry, 1991a). Furthermore, in all such analyses there is the major problem of finding adequate modern human skeletal samples with reliable associated body weights.[5] Because of these problems, various attempts to estimate body mass in early hominids from such structural parameters have arrived at quite different answers, depending on the particular skeletal features and reference samples employed (e.g., see McHenry, 1988, 1991a; Jungers, 1988b).

Superimposed on these problems here is the further complication of KNM-WT 15000's juvenile status. Work in progress on the juvenile sample from Pecos Pueblo (Ruff et al., in prep.) indicates that different skeletal structural properties (joint size, diaphyseal cross-sectional properties, bone length) have different growth trajectories, and that the relationship of each property to body size varies throughout growth to adulthood. Therefore, it is particularly inadvisable to attempt to estimate the body mass of KNM-WT 15000 using adult regression formulae based on these structural properties.

In the present analysis we took a different approach that is fundamentally morphometric rather than biomechanical. That is, we attempted to estimate KNM-WT 15000's body weight by reconstructing the size and shape of his body. The size and shape parameters were stature and biiliac breadth; multiple regressions of body weight on these parameters in modern humans were used as predictive equations.

This approach stems naturally from the re-

TABLE 11.11.

Body weight prediction equations derived from 40 world-wide adult population samples (Ruff, 1991).

REFERENCE GROUP	EQUATION[a]	r	S.E.E.[b]	% S.E.E.
Both sexes (n = 40)	WT = .689(ST) − 53.1	.641	7.1	12.2
	WT = 3.412(BI) − 32.2	.782	5.8	10.0
	*WT = .404(ST) + 2.712(BI) − 78.8	.853	4.9	8.4
Males (n = 21)	WT = .888(ST) − 87.1	.784	5.6	9.3
	WT = 3.884(BI) − 42.5	.794	4.0	6.6
	*WT = .413(ST) + 2.892(BI) − 84.8	.941	3.2	5.3
Females (n = 19)	WT = .659(ST) − 47.0	.451	8.4	15.0
	WT = 3.021(BI) − 24.2	.723	6.5	11.6
	WT = .252(ST) + 2.712(BI) − 55.4	.740	6.5	11.6

a. WT = body weight (kg); ST = stature (cm); BI = bi-iliac breadth (cm); asterisk indicates equation used for KNM-WT 15000.

b. Standard error of estimate.

sults presented earlier in the section on body shape, where the human body was modeled as a cylinder, with stature as the height of the cylinder and bi-iliac breadth as the breadth (also see Ruff, 1991). From this model it follows that body mass should be related to some product of bi-iliac breadth and stature.

Reference samples

To test this prediction, as well as to develop specific weight prediction equations for KNM-WT 15000, we carried out regressions of body weight on stature, bi-iliac breadth, and stature and bi-iliac breadth in several different living samples. These included: (a) a world-wide sample of 40 adult sex/population means—those from among the 71 samples used previously (see Figure 11.4) which had associated body weights; (b) the 32 adult male African sample means listed by Hiernaux (1968) with matched bi-iliac breadth, stature, and body weight; (c) a sample of several hundred Karkar Islanders (New Guinea), including both juveniles and adults, for whom extensive anthropometric data were collected by Harvey (1974; original data kindly made available by Dr. Leslie Aiello); and (d) a small sample of age-specific means for Tanzanian adolescents reported by Hautvast (1971).

TABLE 11.12.

Body weight prediction equations derived from a study of 32 adult male black African samples (Hiernaux, 1968).

EQUATION[a]	r	S.E.E.[b]	% S.E.E.
WT = .637(ST) − 50.0	.826	2.8	5.0
WT = 3.695(BI) − 37.1	.616	3.8	7.0
*WT = .738(ST) − .963(BI) − 42.6	.831	2.8	5.0

a. WT = body weight (kg); ST = stature (cm); BI = bi-iliac breadth (cm); asterisk indicates equation used for KNM-WT 15000.

b. Standard error of estimate.

Least squares (Model I) regression equations for the world-wide, African male, and Karkar Island samples are presented in Tables 11.11 to 11.13, respectively. Results are shown for various subsamples broken down by sex, age (when available), and other factors. Correlations, standard errors of estimate (S.E.E.), and percent standard errors of estimate (%S.E.E.) are given for each equation. Other combinations of predictor variables—bi-iliac breadth squared and stature or the product of bi-iliac breadth squared and stature (see Figure 11.5)—or regressions using log-transformed variables did not improve weight prediction.

The regression equation for the Tanzanian male adolescent sample (Hautvast, 1971) is based on the seven yearly means of stature, bi-iliac breadth, and weight for boys 11 to 17 years old in the Nyakuysa sample (the other sample reported in this study included only smaller boys through 14 years of age). The regression equation is: weight = .240(stature) + 3.991(bi-iliac) − 82.5 (r = .998, S.E.E. = .9 kg, %S.E.E. = 2.1). Although the number of data points is small here, it was thought important to include the results, since this is the only sample of African juveniles for which we could find matched stature, bi-iliac breadth, and body weight data.

Several observations are apparent from the results shown in Tables 11.11–11.13. First, the addition of bi-iliac breadth to stature (i.e., multiple regression equations) significantly improves body weight estimation, except among adult African males. In fact, in the world-wide sample (Table 11.11), bi-iliac breadth *alone* is a better weight predictor than stature alone, and the addition of stature to bi-iliac breadth does not add to predictive accuracy among females. The exception for African males is actually expected, since bi-iliac breadth varies so little (while stature varies greatly) in this group (Figure 11.5), for reasons discussed previously. Second, prediction of body weight from these skeletal dimensions in males is much better than in females (Table 11.11). This is probably due to greater variation in the soft tissue component of body mass among females. Finally, the best predictor equations—multiple regressions using both dimensions among adult males—give weight estimates with S.E.E.'s of only about ±3 kg (±5–6 %S.E.E.). This includes analyses for both sample means (Tables 11.11–11.12) as well as individuals (Table 11.13).

Body weight of KNM-WT 15000

Applications of the equations to KNM-WT 15000 are shown in Table 11.14. Only the multiple regression mixed-sex or male equations (indicated with asterisks in Tables 11.11–11.13) were used, since these gave the best predictions. For the New Guinea data, predictions based on three subsamples are shown: young adult males (18–39 years), all males 8 years or over (maximum age 67 years), and tall male adolescents over 150 cm in stature (mean 156.5 cm). For the first two subsamples only Karkar Islanders were used, since they are closer in general body build to KNM-WT 15000 than the Highland Lufa sample also collected by Harvey (1974). To obtain an adequate number of individuals for the third subsample, adolescents from both Karkar and Lufa were combined.

With the exception of one prediction equation, all body weight estimates for KNM-WT 15000 as a juvenile (i.e., at his time of death) fall between 47.7 and 50.7 kg. All estimates using reference samples that contain juveniles range between 47.7 and 48.7 kg. The higher estimate of 53.3 kg using the African adult male equation is probably due to the relative lack of data in this sample in the juvenile size range of KNM-WT 15000 (only one sample mean bi-iliac breadth of 23 cm or less). The one *juvenile* African sample from Tanzania gives a body weight estimate considerably lower—47.7 kg. The New Guinea reference samples that contain juveniles also give lower estimates of body weight than the New Guinea adult sample. Applying the adult male African equation to the juvenile Tanzanian stature and bi-iliac data results

TABLE 11.13.

Body weight prediction equations derived from a study of Karkar Islanders (Harvey, 1974).

REFERENCE SAMPLE	EQUATION[a]	r	S.E.E.[b]	% S.E.E.
Males, 18–39 yr (n = 130)	*WT = .455(ST) + 1.404(BI) − 54.4	.784	3.3	5.8
Males, 8–67 yr (n = 306)	*WT = .498(ST) + 1.877(BI) − 74.6	.965	4.0	8.7
Males,[c] 11–17 yr, stature >150 cm (n = 43)	*WT = .682(ST) + 1.309(BI) − 90.5	.738	4.0	8.1

a. WT = body weight (kg); ST = stature (cm); BI = bi-iliac breadth (cm); asterisk indicates equation used for KNM-WT 15000.

b. Standard error of estimate.

c. Males from both Karkar and Lufa regions (Harvey, 1974).

TABLE 11.14.

Body weight estimates (kg) for KNM-WT 15000 as a juvenile and as an adult.

REFERENCE SAMPLE USED AS A BASIS OF ESTIMATE[a]	JUVENILE WEIGHT	ADULT WEIGHT
World-wide adult, both sexes	48.2	68.1
World-wide adult males	47.8	68.5
African adult males	53.3	68.3
Karkar young adult males	50.7	67.1
Karkar males over 8 yr	48.2	67.5
New Guinea tall male adolescents	48.7	—
Tanzanian male adolescents[b]	47.7	—
Best estimate	48	68

a. See Tables 11.11–11.13. All prediction equations based on measures of stature and bi-iliac breadth; see Table 11.2 for dimensions of KNM-WT 15000.

b. Hautvast, 1971. See text for prediction equation.

in consistent overestimates of weight up to 17 years of age, while the world-wide adult equations applied to this sample are accurate to within 5% of actual body weight over the entire adolescent age range (12 to 18 years). On the basis of these observations, our best estimate of the body weight of KNM-WT 15000 as a juvenile is about 48 kg (106 lb.).

All estimates of KNM-WT 15000's body weight as an adult fall in a narrow range between 67.1 and 68.5 kg. (Reference samples that did not include adults were not used here.) The average best estimate of his adult weight is therefore about 68 kg (150 lb.). None of the available reference samples contain any data points quite in the stature range of KNM-WT 15000 as an adult (185 cm) —the Karkar individuals range up to 179 cm, the African males to 176.5 cm, the world-wide sample to 175.5 cm. However, our estimate seems reasonable when compared with other very tall, tropical populations for whom stature and body weight have been measured. Perhaps the closest in height are the 279 Shilluk and Dinka males measured by Roberts and Bainbridge (1963), in a study we cited earlier, who averaged 181.5 cm tall (s.d. = 6.1 cm) and weighed an average of 58.2 kg (s.d. = 5.9 kg). Judging from bi-iliac/stature ratios derived from somatological photographs (Roberts and Bainbridge, 1963, plates 1–4), besides the subjects being shorter, their bi-iliac breadths also averaged about 1 cm less than that estimated for KNM-WT 15000 as an adult. Thus, their mean body weight is about in line with that predicted for KNM-WT 15000, although perhaps it is a bit lower.

If KNM-WT 15000 was actually closer to 155 cm than 160 cm at his death, as suggested by some of the previous analyses (see above), this would reduce his estimated juvenile body weight by about 1–2 kg. Since this would also *increase* his juvenile bi-iliac/stature ratio from .144 to .148, and this ratio was used to estimate his adult bi-iliac breadth (Table 11.2), this lower stature as a juvenile would result in an increase in estimated adult body weight of about 1–2.5 kg according to most of the reference equations (although not the African adult male equation). Thus, the effects of this possible reduction in juvenile stature are not great, well within approximate 95% c.i.'s for the weight prediction equations (about ± 6–10 kg, except smaller for Tanzanian juveniles).

Conclusions

A morphometric approach using body height (stature) and width (bi-iliac breadth) to estimate body size produces relatively accurate predictions of body weight in a diverse group of modern population samples. When this approach is applied to KNM-WT 15000, estimated juvenile and adult body weights of about 48 and 68 kg, respectively, are obtained.

Discussion

As noted in its original description (Brown et al., 1985), as well as here and elsewhere in the present volume, KNM-WT 15000, either as a juvenile or extrapolated to adult size, was a large individual, even by modern human standards. As we extend this analysis to broader evolutionary and adaptational issues, a natural question to ask is whether KNM-WT 15000 was extraordinarily large for his own time period and geographic region. That is, how typical was he of other early African *Homo*?

No other early *Homo* specimen approaches KNM-WT 15000 in terms of completeness of preservation, so it is not possible to carry out the kind of analysis we performed here to estimate other body sizes. However, using the same general logic, it is possible to arrive at some reasonable estimates of both stature and body mass in a number of other specimens via a similar morphometric approach.

On the basis of what we showed earlier (also see Ruff, 1991), we first assumed that all early *Homo* were tropically proportioned, i.e., that they had long limbs relative to body size and that their bodies were absolutely narrow (relative to populations from higher latitudes). To estimate stature, we used the same modern reference populations employed for KNM-WT 15000. The femur is by far the best represented of the postcranial elements available for early *Homo*. Thus, the only applicable equations are those of Feldesman and Lundy (1988) for South African Bantu (see Table 11.3). The average of their Models I and II and male and female regression equations were used here.

Femoral bicondylar lengths and estimated statures for nine African early *Homo* specimens, including KNM-WT 15000, are shown in Table 11.15.[6] Three of the specimens (KNM-ER 1472, 1481, and 3728) date from about 1.89 Ma and have traditionally been considered *Homo habilis* or *Homo* sp., while the other six specimens date from 1.7–.7 Ma and have been considered *Homo erectus* (e.g., see Feldesman and Lundy, 1988; McHenry, 1991b). Only specimens with at least half the femoral shaft represented have been included; the least certain femoral length estimate is for KNM-ER 736, which preserves about half of the shaft. All of the length

TABLE 11.15.

Estimated body size of early *Homo* specimens from East Africa.

SPECIMEN	AGE OF FOSSIL (million years)	SEX	BICONDYLAR FEMUR LENGTH (cm)[a]	STATURE (cm)[b]	BODY WEIGHT (kg)[c]
KNM-ER 1472	1.89	?	40.0	149	47
KNM-ER 1481	1.89	?	39.5	147	46
KNM-ER 3728	1.89	?	39	145	45
KNM-ER 736	1.70	?	50	180	62
KNM-ER 1808	1.69	F?	48	173	59
KNM-ER 737	1.60	?	44	160	52
KNM-WT 15000 (adult)	1.53	M	51.7	185	68
OH 34	1.0	?	43	158	51
OH 28	0.7	F	45	163	54
Mean for fossil specimens				162	54
Mean for first 3 fossils (oldest)				147	46
Mean for last 6 fossils (youngest)				170	58
Mean for living Africans[d]				159.0	52.6
Mean for living Africans, pygmies excluded[e]				162.3	55.1
Mean for living Africans of tall stature[f]				166.6	57.0

a. Measured or estimated by authors from original specimens, except OH 34 (Day and Molleson, 1976).

b. Average of Model I and Model II, male and female equations based on study of South African Bantus (Feldesman and Lundy, 1988), except KNM-WT 15000 (present study).

c. Determined from multiple regression of body weight on stature and latitude in living populations (see text), except KNM-WT 15000 (present study).

d. Mean of males and females in 21 sex/population samples (see text).

e. N = 17 sex/population samples.

f. Mean of males and females in 4 populations—Sara (Chad), Tutsi (Zaire and Rwanda), Sahalian (Niger), Sudanian (Niger)—described by Hiernaux, 1975, as having an "elongated" stature.

estimates are ours, except for that of OH 34 (Day and Molleson, 1976).

To derive body weights for these specimens requires some way of factoring in variation in body breadth. As we showed earlier, absolute body (bi-iliac) breadth is highly correlated with latitude, in both living and fossil hominids. Therefore, using our world-wide sample of living human populations (Table 11.11),[7] we derived multiple regression equations of body weight on stature and latitude. The resulting formula (mixed sex) is: weight = .504(stature) + .218(latitude) − 29.1 (r = .833, S.E.E. = 5.0). The correlation and S.E.E. of this equation are almost as good as those of the stature/bi-iliac prediction of body weight (Table 11.11), as expected given the very high correlation of bi-iliac breadth and latitude. An absolute latitude of 3° was used for both Turkana and Olduvai specimens. Although some specimens have previously been assigned tentative sexes (the most certain of which is probably the female assignment of OH 28 because it is associated with a pelvic bone), the mixed-sex formula was used here for all specimens; use of a sex-specific formula makes little difference. The estimate of body weight for KNM-WT 15000 of 68 kg is the adult weight derived here using the bi-iliac/stature equations (Table 11.14). An estimate based on a stature/latitude regression for modern males (world-wide sample) gives an almost identical adult body weight of 67 kg for the specimen.

The mean estimated stature and body weight of the nine early *Homo* specimens in Table 11.15 are 162 cm (5′4″) and 54 kg (119 lb.). There is an increase of about 16% in stature and 26% in body weight between pre– and post–1.7 Ma specimens, with the later *(Homo erectus)* specimens averaging 170 cm (5′7″) and 58 kg (128 lb.). These figures combine, presumably, both males and females, of which the total representations are unknown. Even if it is assumed that most of the specimens are males, the average estimated body size for the *Homo erectus* sample is still large. This is illustrated by listing the mean stature and body weight for the living African samples from our world-wide comparative sample, shown at the bottom of Table 11.15. The mean body size of living Africans (male-female average), even excluding Pygmies, is remarkably similar to that of the early *Homo* sample. If only post–1.7 Ma specimens are considered, they average considerably taller than the mean of living Africans and are actually quite close in stature and estimated body weight to the average of four living "elongated" (Hiernaux, 1975) African populations from just south of the Sahara, also shown in Table 11.15.

Thus, KNM-WT 15000, although still the largest early *Homo* individual yet discovered, is not unusually large in comparison to other early *Homo erectus* from East Africa. There is also no evidence for an overall increase in body size in *Homo* over the last one-and-three-quarter million years. Early East African *Homo erectus* appears most similar in body morphology to tall living African populations who today inhabit relatively open, dry environments. The thermoregulatory advantage of such a physique in this kind of environment has been discussed in detail elsewhere, as have the consequences on obstetrics and lower limb biomechanics of a relatively linear body (Ruff, 1991). The relative constriction of the pelvis with an increase in body size in *Homo,* brought about by thermoregulatory constraints on body breadth, may have contributed to the development of the rotational birth process and infant altriciality in *Homo* (Ruff, 1991, figure 6).

Other early hominids (i.e., *Australopithecus*) were much smaller than *Homo erectus* (Feldesman and Lundy, 1988; Jungers, 1988a; McHenry, 1991a,b). This includes even the purportedly larger males of the same species (McHenry, 1991a,b). Their shorter and thus relatively wider bodies, unlike *Homo erectus,* could have been adapted for either open/dry or closed/wet environments (Ruff, 1991). They also would not have been subject to the same obstetric constraints as *Homo* (Ruff, 1991; Tague and Lovejoy, 1986).

From a methodological standpoint, it is of interest to attempt to estimate body weights of australopithecines using the morphometric approach favored here. One difficulty in such an exercise is the relatively more elliptically shaped pelvis characteristic of australopithecines, at least small early australopithecines (AL 288-1, STS 14; Robinson, 1972; Tague and Lovejoy, 1986). Because of this basic shape difference, which we must assume extended to some extent to the rest of the body, equations based on bi-iliac breadth and stature in living humans will systematically overestimate body weight in these specimens. The amount of error could be calculated by modeling the body as an elliptical rather than a circular cylinder (Figure 11.5) and factoring in an anteroposterior (A-P) as well as mediolateral (M-L) dimension.[8] The formula for volume = mass of an elliptical cylinder is $\pi \times r1 \times r2 \times l$, with l the length (stature) and $r1$ and $r2$ here considered the M-L and A-P radii, respectively. Unfortunately, there are no data available on living populations for external A-P breadth in the pelvic region, and chest A-P breadth, sometimes measured among the living, cannot be estimated accurately in fossils. As a next best approximation, we used internal midplane A-P breadth of the pelvis (Tague, 1989), for which there are data available for both recent humans (ibid.) and at least one early australopithecine, AL 288-1 (Tague and Lovejoy, 1986). Although an internal pelvic measurement, and not directly anteroposterior in orientation, this dimension should give a rough guide to variations in external A-P diameter. The average bi-iliac/A-P midplane breadth ratio in female American blacks, American whites, and Amerindians (average of four groups) is 2.063 (Tague, 1989 and pers. comm.). The same ratio in AL 288-1 is 3.515. The ratio between modern females and AL 288-1 in this shape index—.5871—can be used to correct for differences in the relative size of the second (A-P) radius in the elliptical cylinder formula above. Using this correction factor, AL 288-1's bi-iliac breadth of 25.8 cm (Ruff, 1991), her stature of about 107 cm (Jungers, 1988a), and the female world-wide reference sample bi-iliac/stature weight prediction equation (Table 11.11), we estimate the body weight of AL 288-1 at 23.5 kg.

It is also of interest to compare both this body weight and the weights of the early *Homo* specimens shown in Table 11.15 with those previously estimated from articular and long bone shaft dimensions (Jungers, 1988b; McHenry, 1988, 1991a). The comparison is shown in Table 11.16. Only AL 288-1 was specifically included in each of the analyses; where possible (given preservation of the specimen), body weights of other specimens were calculated from the published prediction equations. Results are shown for both intrahuman and general hominoid (with or without humans) equations, if given.

In general, body weights estimated by the morphometric approach used here are lower than

TABLE 11.16.

Comparison of estimates (kg) in present study with previous estimates of body weight in early hominids.

		ESTIMATES BASED ON FEMORAL SHAFT DIMENSIONS McHenry 1988		ESTIMATES BASED ON ARTICULAR DIMENSIONS[b]		
				McHenry, 1991	Jungers, 1988	
SPECIMEN	PRESENT STUDY[a]	INTRA-HUMAN	INTER-HOMINOID	INTRA-HUMAN	INTER-HOMINOID	NH INTER-HOMINOID
KNM-ER 736	62	70.7	93.3	—	—	—
KNM-ER 737	52	64.8	77.0	—	—	—
KNM-ER 1472	47	51.6	51.1	(49.6)	(67.7)	(80.8)
KNM-ER 1481	46	50.3	48.7	(58.4)	(86.9)	(106.5)
KNM-ER 3728	45	45.5	40.2	—	—	—
AL 288-1	23.5	38.9	29.9	26.7	30.4	30.4

a. See Table 11.15 and text (for AL 288-1).

b. Numbers in parentheses indicate our calculations using published equations. NH = nonhuman.

those estimated from articular or shaft breadths. Interestingly, the only exception is McHenry's (1988) weight estimates for KNM-ER 3728, which are as low as or lower than ours. From personal examination and results of other analyses (Ruff et al., 1993), we have concluded that this femur has almost certainly suffered significant external weathering and loss of periosteal bone; thus, it has smaller than expected external shaft dimensions, resulting in a lower weight estimate. Except for this specimen, McHenry's (1988) weight predictions based on proximal femoral shaft external breadths in a modern human autopsy collection are consistently higher than ours by about 10–15% in three specimens and almost 25% in the fourth. We explain this difference as very likely the result of generally higher mechanical loadings on the skeleton in earlier hominids, which would produce more robust long bones relative to body weight (Ruff et al., 1993). McHenry's (1991a) intrahuman articular (femoral head) predictions are quite close to ours for KNM-ER 1472 but higher for KNM-ER 1481. Applying Jungers' (1988b) hominoid articular equations to these two specimens results in very large body weight estimates, not surprising since nonhuman hominoids have relatively smaller femoral heads than humans (Ruff, 1988) (it should be noted that these equations were developed specifically for australopithecines, not *Homo*).

The results for AL 288-1 are particularly interesting. Our morphometric estimate is closest to McHenry's (1991a) recent estimate of 26.7 kg, which incorporated the positive allometry of the femoral head exhibited by humans (Ruff, 1988) (his earlier estimates did not). Other estimates using both hominoid and human reference samples are all much higher. Even given some error in our method of correcting for differences in pelvic (body) cross-sectional shape, the unknown effects of longer upper limbs and possibly shorter lower limbs relative to trunk size in this specimen (Jungers, 1982; Franciscus and Holliday, in press), as well as error in its stature estimate, still we would favor the lower end of the range of body weight values shown in Table 11.16, perhaps on the order of about 25 kg. This in turn would imply that both lower limb shaft and articular dimensions of AL 288-1 are as large or larger relative to body weight as modern humans and are not like great apes in proportion (Ruff, 1988).

Thus, a morphometric approach to the estimation of body size in earlier hominids may provide some independent evidence useful in evaluating the results of other techniques. Theoretically, this kind of approach could be used to investigate various biomechanical questions, such as the robusticity of particular skeletal elements relative to body mass, since it does not rely on mechanical factors and thus would not involve circular reasoning. We are currently investigating such questions in various structural analyses of the postcranial skeleton of KNM-WT 15000 and other early hominids (Ruff et al., 1993).

Summary

In terms of both relative limb length and body breadth, KNM-WT 15000 had a body proportioned similarly to modern human tropical populations and unlike higher-latitude populations. This is not a result of any special phylogenetic relationship but rather the application of basic thermoregulatory principles. KNM-WT 15000's tall, thin body and long limbs were adaptive for diurnal activity in a hot, relatively dry climate, in an environment that was probably more open than closed. Other contemporary *Homo erectus* from East African were also tall. This morphology appears to represent a fundamental distinction from both earlier and (probably) contemporary australopithecines, who were less specifically adapted for inhabiting open environments. In addition to its environmental implications, the body shape of KNM-WT 15000 indicates that only tropically proportioned modern reference samples should be used in reconstructing his stature and body weight.

Using prediction equations based on adult reference samples in the same general size range as KNM-WT 15000, his stature at the time of his death is calculated at about 160 cm (5′3″), not including a slight downward adjustment for his lower cranial height relative to modern humans. This should be taken as a maximum estimate, however, since there are some other indications that he may actually have been several centimeters shorter than this. Stature estimates here are complicated by the lack of currently available prediction equations for juveniles *per se,* with no modern data at all for juveniles of about the same size and age as KNM-WT 15000.

The adult stature of KNM-WT 15000 can be estimated either from his extrapolated adult long bone lengths or from his juvenile stature extrapolated to adulthood. The first method is favored here, since it avoids the problems in juvenile stature estimation referred to above. Remaining growth in long bone lengths is estimated from several modern samples, living and archeological. From his extrapolated adult skeletal size, and again using appropriately sized modern reference samples, we estimate the adult stature of KNM-WT 15000 at about 185 cm (6′1″). Extrapolations from his estimated juvenile stature yield adult stature estimates several centimeters higher, but we believe these estimates to be less accurate. Either method assumes that growth trajectories (in long bone length or stature) in KNM-WT 15000 were similar to modern human populations.

Body weight of KNM-WT 15000 is estimated using a morphometric rather than biomechanical approach, avoiding difficulties resulting from probable differences in mechanical loading of the skeleton between earlier and recent hominids. Multiple regressions of body weight on stature and bi-iliac breadth are derived for several different living human samples, including tropical adults and juveniles. Application of these equations to KNM-WT 15000 yields body weight estimates of about 48 kg (106 lb.) as a juvenile and 68 kg (150 lb.) as an adult.

KNM-WT 15000, who would have been the largest early hominid known had he lived to adulthood, is not extraordinarily large when compared with other *Homo erectus* from East Africa. The average estimated stature and body weight of six African *Homo erectus* specimens dating to 1.7–.7 Ma is 170 cm (5′7″) and 58 kg (119 lb.). These include, probably, both males and females. Body weights estimated from other skeletal parameters in modern reference samples, such as articular and diaphyseal breadths, are almost uniformly higher than this, probably because of the greater mechanical loading and thus greater robusticity of the postcranial skeleton in earlier *Homo.*

Acknowledgments

We thank Leslie Aiello for the use of her data from New Guinea, Bob Tague for unpublished modern pelvic data, and Holly Smith for very useful discussions. C.R. was funded by NSF grants BNS-8519349 and BNS-8919155.

NOTES

1. Stature continues to increase very slightly to at least age 20 (Trotter and Gleser, 1958), because of continued growth in length of the vertebral column, while all long bone epiphyses, at least in the lower limb, are generally fused by age 18 (Krogman, 1962). Thus, the femur/stature index decreases slightly from 18 to 21+ years (by about .0010 to .0015 units, calculated from data given in Trotter and Gleser, 1958, tables 3 and 8). Some of the variation between the African samples shown in Figure 11.11 may be due to the younger average age of most of the Ugandan samples (see Table 11.3), although one of these (the Nilote sample) included no individuals under 20 years and another (Nilohamite tribesmen) may very well have primarily included individuals older than that.

2. Maresh's (1970) bone lengths were corrected for magnification from information given in Green et al. (1946; also see Gindhart, 1973) and adjusted for a different tube-film distance: femoral lengths over 45 cm were multiplied by .954, femoral lengths under 45 cm by .971, tibial lengths over 36 cm by .969, and tibial lengths under 36 cm by .977. Anderson et al.'s (1963) bone length measurements were converted to maximum lengths using their published illustrations of the method (Anderson et al., 1948, figure 1) and additional information derived from our ongoing study of a sample of juvenile skeletons from Pecos Pueblo (Ruff, unpublished data): their femoral lengths were multiplied by 1.02, tibial lengths by 1.06.

3. One other potential long bone growth study—the radiographic study of British children reported by Tupman (1962)—is not included here, since there are some apparent discrepancies between his results and those reported by all other researchers. Tupman used skeletal rather than chronological ages (using hand-wrist standards), and reported a cessation of growth in length of the femur and tibia in boys of his sample of 15.4–15.5 years. Every other comparable study of which we are aware, cross-sectional or longitudinal, has found a much later cessation of growth of the femur and tibia in boys—about 17 to 18 years (Anderson et al., 1963; Maresh, 1970; Gindhart, 1973). The later age is also consistent with many studies of epiphyseal union in the femur and tibia (Krogman, 1962). Tupman claimed to have used only individuals whose skeletal and chronological ages matched to within 1 year, "and usually to within six months," so that the difference in types of ages used in the studies should not have been an important factor, although this cannot be ruled out. It is possible that since the design of his study—which was basically cross-sectional and did not include large samples of individuals of each sex at each age group (about 10 boys at each year)—introduced some bias. In any event, his calculations of remaining growth in femoral and tibial length to "adulthood" (i.e., 15–16 years), when applied to the average for his 11–12-year-old boys, are some 20–30% smaller than those obtained by all other investigators (percentage changes in length; see Table 11.7). Therefore, until these discrepancies can be resolved, we consider his long bone growth estimates to be unreliable.

4. An additional population sample of Turkana from Kenya (Little et al., 1983; Little and Johnson, 1987) was not included here because it is apparent that this sample is characterized by a negative secular trend in adult stature; in other words, in cross-sectional comparisons stature continues to increase through the fourth or fifth decade (also see Tobias, 1985, table 1). This phenomenon is characteristic of a number of populations throughout the world, including some others in East Africa (Little and Johnson, 1987). Its effect is to artificially inflate the apparent increase in stature from childhood to full "adult" stature. Thus, the Turkana appear to show a relatively large percentage change from 11–12 years to "adulthood," much larger than any of the other samples included in Table 11.10. This negative secular trend may also indicate a recently worsening economic situation in this population, which could explain the delayed adolescent growth spurt observed (Little and Johnson, 1987, figure 3). It seems unlikely that such conditions generally characterized *Homo erectus,* although this cannot be ruled out. (Note that the Australian aborigine sample studied by Abbie, 1967, does not exhibit these characteristics.) In any event, use of this sample as a standard for estimating remaining growth in stature of KNM-WT 15000 would seem inadvisable.

5. We have recently addressed this problem using radiographs of living individuals whose body weights could be ascertained (Ruff et al., 1991). This study highlighted another problem in using such samples for estimating body weight in earlier hominids, how-

ever: modern industrialized populations tend to gain weight during adulthood, but this weight gain is not reflected in any change in articular size. Thus, articular dimensions do not correlate particularly well with body weight and also show a systematic deviation relative to what would be expected for earlier, presumably leaner humans.

6. OH 62, a purported *Homo habilis* (Johanson et al., 1987), is not included here, partly because the preservation of the femur makes it difficult to arrive at an accurate length estimate, and partly because we believe that the grouping of this specimen within the same taxon as the other specimens of similar age listed in Table 11.15 is questionable.

7. The number of sex/population means used here was 39 rather than 40, since one sample population (Apaches) had recently migrated and its original latitude is not known with any precision.

8. Note that this was not necessary in our previous analyses when basic body cross-sectional shape was assumed to be similar in modern reference and fossil samples, a reasonable assumption for KNM-WT 15000 and, by extension, other early *Homo* (also see Ruff, 1991). We did examine the effect of factoring in a body anteroposterior breadth dimension in our weight prediction equations for the modern reference samples. Among Hiernaux's (1968) male African sample, the addition of chest A-P breadth to bi-iliac breadth and stature resulted in slightly better body weight predictions (S.E.E. decreased from 2.71 to 2.24 kg; $n = 28$). In the Karkar Islander males, either young adults or the total sample (8+ years), the inclusion of A-P chest breadth had an even smaller effect, reducing the S.E.E. of weight by less than .2 kg. Chest breadth data were too scanty to test in the world-wide modern human reference sample, although we did note no apparent systematic relationship between chest shape and climate. Since chest breadths cannot be measured with any accuracy in skeletal material, these results cannot be applied directly to fossil hominids. However, they indicate that using only an M-L body breadth dimension in the cylindrical model (Figure 11.5) and in weight predictions for *Homo* probably did not bias the results in any significant way.

CHAPTER 12

THE THORACIC AND LUMBAR VERTEBRAE

BRUCE LATIMER AND CAROL V. WARD

Vertebrae are fragile and rarely preserved in the fossil record. The Nariokotome Homo erectus *skeleton has the most complete vertebral column known for an early hominid. The specimen reveals fundamental adaptations of the spine resulting from cumulative effects of individual segments. With it we may determine whether hominids of 1.5 million years ago had yet attained those features of the human spinal column that are critical to our unique form of posture and locomotion.*

The hominid vertebral column has undergone dramatic reorganization to accommodate the kinematic and kinetic consequences of force transmission during habitually upright posture and locomotion. These functional adjustments and their correlated structural modifications have long been used to differentiate humans from apes (Huxley, 1863; Cunningham, 1886). Unfortunately, evolutionary studies of the human vertebral column have been handicapped by the absence of appropriate fossil material. Vertebrae are relatively fragile and rarely survive as fossils (Brain, 1981). In addition, while the hominid vertebral column is readily identifiable when viewed in its entirety, many of its distinctive characteristics reside in the cumulative effects of its numerous elements. As a result, isolated vertebrae can be less useful in a diagnostic sense.

In this light, it is especially significant that a nearly complete thoracolumbar vertebral column is included in the 1.5-million-year-old *Homo erectus* skeleton, KNM-WT 15000, recovered from the site of Nariokotome, West Lake Turkana (Brown et al., 1985). The spinal column of this remarkable skeleton is well preserved; a minimum of nineteen presacral vertebrae are at least partially represented.

KNM-WT 15000 is estimated to have been roughly equivalent in its dental developmental stage to an 11-year-old modern human (Smith, Chapter 9, this volume). This juvenile developmental status is supported by the absence of secondary center synostosis in the thoracic and lumbar vertebrae. As a juvenile, KNM-WT 15000 presents a unique opportunity to study ontogenetic processes in *Homo erectus*. However, any interpretations based on this single specimen must be viewed as highly tentative, as earlier and later developmental stages of the *H. erectus* vertebral column are almost entirely unknown at present. (An adult seventh cervical vertebra and a first thoracic vertebra, KNM-ER 164 C, were recovered in 1969 from Koobi Fora; see Leakey and Leakey, 1978. These elements were badly damaged and provided little comparative information for the present analysis.) This problem is compounded by the enhanced morphological variability that naturally accompanies ontogeny (Mensforth, 1985; Tanner et al., 1990).

In the following analyses, the KNM-WT 15000 vertebrae are compared with those of juvenile and adult modern humans and with those of modern chimpanzees. This comparative sample provides an appropriate context within which to place KNM-WT 15000, permitting a preliminary assessment of the ontogenetic and evolutionary status of the fossil vertebrae.

Materials and Methods

The following discussion of the KNM-WT 15000 vertebrae adheres to the identification and numbering system used in the accompanying anatomical descriptions (Walker and Leakey, Chapter 7, this volume). Focus is primarily on the thoracic and lumbar vertebrae, as these regions are well represented in the fossil skeleton. Thoracic and lumbar vertebrae were identified using articular facet orientation.

Measurements of the fossils were taken on resin casts and were checked for accuracy against the original specimens. Modern skeletal material used in this study is part of the Hamann-Todd Osteological Collection curated by the Cleveland Museum of Natural History. No skeletons in the comparative sample had any evidence of neoplastic, traumatic or degenerative diseases involving the spine and none had anomalies suggesting congenital or developmental spinal malformation.

The sample of 10 chimpanzees included 5 males and 5 females. All specimens were fully adult, as judged by complete synostosis of all skeletal elements. Comparison of the KNM-WT 15000 vertebrae with those of chimpanzees allows an assessment of the evolutionary development of the *Homo erectus* spinal column.

The modern humans were divided into three subsamples: a group of 30 adults ages 30 to 35 years, each with 17 thoracolumbar vertebrae (12 thoracic and 5 lumbar); a group of 10 adults ages 22 to 60, each with 18 thoracolumbar vertebrae (12 thoracic and 6 lumbar); and a juvenile group of 12 individuals ages 10 to 14 years at time of death (see Table 12.1).

The primary adult human sample was lim-

TABLE 12.1.

Characteristics of the juvenile sample ($N = 12$).

SPECIMEN NUMBER	AGE	SEX
404	11	M
526	11	F
633	13	F
645	12	F
710	10	M
1240	12	F
1441	10	M
1688	10	M
1772	12	F
2118	13	F
2135	14	F
3112	14	M

ited in age distribution to between 30 and 35 years. This procedure avoids potential problems that can occur in younger specimens owing to late fusion of the annular ring and in older specimens to age-progressive deterioration of the vertebral column. The sexes were equally represented.

The second adult human group was selected on the basis of each individual having 6 lumbar vertebrae in addition to 12 thoracic vertebrae. This latter group provides a comparative sample for KNM-WT 15000, which also has 6 lumbar vertebrae (see discussion below). Comparison of the two groups of adult humans also highlights any potential morphological consequences attributable to variations in the number of lumbar elements. Only males are represented in this sample. In the following graphs, data from the 6-lumbar group are presented only where their inclusion assists in the interpretation of the fossil specimen.

Although KNM-WT 15000 and 10 of the modern human adults have 6 lumbar vertebrae, no individual in the modern human juvenile sample had more than 5. An examination of one thousand adult spines in the Hamann-Todd Collection indicated that roughly 4 percent had an additional lumbar vertebra. Thus, it is not surprising that no juvenile individual in the sample possessed 6 lumbar elements.

No effort was made to partition the juvenile or adult comparative samples on the basis of race or sex. Similarly, no assessment was made of potential sex and race differences known to occur in adult spines (Scoles et al., 1988) as well as in vertebral developmental processes (Veldhuizen et al., 1986; Nehme et al., 1980).

The developmental status of each juvenile spine was assessed using several morphological indicators, such as overall size, billowing of end plates, depth and rugosity of radial grooves, and absence of secondary centers of ossification. In neither KNM-WT 15000 nor any of the juvenile human sample was there any evidence of synostosis of the annular ring. Although the juveniles chosen range in age from 10- to 14-years-old, each specimen was selected on the basis of its similarity in developmental status to KNM-WT 15000. Therefore, data obtained from the juvenile specimens are combined and used as sample averages and are not reported as a consecutive chronological series. The 4-year age range of human juveniles that exhibit roughly equivalent stages of development illustrates the extreme variability in modern vertebral ontogenesis. With this in mind, we made no effort to estimate the chronological age of the fossil specimen on the basis of the developmental stage of the vertebral elements. Such estimates using more reliable indicators are made elsewhere in this volume (e.g., Chapter 9).

The division of the modern human sample into juveniles, in whom vertebral growth was in progress, and adults, in whom growth had ceased, allows an appreciation of the relative magnitudes and velocities of growth that are necessary to achieve adult human morphology. It also highlights the differential growth that takes place throughout the vertebral column. Additionally, comparing KNM-WT 15000 with this juvenile sample allows appraisal of morphological similarities between *Homo erectus* and modern human juveniles of a similar ontogenetic stage. Such comparisons can suggest whether *Homo erectus* achieved adult vertebral morphology along a developmental trajectory like that of *Homo sapiens*.

Areal measurements were taken by tracing relevant outlines on flexible, clear acetate overlays placed directly over the vertebral surfaces. This direct method of measurement avoids potential scaling and parallax problems associated with the use of photographs. It also allows accurate conformation to surface topography, providing a reliable method of measuring curved surfaces (e.g., articular facets). Tracings were digitized and areas recorded to the nearest square millimeter. Radiographs were taken using a Hewlett-Packard Cabinet X-ray system, Faxitron series, model A3855A, with Kodak X-Omat TL radiographic film.

Comparing specific vertebral elements is often complicated by the fact that individuals or species may have variable numbers of thoracic or lumbar vertebrae. To circumvent this complication, we identified and compared the individual vertebral elements according to their location relative to the sacrum. Each vertebral element was numbered sequentially from the lumbosacral junction cranially. Thus, the first presacral vertebrae is referred to as number 1, and numbers increase sequentially in a cranial direction. This convention avoids numerical and nomenclatural discrepancies and allows direct comparison of predominant morphological patterns within the thoracolumbar vertebral columns of the various samples. All the graphs, except Figure 12.11, employ this convention. In the accompanying text, individual vertebral elements are also designated by conventional nomenclature.

In the following study, vertebral elements of KNM-WT 15000 are examined for the presence of human-like thoracic and lumbar curvatures. The influence of the juvenile developmental status of KNM-WT 15000 on its vertebral morphology is also examined. This study also addresses potential evolutionary and structural implications of a sixth lumbar vertebra in *Homo erectus*. Finally, the evolutionary development of the hominid vertebral column is briefly discussed in light of the information provided by KNM-WT 15000.

Additional methodologies are more easily understood in the context of their use and are described below. In order to reduce clutter and facilitate interpretation of figures, only species means are presented graphically. Collateral data are provided in associated tables.

Vertebral wedging

The human spine has undergone profound alteration of its structure and function in response to the demands of erect bipedal locomotion. These evolutionary modifications include a suite of pelvic and thoracic adaptations that function to control the orientation of the torso in a sagittal plane during bipedal walking and running.

Postural control of the hominid torso in a sagittal plane is accomplished through several associated anatomical adaptations. These include: (1) Vertical alignment of the sacroiliac and hip joints to permit the body weight vector of the torso to pass through the biacetabular axis. (2) Reduction in iliac height, which lowers the torso's center of mass and results in closer approximation of the sacroiliac joints to the biacetabular axis. This alteration reduces torque about the hip joints such as that produced by the posteriorly directed ground reaction during single limb heel strike in combination with the anteriorly directed momentum of the torso. The diminishment of iliac height also necessitated the oblique orientation of the sacrum (and consequently the lumbosacral articulation) so that the sacral body would not occlude the obstetric canal. This anteroinferior inclination of the cranial plane of the first sacral vertebra, in turn, required the development of the lumbar lordosis. The lumbar lordosis moves the torso's center of mass over the sacroiliac and the hip joints (see number 1), thereby reducing the muscular effort that would otherwise be required for postural control of the torso in the sagittal plane. (3) Invagination of the thoracic vertebral column toward the center of the thoracic cage, which creates an increased moment arm for the epaxial musculature. Associated with this modification is the more dorsal positioning of the costal angles and dorsal prolongation of the ilium. These skeletal alterations increase the moment arms of the erector spinae and the gluteus maximus, respectively, helping to stabilize the torso over the hind limbs (see discussion in Tague and Lovejoy, 1986; Lovejoy, 1988; and Jellema et al., Chapter 13, this volume).

It is clear that the pelvis (see Walker and Ruff, Chapter 10, this volume) and rib cage (see Jellema et al., Chapter 13, this volume) of KNM-WT 15000 had already undergone these striking modifications. The thoracolumbar vertebral column of this individual also demonstrates dramatic adaptive alterations for postural control of the torso.

As noted, orientation of the center of mass of the thorax over the biacetabular axis requires the presence of a lumbar concavity or lordosis. In order to similarly orient the more cranial regions of the vertebral column, additional spinal curvatures are necessary. These take the form of the thoracic kyphosis and the cervical lordosis. Humans are unique among extant mammals in having this series of sinusoidal curves in their vertebral columns. The structural and mechanical correlates of these curvatures are marked and numerous, and they serve to differentiate hominid vertebrae from those of other mammals.

Spinal curvatures are partially a consequence of differential anterior and posterior heights of the intervertebral discs. Wedging of the individual vertebral bodies also contributes to vertebral curvatures (Figure 12.1 and Table 12.2). The method for calculating the degree of wedging for each vertebra, shown in Figure 12.1, follows that of Digiovanni et al. (1989) and Scoles et al. (1991). This method is favored over the use of simple ratios, as it takes into account the anteroposterior

TABLE 12.2.

Vertebral wedging (angle α, in degrees); see Figure 12.1.

	HOMO SAPIENS								
	ADULT WITH 5 LUMBAR VERTEBRAE (*N* = 20)		ADULT WITH 6 LUMBAR VERTEBRAE (*N* = 10)		JUVENILE (*N* = 12)			*PAN TROGLODYTES* (*N* = 10)	
LEVEL	Mean	S.D.	Mean	S.D.	Mean	S.D.	KNM-WT 15000	Mean	S.D.
18			4.06	1.00			5.7		
17	2.81	3.26	1.50	1.51	6.86	1.79	5.3	6.88	2.93
16	− .19	2.27	.12	1.54	4.96	3.91	3.5	8.05	2.86
15	1.03	1.99	.98	2.82	4.04	2.09		5.09	1.98
14	2.23	2.27	1.96	1.68	2.63	2.42	5.2	5.03	2.69
13	2.57	1.70	2.34	2.22	4.69	2.49		4.33	2.69
12	4.63	2.03	2.92	1.36	4.78	1.33	5.8	3.37	1.95
11	3.73	1.41	3.10	.89	5.78	1.37	3.0	3.61	2.22
10	3.23	1.53	2.73	1.27	4.78	1.69		3.96	2.29
9	1.67	1.98	1.26	1.93	3.97	1.88		3.56	2.03
8	1.90	1.58	2.29	1.09	2.96	2.49	4.8	4.08	2.37
7	3.96	1.87	2.88	1.46	3.89	1.76		4.36	1.85
6	4.38	1.97	3.33	.60	3.16	1.81	5.4	5.38	3.00
5	3.00	2.39	2.62	1.00	3.68	1.73	4.7	5.58	2.21
4	− .47	2.04	− .41	1.85	1.49	1.23		5.60	3.36
3	−3.11	2.06	−3.49	1.41	1.36	2.01		4.65	2.62
2	−3.72	2.42	−3.59	1.48	− .99	2.14	− 5.4	3.49	2.71
1	−7.93	2.12	−5.83	.79	−5.45	1.91	−10.2	.48	2.43
0									

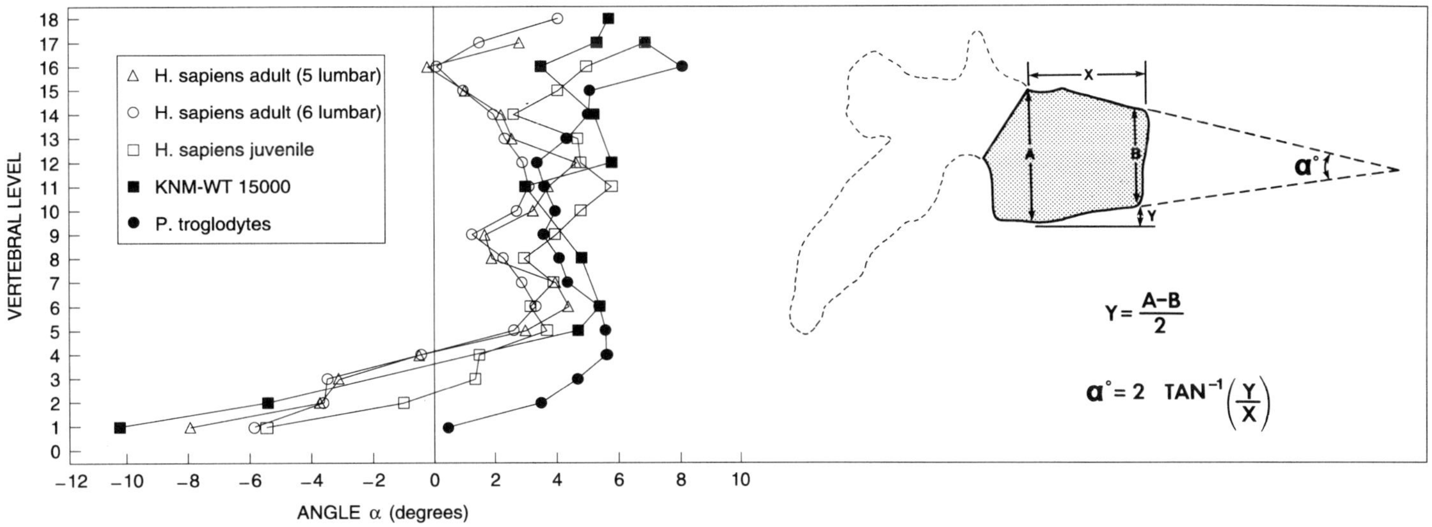

FIGURE 12.1.
Degrees of wedging (angle α) of the thoracolumbar vertebrae in HOMO SAPIENS, PAN TROGLODYTES, *and KNM-WT 15000. Only means are presented graphically; additional details are provided in Table 12.2. Vertebral level refers to the vertebra's sequential position relative to the sacrum: 0 designates the first sacral vertebra and 1 refers to the most caudal lumbar vertebra. The method of computing wedging is illustrated at right. Negative wedging indicates a lordosis and positive wedging a kyphosis. Note that all the hominids share a similar pattern. Also note that the chimpanzees fail to demonstrate a lordosis in their lumbar vertebrae.*

length of the vertebral bodies. Positive values reflect anterior wedging (kyphosis) and negative values reflect posterior wedging (lordosis). A greater degree of wedging (larger angle α) in either direction indicates a smaller radius of curvature of the region in question.

Adult humans have a remarkably consistent pattern of vertebral wedging. The greatest degree of anterior wedging occurs around the first thoracic element (T1), then at the midthoracic level (T6 or T7), and again at the thoracolumbar junction (T12–L1). This pattern creates a unique wedging profile with three peaks of thoracic kyphosis followed by the lumbar lordosis. Alternatively, this pattern could also be described as two thoracic regions that show relatively reduced anterior wedging (T3–T4 and T9–T10) (see Figure 12.1).

No important differences occur between human adults with 5 lumbar vertebrae and those with 6 lumbar vertebrae. The latter group tends to have slightly less lordotic (negative) wedging in their ultimate lumbar elements. As a consequence, the total summed lumbar lordosis is lower than in the 5-lumbar group, as is the associated kyphotic (positive) wedging in their thoracic vertebrae. Despite this slight difference in the magnitude of the curvatures, the 6-lumbar group demonstrates the hominid pattern of 3 kyphotic peaks in the thoracic column followed by a lordotic curve in the lumbar region.

Decreased wedging in this 6-lumbar sample is probably not a direct result of this group having an additional lumbar vertebra. Instead, it is more likely to be a consequence of this particular sample. Although selected randomly with regard to sex, this sample was comprised of 10 male skeletons. Males typically exhibit less lordosis than females (Williams and Warwick, 1980), providing an explanation for the lower magnitudes of curvatures seen in this group.

Relative to adults, human children have greater anterior vertebral wedging and, consequently, greater kyphosis in the upper and middle thoracic regions. They also have less negative wedging, and thus lordosis, in their lumbar vertebrae. The overall picture is one of a flatter profile and a more uniformly kyphotic spine in juveniles. Although the magnitude of wedging differs between adults and children in the lower thoracic and lumbar vertebrae, the pattern is comparable. However, this is not the case in the upper thoracic region, where adults and juveniles differ in a striking manner. In the upper thorax, juveniles display much greater kyphosis in the first through the third thoracic vertebrae.

Clearly, differential growth of the vertebral bodies must occur in order to transform the juvenile condition to an adult one. In this sense, Figure 12.1 is a method of assessing superoinferior growth of the spine when seen in lateral view. The juvenile spine will acquire the adult morphological pattern by differentially adding bone to the anterior margins of the vertebral bodies. This will decrease the kyphosis of its upper and middle thoracic column and simultaneously increase the lordosis of its lumbar vertebrae. This pattern of differential growth is further supported by earlier observations that synostosis of the posterior and lateral margins of the ring apophysis precede anterior fusion, contributing to the augmentation of anterior body height (Scoles et al., 1991). The magnitude of differences between adult and juvenile conditions and the amount of growth which must occur for juveniles to achieve the adult condition are in general agreement with the skeletal measurements of Tulsi (1971).

Importantly, none of the vertebrae included in the juvenile sample have fused apophyseal rings. Much of the dissimilarity between adult and juvenile wedging results from differential superoinferior body thickness provided by the fused annular ring in the adults. Because fusion of the ring apophysis does not occur until late in the second decade or even early in the third decade of life (Bick and Copel, 1950), and because vertebral growth does not cease entirely until the ring fuses (Scoles et al., 1991), adult vertebral conformation is a relatively late ontogenic acquisition. Highly idiosyncratic loading conditions during this extended developmental period explain the marked levels of variation among individuals.

Chimpanzees show much less variability in wedging along the vertebral column than do the hominids. They have a relatively consistent amount of anterior wedging and a nearly uniform radius of curvature throughout the middle and lower thoracic column. Perhaps most importantly, although chimpanzees do have some flattening (increase in radius) in the lumbar region, they fail to develop a lordosis, or negative wedging, in this region. The reduction in lumbar vertebral elements that characterize these animals (3–4 lumbar vertebrae; Keith, 1903; Schultz, 1930) and their lack of posterior vertebral body wedging in the lumbar region militate against their having a hominid-like lumbar lordosis.

KNM-WT 15000 exhibits a demonstrable hominid pattern with 3 distinct kyphotic peaks; one at the top of the thoracic column (T1), one at midthorax (T7), and another at the thoracolumbar junction (L1). This anteriorly concave thoracic profile is followed by an unmistakable and pronounced lumbar lordosis. On the basis of its vertebral profile, KNM-WT 15000 differs dramatically from the chimpanzees. This specimen can be allied functionally and morphologically only with modern humans.

Trunk balance depends upon the thoracic and lumbar curves having approximately equal magnitudes with apices in opposite directions. Any significant alteration in one of the primary curves necessitates an equal but opposite alteration in the other. The high degree of lumbar lordosis in KNM-WT 15000 is counterpoised by a similarly extreme thoracic kyphosis, preserving the position of the upper body over the pelvis.

In light of the ontogenic alterations described above for modern humans, it seems unlikely that KNM-WT 15000 would parallel this pattern and with further maturity decrease its kyphosis while increasing its lordosis. Since alterations like these would potentially unbalance the column, the fossil specimen may be better thought of as representing an adult condition. This suggestion is formed solely on the mechanical necessity of trunk balance in a habitual biped and must be viewed as tentative in the absence of information regarding adult vertebral morphology in *Homo erectus*.

Vertebral wedging patterns in the hominid thoracolumbar spine are markedly dissimilar from those of any other mammal. In combination with the associated modifications in the pelvis, these curvatures are probably among the earliest anatomical alterations for habitual bipedality. The primary reason for these adaptive changes is postural control of the torso in a habitually erect biped. Equivalent vertebral changes are clearly seen in

FIGURE 12.2.
Posterior views of lumbar vertebral columns in P. TROGLODYTES (right) *and* HOMO SAPIENS (left). *Note the caudally progressive increase in the distances between superior facets (black ellipses and shaded areas) in the hominid. This is in contrast to the progressively diminishing distance in the ape. The pyramidal configuration in hominids enables the contiguous facets to imbricate (see arrows) in the position of lumbar extension and allows the development of the lumbar lordosis.*

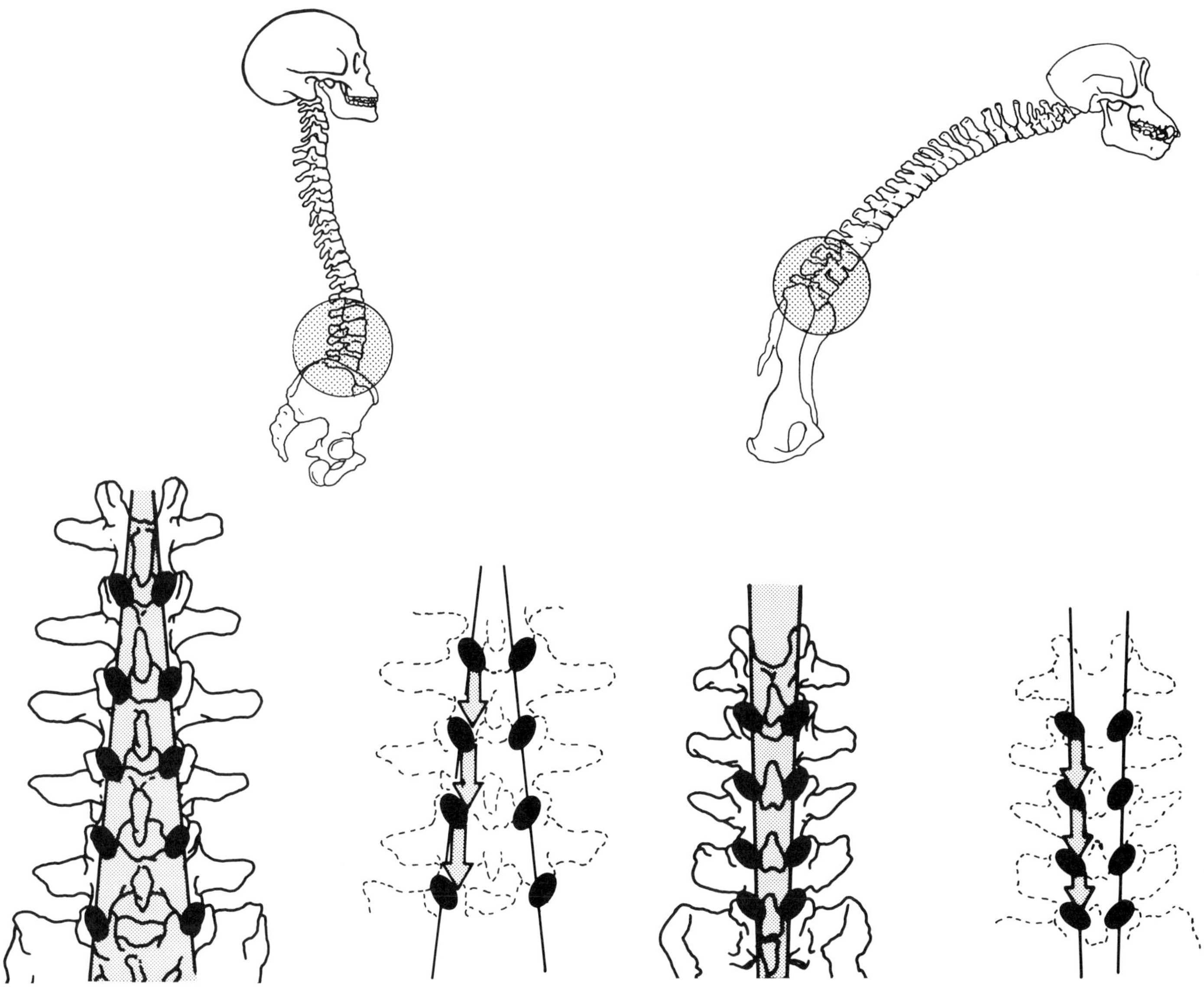

the earlier hominid genus *Australopithecus* (Robinson, 1972; Latimer and Ward, in preparation). Thus, their presence in *Homo erectus* is not surprising.

Interfacet distances

Numerous morphological and functional correlates of the uniquely hominid series of spinal curvatures are easily appreciated. In the lumbar lordosis, posterior portions of the vertebrae converge spatially upon one another because they are dorsal to the axis about which lumbar extension occurs. This convergence has influenced a suite of associated anatomical alterations.

One of these morphologies is seen in dorsal view (Figures 12.2 and 12.3; Table 12.3). Humans exhibit a pyramidal increase in lumbar interfacet distances caudally throughout the lumbar column. This differs dramatically from the condition in other hominoids, wherein the vertebral facets remain equidistant or actually converge caudally (see Figure 12.2).

Without a pyramidal increase in interfacet distance, convergence between adjacent articular facets would result in physical impingement and application of potentially damaging stress on intervening structures, such as the laminae and pars interarticularis. The hominid lumbar lordosis can obtain only if contiguous facets have this pyramidal configuration enabling the facets to imbricate without impinging upon intervening structures (Figure 12.2).

TABLE 12.3.

Interfacet distance (mm); see Figure 12.3.

	HOMO SAPIENS								
	ADULT WITH 5 LUMBAR VERTEBRAE (*N* = 30)		ADULT WITH 6 LUMBAR VERTEBRAE (*N* = 10)		JUVENILE (*N* = 10)			*PAN TROGLODYTES* (*N* = 16)	
LEVEL	Mean	S.D.	Mean	S.D.	Mean	S.D.	KNM-WT 15000	Mean	S.D.
18			37.09	1.15			32.5		
17	36.10	2.58	31.30	2.15	29.72	2.54	26.6	27.23	2.22
16	29.58	1.79	27.25	1.61	25.11	2.09	24.6	24.08	2.12
15	26.02	1.57	24.83	1.47	22.89	2.08		20.58	1.48
14	24.03	1.38	23.46	1.31	20.56	1.82	21.4	18.71	1.55
13	22.62	1.34	23.13	1.41	19.38	1.76		17.90	1.54
12	22.70	1.17	26.78	1.04	19.28	1.88	21.0	17.86	1.55
11	22.55	1.35	24.61	1.32	19.79	2.60	20.8	17.95	1.54
10	23.18	1.65	25.71	2.12	20.17	1.88		18.15	1.31
9	24.41	1.33	26.60	1.95	20.69	1.80	20.5	18.48	1.58
8	25.98	1.65	27.10	1.81	22.54	2.55	22.6	19.41	1.59
7	26.37	1.64	30.00	1.47	21.89	1.73		19.90	1.77
6	27.45	3.48	28.44	2.83	21.58	1.50	26.4	20.90	1.76
5	26.90	3.02	29.38	2.75	24.83	1.66		22.81	2.71
4	27.38	2.06	31.49	2.44	25.60	1.48		24.96	2.33
3	29.58	2.49	32.95	2.80	27.05	2.33	26.0	27.25	2.61
2	32.51	3.88	35.00	2.00	29.78	2.97	31.5	27.02	2.54
1	37.43	3.49	39.15	2.90	32.96	3.26	32.0	25.85	2.59
0	43.76	2.72	41.40	5.10	37.44	2.67	33.5	23.74	2.72

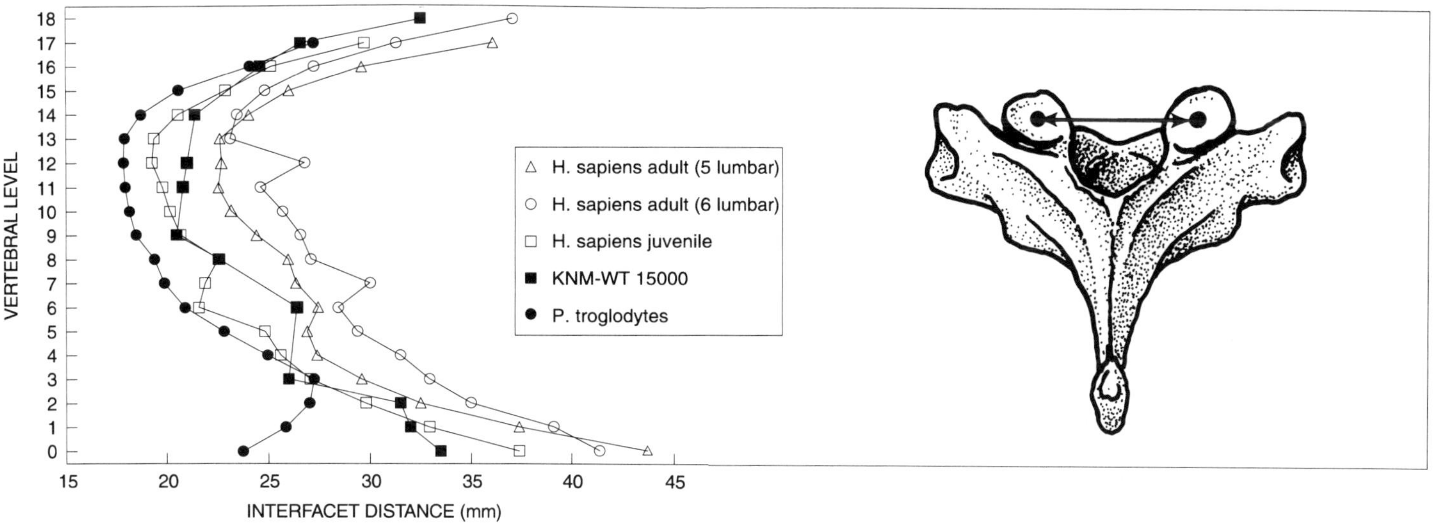

FIGURE 12.3.
Interfacet distances along the thoracolumbar vertebral column. Only means are presented graphically; additional details are provided in Table 12.3. Distances were measured between the centers of the superior articular facets (see illustration). Note that all the hominids show a progressive increase in this distance in the lumbar column. In direct contrast, P. TROGLODYTES *shows a reduction in this dimension. Graphic conventions as in Figure 12.1.*

In the presence of a strong lumbar lordosis, inadequate separation between contiguous superior and inferior lumbar facets can result in these structures pinching the intervening laminar region. This pressure potentially contributes to the developmental sequelae of the uniquely human vertebral condition known as spondylolysis (Latimer, in preparation). This condition involves the physical separation, usually bilateral, of the vertebral arch from the body. Dissolution of the arch occurs through the pars interarticularis, the region where the contiguous facets converge upon one another during extension.

Hominid and chimpanzee vertebral elements demonstrate similar patterns of interfacet dimensions except in the lumbar region. Here, hominids display a marked incremental increase in interfacet distances, in clear contrast to their diminishment in chimpanzees (Figures 12.2 and 12.3; Table 12.3). This difference in lumbar morphology is directly attributable to the lordotic curvature in hominids that makes the pyramidal configuration obligatory.

Given the human-like vertebral wedging pattern in the fossil skeleton (Figure 12.1), it is not surprising that KNM-WT 15000 also shows a pattern of interfacet dimensions that is equivalent to that seen in modern *Homo.* As was the case in vertebral wedging, this pattern is also seen in the earliest hominid genus *Australopithecus* (Latimer and Ward, in preparation). This uniquely hominid condition stands in sharp contrast to that seen in other hominoids, which lack any semblance of these adaptations for a lumbar lordosis.

While sequential increases in the distances between lumbar facets are characteristic of hominids, it is possible that they are simply a structural correlate of an incremental increase in the sizes of the lumbar vertebral bodies. To test this possibility, we examined the ratio formed between the transverse diameters of the vertebral bodies and interfacet distances (Figure 12.4; Table 12.4). It is clear that among hominids, interfacet distances increase faster than do transverse body diameters. The increases in hominid interfacet dimensions (Figure 12.3) are therefore not simply the result of larger lumbar bodies. Moreover, the pattern of incremental areal dimensions in the hominoid thoracolumbar column fails to support the suggestion that hominids have differentially augmented their lumbar vertebral bodies (see below, Figure 12.12 and Table 12.11).

Another feature of the human and the KNM-WT 15000 vertebral columns is also a func-

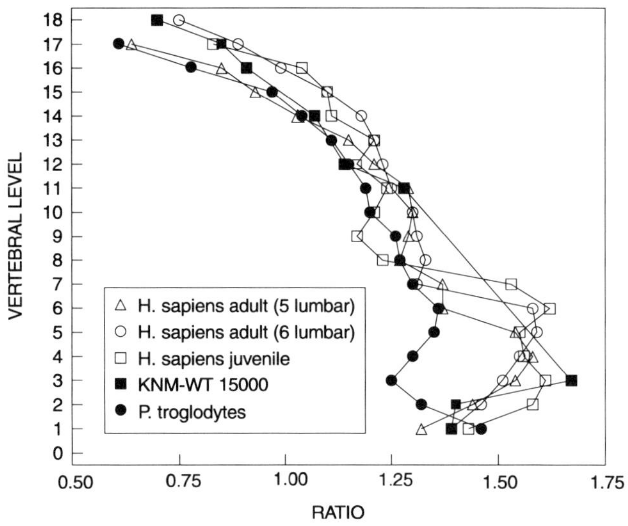

FIGURE 12.4.
Ratio of the transverse diameters of the vertebral bodies to the interfacet distances (see Figure 12.3 and Table 12.3). Only means are presented graphically; additional details are provided in Table 12.4. Note the relative decrease in this ratio in the lumbar columns of the hominids, indicating that the interfacet distances are increasing faster than the transverse diameters of the vertebral bodies. Graphic conventions as in Figure 12.1.

TABLE 12.4.

Ratio of transverse body diameter to interfacet distance; see Figure 12.4.

	HOMO SAPIENS								
	ADULT WITH 5 LUMBAR VERTEBRAE (N = 20)		ADULT WITH 6 LUMBAR VERTEBRAE (N = 10)		JUVENILE (N = 11)			*PAN TROGLODYTES* (N = 10)	
LEVEL	Mean	S.D.	Mean	S.D.	Mean	S.D.	KNM-WT 15000	Mean	S.D.
18			.75	.03			.70		
17	.64	.05	.89	.03	.83	.06	.85	.61	.18
16	.85	.08	.99	.07	1.04	.11	.91	.78	.09
15	.93	.07	1.10	.04	1.10	.08		.97	.10
14	1.03	.06	1.18	.05	1.11	.09	1.07	1.04	.10
13	1.15	.10	1.21	.10	1.21	.11		1.11	.10
12	1.21	.11	1.23	.11	1.17	.07	1.14	1.15	.10
11	1.29	.12	1.25	.18	1.24	.13	1.28	1.19	.09
10	1.30	.13	1.30	.15	1.21	.08		1.20	.11
9	1.29	.12	1.31	.12	1.17	.05		1.26	.12
8	1.27	.07	1.33	.13	1.23	.09		1.27	.12
7	1.37	.09	1.31	.22	1.53	.11		1.30	.14
6	1.37	.11	1.58	.22	1.62	.16		1.36	.10
5	1.54	.17	1.59	.26	1.55	.13		1.35	.15
4	1.58	.10	1.55	.27	1.56	.08		1.30	.12
3	1.54	.11	1.51	.20	1.61	.12	1.67	1.25	.12
2	1.44	.14	1.46	.19	1.58	.15	1.40	1.32	.10
1	1.32	.14	1.39	.13	1.43	.10	1.39	1.46	.13
0									

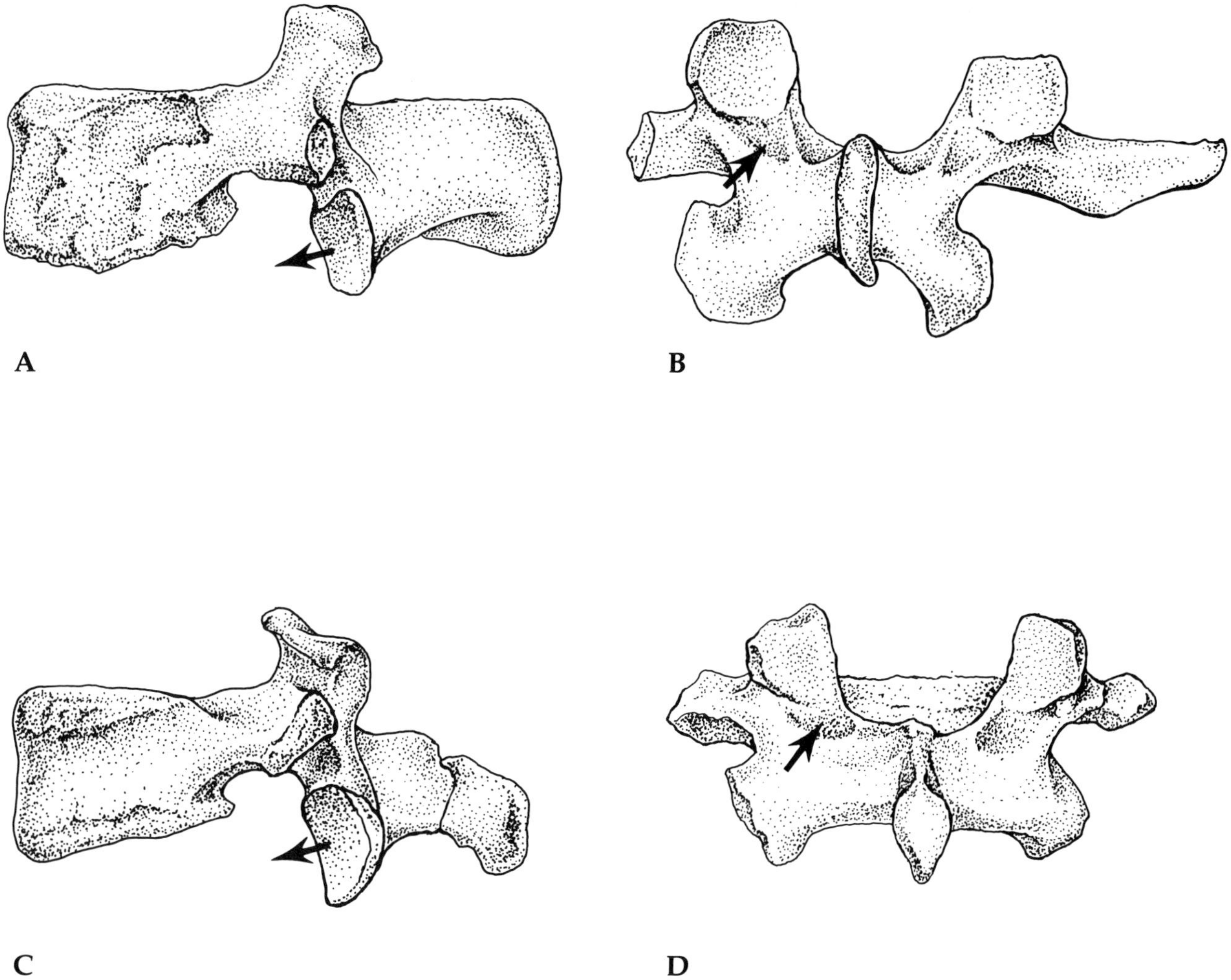

FIGURE 12.5.
Lateral and posterior views of the fifth (A, B) *and sixth* (C, D) *lumbar vertebrae of KNM-WT 15000. In the posterior views* (B, D), *the arrows indicate the nonarticular fossae that accommodate the inferior articular processes of the superjacent vertebrae. In the lateral views* (A, C), *the arrows indicate the anteriorly facing inferior articular facets (see also Figure 12.7 and Table 12.6). Both of these features are related to the lumbar lordosis.*

tion of increasing interfacet distances. This is the striking enlargement of the lower lumbar and sacral vertebral canals (see MacLarnon, Chapter 15, this volume). The human spinal cord usually ends between the first and second lumbar vertebrae (Noback and Harting, 1971). At this level, the size of the canal is considerably greater than its contents (Dommisse, 1975). Caudal to this point, the progressively enlarging canal contains only the cauda equina and spinal meninges. No proposals need be presented to explain this condition other than that canal expansion is the result of selection acting to increase the distances between the articular facets for the reasons given above. The augmented width of the lumbar vertebral canal is simply incidental to this change. This also accounts for the caudally progressive increase in transverse canal width (see MacLarnon, Chapter 15, this volume) and the trefoil cross-section of the lower lumbar and sacral canals. Indeed, if space within the canal were an issue, the triangular cross-section would be a poor option.

Laminar fossae

An additional anatomical characteristic accompanies the physical convergence of the adjoining vertebral arches that is a consequence of the habitually lordotic lumbar column. In posterior view, a small excavated fossa is often present at the medial margin of the base of the superior articular process (Figure 12.5). This shallow, nonarticular hollow accommodates the caudalmost portion of the inferior articular process of the superjacent lumbar

TABLE 12.5.

Inferior process length (mm); see Figure 12.6.

	HOMO SAPIENS						
	ADULT WITH 5 LUMBAR VERTEBRAE (N = 30)		JUVENILE (N = 10)		KNM-WT 15000	*PAN TROGLODYTES* (N = 10)	
LEVEL	Mean	S.D.	Mean	S.D.		Mean	S.D.
18					11.79		
17	19.97	1.92	14.13	2.49	13.28	12.92	.98
16	20.95	1.66	16.11	1.31	13.98	14.40	.90
15	21.17	.85	16.85	1.24		15.24	.55
14	21.03	.98	17.17	1.40	16.60	15.46	1.27
13	21.74	1.25	17.66	1.47		16.30	.67
12	22.69	1.51	18.22	1.68	17.78	16.52	.78
11	23.41	1.66	18.77	1.40	18.38	16.62	1.15
10	23.34	1.38	19.68	1.51		16.78	1.20
9	23.19	1.35	20.41	1.88	20.96	17.52	1.00
8	24.81	1.68	20.53	1.69	22.30	17.70	.28
7	27.67	2.48	22.14	1.96		18.26	.90
6	30.69	2.34	24.77	1.69		20.52	1.42
5	30.10	1.83	26.77	2.30	26.53	21.24	.77
4	30.34	2.13	27.59	1.66	23.71	21.92	.58
3	28.62	2.07	26.85	1.89	22.36	23.20	1.16
2	27.71	1.40	25.25	2.39	21.96	23.16	1.58
1	24.81	1.54	23.42	1.82	18.73	21.58	1.80
0							

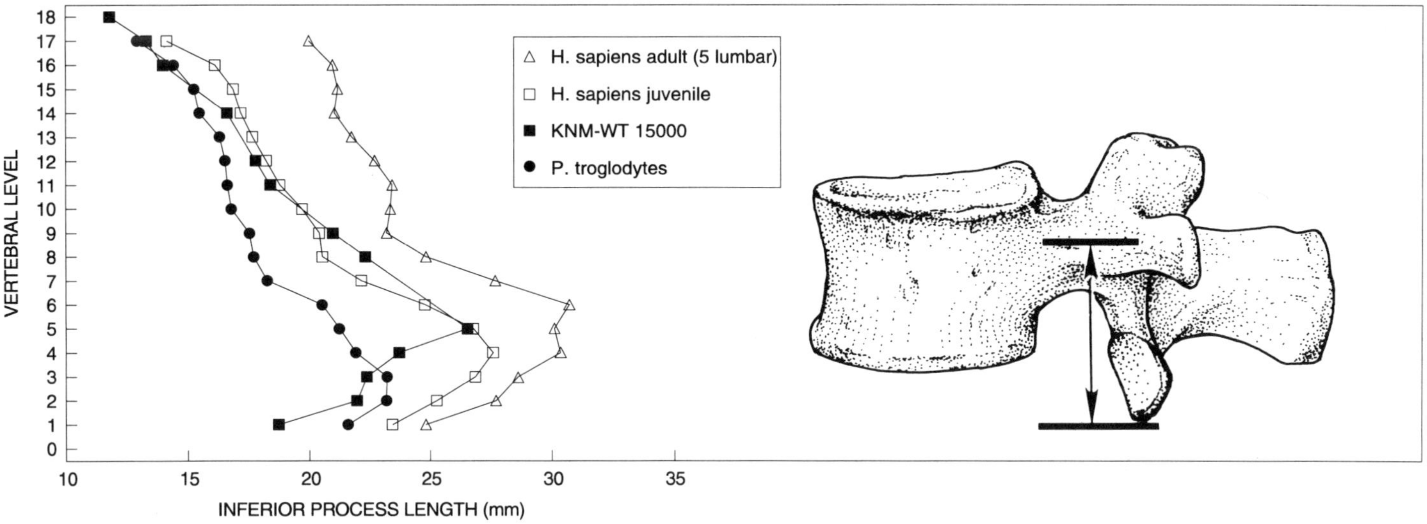

FIGURE 12.6.
Superoinferior lengths of the inferior processes of thoracolumbar vertebrae. Only means are presented graphically; additional details are provided in Table 12.5. Measurements were taken from the center of the pedicle (see illustration) to the most inferiorly projecting point on the inferior process. Note that all the hominids show a similar pattern of process reduction in the lumbar column. Graphic conventions as in Figure 12.1.

vertebra during extreme extension. Although the hominid inferior articular processes are reduced in length and angled posteriorly, they continue to impinge directly upon the laminae (the pars interarticularis) of the adjacent inferior vertebra. This observation lends additional support to the developmental relationship posited between inadequate interfacet distances and the uniquely hominid vertebral condition spondylolysis. Neither a laminar area of contact nor spondylolysis occurs in the vertebrae of great apes. The presence of a nonarticular fossa in modern human and KNM-WT 15000 lumbar vertebrae further demonstrates the habitually lordotic posture in these species.

Length of the Inferior Articular Processes

The hominid lumbar lordosis also necessitates an absolute decrease in the superoinferior lengths of the inferior articular processes. Negative wedging of the lumbar vertebral bodies and intervertebral discs means that if the posterior articular relationships are to remain roughly constant, structures on the concave (posterior) side of the lordotic curve must be either angled posteriorly and/or reduced in length.

In hominids, inferior articular processes of the lumbar vertebrae exhibit both a reduction in length (Figure 12.6; Table 12.5) as well as a greater posterior angulation along the long axis of the process. The hominid pattern reveals a relatively uniform increase in inferior process lengths until the thoracolumbar transition. Here, the profile abruptly reverses direction, reflecting a dimensional reduction in inferior process length throughout the lumbar region (Figure 12.6).

The chimpanzee profile is similar in the thoracic region to that of the hominids. However, chimpanzees lack a marked diminishment in inferior process length in the lumbar column. Interestingly, chimpanzees show no increase in process length of the penultimate lumbar vertebra and only a slight reduction in this dimension in the ultimate vertebra. These features are related to the overall reduction in anterior wedging of these elements.

With regard to its sequential inferior facet length profile, KNM-WT 15000 is clearly allied with modern humans. Indeed, the striking reduction in process length that occurs from the superiormost lumbar element to the last lumbar vertebra is directly attributable to the marked degree of

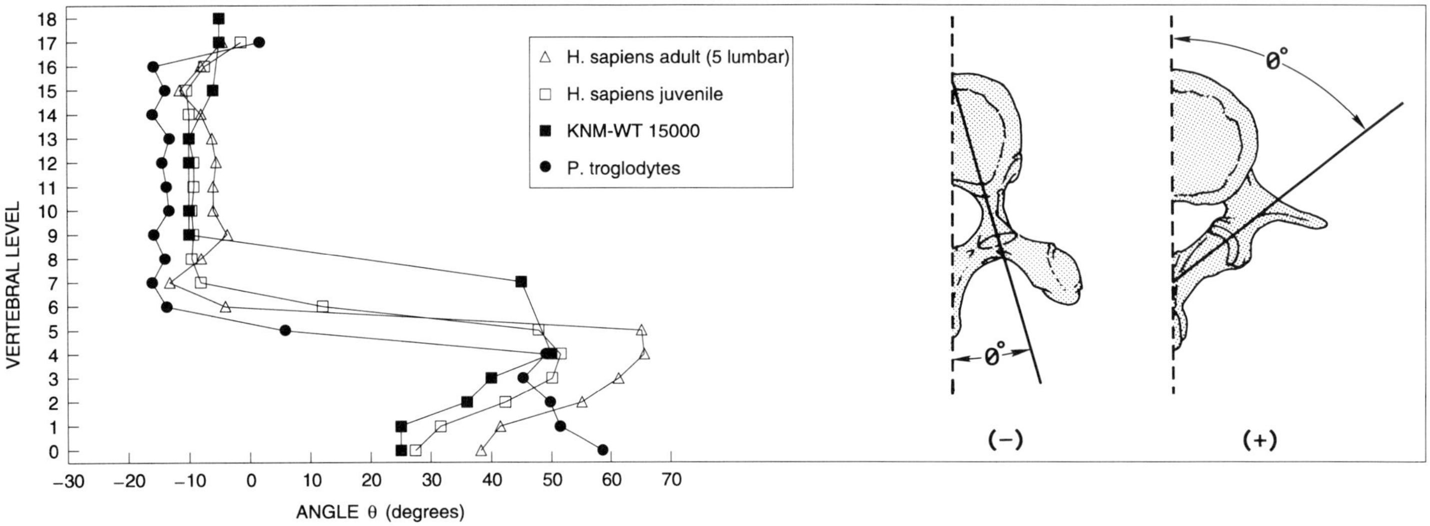

FIGURE 12.7.
Angle θ measuring the orientation of the superior articular facets relative to the sagittal plane. Only means are presented graphically; additional details are provided in Table 12.6. Measurements were taken by directing a normal to the centroid of the superior articular facet and measuring the angle it made to the sagittal plane (see illustration). If the normal was directed anteromedially, the angular measurement was considered negative. If the normal was directed anterolaterally, the angular measurement was considered positive. Zero degrees indicates that the facet was in a coronal plane. Note that angle θ becomes smaller in the lumbar column of the hominids, indicating that the facet orientation was progressively more coronal. In contrast, angle θ increases in the lumbar vertebrae of PAN. *Graphic conventions as in Figure 12.1.*

negative wedging in its lumbar vertebral bodies. The oblique posteroinferior orientation of the inferior articular process in this specimen also is equivalent to the condition in modern humans.

Orientation of Articular Facets

In addition to an increase in transverse interfacet distances and a reduction in inferior articular process lengths, the orientation of the KNM-WT 15000 lumbar articular facets is also associated with a human-like lordotic curvature. As pointed out by Brown et al. (1985), the inferior articular facets of the fossil fifth lumbar vertebra are relatively flat and face almost directly anteriorly. This description applies equally well to the sixth lumbar vertebra (Figure 12.5). Brown et al. (1985) contrasted this orientation to what they considered to be the more usual human condition in which the lumbar articular facets face somewhat laterally.

Examination of a large number of subadult and adult human spines indicates that the conformation and orientation of the KNM-WT 15000 articular facets fall easily within the range of normal human variation (Figure 12.7; Table 12.6). Humans and chimpanzees, however, exhibit marked differences in lumbar facet orientation. The fossil pattern parallels that of modern human juveniles. Hominid lumbar articular facets become increasingly coronally oriented toward the sacrum. This is in direct contrast to the articular facets in chimpanzees, which become more sagittally oriented toward the caudal end of the column.

The reason for this articular orientation in hominids is the role that the facet joints play in resisting shear stresses at the lumbosacral junction. In the normal human, the sacrum and lumbosacral joint tilt anteriorly and inferiorly. This orientation is necessary to minimize parturitional difficulties that would otherwise ensue owing to the projection of the sacrum into the birth canal (see Tague and Lovejoy, 1986). This oblique orientation of the lumbosacral articulation means that any vertical compression force will act to displace the lower lumbar vertebrae (especially the last two, which exhibit the most negative wedging) in an anteroinferior direction. This tendency is resisted by the coronally oriented articular facets.

The extremely low coefficient of friction in healthy synovial joints necessitates that the primary joint reaction vectors remain normal to the articular surface (see Latimer et al., 1987). Lumbar facet joints must become progressively more coronally oriented and larger to resist the increasing tendency for anterior displacement of the lower lumbar vertebral bodies. Therefore, the coronal orientation that characterizes hominid lumbar articular facets is also a correlate of the lumbar lordosis. In light of the accentuated lumbar lordosis in KNM-WT 15000, a marked coronal orientation of the lower lumbar articular facets in this specimen is not unexpected.

An additional observation concerning the orientation of the lumbar articular facets centers on the ontogenesis of these joints. Cihak (1970) pointed out that, at birth, human lumbar articular surfaces are in a coronal plane. Only later do they undergo a process of "sagittalization" that results in a more anteroposterior orientation of these articular surfaces. He further reports that this process is essentially completed by 18 months of age, at which time the joints assume the adult configuration. However, Roaf (1971) suggests that changes in articular orientation may continue until puberty. In the analysis presented here (Figure 12.7; Table 12.6), juvenile humans and KNM-WT 15000 have slightly more coronally oriented articular facets than do adults. This may indicate that, at least in this small sample of juveniles, maturational changes in facet orientation were incomplete.

TABLE 12.6.

Superior facet angle (θ, in degrees); see Figure 12.7.

	HOMO SAPIENS						
	ADULT WITH 5 LUMBAR VERTEBRAE (*N* = 15)		JUVENILE (*N* = 12)			*PAN TROGLODYTES* (*N* = 9)	
LEVEL	Mean	S.D.	Mean	S.D.	KNM-WT 15000	Mean	S.D.
18					−5.0		
17	−4.5	2.17	−1.4	1.67	−5.0	1.7	6.85
16	−8.0	4.74	−7.4	4.81		−15.8	7.24
15	−11.5	4.38	−10.4	5.07	−6.0	−13.9	7.68
14	−8.0	4.30	−10.0	4.83		−16.0	5.01
13	−6.2	3.34	−10.0	4.54	−10.0	−13.2	6.06
12	−5.5	1.80	−9.2	4.91	−10.0	−14.4	5.73
11	−6.0	2.54	−9.2	4.31		−13.7	5.49
10	−6.0	.70	−9.6	4.73	−10.0	−13.3	4.10
9	−3.7	3.49	−9.3	3.62	−10.0	−15.8	3.58
8	−8.0	3.74	−9.6	6.00		−14.0	3.55
7	−13.2	2.94	−8.1	3.98	45.0	−16.1	5.74
6	−4.0	2.89	12.1	7.18		−13.7	7.37
5	65.1	7.28	47.9	6.37		5.8	4.42
4	65.5	6.76	51.5	7.24	50.0	49.1	5.95
3	61.3	5.80	50.1	7.84	40.0	45.3	7.17
2	55.0	6.16	42.3	6.89	36.0	49.8	7.84
1	41.5	7.51	31.6	7.67	25.0	51.4	7.37
0	38.3	6.24	27.4	8.17	25.0	58.6	8.84

Articular areas

Another morphological feature that can be used to compare KNM-WT 15000 with humans and chimpanzees is the incremental change in surface areas of the superior articular facets. In KNM-WT 15000 and modern humans there is a dramatic increase in the surface areas of the lumbar and sacral articular facets (Figure 12.8; Table 12.7). In direct contrast, chimpanzees exhibit a progressive reduction in lumbar and sacral articular areas.

The observed increases in the hominid lumbar column are again a consequence of lumbar lordosis and the oblique orientation of the lumbosacral junction. These two features result in a proportional increase in loading through the articular facets of these elements, largely to resist the shear stresses that would otherwise be imposed on the intervening discs. However, only the facets of vertebrae immediately contiguous with the sacrum are thus affected, owing to the anteroinferior inclination of the lumbosacral junction. The patterns of incremental facet sizes in the thoracic columns are similar for hominids and chimpanzees, indicating that habitually erect bipedality does not result in a relative increase in loading of the posterior elements of the thoracic vertebrae. Moreover, the angular orientation of the articular facets in the thoracic column (angle θ; Figure 12.7) also fails to support the concept of greater loading of the posterior elements resulting from bipedality.

A ratio of articular facet areas to vertebral endplate areas provides another method of assessing differential increases in articular areas within individual vertebrae. Both humans and KNM-WT 15000 have, on average, smaller thoracic articular facets relative to vertebral body sizes than do chimpanzees (Figure 12.9; Table 12.8). In the lower lumbar elements, the chimpanzee and hominid lines converge, indicating that both humans and KNM-WT 15000 have a relative enlargement in lumbar articular facet areas.

Figures 12.8 and 12.9 show several other interesting points. The human adults with 6 lumbar vertebrae have absolutely larger superior articular facets than do humans with 5 lumbar elements. This is not related to the extra vertebra but more likely reflects the fact that all the individuals in the 6-lumbar sample are male.

In absolute surface areas of the articular facets (Figure 12.8; Table 12.7), the juvenile and adult profiles diverge in the thoracic and lower lumbar regions. This means that superior articular facet areas in juveniles are more similar to adults in the middle and upper thoracic regions than they are in the lumbar column, an observation that implies differential growth along the vertebral column. If growth rates remain constant, this indicates that maturation of the lumbar region probably occurs later than in the middle and upper thorax.

Differential growth appears to occur not only in the vertebral column as a whole but also within individual vertebral elements (Figure 12.9). When normalized by vertebral body surface areas, the areas of the juvenile upper and middle thoracic facets are either equal to or relatively larger than those in adults. It seems likely that the posterior portions of the growing thoracic vertebrae, as represented by the superior articular facets, mature to an adult size faster than do the anterior portions.

KNM-WT 15000 falls comfortably within the human ranges for both measures of articular facet area, indicating equivalent load distribution along the vertebral column. It differs quite markedly from the pattern seen in the chimpanzee sample. In agreement with the discussion of vertebral wedging, the sequential maturation of articular dimensions in the fossil vertebral column appears to be more similar to the pattern in adult humans, especially in its lumbar column. That is, in absolute articular dimensions (Figure 12.8) KNM-WT 15000 is somewhat larger in its lumbar column than would be expected from a comparison with the juvenile humans. This observation is likely related to the rather extreme lumbar lordosis in this specimen and the necessity for augmented surface areas in this region.

Spinous processes

Brown et al., (1985) reported that the spinous processes on the KNM-WT 15000 vertebrae were

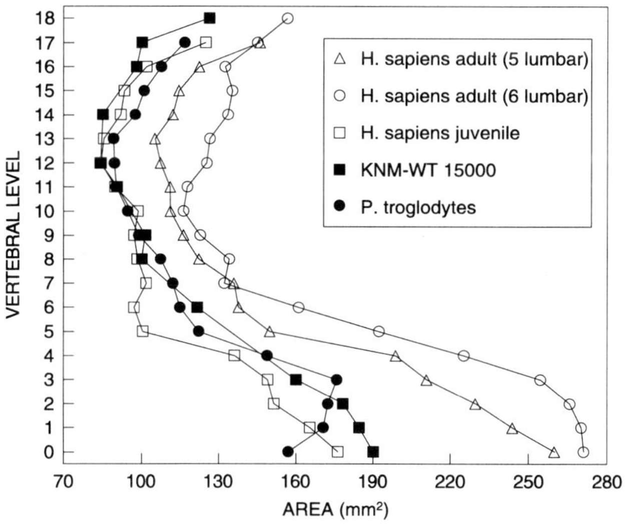

FIGURE 12.8.

Surface areas of the superior articular facets in the thoracolumbar vertebrae. Only means are presented graphically; additional details are provided in Table 12.7. The measurements presented are the averaged areal dimensions for the right and left articular facets. Note the dramatic increase in facet surface areas in the lumbar columns of the hominids. In contrast, PAN *shows a decrease in the lumbar region. Graphic conventions as in Figure 12.1.*

TABLE 12.7.

Superior facet surface area (mm^2); see Figure 12.8.

	HOMO SAPIENS								
	ADULT WITH 5 LUMBAR VERTEBRAE (*N* = 20)		ADULT WITH 6 LUMBAR VERTEBRAE (*N* = 10)		JUVENILE (*N* = 11)			*PAN TROGLODYTES* (*N* = 10)	
LEVEL	Mean	S.D.	Mean	S.D.	Mean	S.D.	KNM-WT 15000	Mean	S.D.
18			156.6	21.50			126.0		
17	146.0	21.08	145.0	23.45	124.9	25.77	100.0	116.7	18.48
16	122.4	26.24	132.5	26.49	101.9	17.17	98.0	107.6	11.89
15	114.5	16.61	135.3	25.49	93.2	12.51		100.9	12.54
14	112.1	19.63	133.5	22.16	91.8	12.51	85.0	97.5	13.91
13	105.2	14.47	126.5	19.62	85.4	10.98		89.1	13.45
12	107.3	10.82	125.3	19.71	84.3	13.36	84.0	89.4	13.53
11	111.1	12.23	117.8	16.89	89.8	13.69	90.5	90.1	8.41
10	111.2	15.00	116.3	19.03	98.6	14.95		94.6	10.48
9	116.3	14.74	122.8	20.11	97.0	12.99	101.5	98.8	13.63
8	122.3	18.60	134.1	19.18	98.4	15.03	100.0	107.4	10.28
7	136.0	27.57	132.2	17.74	101.8	14.46		112.1	12.61
6	137.7	22.54	161.0	46.98	97.0	8.22	121.6	114.9	13.95
5	149.9	21.15	192.3	38.46	100.4	14.99		122.2	21.55
4	198.9	32.25	225.2	36.78	136.3	29.90		148.7	20.44
3	210.8	28.07	254.5	32.35	149.0	22.50	160.0	175.8	19.18
2	229.6	26.32	265.5	36.61	151.4	21.03	178.2	172.4	18.54
1	243.8	37.08	270.0	43.96	165.5	28.48	184.5	170.6	21.82
0	259.9	25.28	271.0	30.52	176.3	25.37	190.0	157.0	20.57

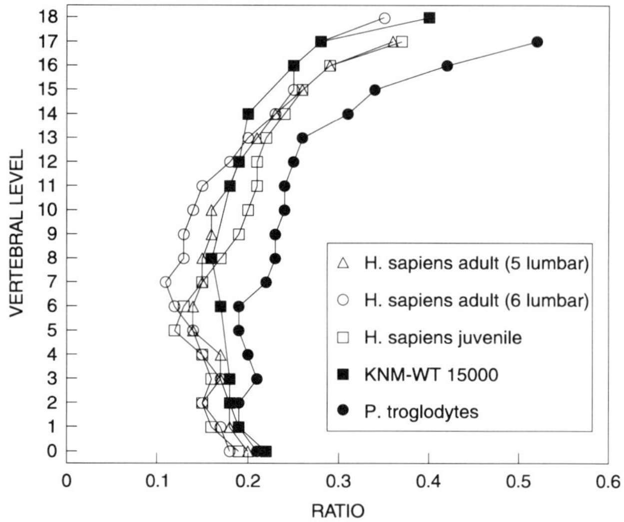

FIGURE 12.9.
Ratio of the averaged surface areas of the right and left superior articular facets (Figure 12.8; Table 12.7) to the surface areas of the superior endplates (Figure 12.12; Table 12.11). Only means are presented graphically; additional details are provided in Table 12.8. Note that when compared to PAN, *hominids have relatively smaller facet areas until the lumbar region. Graphic conventions as in Figure 12.1.*

TABLE 12.8.

Ratio of superior facet area to vertebral body surface area; see Figure 12.9.

	HOMO SAPIENS								
	ADULT WITH 5 LUMBAR VERTEBRAE (*N* = 20)		ADULT WITH 6 LUMBAR VERTEBRAE (*N* = 10)		JUVENILE (*N* = 11)			*PAN TROGLODYTES* (*N* = 10)	
LEVEL	Mean	S.D.	Mean	S.D.	Mean	S.D.	KNM-WT 15000	Mean	S.D.
18			.35	.05			.40		
17	.36	.06	.28	.04	.37	.03	.28	.52	.06
16	.29	.06	.25	.04	.29	.05	.25	.42	.05
15	.26	.03	.25	.02	.26	.02		.34	.04
14	.23	.03	.23	.02	.24	.02	.20	.31	.05
13	.21	.02	.20	.02	.22	.03		.26	.04
12	.19	.03	.18	.02	.21	.02	.19	.25	.04
11	.18	.03	.15	.02	.21	.03	.18	.24	.03
10	.16	.03	.14	.01	.20	.03		.24	.03
9	.16	.02	.13	.01	.19	.03		.23	.04
8	.15	.02	.13	.01	.17	.03	.16	.23	.03
7	.15	.02	.11	.01	.15	.02		.22	.03
6	.14	.03	.12	.03	.13	.01	.17	.19	.02
5	.14	.02	.14	.04	.12	.03		.19	.04
4	.17	.03	.15	.03	.15	.02		.20	.03
3	.17	.03	.17	.02	.16	.02	.18	.21	.03
2	.18	.03	.15	.01	.15	.02	.18	.19	.03
1	.18	.03	.17	.02	.16	.02	.19	.19	.02
0	.20	.03	.18	.03	.19	.02	.22	.21	.02

longer and more horizontally oriented than they usually are in modern humans. They also observed that the laminae of adjacent thoracic vertebrae were less imbricated than is usually the case in modern humans. Comparison of the upper and midthoracic vertebrae in KNM-WT 15000 (T1–3; T5; T7–8) with those of modern humans and chimpanzees reveals that while the KNM-WT 15000 spinous processes are not absolutely longer than those of humans, they are more horizontally oriented. As the horizontal length of the process varies with the cosine of angle β (Figure 12.10; Table 12.9), the processes do, as a consequence, extend somewhat further posteriorly.

The extreme horizontal orientation of the upper thoracic vertebrae in KNM-WT 15000 (T1–3) exceeds even the mean for the chimpanzee sample (Figure 12.10; Table 12.9). The spinous processes of the midthoracic vertebrae (T7–8) are only slightly less horizontal than the chimpanzee average. However, the lumbar spinous processes clearly exhibit a human-like horizontal orientation. Chimpanzee lumbar spinous processes are directed inferiorly in the lumbar region, in contrast to the horizontal orientation seen in hominids.

Two points can be made regarding this unusual juxtaposition of features within the vertebral column of KNM-WT 15000. First, from Figure 12.10 (angle β) it can be seen that the overall profile of the thoracolumbar column of the fossil is, while rather extreme, most like that of modern humans. That is, both adult and juvenile humans and the fossil specimen show two regions of reduced inferior angulation of the processes; one in the upper thorax and the other in the lumbar column. These two areas are separated by the midthorax, wherein the processes are more steeply angled and as a consequence their laminae more strongly imbricated. By contrast, the chimpanzee profile is more uniform and fails to show as great a difference between contiguous regions of the vertebral column.

This difference between hominids and chimpanzees corresponds to the contrasting vertebral wedging patterns in these two groups. In hominids, the two areas with the most horizontally oriented spinous processes are also regions that are within or immediately adjacent to lordotic curvatures. A lordotic profile requires that the processes be more horizontally oriented to permit the posteriorly directed concavity. Great apes, with their more inferiorly directed lumbar spinous processes, cannot achieve the lower lordotic curve because in forming the posteriorly directed concavity the processes would impinge directly upon one another. Similarly, in the uppermost thorax, immediate proximity to the superjacent cervical lordosis also necessitates less inclined processes.

For like reasons, the apex of the thoracic kyphosis is also the region with the most extreme declination of the spinous processes (T6–8). As the kyphotic concavity faces anteriorly, if the spinous processes along the thoracic column are to remain in a relatively consistent vertical plane, then those at the upper and lower extremities of the kyphosis (those areas contiguous to the lordoses) must project further dorsally (more horizontally) while those at the center of the curve must decline at a greater angle. Moreover, if all the spinous processes projected at identical angles relative to their vertebral bodies, the sinusoidal series of spinal curvatures would lead to an increasing divergence of the posteriorly projecting processes at the apex of the kyphosis. In order to maintain equal distances between the ends of the processes, it is necessary that the projection angle of the processes change along the column. From Figure 12.10 it is clear that while both hominids and chimpanzees adhere to this pattern, it is more extreme in the former group owing to the hominid series of curvatures.

Second, KNM-WT 15000 is characterized by a vertebral column that is deeply invaginated into the thoracic cage. This feature is an adaptation for habitually erect bipedality. Invagination of the column increases the moment arm of the epaxial musculature. In so doing, it reduces the energy needed for postural control of the upper body over the pelvis. This configuration also necessitates a more horizontal projection of the thoracic spinous processes in KNM-WT 15000.

Anatomical adjustments associated with the ventral invagination of the column include tighter curvatures (smaller radii) of the rib's vertebral extremities and, as a consequence, a more dorsal positioning of the costal angles. Both of these adaptive alterations in the rib cage had taken place in KNM-WT 15000 (see Jellema et al., Chapter 13, this volume). The rib curvatures in the fossil specimen are especially marked (small radii of curvatures and large included angles; Jellema et al., Chapter 13, this volume) in the upper thorax, the region which also has the most exaggerated horizontal spinous process orientation.

The strong invagination of the thoracic column into the rib cage is further confirmed by measuring the angle that a line through the rib facets on the transverse process and body of the thoracic vertebrae make to the sagittal plane in cranial view (angle Σ, Figure 12.11 and Table 12.10). This measurement was taken directly from superoinferior radiographs of individual thoracic vertebrae upon which the costal articular surfaces had been marked with lead tape (the line through the articular centroids approximates the costal neck axis; this feature and its functional ramifications are discussed further in Jellema et al., Chapter 13, this volume). KNM-WT 15000 and modern humans have thoracic columns that are more deeply invaginated into the thorax than do chimpanzees. It is notable that the fossil specimen exhibits, in both its costal and vertebral morphology, a rather extreme ventral displacement of its thoracic column.

Since the thoracic portions of the erector spinae lie within the gutter on either side of the vertebral column, the dorsal projection of the costal elements and the posteriorly angled vertebral transverse processes necessarily reduce the available space for this postural muscle that is so critical to an erect biped. The horizontally oriented spinous processes have the effect of deepening the

TABLE 12.9.

Spinous process angle (β, in degrees); see Figure 12.10.

	HOMO SAPIENS						
	ADULT WITH 5 LUMBAR VERTEBRAE (*N* = 20)		JUVENILE (*N* = 12)			*PAN TROGLODYTES* (*N* = 10)	
LEVEL	Mean	S.D.	Mean	S.D.	KNM-WT 15000	Mean	S.D.
18					9		
17	32.7	8.97	32.0	8.67	8	24.2	4.94
16	37.1	9.28	37.9	7.63	4	24.8	7.22
15	38.3	9.08	43.5	8.98		27.0	7.18
14	47.6	9.62	49.9	9.58	32	32.5	6.51
13	58.2	8.23	58.2	9.99		34.0	4.58
12	63.5	6.56	62.2	9.42	35	35.0	4.60
11	60.4	6.10	61.1	8.70	35	32.0	2.83
10	57.9	5.58	58.5	6.96	39	32.9	2.88
9	49.4	7.57	56.5	5.37		33.3	3.85
8	41.1	8.81	46.5	5.24		29.1	4.66
7	36.9	8.52	37.4	7.96		24.3	4.58
6	20.0	8.96	26.2	9.96		21.0	5.42
5	15.7	6.15	19.1	8.56		19.3	6.50
4	14.7	4.19	17.1	8.79	14	21.1	6.89
3	16.1	6.58	20.4	8.98		28.1	4.81
2	17.7	5.72	20.7	8.91	9	27.9	4.99
1	19.9	8.44	25.3	9.69	25	30.9	7.80
0							

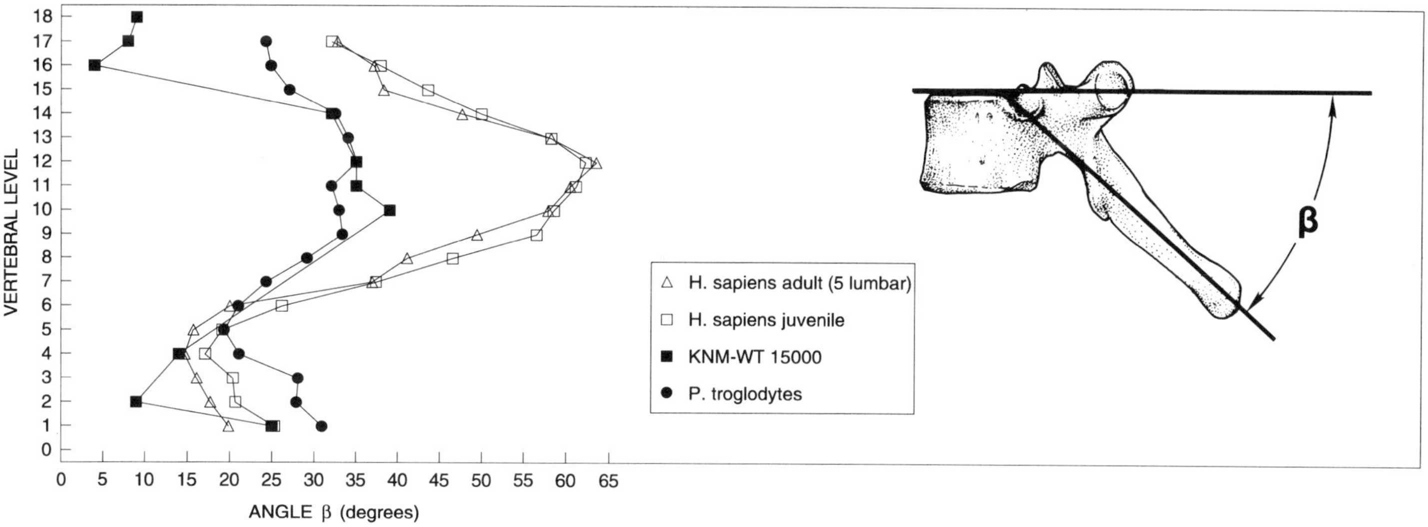

FIGURE 12.10.
The angle (β) that the spinous process makes to the plane of the superior endplate (see illustration). Measurements were taken directly from lateral radiographs. Only means are presented graphically; additional details are provided in Table 12.9. Note that while KNM-WT 15000 and the modern human groups show similar profiles, the fossil specimen is extreme in its upper and midthorax. Graphic conventions as in Figure 12.1.

vertebral gutter (as noted, the horizontal length varies as the cosine of angle β), thereby compensating for what would otherwise be a reduction in surface area for the muscle. This would also increase the moment arm of the associated musculature.

In KNM-WT 15000, the pattern exhibited along the vertebral column is distinct from that in apes and similar to that seen in modern humans. This similarity in the pattern of spinous process orientation indicates that it is directly influenced by vertebral curvatures. The exaggerated angles of the KNM-WT 15000 thoracic spinous processes are a consequence of its correspondingly extreme wedging profile combined with the marked invagination of its thoracic vertebral column into its rib cage.

VERTEBRAL BODY SURFACE AREAS

Modern humans have large vertebrae relative to their body sizes (Schultz, 1953). It is therefore of interest to examine the surface areas of the preserved vertebrae of KNM-WT 15000 to discover if this adaptation preceded modern *Homo sapiens*. However, this process is confounded by the subadult status of the fossil skeleton and the nearly total lack of information concerning the adult *Homo erectus* spine. Figure 12.12 presents sequential body surface areas for the thoracic and lumbar columns of chimpanzees, adult and juvenile humans, and KNM-WT 15000 (see also Table 12.11).

Several important observations can be made regarding vertebral endplate surface areas in these animals. First, all the adult groups exhibit roughly equivalent incremental increases in vertebral areal dimensions. Modern humans do indeed have relatively larger vertebral bodies, and areal augmentation has taken place throughout the column and is not regionally localized. In a metameric series like the vertebral column, such a uniform increase in size along the entire column might be expected and perhaps represents an ontogenic constraint.

Second, human juvenile and adult averages diverge caudally, indicating that the proportionate growth that must occur to transform the subadult vertebra to its fully adult size is similarly divergent. Comparing group averages, only 19% of growth remains to achieve adult areal dimensions in the upper thorax (T1), and about 26% at the ultimate lumbar vertebra. This once again demonstrates the growth differential that occurs along the modern human vertebral column.

Unfortunately, little can be said concerning

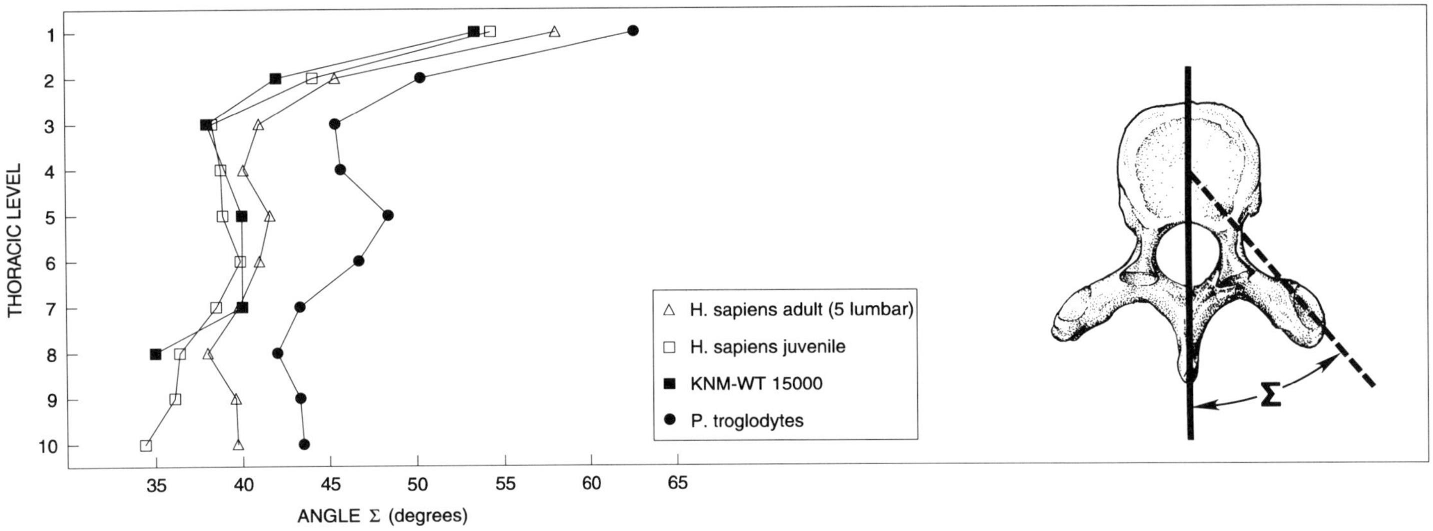

FIGURE 12.11.

The angle Σ formed between a line through the vertebral costal facets and the sagittal plane. Measurements were taken directly from superoinferior radiographs. Only means are presented graphically; additional details are provided in Table 12.10. Note that KNM-WT 15000 and the modern humans demonstrate similar profiles, indicating ventral invagination of the thoracic columns. Thoracic level 1 is the first (uppermost) thoracic vertebra and sequential numbers indicate the caudal progression down the thoracic column.

TABLE 12.10.

Angle Σ (in degrees); see Figure 12.11.

	HOMO SAPIENS					*PAN TROGLODYTES* (*N* = 10)	
	ADULT WITH 5 LUMBAR VERTEBRAE (*N* = 10)		JUVENILE (*N* = 10)				
THORACIC LEVEL	Mean	S.D.	Mean	S.D.	KNM-WT 15000	Mean	S.D.
1	58.0	4.28	54.3	7.32	53.4	62.6	3.72
2	45.4	4.73	44.1	5.50	42.0	50.3	4.42
3	41.0	4.89	38.3	3.46	38.0	45.4	4.80
4	40.1	4.63	38.8	3.21		45.7	5.23
5	41.6	7.65	38.9	3.17	40.0	48.4	4.94
6	41.0	7.26	39.9	3.91		46.7	3.22
7	39.8	5.47	38.5	3.55	40.0	43.3	3.74
8	38.0	4.17	36.4	3.77	35.0	42.0	4.58
9	39.6	6.90	36.1	3.85		43.3	4.85
10	39.7	5.15	34.4	3.38		43.5	3.52

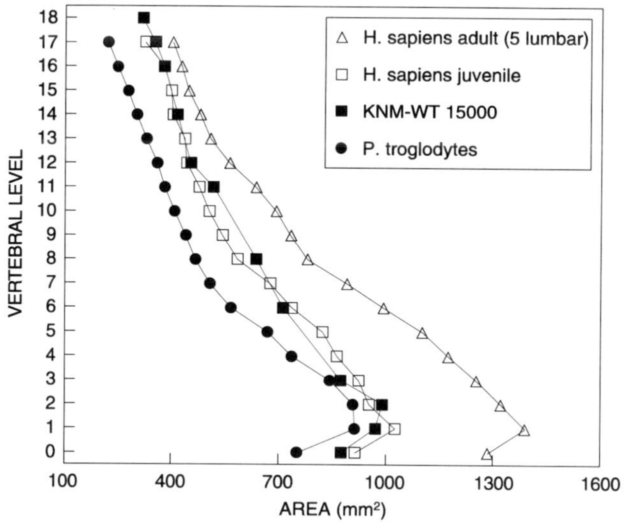

FIGURE 12.12.

Areal dimensions of the superior endplates of the thoraco-lumbar vertebrae. Tracings were taken directly from individual vertebrae. The surface area of the first sacral vertebra of KNM-WT 15000 was estimated from the inferior endplate of its sixth lumbar vertebra. Only means are presented graphically; additional details are provided in Table 12.11. Note the similarity in profiles, indicating that no regional increase in areal dimensions has occurred in hominid evolution. Graphic conventions as in Figure 12.1.

TABLE 12.11.

Vertebral body surface area (mm^2); see Figure 12.12.

	HOMO SAPIENS						
	ADULT WITH 5 LUMBAR VERTEBRAE (*N* = 10)		JUVENILE (*N* = 10)			*PAN TROGLODYTES* (*N* = 10)	
LEVEL	Mean	S.D.	Mean	S.D.	KNM-WT 15000	Mean	S.D.
18					318		
17	403	47.5	326	24.5	353	220	30.7
16	428	61.8	379	38.3	379	248	21.5
15	448	65.4	399	57.5		278	27.8
14	480	70.4	404	53.9	416	302	30.8
13	509	73.6	437	82.0		330	33.1
12	562	85.0	443	66.4	454	360	40.6
11	636	107.1	478	81.5	517	381	44.2
10	692	130.8	506	79.9		409	55.3
9	732	120.5	543	80.3	637	440	59.2
8	779	128.6	585	69.9		467	59.4
7	890	124.6	676	74.9		508	56.9
6	992	129.7	736	89.8	712	567	66.3
5	1101	181.4	822	117.5		669	85.6
4	1174	176.4	862	117.6		736	95.4
3	1253	194.7	923	109.9	873	842	118.5
2	1322	202.7	952	107.9	990	908	120.3
1	1390	173.6	1025	129.6	971	912	140.2
0	1285	203.3	913	104.0	875	752	124.2

the pattern and tempo of vertebral growth in KNM-WT 15000. The absence of an adult *Homo erectus* specimen with which to gauge relative maturation status means that the amount of growth necessary to achieve adult vertebral size cannot be estimated reliably. Nevertheless, it is clear that in pattern and absolute size, the specimen is essentially identical to this sample of 10–14-year-old juvenile humans that it also matches in degree of ontogenic development. Given the strong similarity in sequential patterning of vertebral surface areas (Figure 12.12), growth trajectories of *Homo erectus* would probably have been similar to those of modern humans. However, in view of the large estimated body size in KNM-WT 15000, the magnitude of growth changes in the vertebrae and the subsequent relative dimensions of the adult vertebrae may have been quite different from modern humans.

In order to examine relative sizes of the KNM-WT 15000 vertebral elements it is necessary to compare its vertebral dimensions either to another skeletal variable or to an estimate of body mass. In light of differential growth velocities recognized to occur within areas of the growing skeleton (Tanner et al., 1966) and the confounding influence that these can have on attempts at normalization of vertebral size, we chose to use the body weight estimate of 48 kilograms for KNM-WT 15000 presented by Ruff and Walker (Chapter 11, this volume). Because of the inherent uncertainty of estimating the body weight of extinct species, a range of values (48 ± 5 kg) is depicted in Figure 12.13 and given in Table 12.12.

When viewed in relation to body weight, modern humans have expanded vertebral endplate surface areas (Figure 12.13). Moreover, in their upper thoraxes juvenile humans have relatively larger endplate surface areas than do the adults. This demonstrates that at this particular ontogenic stage, vertebral growth precedes the body weight increases that will subsequently take place during the adolescent period.

By contrast, although chimpanzees show a similar profile, the line is shifted to the left, indicating that as adults they have significantly smaller vertebral areal dimensions. Chimpanzees do not overlap with any of the modern human groups, even at the level of one standard deviation (Table 12.12).

Several observations regarding the relative vertebral areas in KNM-WT 15000 are of interest. First, at the estimated body weight of 48 kilograms, this specimen overlaps the adult human ranges in the upper and midthorax. However, from the lower thoracic region to the lumbar column, the lines progressively diverge, indicating that this individual has relatively smaller areal dimensions to its lumbar elements. Indeed, even in the highly unlikely event that the body weight estimate of 48 kilograms is over 5 kilograms too heavy, the normalized areas of the 4 caudalmost vertebral elements (L4, L5, L6, S1) are barely accommodated within any of the modern human ranges. This divergence in both relative and absolute (see Figure 12.12) surface areas in the lumbar column may indicate that the upper and midthoracic elements of the fossil specimen are slightly ahead in maturation status than are the lumbar vertebrae. This superoinferior gradient in growth along the vertebral column was also noted in several of the previous measurements. Second, in its upper thorax the fossil specimen is even further divergent from modern juveniles, which have, on average, relatively larger vertebral areas than do adults (Figure 12.13; Table 12.12). This occurs even though in absolute areal dimensions (Figure 12.12) KNM-WT 15000 falls within the human juveniles. The large discrepancy in body weights between modern juveniles and KNM-WT 15000 (the juvenile average body weight is 37.8 kilograms) means that the normalized values are much lower in the fossil elements. This dissimilarity implies either a marked difference in ontogenic processes in the fossil or that KNM-WT 15000 underwent similar rates of growth but would have had relatively smaller vertebral surface areas as an adult. Given the strong similarities demonstrated between modern humans and the KNM-WT 15000 vertebrae in overall developmental patterns, the latter suggestion is viewed as more likely.

This is especially true in view of the fact that when compared to modern *Homo*, the earlier hominid genus *Australopithecus* also had relatively small vertebral areas (Latimer and Ward, in preparation). It thus seems likely that KNM-WT 15000

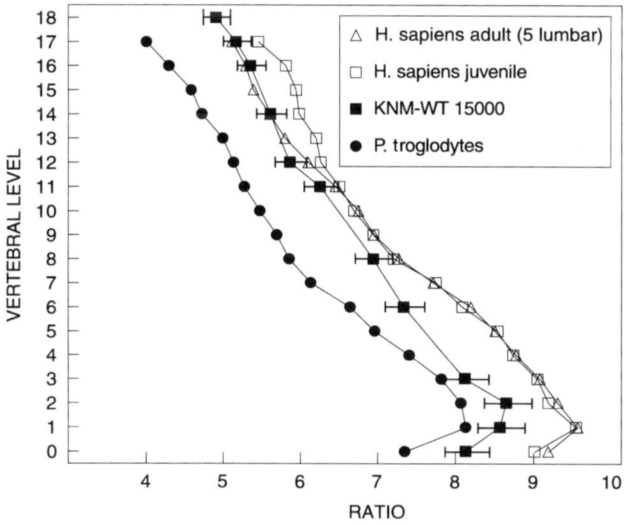

FIGURE 12.13.

Ratio of the square root of the superior endplate areas of the thoracolumbar vertebrae to the cube root of the individual's body weight. Modern human body weights are from the Hamann-Todd Collection; the body weight for KNM-WT 15000 (48 kg) is from Ruff and Walker (Chapter 11, this volume); and the body weights for P. TROGLODYTES *are from Jungers (1985). For modern humans and* PAN, *only means are presented graphically; additional details are provided in Table 12.12. For KNM-WT 15000, the black squares represent the ratios calculated assuming a body weight of 48 kilograms and horizontal lines plot the values for body weight = 43 kg (lines to the right of the squares) and body weight = 53 kg (lines to the left of the squares). Note that in the lumbar vertebrae the fossil specimen falls between the chimpanzee and adult human lines. Graphic conventions as in Figure 12.1.*

TABLE 12.12.

Ratio of the square root of the vertebral surface area to the cube root of body weight; see Figure 12.13.

	HOMO SAPIENS								
	ADULT WITH 5 LUMBAR VERTEBRAE (*N* = 10)		JUVENILE (*N* = 10)		KNM-WT 15000 estimated to weigh—			*PAN TROGLODYTES* (*N* = 10)	
LEVEL	Mean	S.D.	Mean	S.D.	53 kg	48 kg	43 kg	Mean	S.D.
18					4.74	4.90	5.09		
17	5.11	.34	5.45	.12	5.00	5.16	5.36	4.00	.31
16	5.29	.47	5.81	.16	5.18	5.35	5.55	4.29	.25
15	5.39	.48	5.94	.23				4.58	.30
14	5.60	.51	5.98	.22	5.43	5.61	5.82	4.72	.31
13	5.80	.52	6.20	.40				4.99	.32
12	6.10	.52	6.26	.34	5.67	5.86	6.08	5.13	.36
11	6.45	.62	6.50	.38	6.05	6.25	6.49	5.27	.36
10	6.75	.70	6.69	.44				5.47	.42
9	6.95	.63	6.94	.39				5.69	.43
8	7.28	.61	7.21	.33	6.71	6.94	7.20	5.85	.44
7	7.72	.61	7.75	.32				6.13	.43
6	8.20	.57	8.09	.38	7.10	7.34	7.61	6.64	.50
5	8.51	.80	8.54	.47				6.96	.55
4	8.78	.77	8.74	.44				7.41	.56
3	9.08	.82	9.05	.41	7.86	8.12	8.43	7.82	.68
2	9.31	.79	9.19	.35	8.37	8.65	8.98	8.07	.66
1	9.56	.69	9.54	.42	8.29	8.57	8.89	8.13	.71
0	9.19	.77	9.01	.40	7.87	8.13	8.44	7.35	.70

reflects the primitive hominid condition and that modern human-like augmentation of the vertebral areas had yet to occur in *Homo erectus*. This suggestion must, however, remain tentative in that no adult lumbar vertebral elements are available for this species and it is still a possibility that the KNM-WT 15000 specimen would have undergone a subsequent period of dramatic vertebral growth and would have caught up to modern humans. This again seems improbable in that at the estimated adult body weight of 68 kilograms (Ruff and Walker, Chapter 11, this volume), for KNM-WT 15000 to have achieved average modern adult human relative lumbar vertebral areas it would have had to undergo 30–36% of additional growth. By comparison, modern human juveniles need only increase 25–28% to achieve equivalent adult areal dimensions.

Discussion and Summary

Upright, bipedal locomotion places unique mechanical demands on the human vertebral column. In response, the human spine has undergone a series of dramatic alterations of its structure and function from the primitive hominoid condition. Trunk balance over the supporting hind limbs is facilitated by a series of vertebral curvatures. No other extant mammal exhibits an adaptation to trunk balance even remotely similar to that of modern humans.

The structural and mechanical correlates of this adaptation are numerous and can be used to access the functional capabilities of the KNM-WT 15000 vertebral column. In the preceding comparisons of modern humans, chimpanzees, and KNM-WT 15000, the fossil specimen has been shown to be strikingly similar to humans in nearly all significant aspects.

The magnitude and pattern of vertebral wedging in KNM-WT 15000 clearly indicates a lumbar lordosis and thoracic kyphosis fully commensurate with those seen in modern humans. Associated anatomical alterations for bipedality in the pelvis (see Walker and Ruff, Chapter 10, this volume) and the rib cage (see Jellema et al., Chapter 13, this volume) are also present in this specimen.

Associated with these pelvic alterations, especially the reduction in pelvic height, was the necessity that the sacrum and the lumbosacral articulation maintain their oblique orientation so as not to occlude the obstetric canal. This sacral orientation, in turn, demanded the development of the lumbar lordosis (and the associated kyphotic and cervical curvatures) for postural control of the vertically aligned thorax. In all likelihood the pelvic modifications and the vertebral curvatures occurred simultaneously.

Along with the development of the vertebral curvatures in the hominid column was an associated suite of features including alterations in: vertebral body shape; articular facet size and orientation; spinous process orientation; and size and angulation of the lumbar inferior articular processes. All of these alterations had clearly taken place in the KNM-WT 15000 vertebral skeleton.

The completed acquisition of these features in the *Homo erectus* vertebral column provides evidence that the habitual upright bipedality that characterized this species was equivalent to that of modern humans. In view of the fact that similar modifications are also present in the vertebral columns of *Australopithecus* (Latimer and Ward, in preparation), it is clear that this suite of features was among the earliest and most necessary anatomical adaptations for bipedality.

There are nonetheless two fundamental differences between the KNM-WT 15000 vertebrae and those of modern humans, both of which help to illuminate the evolution of the vertebral column in *Homo sapiens*. First, the lumbar column of KNM-WT 15000 is made up of 6 lumbar vertebrae rather than the 5 elements which characterize roughly 96% of modern humans. The only other suitably complete Plio-Pleistocene vertebral skeleton, the *Australopithecus africanus* partial skeleton STS 14 from South Africa, also has 6 lumbar elements (Robinson, 1972, and personal observation). The presence of an additional lumbar element in the only two early hominid vertebral skeletons available suggests that this probably represents the primitive condition for hominids.

Although scant evidence renders interpretation of this transition tentative, we feel that the early development of the lumbar lordosis in hominids very likely included the incorporation of the ultimate thoracic vertebra into the lumbar series. This scenario also explains the reduction in the number of thoracic vertebrae (and ribs) from 13, the primitive hominoid condition (Schultz, 1930), to 12 in hominids.

Modern African apes and *Pongo* have only 3–4 lumbar vertebrae but retain the primitive number of 13 thoracic elements for a total of 16–17 thoracolumbar vertebrae. Early hominids, in contrast, appear to have had 18 thoracolumbar segments (12 thoracic, 6 lumbar), suggesting that the early hominid spinal column may have developed from a numerical sequence similar to that in the extant hylobatids, which is 5 lumbar and 13 thoracic elements. This number is also similar to that found in the early Miocene ape *Proconsul nyanzae* from western Kenya, which had 5–7 lumbar vertebrae (Ward, 1991; Ward et al., 1993). This specimen, KNM-MW 13142, has the only nearly complete lumbar column known for any early Miocene primate. The generalized vertebral count of *Proconsul nyanzae* suggests that the primitive hominoid condition probably was not as specialized as are modern great apes. If the hominid vertebral column did evolve from a more generalized condition than that represented by the extant great apes, then African apes would be convergent upon *Pongo* in their thoracolumbar vertebral count.

Regardless of the pre-hominid evolutionary trajectory, all available evidence suggests that an additional lumbar vertebral element was present in the lumbar columns of early hominids. Habitual bipedality required the suite of pelvic and vertebral modifications already described. The critical adoption of a lumbar lordosis may have been facilitated by the inclusion of the additional vertebra, so that each individual element underwent fewer degrees of negative wedging. Subsequent changes to the more usual condition in modern humans, that of 5 lumbar elements, took place sometime after KNM-WT 15000.

The other possible difference between KNM-WT 15000 and modern *Homo sapiens* is the discrepancy in vertebral endplate surface areas relative to body size. Modern human adults and juveniles have, on average, slightly larger endplate areal dimensions than did KNM-WT 15000, especially in their lumbar elements. It is possible that this observation could be a consequence of a very large error in the body weight estimate of the fossil. Alternatively, it might be the result of a radically different ontogenic trajectory between the vertebral column and the remainder of the postcranial skeleton in the fossil. Neither of these explanations are satisfactory and, in our opinion, it appears as though the KNM-WT 15000 vertebral column, and perhaps that of early *Homo erectus*, was characterized by relatively smaller vertebral bodies than those in modern *Homo*.

As with the presence of an additional lumbar element, small vertebral body size also is shared between *Australopithecus* and *Homo erectus*. These two features seem to distinguish the vertebral columns of KNM-WT 15000 and australopithecines (Latimer and Ward, in preparation) from those of modern *Homo*. It is important to stress, however, that both of those dissimilarities must remain tentative in the absence of additional, and fully mature, *Homo erectus* vertebral fossils.

The vertebral column of KNM-WT 15000 was in both pattern and anatomy similar to those of modern humans. It was definitely characterized by a series of human-like sinusoidal curvatures, as demonstrated by its pattern of vertebral wedging, interfacet distances, inferior articular process morphology, and by the size and orientation of articular facets and spinous processes. This suite of anatomical features occurs only in hominids and serves to underscore the fundamental similarities between this specimen and modern *Homo sapiens*.

Acknowledgments

We wish to thank Robert P. Mensforth, C. Owen Lovejoy, and James C. Ohman for helpful discussion and/or critical reading of earlier versions of this chapter. Art was prepared by K. Dieringer and L. Gudz. Typing and editorial assistance was provided by Tamara Cruise.

We also wish to express our appreciation to Richard Leakey and Alan Walker for the opportunity to study the KNM-WT 15000 specimen.

CHAPTER 13

THE RIB CAGE

LYMAN M. JELLEMA, BRUCE LATIMER, AND ALAN WALKER

Like vertebrae, ribs of early hominids are almost never found. Because we found nearly all the ribs of the Nariokotome hominid, we can reconstruct the thorax of this Homo erectus *specimen with great accuracy. Having done so, we can determine whether these early hominids already had the "barrel-shaped" thorax typical of humans or whether they still retained the anthropoid configuration, or "funnel-shaped" thorax, we now associate with chimpanzees.*

The differences in thoracic shape between humans and apes are dramatic. These anatomical dissimilarities have long been noted (e.g., Keith, 1923; Schultz, 1950, 1961, 1963) and have been used to distinguish humans from other species. The near absence of fossil hominid ribs and thoracic vertebrae has, however, hindered study of the evolutionary transitions that led to the easily recognized distinctions in the living taxa.

The relative dearth of analyses focusing on the evolution of the hominid thoracic cage is directly linked to the fact that these naturally fragile bones are all but missing from the fossil record (Brain, 1981). In addition, it is also a consequence of the metameric arrangement of the thoracic skeleton. That is, the homologous and sequential nature of ribs and thoracic vertebrae often renders isolated or fragmentary bones somewhat difficult to interpret. It is frequently the overall pattern within the complete or nearly complete vertebral column and rib cage that is distinctive, a pattern not easily discernible from individual, disarticulated bones. Once the pattern within the anatomical composite is recognized, however, the contributions of the individual elements can be appreciated.

In view of these analytical problems, it is especially remarkable that a nearly complete thoracic cage was recovered with the KNM-WT 15000 skeleton (details regarding the recovery of this specimen are presented in Walker, Chapter 4, this volume). Portions of ribs 1 through 11 are present, often with both sides available. These ribs combined with the preserved thoracic vertebrae (see Latimer and Ward, Chapter 12, this volume) allow a unique view of an early hominid thoracic skeleton.

The remarkable preservation and completeness of the KNM-WT 15000 skeleton ensures accurate numerical assignation for the individual ribs (see Walker and Leakey, Chapter 7, this volume) as well as an appreciation for the conformation and fundamental properties of the thoracic cage. This specimen, which presents the first opportunity to study the thoracic skeleton in *Homo erectus*, will perhaps shed light on the evolutionary transformations that resulted in the thoracic morphology seen in modern *Homo sapiens*.

The juvenile developmental status of the KNM-WT 15000 skeleton (see Smith, Chapter 9, this volume) is supported by the total absence of secondary center synostosis in the thoracic elements. No effort is made in the present chapter to affix sex or chronological age to the specimen on the basis of its thoracic skeleton. Such determinations are better left to more reliable anatomical indicators (presented elsewhere in this volume—see Smith, Chapter 9; Walker and Ruff, Chapter 10).

Materials and Methods

The extant mammalian skeletal material used in this analysis is from the Hamann-Todd osteological collection and is curated by the Cleveland Museum of Natural History. Because KNM-WT 15000 was subadult, it was necessary to include a juvenile as well as an adult comparative sample. The samples of modern humans, both adult and juvenile, are the same specimens used in the analysis of the KNM-WT 15000 thoracic and lumbar vertebrae (see Latimer and Ward, Chapter 12, this volume). The thirty adult individuals varied between the ages of 30 and 35 years of age at the time of death. The twelve juvenile specimens ranged in age from 10 to 14 years. Ten adult common chimpanzees were also used for purposes of comparison. Sample sizes varied according to specimen preservation and the particular analysis and are noted below. Sexes were balanced and no pathological specimens were included.

The identifications and numbering systems used by Walker and Leakey (Chapter 7, this volume) are adhered to in this analysis. Measurements presented for the fossils were made on resin casts and were checked for accuracy against the original specimens.

Radiographs were taken using a Hewlett-Packard Cabinet X-Ray system, Faxitron series, model A3855A, with Kodak X-0mat TL radiographic film. Rib lengths, areal, and curvature measurements were digitized from radiographs or photographs.

In the analyses presented below, the individual fossil ribs chosen for study vary owing to differences in preservation. If we felt the bone could be reconstructed with reasonable accuracy, we included the reconstruction. All such reconstructions are noted.

In addition, in several of the following graphs and tables only ribs 1–10 are presented. The floating ribs were omitted because the measurements were not comparable or could not be taken on the fossil specimen.

Additional methodologies used in the following analyses are more easily understood in the context of their use and are described below. To reduce clutter in the figures, only species means are graphically presented. Collateral data are given in the associated tables.

In the following study we will examine the shape and conformation of the thoracic cage in KNM-WT 15000 and the potential influence that its juvenile developmental status had on its morphology. We will then discuss the implications that the KNM-WT 15000 specimen has for the evolution of the modern human rib cage.

THORACIC SHAPE

Hominoids are peculiar among modern catarrhines in having a thorax with a significantly greater transverse than sagittal diameter. This distinctive thoracic conformation is related to the distinctive locomotor form practiced by these animals. The broad thorax enhances lateral shoulder mobility by positioning the scapulae dorsally on the thoracic cage, an anatomical orientation in which the glenoids face laterally rather than ventrally. This scapular orientation also alters the lines of action of the shoulder musculature, thereby facilitating powerful forelimb adduction rather than caudally directed retraction.

Although the transversely broad thorax is common to all extant hominoids, differences in thoracic shape and conformation do nevertheless separate African apes and humans. The human thorax is often described as "barrel-shaped," whereas apes are said to have a "funnel-shaped" thorax (see Figure 13.1; Schultz, 1930, 1961, 1969).

In the chimpanzee, the funnel-shaped chest is the consequence of the narrow upper thorax being juxtaposed over a transversely broad pelvis which has coronally oriented iliac blades. To maintain the necessary spatial and anatomical relationships between the wide false pelvis and the inferior thoracic cage, the rib cage flares laterally in its caudal progression down the spine. It is, therefore, the spatial relationship between a transversely broad pelvis and the inferior thoracic cage that results in the so-called funnel-shaped thorax of pongids (see Figure 13.1).

In modern hominids, in contrast, the inferior portion of the thoracic cage is narrower so that the muscles and facia of the lateral abdominal walls preserve the necessary anatomical relationships over ilia that have been markedly reoriented for habitual bipedality. This is to say that the inferior-most ribs are modified to parallel the sigmoid curvature that is characteristic of the hominid iliac crest. In this conformation, the human lower thoracic cage has become noticeably narrowed and more curved. The adaptive modifications in the hominid pelvis include not only the more sagittally oriented anterior portions of the ilia but the dorsal prolongation of the iliac tuberosity as well (see below).

Additional anatomical sequelae associated with the evolutionary transition to bipedality include: (1) a volumetric expansion of the cranial third of the thoracic cage; (2) the invagination of the thoracic vertebral column ventrally into the thoracic cage; and (3) the inferior declination of the ribs. Each of these thoracic adaptations can be read from the ribs and will be used to evaluate the rib cage of KNM-WT 15000.

First, rib curvatures, lengths, and areal measurements are examined to see if KNM-WT 15000 had undergone the changes in shape and volume associated with the hominid thoracic cage. Curvature measurements in combination with vertebral measurements will also indicate whether the vertebral column was positioned ventrally in this specimen.

Second, the ribs and associated vertebrae are examined for evidence of rib declination. Finally, discussion centers on the evolutionary transitions that resulted in the modern hominid thoracic cage.

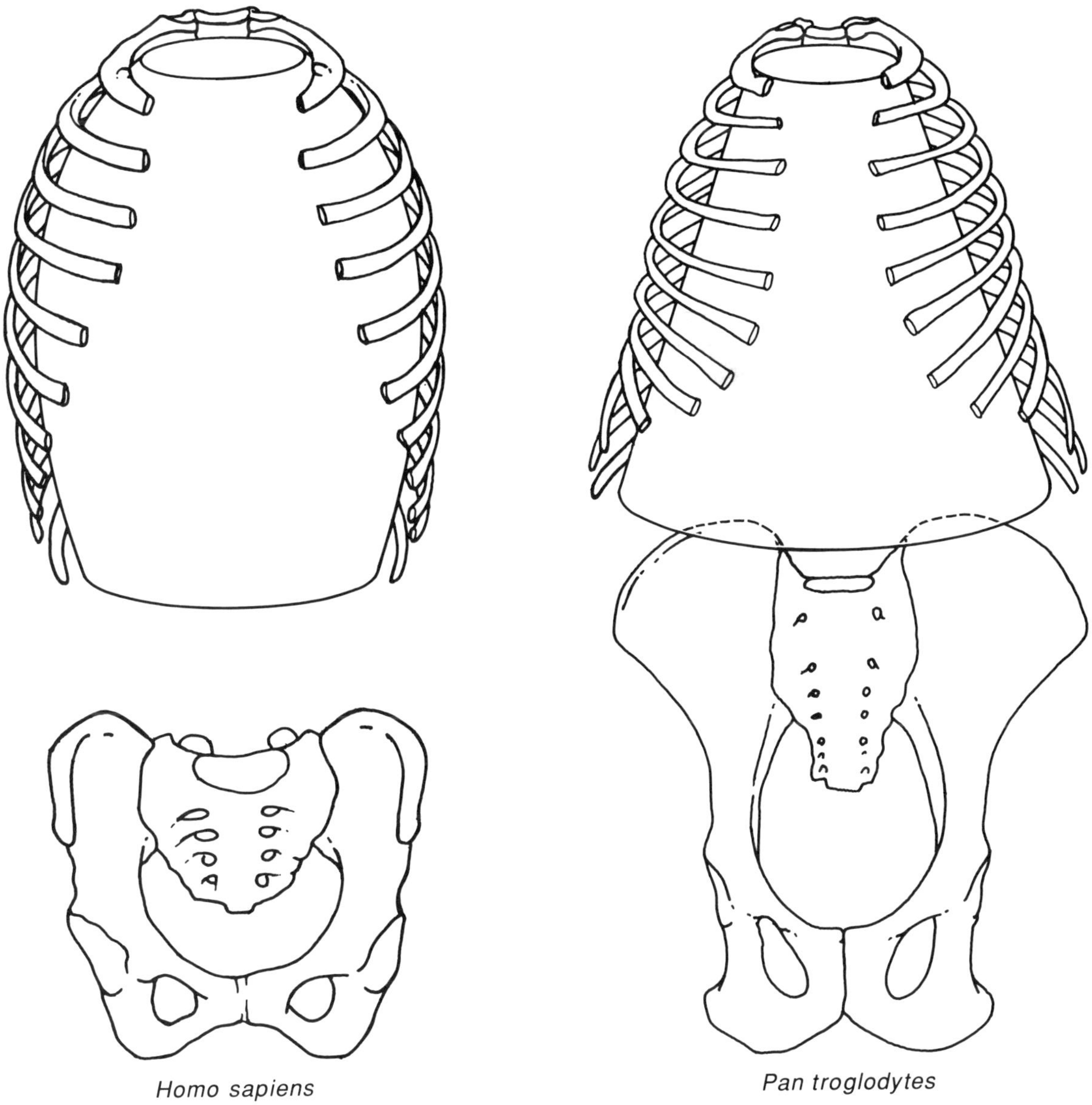

FIGURE 13.1.
A comparison of the thoracic cage and the pelvic girdle in PAN TROGLODYTES *and* HOMO SAPIENS. *The barrel-shaped hominid rib cage is narrowed at the bottom to conform to the reoriented ilia. In contrast, the rib cage in the ape flares in its most inferior segments to conform to the transversely broad false pelvis; it thereby creates a funnel-shaped thorax. The ape and human thoracic shapes are superimposed below. To compensate for the volumetric reduction that accompanied the pelvic transformations in* HOMO, *the cage was elongated and expanded near the thoracic operculum.*

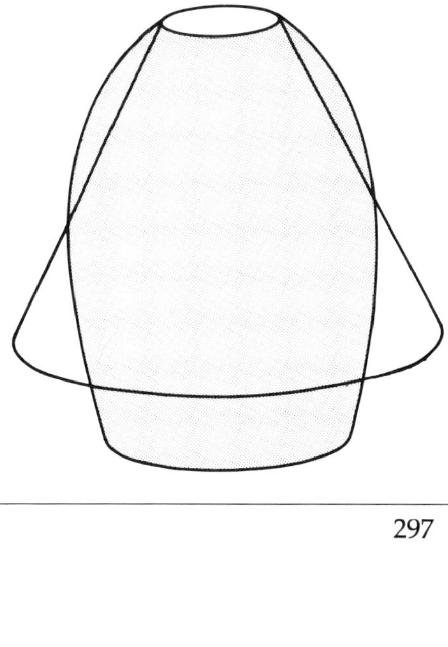

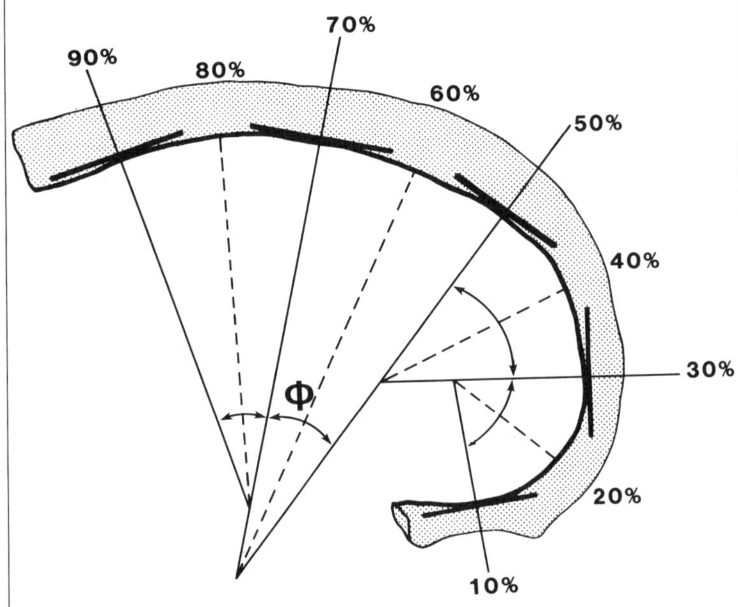

FIGURE 13.2.
An illustration of the method used to define the radii of curvatures and included angles (angle ϕ).

Costal curvatures

Because ribs have complex curvatures with multiple radii, a method of comparing the changing radius of curvature from point to point along the curve was necessary. This was accomplished by digitizing photographs or radiographs of each rib and dividing each bone into 10 arcs of equal lengths (beginning at the rib head and proceeding along the pleural surface)(see Figure 13.2). Tangents were then constructed at the beginning point of every other arc—starting with the arc covering the first 10 percent, then 30 percent, then 50 percent and through to 90—and the normal for each tangent was found. The normals from two consecutive tangents intersect at a point that approximates the instantaneous center of the arc subtending the angle ϕ formed by the two normals.

A line drawn through the point of intersection (the instantaneous center) and bisecting the angle is an instantaneous radius. The angle between the two normals is the included angle. In this manner, a series of instantaneous centers and instantaneous radii were constructed for the points at 20, 40, 60, and 80 percent of the total arc length of each rib. Included angles were also constructed at each instantaneous radius.

TABLE 13.1A.

Normalized radius of curvature at 20% segment of total rib length; see Figure 13.3A, top left.

RIB NUMBER	*HOMO SAPIENS* ADULT ($N = 10$) Mean	s.d.	*HOMO SAPIENS* JUVENILE ($N = 10$) Mean	s.d.	KNM-WT 15000	*PAN TROGLODYTES* ($N = 10$) Mean	s.d.
1	.39	.18	.29	.03	.34	.45	.03
2	.20	.03	.19	.03	.14	.18	.03
3	.18	.02	.19	.03	.14*	.19	.04
4	.17	.02	.19	.02	.16*	.18	.02
5	.16	.01	.22	.04	.15*	.19	.03
6	.15	.01	.19	.03		.18	.04
7	.15	.02	.19	.03	.16*	.18	.03
8	.17	.02	.21	.03	.18*	.21	.04
9	.19	.03	.23	.04		.22	.07
10	.23	.03	.27	.04	.25*	.23	.04

*Measured from reconstructed rib.

Because ribs vary in overall length, for the purpose of comparison the instantaneous radii were normalized by dividing each instantaneous radius by the respective rib's total arc length. In addition, because radius length, even when normalized for size, fails to adequately convey the compass of the subtended arc, the included angle is also presented for each segment.

In Figures 13.3A–B (see also Tables 13.1A–D, 13.2A–D), the normalized radii and included angles are presented for ribs 1–10. This analysis demonstrates that each rib is made up of several systems of curvatures and that its component curvatures can be differentiated and compared by this method.

It is apparent that the normalized radius of curvature measurement usually varies inversely with the included angle measurement. That is, the larger the radius (arc is less curved), the fewer degrees are encompassed within the included angle. Conversely, the arcs with smaller radii (more curvature) generally have larger included angles.

In Figure 13.3A (see also Tables 13.1A and 13.1B) the normalized radii of curvature of the proximal fifth of the ribs (20% segment) fail to separate the four groups clearly. However, examination of the associated included angles reveals that KNM-WT 15000 and modern human adults show a pattern of greater curvature (more degrees of the associated arcs) along the middle thorax (ribs 4–8). Note that the *Homo erectus* specimen shows a rather extreme degree of curvature (short radius and large included angle) in the most proximal portion of ribs 2 and 3. This is also reflected in the morphology of the rib and is discussed below.

This heightened curvature at the proximal fifth of the adult human and *Homo erectus* ribs indicates that the neck of the rib is more sharply flexed toward the rib corpus than it is in the chimpanzee ribs. This flexion is directly associated with the more dorsal orientation of the ribs relative to the thoracic vertebral column (see below) and with the expanded transverse diameter of the upper thorax. The juvenile humans also appear less flexed, suggesting a growth differential in the proximal portions of the rib and an increase in curvature as the bone develops. Also indicated in the data for juvenile humans is the progressive ventral displacement of the thoracic vertebral column that occurs during human growth (see below). Both of these views are supported by the observations that children have a higher thoracic ratio of dorsoventral to transverse diameter than do adults, as well as less ventrally positioned thoracic vertebral columns (Schultz, 1969).

TABLE 13.1B.

Included angle (in degrees) at 20% segment of total rib length; see Figure 13.3A, bottom left.

	HOMO SAPIENS						
	ADULT (*N* = 10)		JUVENILE (*N* = 10)			*PAN TROGLODYTES* (*N* = 10)	
RIB NUMBER	Mean	S.D.	Mean	S.D.	KNM-WT 15000	Mean	S.D.
1	31.98	10.02	38.46	9.57	32.80	32.56	15.35
2	55.60	9.50	59.13	11.28	76.85	62.30	12.06
3	61.98	7.31	59.20	9.15	73.30*	59.67	11.33
4	66.33	9.80	58.97	6.14	65.44*	59.66	5.36
5	67.60	5.99	53.07	10.71	69.63*	57.58	9.55
6	72.39	3.05	58.97	8.97		62.99	11.45
7	72.80	10.42	60.08	10.18	66.49*	59.89	10.08
8	63.47	8.76	54.48	8.94	57.94*	54.81	11.06
9	58.72	9.55	49.31	9.15		53.83	14.69
10	49.77	9.57	42.14	7.66	44.57*	49.49	7.36

*Measured from reconstructed rib.

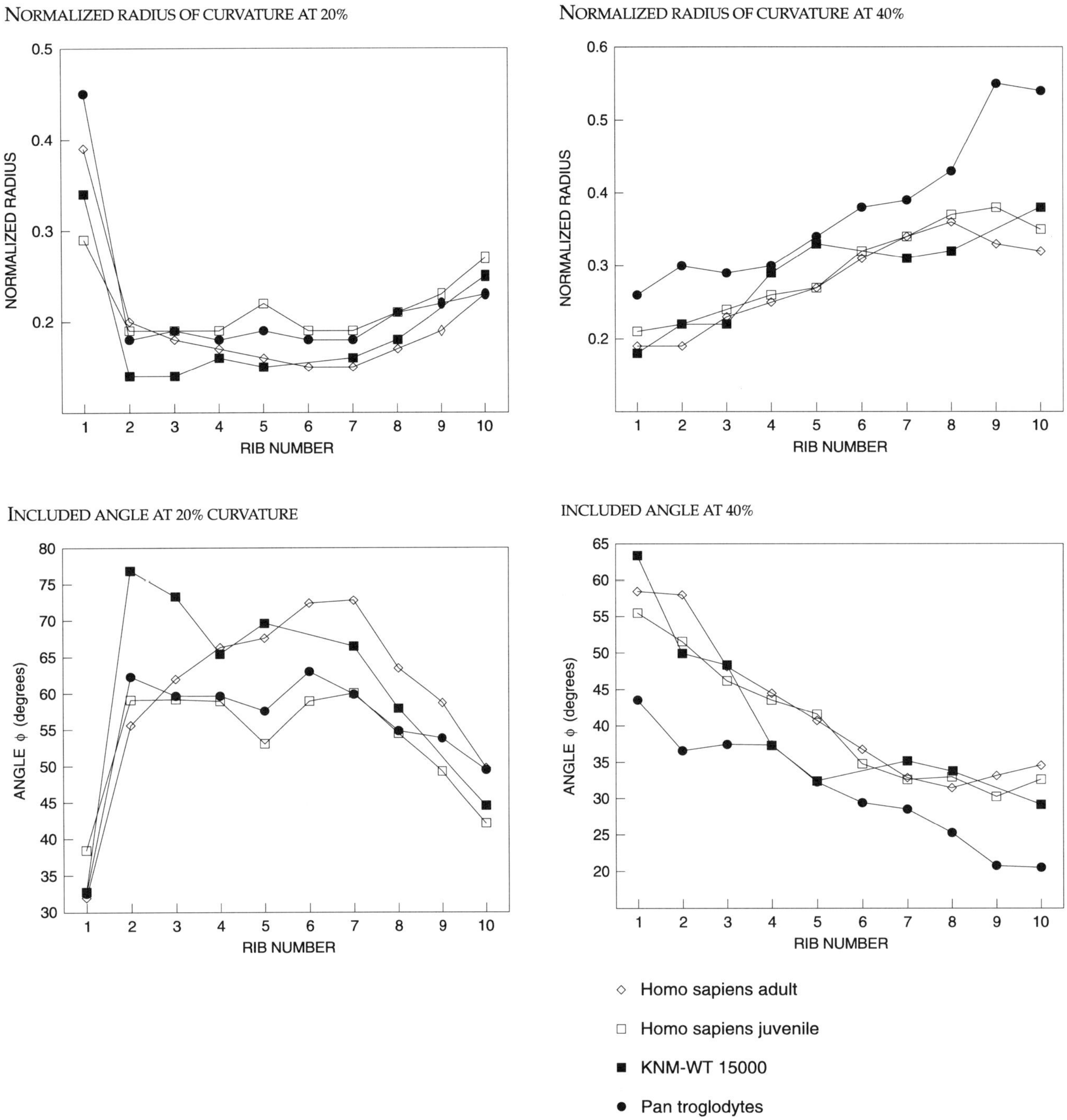

FIGURE 13.3A.

Normalized radii of curvatures and included angles at 20% and 40% of costal arc length for ribs 1–10 (see Tables 13.1A–13.1D). Note the similar patterns in HOMO SAPIENS *and KNM-WT 15000.*

TABLE 13.1C.

Normalized radius of curvature at 40% segment of total rib length; see Figure 13.3A, top right.

RIB NUMBER	*HOMO SAPIENS* ADULT (N = 10) Mean	s.d.	*HOMO SAPIENS* JUVENILE (N = 10) Mean	s.d.	KNM-WT 15000	*PAN TROGLODYTES* (N = 10) Mean	s.d.
1	.19	.02	.21	.03	.18	.26	.05
2	.19	.02	.22	.04	.22	.30	.37
3	.23	.01	.24	.04	.22*	.29	.04
4	.25	.02	.26	.04	.29*	.30	.06
5	.27	.04	.27	.05	.33*	.34	.04
6	.31	.04	.32	.06		.38	.06
7	.34	.07	.34	.05	.31*	.39	.03
8	.36	.08	.37	.17	.32*	.43	.05
9	.33	.04	.38	.07		.55	.13
10	.32	.04	.35	.09	.38*	.54	.09

*Measured from reconstructed rib.

The KNM-WT 15000 specimen demonstrates a pattern most similar to modern human adults. It is, however, extremely curved even for modern humans in its second and third ribs, indicating a significant volumetric expansion in its upper thorax. In contrast to the hominid condition, chimpanzee ribs are overall less curved (longer relative radii and smaller included angles).

At the 40-percent segment (Figure 13.3A, Tables 13.1C and 13.1D) each group shows a similar pattern of a sequential decrease in rib curvature in inferior progression. The chimpanzee ribs are less curved throughout the entire thoracic cage, an observation reflected in the radii of cur-

TABLE 13.1D.

Included angle (in degrees) at 40% segment of total rib length; see Figure 13.3A, bottom right.

RIB NUMBER	*HOMO SAPIENS* ADULT (N = 10) Mean	s.d.	*HOMO SAPIENS* JUVENILE (N = 10) Mean	s.d.	KNM-WT 15000	*PAN TROGLODYTES* (N = 10) Mean	s.d.
1	58.48	7.36	55.50	8.43	63.38	43.55	10.93
2	58.00	7.28	51.57	11.09	49.94	36.58	3.69
3	48.15	4.00	46.19	7.69	48.36*	37.45	4.51
4	44.50	4.44	43.54	6.84	37.30*	37.36	5.85
5	40.72	5.62	41.59	6.00	32.46*	32.30	3.17
6	36.77	5.37	34.79	6.05		29.46	4.76
7	32.90	6.60	32.63	5.37	35.20*	28.59	2.85
8	31.50	5.65	33.01	8.03	33.79*	25.34	3.19
9	33.17	4.42	30.31	5.11		20.79	4.25
10	34.60	5.25	32.66	8.66	29.21*	20.56	4.04

*Measured from reconstructed rib.

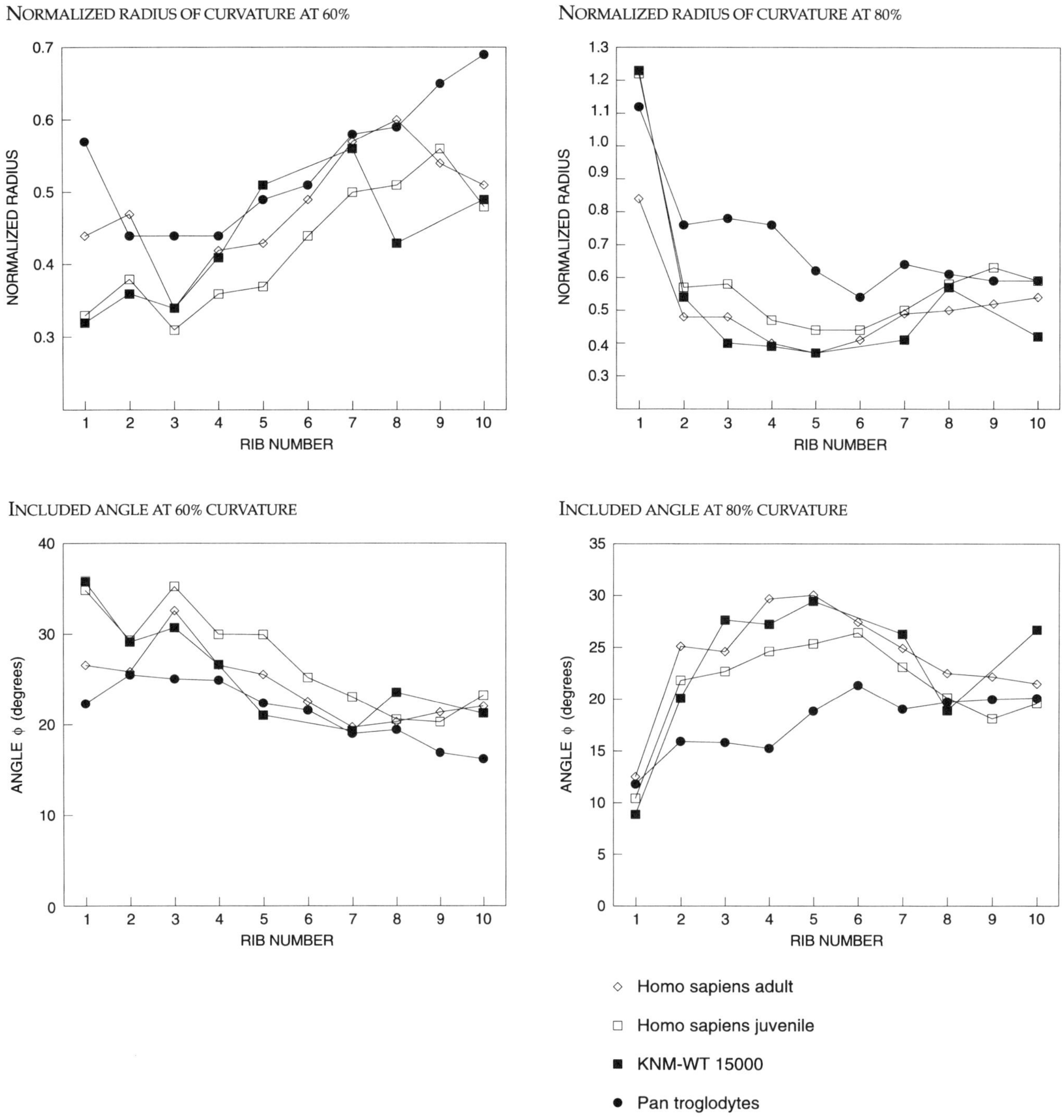

FIGURE 13.3B.
Normalized radii of curvatures and included angles at 60% and 80% of costal arc length for ribs 1–10 (see also Tables 13.2A–13.2D).

TABLE 13.2A.

Normalized radius of curvature at 60% segment of total rib length; see Figure 13.3B, top left.

RIB NUMBER	*HOMO SAPIENS* ADULT (*N* = 10) Mean	S.D.	*HOMO SAPIENS* JUVENILE (*N* = 10) Mean	S.D.	KNM-WT 15000	*PAN TROGLODYTES* (*N* = 10) Mean	S.D.
1	.44	.13	.33	.06	.32	.57	.26
2	.47	.19	.38	.08	.36	.44	.21
3	.34	.54	.31	.03	.34*	.44	.05
4	.42	.07	.36	.03	.41*	.44	.07
5	.43	.06	.37	.05	.51*	.49	.06
6	.49	.08	.44	.07		.51	.07
7	.57	.10	.50	.12	.56*	.58	.09
8	.60	.21	.51	.08	.43*	.59	.14
9	.54	.15	.56	.15		.65	.10
10	.51	.10	.48	.09	.49*	.69	.12

*Measured from reconstructed rib.

vature and included angle measurements. Although KNM-WT 15000 falls on the mean for *Pan* in ribs 4 and 5, these ribs are poorly preserved and undoubtedly have suffered plastic deformation (this is especially true of rib 5, CB; see Walker and Leakey, Chapter 7, this volume). The remainder of the fossil ribs show a human-like pattern at the 40-percent point.

Moving distally to the point which is just beyond midshaft (60% of the rib length; see Figure 13.3B, Tables 13.2A and 13.2B), chimpanzee ribs are on average slightly less curved than are hominid ribs. This is especially true at rib 3, where the chimpanzee shows a clear narrowing of the thorax, and at ribs 9 and 10, where the chimpanzee thorax flares laterally to conform to the broad false pelvis.

TABLE 13.2B.

Included angle (in degrees) at 60% segment of total rib length; see Figure 13.3B, bottom left.

RIB NUMBER	*HOMO SAPIENS* ADULT (*N* = 10) Mean	S.D.	*HOMO SAPIENS* JUVENILE (*N* = 10) Mean	S.D.	KNM-WT 15000	*PAN TROGLODYTES* (*N* = 10) Mean	S.D.
1	26.57	6.81	34.83	7.93	35.80	22.30	9.50
2	25.84	8.21	29.32	5.15	29.13	25.50	4.18
3	32.56	5.61	35.25	3.77	30.70*	25.05	3.47
4	26.56	3.97	29.97	2.64	26.63*	24.89	4.64
5	25.53	3.15	29.92	4.69	21.03*	22.36	2.70
6	22.53	3.85	25.17	3.26		21.58	3.19
7	19.73	3.13	23.05	5.46	19.29*	19.00	3.11
8	20.32	6.33	20.59	3.49	23.54*	19.43	4.19
9	21.37	5.00	20.29	4.40		16.90	2.44
10	22.04	3.83	23.23	4.22	21.24*	16.21	3.30

*Measured from reconstructed rib.

It is interesting that at this point, in contrast to the previous curvature measurements, the ribs of human juveniles are more curved (small radii; larger included angles) than are those of the adults. This differential curvature along the length of the rib again indicates a relative change in torso shape in growing humans as the ribs become progressively wider transversely. KNM-WT 15000 follows the human pattern except for rib 5 (CB), which as noted above is quite clearly less curved than it should be owing to postmortem damage. In addition, rib 8 in the fossil specimen shows an accentuated degree of curvature, even in comparison with those of modern humans. This latter observation may reflect the narrow pelvis in this individual.

At the 80-percent rib segment (Figure 13.3B, Tables 13.2C and 13.2D) the contrast between ape and hominid curvature is more marked than at any other region. This is especially apparent from ribs 2 through 7, where the ape ribs are clearly less curved than are the hominid ribs. KNM-WT 15000 has a pattern essentially identical to that in adult and juvenile humans. It is noteworthy, however, that rib 10 in KNM-WT 15000 also shows extreme curvature, again indicating a definite narrowing of the inferior portions of the cage in order to conform to the narrow false pelvis (see Walker and Ruff, Chapter 10, this volume).

These comparisons of rib curvatures demonstrate that human ribs are relatively more curved than are chimpanzee ribs, primarily along their proximal and distal segments. This enhanced curvature in humans is most extreme in the cranial and inferior portions of the rib cage, areas which contribute to the barrel shape of the thorax (Figure 13.1). The greater curvatures in the proximal region of the adult hominid ribs result from the more ventrally located vertebral column (see below). The greater curvatures in the distal portions of the lower hominid ribs parallel the more sagittally oriented, sigmoid curvature of the hominid ilia. Conversely, the straighter ribs in the apes parallel the broad, flat, coronally oriented ilia and result in a more funnel-shaped thorax. KNM-WT 15000 is human-like in both the individual rib curvatures as well as the patterns of curvatures throughout the thoracic cage. The *Homo erectus* specimen clearly had a barrel-shaped chest. This is not surprising, in light of the narrow human-like pelvis (Walker and Ruff, Chapter 10, this volume) to which the thorax was juxtaposed.

TABLE 13.2C.

Normalized radius of curvature at 80% segment of total rib length; see Figure 13.3B, top right.

	HOMO SAPIENS						
	ADULT (*N* = 10)		JUVENILE (*N* = 10)			*PAN TROGLODYTES* (*N* = 10)	
RIB NUMBER	Mean	S.D.	Mean	S.D.	KNM-WT 15000	Mean	S.D.
1	.84	.32	1.22	.51	1.23	1.12	.54
2	.48	.16	.57	.25	.54	.76	.25
3	.48	.13	.58	.29	.40*	.78	.29
4	.40	.11	.47	.12	.39*	.76	.19
5	.37	.06	.44	.09	.37*	.62	.16
6	.41	.07	.44	.11		.54	.12
7	.49	.18	.50	.13	.41*	.64	.21
8	.50	.08	.58	.17	.57*	.61	.19
9	.52	.15	.63	.13		.59	.16
10	.54	.11	.59	.14	.42*	.59	.19

*Measured from reconstructed rib.

Rib length

A measurement of total rib length taken along the arc of the pleural surface allows comparison of the patterns of sequential lengths among the sample groups (see Figure 13.4 and Table 13.3). The dissimilarities between hominids and apes in absolute rib lengths are most marked between ribs 2 through 5, where the hominids are uniformly longer and, as demonstrated above, more curved. Indeed, in the upper thorax even the human juveniles are absolutely longer than are the adult apes (the lines cross at rib 4), once again indicating the relative expansion of the cranial portion of the hominid thorax. The differences are reversed in the lower thorax, wherein the lower ribs in apes are relatively longer, a factor that, in concert with the relatively lessened curvatures, produces the funnel-shaped chest.

Also apparent is the observation that in rib lengths, the human adult and juvenile mean values are closer to one another in the upper thorax and thereafter diverge inferiorly. This pattern suggests a superoinferior growth differential along the thoracic cage, wherein the cranial portions achieve adult size earlier than do the inferior areas (a pattern also seen in the vertebral column; Latimer and Ward, Chapter 12, this volume). The human ontogenetic pattern in thoracic growth also contributes to the differences in rib declination and thoracic kyphosis that are seen in children and adults (see below).

In Figure 13.4, KNM-WT 15000 shows a sequential pattern that is unequivocally hominid. In view of the growth phenomena noted above in the modern humans, it is somewhat surprising to see that the fossil specimen, although of juvenile developmental status, is essentially identical to the modern human adult averages in rib lengths.

The differences between the hominid and pongid patterns in costal lengths are additionally confirmed in Figure 13.5 (see also Table 13.4). In this exercise, each individual rib length was divided by the length of the associated rib 7. Rib 7 was chosen as a normalizing factor because it is from the middle of the thoracic skeleton and thus provides evidence regarding the relative contribution that each rib makes to the shape of the rib cage.

Again, KNM-WT 15000 and the modern humans, both juveniles and adults, share a pattern

TABLE 13.2D.

Included angle (in degrees) at 80% segment of total rib length; see Figure 13.3B, bottom right.

	HOMO SAPIENS						
	ADULT (*N* = 10)		JUVENILE (*N* = 10)			*PAN TROGLODYTES* (*N* = 10)	
RIB NUMBER	Mean	S.D.	Mean	S.D.	KNM-WT 15000	Mean	S.D.
1	12.49	5.50	10.40	4.13	8.84	11.80	4.80
2	25.11	7.41	21.81	5.89	20.09	15.90	5.11
3	24.59	7.03	22.67	9.47	27.64*	15.80	4.88
4	29.68	8.32	24.62	6.44	27.22*	15.24	3.78
5	30.04	5.12	25.35	4.43	29.44*	18.83	4.14
6	27.43	5.18	26.41	6.47		21.33	4.74
7	24.91	8.72	23.11	5.78	26.29*	19.04	6.02
8	22.50	4.16	20.12	5.44	18.89*	19.70	5.60
9	22.16	4.58	18.11	4.23		19.97	5.71
10	21.48	5.41	19.59	4.06	26.68*	20.07	5.14

*Measured from reconstructed rib.

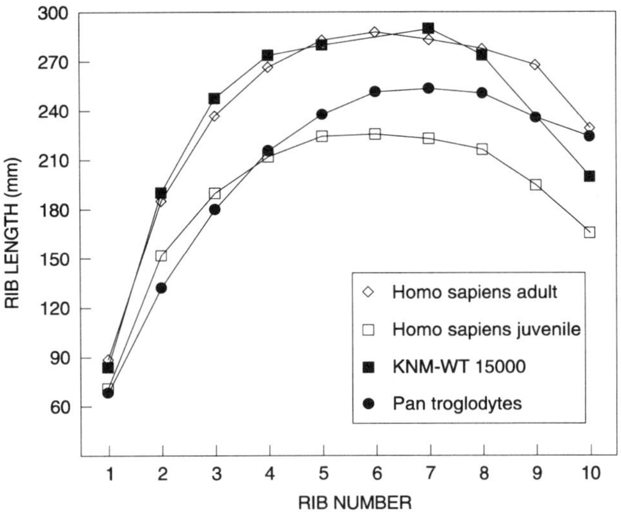

FIGURE 13.4.
Arc lengths measured along the pleural surfaces of the ribs (see Table 13.3). Note the overall similarities in pattern between modern humans and KNM-WT 15000.

of thoracic morphology that is characterized by an expanded upper thorax connected to a narrowed inferior thorax and pelvis. This is contrasted by the pongid condition with its progressive expansion of the thorax over the transversely wide pelvis. It can also be noted from Figure 13.5 that the two most inferior ribs plotted in this analysis of KNM-WT 15000 are relatively shorter than those of either adult or juvenile humans. The length of these ribs is in agreement with the previously mentioned differences in curvatures and again suggests a relatively small circumferential dimension in the inferior-most portion of the fossil rib cage, a feature directly associated with the narrow bi-iliac diameter in this specimen (Ruff and Walker, Chapter 11, this volume).

Thoracic volume

Costal curvatures and arc lengths both show that when compared with the condition in extant African apes, KNM-WT 15000 and modern humans have a relatively expanded upper thorax in combination with a relatively narrow inferior thorax. Both of these features create the barrel-shaped thorax of hominids (Figure 13.1). As noted, the narrow lower thoracic cage is owing to its proximity to the anatomically reorganized hominid pelvis. In view of this, the greater curvature and consequent volumetric augmentation of the cranial half of the hominid thorax likely represents a compensatory response to the diminished inferior

TABLE 13.3.

Rib length (mm); see Figure 13.4.

	HOMO SAPIENS					*PAN TROGLODYTES*	
	ADULT (*N* = 10)		JUVENILE (*N* = 10)			(*N* = 10)	
RIB NUMBER	Mean	S.D.	Mean	S.D.	KNM-WT 15000	Mean	S.D.
1	88.65	8.64	71.18	10.23	84.16	68.45	3.52
2	185.18	13.95	151.85	19.27	190.26	132.20	14.52
3	236.84	17.95	189.76	20.37	247.67*	179.96	16.34
4	266.45	20.69	212.04	17.58	273.84*	215.57	18.89
5	282.89	21.50	224.34	18.12	280.00*	237.66	21.02
6	287.82	25.56	225.63	17.35		251.59	20.77
7	283.25	23.38	222.85	15.56	290.00*	253.49	21.30
8	277.77	26.02	216.29	24.50	273.91*	250.75	16.89
9	267.82	35.36	194.58	16.42		235.90	17.51
10	229.32	30.93	165.57	12.44	199.71*	224.18	16.65

*Measured from reconstructed rib.

thorax. Coincident with this feature of the hominid thorax is the fact that the thorax of the African apes is not simply flared at the bottom but, in addition, is actually narrowed at the top when compared with those of hominids (Figure 13.1).

Mammals have lung volumes that are isometrically equivalent in relation to their body mass (interspecific slope of 1.06; Gehr et al., 1981). It is likely that the differences demonstrated above between the pongid and hominid ribs represent alternative ways of achieving the required thoracic volume in thoraxes of dissimilar shapes. For a comparison of the relevant costal areas between the different taxa, it was necessary to know the functional radius of the rib in question. That is, during ventilation, any point on a rib inscribes an arc as the rib rotates about its neck axis. The neck axis is defined here as a straight line through the centroids of the costal articular surfaces (see Figure 13.6). This line is equivalent to a line piercing the articular centroids of the costal facets on the associated vertebrae (see Figures 13.11 and 13.16).

The orthogonal distance from any point on the rib to the axis of rotation is thus a rotating radius. The maximum costal radius is that point on the rib that corresponds to the greatest orthogonal distance from the neck axis; this point is determined by curvature of the costal body and the relative orientation of the neck axis. It is important to note that although the maximum radius usually touches the distal-most point of the rib's sternal end (except for ribs 2 and 3 in hominids, see below), it need not and usually does not come in contact with the rib's vertebral end (see Figure

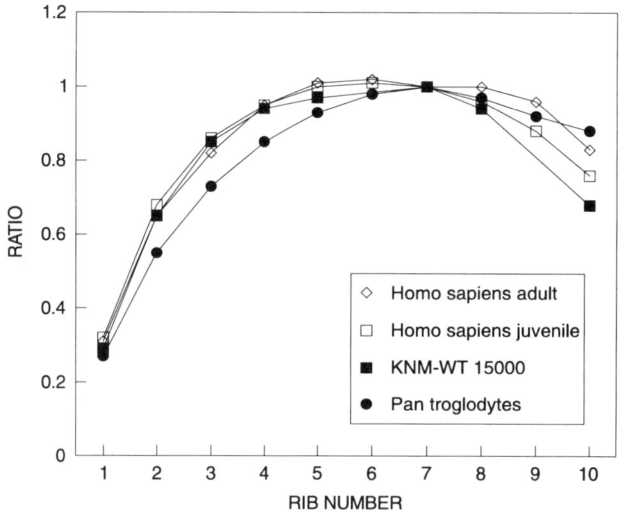

FIGURE 13.5.
Arc length for each individual rib divided by the arc length of rib 7 (see Table 13.4). Note the similar patterns in KNM-WT 15000 and modern humans.

TABLE 13.4.

Ratio of individual rib length to length of rib 7; see Figure 13.5.

RIB NUMBER	*HOMO SAPIENS* ADULT (*N* = 8) Mean	S.D.	*HOMO SAPIENS* JUVENILE (*N* = 8) Mean	S.D.	KNM-WT 15000	*PAN TROGLODYTES* (*N* = 9) Mean	S.D.
1	.31	.02	.32	.03	.29	.27	.01
2	.65	.03	.68	.06	.65	.55	.02
3	.82	.04	.86	.04	.85*	.73	.03
4	.95	.11	.95	.03	.94*	.85	.03
5	1.01	.03	1.00	.04	.97*	.93	.05
6	1.02	.04	1.01	.04		.98	.02
7	1.00		1.00		1.00*	1.00	
8	1.00	.02	.96	.10	.94*	.97	.02
9	.96	.08	.88	.28		.92	.05
10	.83	.07	.76	.05	.68*	.88	.03

*Measured from reconstructed rib.

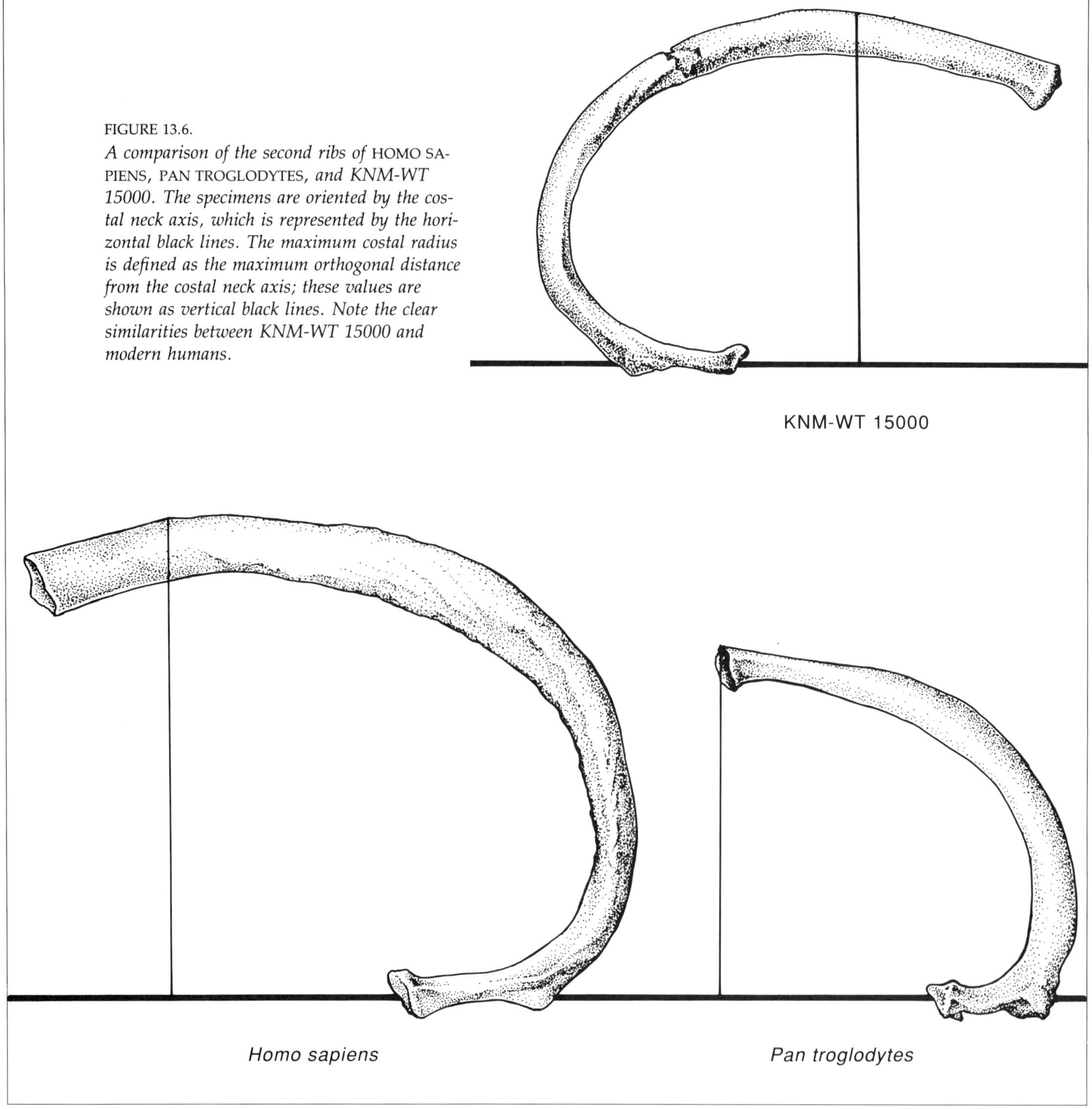

FIGURE 13.6.
A comparison of the second ribs of HOMO SAPIENS, PAN TROGLODYTES, *and KNM-WT 15000. The specimens are oriented by the costal neck axis, which is represented by the horizontal black lines. The maximum costal radius is defined as the maximum orthogonal distance from the costal neck axis; these values are shown as vertical black lines. Note the clear similarities between KNM-WT 15000 and modern humans.*

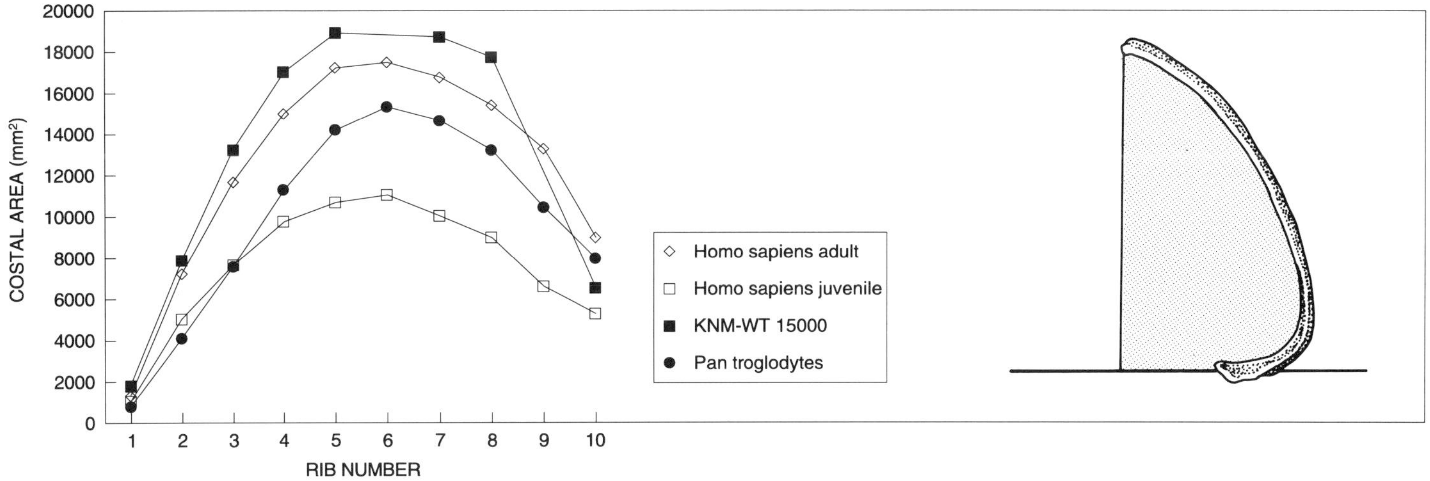

FIGURE 13.7.
Costal areas in human adults and children, chimpanzees, and KNM-WT 15000 (see Table 13.5). Measurements include the area enclosed within the costal neck axis, the maximum costal radius, and the arc of the rib.

13.6). The area encompassed by the maximum costal radius, the arc of the rib's pleural surface, and the neck axis is used in this analysis to compare the functional areas of the stationary ribs.

From Figure 13.7 (and Table 13.5) it can be seen that the patterns of sequential costal areas differ between the hominids and the chimpanzees. Again, the differences between the apes and the hominids are greatest in the upper thorax (ribs 2–5) and then progressively decrease, indicating expansion at opposite ends of the rib cage. The growth differential between the upper and lower thoracic cage is again shown in the human juveniles, wherein the upper thorax (ribs 1–4) more closely approaches the adult size than does the lower thorax.

KNM-WT 15000 has a human-like pattern of sequential costal areas. This is demonstrated even more clearly in Figure 13.8, which graphs the

TABLE 13.5.

Costal area (mm^2); see Figure 13.7.

RIB NUMBER	*HOMO SAPIENS* ADULT (*N* = 10) Mean		*HOMO SAPIENS* JUVENILE (*N* = 10) Mean		KNM-WT 15000	*PAN TROGLODYTES* (*N* = 10) Mean	
	Mean	S.D.	Mean	S.D.		Mean	S.D.
1	1329	306	1045	339	1820	793	210
2	7230	1104	5036	1117	7874	4103	808
3	11671	2201	7640	1861	13218*	7577	1080
4	15010	1982	9777	1825	17033*	11302	1585
5	17231	1918	10702	1958	18921*	14214	1688
6	17492	2770	11060	2275	19207†	15342	2004
7	16772	3168	10045	2283	18719*	14671	2608
8	15418	3657	8995	2477	17732*	13225	3614
9	13281	4000	6617	1199	13044†	10455	1815
10	8988	2206	5289	953	6536*	7968	2657

*Measured from reconstructed rib.
†See caption to Figure 13.7.

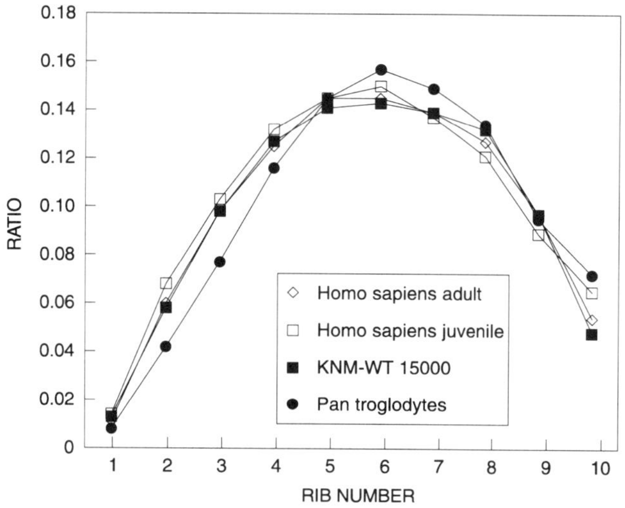

FIGURE 13.8.
Costal area of an individual rib divided by the summed costal areas of ribs 1–10 (using data shown in Figure 13.7 and also Table 13.6). Note that the hominids share an expanded upper thorax relative to the chimpanzee. Areas for the missing ribs in KNM-WT 15000 were estimated.

ratio of the individual rib areas to the total summed areas of rib 1 through 10. This ratio allows an appreciation of the relative contribution that each rib makes to the thoracic cage. (See also Table 13.6.) It is not surprising that the relative contribution of the upper thorax is greater in the hominids than it is in *Pan* (the lines cross at rib 5).

Perhaps the most unexpected outcome of this analysis is the absolutely large area exhibited by the *Homo erectus* specimen. Although only a juvenile, this specimen's costal areas actually exceed the modern human adult averages. Although when normalized by body mass estimates, as in Figure 13.9 and Table 13.7, KNM-WT 15000 falls within the human range, it is nevertheless extreme for the human juvenile sample (assuming a subadult body weight of 48 kg; Ruff and Walker, Chapter 11, this volume). Figure 13.9 suggests that this specimen had a relatively large thoracic volume compared with modern humans of equivalent ontogenetic status. If the estimated adult body weight of 68 kg (Ruff and Walker, Chapter 11, this volume) is used to normalize the thoracic volume of KNM-WT 15000, this specimen is easily within the ranges of the comparative samples. This latter observation may indicate a different ontogenic pattern than that which characterizes modern humans.

Apes and hominids have very similar proportions of thoracic volume (as approximated by rib areas) relative to their respective body sizes. These thoracic dimensions are however, achieved in a slightly different way. The thoracic cage of

TABLE 13.6.

Ratio of individual rib area to sum of all rib areas; see Figure 13.8.

RIB NUMBER	*HOMO SAPIENS* ADULT (*N* = 10) Mean	S.D.	*HOMO SAPIENS* JUVENILE (*N* = 10) Mean	S.D.	KNM-WT 15000	*PAN TROGLODYTES* (*N* = 10) Mean	S.D.
1	.011	.001	.014	.004	.013	.008	.002
2	.060	.009	.068	.012	.058	.042	.008
3	.098	.017	.103	.016	.098*	.077	.009
4	.125	.011	.132	.011	.127*	.116	.014
5	.145	.015	.145	.010	.141*	.145	.012
6	.145	.012	.150	.015	.143*	.157	.013
7	.139	.015	.137	.024	.139*	.149	.018
8	.127	.018	.121	.021	.132*	.134	.025
9	.097	.036	.089	.018	.097*	.095	.034
10	.054	.029	.065	.009	.048*	.072	.030

*Measured from reconstructed rib.

KNM-WT 15000 is essentially identical to that of humans in its overall shape and pattern and, while it does have an absolutely large thoracic volume for its developmental age, when adjusted for body size its volume falls within the range of the modern human sample. Whether this indicates an individual or species-specific growth differential compared with modern humans or an actual augmentation of thoracic dimensions is at present unresolvable.

The preceding analyses have demonstrated that KNM-WT 15000 and modern humans have, when compared with *Pan*, expanded upper thoraxes. This has been shown above in examinations of rib curvatures, rib lengths, and the areas encompassed by the ribs. Expansion of the upper thorax can also be demonstrated in the morphology of the ribs.

The maximum costal radius for rib 2 is illustrated in Figure 13.10 (also see Figure 13.6). This illustration makes clear the effect that the curvature inequalities described above can have on circumscribed areas in the upper thorax. The maximum costal radius in the chimpanzee rib 2 is through a point on the sternal extremity of the bone. In KNM-WT 15000 and in modern humans, the maximum radius of rib 2 (and sometimes rib 3) corresponds to a point which is lateral to the rib's sternal end (see Figure 13.6). This obvious difference highlights the contrasts in shape of the upper thorax discussed above. Moreover, if the human bone is superimposed over the chimpanzee rib (Figure 13.10) so that their two sternal ends coincide spatially, it is apparent that the human neck axis angles more posteriorly relative to the sagittal plane.

This observation is also confirmed by measuring the angle that the rib facets on the transverse process of the thoracic vertebrae make to the sagittal plane in cranial view (angle Σ, see Figure 13.11 and Table 13.8). This measurement was taken directly from superoinferior radiographs of the individual thoracic vertebrae in which the costal articular surfaces had been demarcated with lead tape.

Another measurement confirming the differences in thoracic shapes is a ratio formed between the maximum costal radius and maximum

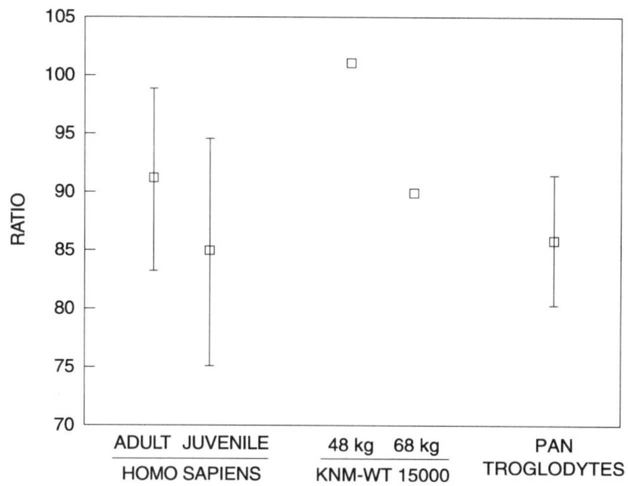

FIGURE 13.9.
The square root of summed costal areas for ribs 1–10 divided by the cube root of body weight. Body weights of humans were taken from records at the Cleveland Museum of Natural History. The body weight for PAN *is an average of male and female estimates from Jungers (1988). The values used for KNM-WT 15000 are estimates of the specimen's juvenile (48 kg) and adult (68 kg) weights from Ruff and Walker (Chapter 11, this volume). The squares on the plot represent means and the bars represent 1 standard deviation. See Table 13.7.*

costal length. The closer the ratio is to one, the nearer the radial length is to the arc length—or, in other words, the straighter the rib. Hominid ribs have a relatively small ratio, and thus greater differences between these two measurements, again indicating more curvature (Figure 13.12 and Table 13.9). It is not surprising that among the hominids, with their expanded upper thorax, the greater curvature is most apparent in ribs 2 through 5. Although the fossil specimen shows what is

TABLE 13.7.

Ratio of the square root of the sum of all rib areas to the cube root of body weight; see Figure 13.9.

	MEAN	S.D.
Homo sapiens adults	90.95	7.84
Pan troglodytes	85.66	5.54
Homo sapiens children	84.82	9.64
KNM-WT 15000 (48 kg)	100.76*	
KNM-WT 15000 (68 kg)	89.71*	

*Measured from reconstructed rib.

clearly a hominid pattern, it has an extreme curvature even for modern humans in its upper thorax (ribs 2–4).

Both KNM-WT 15000 and modern humans have, on average, a more posteriorly angled neck axis than do chimpanzees and this can be demonstrated in the rib as well as the associated vertebrae. When combined with the hominid curvatures in the proximal portions of the ribs, especially ribs 2–5 (see above), the posteriorly angled neck axis results in the more dorsally positioned costal angle relative to the thoracic vertebrae, a slight increase in volume (Figure 13.13), and a ventrally positioned thoracic vertebral column (see below). Simultaneously, this alters the orientation and direction of the costal radii, which, as described above, are orthogonal to the neck axis. Hominids have somewhat more laterally directed costal radii (i.e., more posteriorly directed rib axes) than do apes, a feature that contributes to their relatively greater transverse diameter in the upper thorax. This observation may also help explain the relatively broader sterna in humans than in chimpanzees.

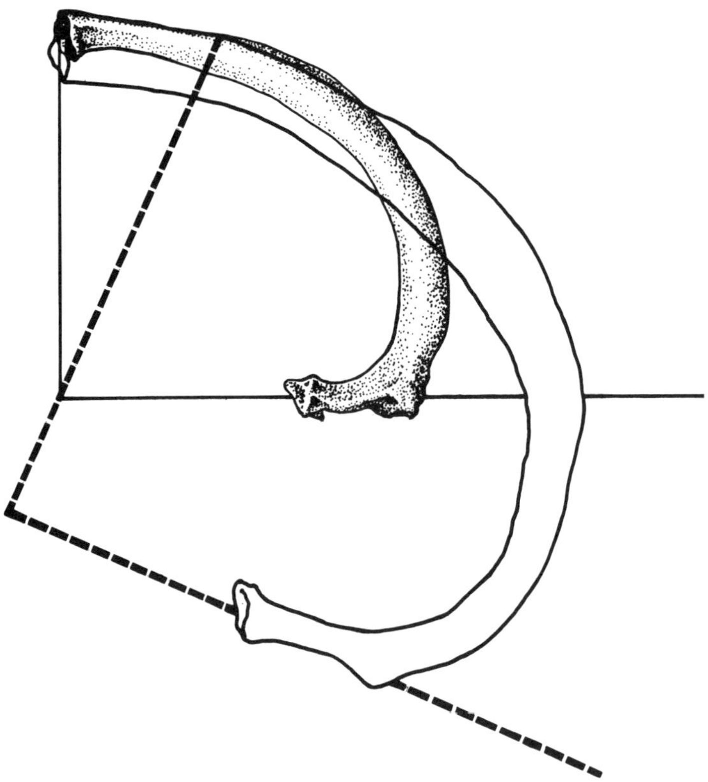

FIGURE 13.10.
A human second rib (outline) *superimposed on a chimpanzee second rib* (stippled). *Both are oriented to their respective sternal ends. Note that in humans the costal neck axis is more posteriorly directed.*

FIGURE 13.11.
The angle Σ between a line through the costal articular facets and the sagittal plane in superior view. Measurements were taken directly from radiographs. Note that PAN *shows greater angulation than do the hominids. See also Table 13.8.*

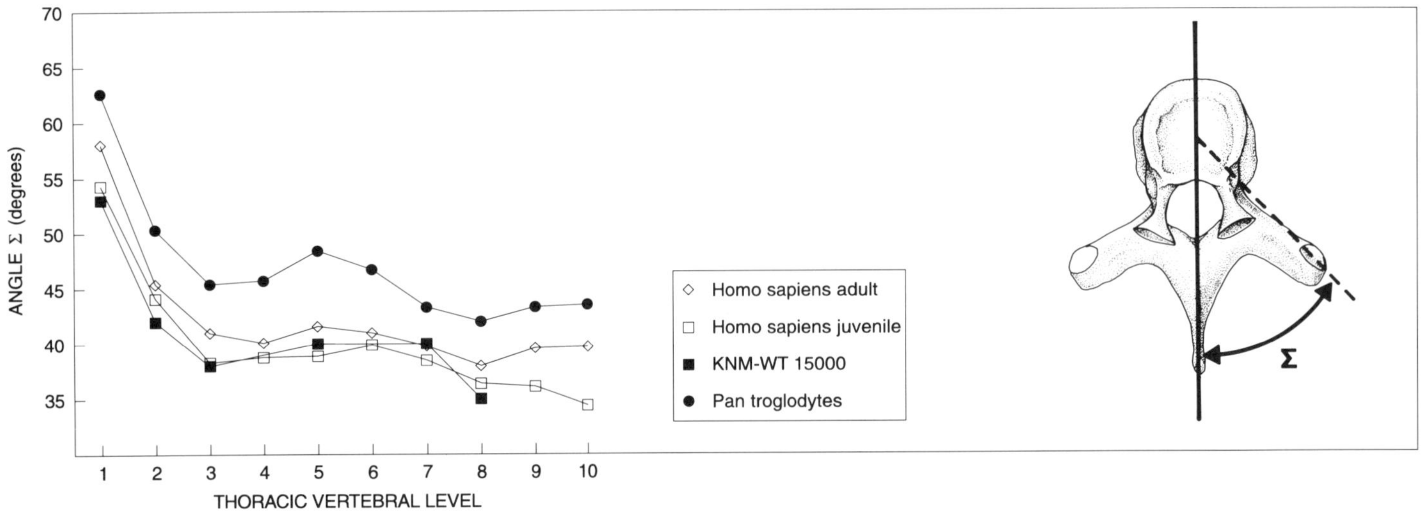

Thoracic cage shape and sequential patterns of rib morphology discriminate extant pongids from hominids. Judging from the only specimen available that is complete enough to permit accurate comparison, *Homo erectus* was essentially identical to modern human adults, which is to say that this specimen had a barrel-shaped, modern human–like thorax.

Although modern hominoids have equivalent chest volumes when corrected for body size, these volumetric equalities are achieved through quite different shapes. Pongids with their funnel-shaped chests are narrow at the top of the rib cage and become progressively wider in order to conform to the transversely broad pelvis. Humans and *Homo erectus,* having undergone the radical reorganization of the pelvis for habitual bipedality, have narrowed their inferior thoracic cages thereby preserving anatomical continuity between the lower thorax and the false pelvis. The diminishment of the thoracic volume that accompanied this evolutionary transition was partially compensated by an increase in the transverse diameter in the upper thorax and by the more dorsal position of the proximal ribs (Figure 13.13). This modification is reflected in rib curvatures, relative rib lengths, and associated costal areas. Some additional pulmonary space is gained by angling the costal neck axis more posteriorly, although this latter alteration probably has more to do with invagination of the spine into the thorax for postural purposes than it does with augmentation of thoracic volume (see below).

Position of the Thoracic Vertebral Column

Schultz (1961, 1969) pointed out that while all extant hominoids are characterized by a ventrally placed thoracic spine this feature is carried to an extreme in modern humans. The location of the thoracic spine does, moreover, change during human development, such that human infants possess a less ventrally oriented thoracic vertebral column than do adults (Schultz, 1961, 1969). In that this marked ventral position of the thoracic spine is unique to adult *Homo sapiens* and is achieved during growth, its condition in the fossil skeleton is of some interest.

As noted above, human (and *H. erectus*) ribs and vertebrae have a more posteriorly angled neck axis than is the case in *Pan* (Figures 13.10 and

TABLE 13.8.

Angle Σ (degrees); see Figure 13.11.

RIB NUMBER	*HOMO SAPIENS* ADULT (*N* = 10) Mean	s.d.	*HOMO SAPIENS* JUVENILE (*N* = 10) Mean	s.d.	KNM-WT 15000	*PAN TROGLODYTES* (*N* = 10) Mean	s.d.
1	58.0	4.28	54.3	7.32	53.0	62.6	3.72
2	45.4	4.73	44.1	5.50	42.0	50.3	4.42
3	41.0	4.89	38.3	3.46	38.0	45.4	4.80
4	40.1	4.63	38.8	3.21		45.7	5.23
5	41.6	7.65	38.9	3.17	40.0	48.4	4.94
6	41.0	7.26	39.9	3.91		46.7	3.22
7	39.8	5.47	38.5	3.55	40.0	43.3	3.74
8	38.0	4.17	36.4	3.77	35.0	42.0	4.58
9	39.6	6.90	36.1	3.85		43.3	4.85
10	39.7	5.51	34.4	3.38		43.5	3.52

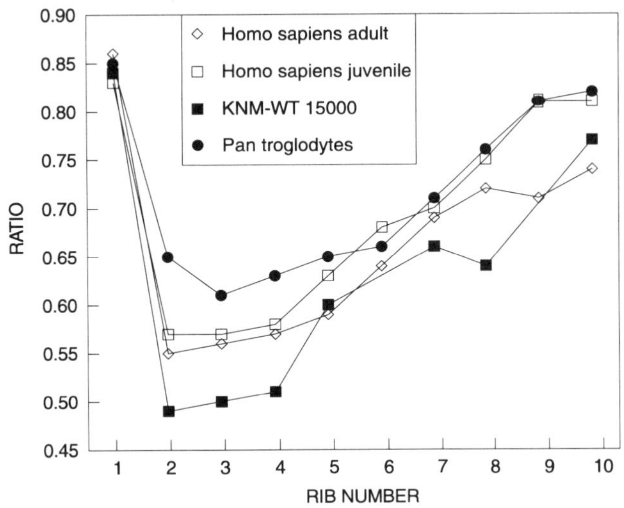

FIGURE 13.12.

Ratio of the maximum costal radius to the costal arc length (see Table 13.9). The straighter the rib, the closer the ratio is to 1. The humans and KNM-WT 15000 have ribs that are more curved, especially in the upper thorax.

13.11). Associated with this observation is the fact that hominid ribs are, on average, more curved in their proximal portions, such that the posterior wall of the thorax extends further dorsally beyond the bodies of the thoracic vertebrae (see Figure 13.13). This evolutionary modification provides a slight increase in thoracic volume. The primary adaptive significance of this alteration, however, is as a postural mechanism to help balance the thorax over the pelvis.

In hominids, habitual bipedality has resulted in the torso being placed in a precarious position over the pelvis. Latimer and Ward (Chapter 12, this volume) discuss some of these postural adaptations in the vertebral column. During active bipedal locomotion, muscular forces are solely responsible for maintaining the trunk's vertical orientation. During single-limb heel strike, with its posteriorly directed ground reaction, the anteriorly directed momentum of the upper body is resisted by a group of dorsally positioned muscles linking the pelvis to the upper trunk. Lovejoy's (1988) discussion of gluteus maximus (also see Tague and Lovejoy, 1986) and its role in resisting trunk momentum during hominid locomotion provides an example of this important muscular function. Clearly, the epaxial spinal musculature also contributes significantly to the maintenance of trunk verticality, and any skeletal modification that positioned these muscles further posteriorly from the transverse axis of flexion aids in this critical role. This result is obtained through the ventral invagination of the hominid thoracic vertebral column into the thoracic cage itself.

The dramatic mechanical consequences of

TABLE 13.9.

Ratio of costal radius to rib length; see Figure 13.12.

RIB NUMBER	*HOMO SAPIENS* ADULT (*N* = 10) Mean	S.D.	*HOMO SAPIENS* JUVENILE (*N* = 10) Mean	S.D.	KNM-WT 15000	*PAN TROGLODYTES* (*N* = 10) Mean	S.D.
1	.86	.06	.83	.06	.84	.85	.05
2	.55	.04	.57	.09	.49	.65	.06
3	.56	.05	.57	.07	.50*	.61	.05
4	.57	.05	.58	.07	.51*	.63	.06
5	.59	.04	.63	.07	.60*	.65	.05
6	.64	.05	.68	.08		.66	.07
7	.69	.05	.70	.07	.66*	.71	.06
8	.72	.05	.75	.08	.64*	.76	.06
9	.71	.06	.81	.05		.81	.05
10	.74	.04	.81	.05	.77*	.82	.03

*Measured from reconstructed rib.

these evolutionary modifications in hominids are shown in the alterations in rib curvatures, vertebral column orientation, and the dorsal prolongation of the iliac tuberosity, which together nearly double the moment arm of the hominid erector spinae relative to that of apes (Figure 13.13). At the same time, the more ventral placement of the hominid column (as well as the transversely broad thorax) proportionally reduces the moment arm of the center of mass of the upper body, further enhancing the epaxial muscle's ability to resist the trunk's anteriorly directed momentum during walking and running.

In addition, moving the hominid vertebral column closer to the centroid of the thoracic cylinder, and thus nearer to the center of mass, also minimizes potentially damaging eccentric loads on the individual vertebrae and their intervertebral discs. Such eccentric loading is especially problematic given the unique cyclic rotation that takes place in the hominid vertebral column during locomotion (the thorax is 180° out of phase with the pelvis) and the potentially damaging shear stresses this imposes on the intervertebral discs. Finally, the greater curvature already shown to take place in the proximal portions of the hominid ribs increases the relative length of the dorsal rib segments and thereby significantly enhances the ability of the erector spinae muscles to modulate transverse and anteroposterior flexion as well as rotatory movements of the trunk.

The invagination of the vertebral column

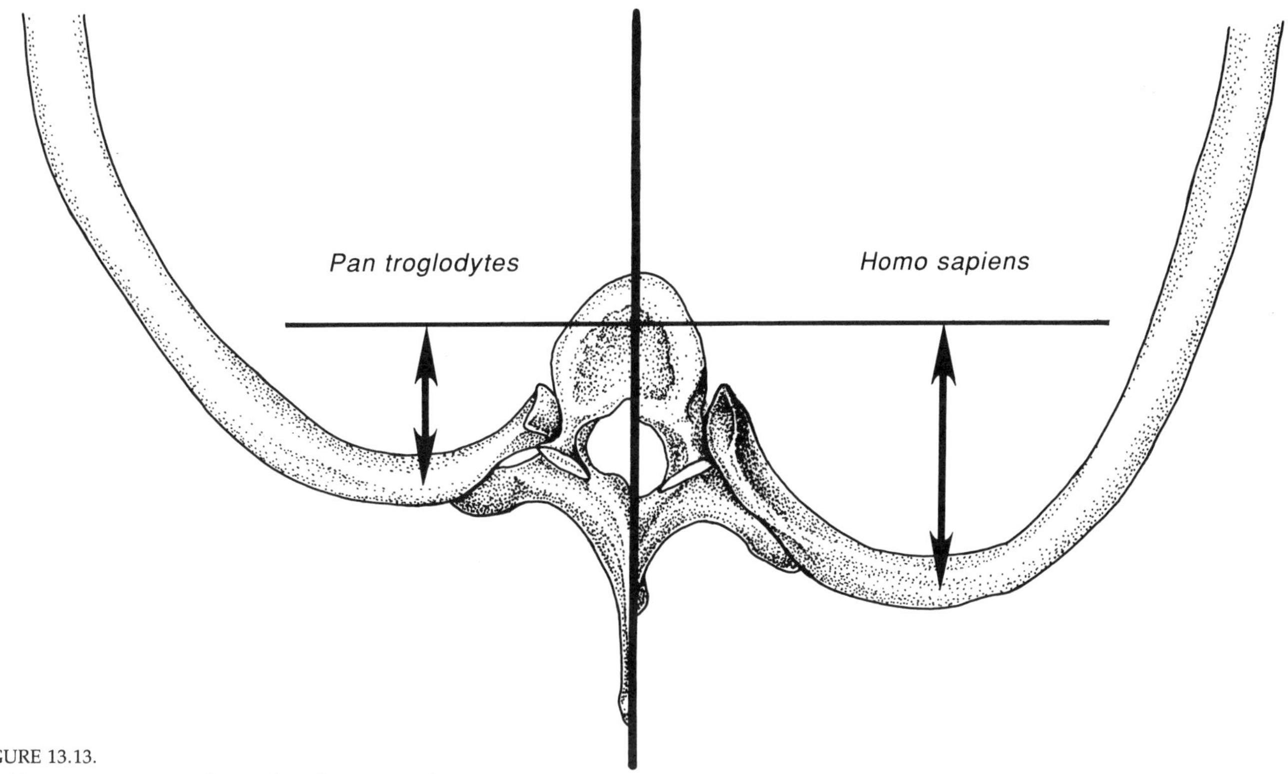

FIGURE 13.13.
A chimpanzee (left) *and a modern human* (right) *seventh thoracic vertebra with associated ribs. Note the more dorsal positioning of the human rib and the greater moment arm* (arrows) *this produces about the center of the vertebral body. The orientation of the human rib also increases potential thoracic volume.*

ventrally into the thorax as well as the related rib modifications are also associated with the lordotic curve in the lumbar column and the dorsal extension and marked expansion of the iliac tuberosity in hominids. Latimer and Ward (Chapter 12, this volume) examine the lumbar lordosis in KNM-WT 15000 and Walker and Ruff (Chapter 10, this volume) show the marked similarities between this specimen and modern humans in iliac tuberosity morphology. These skeletal modifications provide enhanced area and a greater moment arm for the erector spinae (and gluteus maximus; see Tague and Lovejoy, 1986; Lovejoy, 1988), thereby augmenting the ability of the muscles to control trunk movements over the pelvis. These adaptive mechanisms are unique to bipedal hominids and are reflected in the morphology of their thoracic and pelvic skeletons. In all features reflecting these mechanisms, the KNM-WT 15000 thoracic skeleton is essentially identical to that of modern humans.

Declination of the Ribs

In modern humans the lateral and distal segments of the ribs assume a more oblique orientation than they do in African apes. This declination of the ribs increases from childhood to adulthood and is associated with the inferior descent of the shoulders as well as the inferior displacement of the thoracic operculum (Todd, 1912; Ohman, 1984, 1986). Concurrent with the increasing declination of the ribs in humans are the progressive diminishment of the anteroposterior diameter relative to the transverse breadth of the thorax (Schultz, 1969) and the superoinferior elongation of the thoracic cage (see Figure 13.1).

Additional consequences of the inferior declination in human ribs are the "twisting" along the long axis of the rib and the flattening of the rib corpus. Because the declination, twisting, and flattening of the ribs is extreme in adult humans and undergoes modification during growth, these features were also looked for in the *Homo erectus* specimen.

The angular orientation of the ribs relative to the vertebral column was determined by articulating the individual ribs with their associated vertebrae and measuring the angle (α) formed between the cranial surface of the respective vertebra and the lateral segment of the rib (see Figure 13.14 and Table 13.10). If, as in the case of several ribs of KNM-WT 15000, there were no associated vertebrae with which to articulate the ribs, the planar pleural surface of the rib head was used to orient the specimens into appropriate anatomical position. In this latter procedure the plane formed by the cranial endplate of the vertebra was estimated by drawing a transverse plane orthogonal to the rib head's pleural surface. Several ribs were measured with their associated vertebrae and then remeasured without their vertebrae, but similar results were obtained.

As an additional indicator of the inferior displacement of the rib, the orthogonal distance from the transverse plane formed by the cranial surface of the vertebral body to the distal most point on the rib was also measured (see Figure 13.15 and Table 13.11). Because this latter measurement necessarily varies with the length of the rib (longer ribs would have greater vertical displacement only by virtue of being longer), it was normalized by the maximum costal radius.

In Figure 13.14, modern human adults are shown to have somewhat greater costal angles than do juveniles, an observation that is in concert with the earlier observation that the ribs are more horizontally oriented in children than in adults and that rib declination increases during growth (Keith, 1903, 1923; Todd, 1912; Schultz, 1963). The reasons for the increase in rib declivity reside in the patterns of rib growth and in the ontogenetic differences in the kyphotic curvature of the thoracic vertebral column.

First, not only do modern human adults have larger costal angles (angle α), as measured in Figure 13.14, they also have greater normalized superoinferior distances between the rib heads and sternal ends (Figure 13.15). This means not only that adult ribs are longer and more angled but also that growth on the rib's sternal extremity proceeds along a spiral trajectory in an inferomedial direction such that the superoinferior distance increases faster than the maximum costal radius. This differential growth and increasing angulation

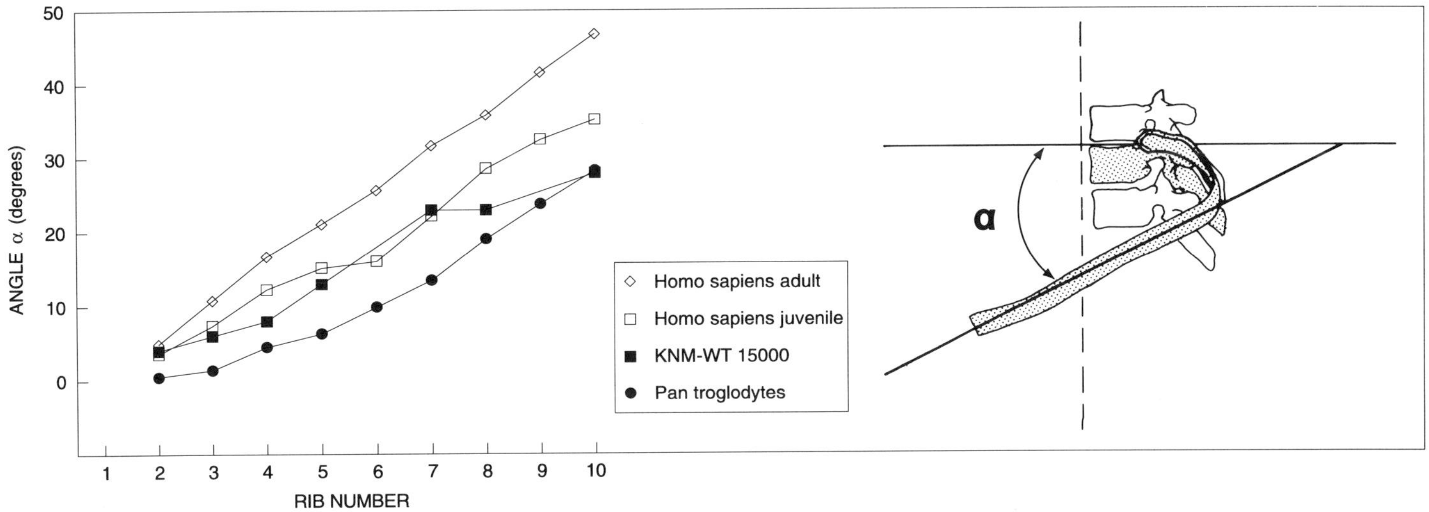

FIGURE 13.14.
The angle α that the lateral segment of the rib makes to the transverse plane of the superior vertebral endplate. Note the difference between juvenile and adult humans and the similarity between KNM-WT 15000 and juvenile humans. See also Table 13.10.

at the sternal end of the rib also corresponds to the change from a more cylindrical thoracic cross-section in children to the transversely broad thorax in adults, a change noted earlier in the discussion of costal curvatures. When measurements are made relative to the associated vertebrae, children not only have slightly less angled ribs, they also have shorter, less curved ribs, a combination which heightens the appearance of a more rounded thorax with less rib declination.

Second, because the costal margins must remain parallel (so as not to impinge on one another during ventilatory movements), any decrease in the kyphosis of the lower thoracic column would increase the declination of the lower ribs and as a consequence would influence the entire thoracic cage. Thus, the decrease in thoracic kyphosis that naturally occurs in the inferior half of

TABLE 13.10.

Angle α (degrees); see Figure 13.14.

RIB NUMBER	*HOMO SAPIENS* ADULT (*N* = 10) Mean	S.D.	*HOMO SAPIENS* JUVENILE (*N* = 9) Mean	S.D.	KNM-WT 15000	*PAN TROGLODYTES* (*N* = 10) Mean	S.D.
1							
2	4.90	3.11	3.56	3.24	4.00	.50	1.50
3	10.80	3.28	7.40	3.12	6.00	1.40	1.56
4	16.70	6.54	12.30	4.78	8.00*	4.50	1.69
5	21.10	6.92	15.20	4.85	13.00*	6.30	2.00
6	25.70	7.34	16.10	7.99		9.90	4.23
7	31.70	8.11	22.20	6.38	23.00*	13.50	5.00
8	35.80	8.54	28.60	8.29	23.00*	19.10	6.09
9	41.60	8.48	32.50	7.35		23.80	5.71
10	46.78	7.22	35.20	6.96	28.00*	28.30	5.31

*Measured from reconstructed rib.

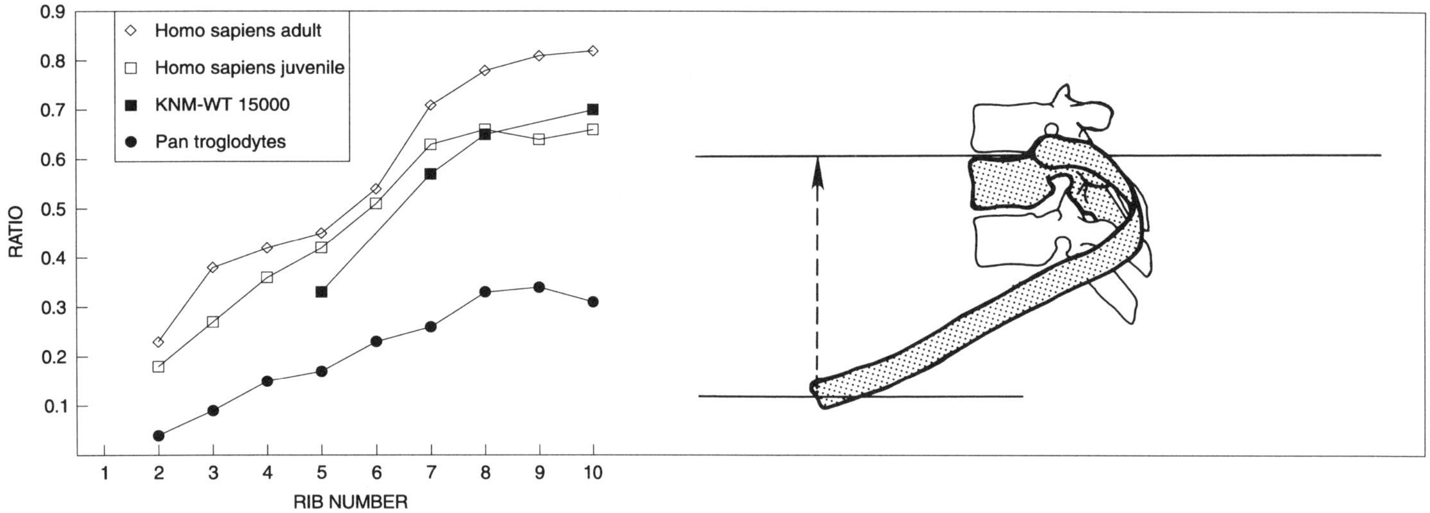

FIGURE 13.15.
The ratio of the superoinferior distance (the orthogonal distance from the transverse plane of the superior surface of the vertebra to the most distal point on the rib) to the maximum costal radius (see also Table 13.11). Note that KNM-WT 15000 is similar to human juveniles.

the thoracic column during growth (Latimer and Ward, this volume; Scoles et al., 1991) necessarily results in and contributes to the increase in rib declination in the adult. The less abrupt inclination and cylindrical shape of the thoracic cage in human children not only signals an actual difference in the angle that a rib makes to its respective vertebrae (as measured in Figure 13.14) but is also a manifestation of naturally occurring growth phenomena in the human thorax.

It is apparent that the ribs of the chimpanzee are uniformly less obliquely inclined than are either human group. This is understandable when one considers the anatomical context of the superoinferiorly elongated iliac blades and the short-

TABLE 13.11.

Ratio of superoinferior distance to costal radius; see Figure 13.15.

RIB NUMBER	*HOMO SAPIENS* ADULT (*N* = 8) Mean	S.D.	*HOMO SAPIENS* JUVENILE (*N* = 7) Mean	S.D.	KNM-WT 15000	*PAN TROGLODYTES* (*N* = 8) Mean	S.D.
1							
2	.23	.05	.18	.08		.04	.05
3	.38	.09	.27	.06		.09	.05
4	.42	.08	.36	.05		.15	.04
5	.45	.08	.42	.09	.33*	.17	.03
6	.54	.07	.51	.03		.23	.07
7	.71	.10	.63	.08	.57*	.26	.06
8	.78	.06	.66	.05	.65*	.33	.08
9	.81	.05	.64	.04		.34	.10
10	.82	.04	.66	.10	.70*	.31	.06

*Maximum radius as determined from reconstructed rib.

ened lumbar vertebral column in these animals, adaptations that quite simply do not provide enough space to permit human-like rib declination.

Rib declination in KNM-WT 15000 is most similar to modern human juveniles in both pattern and degree. The two exceptions are the fossil ribs 8 and 10, which are closer to the average costal angles in *Pan*. However, reference to Figure 13.15 shows that the *Homo erectus* specimen is again closer to human children in the superoinferior distances between the proximal and distal rib extremities even in ribs 8 and 10. The explanation for this apparent discrepancy in the graphic positions of these particular ribs resides in their extreme curvatures (Figure 13.3A and 13.3B, Tables 13.1A–D and 13.2A–D). With even minimal declination, ribs that have a high degree of curvature, like those of ribs 8 and 10 in KNM-WT 15000, will as a consequence also show a greater superoinferior distance between the two extremities. Therefore, although the fossil ribs 8 and 10 are less inclined on average to their respective vertebrae, the strong curvatures of these ribs counteract these differences and place these specimens within the ranges of modern human juveniles.

Additionally, the superoinferior distance between the transverse plane of the vertebral endplate and the sternal rib end increases with the angle that the maximum costal radius makes to the centroidal axis of the thorax. Indeed, the superoinferior distance measurement is a better indicator of the angle that the maximum costal radius makes to the vertebral column than is the costal angle as measured in Figure 13.14 because it takes the curvature and inferior inclination of its sternal extremity into account. Because the magnitude of costal elevation during inspiration is proportional to the maximum costal radius and the elevation angle of the rib (Jordanoglou, 1970; Latimer and Ohman, in preparation), the measurements of the normalized superoinferior distance (Figure 13.15) provide evidence not only of structural similarity but of functional equivalence between *Homo erectus* and modern humans as well.

One final observation related to rib angulation is the inclination that the rib neck axis makes to the sagittal plane (angle β; Figure 13.16 and Table 13.12) in anterior view. This measurement was accomplished by circumscribing the costal facets on the vertebrae with lead markers and radiographing the vertebrae in a ventral-to-dorsal direction. Throughout the entire thoracic column, the angle formed between the mid-sagittal plane and a line through the centers of the vertebral facets is greater in the humans than in the apes.

This difference means that the transverse processes of the vertebrae and the associated costal facets are more superiorly oriented in the humans. Interestingly, although thoracic vertebrae 1, 2, 3, and 5 fall within the human ranges, the only other vertebrae that could be measured (T7 and T8) in the *Homo erectus* skeleton have an estimated angular orientation most like *Pan*. A consequence of this angular difference in the costal neck axis is the influence it has upon the orientation of the rib corpus. If the proximal rib facet on the vertebral body is unchanged and only the transverse process is elevated (thereby increasing the angle), the sternal extremity of the rib corpus will naturally decline further laterally and inferiorly. Indeed, were it not for the oblique superolateral growth trajectory of the transverse process and the concomitant relative elevation in its lateral costal facets, the ontogenetic declination of human ribs would result in the sternal extremities more closely approximating the mid-sagittal plane. Therefore, the greater neck axis angle relative to the sagittal plane (angle β) in the hominid vertebrae is also indicative of a more oblique rib orientation.

As noted, the angular orientation of KNM-WT 15000 is closer to the average of the chimpanzees in two vertebrae (T7 and T8), perhaps indicating somewhat less rib declination than occurs on average in humans at these particular vertebrae. This, however, is unlikely given the angular orientation of the upper ribs (Figure 13.14) and transverse processes of the other vertebrae (Figure 13.16) and the requirement that the rib margins remain essentially parallel. An alternative explanation is that the inferior displacement of the lowest ribs in KNM-WT 15000 specimen occurred in the rib corpus rather than in the costo-vertebral joints. This latter idea is supported by the fact that

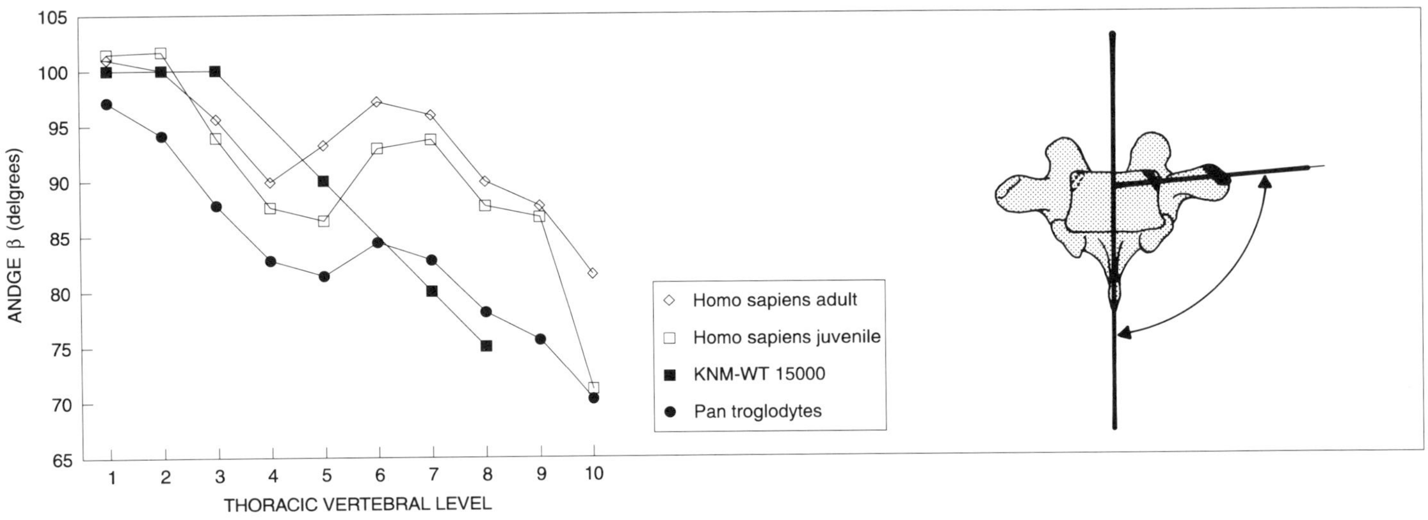

FIGURE 13.16.
The angle β that the costal neck makes to the sagittal plane in anterior view (see also Table 13.12). The measurements were taken directly from radiographs.

the superoinferior displacement of the fossil ribs 7 and 8 (Figure 13.15) falls in the range of human juveniles, indicating that the ribs' inferior displacement is achieved in the body of the ribs and not in the bone's proximal segments.

Overall, rib declination in KNM-WT 15000 is most similar to that in modern human juveniles. Moreover, given the likelihood that with additional growth its thoracic kyphosis would have reduced (Latimer and Ward, Chapter 12, this volume) and the transverse processes of the thoracic vertebrae would have increased their angular orientation, the ribs of this specimen could be expected to parallel a modern human growth trajectory and probably would have become even more obliquely angled.

TABLE 13.12.

Angle β (degrees); see Figure 13.16.

	HOMO SAPIENS						
	ADULT (*N* = 10)		JUVENILE (*N* = 10)			*PAN TROGLODYTES* (*N* = 10)	
RIB NUMBER	Mean	S.D.	Mean	S.D.	KNM-WT 15000	Mean	S.D.
1	101.0	7.79	101.5	5.95	100.0	97.1	6.94
2	100.0	6.09	101.7	3.19	100.0	94.1	6.41
3	95.6	5.96	93.9	8.15	100.0	87.8	5.82
4	89.9	5.46	87.6	5.67		82.8	4.55
5	93.2	9.93	86.4	11.57	90.0	81.4	5.88
6	97.1	4.78	92.9	8.54		84.4	7.99
7	95.9	6.36	93.7	8.53	80.0	82.8	5.26
8	89.9	8.57	87.7	6.73	75.0	78.1	5.08
9	87.7	11.47	86.7	8.03		75.6	7.25
10	81.5	13.11	71.1	9.10		70.2	7.16

Rib torsion

A structural consequence of declination in the hominid rib cage is the twisting that occurs along the long axis of the rib corpus. This spiraling along the rib shaft increases inferiorly in humans as a consequence of an obliquely oriented rib having to conform to the narrowing of the inferior half of the thoracic cage (Figure 13.17).

For a comparison of torsion in hominoid ribs, a measurement was made of the angle that the sternal end of each rib made to the vertebral column: each rib was articulated with its associated vertebra and a thin steel post was affixed parallel to the long axis of its sternal end. Sharp shadow tracings were projected against an angular diagram, allowing measurement of the angle formed by the steel post in the midsagittal plane. If the rib's sternal cross-section inclined superoposteriorly toward the cranial end of the thorax, it was given a positive angular measurement. In contrast, if the long axis of the sternal end of the rib inclined toward the pelvis, it was assigned a negative angular measurement. The torsion of the rib would be zero if the long axis of the sternal end was parallel to the midsagittal plane (Figure 13.18 and Table 13.13).

It can be seen from Figure 13.18 that as they progress inferiorly the chimpanzee ribs increase slightly in the amount of twisting along the corpus; this torsion is always in the positive direction. Similarly, in the humans and KNM-WT 15000, in the upper half of the rib cage the ribs measure positively, which is to say that the sternal ends of the bones angle superiorly and dorsally. In direct contrast to the condition in pongids, however, at midthorax there is an abrupt reversal in the direction of the rib torsion and the sternal rib ends incline negatively—that is, inferiorly and dorsally. This dramatic change is owing to the tapering of the hominid thoracic cage as well as to the greater declination of the lower ribs. It is interesting to note that the fossil specimen and the juvenile modern humans show the changeover in rib torsion at rib 8 whereas the adults show it at rib 7. This again is indicative of an ontogenetic change in thoracic shape wherein juvenile thoracic cages continue to flare laterally one segment further inferiorly than do those of adults; it also shows that KNM-WT 15000 is following the modern human pattern.

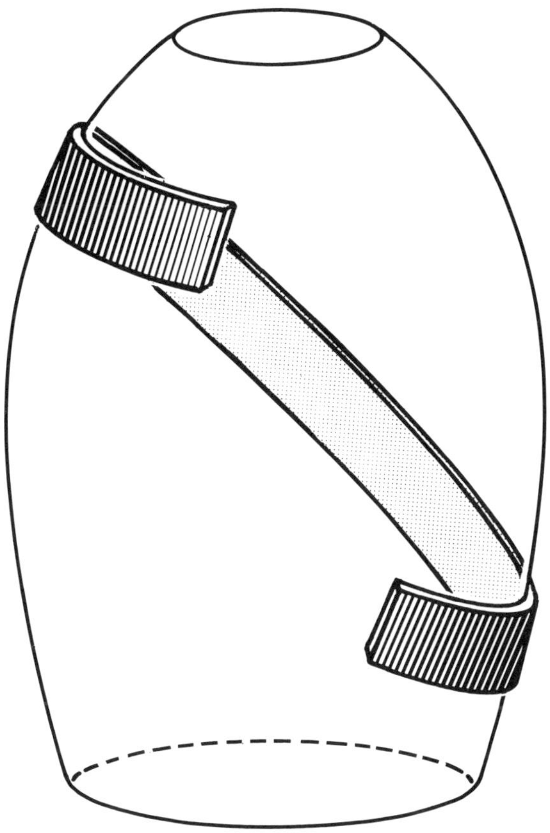

FIGURE 13.17.
A diagrammatic representation of the structural reasons for costal torsion in hominid ribs. The spiraling along the long axis of the rib is a consequence of the angle that the rib makes to the long axis of the thorax. The ribs that are most obliquely oriented also have the most torsion.

Cross-sectional shape of the ribs

Another feature that is related to the declination of the ribs is the elliptical or flattened cross-section that characterizes the rib corpus in hominids. Figure 13.19 (see also Table 13.14) illustrates the dif-

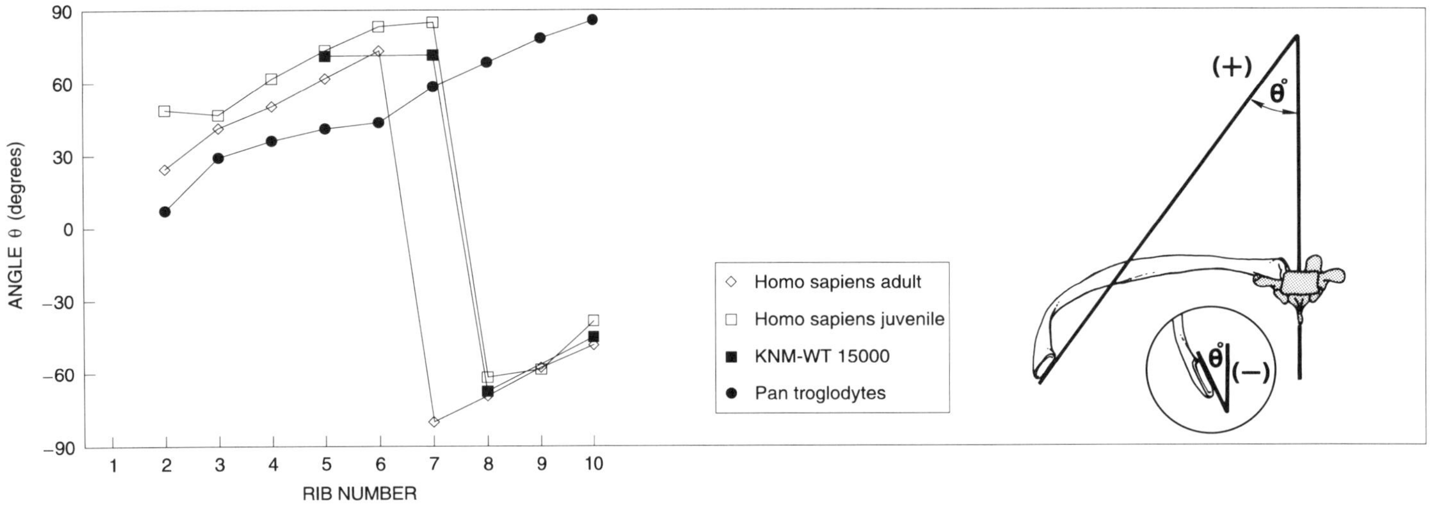

FIGURE 13.18.
The measurement of rib torsion or spiral twisting that takes place along the length of the rib corpus (angle θ). It is clear that the pattern in the humans and KNM-WT 15000 differs markedly from that in the chimpanzees. Because of measurement difficulty, rib 1 is not included. See also Table 13.13.

ferences in the ratio formed between the minimum and maximum diameters taken at mid-shaft of the rib corpus. It is apparent that chimpanzees have a larger ratio, which is to say that the two dimensions are more nearly equal and that the ribs are on average rounder than they are in modern humans. The rib profiles of KNM-WT 15000 are both in overall pattern and in individual rib measurements very similar to the human profiles.

The adaptive reasons for modifying rib cross-sections are undoubtedly complex and vary across taxa (Jenkins, 1970). There is also a great deal of overlap in comparisons of this feature in modern pongids and hominids. Nevertheless, the basic trend is that the apes have ribs that vary less in their cross-sectional dimensions than do humans (Figure 13.19).

TABLE 13.13.

Angle θ (degrees); see Figure 13.18.

	HOMO SAPIENS						
	ADULT (*N* = 8)		JUVENILE (*N* = 7)			*PAN TROGLODYTES* (*N* = 8)	
RIB NUMBER	Mean	S.D.	Mean	S.D.	KNM-WT 15000	Mean	S.D.
1							
2	24.40	9.77	48.67	12.39		7.20	10.85
3	41.20	10.85	46.67	10.27		29.20	8.08
4	50.20	4.66	61.67	6.24		36.00	7.04
5	61.80	6.58	73.33	10.40	71.00	41.00	7.35
6	73.25	10.96	83.33	9.43		43.60	7.96
7	−80.00	1.87	85.00	5.00	71.50	58.40	9.00
8	−69.40	8.87	−61.67	14.34	−67.50	68.40	12.34
9	−57.60	4.59	−58.33	6.24		78.40	10.01
10	−48.20	6.62	−38.33	6.24	−45.00	85.80	5.49

The greater degree of flattening in the hominid lower ribs demonstrated in Figure 13.19 can perhaps be best understood as a structural solution to the functional necessity of maintaining an adequate planar surface on the pleural aspect of the rib corpus as it twists and inclines around an ellipsoid. Put another way, a flattened, ribbon-like rib can obliquely warp and twist around a barrel-shaped ellipsoid like the hominid thorax while simultaneously preserving congruous contact with the pleural surface (see Figure 13.17).

In addition, during ventilatory rib elevation in a vertical and sagittal direction, it is possible to keep the internal surfaces of a flattened rib flush against the curved pleural surface, thereby avoiding separation and reduced contact between the two opposing surfaces. A flat rib, however, would be less effective during inspiratory elevation in a transverse and lateral direction because the axial rotation that accompanies this type of rib motion would significantly reduce costal-pleural contact area by twisting the internal rib surface such that it would no longer be coplanar to the pleural surface. In this latter instance, a more circular rib cross-section would be superior both in its resistance to the mechanical stresses placed upon it during elevation and in the preservation of a uniform amount of contact area between pleural and costal surfaces. Also, during elevation of the ribs in a transverse and lateral direction, a cylindrical rib would present an opposing contact margin

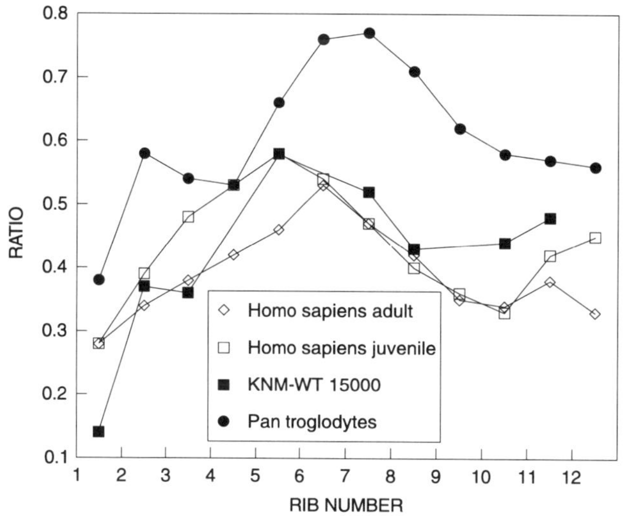

FIGURE 13.19.
The ratio of minimum to maximum costal thickness at midshaft. The closer this ratio is to 1, the rounder the rib is at midshaft. See also Table 13.14.

TABLE 13.14.

Ratio of minimum to maximum costal thickness at midshaft; see Figure 13.19.

RIB NUMBER	*HOMO SAPIENS* ADULT (*N* = 10) Mean	S.D.	JUVENILE (*N* = 10) Mean	S.D.	KNM-WT 15000	*PAN TROGLODYTES* (*N* = 10) Mean	S.D.
1	.28	.05	.28	.07	.14	.38	.05
2	.34	.06	.39	.04	.37	.58	.01
3	.38	.02	.48	.05	.36	.54	.07
4	.42	.04	.53	.13		.53	.05
5	.46	.03	.58	.09	.58	.66	.11
6	.53	.08	.54	.10		.76	.07
7	.47	.11	.47	.07	.52	.77	.05
8	.42	.03	.40	.06	.43	.71	.08
9	.35	.06	.36	.05		.62	.06
10	.34	.04	.33	.06	.44	.58	.11
11	.38	.09	.42	.09	.48	.57	.07
12	.33	.06	.45	.12		.56	.09

with a nearly constant radius. In contrast, if similarly elevated, a flat, hominid-like rib would angle its sharper margin (significantly smaller radius) toward the pleural surface, creating a stepped, nonuniform pleural contact area. In a barrel-shaped thorax with a reduced radius in its inferior portions, a stepped costal-pleural contact would result in an obstruction to further rib elevation especially, in the inferior-most ribs.

Factors influencing costal movement and the changes in thoracic diameters include the maximum costal radius, the angles of rib declination, and the angles that affect the orientation of the neck axis within the rib cage (Jordanoglou, 1970; Latimer and Ohman, in preparation). Each of these factors has been shown to differ between humans and chimpanzees and can be expected to influence the mechanical behavior of the rib cage as well as the cross-sectional profiles of the individual ribs. Based upon the skeletal measurements presented above, preliminary vector analysis of costal movements in humans and pongids suggests that rib cross-sectional shape differences correspond to functional differences in the rib cage and that hominids are characterized by a greater relative magnitude of costal elevation in the anterior and vertical direction. Apes, in contrast, when undergoing equivalent angles of costal elevation have vectors directed in more horizontal and transverse directions. Therefore, the flattened rib cross-sections in humans and in KNM-WT 15000 can be attributed to the alterations in rib mechanics that occurred as a direct consequence of the evolutionary modifications in thoracic shape. As a result of the evolutionary change to a barrel-shaped thorax, human costal movements are more vertical and sagittal in direction. In contrast, chimpanzees with their funnel-shaped rib cages have costal movements that are directed primarily in a transverse and lateral direction. These alterations in costal mechanics resulted in the flat and round rib profiles, respectively.

It is important to emphasize, however, that these differences in costal movement do not imply any functional advantage; rather, they are ways of preserving equivalent pulmonary function in thoraxes of markedly dissimilar shapes. With regard to thoracic shape, rib declination, rib torsion, and cross-sectional shape, KNM-WT 15000 falls within modern human ranges, indicating not only structural but mechanical equivalence as well.

Discussion

Our review of the anatomical features of the KNM-WT 15000 ribs demonstrates that the thoracic cage of this specimen was in almost all respects indistinguishable from those of modern humans. Although subadult, this individual clearly had a barrel-shaped thorax, a ventrally positioned thoracic vertebral column, and ribs that decline inferiorly; these three features are directly associated with bipedality and distinguish hominids from modern pongids.

As discussed, the reduction in circumferential dimensions of the inferior portion of the thoracic cage in hominids is a result of the immediate juxtaposition of the rib cage to the dramatically altered bipedal pelvis. That is, the repositioning of the hominid ilia into a more sagittal orientation and the reconfiguration of the iliac crest into a sigmoid curve necessitated the tapering and increase in curvature (smaller radius and larger included angle) of the lower thoracic cage. It is significant that this narrowing of the radial dimensions of the lower rib cage is especially marked in KNM-WT 15000. This is in agreement with the narrow bi-iliac breadth presented in Ruff and Walker (Chapter 11, this volume) and highlights the structural relationship posited here between the upper and lower torso. Also associated with this marked tapering of the lower part of the thoracic cage in KNM-WT 15000 is the compensatory increase in the upper thorax.

As noted above, the expansion of the upper thorax in this specimen is extreme even for this sample of modern humans. This again is in concurrence with the narrow bi-iliac breadth, which causes a volumetric reduction in the lower rib cage that, in turn, leads to a compensatory volumetric expansion of the cranial portion of the cage.

It is important to note that while these alterations in thoracic shape can be used to discriminate between pongids and hominids (and to a lesser extent early from later hominids), they do not imply like differences in pulmonary function. The pulmonary system was not the primary target for natural selection. Rather, it was the locomotory system that was under direct selectional pressure. The thoracic cage paralleled the critical shape alterations in the pelvis in order to preserve the structural and spatial continuity between the upper and lower torso. Therefore, the majority of the modifications noted in this summary should be viewed as secondary changes and simply a way of maintaining pulmonary space and function in the face of dramatic changes in the locomotory system. There is thus no need to advance arguments about selection pressures on enhanced ventilatory capabilities in extinct or extant hominids, especially if they are founded solely on morphological changes in the ribs and the shape of the thoracic cage.

Reference to an earlier hominid, the Pliocene skeleton AL 288-1 ("Lucy"), provides additional evidence bearing on thoracic cage shape as well as on the evolutionary transitions that lead to modern human thoracic morphology. Although clearly habitually bipedal, this diminutive specimen (and the rest of her species, *Australopithecus afarensis*) was, probably, characterized by a thoracic cage that became progressively broader in its inferior parts. "Lucy" therefore did not have a modern human barrel-shaped rib cage; this suggestion is based upon the exceptionally broad pelvis (see Tague and Lovejoy, 1986, and Ruff, 1991 for a discussion of pelvic breadth) to which the inferior ribs had to conform. The failure of "Lucy's" rib cage to taper in its inferior segments in no way signals a change in ventilatory or locomotory capabilities; rather, it indicates a wide bi-iliac breadth.

As "Lucy" was bipedal, it is clear that there is no necessary functional alliance between bipedality and a barrel-shaped thorax (Ohman, 1984, 1986). Subsequent evolutionary modifications in the hominid false pelvis related to obstetric and visceral accommodation (Tague and Lovejoy, 1986) resulted in the alterations (seen in later hominids) to a relatively constricted inferior rib cage.

Our review of the anatomical features of the KNM-WT 15000 thoracic cage demonstrates that the morphological transitions leading to modern humans were essentially completed by 1.5 million years ago (Brown et al., 1985). These changes include the barrel-shaped thorax, the ventrally invaginated thoracic column, and the inferiorly declined ribs.

All of these features are present in the KNM-WT 15000 skeleton, largely in modern human form. The slight differences that do occur between modern humans and this specimen are a consequence of its narrow pelvic breadth and are not obviously related to pulmonary function.

Acknowledgments

We wish to thank Tamara Cruise, Robert P. Mensforth, C. Owen Lovejoy and James C. Ohman for valuable comments and discussion during the preparation of this chapter. Art work was done by K. Dieringer and L. Gudz.

Two of us (LMJ and BL) especially wish to thank Alan Walker and Richard Leakey for the opportunity to study the ribs of the KNM-WT 15000 specimen.

CHAPTER 14

THE ENDOCAST

DAVID BEGUN AND ALAN WALKER

One of the major evolutionary adaptations that distinguish the human species is our relatively large brain. The Nariokotome skeleton is the first early hominid discovered to have a skull complete enough to estimate cranial capacity associated with a postcranial skeleton complete enough to estimate body size accurately. With this specimen we can test previous estimates of brain size relative to body size based on much less complete remains. The new estimates reveal an interesting pattern of brain-size change from nonhuman hominoids to Homo sapiens.

Endocasts of African fossil hominids have been a focus of attention and controversy in paleoanthropology for as long as there has been a record of fossil hominids in Africa. Dart (1925) claimed to recognize a human-like quality of the Taung endocast, despite its chimp-like size. This important characteristic attested to the hominid status of the otherwise "primitive" form he called *Australopithecus.* Since then, much literature has been devoted to various aspects of hominid endocasts as additional specimens, both natural and artificial, became available, first in South Africa and more recently in East Africa.

There have been several foci of this attention. The first and most controversial concerns precise determination of sulcal patterns in endocasts. Different interpretations of the positions of various sulci have led to substantially different perspectives on the organization of fossil hominid brains, which have for the most part been considered within the context of a unilinear sequence from great apes to humans. We choose to avoid this contentious domain completely by describing only the most obvious sulci (which are, in any case, only apparent anteriorly on KNM-WT 15000) and by ignoring the indistinct and therefore debatable sulcal patterns of the lateral and posterior portions of the endocast. Other indicators of overall cerebral shape and the degree of development of particular portions of the endocast will receive more attention here. The KNM-WT 15000 endocast is particularly useful in this regard because it is relatively undistorted and provides clear evidence of the morphology of various cerebral and cerebellar components.

A second area of much concern in hominid paleoneurology involves the size of endocasts and their relationship to body size. Two difficulties have plagued this area of research since the beginning. The first involves the accurate measurement of endocranial volume, especially of partial endocasts or of endocasts made from extensively reconstructed crania. The second, much greater, difficulty involves the calculation of relative brain size. The KNM-WT 15000 specimen gives us few problems in overcoming the first difficulty, since it is largely complete. A few parietal fragments are missing, but their presence on the other side and the relatively smooth contours of this portion of the specimen make these deficiencies relatively minor. A slightly bigger problem concerns the basal aspect of the specimen, which is damaged in the areas of the temporal and inferior frontal lobes. However, large portions of these areas are preserved, and the missing sections have been reconstructed using data on the available anatomy in conjunction with comparisons to other specimens more complete in these regions, such as KNM-ER 3733 and KNM-ER 3883. When it comes to estimating relative brain size, KNM-WT 15000 is, as far as we know, the only Pliocene or early Pleistocene hominid with a reasonably complete braincase that is also associated with enough of the postcranial skeleton such that a reliable estimate of body weight can be made. Body weight estimates based on dental measurements, especially in primitive hominids (which are clearly different in their relative tooth sizes from living hominoids), cannot be considered reliable without supporting evidence from the postcrania through which the forces generated by the body's mass and muscle activity are actually being transmitted. KNM-WT 15000 is the first specimen to allow a relatively straightforward calculation of relative brain size in a primitive, non-*Homo sapiens* hominid. However, even this nearly ideal specimen is not without some problems (see Smith, Chapter 9, and Ruff and Walker, Chapter 11, this volume).

A third focus of attention in hominid paleoneurology concerns the pattern of blood flow to and from the brain, the meninges, and skull. This endocast, unlike many others from East Africa, preserves numerous details of both the meningeal and venous sinus systems, which can be described with some confidence.

Finally, much of the literature in hominid paleoneurology has been devoted to speculation about the function of the primitive hominid brain, the factors that can account for its expansion, and the pattern of evolution implied by so-called trends in the evolution of the hominid brain. With the exception of a brief consideration of possible functional implications of the KNM-WT 15000 endocast, these issues will not concern us here. Most

TABLE 14.1.

Estimates of endocranial volume (cc) for KNM-WT 15000; the volumes of water displaced by each specimen were measured separately.

	N	MEAN	RANGE
Restoration 1	6	881.2	873.0–896.2
Restoration 2	10	879	872–888

TABLE 14.2.

Endocasts used in this analysis.

EAST AFRICA	CHINA
KNM-ER 1470	Locus DI (Skull 2)
KNM-ER 1805	Locus EI (Skull 3)
KNM-ER 1813	Locus LI (Skull 10)
KNM-ER 3732*	Locus LIII (Skull 12)
KNM-ER 3733	
KNM-ER 3883	
KNM-ER 13750	
KNM-WT 17000	
OH 5	

*This specimen is fragmentary and was not included in the metric analysis.

can be addressed, not to mention resolved, only when more specimens such as KNM-WT 15000 are recovered.

Materials and Methods

The calvaria was recovered from the disturbed surface sediment and desert lag at Nariokotome. Only the facial skeleton was seen *in situ* and that was embedded in the roots of an *Acacia* tree. The individual skull bones were assembled in the field by gluing together many fragments with Durofix (Rawlplug Ltd). Most of the nearly 70 pieces did not have much adhering matrix, but any still present was removed in the laboratory using pneumatic Airscribes (Chicago Pneumatic Corp.). The individual bones were then reassembled into the calvaria and the slight distortion in the curvatures of the fossilized bones were corrected for as much as possible by adjusting the glue joins. It is important to note that the positions of the temporal bones were fixed by aligning them on the mandibular condyles, while at the same time the facial skeleton and frontal bones were being aligned by

TABLE 14.3.

Measurements (mm) of fossil hominid endocasts.

MEASUREMENT*	KNM-WT 15000	KNM-ER 3733	KNM-ER 3883	KNM-ER 1470	KNM-ER 1805	KNM-ER 1813	KNM-ER 13750	KNM-WT 17000	OH 5	DI (Skull 2)	EI (Skull 3)	LI (Skull 10)	LIII (Skull 12)
Length (1)	158	143	150	136	135	127	128	119	130	165	157	171	168
Breadth (2)	116	120	122	116	100	97	100	95	104	115	120	128	125
Height (3)	100	92	92	84	81	81	—	71	77	106	102	108	106
Frontal breadth (4)	88	92	91	87	84	74	76	78	79	95	91	106	99
Cerebral height (5)	107	98	98	98	91	88	—	75	88	113	102	114	113
Frontal height (6)	66	76	68	62	60	65	60	55	62	75	70	77	75
Frontal chord (7)	75	73	68	60	-	68	65	62	70	75	81	91	89
Parietal chord (8)	93	83	82	87	74	76	76	68	77	99	87	94	87
Occipital breadth (9)	103	100	110	103	90	82	86	86	91	92	89	93	93

*Numbers refer to dimensions illustrated in Figure 14.1.

placing the upper and lower teeth in centric occlusion. Walker and Leakey (Chapter 6, this volume) give details of preservation and descriptions of individual bones.

A silicone rubber mold of the endocranial cavity was then prepared. Gaps in the cranial vault were filled with modeling clay before several layers of silicone rubber (Dow Corning RTV) were applied to the endocranial surface. This hollow mold was then filled with plaster of paris through the hole in the cranial base where part of the sphenoid body was missing. The plaster was to stabilize the rubber mold so as to prevent distortion when the bones were removed. The glue joins between the bones were then dissolved with acetone and the individual bones, or large parts of them, were removed from the mold. "Feathers" of rubber that were themselves molds of gaps between the bones were trimmed from the endocranial mold. After being coated with a thin soap solution, the mold itself was then used to make a two-part mold, using the same silicone rubber and large plaster-of-paris keepers (see Smith and Latimer, 1989, for details). Several plaster-of-paris copies of the endocast were then made from this final mold. Missing parts of the endocast were carved from the extra plaster around the base, and the extra plaster pieces filling spaces where pieces of vault were missing were pared down so that their edges and curvatures were congruent with the surrounding endocranial cast surface.

An initial endocranial reconstruction was made shortly after the cranium was reconstructed (Brown et al., 1985). Two unreconstructed copies of the endocast were taken later by each present author to be reconstructed independently. The reconstruction involved restoration of the medial aspects of both temporal lobes, the medial section of the inferior aspects of the frontal lobes and the olfactory area, and the midline basicerebral structures from the anterior edge of the spheno-occipital to the optic chiasma. All of these areas were partially preserved on the unreconstructed endocast. Restoration involved relatively minor alterations that were straightforward and relied on comparisons and mirror imaging. Each separately reconstructed specimen was coated with plastic varnish and measured by the volumetric displacement method described in Tobias (1971). The volumetric data upon which the final capacity was based are given in Table 14.1. The comparative sample used in this analysis is listed in Table 14.2.

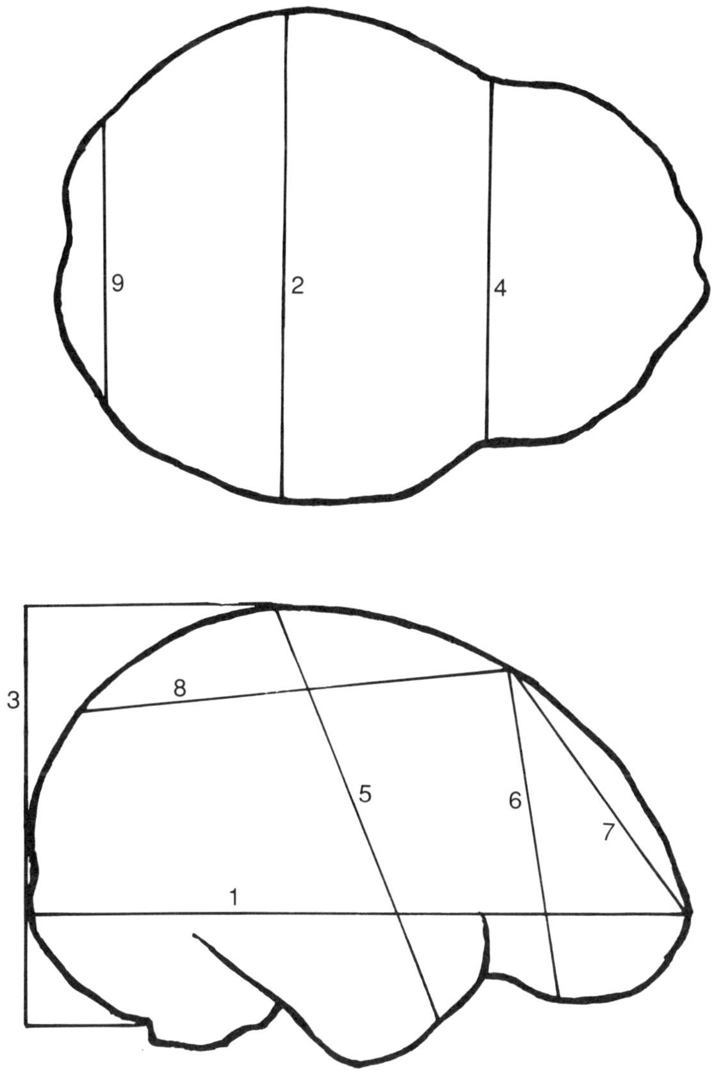

FIGURE 14.1.
Superior (top) *and right lateral* (bottom) *views of an endocast of KNM-WT 15000 showing measurements used in study (see Table 14.3).*

The unaltered, intact surface of the endocast was then compared in detail to the specimens listed in Table 14.2. Measurements (see Figure 14.1 and Table 14.3) were taken with electronic sliding calipers and recorded to the nearest millimeter.

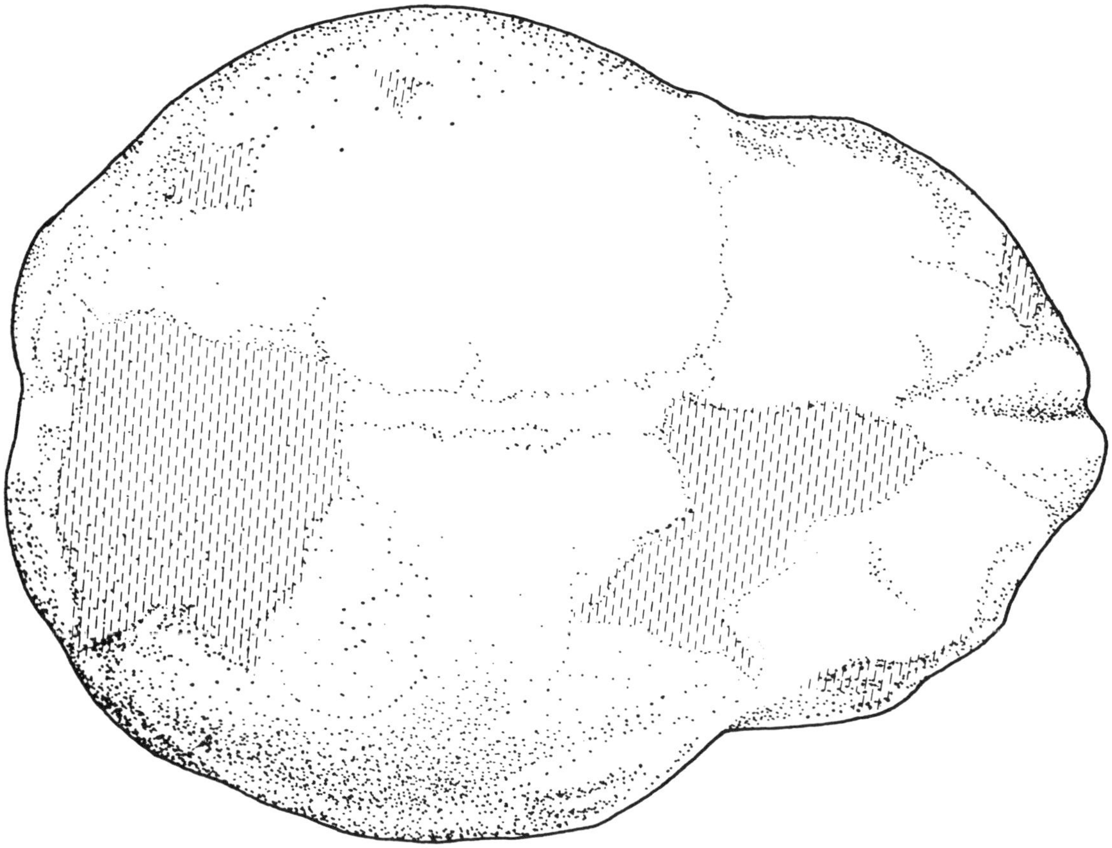

FIGURE 14.2.
Superior view of endocast.

Description of the Reconstructed Endocasts

Figures 14.2 to 14.7 show standard views of Reconstruction #1. The only major differences between the two reconstructions are mostly in the rostral and sella turcica regions. These differences in modeling in areas where parts were missing did not cause any appreciable difference in cranial capacity between the two reconstructions. The KNM-WT 15000 endocast shows slight asymmetry and distortion due to plastic deformation of the cranium during fossilization. Viewed from behind, the superior part of the endocast is clearly displaced to the left. The endocast is metrically and morphologically similar to other *Homo erectus* endocasts from the Turkana basin, such as KNM-ER 3733 and KNM-ER 3883, though there are some minor differences. The specimen is relatively and

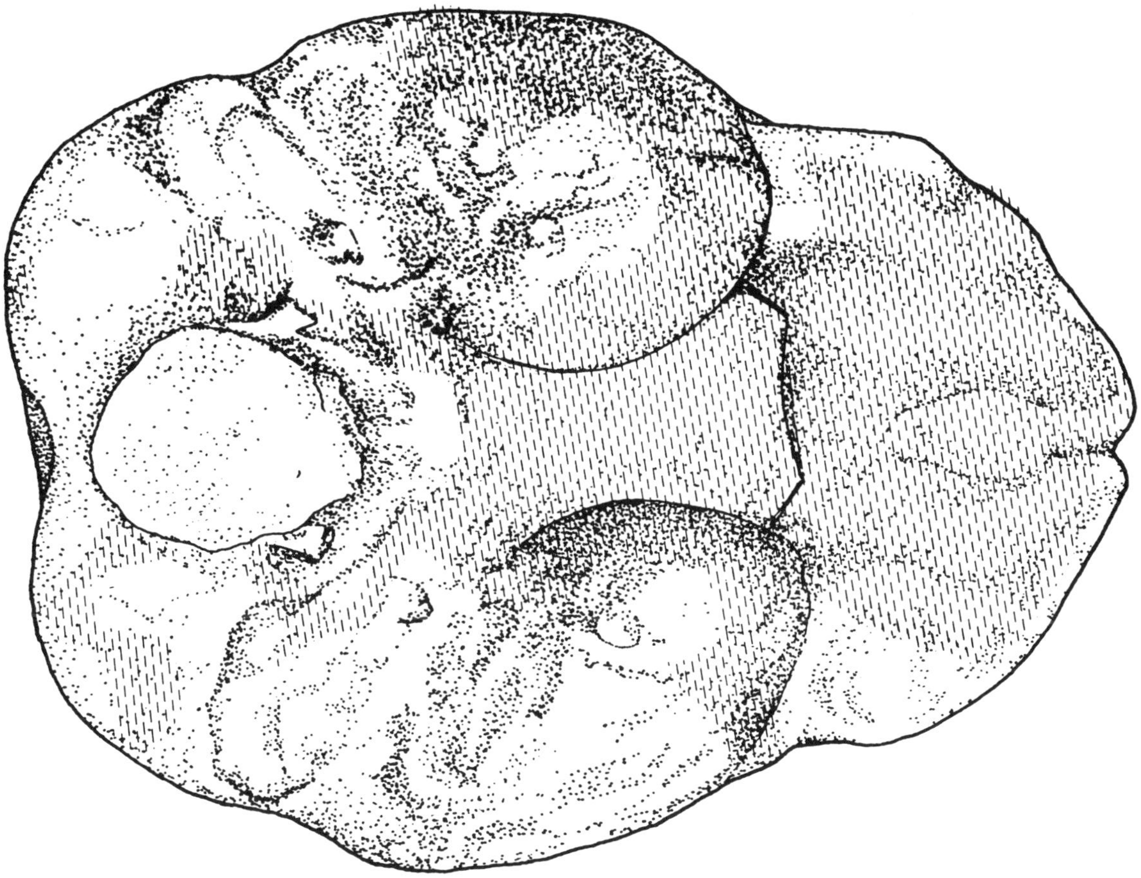

FIGURE 14.3.
Inferior view of endocast.

absolutely long and narrow compared with other *Homo erectus* from Kenya. It is also comparatively high, measured either from the cerebellum or from the temporal lobes, though this may be attributed to the fact that it was a juvenile. KNM-WT 15000 has very distinct and well-developed right fronto-petalia (LeMay, 1976) and fairly strongly bulging parietal lobes as well, compared with both East Turkana specimens, which have more smoothly rounded parietal contours. The frontal lobes are much longer, lower, and more strongly projecting anteriorly, and they are considerably less broad than those of KNM-ER 3733 and 3883, both of which have very short, broad frontal lobes (Table 14.3). The frontal bone of KNM-WT 15000 has a frontal sinus well developed on the left side and poorly developed on the right, and this may have affected the conformation of the frontal poles in KNM-WT 15000.

The orbital surfaces of the frontal lobes are

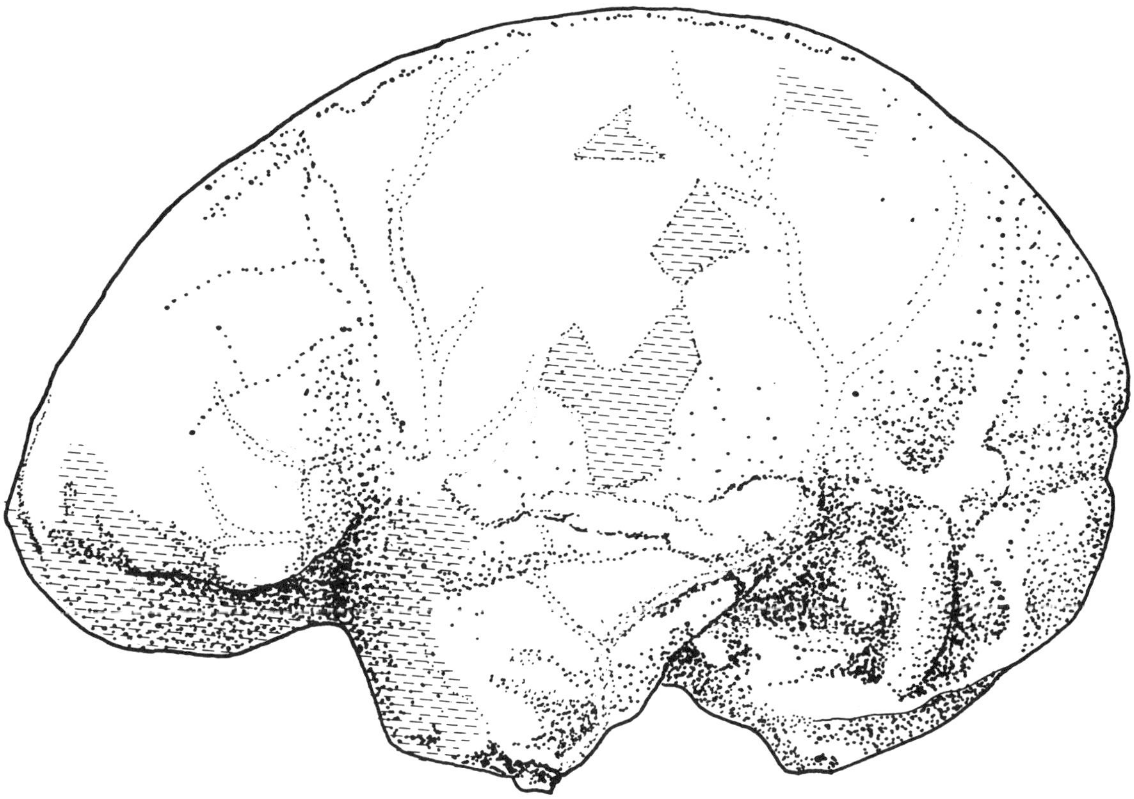

FIGURE 14.4.
Left lateral view of endocast.

not preserved except for strips about 1–2 centimeters broad on their lateral parts. These strips are relatively smooth and flat ventrally. The posterior-most portion of the inferior aspects of the frontal lobes end at deep, incised depressions to accept the lesser wings of the sphenoid bone. Just anterior to the depression on the left side only, a relatively broad area protrudes inferiorly and laterally. In lateral view, from posterior to anterior, it consists of a shallow groove immediately superior to the superior end of the sphenoidal lesser wing depression, a low convex area anterior to this, a second shallow groove, and a relatively large inferiorly and laterally protruding convex area. The posterior groove probably corresponds either to the central or the precentral sulcus and the more anterior groove either to the precentral or to the vertical limb of the arcuate sulcus. The protruding convexities on the left side of KNM-WT 15000 correspond to the strongly bulging area

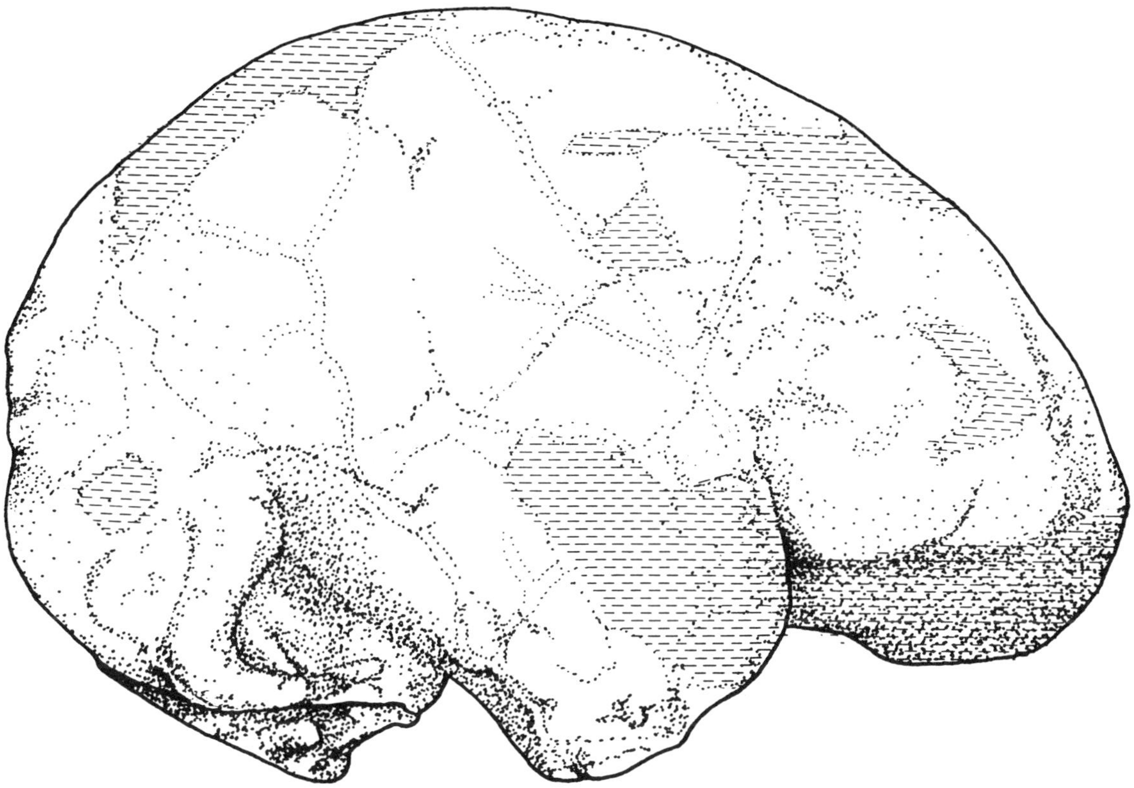

FIGURE 14.5.
Right lateral view of endocast.

of the inferior frontal gyrus near the anterior end of the lateral cerebral sulcus, otherwise known as Brodmann's area 44 and part of 45, or "Broca's" area (Williams et al., 1989). It is significantly larger and more projecting, both ventrally and laterally, on the left side than on the right. The corresponding area on the right side appears deflated and has more strongly marked sulci around it than its counterpart on the left. These convolutions are shorter anteroposteriorly and appear more projecting because of the deeper sulci on the right side. On the left side the gyri appear inflated, with the sulci more "filled in" and thus less distinct.

The temporal lobes are large and strongly projecting anteriorly and ventrally on KNM-WT 15000. Their anterior edges are broad and rounded. Laterally they are convex anteroposteriorly and dorsoventrally, the surfaces facing slightly ventrally. The parietal lobes are large and laterally bulging. In contrast to the frontals, they

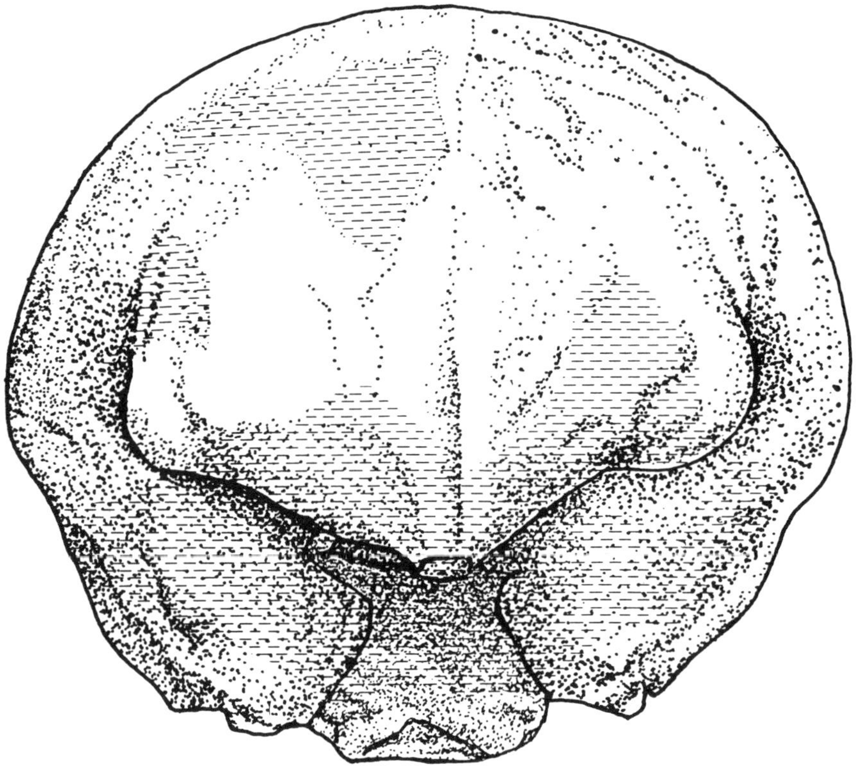

FIGURE 14.6.
Anterior view of endocast.

are very high, their maximum height being greater than in the other East African specimens included in this analysis (Table 14.3). Maximum height of the parietals occurs midway along their length. The occipital lobes are relatively small compared with the parietals, though their anterior boundaries are difficult to see. Whether or not there was an occipito-petalia cannot be determined because of the slight, but obvious, distortion in this region. The cerebellar hemispheres posterior to the occipital lobes are also large relative to the latter. The temporal, parietal, and occipital lobes all show some slight convolutional detail, but this is insufficiently distinct to warrant conclusive identification of specific sulci.

Some basicerebral morphology of KNM-WT 15000 is preserved in detail. A cast of the middle meningeal vessels passing through the foramen spinosum is visible on the inferior surface of the temporal lobe on the right side. The medial edges

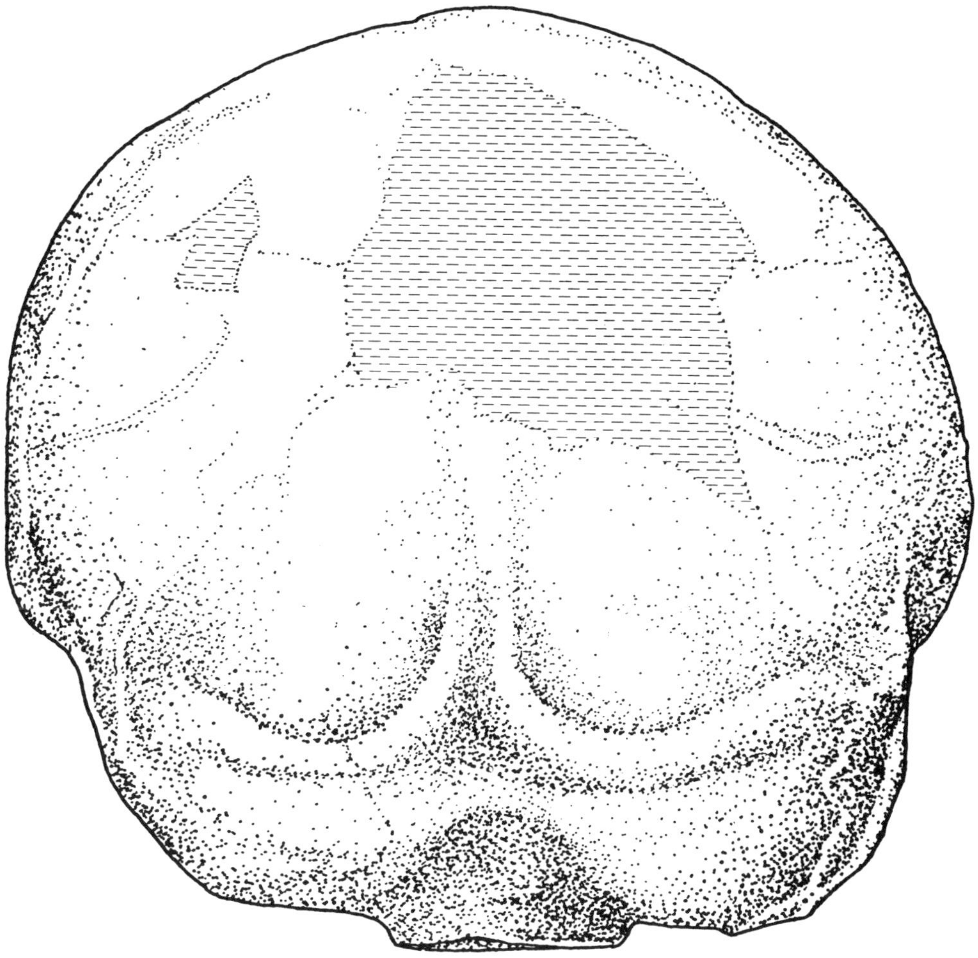

FIGURE 14.7.
Posterior view of endocast.

of the temporal lobes, just anterior and medial to the foramen spinosum, preserve casts of the foramen ovale. The posterior portions of the temporal lobes are marked by the broad, V-shaped valley representing the petrous portions of the temporal bones. Protruding from the posterior walls of the valley are casts of the internal acoustic meati on both sides. Just distal to these structures on both sides is a cast of the jugular foramen and associated structures continuous with the distal end of the sigmoid sinuses. More medially, casts of the hypoglossal canals project anteriorly and laterally from the walls of a large foramen magnum, whose border is preserved intact. The cast within the borders shows that the medulla was oval in cross-section with the long axis placed anteroposteriorly. The midline basicranial structures from the anterior edge of the foramen magnum to the olfactory structures anteriorly are represented only by a short piece of basioccipital surface.

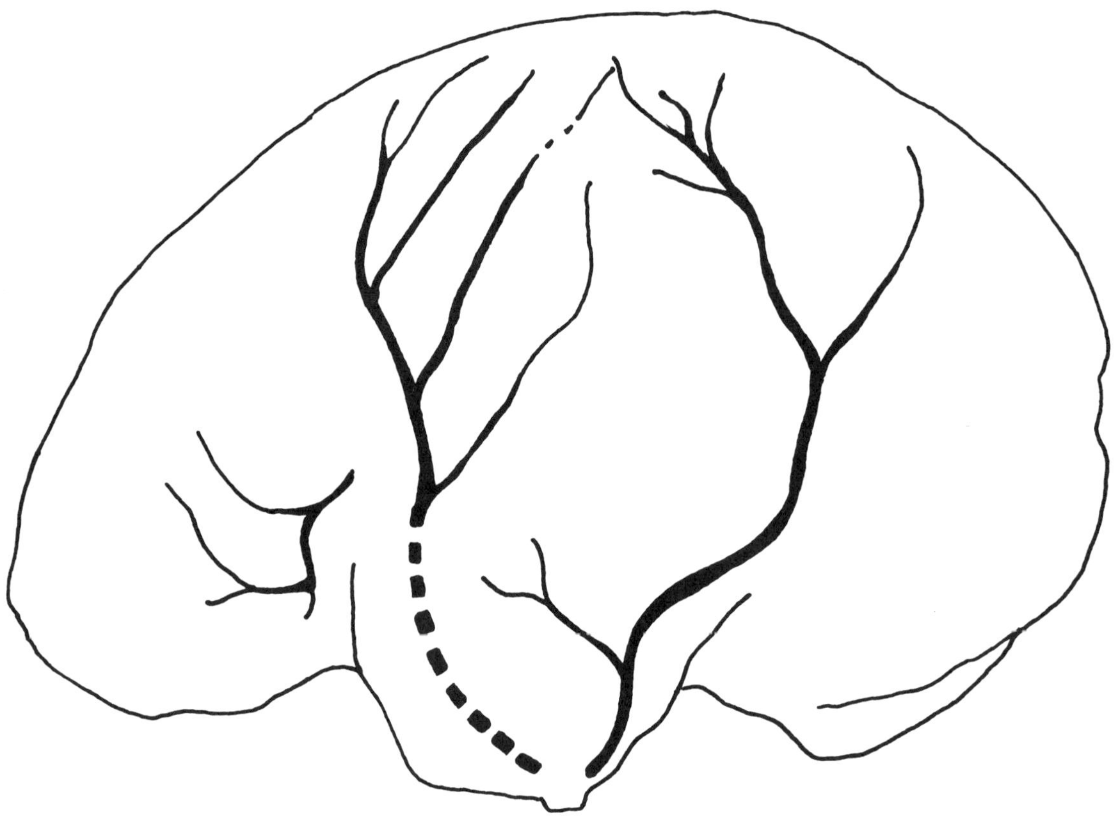

FIGURE 14.8.
Left lateral view of endocast showing meningeal vessel grooves.

Meningeal vessels

The pattern of meningeal vessels on KNM-WT 15000 is illustrated in Figures 14.8 and 14.9. Despite the individual's young age, the vessels have been very well preserved compared with most other East African endocasts. On the left side the middle meningeal vessels can be traced from close to the foramen spinosum about halfway up the lateral surface of the temporal lobe. The vascular cast divides into a large posterior division and a relatively smaller anterior one. The anterior division runs anteriorly toward the squamosal suture but fades 1–2 centimeters from the middle meningeal trunk. It can be picked up again just dorsal to the squamosal suture and posterior to the coronal suture, where it divides into an anterior branch, which crosses the coronal suture, a middle branch, which runs roughly parallel to the coronal suture, and a posterior branch, running dorsally and posteriorly. The posterior branch can be traced for about 2 centimeters. The middle branch further divides into one branch leading dorsally and posteriorly and a second continuing dorsally along the

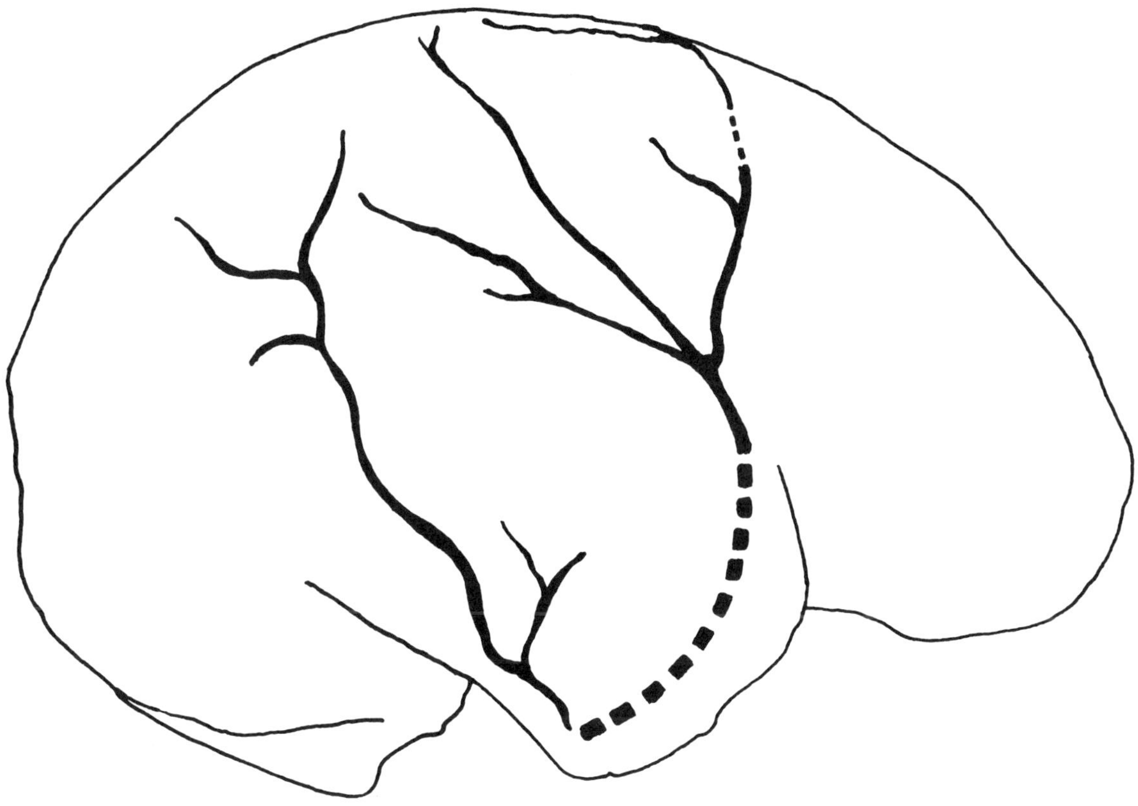

FIGURE 14.9.
Right lateral views of endocast showing meningeal vessel grooves.

coronal suture. Both of these branches fade out within 1 or 2 centimeters. The anterior branch runs anteriorly from the coronal suture and then divides into a small twig, which runs directly posterior to the most prominent portion of the bulge in the third inferior frontal convolution (Broca's cap), and a larger twig, which runs over the bulge and then divides into two smaller branches, one proceeding dorsally and the other more anteriorly.

The posterior division of the grooves representing the meningeal vessels runs posteriorly along the temporal lobe and is lost for about 1 centimeter in an area corresponding to the posterior portion of the squamosal suture. Dorsal to this it divides into a posterior branch, running parallel to the lambdoid suture, and an anterior branch, running dorsally and slightly anteriorly. This branch is the most highly ramified of all the meningeal traces. Two short branches diverge anteriorly, running toward the branches of the anterior division, though no clear anastomosis is preserved. Dorsal to these, two more branches diverge, both directed dorsally but one more posteriorly. Close to the sagittal suture another branch appears running parallel to the more dorsal branch just described. The origin of this segment is not apparent.

On the right side the traces of the meningeal vessels are not as apparent as on the left. The cast of the trunk of the middle meningeal vessels is visible inferiorly on the lateral surface of the temporal lobe. It divides into anterior and posterior divisions, as on the left side. The anterior division runs anteriorly and slightly dorsally for about a centimeter and is then lost in a relatively broad area of missing bone of the temporal squama in the area of the squamosal suture. The anterior division reappears dorsal to the missing area and anterior to the coronal suture, where it divides into a branch leading anteriorly and another that crosses the coronal suture and ramifies. The anterior branch runs directly anteriorly for several centimeters and then disappears without apparent further ramification. The posterior branch ramifies posterior to the coronal suture. One branch follows the coronal suture dorsally. A second branch runs dorso-posteriorly and branches into two smaller twigs that fade away after about 1–2 centimeters. A third branch runs more posteriorly still and contributes to an anastomosis with a branch of the posterior division.

The posterior division of the meningeal vessels on the right side runs along a smoothly curved line dorsally and posteriorly across the temporal lobe toward the squamosal suture. About 1 centimeter dorsal to the point at which it crosses the suture, it branches into an occipital branch, which heads directly for the lambdoid suture, and a dorsally directed branch, which divides into anterior and posterior twigs. The anterior twig runs dorsally and slightly anteriorly and continues as an anastomotic twig to the anterior division.

Venous sinuses

Traces of the cerebral venous drainage system of the brain of KNM-WT 15000 preserve additional details of the pattern of blood flow in this individual. Transverse sinuses are very strongly marked from their inception at the confluence of sinuses, near inion. The right transverse sinus is considerably larger than the left. This is the common pattern in modern humans and is related to the fact that the smaller straight sinus usually runs to the left while the larger superior sagittal sinus drains into the right transverse sinus (Browder and Kaplan, 1976). The transverse sinuses cross the posterior portion of the endocranial cast between the occipital lobes and the cerebellum as far as the lateral-most portions of the cerebellar hemispheres, at which point they drain into very large sigmoid sinuses. These curve ventrally and then swing forward toward the jugular foramen. Unlike the transverse sinuses, the sigmoid sinuses are about equal in caliber. Conroy (1982) made a survey of living primates as well as of the literature on the asymmetry of cerebral venous flow. He points out that the dominance of the right side of the venous pathway has long been a well-known feature of humans. But he also points out that while several different functional explanations have been proposed since 1877, the extent to which venous asymmetry is present in some nonhuman anthropoids has been overlooked. However, it does seem that the confluence of sinuses is more symmetrical in modern great apes (Kimbel, 1984) and that the asymmetrical condition is the derived one.

The region between the confluence of sinuses and the posterior border of the foramen magnum is well preserved, as are the lateral borders of the foramen magnum, yet there is no indication of the development of an occipital or a marginal sinus as seen in some australopithecines and occasionally in *Homo* (Tobias, 1967; Kimbel, 1984).

Comparative studies

East Africa

The KNM-WT 15000 endocranium is very similar to two other *Homo erectus* endocasts from the Turkana basin, KNM-ER 3733 and 3883. However, endocasts produced from these crania differ in several ways from the KNM-WT 15000 specimen. The calvaria of KNM-ER 3883 is distorted somewhat and appears flatter than it was in life. KNM-ER 3883 and 3733 both have relatively broader frontal lobes with broader and more rounded poles. KNM-ER 3883 has relatively flat frontal lobes viewed laterally, like those of KNM-WT 15000, while those of KNM-ER 3733 are more vertically oriented. The orbital surfaces of the frontal lobes are preserved on KNM-ER 3883. They are flat to slightly concave. Interorbitally, an elongated, relatively broad rostral region projects slightly beyond the orbital plane from just posterior to the frontal poles to the optic chiasma. This region is not preserved on KNM-ER 3733. Reconstruction #2 of KNM-WT 15000 in this area was based in large part on KNM-ER 3883. KNM-ER 3883 has a clearly defined right fronto-petalia, the right lobe being broader and slightly more projecting than the left. On KNM-ER 3733 the asymmetry between the lobes is not so pronounced. On KNM-WT 15000 the asymmetry is strong but different in the sense that the more prominent right lobe is much more anteriorly projected but only slightly broader than the left. As noted above, this may be due to the unequal development of the frontal sinuses.

Broca's caps in KNM-ER 3733 and 3883 are less well preserved than on KNM-WT 15000 but are more rounded and project laterally and inferiorly. Sulci around Broca's cap are more strongly marked on the right sides of all three specimens, the left sides having the appearance of having been "filled in" to a certain extent by cortical expansion.

Both KNM-ER 3733 and 3883 share with KNM-WT 15000 large, rounded, anteriorly and inferiorly projecting temporal lobes. The parietal lobes bulge bilaterally in all three specimens but seem to be largest on KNM-ER 3733, which has nearly vertical parietals laterally. The maximum parietal height is slightly more anterior on KNM-ER 3733 and 3883 than on KNM-WT 15000, though the absolute height is greatest on KNM-WT 15000. The occipital lobes of KNM-ER 3733 and 3883 appear to be somewhat larger relative to the cerebellum than on KNM-WT 15000, and both have well-developed left occipito-petalia.

A number of endocasts from the Turkana Basin can be compared to KNM-WT 15000. These include KNM-ER 1470, 1805, and 1813, all of which are usually attributed to early *Homo* (Chamberlain and Wood, 1987).

The frontal lobes of KNM-ER 1470 are smaller but are also relatively short, broad, and high compared with those of KNM-WT 15000. On KNM-ER 1813 they are very similar in shape to those of KNM-WT 15000, that is, they are elongated, low, and mediolaterally constricted and have prominent poles. KNM-ER 1813 also has a well-marked right fronto-petalia, though it is less well marked than that on KNM-WT 15000. KNM-ER 1805 is intermediate in frontal lobe morphology between the low, narrow pattern of KNM-WT 15000 and KNM-ER 1813 and the taller, broader pattern of KNM-ER 1470, 3733, and 3883. The frontal poles are damaged on this specimen, so that a petalia is not visible. Inferiorly and posteriorly, the frontal lobes of KNM-ER 1470 are asymmetrical, with the left third inferior frontal convolution and Broca's area being larger and more inflated than the equivalent part on the right (Falk, 1983; Holloway, 1983; Tobias, 1983). The area, however, is noticeably less protuberant than on KNM-WT 15000. KNM-ER 1813 is damaged in this area. KNM-ER 1805 is more complete and shows no clear-cut evidence of an asymmetry (Falk, 1983).

The temporal lobes on KNM-ER 1470 are large, rounded, and anteriorly and inferiorly projecting, as on the early *Homo erectus* grade specimens described above. On KNM-ER 1813, they are relatively smaller, narrower, and less projected, and on KNM-ER 1805 they are intermediate in size and shape. The parietal lobes of KNM-ER 1470 are high, as in *Homo erectus* generally, and their maximum height is posteriorly placed, as on KNM-WT 15000, compared with KNM-ER 3733 and 3883. The occipital lobes are relatively strongly projected and broad in 1470, with a moderate petalia on the left side. On both KNM-ER 1805 and 1813 the occipital lobes are smaller and less projecting, without clear asymmetries.

The pattern of meningeal vessels is difficult to make out on KNM-ER 3733 (but see Saban, 1986). This cranium was filled with extremely hard matrix when found. The separation between bone and matrix proved to be poor and so the internal surface of the cranium is abraded and crossed by the numerous breaks and fissures which can be seen on the outside of the cranium. The vessels are somewhat more visible on KNM-ER 3883, though they are less well defined than on KNM-WT 15000. On the left side there appears to be a well-defined anterior division proceeding toward the posterior inferior portion of the frontal lobe, and a larger posterior division that ramifies as on KNM-WT 15000 beyond the squamous suture. There is no clear-cut evidence of anastomotic branches, but this is probably the result of relatively poor preservation rather than a demonstrable absence of these structures.

The pattern of meningeal vessels is visible on KNM-ER 1470. On the right side, the anterior division is relatively small and difficult to see. It can be traced to a single branch in the vicinity of the left third inferior frontal convolution. The posterior division is similar to KNM-WT 15000 in that it ramifies into a number of large branches heading dorsally and posteriorly mid-parietally and toward lambda, the occipital lobes, and the cerebellum. On the left side, the anterior division is better developed, with large branches running along the coronal suture and toward bregma.

On KNM-ER 1805 the meningeal pattern is also relatively well marked. As on KNM-ER 1470, and in contrast to KNM-WT 15000, the anterior division is relatively small and poorly ramified. The posterior division is also less branching than on either KNM-ER 1470 or KNM-WT 15000, dividing mid-parietally into a number of branches running between bregma and lambda.

On KNM-ER 1813 the meningeal pattern is partially preserved on the left side. It differs from the others in having a very posteriorly positioned division that runs along the posterior edge of the temporal lobe, giving off several small branches near the transverse sinus. An anterior branch runs supero-anteriorly across the temporal lobe, crossing the squamosal suture and then ramifying into several branches heading dorsally and posteriorly on the parietal lobe. No branches can be traced crossing the coronal suture.

The main difference between KNM-WT 15000 and KNM-ER 1805 and 1813 is in the development and distribution of branches of the anterior division of the meningeal vessels. In KNM-WT 15000, the anterior division is large and sends branches both along the coronal suture and to the frontal lobes. It thus covers the frontal lobes and the anterior portion of the parietals. In KNM-ER 1805 and 1813, the anterior division is poorly developed anterior to the coronal suture and covers most of the parietal lobes. The posterior division is relatively small and is confined to the most posterior portions of the parietal and to the occipital lobes. In KNM-WT 15000, the posterior division ramifies both posteriorly and anteriorly on the parietals and sends one or more anastomotic branches to the anterior division.

It should be noted that the homologies of the branches of the meningeal vessels in different endocasts are not clearly established. The vessels discussed here are named on the basis of their relative position on the endocasts. They may or may not be homologous among specimens. It is clear, however, that there are differences in the pattern of meningeal blood flow between KNM-WT 15000 and smaller specimens such as KNM-ER 1805 and 1813. The pattern is intermediate in KNM-ER 1470, with its somewhat better-developed anterior division, especially on the left side, and a posterior division that covers more of the parietal lobes. Transverse sinuses are visible on KNM-ER 3883 and sigmoid sinuses on both KNM-ER 3733 and 3883. On KNM-ER 3883, the transverse sinus on the right side is larger than the left, as in KNM-WT 15000. The sigmoid sinuses are similar on all three specimens. None of the specimens show any sign of occipital or marginal sinuses.

China

The endocasts of *Homo erectus* crania from Zhoukoudian were compared with that of KNM-WT 15000 using University Museum replicas (Casting Program, University Museum, University of Pennsylvania). These casts are of good quality, with adequate detail for general comparisons. Areas of reconstruction were clearly marked. The reconstructions and dimensions closely match those provided in Weidenreich (1943). The Zhoukoudian endocasts are all quite similar in overall morphology to KNM-WT 15000, with some important and consistent differences. Of the four endocasts compared with KNM-WT 15000, three (Skulls 2, 3, and 12) are similar in size (915–1030 cc) while Skull 10, at 1225 cc, is substantially larger. Despite the size range, all the specimens are relatively narrow, compared with their height and length, as in KNM-WT 15000 (see Figures 14.10 and 14.11). All preserve the frontal poles intact, and none show any substantial development of fronto-petalia. Two of the largest specimens (Skulls 10 and 12) show development of occipito-petalia while the smallest one (Skull 3) shows none. Skull 2 has an estimated cranial capacity equal to that of Skull 12, but its occipito-petalia is not preserved.

All the specimens have moderately to strongly developed, broad sagittal keels, especially along the parietal portions of the endocasts. To either side of these keels, the parietal surfaces are slightly depressed and are oriented at a relatively acute angle to the mid-sagittal plane. This shape is almost certainly a reflection of the strongly keeled morphology of the Zhoukoudian crania; it is not characteristic of any other specimens included in this analysis. Skulls 2 and 12 have this keel continued mid-sagittally between the frontal lobes and have broad, well-marked depressions to either side of the keel just anterior to the coronal suture. The frontal keels on Skulls 2 and 10 are more rounded and the depressions less deep. In all four specimens, the frontal keel ends anteriorly as a broad bulge facing anteriorly and superiorly. The three smaller specimens all have long, low frontal lobes, though the smallest (Skull 3) is closest in morphology in this area to KNM-WT 15000. The frontal lobes of the largest specimen, Skull 12, are higher and broader relative to length.

The parietal regions of all the specimens from Zhoukoudian examined here are broadest bilaterally in the region adjacent to the temporal lobes. The parietal surfaces are flat and oriented laterally and slightly superiorly, whereas in KNM-WT 15000, this area is rounded and oriented directly laterally. Roughly mid-parietally, in lateral view, the parietal surfaces angle in toward the parietal keel; they are relatively flat and face supero-laterally. This shape again is directly related to the keeled shape of these crania. In KNM-WT 15000, the parietal arc from the squamosal suture to the mid-sagittal plane is smoothly rounded, and the maximum breadth of the parietal region occurs more superiorly on the lobes.

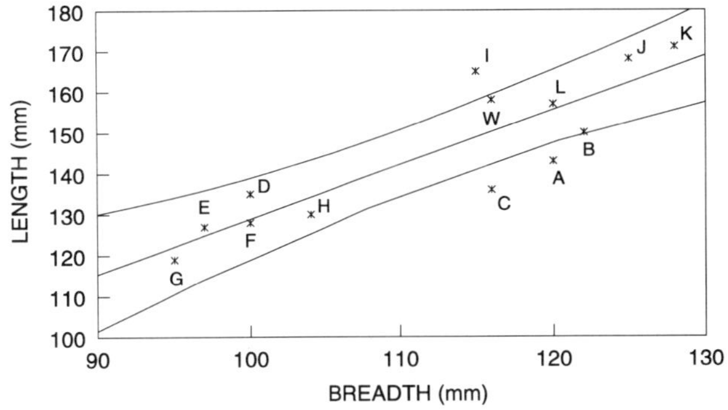

FIGURE 14.10.
Bivariate plot, least-squares linear regression, and 90% confidence interval of length and breadth (mm) of fossil endocasts; r = 0.87; slope = 1.335.

A = KNM-ER 3733	*H = OH 5*
B = KNM-ER 3883	*I = Skull 2 (Zhoukoudian)*
C = KNM-ER 1470	*J = Skull 3 (Zhoukoudian)*
D = KNM-ER 1805	*K = Skull 10 (Zhoukoudian)*
E = KNM-ER 1813	*L = Skull 12 (Zhoukoudian)*
F = KNM-ER 13750	*W = KNM-WT 15000*
G = KNM-WT 17000	

The temporal lobes in the Chinese specimens are large and bilaterally flared, and in lateral view they face laterally such that the broadest portion of the entire endocast occurs near the middle of the temporal squama of the crania. In KNM-WT 15000, the temporal lobes in lateral view are smaller, less flared, and oriented infero-laterally such that they appear tucked under the parietal lobes. Parts of the basal aspects of the temporal

FIGURE 14.11.
Bivariate plot, least-squares linear regression, and 90% confidence interval of length and height (mm) of fossil endocasts. Abbreviations as in Figure 14.10; r = 0.986; slope =1.351.

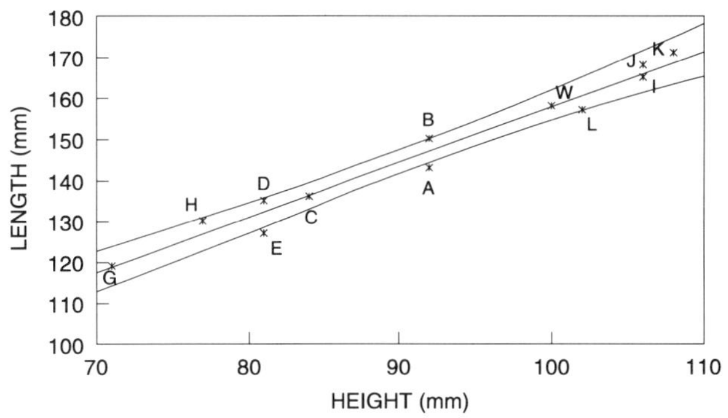

lobes are preserved only on Skulls 2 and 3 and are flat and oriented inferiorly. In KNM-WT 15000, this region on either side is more rounded and faces more infero-laterally.

The occipital lobes of the Zhoukoudian specimens are dorso-ventrally flattened, broad, and strongly projected posteriorly, probably in accordance with the morphology of the occipital squama in these crania. The cerebellar lobes are poorly preserved but can be seen on all specimens to be situated well anterior to the occipital lobes. The occipital lobes thus project posteriorly well beyond the cerebellar surfaces. In KNM-WT 15000, the cerebellum is not tucked under the occipital lobes but projects posteriorly almost as much as the occipital lobes.

The meningeal blood vessels are variably preserved on the endocasts from Zhoukoudian. The cast of the groove for the middle meningeal vessels is visible and originates about midway along the posterior edge of the temporal lobe in basal view in Skulls 2, 3, and 12. On Skulls 3 and 12, a vascular cast runs anteriorly from this point across the temporal lobe to a point close to the anterior edge of the lobe in lateral view. This is in contrast to the configuration in KNM-WT 15000, where the vascular cast branches into its major divisions shortly after emerging from the foramen. On Skull 2 the vascular cast reaches halfway across the infero-lateral surface of the temporal lobe before its course becomes obscured by damage in the area. On Skull 12, a large branch of the meningeal vessels diverges superiorly and posteriorly about midway along the vascular cast running between the anterior and posterior edges of the temporal lobe in lateral view. On Skull 3, the branch occurs closer to the anterior edge of the temporal lobe, probably within the lateral cerebral sulcus. The vascular cast continuing anteriorly from the point where the meningeal vessels diverge (relatively anterior on Skull 3 and posterior on Skull 12) may represent a venous anastomosis between the sphenoparietal sinus and the middle meningeal vein. Where the bifurcation occurs more posteriorly (on Skull 12), the branch from the sphenoparietal sinus is very large, and it continues toward the anterior edge of the temporal lobe and disappears into the lateral cerebral sulcus. On Skull 3, the sphenoparietal branch is smaller and shorter, and its connection to the meningeal vein appears to consist of a number of smaller branches.

From the posterosuperior end of the lateral cerebral sulcus on all four specimens, a large branch ascends along the coronal suture. It has no apparent connection to the meningeal branches just described but was probably connected to the temporal or sphenoidal portion of the meningeal vessels in the region of the parietal's sphenoidal angle. On Skulls 3 and 12, several small branches can be traced from the coronal or middle branch anteriorly across the coronal suture toward areas 44 and 45. On Skulls 2 and 10, the relevant regions are not preserved. The posterior division is strongly ramified in all specimens and sends large branches toward the parietal and occipital lobes. On Skulls 2, 10, and 12, where they are clearly visible, the branching pattern of the posterior division is more complex than in KNM-WT 15000. On all three of these specimens, there are clearly defined anastomotic branches between anterior and posterior divisions in the area of the parietal lobes.

The venous sinus systems on the Zhoukoudian specimens are all quite similar to one another. Where preserved, on Skulls 3, 10, and 12, the transverse sinuses are large and right-side dominant, though the difference in caliber between right and left sides on Skull 3 is minor. The right transverse sinus on all three specimens is directly continuous with a strongly marked superior sagittal sinus running between the projected occipital poles and along the more prominent part of the sagittal keel. The superior sagittal sinus is visible on Skull 2 as well, between the parietal lobes on the sagittal keel. On Skulls 2 and 3, the superior sagittal sinus can be traced anteriorly as far as bregma, and on Skull 3 all the way to the gap between the anterior portions of the frontal lobes. On KNM-WT 15000, the superior sagittal sinus cannot be reliably detected, even just anterior to the confluence of sinuses where it is best developed on the Zhoukoudian specimens. The sigmoid sinuses are preserved only on both sides on Skull 3, where they are large and of roughly equal caliber.

In summary, the morphology of the endocasts of *Homo erectus* from Zhoukoudian is similar in its major features to that of KNM-WT 15000. Detailed differences include the development of mid-sagittal keels, expanded temporal lobes, projected and broad occipital lobes, and retracted cerebellar hemispheres, all characteristic of the Chinese specimens and not found in the East African sample. The Chinese endocasts can also be distinguished from the East African material by the development of the superior sagittal sinus, the apparently well-developed connection between the sphenoparietal sinus and the middle meningeal vein, and the more complex pattern in the posterior division of the middle meningeal vessels.

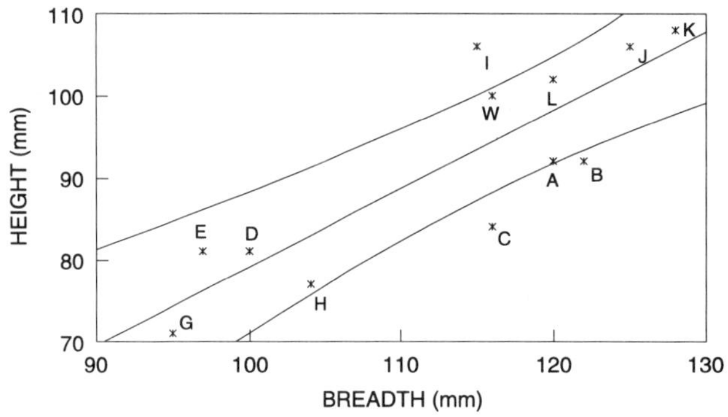

FIGURE 14.12.
Bivariate plot, least-squares linear regression, and 90% confidence interval of height and breadth (mm) of fossil endocasts. Abbreviations as in Figure 14.10; r = 0.847; slope = 0.957.

METRIC COMPARISONS

Nine standardized measurements, as illustrated in Figure 14.1, were taken on the KNM-WT 15000 endocast and on the comparative series. These measurements were chosen so as to be easily repeatable and to measure the major portions of the endocasts. Figure 14.10 is a bivariate plot of length on breadth. It shows a general size trend, with the smaller specimens falling toward the left side of the graph. KNM-WT 15000 (*W* on the plot) falls among the larger endocasts, just at the upper 90% confidence limit of the regression line. It is slightly longer than the regression predicts, falling above all values except Skull 2 from Zhoukoudian. Skull 2, however, is the least satisfactory of the Chinese endocasts included here, because the temporal regions are mainly reconstructed and the maximum breadth measurement is suspect. Despite this, KNM-WT 15000 falls generally in the group that includes the two other early *erectus* grade specimens from Kenya and close to the Chinese specimens. Figure 14.11 shows a tighter correlation between length and height than exists between length and breadth. In this case, KNM-WT 15000 falls exactly on the regression line, in the middle of the *erectus* grade specimens, between the smaller Kenyan specimens and the larger specimens from China. The last bivariate plot shows KNM-WT 15000 again at the upper 90% confidence limit in height relative to breadth (Figure 14.12). In this respect, it is narrow like the Chinese specimens but unlike the other, large Kenyan specimens, which are wide. In all these plots, KNM-WT 15000 falls very close to or within the 90% confidence intervals of the regression line; it is closest to the Chinese specimens and next closest to the other *erectus* grade specimens from Kenya.

Metric comparisons of endocast shape among different groups of hominids are somewhat

FIGURE 14.13.
Means and ranges of the index of length/breadth in fossil hominid endocasts.
WT =KNM-WT 15000
Hsp =KNM-ER 1470, KNM-ER 1805, KNM-ER 1813
Hep = Zhoukoudian skulls 2, 3, 10, and 12
Hea = KNM-ER 3733, KNM-ER 3883
Aust = KNM-ER 13750, KNM-WT 17000, OH 5

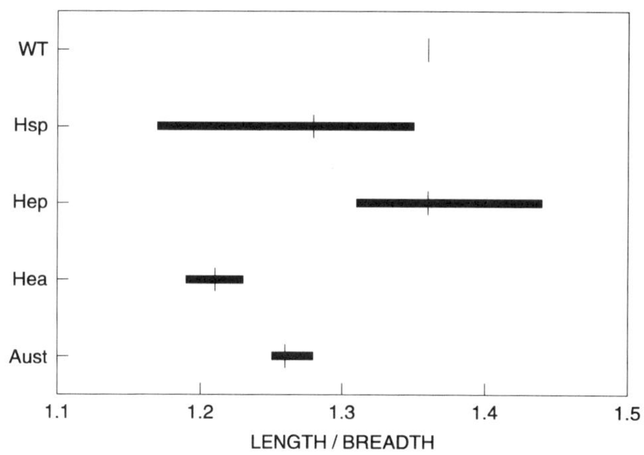

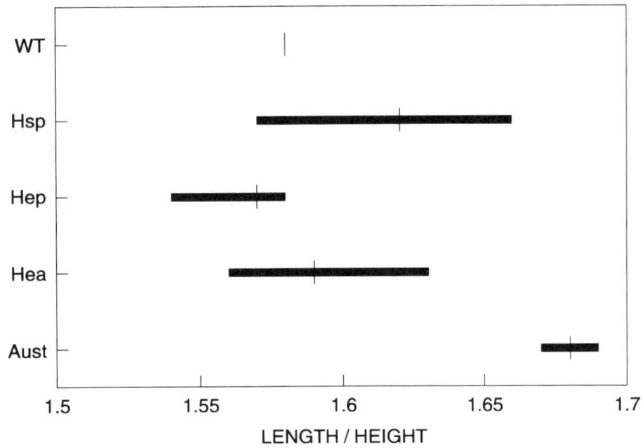

FIGURE 14.14.
Means and ranges of the index of length/height of fossil hominid endocasts. Abbreviations as in Figure 14.13.

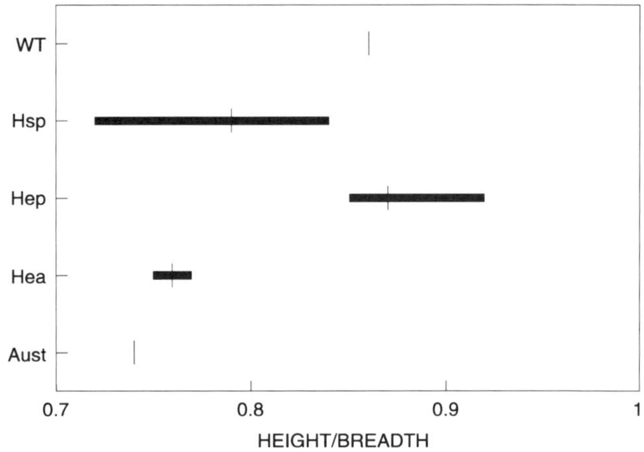

FIGURE 14.15.
Means and ranges of the index of height/breadth of fossil hominid endocasts. Abbreviations as in Figure 14.13.

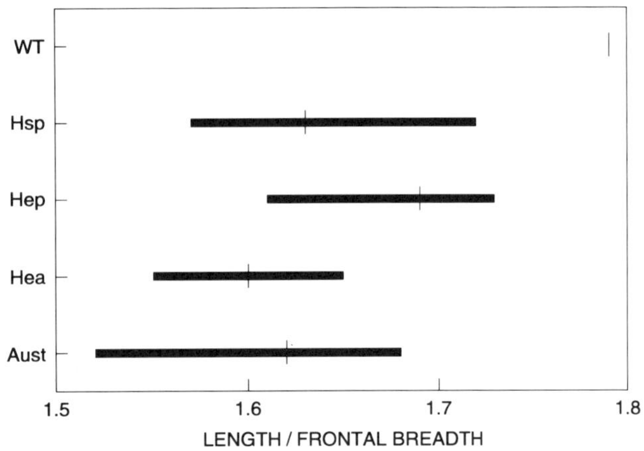

FIGURE 14.16.
Means and ranges of the index of length/frontal breadth of fossil hominid endocasts. Abbreviations as in Figure 14.13.

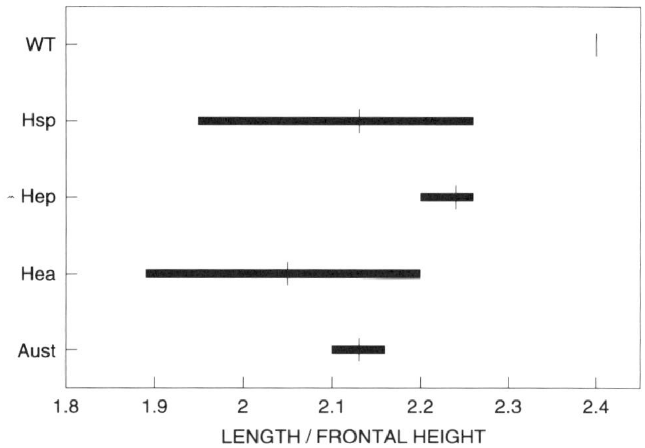

FIGURE 14.17.
Means and ranges of the index of length/frontal height of fossil hominid endocasts. Abbreviations as in Figure 14.13.

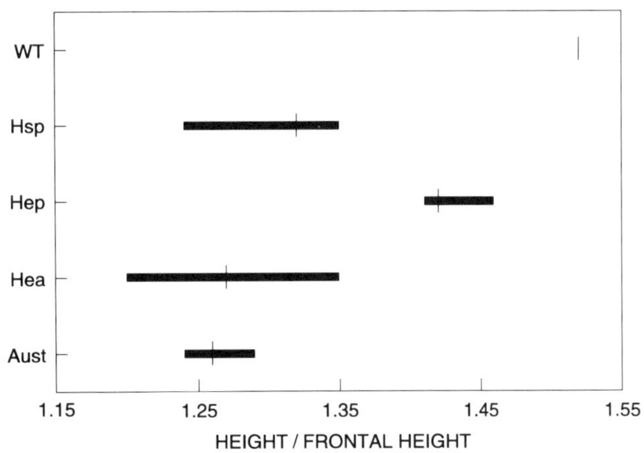

FIGURE 14.18.
Means and ranges of the index of height/frontal height of fossil hominid endocasts. Abbreviations as in Figure 14.13.

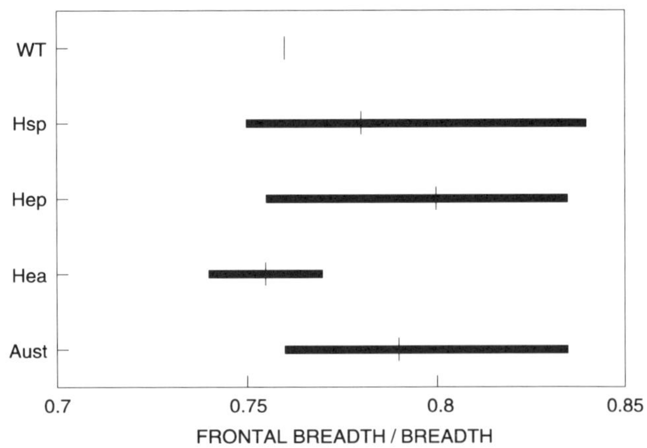

FIGURE 14.19.
Means and ranges of the index of frontal breadth/breadth of fossil hominid endocasts. Abbreviations as in Figure 14.13.

easier to see in mean/range plots of a number of indices (Figures 14.13 to 14.22). In Figure 14.13, KNM-WT 15000 is just above the range of three values for archaic *Homo* and close to the mean for *Homo erectus* from Zhoukoudian. It is longer relative to breadth than the other African early *erectus* grade hominids. In length relative to height (Figure 14.14), all three *Homo erectus* grade groups fall close together, while the archaic *Homo* group is slightly longer relative to height, though the overlap in ranges is substantial. Figure 14.15, which shows the ranges of the height/breadth index, again groups KNM-WT 15000 with *Homo erectus* from China. Thus, these specimens are long and high relative to breadth and similar in length relative to height.

KNM-WT 15000 has a low, narrow frontal lobe relative to overall length, breadth, and height, as indicated in Figures 14.16 through 14.20. Figures 14.16 and 14.17 plot the ranges of length relative to frontal breadth (Figure 14.16) and height (Figure 14.17). In both indices, the means of the sample from China are closer to KNM-WT 15000 but they overlap the ranges of both *Homo* samples from East Africa, while KNM-WT 15000 falls well outside the range of any specimen. In Figure 14.18, overall cerebral height is great relative to frontal height in KNM-WT 15000 and to a lesser extent in the Zhoukoudian sample, while frontal breadth relative to overall breadth (Figure 14.19) is low, though within the ranges of the other groups and close to the means for the East African *Homo erectus* grade specimens. This is true in spite of the low overall breadth in KNM-WT 15000 (Figures 14.13–14.15). Frontal breadth in KNM-WT 15000 is also low compared with cerebral height (Figure 14.20), and again the Zhoukoudian sample falls closest to KNM-WT 15000.

KNM-WT 15000 is most similar to all the African groups in length (Figure 14.21) and breadth (Figure 14.22) relative to occipital breadth; it is distinguished in both these indices from the Zhoukoudian sample, which has substantially larger relative occipital breadth.

In summary, these indices indicate that KNM-WT 15000, and to a lesser extent the sample from Zhoukoudian, had small frontal lobes and narrow endocasts overall compared with other

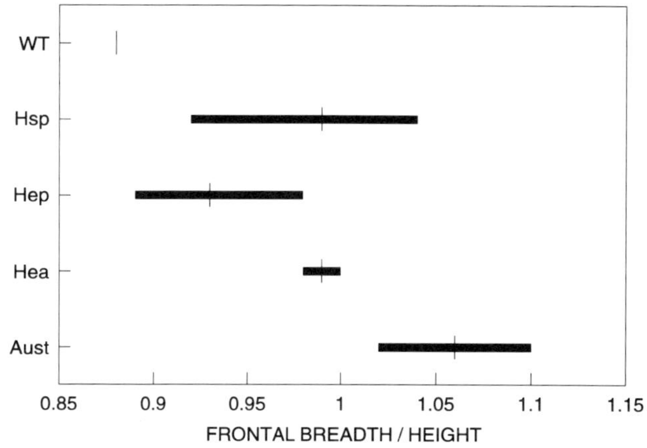

FIGURE 14.20.
Means and ranges of the index of frontal breadth/height of fossil hominid endocasts. Abbreviations as in Figure 14.13.

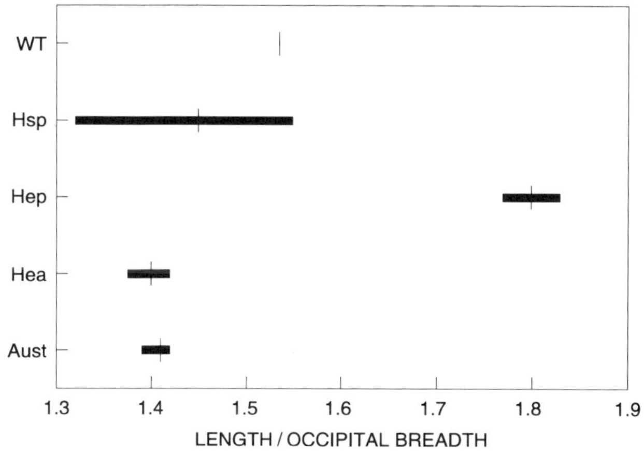

FIGURE 14.21.
Means and ranges of the index of length/occipital breadth of fossil hominid endocasts. Abbreviations as in Figure 14.13.

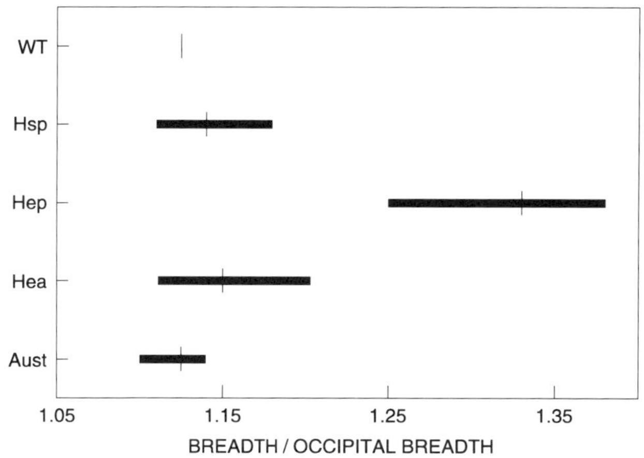

FIGURE 14.22.
Means and ranges of the index of breadth/occipital breadth of fossil hominid endocasts. Abbreviations as in Figure 14.13.

East African *Homo erectus* grade endocasts and archaic *Homo. Homo erectus* (from Zhoukoudian) has relatively larger occipital lobes than all other specimens considered here.

In contrast, KNM-WT 15000 falls close to the means for both *erectus* grade groups from China and East Africa in the length/height index (Figure 14.14), while the archaic *Homo* group has relatively longer, or more shallow, endocasts.

Figure 14.23 shows a scatterplot of two factors resulting from a principal components analysis using the data from Table 14.3. Factors 1 and 2 together explain about 90% of the variation in these data. KNM-WT 15000 (W) groups most closely with KNM-ER 3733 (A) and falls almost equidistant between KNM-ER 1470 (C) and KNM-ER 3883 (B) on the one hand and the Chinese *erectus* (I–L) on the other. KNM-WT 15000 is widely removed from the three other early hominid endocasts on the plot. Overall, then, the endocast of KNM-WT 15000 seems to be closest in size and shape to other *erectus* grade hominids, but it is as much like the Chinese as the East African hominids, despite the substantial differences in geological age and geographical distribution.

Brain size

The volume of the KNM-WT 15000 endocast is estimated to be about 880 cc (Table 14.2). This is slightly greater than the volumes obtained by Holloway (1983) for the morphologically very similar specimens KNM-ER 3733 and 3883, and lower than the estimate for OH 9 (Table 14.4). It must be noted that the value for KNM-ER 3883 is an underestimate (see Walker and Leakey, Chapter 6, this volume). Tobias (1971), in discussing the cranial capacity of OH 7, suggested a correction factor of about 3.5% to compensate for the fact that the specimen was not fully mature (its second molars were erupting at death). This correction is based on average values for percentage of completed cranial capacity growth at this dental age, regardless of actual chronological age, in great apes and humans (Tobias, 1971). If the same correction is applied to the KNM-WT 15000 endocast, which is also from an individual with recently erupted second molars, the revised endocranial volume is about 909 cc. This is about the same as the means for *Homo erectus* from East Africa and close to that of *Homo erectus* from Indonesia (Table 14.4). It should be noted, however, that there is extensive overlap in the ranges of variation and in standard deviations among the three samples of *Homo erectus* listed in Table 14.4. The uncorrected value for KNM-WT 15000 and the range of values obtained from each estimate of both reconstructions (Table 14.2) are all well within one standard deviation of both the Indonesian and East African samples. It can be concluded, then, that the endocranial volume of KNM-WT 15000 was fairly typical for *Homo erectus*.

As noted above, KNM-WT 15000 is unique in its association of a well-preserved cranium with a postcranium sufficiently complete to provide a reliable basis for estimating body size. Despite the fact that brain size to body size relationships have been widely discussed in the paleoanthropological literature, prior to the discovery of KNM-WT 15000, no pre–Upper Pleistocene specimen provides data from which relative brain size can be estimated with any accuracy. It is therefore necessary to compare the evidence for relative brain size in KNM-WT 15000 to previous interpretations of relative brain sizes that were based on comparisons of specimens from different individuals and even, in some cases, from different localities.

Estimates of body weight in *Homo erectus* are few in number because, until recently, the postcranial skeleton of this taxon has been very poorly known. Frequently cited estimates range from about 50 to 53 kg (Hofman, 1983; Martin, 1983, 1989; McHenry, 1982; Rightmire, 1986, 1990; Pilbeam and Gould, 1976), almost without exception based on Weidenreich's (1941) reconstructions of femora from Zhoukoudian. Weidenreich used reconstructed femoral length to estimate body weight. The femoral pieces from Zhoukoudian are very fragmentary, and none preserves any portion of the articular surfaces. Although Weidenreich's femoral reconstructions are relatively short, the shafts themselves are quite robust, especially transversely. The estimated midshaft transverse diameter of the Zhoukoudian specimens is very large, only slightly smaller that the transversely massive femoral shaft OH 28, usually also attributed to *Homo erectus* (Day, 1971). Recent work on estimating body weight has demonstrated that the cross-sectional properties of the femur in primates are strongly correlated with body weight, even in *Homo,* despite fundamental differences in patterns of limb loading (e.g., Ruff, 1990). Ruff (personal communication) has used a variety of techniques to arrive at an estimate of body weight for the fragmentary *Homo erectus* specimens from China. He reports that a reasonable estimate is 55 to 60 kg. Chapter 11 gives the methods and estimates of the weight at death and projected adult weight of KNM-WT 15000. With the accurate estimates of adult body weight at 68 kg (Ruff and Walker, 1991) and adult brain weight at 909 grams, the relative size of the KNM-WT 15000 brain can be estimated.

There are a number of techniques for calculating relative brain size, but perhaps the most widely cited is encephalization quotient (Jerison, 1973; Holloway, 1980, 1982; McHenry, 1982; Hofman, 1983; Radinsky, 1982; Falk, 1980), or its equivalent, the index of progression (Bauchot and Stephan, 1969). As Radinsky (1982), Holloway (1982), Martin (1983, 1989), Pagel and Harvey

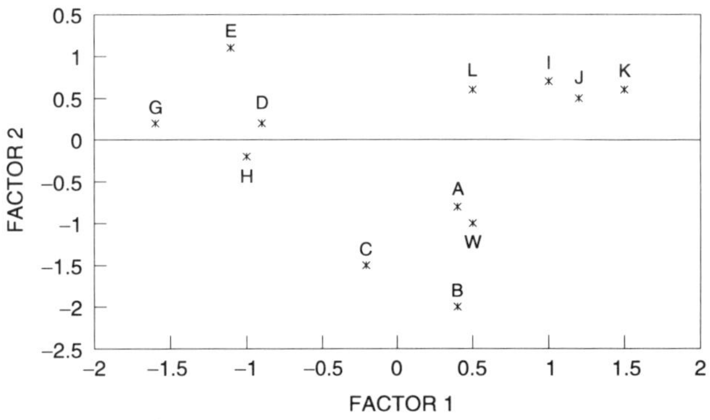

FIGURE 14.23.

Principal components analysis of the data from Table 14.3. Cord data were excluded from the final plot to include KNM-ER 1805, for which these dimensions are not available. When cord data are included, the association of specimens is identical but the first two factors explain slightly less of the variance. The cord data did not distinguish reliably among the endocasts in the bivariate analyses. Factor 1 explains 79.4% of the variance, and primarily reflects the influence of length and the height measurements, with somewhat less influence exerted by breadth and the frontal dimensions. Factor 2 explains 13.8% of the variance and primarily reflects the influence of occipital breadth. Abbreviations as in Figure 14.10.

TABLE 14.4.

Endocranial volumes for *Homo erectus* specimens.

SPECIMEN	*N*		VOLUME (cc)	GEOLOGIC AGE[a] (million years)
KNM-WT 15000	1		909 (adult estimate)	1.53
East Africa[b] (Holloway, 1983)	3	mean	910	1.7–1.2
		range	804–1067	
		S.D.	140.9	
Indonesia[c] (Holloway, 1983)	6	mean	931	1.2–0.7
		range	815–1059	
		S.D.	92.1	
China (Weidenreich, 1943)	9	mean	1001.1	0.9–0.5
		range	780–1225	
		S.D.	137.4	
All *Homo erectus*		mean	959.2	
		S.D.	121.2	

a. Dates from Feibel et al., 1989; Pope and Cronin, 1984; Hay, 1976; Liu and Ding, 1984; and Wu, 1985.

b. Excludes OH 12, the endocast volume of which is considered unreliable (Holloway, 1983).

c. Excludes Solo specimens.

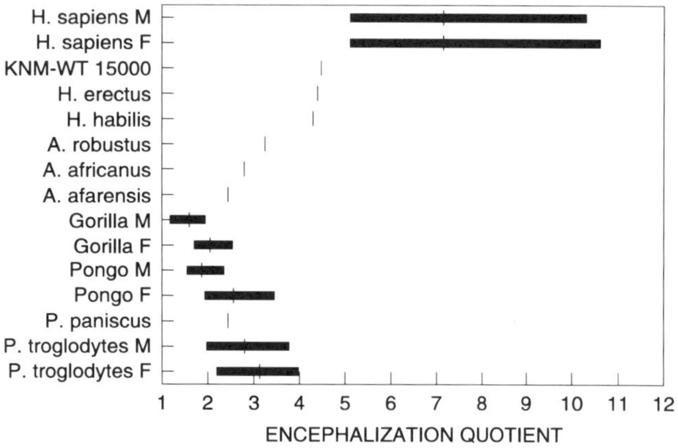

FIGURE 14.24.
E.Q. means and ranges for fossil and living hominoids, combined-sex samples unless otherwise indicated. Data from Table 14.5; E.Q. values determined by formula 1.

(1988), and Harvey (1988) note, the choice of formulas used to calculate E.Q. is somewhat arbitrary. It is important to be aware that each formula for calculating E.Q. is based on slightly to widely varying data sets and assumptions and that, for the most part, each formula produces a different E.Q. for a given set of new data. Deacon (1990) has recently called attention to several problems which he sees plaguing studies of brain-size evolution, and we are sympathetic to his point of view. He is critical of the use of encephalization, regardless of the way it is calculated, as an explanation for brain/body size allometry. We would

FIGURE 14.25.
Mean E.Q. values for extant great apes, humans, and fossil hominids using four different formulas (see Table 14.5).

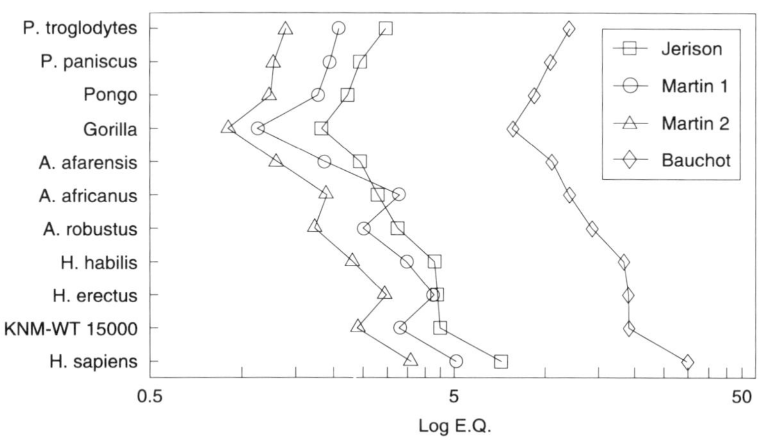

caution, however, against throwing the baby out with the bath water. Though Deacon (1990) is probably correct in pointing to the inadequacy of encephalization as an explanation, it remains a very useful descriptive tool. Encephalization does distinguish many fossil hominids from one another, and it distinguishes many fossil hominids from living great apes. The analysis of encephalization in fossil hominids reveals a pattern of change that can be described in some detail. In the discussion that follows, encephalization is treated as a continuously variable character, the distribution of which reveals a pattern of change, rather than as a process that accounts for these changes. The processes, both intrinsic and external, that have accounted for the fact of changes in encephalization in hominid evolution are for the most part beyond the scope of this analysis.

Four formulas are used to calculate E.Q. in Table 14.5 and Figures 14.24 and 14.25. Formulas 1 and 3 (Jerison, 1973; Martin, 1983, 1989) are the result of regressions of body weight and brain weight in placental mammals generally. Formula 2 (Bauchot and Stephan, 1969) is based on basal insectivores, which may account for its lower slope, and does account for the much larger departures from predicted brain sizes reflected in the larger E.Q.'s for all taxa. Formula 4 (Martin, 1983, 1989) is based on the regression of body weight and brain size in anthropoids. The most important difference in these formulas is the slope of the regression line, which is smaller in the taxonomically more restricted group (formula 4) and largest in formula 3 (Martin, 1983, 1989). A slope of about 0.75 across mammalian orders is currently considered the best estimate (Bauchot, 1978; Hofman, 1983; Martin, 1981, 1983, 1989; Martin and Harvey, 1985; Pagel and Harvey, 1988), as opposed to 0.67 (Jerison, 1973; Pilbeam and Gould, 1974). McHenry (1988) has suggested that regressions with lower slopes (derived from lower-level taxonomic units) are perhaps more relevant when analyzing relative brain size in closely related forms. However, Pagel and Harvey (1988) have shown that the reduction of the slope of regression lines in taxonomically more restricted groups results in most instances from error variation, and thus represents a statistical artefact. This may account for the lower slope in the basal insectivore and anthropoid regressions. Though it remains unclear what slope is best, very low slopes from low-level taxa (such as the 0.23 value proposed by McHenry, 1988) are probably suspect, and they are not included in this analysis.

McHenry (1988) calculates E.Q.'s for *Homo erectus* and other fossil hominids using new estimates of body size based on femoral shaft cross-sectional areas. His estimates for *Homo erectus* size are larger than previous estimates, and he suggests that on average *Homo erectus* may have been even larger. This is due to the fact that he chose to include a very small femoral specimen (KNM-ER 738) and exclude a much larger one (KNM-ER 736), though he gave no morphological criteria for such a taxonomic decision (a fact McHenry implicitly recognizes). Day (1986) considers KNM-ER 736 to be either *Australopithecus* or *Homo* and KNM-ER 738 to be *Australopithecus*. Jungers (1988) also places KNM-ER 738 in *Australopithecus (Paranthropus)* and estimates body weight in this individual at 54.8 kg, on the basis of femoral head dimensions. KNM-ER 736 is like several early *Homo* femoral shafts in being platymeric.

As can be seen in Figure 14.25, the comparison of mean E.Q. values is very similar from one taxon to the next regardless of the index used. The

TABLE 14.5.

Relative brain sizes (E.Q.) in some living and fossil hominid species.

HOMINID	E.Q. AS DETERMINED BY FORMULA— 1 (Jerison)	2 (Bauchot)	3 (Martin 1)	4 (Martin 2)
KNM-WT 15000 (68 kg)	4.48[a]	19.12[a]	3.28	2.39
Pan paniscus	2.44[b]	10.39[b]	1.94	1.27
P. troglodytes (♀ range/mean)	2.19–3.99/3.12[c]	9.00–15.97/12.65[c]	2.08	1.39
P. troglodytes (♂ range/mean)	1.97–3.78/2.80[c]	8.21–15.15/11.45[c]	—	—
Gorilla gorilla (♀ range/mean)	1.70–2.55/2.05[c]	7.21–10.58/8.59[c]	1.12	0.90
Gorilla gorilla (♂ range/mean)	1.17–1.96/1.60[c]	5.16–8.33/6.91[c]	—	—
Pongo pygmaeus (♀ range/mean)	1.93–3.46/2.56[c]	7.92–13.83/10.42[c]	1.77	1.24
Pongo pygmaeus (♂ range/mean)	1.54–2.36/1.87[c]	6.51–9.78/7.84[c]	—	—
A. afarensis	2.44[d]	10.52[e]	1.87[e]	1.30[e]
A. africanus	2.79[d]	11.98[e]	3.29	1.90
A. robustus	3.24[d]	14.25[e]	2.50[e]	1.73[e]
Homo habilis	4.30[a,d]	18.40[a,e]	3.47	2.32
Homo erectus	4.40[a,d]	19.09[a,e]	4.29	2.98
Homo sapiens (modern)	7.15[a,c]	29.80[a,e]	5.08	3.60

Formula 1[f]: E.Q. = observed cranial capacity/0.12 $BW^{0.67}$

Formula 2[f]: E.Q. = observed cranial capacity/0.0429 $BW^{0.63}$

Formula 3[f]: E.Q. = observed cranial capacity/$10^{0.76 \log_{10} BW + 1.77}$

Formula 4[f]: E.Q. = observed cranial capacity/$10^{0.60 \log_{10} BW + 2.68}$

a. Above the ranges of E.Q.'s for nonhuman anthropoids (Holloway, 1982).

b. Body weights and cranial capacities from Martin, 1983.

c. From Holloway, 1980.

d. From McHenry, 1988. The values for all but *Homo habilis* are substantially lower than previous estimates (McHenry, 1982), because of new, heavier body weight estimates for these forms. In contrast, *Homo habilis* has dropped 7.5 kg and now has a slightly higher E.Q. Even so, McHenry's new estimates are lighter than those of Jungers, 1988, which would yield somewhat smaller E.Q.'s for australopithecines.

e. Body weights and cranial capacities from McHenry, 1988. *A. africanus* has absolutely higher E.Q.'s according to the Martin equations because Martin (1983, 1989) uses a lower body weight than McHenry.

f. E.Q. = encephalization quotient; *BW* = body weight (kg). Formula 1 from Jerison, 1973; formula 2 from Bauchot and Stephan, 1969; formulas 3 and 4 from Martin. 1983, 1989.

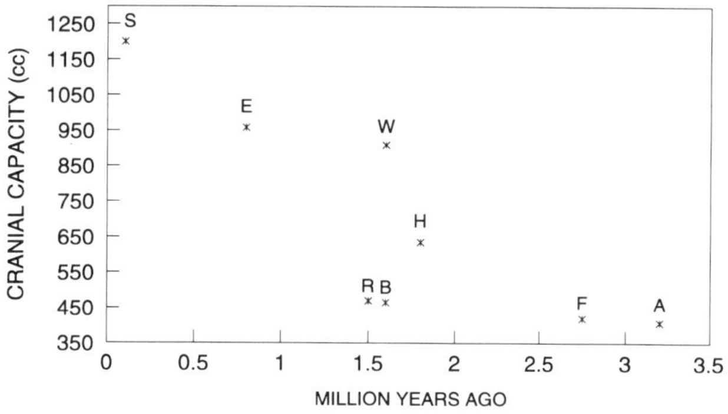

FIGURE 14.26.

Bivariate plot of cranial capacity (in cubic centimeters) and approximate midpoint in the temporal ranges (million years before present) of fossil hominids. The apparent difference between the "relative" brain size of HOMO HABILIS *and* HOMO ERECTUS *disappears when the values are adjusted (by calculating the E.Q.; see Figure 14.27).*

A = AUSTRALOPITHECUS AFARENSIS
B = AUSTRALOPITHECUS BOISEI
E = HOMO ERECTUS
F = AUSTRALOPITHECUS AFRICANUS
H = HOMO HABILIS
R = AUSTRALOPITHECUS ROBUSTUS
S = HOMO SAPIENS
W = *KNM-WT 15000*

only exception appears to be for the relationship between *A. robustus* and *A. africanus,* where *A. africanus* appears to have a relatively larger brain than *A. robustus* according to Martin's data. However, this is an artefact of the choice of body weight estimates. Martin's estimate for *A. africanus* body weight is considerably lower than more recent estimates (McHenry, 1988; Jungers, 1988), which has the effect of elevating relative brain size. This discrepancy illustrates the problem of comparing E.Q. values for fossil and even recent taxa without considering both the formula used to calculate the index and the data used to estimate body weight and cranial capacity. Apart from this minor inconsistency, the pattern is consistent. By any formula, modern non-human hominoid E.Q. means are inversely proportional to their body size means, though there is extensive overlap in the ranges of variation. All three australopithecines are close to the mean for chimps, though, in contrast to the living great apes, their E.Q.'s are directly proportional to their estimated body size. Australopithecine E.Q.'s are all very similar and are probably not significantly different from one another, given the variability in E.Q.'s measured in modern hominoids (Holloway, 1980). E.Q.'s for Plio-Pleistocene *Homo* are higher than in australopithecines but, interestingly, there does not appear to be a difference in encephalization between *Homo habilis* and *Homo erectus.* The values for *Homo erectus* are based on data from McHenry (1988) and Martin (1983, 1989) and do not include the data from KNM-WT 15000. In sum, the E.Q. estimates for KNM-WT 15000, the only *Homo erectus* specimen with associated cranial and postcranial parts complete enough to provide brain and body weight estimates, tend to confirm previous E.Q. estimates based on isolated specimens.

The results for *Homo habilis* are difficult to interpret. There may be a bias favoring higher E.Q.'s for *Homo habilis.* Most of the specimens included in McHenry's *Homo habilis* group, including all of the larger postcranials, are unassociated cranial and postcranial specimens from East Turkana and may belong to one or more taxa other than *Homo habilis.* If *Homo habilis* in fact was closer in mean body size to OH 62, the smallest individual included in McHenry's *Homo habilis* group, then the E.Q. for this taxon may be higher. On the other hand, if *Homo habilis* were closer in mean brain size to the specimens from Olduvai and body size close to McHenry's estimate, then mean E.Q. may have been somewhat lower. However, given the fact that the endocranial volumes of *Homo habilis* from Olduvai are larger than in australopithecines (Holloway, 1982; Tobias, 1971), the E.Q.'s of *Homo habilis* are likely to have been within the range of variation of *Homo erectus,* the E.Q.'s of which are not much larger than those of australopithecines.

It is very important to remember that species E.Q.'s are very variable both within and between sexes (Table 14.5, Figure 14.25). Using species means made by averaging average male and female body weight and cranial capacity, though perhaps they are comparable to samples from the fossil record, has the effect of masking marked sexual dimorphism in E.Q. in living great apes. The Jerison E.Q. formula, for instance, results in average male E.Q.'s of only 73% of the female average in *Pongo,* 78% in *Gorilla,* and 90% in *Pan* (Table 14.5). Ranges between the sexes are very large, with individuals (*Pongo* females) up to 225%

more encephalized than others (*Pongo* males). Even ranges within sexes are large, with some male *Pan* being nearly twice as encephalized as others (Table 14.5).

The pattern of change in relative brain size through time, when body size, brain size, and ranges of variation are considered together, differs from the pattern of more or less continuous increases in cranial capacity from *Australopithecus* through *Homo habilis-erectus-sapiens* that is observed when brain size is considered alone (Holloway, 1982; Falk, 1987; Figures 14.26 and 14.27). There is no demonstrable change in E.Q. from *H. habilis* to *H. erectus* (Figures 14.24 and 14.27), despite a substantial increase in cranial capacity (Figure 14.26), which is thus misleading when considered alone. In addition, the pattern of change seems to differ depending both on whether E.Q. or actual volumes are used.

It is also interesting to note the spread of data within *Homo erectus,* as represented by KNM-WT 15000 *(W)* and other *Homo erectus (E).* Figure 14.28 shows the variation of endocranial volumes in *Homo erectus* through time. Variability is high throughout the temporal range. (We have not analyzed this data set by formal regression methods because the number of data points is small and the distribution along the *X*-axis is too regular.) In contrast, Rightmire (1985), who provides a linear regression of a similar data set, concludes that there might be a single, slight trend toward increasing brain size but that the slope of the regression line is not significantly different from zero (no change). The results presented here may be related to missing data between the bulk of the Asian and African samples, or they may be a reflection of some independence in brain size evolution in *Homo erectus* from Africa on the one hand and in *Homo erectus* from Asia on the other. The data do not obviously support either an interpretation of gradual evolutionary change or of punctuated change in brain size in this taxon.

The ranges of variation in E.Q. for early *Homo* are likely to have overlapped with the ranges of variation in *Homo sapiens* (Figure 14.24), especially if fossils commonly attributed to "archaic *Homo sapiens*" are included in the *Homo sapiens* range. There does appear to be some increase in

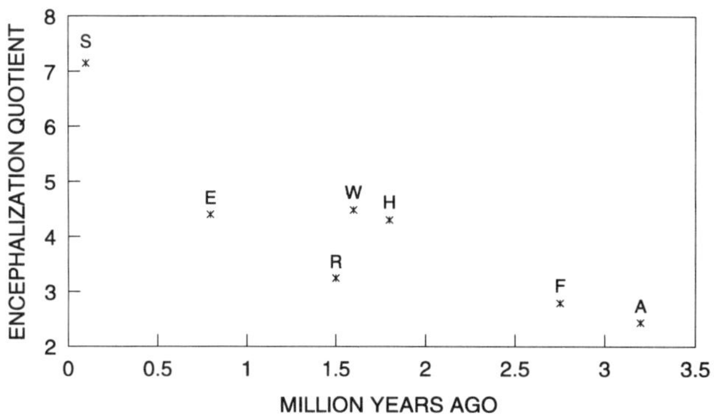

FIGURE 14.27.
Bivariate plot of E.Q. and approximate midpoint in the temporal ranges (million years before present) of fossil hominids. Abbreviations as in Figure 14.26.

E.Q. between *Australopithecus* and *Homo habilis,* though there was probably overlap in their ranges, and between the ranges for *Homo habilis* and modern great apes. According to the Jerison regression, *Homo habilis,* with an average cranial capacity of 642 cc, would have had to have been over 60 kg on average to have had average E.Q.'s as small as the mean in the australopithecines with the largest E.Q.'s. KNM-WT 15000 would have had to have weighed over 110 kg to achieve the same effect.

When both brain size and body size are considered together, trends toward increasing

FIGURE 14.28.
Bivariate plot of endocranial volumes (cubic centimeters) and approximate geological age (million years before present) of HOMO ERECTUS *specimens from East Africa (OH 9, KNM-ER 3883, KNM-ER 3733, KNM-WT 15000), China (Lantian, Zhoukoudian), and Indonesia. Data from Table 14.4 and references cited therein.*

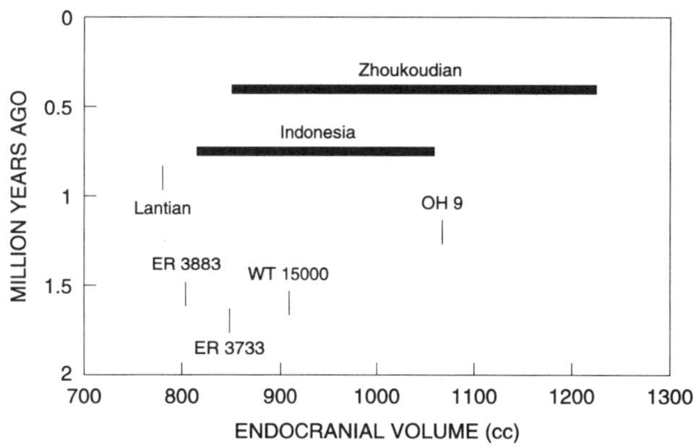

brain size are less marked than when brain size is considered alone. However, a somewhat different trend in brain size change does emerge from this analysis. Among living genera of great apes, increasing body size is associated with decreasing E.Q., that is, decreasing amounts to which larger hominoid taxa depart from an expected relative brain size for mammals (Jerison, 1973; Martin, 1983, 1989), basal insectivores (Bauchot and Stephan, 1969), or anthropoids (Martin, 1983, 1989) of the same size. Within *Australopithecus*, however, the larger (and later) species have larger mean E.Q.'s. Between *Australopithecus* and *Homo*, the latter, which are on average larger, also have larger E.Q.'s. Within the genus *Pan*, *P. troglodytes*, which on average is larger, also has a mean E.Q. larger than *P. paniscus*. However, the E.Q. of *Pan paniscus* is based on a very small number of specimens and may not be representative of the taxon. In *Homo*, despite possibly substantial increases in body weight from *H. habilis* to *H. erectus*, there is relatively little or no change in E.Q.

It is possible that the increasing E.Q.'s of increasingly large and later australopithecines is an artefact resulting from very small sample sizes, but it could also be that the pattern from *Australopithecus* through to modern *Homo* is real and that it represents a departure from the pattern seen in living hominoids, as previously noted, using different criteria, by Pilbeam and Gould (1976). The pattern of change in relative brain size seen among great apes is predictable from the fact that brain size scales negatively with respect to body size (Pilbeam and Gould, 1976; Martin, 1983). Relatively larger great apes should have relative brain sizes that depart less from more general mammalian or primate baselines. That relative brain size in *Homo* and possibly in *Australopithecus* expands with increased body size probably indicates strong selective pressure for larger relative brain size despite increasing body size. Brain expansion therefore cannot be seen as a consequence of body size increases. Even the apparent absence of relative brain expansion between *Homo erectus* and *Homo habilis* may represent evidence of selection for increasing brain size, given the fact that *Homo erectus* is bigger than *Homo habilis*, probably, and could be expected to have had both a relatively smaller brain and a relatively smaller E.Q., following the hominoid pattern. Incidentally, the hypothesis of selection for increased brain size is testable, with more complete fossils of early *Homo* with associated brain and body sizes, though interpretations of the precise mechanisms or set of selection pressures will probably remain speculative.

Asymmetry and Handedness

The development of X-ray computerized tomography (CT) has led to an easy and accurate assessment of cranial asymmetries. Some of these, such as asymmetry of the ventricles, are not of much interest for paleontologists, but hemispheric asymmetry has lately assumed some importance.

Following the observations of anatomists who for nearly a century had pointed out that human cerebral hemispheres are often asymmetrical, LeMay (1976) used CT to look at the brains of living subjects. Pooling the separate sex samples (Table 14.6 lists the samples separately), LeMay found that 62.5% of 264 brains had symmetrical frontal projections (fronto-petalia) while 27.7% had right fronto-petalia and only 9.8% had left fronto-petalia. Symmetrical occipital projections (occipito-petalia) were found in 22.7%, left occipito-petalia in 66.5%, and right occipito-petalia in 10.9% of 331 brains. Symmetrical petalias occurred at similar frequencies in right- and left-handed patients, while asymmetrical petalia, though less striking in their appearance among left-handed people, differed in frequency between the groups. Right fronto-petalia were over four times more frequent than left fronto-petalia among right-handers, while left and right fronto-petalia were very similar in frequency (20.7% and 19%, respectively) among left-handers. Occipito-petalias were more similar between the two groups. Although in both groups left occipito-petalia occurred most often, right occipito-petalia occurred about twice as frequently among left-handers. However, the sample sizes for left-handers are only 23% to 28% of those for right-handers, which may account for some of the differences in frequencies. Nevertheless, these data tend to confirm early studies that called attention to the common co-occurrence of left occipito- and right fronto-petalia.

Eliot Smith (1908) speculated that occipital asymmetries were associated with handedness, but LeMay (1976) found that the width of the frontal and occipital is more closely related to handedness than was the length or projection. Seventy percent of right-handers have wider right hemispheres in their frontal portions, and 64% have wider left hemispheres in their occipital portions. Left-handers are less asymmetrical, with a maximum of 46% having some identifiable asymmetry (wider right occipital portion). KNM-WT 15000, with right fronto-petalia, shares a condition found twice as often among right-handers as among left-handers (30.1% and 19%, respectively). The probably nearly symmetrical occipito-petalia seen on KNM-WT 15000 (uncertain because of slight distortion) occurs in roughly equal frequencies among right- and left-handers (22.3% and 24.2%, respectively) but is slightly more common in right-handed males than in left-handed males (Table 14.6). KNM-WT 15000 has a slightly wider right hemisphere measured in the area of the frontal, though probably not in exactly the same plane as in LeMay (1976). Digitized video images of Reconstruction #2 viewed dorsally and ventrally correspond approximately to the transverse plane incorporating the largest surface areas of the frontal lobes (roughly parallel and just dorsal to the orbital surfaces). The left frontal surface area was 87.6% of the right in dorsal view and 87.2% of the right in ventral view. The breadth maxima were between 7.1 and 8.6% smaller on the left side as well. The right occipital may also have been wider, but this measurement is more difficult to make in an endocast and is also subject to some uncertainty due to distortion. Wider right frontals occurred in 70% of the right-handers and 39% of the left-handers, while wider right occipitals occurred in only 16.4% of right-handers (compared with 64% with wider left occipitals) and 46% of left-handers (compared with 22% with wider left occipitals) (LeMay, 1976). Thus, KNM-WT 15000's wider frontal is both more common in right-handers than left-handers and is the most common state among right-handers, while its right fronto-petalia is also more common in right-handers but is not the most common character state (equal fronto-petalia is). The possibly equal occipito-petalia of KNM-WT 15000 is more often found among male right-handers than male left-handers, though left occipito-petalia is most common among right-handers. All in all, if the relationship between hemispheric asymmetry and handedness in *Homo erectus* was similar to that seen in modern humans, then the endocast

TABLE 14.6.

Cerebral hemispheric asymmetries in modern humans (percent of group) and in KNM-WT 15000.

	LEFT-HANDED MALES		LEFT-HANDED FEMALES		RIGHT-HANDED MALES		RIGHT-HANDED FEMALES		KNM-WT 15000
	%	*n*	%	*n*	%	*n*	%	*n*	
Wider right frontal	46.7	14	26.3	5	65.9*	54	73.9	68	yes
Right fronto-petalia	18.2	6	20.0	5	37.4*	37	23.4	25	yes
Equal occipito-petalia	20.0	7	29.6	8	24.8*	33	19.9	27	yes
Left fronto-petalia	24.2	8	16.0	4	9.1	9	4.7	5	no
Equal fronto-petalia	57.6	19	64.0	16	53.4	53	72.0	77	no
Wider left frontal	16.7	5	26.3	5	13.4	11	4.3	4	no
Equal frontal width	36.7	11	47.4	9	20.7	17	21.7	20	no
Left occipito-petalia	62.9	22	48.1	13	67.7	90	69.9	95	no
Right occipito-petalia	17.1	6	22.2	6	7.5	10	10.3	14	no

Note: Modern human data from LeMay, 1976.

*The characters shared with WT-15000, when present, are found most frequently in right-handed males

evidence indicates that KNM-WT 15000 was more likely to have been right-handed (Table 14.6).

Holloway and De La Coste-Lareymondie (1982) have made the most extensive study of endocasts of great apes and fossil hominids. They conclude that human and fossil hominid petalial patterns are similar (indeed they are not significantly different from each other) and, further, that they are significantly different from those of great apes. They suggest, but say they cannot prove, that the common petalial pattern in hominids means that most were right-handed, which is also suggested by the analysis presented here and by Toth (1985) in his work on the archeological evidence.

In their speculative discussion, Holloway and De La Coste-Lareymondie (1982) point out that behaviors potentially important to early hominids, such as increased communicative skills, life in relatively permanent home bases, reliance on stone tools made to standardized patterns, and development of throwing abilities would together provide appropriate selection pressures to produce cerebral asymmetries. Note, however, that they found the hominid petalial pattern was present in their sample of earlier, small-brained hominids too. In fact, the earliest hominids known probably had this petalial condition, since robust australopithecines from both east and south Africa had it and there are even hints of it in the earlier sample from Hadar. That these were tool-using, tool-throwing hominids that regularly communicated with each other in a social setting is not a widely shared view. It is not very clear to what extent these skills were developed in *Homo erectus* or to what extent these skills are related to cerebral asymmetries. Finally, similar asymmetries in petalia and hemisphere size are also found in a number of different non-human primates, though less frequently and usually less well developed (LeMay, 1976).

In any case, for whatever reasons, it seems that hominids had cerebral asymmetries and that these developed much earlier than the first appearance of well-developed Broca's caps (see below), which are not clearly present in any specimen until early *Homo* (Falk, 1983, 1987; Holloway, 1983). That generalized cerebral asymmetries and a specific asymmetry such as the development of Broca's cap are uncoupled in the fossil record suggests that the functional significance of these patterns is complex. The function of general hemispheric asymmetries in *Australopithecus* probably differed from that in modern humans, at least in detail, because they are not associated with more specific asymmetries (like Broca's cap) that seem to work together with general hemispheric asymmetries in generating certain characteristics of human neural function. It seems more likely that cerebral asymmetries in fossil hominids may be functionally similar to modern human asymmetries only at a very basic level, involving primary functions such as motor sequencing or task specialization, rather than at the level of specific behaviors such as throwing or language production (see below).

Signs of Language Capability

Several authors have pointed out that some early hominid endocasts have well-developed areas corresponding to those involved with the control of speech in humans (including this analysis, see above). Tobias (1975, 1981, 1983) has claimed that both Olduvai Hominid 24 and KNM-ER 1470 have "striking development of the inferior parietal lobule . . . together with the strong development of the posterior part of the inferior frontal convolution." Fossil hominid crania are often distorted in some way and so claims for strong development of Wernicke's area must be accepted with caution. Olduvai Hominid 24 was squashed flat when found (Leakey, Clarke, and Leakey, 1971), and the region of the skull overlying the region of the left inferior parietal lobule is not even preserved. KNM-ER 1470 is also distorted (Walker, 1980), but more subtly than is Olduvai Hominid 24. When it comes to recognizing a developed Broca's cap, however, there is little ambiguity. As Tobias (1975) reported, there are only hints of such a development in *Australopithecus*. Falk (1983), without mentioning Broca's cap in particular, considers the South African *Australopithecus* specimens to be like great apes in their frontal lobes. KNM-ER 1470 does have a well-developed Broca's cap (Tobias, 1975; Falk, 1983; Holloway, 1983), but it must also be remembered, as Holloway (1983) has pointed out, that Broca's caps can be found on chimpanzee endocasts. Both Tobias (1983) and Holloway (1983) see the presence of Broca's cap as indicating that there was the possibility for speech in *Homo habilis* as represented by KNM-ER 1470. They both are equally cautious in saying that the evidence of Broca's caps does not prove that language was spoken by early hominids 2 million years ago, but that the evidence is "strongly suggestive" (Tobias) and "corroborative" (Holloway) of some sort of speech (see above).

The work that established that certain regions of the human cerebral cortex were involved in the process of language and speaking and understanding speech was based on patients with traumatic or vascular lesions. This technique still has much to offer (Caramazza and Hillis, 1991), but spectacular confirmation of the importance of areas such as those of Wernicke and Broca has come from Marcus Raichle's group at Washington University. This group used positron emission tomography to visualize the regions of the brain involved in language (Peterson et al., 1988, 1989). This method measures the blood flow to regions of the brain in normal, conscious individuals while they perform tasks involving language. Blood flow is measured rather than neuronal activity because it is easier to do so and because the response by the brain's vasculature to increased local neuronal activity is practically immediate.

The researchers found that speech production activated regions of the left inferior frontal lobe around Broca's area. They concluded, however, that rather than being associated specifically with linguistic or speech functions, these regions are more plausibly associated with some higher-level control of motor programming or action. They cite two reasons for coming to this conclusion. First, simple tongue movements and actual (and also imagined!) hand movements also activated these areas and, second, recent reviews showed that lesions confined exactly to Broca's area produce only motor or praxis defects and not the full-blown symptoms of Broca's aphasia. Much larger lesions in the same area are needed to produce those. So it seems that, as a result of this exciting new work, the development of a Broca's cap may not, after all, be associated exclusively with language but rather with motor programming or praxis. At the most, it suggests that selection for a reorganizing of this region of the cerebral cortex, if it did occur in early hominid populations about 2 million years ago, was only partly or perhaps indirectly related to language capability. And of course, as Tobias and Holloway have pointed out, it is probably no coincidence that this apparent selection occurred at the very time of the development of the first stone tool culture.

Development

Martin (1983) discusses the relationship between adult and neonate brain weights in mammals and provides a regression of these variables in primates (Martin, 1983, p. 19, figure 7). Humans fall substantially above the regression line. This is a reflection of the well-established fact that human postnatal brain growth is relatively greater than in non-human primates due to continued growth at fetal rates (Portmann, 1941; Leutenegger, 1982; Martin, 1983, 1989; Passingham, 1985). The adult brain size of KNM-WT 15000 can be used to predict a value of about 388 grams for its neonatal brain weight (calculated using formula 13 in Martin, 1983, p. 18). The value of 388 grams is well within the range of modern human full-term neonates, despite the KNM-WT 15000 adult cranial capacity being much smaller than those of modern humans. This effect is due to the fact that humans have much larger adult brain sizes than predicted by a regression based on primate values, because of the expanded period of brain growth in humans. Yet the pelvic inlet in KNM-WT 15000 is considerably smaller in absolute dimensions than it is in adult modern human males or females, even those of smaller body size (see Walker and Ruff, Chapter 10, this volume). Pelvic inlet size is not very sexually dimorphic in humans (Martin, 1928). In addition, females of the same population from which KNM-WT 15000 came must have been subjected to the same climate-induced pressures for reduction in transverse dimensions of the body suggested by Ruff (1991, and Chapter 11, this volume) to account for the narrow body build of KNM-WT 15000. Thus, we can conclude that a human-sized neonate head could not have passed through a birth canal the size of that of a female conspecific of KNM-WT 15000. If the size of a *Homo erectus* neonate's brain is smaller than would be predicted by comparison with a large sample of primates, then it is similar in this regard to modern humans. Humans are born with brains about a quarter the size of adult brains, whereas among primates as a whole the neonate brain size is about half that of the adults (Passingham, 1985). The comparatively small estimated size of the neonate brain of *Homo erectus* suggests that there must have been both a reduction in neonatal brain growth advancement and a period of accelerated brain growth relative to non-human primate postnatal norms in *Homo erectus,* as is found in *Homo sapiens,* or that brain growth continued for relatively longer postnatally in *Homo erectus,* which seems less likely given other developmental evidence (see Smith, Chapter 9, this volume).

The pattern of relatively incomplete brain growth at birth coupled with a period of accelerated brain growth in *Homo erectus* infants implies a secondary altriciality (Portmann, 1941; Martin, 1989). "Primary" altriciality, in which infants are born at a comparatively poor level of development, is usually associated in mammals with relatively large litters, small neonates, and small adult brain sizes (Martin, 1983). On the other hand, precocial mammalian neonates are usually born in small litters, are usually relatively large and well developed at birth, and grow into adults with relatively large brains. The concept of secondary altriciality (Portmann, 1941) is a recognition of the fact that although human neonates have many characteristics of precocial neonates (small litters, very large neonates, big adult brains), they also have characters usually associated with altriciality (incomplete development). In most mammals, altricial neonates never "catch up" to their precocial counterparts in adult brain size, despite the often sheltered conditions of their early postnatal lives (Martin, 1983). This is not the case for humans. Human neonates are not born "behind" other precocial primates. Human neonate brain size relative to body size falls very close to the regression line for primates generally (Martin, 1983), but human brains continue to develop at embryonic rates for a year after birth (Martin, 1983). Human neonate brains are relatively small compared with adult humans but not compared with primate neonates. Human neonates are comparatively helpless not because they are born with small brains but because they have a year's worth of growth left before they have human brains that can function at a level equivalent to other primate neonate brains. That is, humans have the equivalent of about a 21-month gestation period (Portmann, 1941; Martin, 1983). The prolonged period of infant dependency that is a direct result of this postnatal neurologic gestation strongly suggests a period of intensive maternal or parental investment and a development of the social and cognitive strategies that follow from it.

Gould (1977) suggests that neoteny as a process may account for the expanded rate and period of human postnatal brain growth. Human neonates are said to retain fetal growth rates later into ontogeny. On the other hand, this does not account for the fact that human neonates are already very large compared with maternal body weight (Martin, 1983). It is possible that growth and development is accelerated rather than delayed in human fetuses, producing giant fetuses proportionally similar, at least in brain to body size, to other primates. Birth may occur earlier than would be expected on the basis of degree of neurologic development because of constraints imposed on mothers related to the metabolic demands of huge fetuses and the mechanical problems associated with birthing giant neonates (Deacon, 1990). The possible advantages that may be associated with extending neonatal brain growth rates beyond birth may be secondary to these more structural influences (Deacon, 1990). Accelerated rates of embryological growth and development coupled with neotenic retention of these rates postnatally may together be necessary to achieve the adult brain sizes characteristic, for whatever reason, of modern humans and *Homo erectus.*

Summary and Conclusions

The remarkable degree of completeness of KNM-WT 15000 allows for a thorough examination of a number of phenomena related to the evolution of the brain in humans. Morphologically, endocasts of *Homo erectus* appear to cluster together in most dimensions, indices, multivariate analyses, and discrete anatomical characters, suggesting a single taxonomic unit (though perhaps above the species level) distinct from early *Homo.* There is no easily identified temporal component to variation in *Homo erectus* brains, with the earliest specimen (KNM-WT 15000) in fact sharing affinities with both early (East African) and late (Chinese) *Homo erectus.*

For the first time, we can confidently talk about degree of encephalization in a pre-*sapiens* fossil hominid, though we still do not know why hominids are encephalized. Though actual E.Q.'s are relatively small in australopithecines and overlap with the ranges of variation of extant great apes, they appear to increase over time despite increases in overall body size. E.Q.'s increase more substantially with early *Homo,* though there is no demonstrable difference between early *Homo* and *Homo erectus.* More substantial increases in E.Q., and more obvious cerebral asymmetries (petalia, Broca's cap) occur with the earliest *Homo,* along with the earliest appearance of unambiguous stone tools. That enlargement of Broca's cap, or of the area in its immediate vicinity, should occur at the same time that stone tool assemblages with complex typologies appear in large numbers is suggestive of some sort of causal linkage. This is especially true given recent findings of the involvement of Broca's cap in hand movements and motor programming and earlier observations linking regions adjacent to Broca's to sequencing motor control in the hand (Falk, 1987; Passingham, 1981). Elaborate control over hand movements and their sequence, and the repeatability of these sequences, has obvious implications for the manufacture of standardized stone tools whether or not a concept or image of a tool type was involved (Holloway, 1969).

It is also probably no coincidence that Broca's cap is involved in movements of the tongue and that it lies adjacent to the inferior portion of the somatomotor area that is involved in movements of the vocal apparatus (muscles of facial expression and mastication, the tongue, and muscles of the pharynx and larynx). Hand movement is the link between lithic assemblages and Broca's area. The development of elaborate control of the hand, as evidenced by lithic assemblages and possibly by the presence of Broca's cap in early *Homo,* may have occurred at the same time as the development of neural control mechanisms in the same part of the brain that are today associated with human language production. It may be that the language function of this region was co-opted much later, after the appearance of other structures necessary for the production of speech (other cortical areas, possible changes in upper respiratory tract configuration, etc.). It may be that language occurred first. There is no way to know at present.

Finally, for the first time we have more direct evidence for a period of delayed maturation and a retention of fetal growth rates, at least for the brain, into early postnatal ontogeny in *Homo erectus.* This is in basic agreement with the conclusions of Mann (1975) based on dental maturation, though he was talking about *Australopithecus.* It is not clear to what extent the australopithecines may have shared this pattern of growth and development with *Homo sapiens* (Smith, 1986; Mann et al., 1987), but it does seem to have been present by *Homo erectus* times.

It is a paradox that reconstructions of the most interesting aspects of fossil hominid biology require the greatest number of assumptions and speculations. For most fossil specimens the number of assumptions required to generate plausible speculations are very numerous because even beautifully preserved crania are biologically incomplete in isolation. KNM-WT 15000 allows us to test many previous assumptions regarding relative brain size and developmental biology because it is so much more complete. Some previous assumptions turn out to have been well founded. For example, hypotheses about relative brain size in

Homo erectus (McHenry, 1988) or developmental delay in fossil hominids (Mann, 1975; Gould, 1977) have not been falsified by this specimen, the only one currently known that can be used to test them. Others, notably those about increased encephalization from *Homo habilis* to *Homo erectus* (Falk, 1987; Martin, 1983; Tobias, 1971), are less well accommodated by the evidence presented here. In the final analysis, we feel that we know a great deal more about the paleoneurology of *Homo erectus* than was possible before the discovery of KNM-WT 15000. Much remains to be understood, but we hope that more complete specimens from other Plio-Pleistocene hominids will serve to test the hypotheses presented here further and to develop new hypotheses concerning the patterns and processes of hominid brain evolution.

Acknowledgments

We thank the Government of Kenya and the Governors of the National Museums of Kenya. D.B. was funded by a Postdoctoral Fellowship, Department of Cell Biology and Anatomy, The Johns Hopkins University, and by a grant from NSERC. A.W. received fellowships from the John Simon Guggenheim Memorial Foundation and the John D. and Catherine T. MacArthur Foundation. We thank B. Brown for the illustrations of Reconstruction #1 and K. L. Rafferty for helping with the volumetric determinations.

CHAPTER 15

THE VERTEBRAL CANAL

ANN MACLARNON

Because the vertebral column of the Nariokotome skeleton is nearly complete, the size and shape of the vertebral canal can be determined and its features compared with those of vertebral canals in other primates. Human canals exhibit anatomical peculiarities not attributable to body size. Early Homo erectus *seems to have had some of these human features and yet not others. The reasons for this are at present speculative, but may have a direct bearing on the origin of some distinctive human traits.*

In their original description of the *Homo erectus* skeleton, KNM-WT 15000, Brown et al. (1985) describe the vertebral canal through the lower cervical and thoracic vertebrae as being relatively small compared with that of modern humans, while the lumbar canal is of about the same relative size. This, it has been pointed out (Leakey and Walker, 1987), implies that the fossil species also had a relatively small spinal cord compared with that of modern humans, as might have been expected in view of its relatively small brain, the other part of the highly integrated central nervous system.

In an earlier study (MacLarnon, 1987) of the variation of vertebral canal and spinal cord dimensions in modern and fossil primates, a number of differences were identified in the cord and canal of humans compared with those of other modern primate species, mainly in the lumbar and sacral regions, although in general results showed little variation in cord and canal dimensions among all modern primates. Analysis of a wider sample of modern mammals also showed that there is little variation in total spinal cord size relative to body size across the whole class, and that the relative total spinal cord size of modern humans is typical for modern mammals. (Some of these analyses are presented here.) In the same study, however, analysis of the relative vertebral canal dimensions of two Eocene primate species showed that they had relatively small cervical and thoracic canals compared with modern primates while their lumbar canals were of similar relative size. It was shown that the differences in the cranial canal dimensions between the fossil and modern primate species were great enough that a spinal cord of modern relative dimensions could not have fitted through the canal of at least some Eocene primate species.

In the light of this previous study of mammal spinal cords and primate cords and canals, the findings of Brown et al. (1985) were unexpected in suggesting a relatively small lower cervical and thoracic spinal cord in a hominid fossil compared with that of modern humans. The previous comparisons, both across a wide range of modern mammals and within modern primates, had indicated very little variation in cord dimensions between modern humans and other species, and that what differences there are in the canal dimensions of humans are most marked in the lumbar and sacral regions and not the thoracic region. During primate evolution as a whole, the cervical and thoracic cord apparently increased, but the similarities between all modern mammals, and particularly between all modern primates, imply that what evolutionary increase in cord dimensions took place, took place much earlier than the last few million years of separate hominid evolution.

Here the relative dimensions of the vertebral canal of KNM-WT 15000 are investigated further and compared with those of a wider range of modern primate species and with an extended sample of modern humans, including juveniles. KNM-WT 15000 is estimated (Smith, Chapter 9, this volume) to have been at an equivalent stage of development to an 11-year-old modern human, and so it is possible that its apparently small vertebral canal dimensions are the result of its juvenile status.

Materials and Methods

Vertebral canal data

For a previous study (MacLarnon, 1987), measurements of the vertebral canal dimensions of 45 adult modern primates were taken, comprising 38 different species (7 lemur species, 6 lorises, 9 New World monkeys, 10 Old World monkeys, 5 apes, and humans), from the collections of The Natural History Museum, London, and the Department of Anthropology, University College London. This sample was increased for the present study. A further 7 ape specimens from The Natural History Museum, London, were measured, giving a total of 2 gibbons, 2 orangutans, 4 chimpanzees, and 4 gorillas. Sixteen modern human specimens from the Spitalfields' collection at The Natural History Museum, London, were measured; these specimens are of known age and hence provided a cross-sectional sample for analysis of the growth of canal dimensions. They ranged in age from 1 year 3 months, when the neural arches and centra of vertebrae are not yet fused, to 37 years, and included 9 specimens under the age of 18 years.

In addition, measurements of the vertebral canal dimensions of modern humans, both juvenile and adult, were collected from a range of published sources. These are mostly from the cervical and lumbar regions of the vertebral column. The studies were usually undertaken to provide baseline data from normal individuals for the interpretation and investigation of the influence of canal dimensions on various clinical conditions. Many of these studies involved measurements from radiographs, one used computerized tomography, and only a few took measurements from skeletons. In only a limited number of the papers (Larsen, 1981; Larsen and Smith, 1981) were the radiographic measurements clearly corrected for magnification errors. In particular, for juveniles, Larsen and Smith (1981) show that previous authors overestimated the growth of vertebral dimensions because they did not correct for the differential magnification of the dimensions of larger and smaller subjects. In addition, the definitions of canal height and width used in previous studies are not always identical with those of the present study, nor even clearly explained. For all these reasons, much of the published data is not absolutely directly comparable with the data collected for the present study. However, it provides a large and wide-ranging sample of modern human data. For this reason, all available published data are included in analyses here, in spite of the problems of direct comparability.

Two dimensions of the vertebral canal are analyzed here. These are the height (dorso-ventral diameter) and the width (transverse diameter) of the vertebral canal in each vertebra. The height and width of the canal are defined relative to the co-ordinate system shown in Figure 15.1A. They are the maximum height and width of the silhouette of the canal as seen from its cranial end (see Figure 15.1B). For the original study (MacLarnon, 1987), measurements were taken using a Reflex microscope or Reflex metrograph (see MacLarnon, 1989). The extra specimens added here, including KNM-WT 15000 itself, were measured using vernier calipers, as only a simpler set of measurements and larger species were involved. Both the canal height and width were measured with the cranial end of the vertebra facing the measurer.

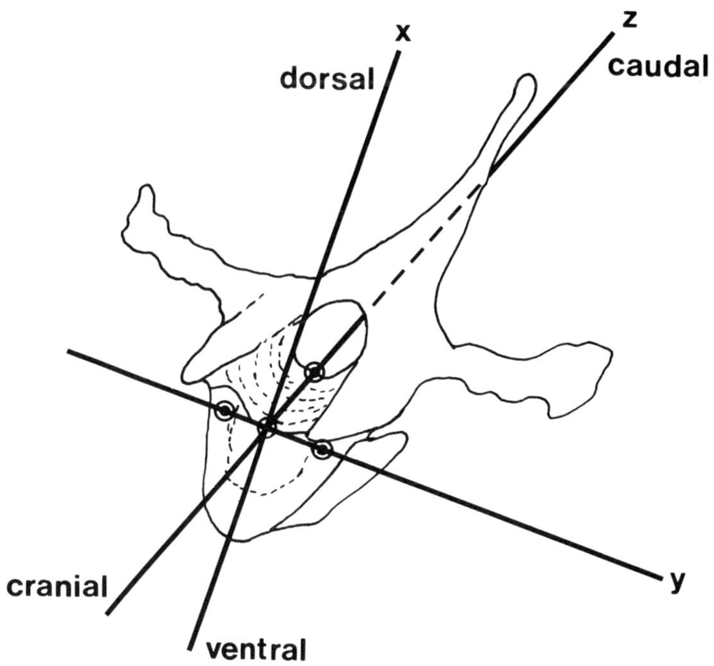

FIGURE 15.1A.
The coordinate system used in measuring the dimensions of the vertebral canal.

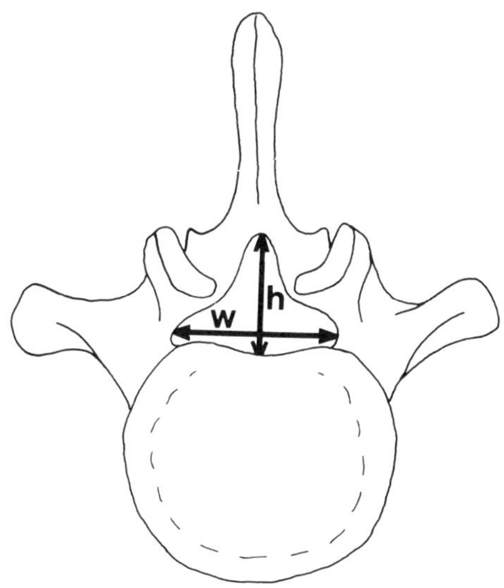

FIGURE 15.1B.
Cranial view of a vertebra demonstrating the measurements taken of the vertebral canal height (h) *and width* (w). *(After Twomey and Taylor, 1988.)*

TABLE 15.1.

Vertebral canal measurements (mm) for KNM-WT 15000.

VERTEBRA	IDENTIFICATION SUFFIX	CANAL HEIGHT	CANAL WIDTH
cervical 7	R	10.5	21.6
thoracic 1	S	10.5	18.7
thoracic 2	T	11.9	14.9
thoracic 3	U	12.2	14.0
thoracic 5	CA	13.1	13.1
thoracic 7	W	13.3	13.2
thoracic 8	V	13.2	12.2
thoracic 9	BI		
thoracic 10	X	13.2	11.9
thoracic 11	Y	13.6	12.9
lumbar 1	AR and BA		
lumbar 2	AV and AA	12.8	17.2
lumbar 3	Z	12.2	17.2
lumbar 4	AB	12.6	18.5
lumbar 5	BM	14.2	20.0
lumbar 6	AC	15.8	21.6
sacral 1	AD		28.0*
sacral 2	BB		>30.0
sacral 3	BC	6.4	30.0*
sacral 4	AE	4.9	16.9
sacral 5	AF		12.8

*Estimates.

For measurement of the canal height, one of the caliper pointers was placed along the "floor" of the canal, the dorsal surface of the vertebral body, in the mid-sagittal plane (i.e., along the *z*-axis). The second pointer was then opened out until it just touched the "roof" of the canal, the dorsal canal surface, again in the mid-sagittal plane. For measurement of the canal width, the caliper pointers were placed inside the vertebral canal, protruding at either end. They were then opened out to touch the "sides" of the canal, its lateral surfaces, so as to measure the maximum diameter of the canal parallel to the *y*-axis, at its minimum size along the *z*-axis. As well as the dimensions of the vertebral canal itself, the cranio-caudal length of the centrum of each vertebra was also measured between the points where the mid-sagittal plane cuts the dorsal surface of the centrum. For axis this was measured excluding the odontoid peg.

Measurement of some of the dimensions of atlas, axis, and the sacrum required special definition. The canal height through atlas was measured with the vertebra in articulation with axis; the odontoid peg therefore formed the ventral surface of the canal through atlas. The canal height through axis was measured from the caudal end of the vertebra to avoid interference from the odontoid peg and the centrum length of axis was measured excluding the odontoid peg. The canal dimensions of the first sacral vertebra were measured as for more cranial vertebrae. In adult specimens, with fused sacra, the dimensions of the canal through the last sacral vertebra in which the neural arch was complete enough were measured from the caudal end of the sacrum. In some instances, the canal width could be measured in a more caudal vertebra than it was possible to measure the canal height. For juvenile specimens, sacral canal dimensions were measured for all unfused sacral vertebrae as for more cranial regions of the vertebral column.

The vertebrae of KNM-WT 15000 were measured using the definitions described above. The first six cervical vertebrae are missing, but most of the rest of the vertebral column is complete and in good enough condition for measurement of the vertebral canal. Details of the vertebrae for which canal measurements were possible are given in Table 15.1. The sacral vertebrae are unfused in this juvenile specimen and so canal measurements were taken for individual sacral vertebrae, although S1–S3 are incomplete and somewhat damaged; their canal widths are estimates and the first two sacral canal heights could not be measured.

Spinal cord data

For a previous study (MacLarnon, 1987), measurements were collected of the gross dimensions of the spinal cord and its major constituent tissues for a series of thirteen non-human primate species, one specimen per species. All of these specimens were also included in the sample of vertebral columns measured as described above. For eight of these specimens, the preservation of the cord was completely successful and well controlled according to a method using 10% Formol-saline devised to minimize shrinkage while preserving the tissue adequately. Complete sets of measurements were possible for these 8 specimens; some measurements, including spinal cord weight, were possible for all 13 specimens. In outline, after preservation of the whole cord within the vertebral column, the vertebrae were individually separated, segmenting the cord transversely at the midpoint of each intervertebral disc. (The segmentation of the cord was therefore defined according to its relationship to the segmentation of the vertebral column and not, as is usual, according to the position of origin of the roots of the spinal nerves.) Each spinal cord segment was weighed and then sections were taken from the cranial end. Photographs of the sections were used to measure the height (maximum dorso-ventral diameter), width (maximum transverse diameter), and cross-sectional area of the total spinal cord, and the cross-sectional areas of the grey matter and white matter for each segment of the spinal cord (for further details see MacLarnon, 1987).

Some additional spinal cord weight data were collected from published sources. Only weights of cords weighed fresh were included; no studies used weak Formol-saline and other preservatives may not give comparable results. (See Figure 15.2.)

Body weight data

For the specimens whose spinal cords and vertebral columns were dissected for MacLarnon (1987), the actual body weights of the specimens are used in analyses here if they fall within the range for wild-caught specimens of the species of the same sex (MacLarnon and Martin, unpublished data collection). For the remainder of these and other modern species, the body weights used in analyses are the species' averages for the appropriate sex, calculated from wild-caught specimens wherever possible (MacLarnon and Martin, unpublished data collection).

For KNM-WT 15000, a body weight of 68 kg is used. This is Ruff and Walker's estimate (Chapter 11, this volume) of the body weight the individual would have reached as an adult. As discussed below, in modern humans the vertebral canal has reached adult dimensions by the stage of development equivalent to that of KNM-WT 15000. Therefore, the canal dimensions of the fossil can be analyzed as if the fossil were adult, and their relative sizes can be compared with those of adults of other species using the adult body weight estimate. In fact, using Ruff and Walker's estimate of the actual body weight of the juvenile individual, 48 kg, would make little difference to the analyses carried out here; and neither the use of the juvenile body weight nor the use of other recent estimates for the body weight of *Homo erectus*, such as the value of 58.6 kg estimated by McHenry (1988), would make any difference to the overall conclusions.

FIGURE 15.2.
Log-log plot of spinal cord weight against body weight for mammal species (data from Donaldson and Hatai, 1911; Hopf and Claussen, 1971; Krompecher and Lipak, 1966; Latimer, 1950; Latimer and Sawin, 1955 and 1957; MacLarnon, 1987; Ridgway et al., 1966). The best-fit line drawn in is the major axis for the whole sample.

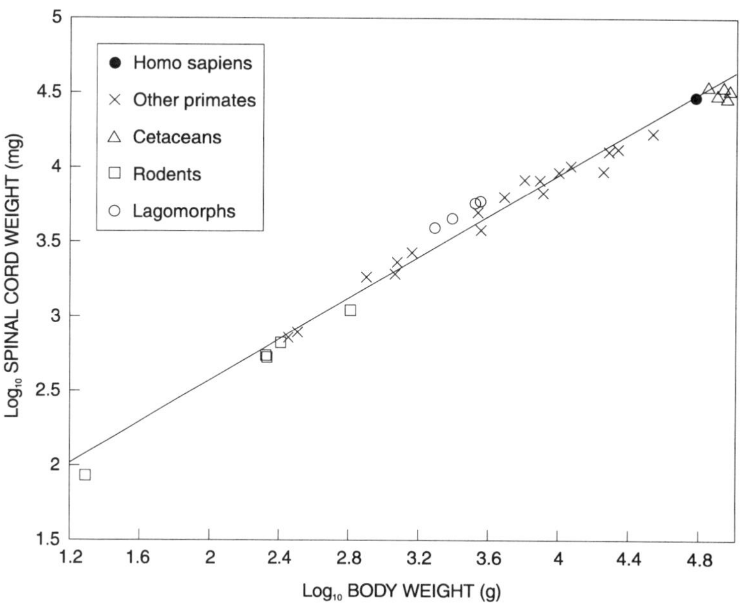

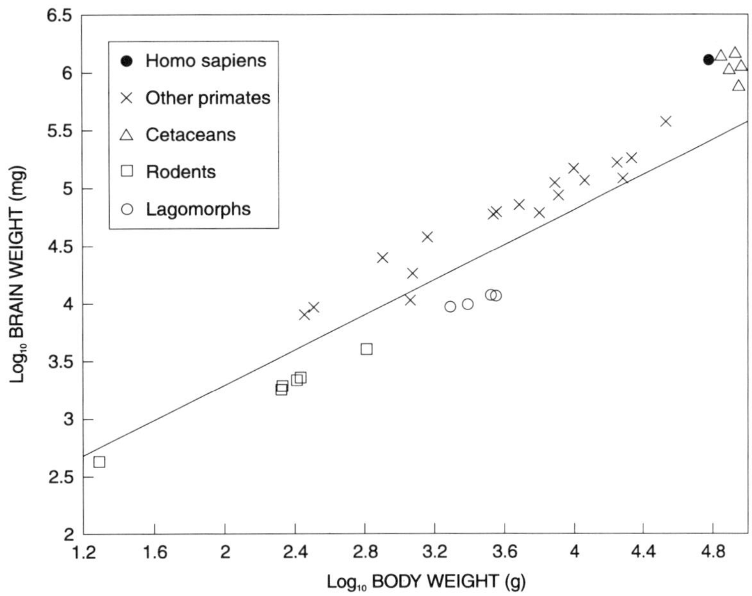

FIGURE 15.3.
Log-log plot of brain weight against body weight for the same mammal sample as in Figure 15.2. The best-fit line drawn in is the major axis for a sample of 309 mammal species from Martin (1981).

Results

Spinal cord dimensions

Figure 15.2 is a log-log plot of spinal cord weight against body weight for the total collected sample of mammal species, which comprises primates, cetaceans, rodents, and lagomorphs. The correlation coefficient for the distribution is high ($r = 0.99$), even for an allometric plot over a wide body size range. There is apparently very little variation in the weight of the spinal cord relative to body weight among modern mammals. The human point lies almost exactly on the best-fit major axis drawn in, indicating that there is nothing particularly special about the relative size of the human spinal cord compared with other modern mammals. These results are in sharp contrast to the amount of variation in relative brain weight among modern mammals. Figure 15.3 is a log-log plot of brain weight against body weight for the same set of mammal species as in Figure 15.2. The line drawn in is the major axis for the relationship derived for a much larger sample of mammal species (Martin, 1981). There is considerably more scatter of the species' points about the line in Figure 15.3 than in Figure 15.2. Relative brain weight is much more variable among mammals than is relative spinal cord weight. The human point lies well above the line in Figure 15.3; modern human brain weight is more than five times bigger than expected for a typical mammal of the same body weight (equation from Martin, 1981).

In Figure 15.4, parts A–G, profiles of the cross-sectional areas of the spinal cord and vertebral canal along the length of the vertebral column are plotted for *Homo sapiens,* and profiles of the cross-sectional areas, heights (dorso-ventral), and widths (transverse) of the cord and canal are plotted for *Pan troglodytes* and *Aotus trivirgatus.* The human plot is a composite of data for the human spinal cord segmented according to the points of origin of the spinal nerves, reported by Donaldson and Davis (1903), and data for the vertebral column measured for MacLarnon (1987) as described above. The plots for *Pan* and *Aotus* are examples using the cord and canal data collected from the same specimens for MacLarnon (1987). Vertebral canal dimensions are plotted on the assumption that they came from the midpoint along the length of each vertebra. (Because the human data come from two different sources, using somewhat different segmentation methods, the cord and canal profiles cannot be absolutely aligned either at their cranial-most ends or along the length of the vertebral column. The segment lengths for the vertebrae are underestimated relative to those of the spinal cord segments because they do not include the intervertebral discs.) These profile plots are examples demonstrating how closely the dimensions of the vertebral canal match those of the spinal cord in primate species. In general terms, all three dimensions of both cord and canal are large at the cranial end of the vertebral column; beyond this they decrease a little, before increasing for the cervical enlargement; they decrease again for the thoracic or interenlargement region; and increase for the lumbar enlargement; caudal to the lumbar enlargement, the cord dimensions decrease much more rapidly than the canal dimensions, which in most primate species decrease con-

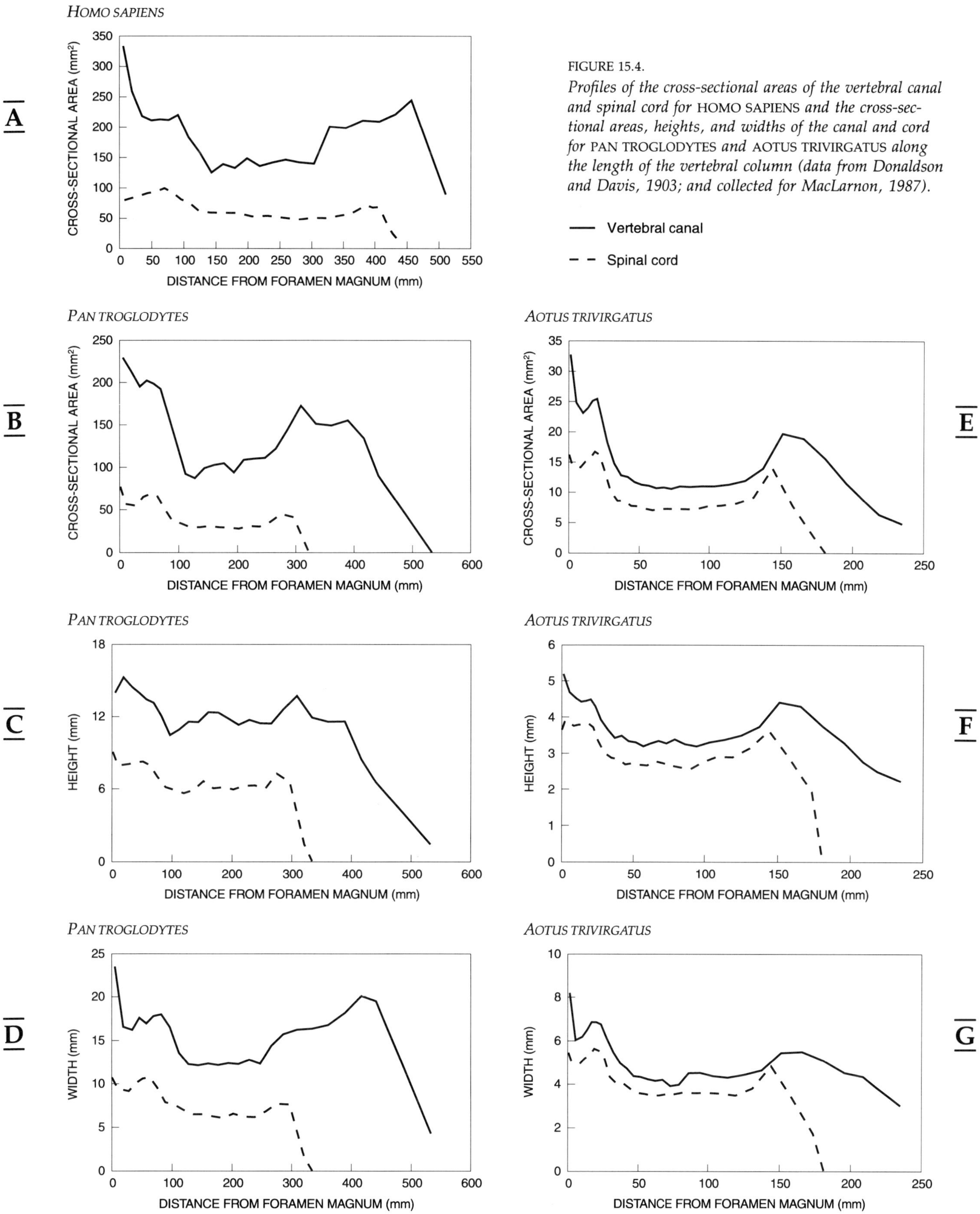

FIGURE 15.4.
Profiles of the cross-sectional areas of the vertebral canal and spinal cord for HOMO SAPIENS *and the cross-sectional areas, heights, and widths of the canal and cord for* PAN TROGLODYTES *and* AOTUS TRIVIRGATUS *along the length of the vertebral column (data from Donaldson and Davis, 1903; and collected for MacLarnon, 1987).*

tinually to the caudal end of the sacrum. In the apes, and especially humans, the canal is particularly large caudal to the lumbar enlargement of the cord. The cord and canal dimensions match quite well: for the sample of eleven species with adequate data measured for MacLarnon (1987), the maximum cross-sectional areas and widths of the cord and canal at their respective cervical enlargements were always within one vertebra of each other, and their cross-sectional areas, heights, and widths at the lumbar enlargement were almost always within one vertebra of each other.

The "double-humped" profiles of the cord and canal dimensions reflect the swellings of the cord in the cervical and lumbar regions for the innervation of the fore- and hindlimbs. The increase of cord and canal dimensions cranially reflects the frontal accumulation of the cord, that is the accumulation of nerve fibers connecting the cord and brain. While some reflex mechanisms are entirely controlled by the postcranial nervous system, the cord is to a great extent a messenger trunk for nervous impulses to and from the brain, carrying sensory information gathered in the postcrania to the interpretive centers and motor commands from the brain to the postcranial muscle fibers.

From the spinal cord, spinal nerves pass out between the vertebrae to more superficial sensory cells and muscle fibers. In early fetal stages, the spinal cord is the same length as the vertebral canal, but afterward the vertebral column outgrows the spinal cord. The caudal end of the cord therefore gradually migrates up the vertebral canal, its cranial end being fixed at the junction with the brain. In all the specimens examined for MacLarnon (1987), the spinal cord ended presacrally, though the exact position varied from the fourth last to the last presacral vertebra. In humans, the spinal cord usually ends between the first two lumbar vertebrae. Caudal to the end of the spinal cord, the vertebral canal contains the cauda equina, the bundle of roots of spinal nerves with more caudal exit points; in other words, the lower lumbar and entire sacral vertebral canal segments do not contain spinal cord tissue. The end point of the spinal cord leaves no mark on the bony canal. The dural sheath surrounding the cord also envelops the cauda equina and is attached to the vertebral canal only via the filum terminale at its caudal end (e.g., to the coccyx in humans). On average, for the specimens investigated in MacLarnon (1987), the spinal cord ended in the second vertebra caudal to the vertebra with the largest lumbar canal cross-sectional area. From the limited evidence available, it appeared to end relatively abruptly after the lumbar enlargement in tailless species (*Pan* and *Homo sapiens*).

More quantitative analyses of the goodness of fit of spinal cord and vertebral canal dimensions were also carried out in MacLarnon (1987). Canal dimensions scale to body weight with higher allometric exponents than the corresponding cord dimensions, and so simple ratios of measures of the two are not meaningful. However, the relative sizes of cord and canal dimensions, once variation in body size has been taken into account adequately, were shown to be positively correlated, including cross-sectional areas, heights, and widths of the cord and canal at the cervical and lumbar enlargements and in the thoracic region, as well as measures of the total canal volume and spinal cord weight. It is therefore possible to make reasonably good estimates of the spinal cord dimensions of primates from their vertebral canal dimensions. *Homo sapiens* fits the relationships between cord and canal dimensions for other primates in most cases. However, because of the great expansion of the human vertebral canal in the lumbar region caudal to the end of the spinal cord, maximum lumbar dimensions of the cord and canal are not well matched for humans at this point. A more cranial first high point in the dimensions of the lumbar canal of humans seems to correspond better to the lumbar enlargements of non-human primates.

As well as the total spinal cord tissue being distributed with a "double-humped" profile along the length of the vertebral column, the two major tissue divisions of the cord, the grey matter and the white matter, are also similarly distributed. In Figure 15.5, the profiles of the cross-sectional areas of the grey and white matter along the cord are illustrated for the same species as before, *Homo sapiens*, *Pan troglodytes*, and *Aotus trivirgatus*. Frontal accumulation is more clearly evident for the nerve fibers of the white matter. Both the nerve fibers and the nerve cell bodies of the grey matter increase in volume in the cervical and lumbar enlargements, providing innervation for the fore- and hindlimbs, respectively. The more abrupt ending of the cord tissue of the two tailless species, caudal to their lumbar enlargements, compared with the long-tailed *Aotus*, can also be distinguished.

EFFECTS OF AGE ON VERTEBRAL CANAL DIMENSIONS

Figure 15.6, parts A–J, are plots of various vetebral canal dimensions against age for the Spitalfields' sample of modern humans measured for the present study. Some cervical and sacral dimensions could not be measured on the youngest specimens because the vertebrae are unfused, and the neural arches and centra could not be paired or are missing. The canal heights and widths at the cervical and lumbar enlargements are the largest values for these dimensions for each specimen in cervical and lumbar vertebrae, respectively. The minimum thoracic dimensions are the smallest dimensions from thoracic vertebra. The cranial sacral dimensions are the canal dimensions through S1. The caudal sacral canal height is the canal height of the last sacral vertebra with a complete neural arch. The caudal sacral canal width is the canal width of the last sacral vertebra with laminae forming the lateral sides of the neural arch. In some specimens, therefore, the caudal sacral canal width is defined as being from a more caudal vertebra than the caudal sacral canal height.

The results shown in Figure 15.6 are discussed below together with similar data from previous studies. There is quite a lot of variation among published studies in conclusions about the age at which human vertebral canal dimensions reach their adult sizes. Some of the variation may well be due to the use of uncorrected measurements from radiographs, because of the variation in size of younger and older children (Larsen and Smith, 1981). While the definitions of canal dimensions varied in the published studies, it is presumed that the growth patterns of canal heights and widths, however they are defined, are similar.

Figure 15.6A suggests that for modern humans, the height of the vertebral canal at the cervical enlargement does not increase significantly from about 2 years of age. However, Figure 15.6B

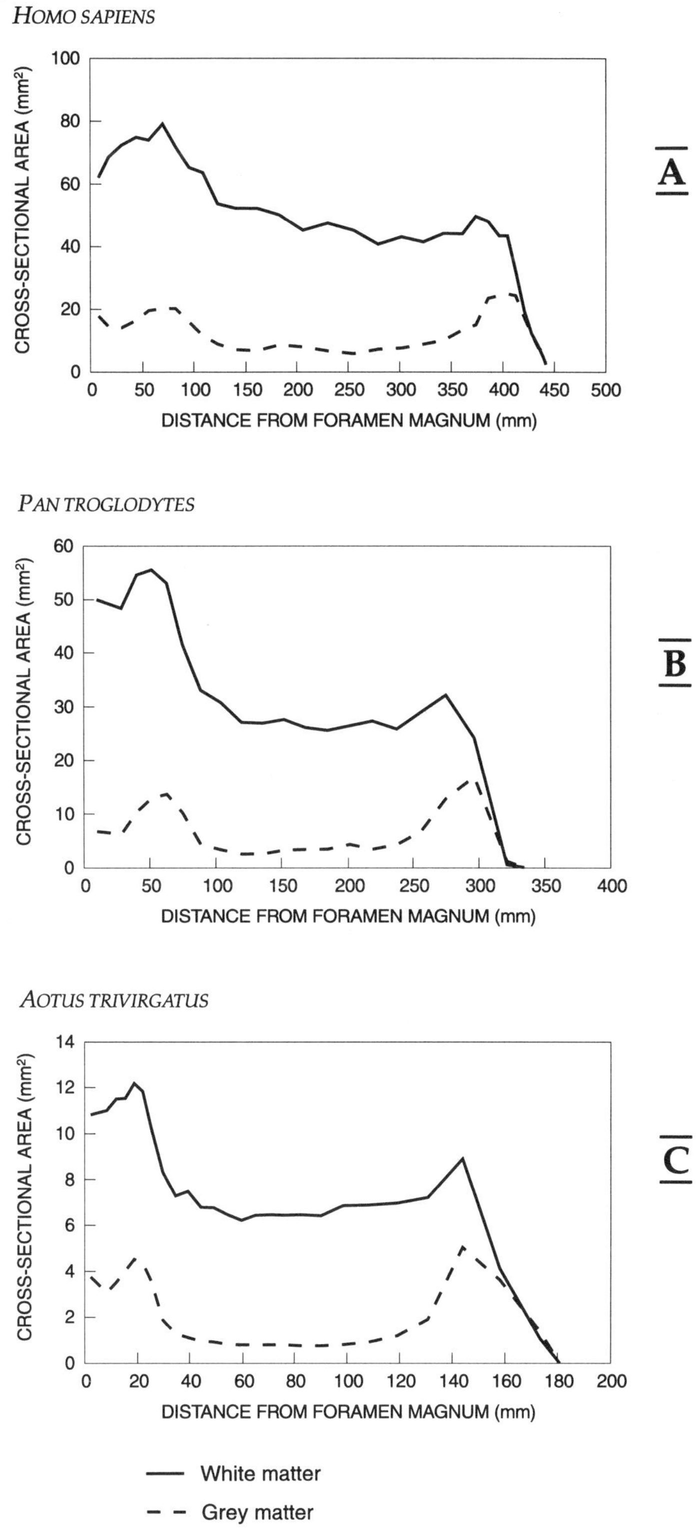

FIGURE 15.5.
Profiles of the cross-sectional areas of the white matter and grey matter of the spinal cord along the length of the vertebral column for HOMO SAPIENS, PAN TROGLODYTES, *and* AOTUS TRIVIRGATUS *(data from Donaldson and Davis, 1903; and collected for MacLarnon, 1987).*

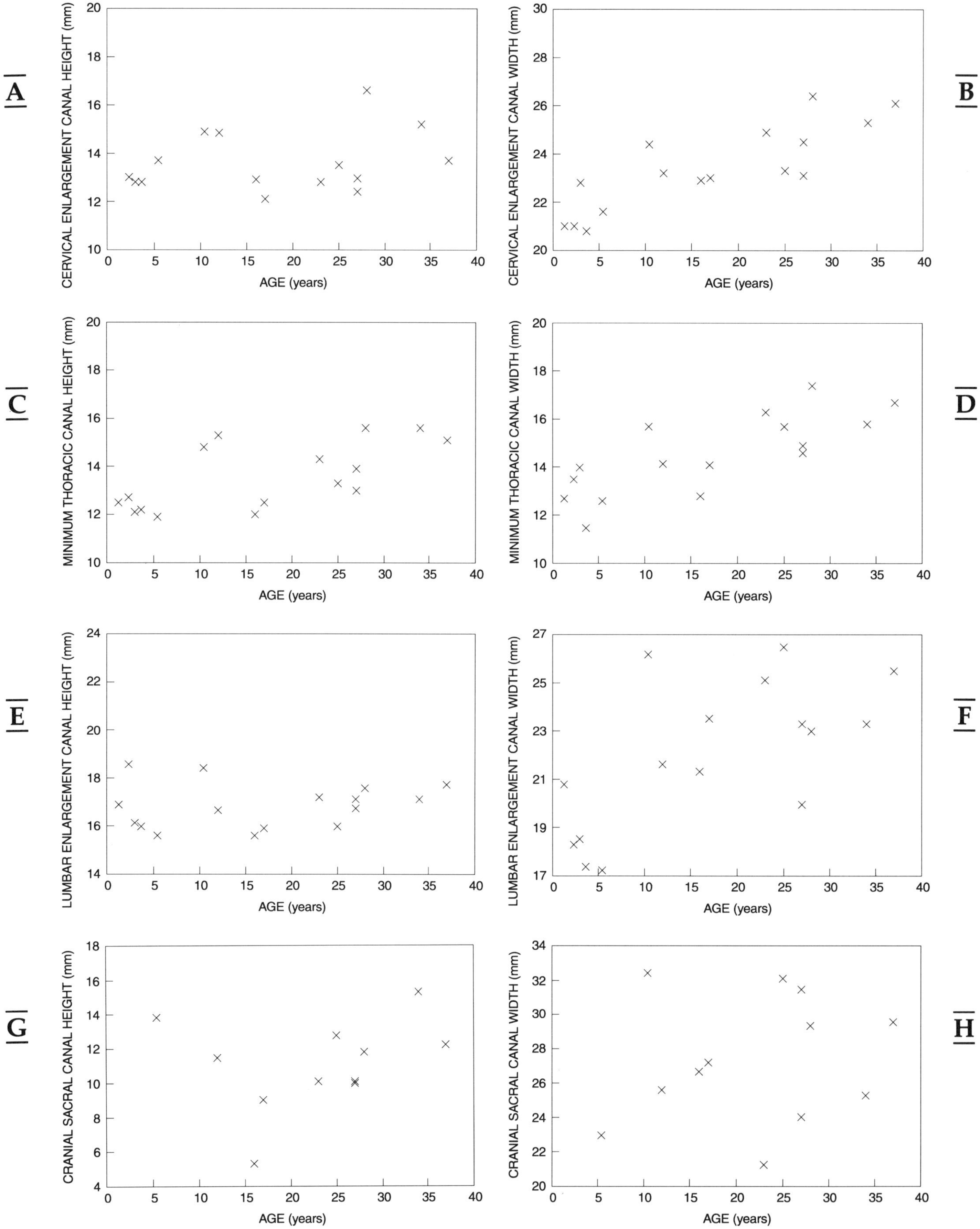
A
CERVICAL ENLARGEMENT CANAL HEIGHT (mm)
AGE (years)
B
CERVICAL ENLARGEMENT CANAL WIDTH (mm)
AGE (years)
C
MINIMUM THORACIC CANAL HEIGHT (mm)
AGE (years)
D
MINIMUM THORACIC CANAL WIDTH (mm)
AGE (years)
E
LUMBAR ENLARGEMENT CANAL HEIGHT (mm)
AGE (years)
F
LUMBAR ENLARGEMENT CANAL WIDTH (mm)
AGE (years)
G
CRANIAL SACRAL CANAL HEIGHT (mm)
AGE (years)
H
CRANIAL SACRAL CANAL WIDTH (mm)
AGE (years)

I

J

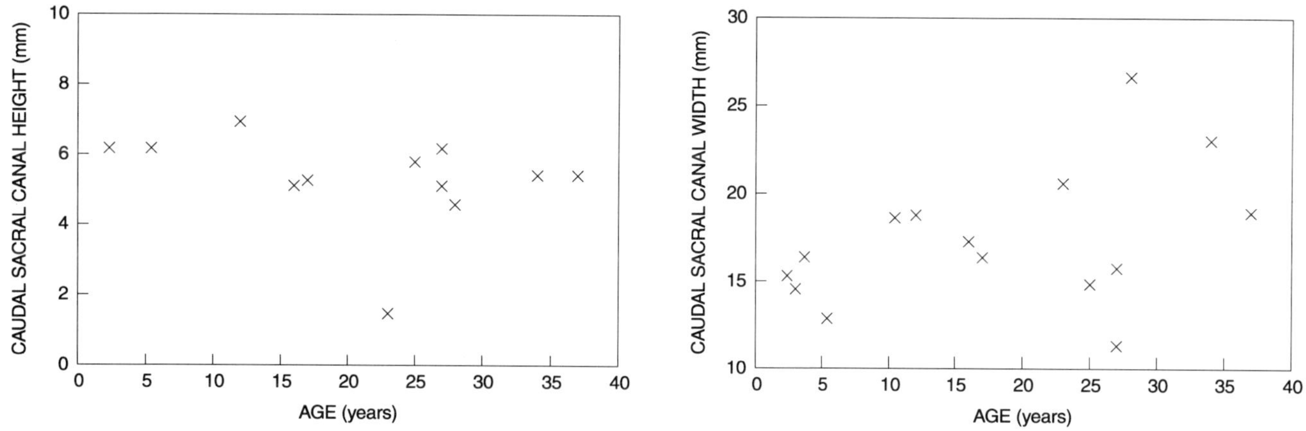

FIGURE 15.6.
Plots of vertebral canal dimensions against age for a sample of modern humans (Spitalfields' collection).

suggests that the corresponding canal width increases a little, at least until about 10 years of age. The results of relevant published studies, all of which were based on radiographs, range from suggesting that cervical canal height reaches its adult size at 3–6 years (Wholey et al., 1958, in Yousefzadeh et al., 1982), or by 6 years for the lower cervical region (which would encompass the cervical enlargement; Yousefzadeh et al., 1982), to suggesting that cervical canal height continues to increase until at least 11–14 years (Markuske, 1977), or even to 18 years (Hinck et al., 1962). However, the reported increases in cervical canal height are never very large; the average increase is about 15% from 3 years to adult size according to the radiographic data of Hinck et al. (1962), and even this may be exaggerated through lack of correction for the size differences of subjects, as the limited skeletal data set presented in Figure 15.6 suggests.

There have been fewer published studies of the growth of cervical canal width in modern humans; again they all involve radiographic measurements. Hinck et al. (1966) show that there is an increase in cervical canal width until 13–14 years, with only slight increase after that. The data of Haworth and Keillor (1962) suggest that by 10 years of age, cervical canal widths are within the lower end of the adult range. From the published radiographic data, the increases in cervical canal width appear to be greater than those for cervical canal height, averaging about 50% between 2 years and adult size, but the present skeletal data suggest that this may be an exaggeration and that an increase of about 20% may be more accurate.

Figures 15.6C and D suggest that in the thoracic region there is a slight increase in thoracic canal height in modern humans from early childhood and a more clearly defined increase, though still small, in thoracic canal width. The former appears to reach adult size by 10 years; the latter may continue increasing through adolescence. The few relevant published studies, which are all radiographic, suggest that both thoracic canal dimensions grow to within adult size ranges by 10 years, with only slight increases on average after that. Their data imply an average increase of about 20% in these dimensions from 2 years to adulthood (Hinck et al., 1962 and 1966; Haworth and Keillor, 1962), which is similar to that suggested by the present skeletal data.

Figures 15.6E and F suggest that there is no growth in lumbar canal height in modern humans from 2 years to adulthood but that there is some early growth in lumbar canal width, which reaches adult size by about 10 years. Published studies of the growth of lumbar canal height also suggest no significant growth in lumbar canal height from at least 3 years, whether or not the radiographic measurements were corrected for variable magnification (Hinck et al., 1965; Larsen and Smith, 1981). Published studies also agree that there is some growth in lumbar canal width after fusion. The ages they suggest for its completion vary from about 6–8 years (Larsen, 1981) to 9–10 years (Hinck et al., 1966). Both published sources and the present data suggest an average increase of about 20–25% in the lumbar canal width between 2–3 years and adult size.

Figures 15.6G and H imply that there is no growth in cranial sacral canal height in modern humans from about 6 years of age and no growth in cranial sacral canal width from 10 years, although, because of the poor data from this part of the canal, the age thresholds cannot be stated with any great confidence. The one available set of relevant published data, the radiographic data set of Haworth and Keillor (1962), implies that cranial sacral canal width definitely reaches lower adult levels by 10 years, perhaps even by 4.5 years. The published data suggest an average increase in cranial sacral canal width of about 20% from 2 years to adult size.

Figures 15.6I and J suggest that caudal sacral canal height does not grow in modern humans from at least 2 years of age but that caudal sacral canal width may grow a little on average, although specimens from 2 years of age have widths within the very wide adult range. The radiographic data of Haworth and Keillor (1962) suggest that average adult sizes of the cranial sacral canal width are reached by 10 years and that there is an increase of about 15% from 2 years to adult size.

The descriptions above of the growth of the vertebral canal dimensions of modern humans are largely based on average dimensions for age groups from the various published studies. It is clear from the wide scatter of points on the parts of Figure 15.6, which never conform closely to a simple growth curve, that there is considerable variation in dimensions within age bands, compared with the total variation in dimensions throughout the growth period. This is also demonstrated by the published data, where standard deviations or other measures of the variation in dimensions within age groups are given. There is usually considerable overlap between the ranges of even one standard deviation either side of the mean, for all age groups from 2–3 years to adulthood, for the dimensions of the canal through any vertebra. That is, the average growth of the dimensions of the vertebral canal in modern humans is not very significant compared with the amount of variation in dimensions within a single age group, including adults.

FIGURE 15.7.
Profiles of the height and width of the vertebral canal of KNM-WT 15000 along the vertebral column. Dashed lines join measured points where vertebrae in between are missing or where the measurements were not possible. See Table 15.1 for data.

HOMO ERECTUS
KNM-WT 15000

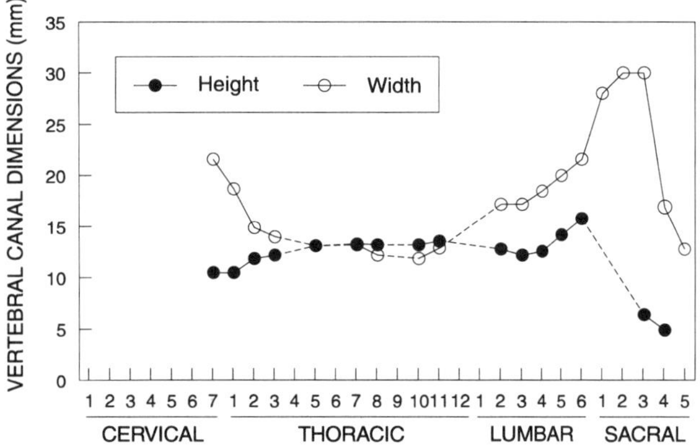

Vertebral canal profiles

In Table 15.1, the heights and widths of the vertebral canal measured for KNM-WT 15000 are presented. In Figure 15.7, these are plotted in a profile diagram showing variation in the dimensions of the vertebral canal along the length of the vertebral column. In Figures 15.8 and 15.9, profiles of the vertebral canal dimensions for adult modern humans and apes are given for comparison.

Of the cervical vertebrae, only the seventh is known for KNM-WT 15000. From C7, the height of the vertebral canal of the fossil increases into the early part of the thoracic region. In most modern primates (see Figures 15.8 and 15.9 and MacLarnon, 1987, for non-hominoids), the height of the vertebral canal decreases from C7 into the thoracic region, showing signs of a cervical enlargement of the canal height even in the last cervical vertebra. However, in a few apes and many humans there is an increase, as for KNM-WT 15000: for example, one orangutan and two gorillas in Figure 15.9 and all the humans in Figure 15.8. In humans, at least, this increase in the cervical canal height through C7 and the cranial-most thoracic region into the thoracic plateau is associated with the lack of an obvious cervical enlargement in the height dimension of the cervical canal, suggesting that KNM-WT 15000 conformed to this common, if not exclusive, human characteristic.

The width of the vertebral canal through C7 of KNM-WT 15000 is markedly larger than for most of the thoracic region. The canal width decreases through T1 and T2 into the flatter profile or plateau of the interenlargement region. This is the caudal end of the cervical enlargement of the canal width of the fossil. Its position and its apparently much clearer definition than any cervical enlargment of the canal height are typical of all modern primates (see Figures 15.8 and 15.9 and MacLarnon, 1987, for non-hominoids).

Over half the thoracic vertebrae of KNM-WT 15000 are known. The profiles of the canal dimensions of the fossil shown in Figure 15.7 are generally flat in this region, caudal to some increase and decrease, respectively, in the canal heights and widths of the first couple of thoracic vertebrae. More or less flat thoracic canal profiles are typical of all modern primates (see Figures 15.8 and 15.9 and MacLarnon, 1987, for non-hominoids).

Partial or complete vertebrae are known for KNM-WT 15000 from L2 to S5. It had six lumbar vertebrae in total, an unusual but not unknown number among modern humans, who normally have five; the great apes usually have four lumbar vertebrae with individuals varying between three and five (Schultz, 1961). L1 is missing and the neural arches of S1–S3 are incomplete, so that canal measurements for these vertebrae are either estimates or not possible, but otherwise complete profiles for the fossil of the canal heights and widths through the lumbar and sacral regions can be drawn. The canal height of KNM-WT 15000 decreases from L2 to L3 and then increases continuously to L6. From the profiles in Figure 15.9, it can be seen that the modern human lumbar canal height profile is quite variable. In the different specimens illustrated, the lumbar canal height decreases throughout the whole lumbar region, decreases through most of the lumbar vertebrae (increasing only in the last one), or decreases in the first lumbar vertebra and then increases to the last (as for KNM-WT 15000), or has a more complex profile, increasing, decreasing, and then increasing again. Previous studies of the dimensions of the vertebral canal in modern humans usually present the data or profile diagrams only for sample average lumbar canal dimensions, which limits their usefulness for comparison with a single individual. However, for about three-quarters of the reported samples (which comprise South African blacks and whites, Americans, Anglo-Saxons and Romans, Italians, Indians, and European Australians), the average lumbar canal height decreases in the cranial-most vertebrae and then increases in at least the last lumbar vertebra (Eisenstein, 1976 and 1977; Hinck et al., 1965; Porter and Pavitt, 1987; Postacchini et al., 1983; Twomey and Taylor, 1988). In other modern primates (see Figure 15.8 and MacLarnon, 1987, for non-hominoids), the lumbar canal height most commonly decreases steadily through the lumbar region and, even if it increases a little between adjacent vertebrae, the general trend through the region is always a decreasing canal height. The lumbar canal height profile of KNM-WT 15000 therefore conforms to a pattern seen only in humans among modern primates. Modern humans are quite variable in this characteristic, but the commonest population average profile is also as for KNM-WT 15000.

In all the groups investigated here, the lumbar canal width varies much more throughout the lumbar region than the lumbar canal height, as is also generally true of the canal width compared with the canal height in other regions of the canal. The width of the vertebral canal of KNM-WT 15000 in the lumbar region increases from L3 to L6 and then increases greatly from L6 to S1 (this last measurement is an estimate). The profile of the lumbar canal width in modern humans is quite variable, as shown in Figure 15.9; sometimes it is quite flat, sometimes there is a humped lumbar enlargment, and sometimes it rises steadily. What is common to all human specimens measured here, however, is an increase in the canal width between the last two lumbar vertebrae and then a usually marked increase from the last lumbar vertebra to S1 (see Figure 15.9). In most other modern primate species, there is a lumbar enlargement with a humped profile and then the canal width decreases into the sacrum (see Figure 15.8 and MacLarnon, 1987, for non-hominoids). A rising profile of the lumbar canal width, continuing into the sacrum, is found among non-human primates only in species with a prehensile tail (MacLarnon, 1987), and in these species it is not as marked as in many modern humans and KNM-WT 15000.

HOMO SAPIENS

Spitalfields no. 2327—23 yr, F

A

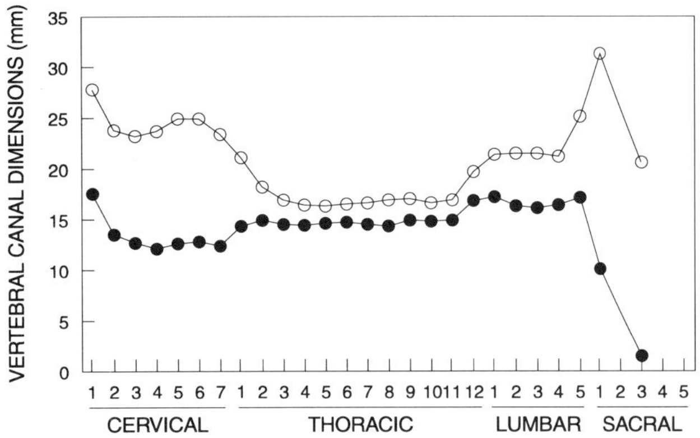

HOMO SAPIENS

Spitalfields no. 2142—27 yr, F

D

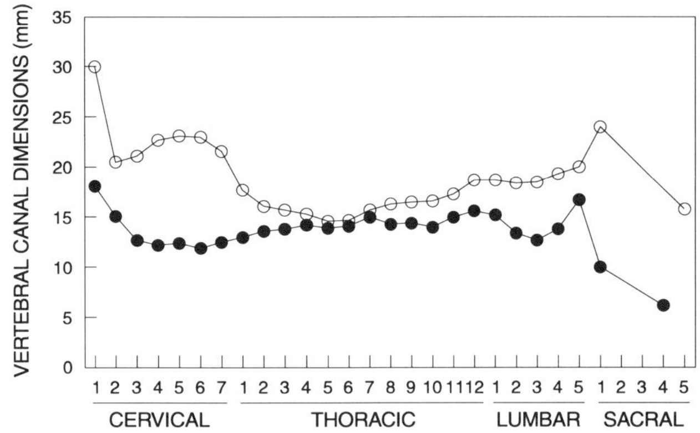

HOMO SAPIENS

Spitalfields no. 2720—25 yr, F

B

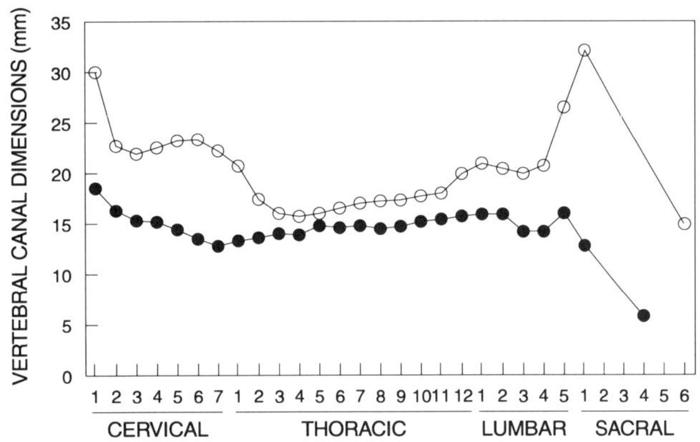

HOMO SAPIENS

Spitalfields no. 2843—28 yr, F

E

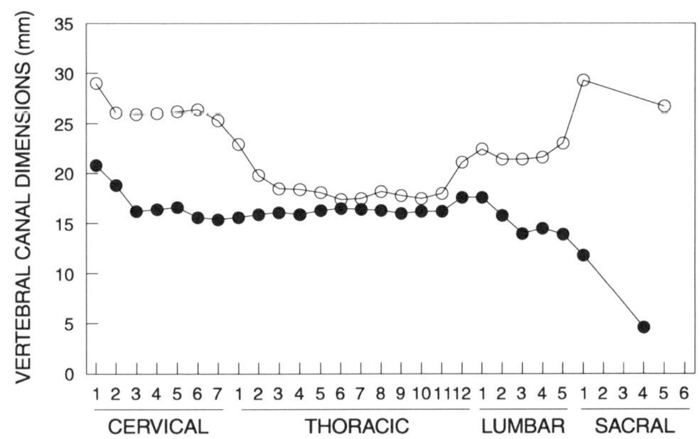

HOMO SAPIENS

Spitalfields no. 2872—27 yr, F

C

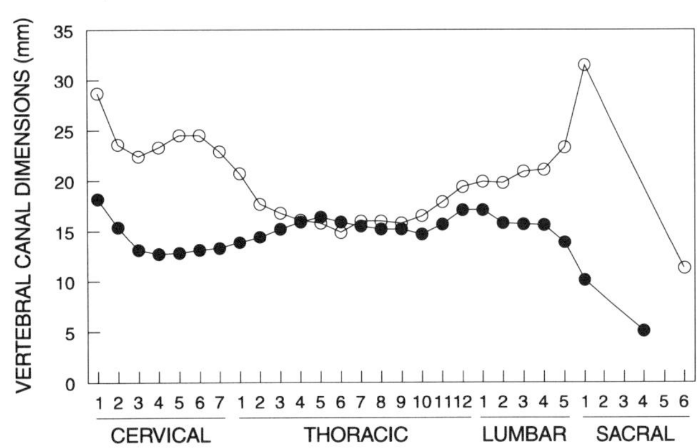

HOMO SAPIENS

Spitalfields no. 2254—34 yr, M

F

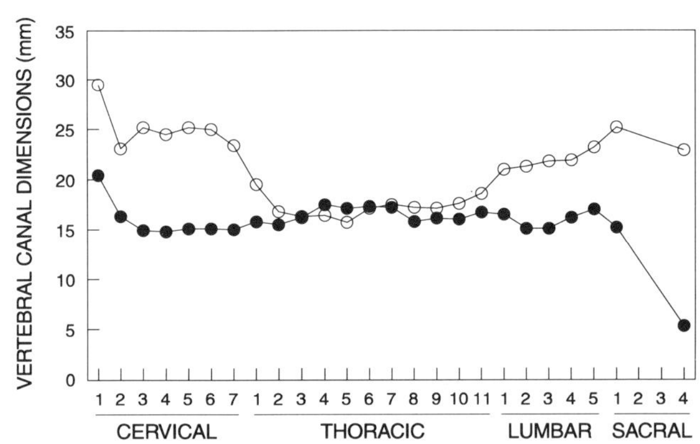

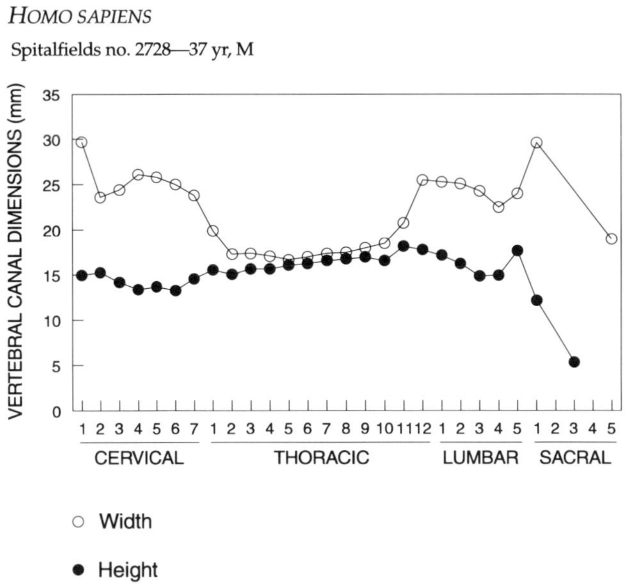

FIGURE 15.8.
Profiles of the height and width of the vertebral canal along the vertebral column for adult modern humans (Spitalfields' collection). M, *male;* F, *female.*

Most previous studies of the lumbar canal width of modern humans again usually present data or profile diagrams only for sample average measurements. However, in one study by Piera et al. (1988), of radiographic measurements of a sample of 215 adult Spanish vertebral columns, the lumbar canal was found to widen continuously from L1 to L5 in 76% of males and 80% of females, and to widen throughout the region between all except one pair of adjacent vertebrae in a further 21% of males and 18% of females. Of the samples from studies presenting only sample average data (comprising Nigerians, Saudis, South African blacks and whites, Americans, Anglo-Saxons and Romans, European Australians, Italians, and Indians), two-thirds have increasing lumbar canal widths throughout the lumbar region, or increases in widths with the exception of equal values for two adjacent vertebrae (Amonoo-Kuofi, 1982; Amonoo-Kuofi et al., 1990; Eisenstein, 1976 and 1977; Haworth and Keillor, 1962; Hinck et al., 1966; Porter and Pavitt, 1987; Postacchini et al., 1983; Twomey and Taylor, 1988). The increase is often most marked between the last two lumbar vertebrae. KNM-WT 15000 therefore has a typical human profile of the lumbar canal width, distinctive from those of most other modern primates, rising throughout the region, including a marked increase in the canal width between the last lumbar vertebra and S1.

Modern humans and KNM-WT 15000 therefore share a combination of quite distinctive profiles of both the height and width of the lumbar vertebral canal, continuing into the cranial sacrum for the canal width. The two dimensions of the canal can be combined to estimate its cross-sectional area using the ellipse formula ($\pi/4 \times$ height $\times$ width). For the majority of modern human specimens measured here, and for KNM-WT 15000, the maximum estimated cross-sectional area of the lumbar and sacral vertebral canal occurs in the last lumbar vertebra; in a few modern humans, by a small margin, it falls in T12 or L1. For the ape specimens measured here, and for all the other non-hominoid modern primates measured for MacLarnon (1987), except for a single orangutan, the maximum cross-sectional area of the vertebrae, caudal to the cervical enlargement, always occurs in the caudal thoracic or more cranial lumbar regions. Swelling of the caudal lumbar and cranial sacral vertebral canal is unique to hominids.

In the sacrum, both the height and width of the vertebral canal of KNM-WT 15000 decrease between S1 and the last sacral vertebrae in which they are measureable, namely S4 and S5, respectively. The canal width seems to increase a little in the first three sacral vertebrae and then to decrease rapidly to S4 and S5, but as the first three measurements are estimates from incomplete vertebrae, this cannot be stated with great certainty.

A

Hylobates concolor

PA 1004, F

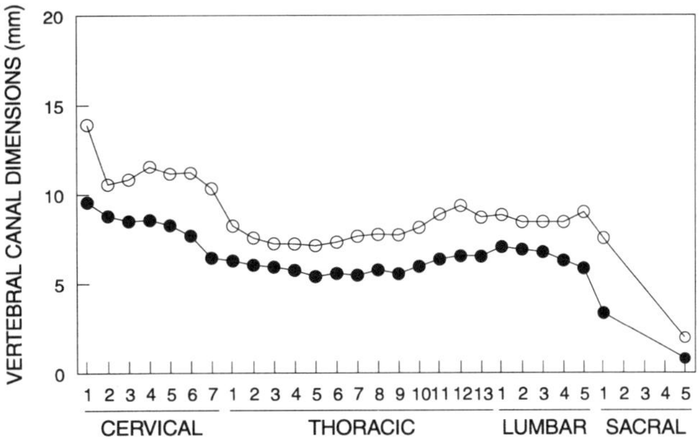

B

Hylobates hoolock

BNMH 1850.11.22.56, U

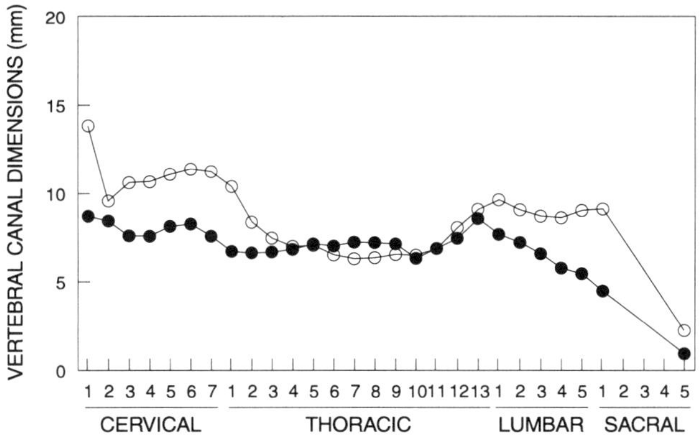

C

Pan troglodytes

BMNH 1976.437, F

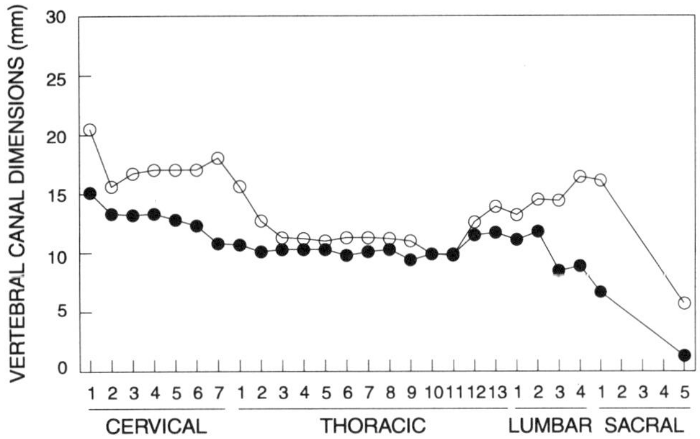

D

Pan troglodytes

PA 357, F

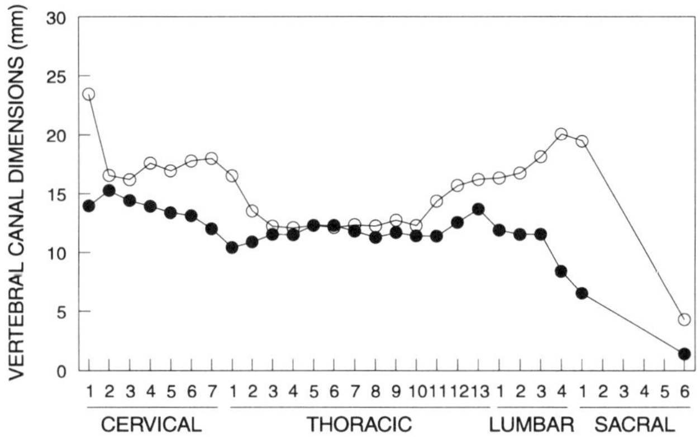

E

Pan troglodytes

BMNH 1924.8.6.1, M

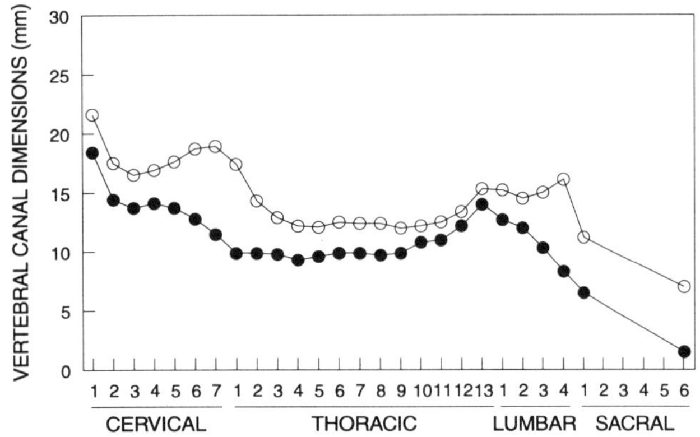

F

Pan troglodytes

BMNH 1901.8.9.84, M

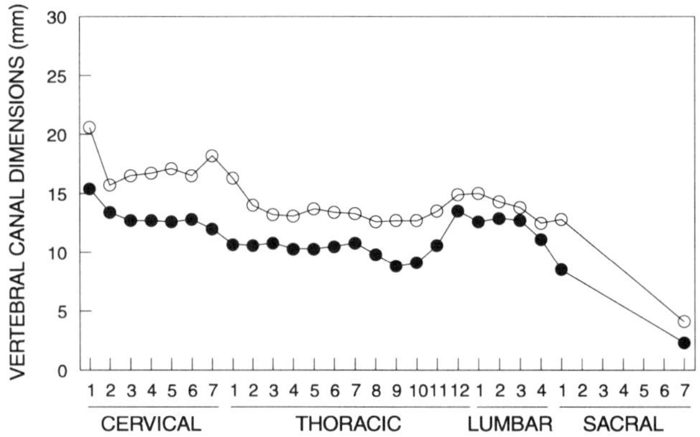

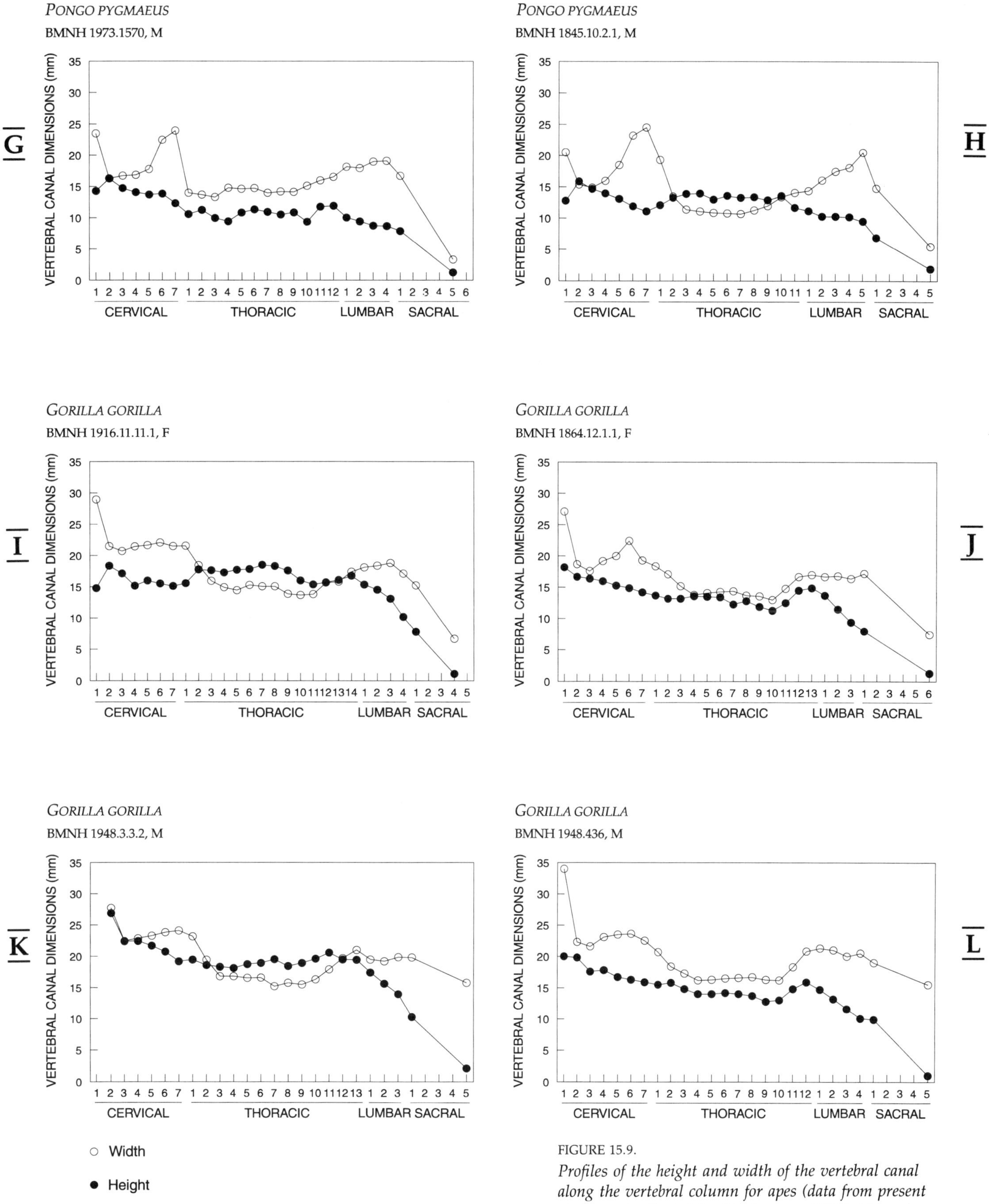

FIGURE 15.9.
Profiles of the height and width of the vertebral canal along the vertebral column for apes (data from present study; and collected for MacLarnon, 1987). M, *male;* F, *female;* U, *sex unknown.*

A B C D E F

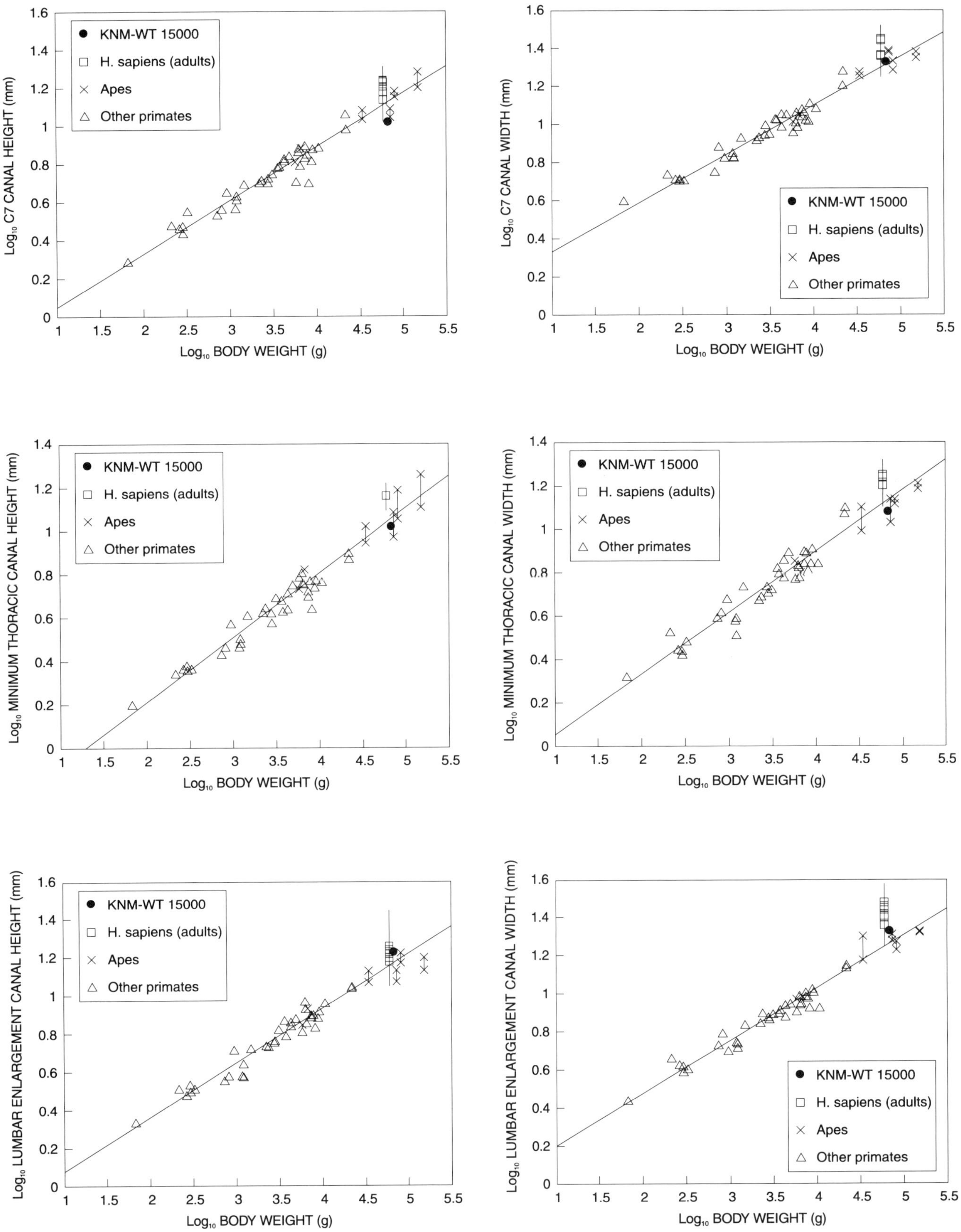
KNM-WT 15000
H. sapiens (adults)
Apes
Other primates
Log10 C7 CANAL HEIGHT (mm)
Log10 C7 CANAL WIDTH (mm)
Log10 MINIMUM THORACIC CANAL HEIGHT (mm)
Log10 MINIMUM THORACIC CANAL WIDTH (mm)
Log10 LUMBAR ENLARGEMENT CANAL HEIGHT (mm)
Log10 LUMBAR ENLARGEMENT CANAL WIDTH (mm)
Log10 BODY WEIGHT (g)

G

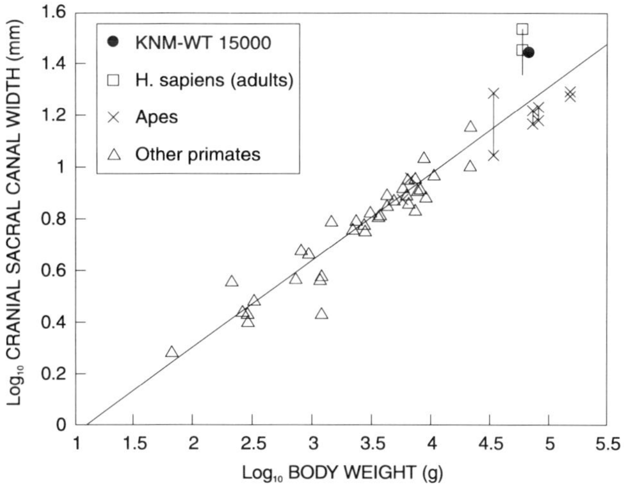

FIGURE 15.10.

Log-log plots of vertebral canal dimensions against body weight for modern primates, including humans, and KNM-WT 15000. Maximum and minimum measurements are plotted for one monkey species and great ape species for which two or more specimens were measured. For modern humans, the individual points plotted are the sample averages for adults from a range of studies (see Table 15.2); the ranges drawn in are the maximum ranges of two standard deviations either side of any of the sample means (data from present study; collected for MacLarnon, 1987; and published sources, see Table 15.2). The best-fit lines drawn in are the major axes for the original sample used in MacLarnon (1987) of 38 modern primate species, including 3 with both male and female representatives (total n = 41). For the plots of caudal sacral dimensions, the major axes were calculated for species with long tails only (20+ caudal vertebrae).

H

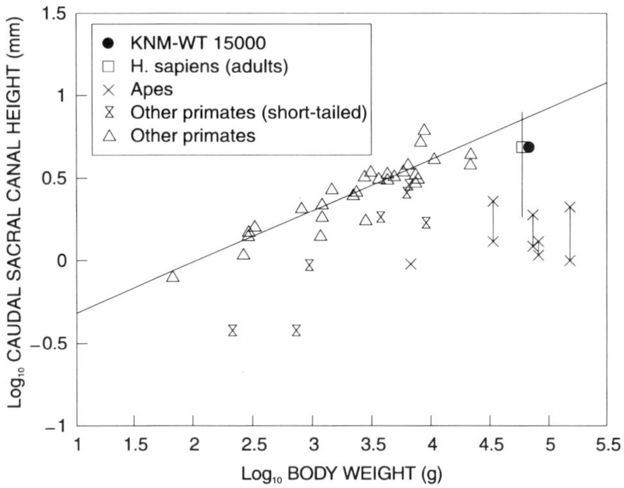

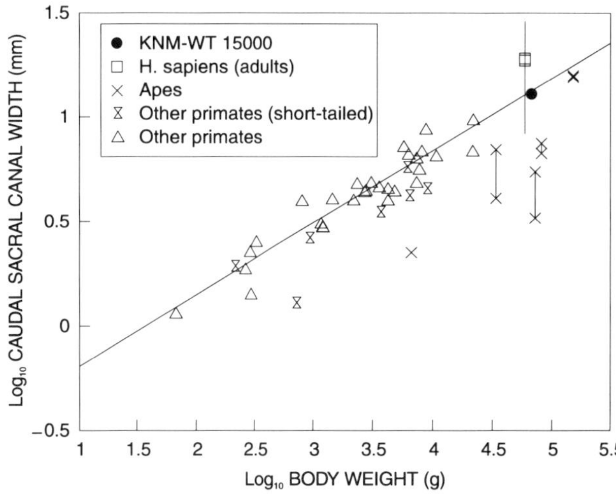

I

Data from the juvenile modern humans measured for the present study show that a rising canal width profile in the cranial sacral canal is not unknown in modern humans, though it is not common. A decrease in sacral canal dimensions between the first and the last vertebrae is the normal pattern for all modern primates, except for some specimens from species with a prehensile tail (see MacLarnon, 1987). The sacral canal generally narrows more in species with short tails or no tail. In the tailless apes, the sacral canal narrows to a very small caudal opening (see Figure 15.9). Tailless modern humans, however, break the pattern of association between tail length and narrowing of the sacral canal. The human sacral canal does not narrow to the very small caudal dimensions of the other hominoids (see Figures 15.8 and 15.9). The sacral canal profile of KNM-WT 15000 is like that of modern humans and unlike that of other tailless modern primates; it narrows from its cranial to its caudal end, but its caudal opening is not particularly small.

Vertebral Canal Dimensions

Figures 15.10 and 15.11 are log-log plots of a series of dimensions of the vertebral canal against body weight for modern primates, including extended

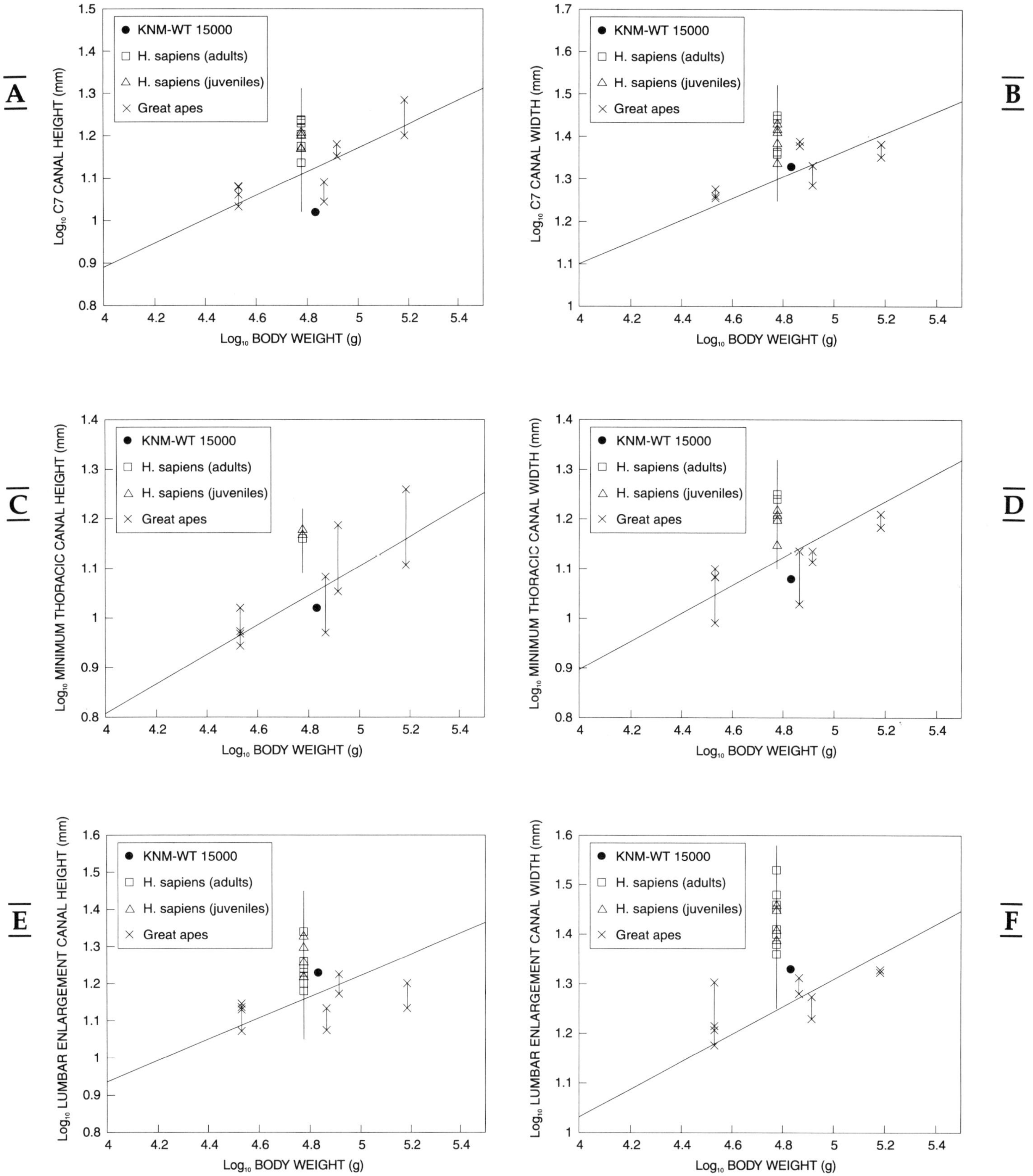

A
KNM-WT 15000
H. sapiens (adults)
H. sapiens (juveniles)
Great apes
Log10 C7 CANAL HEIGHT (mm)
Log10 BODY WEIGHT (g)
B
KNM-WT 15000
H. sapiens (adults)
H. sapiens (juveniles)
Great apes
Log10 C7 CANAL WIDTH (mm)
Log10 BODY WEIGHT (g)
C
KNM-WT 15000
H. sapiens (adults)
H. sapiens (juveniles)
Great apes
Log10 MINIMUM THORACIC CANAL HEIGHT (mm)
Log10 BODY WEIGHT (g)
D
KNM-WT 15000
H. sapiens (adults)
H. sapiens (juveniles)
Great apes
Log10 MINIMUM THORACIC CANAL WIDTH (mm)
Log10 BODY WEIGHT (g)
E
KNM-WT 15000
H. sapiens (adults)
H. sapiens (juveniles)
Great apes
Log10 LUMBAR ENLARGEMENT CANAL HEIGHT (mm)
Log10 BODY WEIGHT (g)
F
KNM-WT 15000
H. sapiens (adults)
H. sapiens (juveniles)
Great apes
Log10 LUMBAR ENLARGEMENT CANAL WIDTH (mm)
Log10 BODY WEIGHT (g)

G

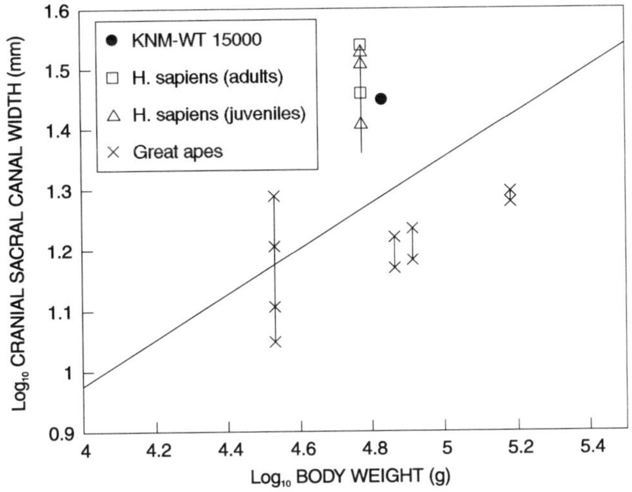

FIGURE 15.11.
Log-log plots of vertebral canal dimensions against body weight for modern great apes (chimpanzees, orang-utans, and female and male gorillas), modern humans, and KNM-WT 15000. All specimens measured are plotted for the great apes. For modern humans, the individual points plotted are the sample averages for adults and the sample averages and individual specimens for juveniles of around 11 years of age, from a range of studies (see Table 15.2); the ranges drawn in are the maximum ranges of two standard deviations either side of any of the sample means. On the plots of caudal sacral canal dimensions, the total range for canal measurements of modern humans (Spitalfields' collection) taken in the same vertebra as for KNM-WT 15000 is also plotted (data from present study; collected for MacLarnon, 1987; and published sources as Table 15.2). The best-fit lines drawn in are the same as for Figure 15.10.

H

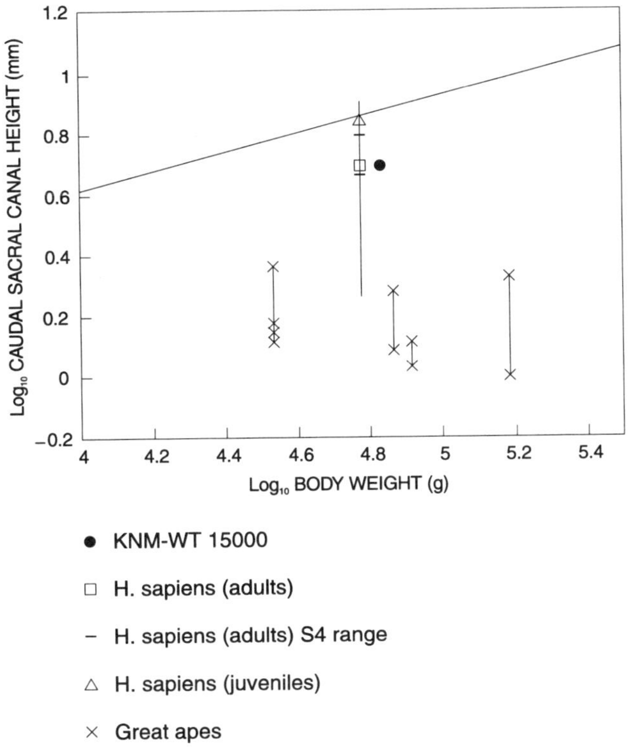

I

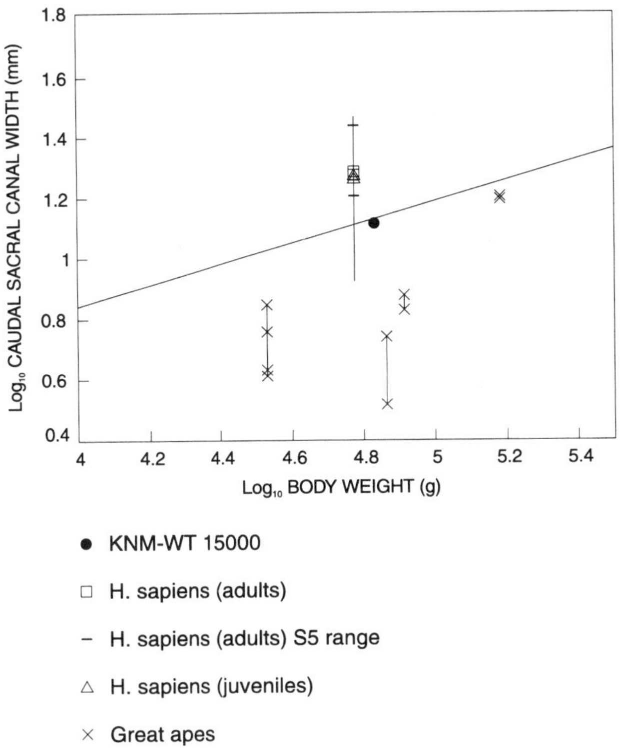

samples of one monkey species, apes, and humans, plus KNM-WT 15000. In each case, the line plotted in is the major axis calculated for the original sample used in MacLarnon (1987) of 38 modern primate species, including three with both male and female representatives (total $n = 41$). For the plots of caudal sacral dimensions, the major axes were calculated for species with long tails only, defined as those with at least 20 caudal vertebrae (i.e., excluding the hominoids and short-tailed species identified on Figures 15.10 and 15.11.)

Table 15.2 is a summary of the modern human data plotted on Figures 15.10 and 15.11, both from the present study and previous published studies, giving details of the samples and methods of measurement. Details of the modern human points and ranges plotted on Figures 15.10 and 15.11, and of the ranges shown for the extended samples of apes and one monkey species, are described in the figure legends.

The canal dimensions were chosen on the basis of the available measurements for KNM-WT 15000 and of which points along the vertebral canal could be defined comparably for all primate species. These are specific vertebrae in the cases of C7 and S1, the cranial sacrum, but the remainder are defined on the basis of landmarks in the total profiles of vertebral canal dimensions. While all primates, as almost all mammals, have 7 cervical vertebrae, the numbers of divisions and the relative lengths of the other regions of the vertebral column vary considerably (Schultz, 1961). This variation in the segmentation of the vertebral canal is related to mechanical properties of the vertebral column and not to the functioning of the contents of the vertebral canal, including the spinal cord. The vertebral canal dimensions are of interest here because of their relationship to the dimensions of the canal's contents, and so it is the total profile of the canal which is of interest, not its segmentation. A comparison of the dimensions of, for example, the canal through L3, might be a comparison of the dimensions of the lumbar enlargement of the canal for some species with the dimensions of the canal cranial or caudal to the lumbar enlargement in others. The external, mechanical reasons for the lumbar form of the articulation processes of certain vertebrae, which define the designation of a particular vertebra as L3, are unrelated to the internal swelling and narrowing of the vertebral canal.

The dimensions of the vertebral canal plotted in Figures 15.10 and 15.11, parts C–I, are defined in the same way as those plotted in Figure 15.6, except for the dimensions of the vertebral canal at the lumbar enlargement in non-hominids. For these species, the dimensions of the canal at the lumbar enlargement are defined as the maximum dimensions for each specimen of the canal height and width between the thoracic minima and the sacrum. They usually occur in lumbar vertebrae, but occasionally they occur in a thoracic vertebra, usually the last one. Only for hominids were the dimensions of the lumbar enlargement defined to be in a lumbar vertebra. This enables the inclusion of many published sources which are limited to the lumbar region, and makes comparison with KNM-WT 15000, which is missing T12, easier. For modern humans, unlike some other modern primates, the difference between the dimensions of the lumbar enlargement according to the two definitions is never more than very slight (see, for example, Spitalfields no. 2728, Figure 15.8G), and use of the restrictive definition of the position of the lumbar enlargement for humans makes negligible differences to the relevant plots of Figures 15.10 and 15.11, and no difference to any of the conclusions reached.

Figures 15.10 and 15.11, parts A, are log-log plots of the height of the cervical canal through C7 against body weight. All the species' points lie within a fairly narrow range of the best-fit line. The modern human range straddles the line, although all of the adult sample averages and the juvenile points and sample averages actually lie somewhat above the line. KNM-WT 15000 lies quite far below the line, indicating that its relative canal height through C7 is at the lower end of the modern primate range, and at the very bottom of the modern human and ape ranges. However, it is not outstanding.

On Figures 15.10 and 15.11, parts B, the range of variation in the relative width of the vertebral canal through C7, represented by the scatter of the species' points about the best-fit line, is even narrower than for the relative height of the canal at this point along the vertebral column. The modern human and ape ranges again straddle the best-fit line, and again all the individual human points for both adults and juveniles fall above the line. Some of the human points lie further above the line than those of any other modern primate, but it is notable that these are all from apparently uncorrected measurements taken from radiographs. KNM-WT 15000 lies above the line, indicating that its relative canal width through C7 is within the upper half of the modern primate range and clearly within the modern human range.

On Figures 15.10 and 15.11, parts C, the modern primates show little variation in the relative minimum height of the vertebral canal in the thoracic or interenlargement region. The male and female gorilla ranges are wide but still straddle the best-fit line. The modern human range lies completely above the best-fit line, overlapping, in distance above the line, only with a very few other species and the upper range of the gorillas. The one adult human sample average and two juvenile points, all from skeletal measurements, lie as far or further above the line than any other modern species. Taken together, the evidence from the human range and points suggests that modern humans have a relatively large minimum thoracic canal height compared with other modern species. KNM-WT 15000 lies a little below the best-fit line; its relative minimum thoracic canal height is well within the range typical of modern primates, showing no sign of the enlargement identified for modern humans.

Modern primates show little variation in the relative minimum width of the vertebral canal in the thoracic or interenlargement region, as illustrated in Figures 15.10 and 15.11, parts D. Among the apes, most have somewhat small relative canal widths at this point, except for some of the chimpanzees. Almost the whole modern human range lies above the best-fit line and all of the individual points, apart from that for the 12-year-old Spitalfields' child, lie about as far or further above the line than any of the other modern species, although again the largest measurements are all from radiographic data. The relative minimum thoracic canal width of modern humans is large to very large compared with that of other primates. KNM-WT 15000 lies a little below the best-fit line; its minimum thoracic canal width is of a typical relative size for a modern primate, and it does not have the large relative width of modern humans.

Figures 15.10 and 15.11, parts E, illustrate the narrow range of variation in the relative height of the vertebral canal at the lumbar enlargement among modern primates. Some of the ape ranges straddle the best-fit line; others fall completely below. The very wide modern human range also straddles the line, but most of it falls above, as do all the individual human points. Several of these points, both adult and juvenile, including some from skeletal data (e.g., South African whites, a nineteeth-century sample, and the 10-year-old Spitalfields' child), lie further above the line than any of the other modern primates. Modern humans have large relative heights of the canal lumbar enlargement. KNM-WT 15000 lies as far above the best-fit line as the most extreme non-human modern primate. This indicates that it has a relatively large lumbar enlargment canal height compared with non-human modern primates, within the range of modern humans.

The distribution of non-human points about the best-fit line on Figures 15.10 and 15.11, parts F, is tight. There is very little variation among these species in the relative width of the vertebral canal at the lumbar enlargement. The whole of the modern human range lies above the best-fit line, and the majority of points, both adult and juvenile, including several from skeletal data (a range of samples of European origin, two South African black samples, a nineteenth-century sample, an Anglo-Saxon sample, a Roman sample, and the 10-year-old Spitalfields' child), lie further above the line than the points for any other modern species. Modern humans have relatively large maximum lumbar canal widths, usually located in the last lumbar vertebra. KNM-WT 15000 also lies above the best-fit line on Figures 15.10 and 15.11, parts F, within the modern human range of distance above the line, and further above the line than all other modern specimens except for one chimpanzee. This indicates that the fossil had an enlarged relative maximum lumbar canal width (from the last lumbar vertebra) compared with

TABLE 15.2.

Vertebral canal measurements for *Homo sapiens*.

CANAL DIMENSION	SOURCE	METHOD[a]	SAMPLE[b]	*n*	MEASUREMENTS (mm) Average	Range[c]
C7 canal height	Present study	S	Spitalfields, adults	7	13.7	11.1–16.3
		S	Spitalfields, 10y 5m	1	14.9	
		S	Spitalfields, 12y	1	14.85	
	Gupta et al. (1982)	R	Indian adults	300	15.98	13.0–20.5
	Wolf et al. (1956)[d]	R	European adults	200	17	
	Payne and Spillane (1957)[d]	R	European adults	—	17.2	
	Burrows (1963)[d]	R	—	300	17.3	
	Bhalla et al. (1977)[d]	R	Indian adults	30	17.2	
	Markuske (1977)	R	East Germans, 7–10y	40	16.0	14.2–17.8
		R	East Germans, 11–14y	40	16.25	14.45–18.05
	Stanley et al. (1986)	T	American adults	52	15.0	10.5–19.4
C7 canal width	Present study	S	Spitalfields, adults	7	23.2	20.8–25.6
		S	Spitalfields, 10y 5m	1	24.4	
		S	Spitalfields, 12y	1	21.9	
	Haworth and Keillor (1962)	R	American adults	100	27.7	
		R	Americans, 8–14y	33	27.1	
	Hinck et al. (1966)	R	American whites, adults	121	28.2	23.1–33.3
		R	American whites, 9–10y	353[e]	26.3	22.0–30.5
		R	American whites, 11–12y	353[e]	25.9	20.2–31.6
	Stanley et al. (1986)	T	American adults	52	22.9	17.7–28.1
Minimum thoracic canal height	Present study	S	Spitalfields, adults	7	14.4	12.3–16.5
		S	Spitalfields, 10y 5m	1	14.8	
		S	Spitalfields, 12y	1	15.3	
Minimum thoracic canal width	Present study	S	Spitalfields adults	7	15.9	13.9–17.9
		S	Spitalfields, 10y 5m	1	15.7	
		S	Spitalfields, 12y	1	14.2	
	Haworth and Keillor (1962)	R	American adults	100	17.6	
		R	Americans, 8–14y	33	16.6	
	Hinck et al. (1966)	R	American whites, adults	121	17.4	13.9–20.8
		R	American whites, 9–10y	353[e]	16.4	13.7–19.1
		R	American whites, 11–12y	353[e]	16.6	12.7–20.6
Lumbar enlargement canal height	Present study	S	Spitalfields, adults	7	17.1	16.0–18.2
		S	Spitalfields, 10y 5m	1	18.4	
		S	Spitalfields, 12y	1	16.65	
	Berry et al. (1987)	S	American Caucasian adults	30	17.3	11.5–23.1
	Eisenstein (1976)	S	South African Caucasian adults	113	18	
		S	Zulu adults	162	16.5	
	Eisenstein (1977)	S	Sotho adults	168	16	
	Hinck et al. (1965)	R	American whites, adults	121	22.0	16.7–28.0
		R	American whites, 9–10y	12	20.0	16.0–23.6
		R	American whites, 11–12y	16	21.2	15.0–24.6
	Larsen and Smith (1981)	R	Norwegians, 9–11y	28	16.6	
		R	Norwegians, 12–14y	25	16.7	
	Huizinga et al. (1951)[f]	S	Nineteenth-century, adults	—	18.0	
	Sand (1970)[f]	S	Middle Ages adults	—	16.0	
	Porter and Pavitt (1987)	S	Anglo-Saxon adults	839[g]	15.23	13.30–17.14[h]
		S	Roman adults	839[g]	15.85	12.68–18.29[h]
	Twomey and Taylor (1988)	S	European Australian adults	96	16.6	11.1–22.4

TABLE 15.2 *(continued).*

Vertebral canal measurements for *Homo sapiens* *(continued).*

CANAL DIMENSION	SOURCE	METHOD[a]	SAMPLE[b]	*n*	MEASUREMENTS (mm)	
					Average	Range[c]
Lumbar enlargement canal width	Present study	S	Spitalfields, adults	7	23.8	19.6–28.0
		S	Spitalfields, 10y 5m	1	26.2	
		S	Spitalfields, 12y	1	21.6	
	Amonoo-Kuofi (1982)	R	Nigerian adults	290	28.6	24.1–33.3
	Amonoo-Kuofi et al. (1990)	R	Saudi adults	340	30.0	24.4–36.1
	Berry et al. (1987)	S	American Caucasian adults	30	26.0	21.0–31.0
	Eisenstein (1976)	S	South African Caucasian adults	113	26	
		S	Zulu adults	162	25	
	Eisenstein (1977)	S	Sotho adults	168	25	
	Haworth and Keillor (1962)	R	American adults	100	30.3	
		R	Americans, 8–14y	33	28.6	
	Hinck et al. (1966)	R	American whites, adults	121	29.9	21.6–38.1
		R	American whites, 9–10y	353[e]	28.3	22.8–33.8
		R	American whites, 11–12y	353[e]	28.7	22.3–35.1
	Larsen (1981)	R	Norwegians, 9–11y	28	24.6	
		R	Norwegians, 12–14y	25	25.6	
	Huizinga et al. (1951)[i]	S	Nineteenth-century adults	—	25.8	
	Sand (1970)[i]	S	Middle Ages adults	—	25.0	
	Piera et al. (1988)	R	Spanish adults	215	33.85	26.01–41.45
	Porter and Pavitt (1987)	S	Anglo-Saxon adults	839[g]	25.55	21.67–29.43[h]
		S	Roman adults	839[g]	25.66	22.09–29.23[h]
	Postacchini et al. (1983)	S	Italian adults	63	25.9	21.1–30.7
		S	Indian adults		22.8	17.6–28.0
	Twomey and Taylor (1988)	S	European Australian adults	96	28.4	22.2–35.8
Cranial sacral canal width	Present study	S	Spitalfields, adults	7	29.0	22.7–32.2
		S	Spitalfields, 10y 5m	1	32.4	
		S	Spitalfields, 12y	1	25.6	
	Haworth and Keillor (1962)	R	American adults	100	34.7	
		R	Americans, 8–14y	33	33.6	
Caudal sacral canal height	Present study	S	Spitalfields, adults	7	4.9	1.8–8.0
		S	Spitalfields, 12y	1	6.9	
Caudal sacral canal width	Present study	S	Spitalfields, adults	7	18.7	8.3–29.1
		S	Spitalfields, 10y 5m	1	18.6	
		S	Spitalfields, 12y	1	18.7	
	Haworth and Keillor (1962)	R	American adults	100	19.2	
		R	Americans, 8–14y	33	18.0	

a. Measurement methods: R = radiographic; S = skeletal; T = computed tomography.

b. Male and female samples were combined where necessary. Spitalfields = Spitalfields' collection at The Natural History Museum, London; y = years; m = months of age.

c. Ranges equal to the average plus and minus two standard deviations; for combined samples, the maximum range of two standard deviations on either side of the individual sample averages is given.

d. Cited in Gupta et al., 1982.

e. For Hinck et al., 1966, *n* = 353 for a combined sample of 3- to 18-year-olds.

f. Cited in Larsen and Smith, 1981.

g. For Porter and Pavitt, 1987, *n* = 839 vertebrae for a combined sample of Anglo-Saxons and Romans.

h. For Porter and Pavitt, 1987, range equals the tenth to the nintieth percentiles.

i. Cited in Larsen, 1981.

non-human modern primates, just within the range of modern humans.

There is more variation among modern primates in the relative size of the cranial sacral canal width than there is for any of the more cranial canal dimensions analyzed, as illustrated by the wider scatter of points about the best-fit line on Figures 15.10 and 15.11, parts G, compared with previous graphs. This is true of the non-human primates on their own, but the most outlying modern species is *Homo sapiens.* Almost the complete range for modern humans, and all the individual points, both adult and juvenile, lie further above the line than the points for all other modern species. Modern humans have an enlarged relative cranial sacral canal width. KNM-WT 15000 also lies much further above the line than any modern non-human primate, within the range of distance from the line of modern humans; the fossil also has an enlarged relative cranial sacral canal width.

There is considerably more scatter of the species' points on Figures 15.10 and 15.11, parts H and I, than for any of the previous graphs; the relative dimensions of the caudal end of the sacral vertebral canal are much more variable than those of any more cranial part of the canal, including the cranial sacrum. Much of this variation is associated with variation in tail length. In most species with short tails or no tail, the sacral canal narrows much more from cranial to caudal ends than in species with long tails. The best-fit lines drawn in describes the relationship between the dimensions of the caudal sacral canal and body weight for long-tailed (20+ caudal vertebrae) species. Most of the points for short-tailed species or those with no tail (hominoids) fall well below the lines; they have relatively small caudal sacral canal dimensions. Modern humans, however, are an exception to this general picture. On Figures 15.10 and 15.11, parts H, the range for the caudal sacral canal height of modern humans straddles the best-fit line, though mostly falling below it, the range of distance from the line being similar to that for short-tailed species. This is unlike the points for any other tailless species. The single sample average point and single juvenile point for modern humans lie a little below the line, well above the ranges for the apes. Modern humans have unusually large caudal sacral canal heights for a tailless species, similar to those of short-tailed species. KNM-WT 15000 also lies only a short distance below the line, in a very similar relative position to the modern human adult average. The caudal sacral canal height of KNM-WT 15000 was measured in S4. The range of measurements for modern humans in which the caudal sacral canal height was also measured in S4 is shown on Figure 15.11H. KNM-WT 15000 lies within the range of distance from the line of these points. The fossil had an enlarged relative caudal sacral canal height for a tailless species, like that of modern humans.

On Figures 15.10 and 15.11, parts I, the range of values for the caudal sacral canal width of modern humans straddles the best-fit line for long-tailed modern primates, but all of the points for human sample averages and individual juveniles lie further above the line than those for any other modern species. Modern humans have an enlarged relative caudal sacral canal width compared with all other modern primates, very different from the typically small relative width of tailless species. KNM-WT 15000 lies just below the best-fit line on these figures. The measurement of the caudal sacral canal width for the fossil was from S5. The range of measurements taken in S5 of the caudal sacral canal width for modern humans is also shown on Figure 15.11I. KNM-WT 15000 lies quite a long way below the lower limit of this range. KNM-WT 15000 had a very large relative caudal sacral canal width compared with most modern non-human primates, especially compared with tailless species, but it was not quite as large as for modern humans with similar sacra.

THORACIC SPINAL CORD

The enlarged parts of the modern human vertebral canal identified in the previous section fall into two groups according to their neural contents: the thoracic canal contains spinal cord; the lower lumbar and sacral canals are beyond the end of the spinal cord and contain the cauda equina. The enlarged thoracic canal of modern humans suggests the possibility of a concomitant enlargement of the spinal cord.

Figure 15.12 is a log-log plot of the minimum cross-sectional area of the spinal cord in the thoracic region for the primate specimens dissected for MacLarnon (1987) for which the tissue preservation was satisfactory, plus comparable human data from Donaldson and Davis (1903), versus body weight. The best-fit line drawn in is the major axis calculated for the non-human species. The human point lies about as far above the line as the most outlying non-human primate point, indicating that humans have quite a large relative spinal cord cross-sectional area at this point along the vertebral column, though not outstanding. Figure 15.12, parts B and C, are similar plots to Figure 15.12A for the white matter and grey matter cross-sectional areas of the cord. The human point lies close to the best-fit line for non-human primates on the white matter plot, indicating that humans have a relative cross-sectional area of the thoracic white matter typical for a primate. However, on the grey matter plot, the human point lies quite a lot further above the best-fit line for non-human species than any of the other species' points. Humans have a larger relative grey matter cross-sectional area in the thoracic region than any other primate species, despite their relative total spinal cord cross-sectional area not being especially large.

Discussion

In analyzing the vertebral canal dimensions of the *Homo erectus* boy skeleton, KNM-WT 15000, the first consideration is whether or not they are affected by its juvenile status. Evidence from the present study, and previous published studies, shows that the average height (i.e., the transverse diameter) and width (i.e., the dorso-ventral di-

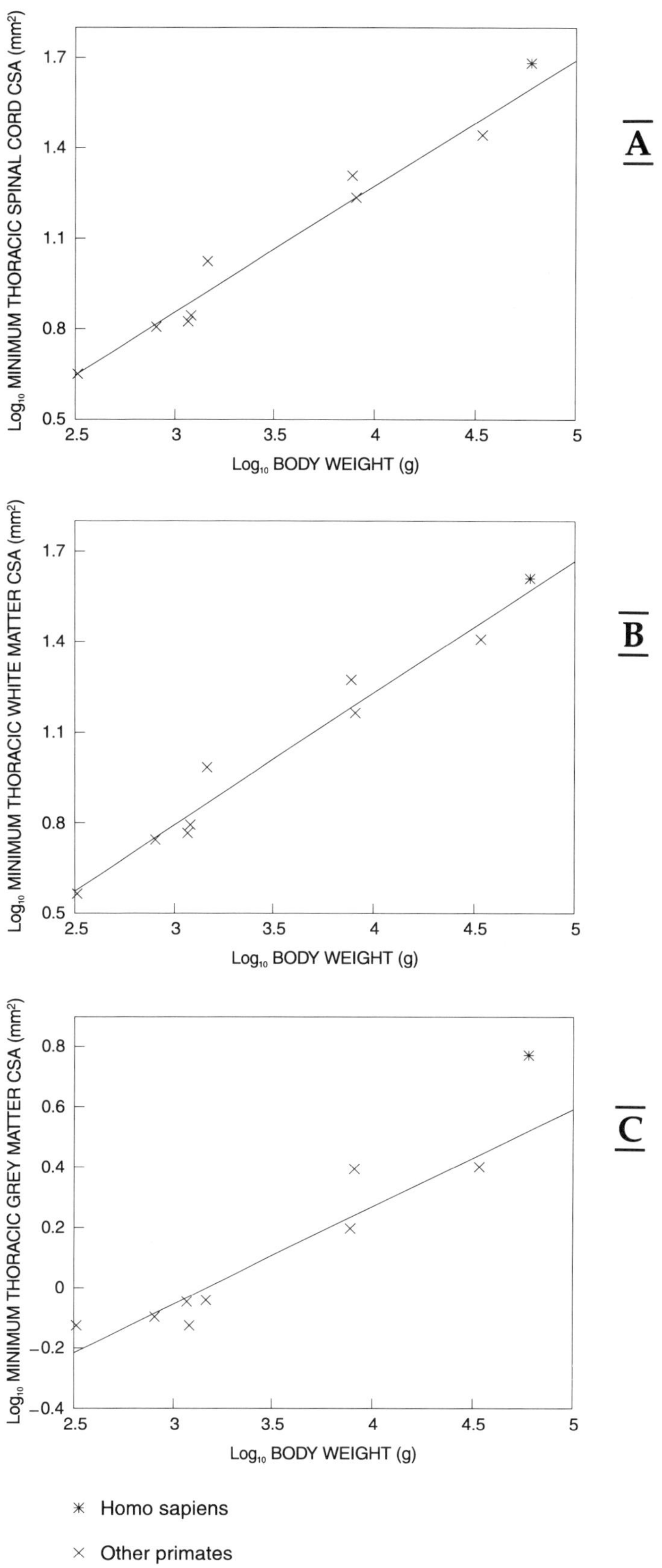

FIGURE 15.12.
Log-log plots of the minimum cross-sectional areas of the total spinal cord and of the white matter and the grey matter of the spinal cord in the thoracic region of the vertebral column against body weight for primates (data from Donaldson and Davis, 1903; and collected for MacLarnon, 1987). The best-fit lines drawn in are the major axes for the nonhuman species.

ameter) dimensions of the modern human vertebral canal increase by as much as 25% from the age of 2–3 years, when the vertebrae begin to fuse, to adulthood. Adult dimensions are generally reached earlier in canal heights than canal widths, and earlier in postcervical regions than in the cervical region. After the age of about 11 years, the equivalent of the stage of development of KNM-WT 15000 (Smith, Chapter 9, this volume), there is no clearcut evidence of growth of either the canal height or width in any region of the vertebral column. In any case, the average increase in canal dimensions throughout the entire growth period is generally insignificant in comparison with the variation in each of the canal dimensions within individual age groups. From this survey, there is no evidence that the juvenile status of KNM-WT 15000 would explain any significant differences between its relative vertebral canal dimensions and those of modern human adults, especially as the fossil is missing the first six vertebrae from the cervical region, the part of the vertebral canal that attains adult size last. Analyses of individual dimensions of the vertebral canal relative to body weight, comparing both juvenile and adult modern humans with adults of other modern primate species, confirm this conclusion.

From the known vertebrae, it is possible to ascertain that the vertebral canal of KNM-WT 15000 has, in general terms, a profile typical of all primates, and indeed of other four-legged vertebrates (see, for example, Ariëns Kappers et al., 1936) along the length of the vertebral column. The canal swells in the cervical and lumbar regions where the spinal cord expands to provide innervation for the fore- and hindlimbs, narrows in the interenlargement region where comparatively little innervation is required, and tapers at the caudal end as the cauda equina peters out, beyond the end of the spinal cord itself. However, at a more detailed level, the vertebral canal of the fossil shows some characteristics typical of, or even unique to, *Homo sapiens* among modern primates, both in its shape and size.

Although in the cervical region only C7 is known for KNM-WT 15000, the canal dimensions of this single vertebra and those of the first few thoracic vertebrae provide sufficient evidence to suggest that the cervical enlargement of the fossil canal largely involved swelling of the canal width and not of the canal height. While not unique to *Homo sapiens* among modern primates, this form of the cervical enlargement of the canal is common among modern humans and uncommon among other species. It results in no difference in the relative size of the total cross-sectional area of the cervical enlargement of the canal between modern humans and other primate species (MacLarnon, 1987). The possible functional significance of such a shape difference is not clear, but it is of interest that *Homo erectus* apparently shared a typical modern human trend away from the common primate shape of the cervical enlargement of the vertebral canal.

The size of the vertebral canal through C7 relative to body weight for KNM-WT 15000 falls within the narrow range of variation found among modern primates, including humans. Both the relative height and width of the canal through C7 in modern humans fall at the upper end of the range for modern primates. The relative canal height of KNM-WT 15000 falls at the lower end of the modern primate range, though just within the human range; the relative canal width of the fossil falls well within the modern human range. The analyses presented here therefore do not give strong support for Brown et al.'s (1985) statement that the cervical canal of KNM-WT 15000 was relatively small compared with that of modern humans, nor their suggestion that it had a smaller relative spinal cord size at this point. However, more cervical vertebrae, and particularly more cranial cervical vertebrae, are needed before firm statements can be made regarding the relative size of the cervical vertebral canal of *Homo erectus*, and hence the relative size of its cervical spinal cord.

Just over half of the thoracic vertebrae are known for KNM-WT 15000, spread throughout the region. After decreasing a little in width and increasing a little in height in the first two thoracic vertebrae, from the third to at least the eleventh thoracic vertebra the thoracic vertebral canal of the fossil maintains fairly constant size and proportions. This is typical of all modern primates, including humans. The relative size of the thoracic vertebral canal in KNM-WT 15000, as measured by the minimum canal height and width in the region, is also right in the middle of the range for modern non-human primates. In this it is unlike modern humans, which have larger relative thoracic canals than all other modern primates. The expansion of the human thoracic vertebral canal therefore must have taken place after the time of *Homo erectus.*

Relative dimensions of the vertebral canal are in broad terms well correlated with the relative dimensions of the spinal cord contained within it among modern primates, both the sizes of the total cord and canal and their dimensions at particular points along the length of the vertebral column. Relative cord size is apparently a major determinant of relative canal size. However, relative cord and canal dimensions are not perfectly correlated, and not all details are matched (MacLarnon, 1987). Analyses presented here show that the enlarged thoracic vertebral canal of modern humans is not associated with an enlargement of the relative girth of the total spinal cord contained within it, compared with other primates. The relative amount of white matter in the modern human thoracic cord is also no bigger than expected for a typical modern primate, but the relative amount of grey matter at this point is larger than for all the non-human species examined. That is, modern humans have the expected amount or number of nerve fibers passing through the thoracic spinal cord, but an increased amount or number of nerve cell bodies within the region, compared with other primates. The bulk of nerve cell bodies in a region of the spinal cord is related to the level of innervation provided by the spinal nerves whose roots arise from that part of the cord. The bulk of these spinal nerve roots presumably increases in line with the increase of the corresponding grey matter, hence requiring an expansion of the girth of the vertebral canal.

The spinal cord within the thoracic region of the vertebral canal broadly gives rise to the thoracic spinal nerves. (Because of the differential growth of the cord and canal described above, a few of the cranial-most thoracic rootlets of the thoracic spinal nerves arise from the cord within the cervical vertebral canal, and the caudal-most thoracic vertebral canal contains cord from which some rootlets of lumbar spinal nerves originate. However, as the minimum vertebral dimensions used here as the measure of thoracic canal size usually occur somewhere in the mid-thoracic canal, and not at its cranial or caudal ends, the enlarged grey matter of the spinal cord of humans is taken to correspond to the part of the cord giving rise to the thoracic spinal nerves.) The thoracic spinal nerves, apart from the first and sometimes the second, innervate the muscles of the walls of the thorax and the overlying skin, as well as the abdominal muscles; they do not innervate the muscles of the limbs. Therefore any increased innervation by the thoracic spinal nerves in humans compared with other primates must be in the mid- or lower trunk regions.

So, modern humans have an enlarged relative thoracic vertebral canal compared with other modern primates, which might be associated with their enlarged relative bulk of grey matter in the thoracic spinal cord, through an increased bulk of associated thoracic spinal nerves. *Homo erectus* does not have an enlarged thoracic canal and it therefore, presumably, had neither an increased bulk of thoracic canal grey matter nor enlarged thoracic spinal nerves compared with other primates. Two possible functional reasons can be suggested for the implied increase in innervation of the mid- or lower trunk region in *Homo sapiens* compared with *Homo erectus:* it could be associated with increased muscular movement or control of the trunk, or it could be associated with increased muscular control of breathing, which involves both intercostal and abdominal muscles. The former could be connected with an improvement in the efficiency of bipedalism; the latter could be connected with the development of speech (Gwen Hewitt suggested the link with increased control of breathing and speech development).

The second to sixth lumbar vertebrae are known for KNM-WT 15000, plus, though somewhat damaged, five sacral vertebrae. In these regions, the vertebral canal of modern humans is distinctively different from that of other modern primates, and in these characteristics KNM-WT 15000 is broadly similar to modern humans. In general terms, the caudal lumbar and sacral canals of the hominids are expanded compared with other primates, the latter particularly in comparison with the sacral canal of other tailless and short-tailed species.

The lumbar enlargement of the spinal cord occurs in the caudal-most part of the thoracic vertebral canal and the cranial part of the lumbar canal in primate species, and this is usually associated with the lumbar enlargement of the vertebral canal, evident in both its height and width. Caudal to this, the spinal cord ends fairly quickly and the vertebral canal usually tapers into the caudal lumbar and sacral regions, where it carries the remaining spinal nerve roots of the cauda equina. Rarely among modern primates, except in humans, does the vertebral canal height, and particularly the vertebral canal width, continue increasing through many vertebrae caudal to the end of the spinal cord in the lumbar region and into the first sacral vertebra. In no modern species, except *Homo sapiens* and possibly some New World monkeys with prehensile tails, is there a very marked increase of the canal width between the last lumbar vertebra and the first sacral vertebra. Together, the increases in both canal dimensions in the lower lumbar region in modern humans result in the last lumbar vertebrae commonly having the largest cross-sectional area of all the postcervical vertebrae; in other modern primates, with the exception of only the occasional specimen, the maximum cross-sectional area of the postcervical canal falls in a caudal thoracic or cranial lumbar vertebra. KNM-WT 15000 shares all these characteristics of the modern human lumbar and cranial sacral canal. The unusual, increasing bore of the vertebral canal in the lower lumbar and cranial sacral canal of the modern and fossil hominids is reflected in their having larger relative maximum lumbar enlargement canal heights and widths, and larger cranial sacral canal widths (the cranial sacral canal height of KNM-WT 15000 is unknown), than all non-human modern primates.

Within the sacrum itself, the vertebral canal narrows caudally in virtually all modern primate specimens. It narrows more in species with shorter tails and even more in most tailless species in which it tapers to a tiny caudal opening. This reflects variation in the relative importance, and therefore size, of the spinal nerves innervating the tail. Modern humans and KNM-WT 15000 have exceptionally large openings of the caudal sacral canal not only for tailless species but also, in the transverse direction, compared with many long-tailed modern non-human primate species as well. The relative caudal sacral canal heights of the hominids are within the lower range for long-tailed modern primates, distinctly different from the very small relative canal heights of the tailless modern apes. Their relative caudal sacral canal widths are even more outstanding: modern humans have relative caudal sacral canal widths which are usually larger than those of any other modern primate, including species with prehensile tails; the relative caudal sacral canal width of KNM-WT 15000 is within the lower range of modern humans and above the average for long-tailed modern primates.

The swelling of the lower lumbar and sacral parts of the vertebral canal in modern humans occurs caudal to the end of the spinal cord, which usually ends between the first two lumbar vertebrae. In other modern primates, it is possible to make a reasonable estimate of where the end point of the spinal cord occurred from the dimensions of the vertebral canal—namely, in the second vertebra caudal to the one at the peak of the lumbar enlargement of the canal (which may be a lumbar or thoracic vertebra). For some modern human specimens, there is a first peak of the cross-sectional area of the lumbar canal in the caudal-most thoracic or first lumbar vertebrae, but the actual maximum lumbar canal size usually occurs in the last lumbar vertebra. Hence, the position of the end point of the modern human spinal cord cannot be estimated accurately from canal dimensions, at least not in the same way as for other primate species. The maximum lumbar canal dimensions of KNM-WT 15000 also occur in the last lumbar vertebra. By the estimation method devised for modern non-human primates, this would imply that the spinal cord of the fossil ended in S2. In all modern primates, the spinal cord ends presacrally. It therefore seems highly probable that the spinal cord of *Homo erectus* ended in the lumbar region, and that its vertebral canal swells beyond the end of the cord as in modern humans. The spinal cord ended in L1 in a chimpanzee (Mac-

Larnon, 1987), and it can therefore be suggested that the spinal cord of *Homo erectus* ended in L1 or L2, as for the two modern hominoids for which data are available.

Caudal to the end of the spinal cord, the vertebral canal contains two major structures, the cauda equina and the internal vertebral venous plexus. The cauda equina of humans comprises the lower lumbar, sacral, and coccygeal spinal nerve roots. The pairs of ventral and dorsal roots combine to form the spinal nerves at the spinal ganglia, which also contain the nerve cell bodies of the sensory dorsal fibers. The ganglia lie in the intervertebral foramina throughout most of the body, but they lie inside the vertebral canal in the sacrum. All the spinal nerves exit from the vertebral canal through their respective intervertebral foramina. The lower lumbar, sacral, and coccygeal spinal nerves primarily innervate the muscles of the hindlimbs and the overlying skin of the lower abdomen and hindlimbs.

The internal vertebral venous plexus is a complex of four interconnected veins which run throughout the length of the vertebral canal. This large, valveless plexus is also connected to many other veins of the body, forming a passage for the flow of venous blood from areas of higher to lower pressure as well as the drainage system for the structures of the vertebral column. In the lower body, it provides an alternative route to the inferior vena cava for the return of blood to the heart. It can be distended greatly by an unusually large flow of blood.

Two suggestions based on the contents of the vertebral canal can be made to explain the evolutionary increase in the size of the lower lumbar and sacral canals in *Homo erectus* and *Homo sapiens*. First, it may reflect an increase in the size of the spinal nerve roots of the cauda equina and the sacral spinal ganglia, associated with increased innervation of the lower body in connection with the development of bipedalism. Second, the internal vertebral venous plexus may have been expanded, which could also be connected with the development of bipedalism through alteration in the pattern of venous blood flow with a habitual upright stance. The latter explanation seems more probable, as the lumbar enlargement of the spinal cord in modern humans, which gives rise to the lumbar spinal nerves, is of typical relative size for a modern primate (MacLarnon, 1987).

However, the expansion of the lower lumbar and sacral vertebral canal in *Homo erectus* and *Homo sapiens* may not be caused by factors intrinsic to the canal itself. Latimer and Ward (Chapter 12, this volume) suggest that the widening of this part of the canal is the indirect consequence of the lumbar lordosis of these bipedal species. The lordosis requires the distance between the pairs of vertebral articular facets to increase caudally through the lumbar region and into the cranial sacrum, contrasting with the constant or even craniocaudally decreasing distance found in the apes. Latimer and Ward suggest that the craniocaudal widening of the interfacet distance in the hominids in turn causes the craniocaudal widening of their lumbar and cranial sacral vertebral canals. At least in the more cranial regions of the vertebral column, however, the dimensions of the canal are quite well correlated with the swellings and narrowings of its major content in these regions, namely the spinal cord. It is to be expected that the size of the canal in the lower lumbar and sacral regions also reflects the size of its contents. Therefore, even if the size of the lower lumbar and cranial sacral canal is primarily influenced by external factors, apparently unlike more cranial regions, its enlargement still provides the space for an enlarged internal vertebral venous plexus. However, whether external or internal factors are the cause of the enlarged lumbar and sacral vertebral canals of the hominids, all the proposed explanations imply that the unique canal profile they share in these regions is a mark of their upright stance and bipedality.

The evidence presented here does not suggest that the spinal cord of *Homo erectus* was very different in total size from that of *Homo sapiens*. The suggested increase in the amount of thoracic grey matter in the cord of *Homo sapiens* makes an insignificant difference to the total size of the spinal cord. This is in contrast to the significant difference between the relative brain sizes of the two hominids. The brain size of KNM-WT 15000 is estimated to have been 878 cc (Begun and Walker, Chapter 14, this volume). Brain size matures early in development in all mammals. It reaches adult size at about 6–7 years in *Homo sapiens,* which is comparatively late in the development process (Martin, 1983). As KNM-WT 15000 is estimated to have been at a stage of development equivalent to an 11-year-old modern human (Smith, Chapter 9, this volume), it is reasonable to assume that it had already acquired its adult brain capacity. On the basis of the estimated brain and body sizes of the fossil, its relative brain size would have been just over three times that expected for a modern mammal, compared with a relative brain size for *Homo sapiens* of over five times the expected capacity (equation from Martin, 1981).

So, while there was a considerable increase in relative brain size between *Homo erectus* and *Homo sapiens,* there was no significant concomitant increase in the total size of the spinal cord. This is in keeping with the lack of significant variation in relative spinal cord size among modern mammals, which show considerable variation in relative brain size. Although there is evidence that relative total spinal cord size increased from the earliest primates of the Eocene to present-day forms, major increases in brain size must have taken place in mammalian evolution, including the hominid lineage, independent of any major increase in relative spinal cord size. The evolutionary increase in brain size, including that between *Homo erectus* and *Homo sapiens,* therefore apparently involved almost exclusively increase in the capacity for information processing and integration, rather than any major increase in information flow to and from the postcranial division of the central nervous system. The one notable exception may have been an increase in the innervation of the thoracic and abdominal regions, which may have been associated with increased control of trunk position or an increase in the control of breathing, perhaps related to the evolution of speech.

Acknowledgments

I should like to thank Richard Leakey and Alan Walker for the opportunity to work on KNM-WT 15000. Measurements of the vertebrae were kindly taken by Mark Teaford and Alan Walker. Access to other skeletal material was given by Paula Jenkins and Theya Molleson. Help with the collection of spinal cord data was given by many people, particularly Alan Ainsworth and Mary Brancker. Joanna Burgess produced most of the diagrams. Gwen Hewitt provided very helpful comments. Bob Martin gave me considerable advice and encouragement during the work which provided the basis for this study. I am very grateful to all of these people.

CHAPTER 16

A Morphometric Study of Facial Growth

Joan T. Richtsmeier and Alan Walker

With the discovery of the first juvenile Homo erectus *facial skeleton at Nariokotome, we took the opportunity to examine the patterns of growth of the face. Several models of facial growth were devised and quantified. We used three-dimensional morphometric techniques to characterize growth in a cross-sectional series of humans and chimpanzees. We then used computer simulation to suggest a pattern of facial growth for* Homo erectus *by producing an adult male* Homo erectus *face using various models and checked this growth pattern against the human and chimpanzee patterns.*

In the past few years paleontologists have become increasingly interested in issues relating to growth and development. This is due partly to the attention that the concept of heterochrony has received as a mechanism for evolutionary change (e.g., Gould, 1966, 1977, 1988; Shea, 1983, 1990; McKinney, 1988). That small manipulations in well-established growth patterns might lead to significant changes in morphologies is an idea that begins to bridge genetic mechanisms of evolution with concepts of growth and development (see Atchley, 1987; Atchley and Hall, 1991; Slatkin, 1987).

The concept of heterochrony deals primarily with changes in *rate* or *timing* of growth. Gould (1988) attributes the prevailing and intense interest in heterochrony by evolutionary biologists (e.g., Bonner, 1982; Atchley, 1987; Raff and Raff, 1987; McKinney, 1988; Strauss, 1990; Raff, 1990; Atchley and Hall, 1991) to the pursuit of information regarding behavior and life history on the one hand and advances in the study of cellular processes and the molecular genetics of development on the other. Put simply, this interest reflects a renewed debate concerning the respective roles of internal versus external influences upon the production of evolutionary change (Gould, 1988).

In this study, we focus on a feature of the growth process that is useful in the study of phylogenetic relationships from an ontogenetic perspective. Here, we consider *patterns* of morphological change through time with advancing age of the specimen. A *growth pattern*, as defined here, consists of the physical interpretation of the geometric changes in morphology due to growth. We summarize facial growth patterns by measuring the direction and magnitude of change at biological loci on the facial skeleton using finite-element scaling analysis and Euclidean distance matrix analysis.

It is not currently known whether modern *Homo sapiens* or one of the great apes serves as a more appropriate model for studying growth in *Homo erectus*. When *Pan* and humans are compared at homologous dental ages, it is obvious that chimps have completed more of their growth than have humans (Smith, Chapter 9, this volume). A feature-by-feature analysis of KNM-WT 15000 indicates that some bones of *Homo erectus* are remarkably close to their counterparts in modern humans, but others are different (Leakey and Walker, 1989). We suggest that growth patterns of specific skeletal systems can also be used as features in phylogenetic analyses. This approach follows de Queiroz's (1985) proposal that ontogenetic transformations be used as characters in systematic analysis.

For example, Martin's (1983) discussion of the rates of brain growth in humans and apes prompted Leakey and Walker's (1989) discussion of the attainment by *Homo erectus* of growth mechanisms that produce secondary altricial young, a uniquely human feature. Assuming growth patterns include phylogenetic information, then it is useful to determine if the facial growth pattern of *Homo erectus* more closely resembles that of modern humans or that of the great apes. We conduct an analysis of facial growth in two modern species, *Pan troglodytes* and *Homo sapiens*, in the hope that comparison of the patterns will provide new information for placing *Homo erectus* within a proper phylogenetic context.

We study facial growth in these species to examine similarities and differences in growth patterns, to suggest generalized models of sexual dimorphism of facial growth that can be further tested and applied to other fossils, and to formulate some hypotheses about the links between species-specific patterns of facial growth and phylogenetic relationships among groups. Although our analysis will focus only on growth patterns, we assume a genetic basis for these morphological trajectories (see Atchley, 1987; Slatkin, 1987; Atchley and Hall, 1991). We also assume that similarities in morphogenetic patterns denote genetic relationships.

In this chapter we use three-dimensional morphometric techniques to characterize the trajectories of adolescent facial growth in modern *Homo sapiens*, *Pan troglodytes*, and specimens of *Homo erectus* recovered from Lake Turkana, Kenya. This is done in the two modern species by comparing samples of subadult individuals with those of adults of both sexes. To characterize growth in *Homo erectus*, the face of KNM-WT 15000 is compared first with that of KNM-ER 3733 and then with computer-simulated male *Homo erectus* faces. The morphometric methods give the direction and magnitude of changes in facial morphology during growth at specific landmarks and determine the change in form required to produce the older from the younger morphology.

Methods and materials

Two methods were used in this study of facial growth patterns: finite-element scaling analysis (FESA) and Euclidean distance matrix analysis (EDMA). The specifics of FESA have been discussed in numerous publications (e.g., Richtsmeier and Cheverud, 1986; Cheverud and Richtsmeier, 1986; Skalak et al., 1986; Lozanoff and Diewert, 1986; Richtsmeier, 1990). Our treatment will therefore be brief and will focus only on aspects of the method pertinent to this application. EDMA has been introduced with thorough explanation elsewhere (Lele, 1991; Richtsmeier and Lele, 1990; Lele and Richtsmeier, 1991, 1992). Only an abbreviated description is given here.

Finite-element scaling analysis (FESA)

This method, based on engineering principles, was introduced for the study of biological form by Lewis et al. (1980). A biological object is divided into smaller geometric units called finite elements. Each vertex of a finite element is a biological landmark. The location of each landmark is defined by three-dimensional coordinates. Hexahedra, tetrahedra, and wedges are three-dimensional element types used to model the biological form according to meaningful osseous components. In this study we modeled the primate face as 10 contiguous finite elements (tetrahedra and wedges) using 18 landmarks (see Tables 16.1 and 16.2 and Figure 16.1). FESA compares forms in order to determine the amount of change required to produce an altered, target morphology from an original morphology.

In this study we compared the original, immature average face to the average male and female adult target faces within each species. (For computation of average morphologies, see Lele and Richtsmeier, 1991.) We interpret the change required to produce the older from the younger morphology as growth.

The difference between forms is estimated from the location of landmarks at the vertices of elements and the connectivities of these landmarks in the construction of the finite-element model. The measurements produced are expressed in terms of strain (see Lewis et al., 1980) local to each of the landmarks used to define the element. This means that information from surrounding landmarks is used to produce local measures of form change and that the design of elements is a critical part of FESA (see Richtsmeier et al., 1990a,b; Richtsmeier, 1991).

FESA provides information about the directions and magnitude of form change for each landmark. This information can be rotated to a coordinate system consistent with anatomical directions for ease of interpretation. Importantly, rotation of the object to a different coordinate system does not change the directions or magnitudes of change at each landmark as they relate to other landmarks or to the entire object.

Graphic displays of the direction of form change local to landmarks are used in this study to summarize growth patterns of the facial skeleton. There is a standard way to interpret these displays, as shown in Figure 16.2. A sphere centered around the landmark is used as a scale representing a condition of no change between the original and target form. Change in form at each landmark is represented as an ellipsoid superimposed on the sphere. If that ellipsoid remains relatively spherical with an increase or decrease in radial length, little change in shape has occurred. In this case, form difference between the original and target forms is due only to change in size. The degree to which the ellipsoid is elongated indicates the magnitude of shape change. For example, the landmark shown in Figure 16.2 shows a great deal of change oriented obliquely along the X-axis but relatively less along the Z- and Y-axes. The orientation of the ellipsoid gives the direction of change, while the length of the major and minor radii indicate the magnitude of the change. To interpret the graphics presented here (Morris,

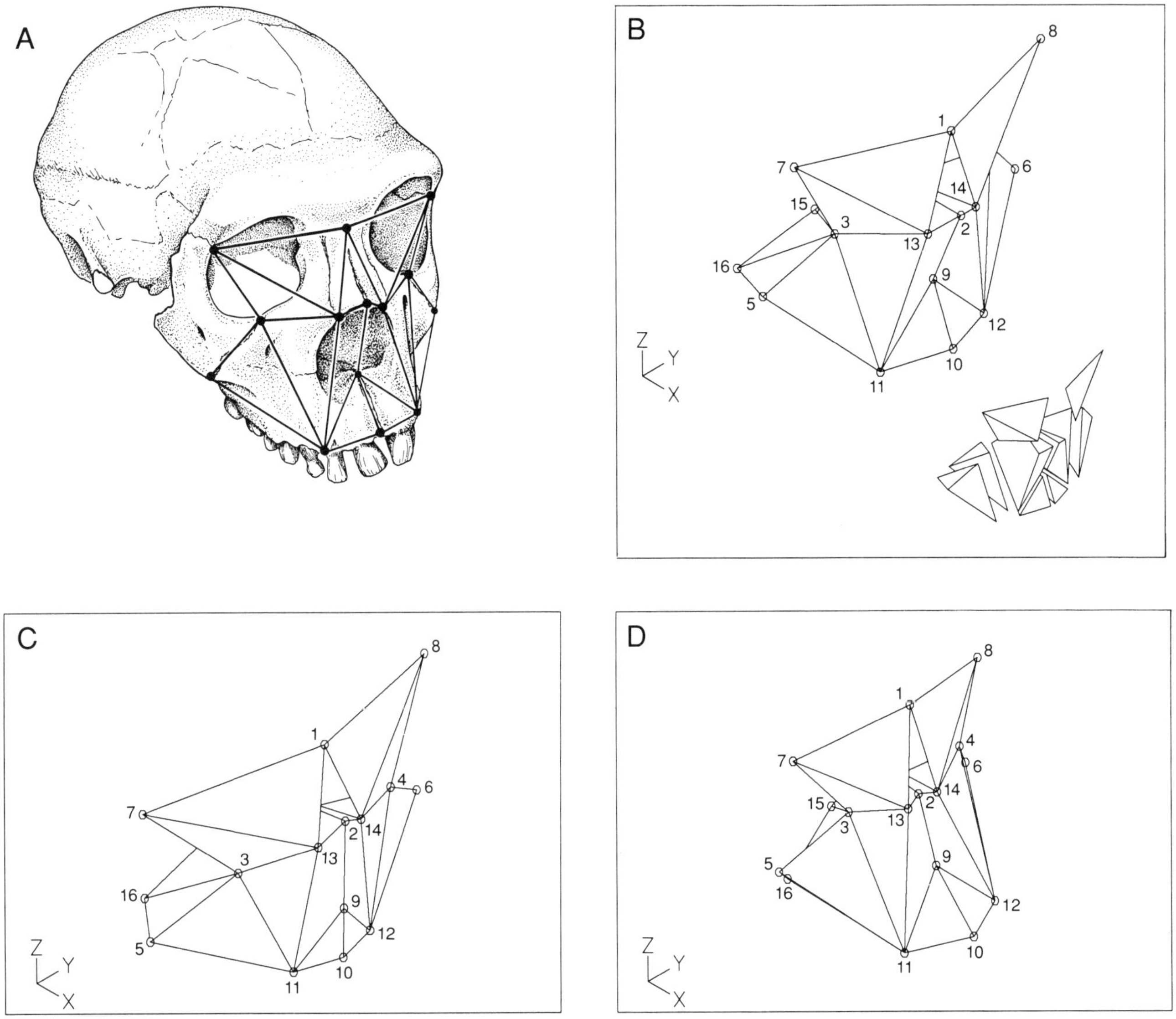

FIGURE 16.1.
Oblique views of the finite-element model of the facial skeleton of KNM-WT 15000 (B), *modern* HOMO SAPIENS (C), *and* PAN (D) *based on the location of eighteen landmarks (see Table 16.1). In panel* B *an exploded version of the finite-element design is shown to aid visualization of the three-dimensional nature of the model. Lists of the landmarks included in each element are provided in Table 16.1 and the traces of the element design are given in Table 16.2. The computer-generated oblique views* (B, C, D) *distort the objects so that some landmarks and/or surfaces that actually exist behind another surface are visible. A more realistic view of the finite element model is shown superimposed over KNM-WT 15000 (panel* A*).*

1991), several views are considered collectively, since our graphics project the ellipsoids calculated in three dimensions onto a two-dimensional plane (Figure 16.2).

EUCLIDEAN DISTANCE MATRIX ANALYSIS (EDMA)

EDMA was developed for the comparison of forms in two or three dimensions (Lele, 1991; Lele and Richtsmeier, 1991, 1992). Using landmark coordinates as raw data, we can describe a three-dimensional object using distances computed between all possible pairs of landmarks. The matrix of these distances is called a Euclidean distance matrix, or *form matrix* (FM). We say that two objects have the same form if their FMs are identical. Two objects have the same shape if the FM describing one form is a constant multiple of the FM describing the second form.

When samples of forms are compared, an average form is first calculated for each sample (see Lele and Richtsmeier, 1991; Lele, 1991). An FM is calculated for the average form for both samples. The two FMs are compared by computing ratios of like linear distances. This matrix of ratios of like linear distances is called a *form difference matrix* (FDM). An FDM for K landmarks includes $K(K - 1)/2$ ratios (in our case there are 18 landmarks and consequently 153 linear distances). Only results of the analysis of the right side of the face (including midline points) are reported here (number of landmarks = 11), in order to reduce the size of the FDM. Analyses of the left and right sides of our sample faces were in agreement.

An FDM presents the differences in form between two objects as a summary of distance-by-distance comparisons. However, the method analyzes all possible linear distances simultaneously. This is distinct from multivariate analysis of a composite of linear distances or the independent analysis of each linear distance, because it allows analysis of a morphologically integrated form in three-dimensional space instead of analysis of a series of geometrically unrelated linear distances in one dimension or in n-dimensional statistical space.

In this study, FDMs were calculated for the comparison of subadult with adult specimens of both sexes within a species. These FDMs were interpreted as representative of the morphological changes occurring due to growth of the face for each sex during the age interval considered. Each ratio therefore indicates the percent increase or decrease in a particular linear distance due to growth during the interval under investigation. It might be expected that all ratios in an FDM representing growth should be greater than 1, but this is not always the case. Decreases in specific linear distances are common in growth (see for example Corner and Richtsmeier, 1991, 1992; Schultz, 1962). They usually indicate changes in

TABLE 16.1.

Landmarks used in analysis listed by number as referenced in text and figures. For incomplete fossils the area was reconstructed by A.W., and landmarks were located on the reconstruction by J.T.R. and A.W.

1	nasion[a]
2	nasale[a]
3	zygomaxillare superior, right
4	zygomaxillare superior, left
5	zygomaxillare inferior, right
6	zygomaxillare inferior, left
7	fronto-zygomatic junction, right[b]
8	fronto-zygomatic junction, left[b]
9	anterior nasal spine
10	intradentale superior[b]
11	premaxillary-maxillary junction at alveolus, right
12	premaxillary-maxillary junction at alveolus, left
13	distal naso-maxillary junction, right
14	distal naso-maxillary junction, left
15	posterior nasal spine[a,b]
16	point on alveolus lateral to M^2, right[a,b]
17	point on alveolus lateral to M^2, left[b]
18	point midway between nasion and posterior nasale spine

a. Landmark missing from KNM-WT 15000.
b. Landmark missing from KNM-ER 3733.

TABLE 16.2.

Element design used in finite element scaling analysis. A landmark trace is the sequence of landmarks (coded by number as listed in Table 16.1) that specify an element. Elements made of four landmarks are tetrahedra. Elements made of six landmarks are wedges. See Figure 16.1 for a graphic display of element design.

ELEMENT	LANDMARK TRACE					
1	3	16	11	5		
2	17	4	12	6		
3	11	9	15	13	2	18
4	9	12	15	2	14	18
5	15	11	10	9		
6	15	10	12	9		
7	15	16	11	3		
8	17	15	12	4		
9	1	7	3	13		
10	8	1	4	14		

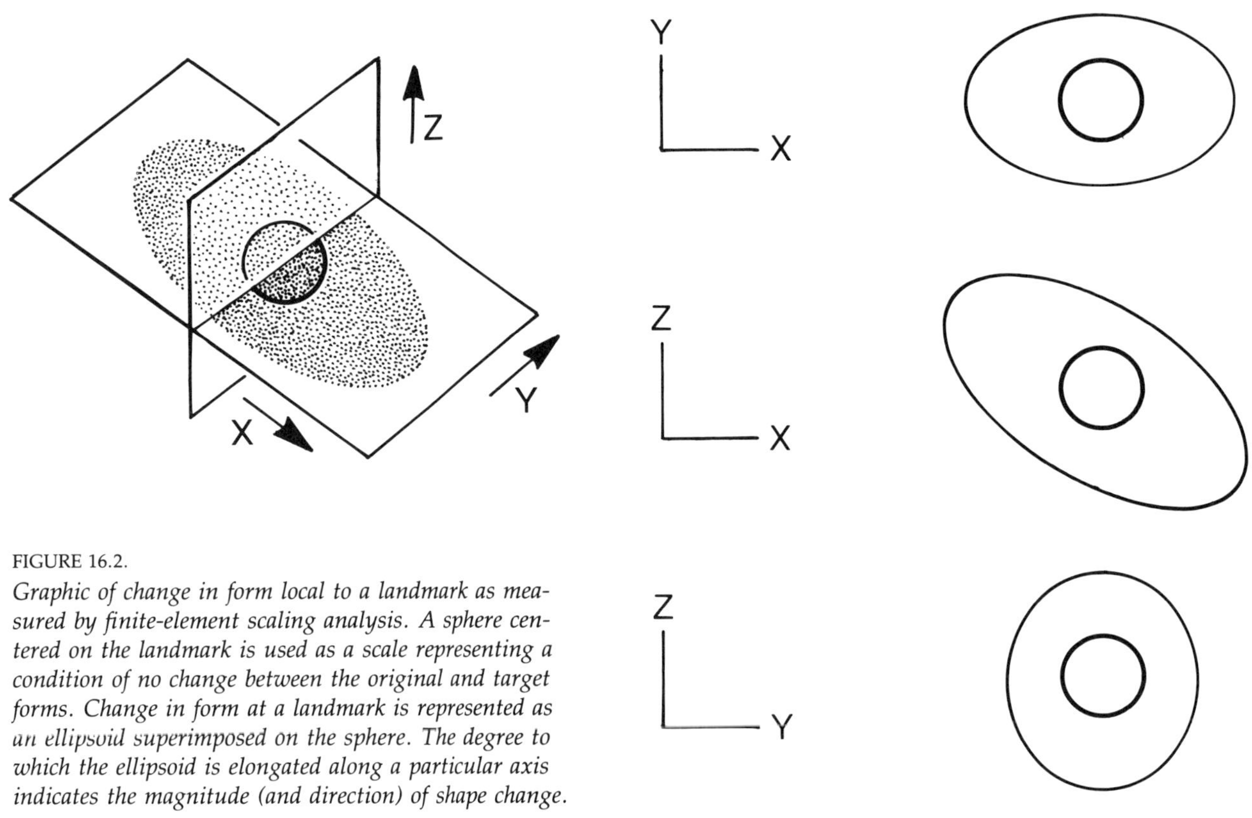

FIGURE 16.2.
Graphic of change in form local to a landmark as measured by finite-element scaling analysis. A sphere centered on the landmark is used as a scale representing a condition of no change between the original and target forms. Change in form at a landmark is represented as an ellipsoid superimposed on the sphere. The degree to which the ellipsoid is elongated along a particular axis indicates the magnitude (and direction) of shape change.

angles between bony structures, changing curvature of a surface, or rotation of parts of bones.

FDMs for the various species-specific growth analyses were compared by sorting the ratios of each FDM from smallest to largest. When arranged in this fashion, the ratios at one end of the FDM describe those linear distances that have changed the least with growth while the other end of the matrix contains the linear distances that have changed the most. Differences in growth patterns are described by comparing the sequence of the landmark pairs and the magnitude of the ratios between analyses.

SPECIMENS AND DATA COLLECTION

The pattern of tooth maturation seen in KNM-WT 15000 is described and some of the implications are discussed by Smith (Chapter 9, this volume). KNM-WT 15000 displays a pattern of dental eruption somewhat different from that seen in modern people, but it is a known human pattern (Garn and Koski, 1957). Smith (Chapter 9) has assigned an age of 11 years to KNM-WT 15000 on the basis of a human pattern of tooth emergence. Modern human females and males experience the onset of the adolescent growth spurt at 10 and 12 years, respectively. If early hominids developed as modern humans do, then the growth pattern exhibited by comparing KNM-WT 15000 to KNM-ER 3733 would include an adolescent growth spurt. If early hominids developed as great apes do, however, then KNM-WT 15000 is 7 years old (Smith, Chapter 9). Watts (1985) estimates the adolescent spurt to begin at approximately 8 years in female chimps and 10 years in male chimps. Thus, if *Homo erectus* ages as *Pan* does, an adolescent spurt would also be included in the growth interval that we study.

We note, however, that the adolescent spurt of *Pan* is of lesser magnitude than that of humans (Watts and Gavan, 1982).

We looked for specimens of modern *Homo sapiens* among the skeletal collections of the Smithsonian Institution National Museum of Natural History according to dental pattern. We chose specimens that shared the dental maturation pattern of KNM-WT 15000 as closely as possible from the Eskimo populations housed in the Division of Anthropology. A sample of North American Arikara Indians was also inspected for individuals that showed the dental eruption pattern of KNM-ER 15000, but none was found. The Eskimo face shows evidence of cold adaptation and might be considered an extreme morphology. This is of little consequence since *growth pattern* is the focus of this study rather than *morphology* and we assume a generalized similarity in facial growth pattern among all modern humans.

Five immature Eskimo specimens with dental patterns matching that of KNM-WT 15000 but of unknown sex make up the immature modern human sample. Four adult human males and four adult human females (aged by the presence of all permanent dentition and sexed by skeletal criteria or, in some cases, on the basis of remaining soft tissue on the original remains) were chosen as the adult modern human samples.

Four immature *Pan troglodytes* were selected using the same dental criteria from the collections of the Smithsonian Institution National Museum of Natural History (N = 2, both males) and the Cleveland Museum of Natural History (N = 2, one male, one unknown sex but X-ray of this specimen shows very large canines within the crypts). Five adult male and four adult female *Pan troglodytes* make up the adult chimpanzee samples. *Pan* individuals were identified as adults if the complete permanent dentition was in place. All *Pan* specimens were wild-shot, and sex identification was made at the time of collection.

From casts of KNM-WT 15000 and KNM-ER 3733, seventeen landmarks (Table 16.1) were identified that could be reliably located either directly on the casts or on reconstructed portions of the fossil faces. An additional landmark was constructed for FESA (see Tables 16.1 and 16.2). Necessary reconstructions of the fossils were done by A.W. while both authors decided on location of the landmarks. Digitizing of specimens was done by J.T.R. using the 3Space digitizer (Polhemus Navigation). Each fossil cast was digitized three times without removing the specimen from the digitizing table in order to reduce measurement error (see Corner et al., 1992). All *Pan* and Eskimo crania were digitized twice. An average form was computed for each individual from the separate digitization episodes using a Generalized Procrustes algorithm (Gower, 1987). These average forms were used as individuals in growth analyses.

Simulating an Adult Male *Homo erectus* Facial Morphology

In order to study growth in *Homo erectus* males, it was necessary to simulate an adult male *Homo erectus* face. There are many ways to simulate this facial morphology. We do not want the growth patterns defined for either of our two comparative samples to predetermine results of our analysis of *Homo erectus* male facial growth. Consequently, we used patterns of adult sexual dimorphism calculated for the two comparative species to simulate the male *Homo erectus* face. The simulation process (Table 16.3) will be described using the pattern of facial sexual dimorphism in *Pan*. An alternate male adult *Homo erectus* face was simulated using the pattern of facial sexual dimorphism calculated for the modern *Homo sapiens* sample. Growth analyses were performed using both simulated *Homo erectus* male forms.

To simulate the male *Homo erectus* face, male and female samples of adult *Pan* were first compared using EDMA. Linear distances calculated for the male form were used as the numerator while linear distances of the female form were the denominator (male/female; see Table 16.3). Second, the three-dimensional coordinates of the landmarks collected from the face of KNM-ER 3733 were translated into FM format. Each element of the FM, d_{ij}, represents the linear distance between landmarks i and j on the face of KNM-ER 3733.

Third, each of the linear distances in the FM representing KNM-ER 3733 was scaled by the ratio for the same linear distance from the FDM calculated for the comparison of male and female adult *Pan* facial morphologies (step 1). The resulting matrix, which we refer to as the *simulated form matrix* (SFM), represents the linear distances between all possible pairs of landmarks in the simulated form. This matrix can be used to calculate the three-dimensional coordinates of landmark locations in the simulated form.

To calculate the three-dimensional coordinates for K landmarks, the SFM is written as a symmetric matrix of dimension $K \times K$ consisting of distances between all possible pairs of landmarks. From this matrix the centered inner product matrix is calculated. The spectral decomposition of the centered inner product matrix consists of K eigenvectors and K eigenvalues, but only the first three eigenvectors are needed to calculate landmark locations. The first three eigenvectors are multiplied by the square root of the corresponding eigenvalues to give the three-dimensional coordinates of the K landmarks for the simulated form.

Two adult male *Homo erectus* faces were simulated by scaling the digitized face of KNM-ER 3733 by the sexual dimorphism pattern calculated for *Pan* and for modern *Homo sapiens* following procedures outlined above and summarized in Table 16.3. This method of simulating a facial morphology does not simply scale a form by a single number, which would produce a face that is of the same shape but larger or smaller in size. By using a complete and specialized suite of scaling values (the FDM), the data are scaled according to an integrated morphological pattern with a precise biological meaning. The simulated form produced represents a morphological assessment based on an explicit biological hypothesis.

The underlying hypothesis for the form produced using the sexual dimorphism pattern of *Pan* is that facial sexual dimorphism of *Homo erectus* is like that of *Pan*. The competing hypothesis em-

TABLE 16.3.

The calculations used to simulate the *Homo erectus* male faces. First (column A), a measure of sexual dimorphism is calculated using EDMA for a modern species; this creates a form difference matrix (FDM). Second (column B), the three-dimensional coordinates of landmark locations for KNM-ER 3733 are put into form matrix (FM) format. Third, the FM representing KNM-ER 3733 (B) is multiplied by the FDM representing sexual dimorphism for a modern species (A). The resulting matrix (column C) contains the dimensions of the linear distances between all possible combinations of landmarks in the simulated form. From this overdetermined set of linear distances, the three-dimensional coordinates of landmark locations for the simulated form are calculated.

(A) SEXUAL DIMORPHISM IN KNOWN SPECIES FDM	×	(B) MORPHOLOGY OF KNM-ER 3733 FM	=	(C) SIMULATED MORPHOLOGY OF ADULT MALE *HOMO ERECTUS* SFM
$\frac{♂ d_{1,2}}{♀ d_{1,2}}$	×	$d_{1,2}$	=	$d_{1,2}$
$\frac{♂ d_{1,3}}{♀ d_{1,3}}$	×	$d_{1,3}$	=	$d_{1,3}$
$\frac{♂ d_{1,4}}{♀ d_{1,4}}$	×	$d_{1,4}$	=	$d_{1,4}$
⋮		⋮		⋮
$\frac{♂ d_{i,j}}{♀ d_{i,j}}$	×	$d_{i,j}$	=	$d_{i,j}$

TABLE 16.4.

Form difference matrices (FDMs) for various growth comparisons as described in text. The entire FDM for each comparison contains 55 linear distances. Ratios have been sorted from smallest to largest. Shown here are the bottom and top extremes of the FDMs. The ratio is defined as the quantity calculated when the linear distance in the adult form is divided by the like linear distance in the subadult form. Landmarks specify endpoints of the linear distances compared.

ESKIMO				*PAN*			
MALE		FEMALE		MALE		FEMALE	
Ratio	Landmarks	Ratio	Landmarks	Ratio	Landmarks	Ratio	Landmarks
1.053	16 11	1.017	7 1	0.969	3 1	0.919	2 1
1.067	16 10	1.021	16 11	0.988	7 3	0.952	9 1
1.076	7 1	1.060	7 3	1.016	2 1	0.955	13 1
1.099	3 1	1.070	3 1	1.020	7 5	0.959	13 9
1.104	7 3	1.071	16 10	1.034	5 1	0.965	7 5
1.110	11 10	1.081	7 2	1.035	13 5	0.968	3 1
1.118	16 9	1.090	13 7	1.035	3 2	0.982	15 1
1.172	13 7	1.110	16 9	1.037	11 10	0.985	16 11
1.178	7 2	1.113	13 3	1.040	13 7	0.998	9 2
1.180	13 3	1.121	3 2	1.043	13 1	1.001	7 3
⋮		⋮		⋮		⋮	
1.348	13 1	1.247	10 3	1.207	15 10	1.122	15 5
1.350	15 2	1.251	11 9	1.207	13 10	1.127	10 5
1.365	16 3	1.251	15 11	1.208	11 2	1.130	13 10
1.367	10 3	1.252	15 10	1.211	10 3	1.140	10 3
1.375	10 9	1.268	16 3	1.219	10 2	1.142	10 2
1.384	16 7	1.286	10 9	1.226	10 9	1.151	11 5
1.451	5 3	1.289	11 3	1.230	11 5	1.154	15 11
1.461	2 1	1.290	5 3	1.267	11 3	1.163	15 10
1.484	11 3	1.306	16 7	1.268	15 11	1.276	10 9
1.490	16 15	1.328	16 15	1.334	16 15	1.328	16 15

bodied in the second simulated form is that *Homo erectus* facial sexual dimorphism is like that of modern *Homo sapiens*. We have no way of directly testing these hypotheses. The proper test awaits the discovery of male early African *Homo erectus* fossil faces. Instead, we will compare KNM-ER 15000 to the simulated adult male *Homo erectus* faces in order to make generalizations regarding primate facial growth patterns.

Results

Eskimo facial growth

Males. The form difference matrix representing facial growth patterns of the male Eskimo (Table 16.4) indicates that all linear distances increase in size during growth from adolescence to adulthood (i.e., all ratios are greater than 1). Minimal change is seen along the maxillary alveolus (from landmark 16 to 10 and to 11). Upper facial width, especially around the orbits (as measured from landmark 1 to 7 and 1 to 3; and from 7 to 3), also experiences only a slight increase (7% to 10%) during this period. Those linear distances showing the greatest increase (from 35% to 49% increase) during adolescent growth include posterior palatal width (from landmark 16 to 15) and all aspects of facial height (from 11 to 3; 5 to 3; 10 to 3; 1 to 2; 16 to 7; 10 to 9, 16 to 3; and 15 to 2).

Females. Comparison of the sequence of landmark pairs in the FDM representing male growth with the FDM representing female growth indicates that the pattern of female facial growth is very similar to that of males. Nine of the 10 linear distances at the minimum end of the female growth FDM are found in the same end of the male growth FDM. Seven of the ten linear distances at the maximum extreme of the female growth FDM are also found in the maximal extreme of the FDM that describes male growth. However, female magnitudes of growth are reduced as compared with those of males. As in males, females experience little mediolateral growth of the orbits and minimal anteroposterior growth along the alveolus. Magnitudes of female facial growth indicate that posterior palatal width and facial height experience the greatest increases. In females, the distance measured from front to back of the palate (from 15 to 10 and from 15 to 11) show marked growth. In males these linear distances are not among those experiencing maximal change.

Generalized Eskimo growth pattern. The FESA graphic of growth in male Eskimos (Figure 16.3) indicates that the majority of change at frontozygomatic junction and nasion occurs along the anteroposterior axis. The same landmarks show less change along the mediolateral axis, confirming the lack of mediolateral growth across the orbits as described by EDMA. Other landmarks of the upper face (2, 13, 3) show a relatively high amount of growth along the anteroposterior axis coupled with increases along the superoinferior axis and reduction or maintenance of the mediolateral dimension. Growth local to landmarks on the lower face and palate (5, 11, 15, 16) is characterized by a large amount of change along the superoinferior axis and little if any change along the mediolateral and anteroposterior axes. A lateral flare of the zygomatic bones (3, 5) is accomplished by local growth patterns. Growth patterns local to the point on the alveolus at M2 result in a widening of the posterior palate and zygomatic region. Magnitudes of growth are similar along the three major axes at the anterior nasal spine (9) and intradentale superior (10), indicating that growth local to these landmarks is characterized as size change accompanied by very little shape change.

Sexual dimorphism: morphology and growth. Male and female Eskimo faces differ predominantly along the superoinferior axis but there is some dimorphism in width of the face. Superoinferior and mediolateral dimensions of the zygomatic bone are particularly dimorphic, as are distances that together describe the superoinferior dimension of the nasomaxillary complex (1 with 2, with 9, with 11, and with 13; and 11 with 13).

As noted above, patterns of facial growth are very similar in males and females. Males (Figure 16.3) experience an increased magnitude in form change as compared with females (Figure 16.4) for the same age interval, but the directions of growth are generally similar between the sexes. Orientations of growth differ at landmarks 5, 6, 13, and 14. At these landmarks males experience an increase along the mediolateral axis. Females, on the other hand, experience a maintenance or reduction in dimensions measured along this axis at these landmarks. Therefore, facial dimorphism at most landmarks is due to differences in magnitudes of growth (males experiencing greater *magnitudes* of change along the same directions noted in female growth), but sexual dimorphism in *directions* of growth underlies the differences in morphology noted at specific points on the nasal complex and inferolateral margin of the zygoma.

Chimpanzee facial growth

Males. The FDM representing male facial growth in chimpanzees includes some ratios that are less than 1, indicating an actual decrease in that particular linear distance with growth (Table

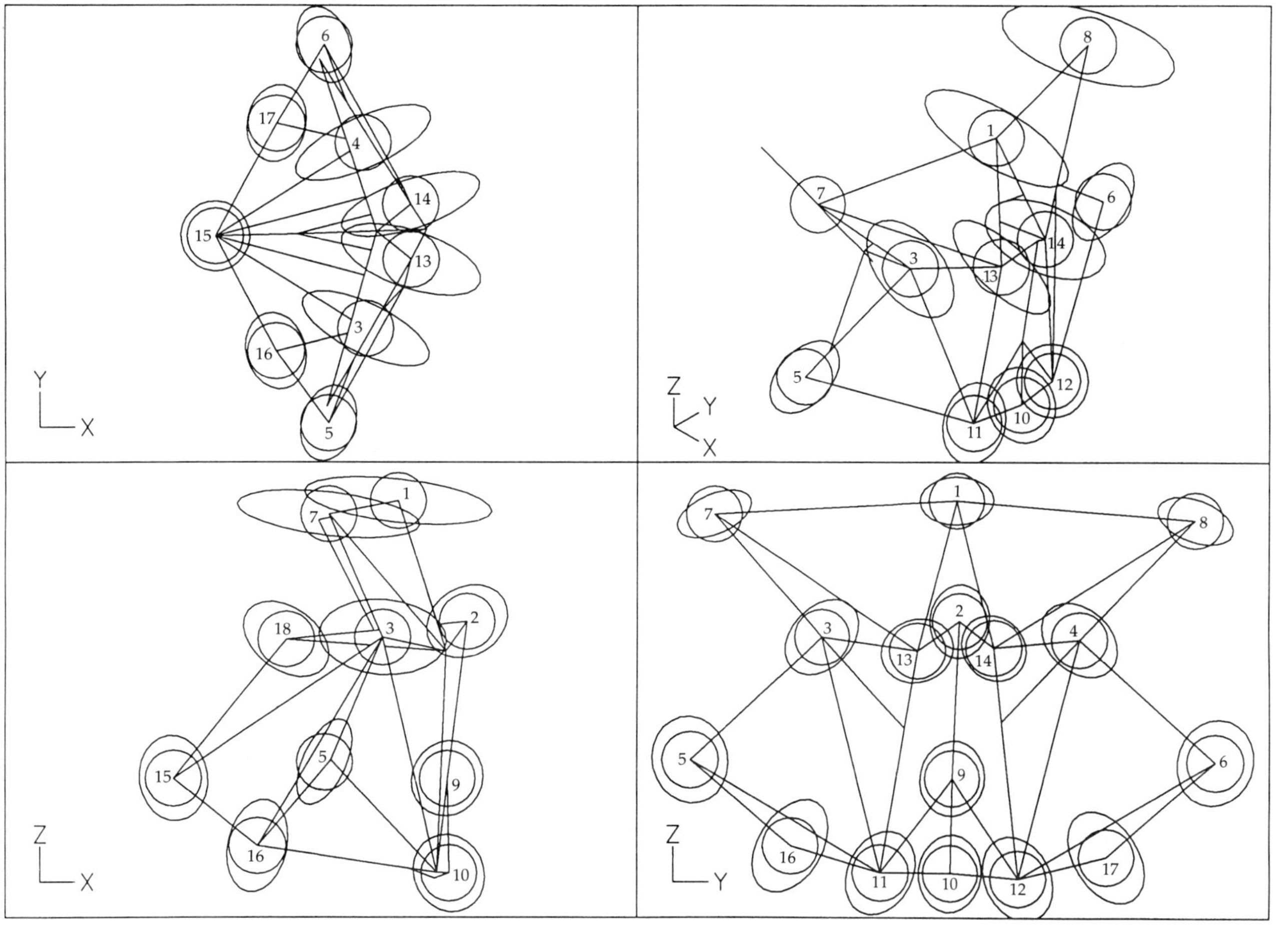

FIGURE 16.3.

FESA graphic (Morris, 1991) of male Eskimo adolescent facial growth pattern calculated by comparing the adolescent Eskimo face with the adult male face. Although growth is studied in three dimensions, graphic displays are collapsed onto a two-dimensional plan and several views are provided. The upper right quadrant is the oblique view (as shown in Figure 16.1). The upper left quadrant shows a superior view of growth of the face. The lower left quadrant shows a lateral view of the face. The lower right shows an anterior view of the face. The positive x *direction runs from posterior to anterior, the positive* y *direction runs from right to left, and the positive* z *direction runs superiorly. These orientations are constant for all FESA graphics presented (Figures 16.3–16.9), and in all graphics ellipses are plotted on the face of the older form. To simplify these drawings and avoid superimposition of ellipses, certain ellipses were excluded from the plots.*

16.4). The male chimpanzee face experiences reduction or maintenance of facial width as measured from the upper midline points (1, 2, and 13) to points on the orbital rim and zygoma (3, 5, and 7). Certain linear distances indicative of superoinferior growth of the upper face (1 to 2, 7 to 5) are also included among the dimensions that experience only small changes with growth.

There is a marked increase (33%) in width of the posterior palate (landmark 16 to 15) during growth. Anteroposterior lengthening of the palate (landmark 15 to 11, 27% increase; 15 to 10, 21% increase) is evident during chimpanzee facial growth. In addition, certain linear distances (from 11 to 3, 27% increase; 11 to 5, 23% increase) indicate the increasing prognathism of the chimpanzee face with aging. Other dimensions that experience maximal change indicate increased prognathism and include distances from intradentale superior to anterior nasal spine, to nasale, and to zygomax-

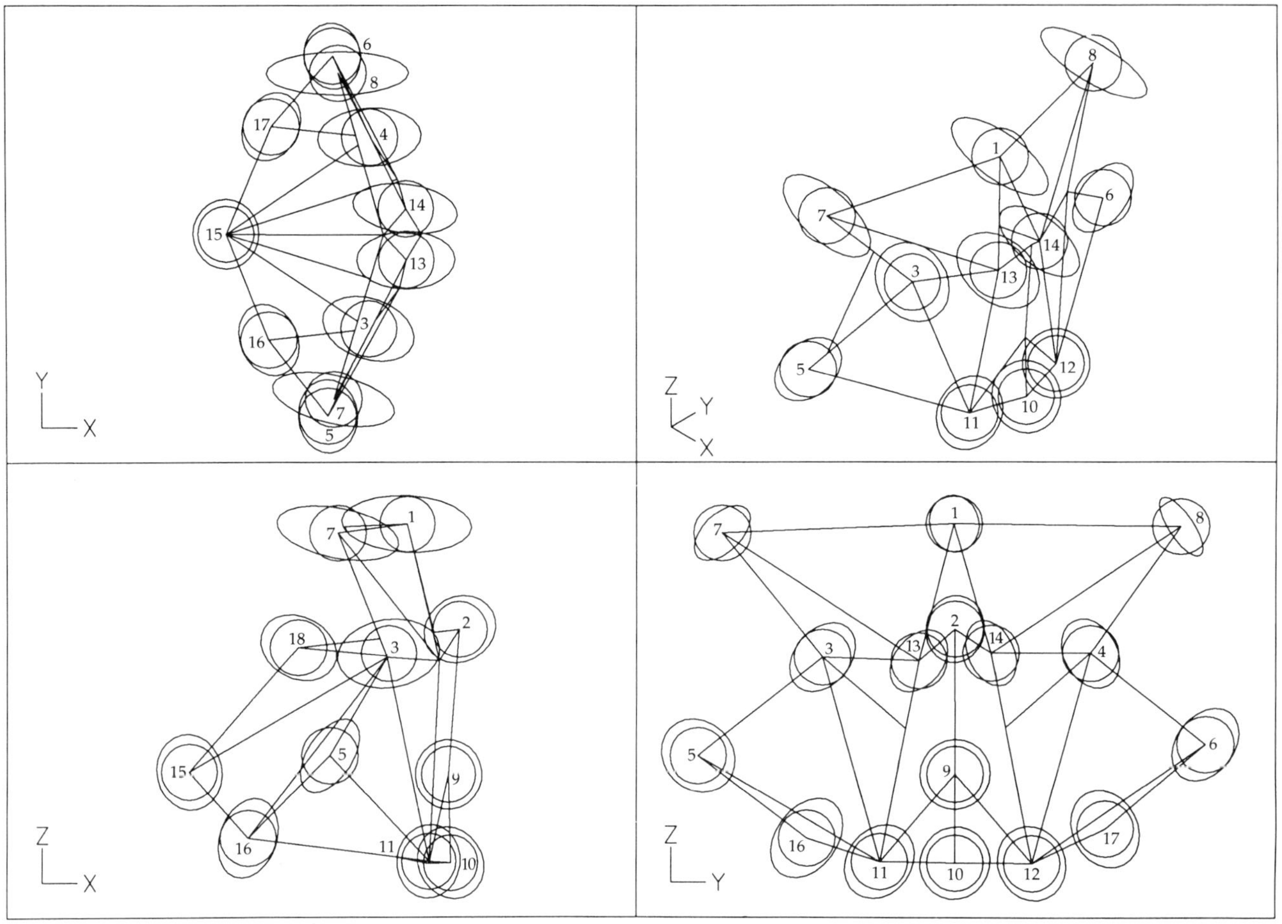

FIGURE 16.4.
FESA graphic of female Eskimo adolescent facial growth produced by comparing the adolescent Eskimo facial form with the adult female facial form.

illary superior; from premaxillary-maxillary intersection to nasale; and from distal nasomaxillary intersection to intradentale superior. Many of these linear distances showed maximal change in our depiction of the Eskimo pattern of facial growth, but in the modern *Homo sapiens* face, increase in these linear distances results in increased facial height as opposed to increased prognathism.

Females. The FDM representing female facial growth in chimpanzees includes several ratios less than 1. The growth pattern for female *Pan*, as described by the extremes of the FDM, is similar to that calculated for males. Females show reduction or maintenance of dimensions measured from nasion to other upper facial landmarks (nasale, anterior nasal spine, distal nasomaxillary intersection, zygomaxillary superior, posterior nasal spine), indicating no positive change in upper facial height during this interval. As in males, females experience great increases (12% to 33%) in palatal width and length. Lower facial dimensions experience from 13% to 15% increase, with growth resulting in pronounced prognathism. This is evidenced by changes at distances measured from intradentale superior to upper facial landmarks (2, 3, 5, 13) and from premaxillary-maxillary intersection to zygomaxillare inferior.

Generalized growth pattern. In *Pan*, as in modern *Homo sapiens*, the growth pattern of the adolescent upper face is characterized by a disproportionate amount of change along the anteroposterior axis, while growth of the lower face and palate occurs primarily along the superoinferior axis. This superoinferior component of lower facial growth is obliquely oriented along an axis stretching from posterosuperior to anteroinferior, suggesting elongation of the lower face and protrusion of the snout during adolescent growth. Little growth occurs directly along either the mediolateral or anteroposterior axis of the lower face. The upper face of *Pan* shows a greater degree of growth along the mediolateral axis than seen in *Homo erectus.* This is especially evident in males. Landmarks surrounding the orbit (7, 1, 3) experience a mediolateral flare oriented superolaterally. This pattern is not seen in growth of the modern *Homo sapiens* face. Some shape change is noted at the anterior nasal spine and intradentale superior, although the magnitude of the size change is the more obvious component of growth noted local to these landmarks.

SEXUAL DIMORPHISM: MORPHOLOGY AND GROWTH. Male and female chimp faces differ predominantly along an axis oriented from superoposterior to anteroinferior. This results in a male chimp face that is more prognathic than the female face. Male chimps also have a longer palate as measured along the anteroposterior axis (from point 11 to 16 and 10 to 17). This distance incorporates the canine. Local dimorphism is predominantly due to the larger male canine. There is no marked facial dimorphism along the mediolateral axis.

Most of the differences in growth patterns between the two sexes appear to relate to differing growth magnitudes. Directions of growth at most landmarks are very similar between the sexes in *Pan* (Figures 16.5 and 16.6).

HOMO ERECTUS FACIAL GROWTH

FEMALE PATTERN: KNM-WT 15000 COMPARED WITH KNM-ER 3733

Like the growth pattern observed for *Pan,* facial growth in *Homo erectus* includes reduction of certain distances (Table 16.5). Many of the linear distances that are reduced during growth involve the posterior nasal spine. We note that the palate of KNM-ER 3733 is highly arched. The superior position of the posterior nasal spine in the adult female specimen may account for the reduction of the following linear distances during growth: posterior nasal spine to anterior nasal spine (9), to nasale (2), to distal nasomaxillary intersection (13), and to nasion (1). Other measures that reduce during growth include the distance from intradentale superior to premaxillary-maxillary intersection, and the distance from the point on alveolus at M2 to intradentale superior, to anterior nasal spine, and to premaxillary-maxillary intersection. This may signal a downward rotation of the premaxilla during growth with maintenance of the length of the postcanine tooth row. Alternatively, male canine roots may be bigger and this could account for substantial enlargement of the adolescent maxillary alveolar length in KNM-WT 15000.

Four of the linear distances at the maximum extreme of the FDM representing growth of *Homo erectus* are found in the maximal extremes of the FDMs representing facial growth in chimpanzees and modern humans. All of these linear distances involve intradentale superior (10 with 2, with 3, with 13, and with 9) and represent aspects of midline facial height (although characterized as prognathism in the chimpanzee). A few of the linear distances at the maximal extreme of the *Homo erectus* FDM suggest increasing width of the face (7 to 2; 13 to 5; 7 to 1), but the majority connote increases in additional aspects of lateral and midline facial height (5 to 3; 16 to 3; 16 to 7; 7 to 5; 13 to 9 and to 10).

MALE PATTERN: KNM-WT 15000 COMPARED WITH SIMULATED FORMS

ESKIMO-BASED SIMULATED FORM. The graphic representation of this comparison of KNM-WT 15000 with a simulated adult male *Homo erectus* is presented in Figure 16.7. Many of the linear distances that show reduction or maintenance in size in this comparison involve landmarks of the posterior palate (15 and 16; Table 16.5). Little or no change occurs in premaxillary width (10 with 11), palatal length (9 with 15), or alveolar length (16 with 9, with 10, and with 11). Those linear distances that experience the greatest magnitudes of change tend

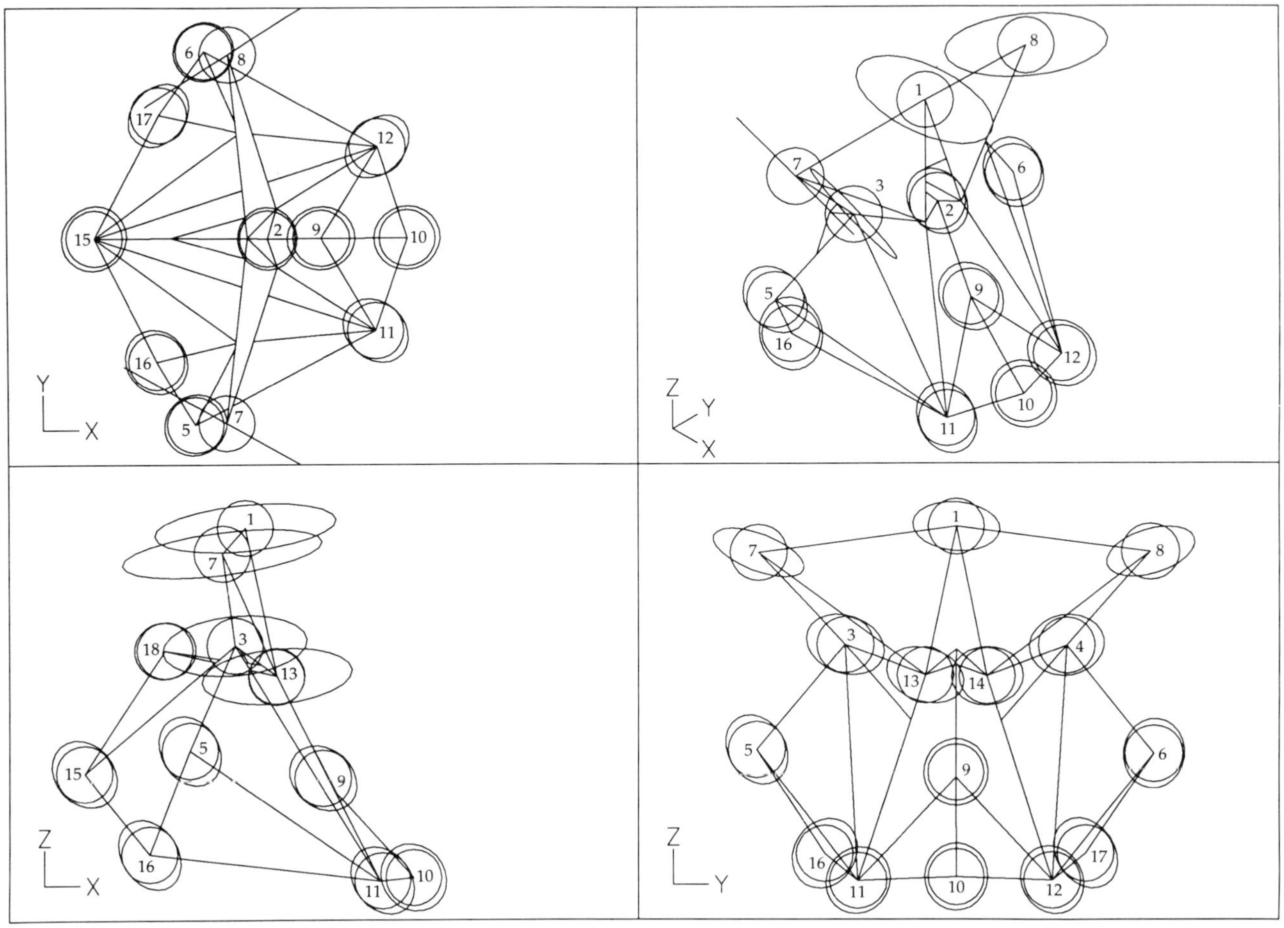

FIGURE 16.5.
FESA graphic of male chimpanzee adolescent facial growth pattern produced by comparing the adolescent chimpanzee facial form with the adult male facial form.

to increase midline facial height (9 with 10; 10 with 13; 3 with 11; 9 with 13; 3 with 10), lateral facial height (3 with 5; 3 with 16; 3 with 11; 6 with 7; 5 with 7), orbital width (1 with 7, 2 with 7), and palatal width (15 with 16).

CHIMPANZEE-BASED SIMULATED FORM. Distances measured from the posterior nasal spine to anterior midline points (9, 2, 13, 10) show reduction or maintenance in this comparison of KNM-WT 15000 with a simulated adult male *Homo erectus* face. Other measures that do not display increased length include dimensions of the nasal bones (13 with 1, 13 with 2), the orbits (3 with 1; 3 with 7), and the alveolus (10 with 11 and 10 with 16).

Midline facial height, especially local to the nasomaxillary complex, experiences the greatest magnitudes of change for this comparison (13 with 9, with 10, and with 11; 9 with 2 and with 3; 10 with 2, with 3, and with 9). Lateral facial height also undergoes substantial enlargement (distances from 3 to 5; 3 to 16; 3 to 11; and 7 to 16 experience from 23% to 38% increases in length). The predominance of zygomaxillare superior in many of these measures and the large magnitude of change in the linear distances that cross the face obliquely from zygomaxillare superior to anterior nasale spine and intradentale superior suggest a large degree of widening of this simulated male *Homo erectus* when compared with KNM-WT 15000.

COMPARISON OF THE TWO GROWTH PATTERNS. The growth patterns calculated for the two simulated forms are similar (Table 16.5). This is most likely due to features of cranial sexual dimorphism held

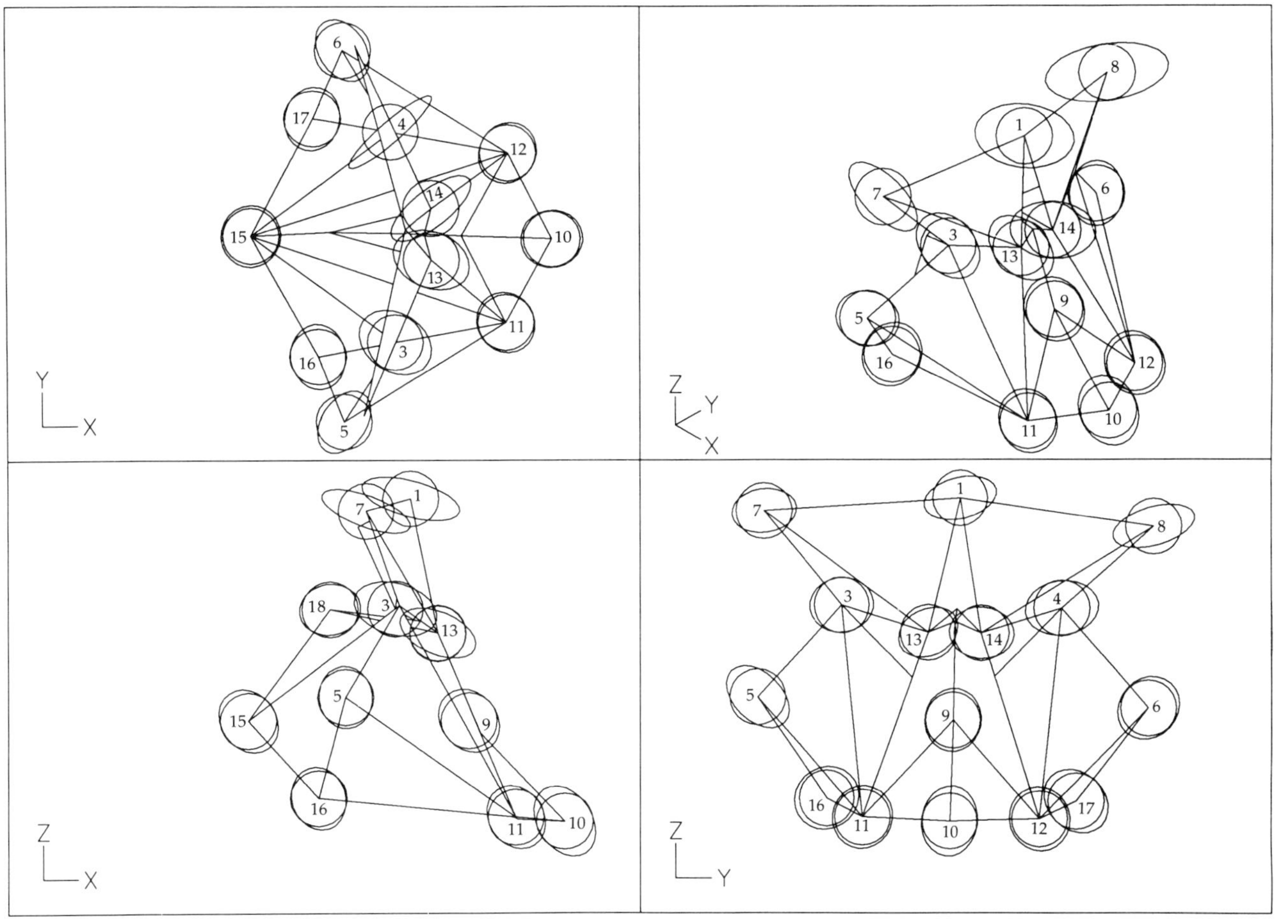

FIGURE 16.6.
FESA graphic of female chimpanzee adolescent facial growth pattern produced by comparing the adolescent chimpanzee facial form with the adult female facial form.

in common by *Pan* and *Homo* (Wood et al., 1991). Nine of 10 linear distances found at the maximum extreme of the FDM describing differences between KNM-WT 15000 and the adult male form simulated using the *Pan* sexual dimorphism pattern are also found in the FDM describing differences between KNM-WT 15000 and the alternate simulated form. There is also correspondence between the linear distances found in the minimum extreme of these FDMs. Likewise the range of magnitudes of change are similar between the two comparisons. One outstanding difference in the two patterns of change is found in the distance measured from zygomaxillare inferior to zygomaxillare superior. The form simulated using the sexual dimorphism pattern of *Pan* shows an increase of 88% in this distance whereas the form simulated using the sexual dimorphism pattern of the Eskimo shows a 47% increase in this dimension.

Directions of change are similar for the analyses using the two simulated forms when viewed on the *Z-Y* plane (Figures 16.8 and 16.9). Comparison of the two analyses show obvious differences in directions of change at landmarks 1, 7, and 8 projected onto the *Z-X* plane. Change at these landmarks is directed almost exactly along the *X*-axis when the adult form is simulated using the sexual dimorphism pattern of *Pan*. If the other simulated form is used, change at nasion and fronto-zygomatic junction is oriented along an axis running from posterosuperior to anteroinferior. Comparison of these patterns of change projected onto the *Y-X* plane shows differences in magnitude, and direction at landmarks 5, 6, 7, and 8, indicating differences in change local to the zygoma using both forms. Transformation of the zygoma as depicted by comparison of KNM-WT 15000 with the *Homo erectus* male simulation based on the Eskimo pattern of sexual dimorphism indicates a lateral expansion and anteroposterior compression of the lower zygoma (points 5 and 6). There is little change local to zygomaxillare inferior if the other simulated form is used for comparison. The orbital rim of the zygoma (points 3, 4, 7, 8) shows an expansion along the anteroposterior axis in both growth simulations, although the exact directions vary between analysis. Other, more subtle differences can be noted in the transformation patterns using the two simulated forms (see Figures 16.8 and 16.9), but the similarities are more striking than the differences.

GENERALIZED HOMO ERECTUS *FACIAL GROWTH PATTERN.* We have compared KNM-WT 15000 with KNM-ER 3733 and with two adult male *Homo erectus* faces simulated using models of sexual dimorphism from modern primate species. We do not pretend to have modeled actual growth of the *Homo erectus* facial skeleton in detail, but instead have described the geometric transformations suggestive of patterns of facial growth. A pattern of

TABLE 16.5.

Form difference matrices (FDMs) for growth analysis of *Homo erectus*. The entire FDM for each comparison contains 55 linear distances. Ratios have been sorted from smallest to largest. Shown here are the bottom and top extremes of the FDMs. The ratio is defined as the quantity calculated when the linear distances in the adult forms are divided by linear distances in the subadult form. Landmarks specify the endpoints of the linear distances compared.

KNM-WT 15000 COMPARED WITH KNM-ER 3733		KNM-WT 15000 COMPARED WITH *HOMO ERECTUS* MALE FACE SIMULATED FROM ESKIMO PATTERN OF SEX DIMORPHISM		KNM-WT 15000 COMPARED WITH *HOMO ERECTUS* MALE FACE SIMULATED FROM *PAN* PATTERN OF SEX DIMORPHISM	
Ratio	Landmarks	Ratio	Landmarks	Ratio	Landmarks
0.842	13 1	0.843	11 10	0.902	15 9
0.842	15 9	0.887	15 9	0.921	13 2
0.873	13 2	0.895	16 10	0.923	13 1
0.901	16 10	0.944	16 9	0.930	15 2
0.904	11 10	0.946	13 1	0.949	15 13
0.916	15 2	0.960	16 11	0.951	3 1
0.922	15 13	0.988	3 1	0.952	15 5
0.934	16 9	0.992	15 2	0.969	16 10
0.939	16 11	0.997	15 5	0.971	7 3
0.948	15 1	0.997	13 2	0.980	11 10
⋮		⋮		⋮	
1.113	10 2	1.219	7 5	1.176	10 9
1.116	13 9	1.222	10 3	1.185	10 2
1.117	10 3	1.226	7 1	1.187	10 3
1.120	13 5	1.237	16 7	1.230	13 11
1.160	7 1	1.253	13 9	1.236	16 7
1.170	16 7	1.264	11 3	1.254	13 10
1.171	13 10	1.272	13 10	1.256	11 3
1.192	16 3	1.279	16 3	1.270	16 3
1.243	10 9	1.305	10 9	1.283	13 9
1.275	5 3	1.473	5 3	1.380	5 3

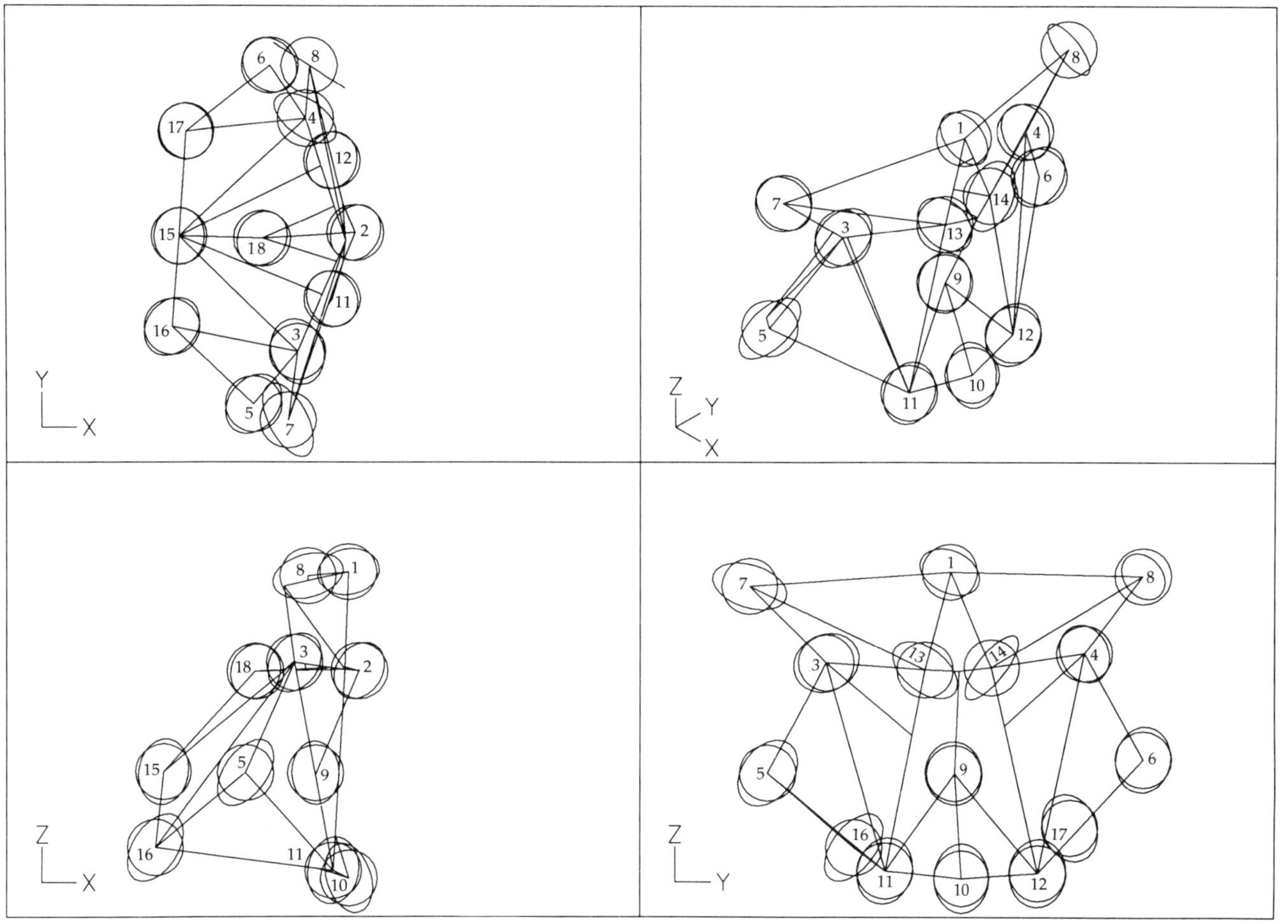

FIGURE 16.7.
FESA graphic of female HOMO ERECTUS *adolescent facial growth pattern produced by comparing KNM-WT 15000 with KNM-ER 3733.*

growth for *Homo erectus* described in this way can be informative when discussed in the context of growth patterns determined for our comparative samples.

Although not as obvious as in the growth patterns of Eskimo or *Pan,* facial growth in *Homo erectus* can be characterized by an upper facial pattern and a lower facial pattern. The major axes of upper facial growth are anteroposterior and mediolateral, while lower facial growth occurs primarily along the superoinferior axis. Magnitudes of upper facial growth in *Homo erectus* are markedly reduced as compared with growth in either of the contemporary species. The pattern of mediolateral growth seen in the *Homo erectus* upper face does not resemble either of the contemporary species closely, but there are similarities in local growth patterns among the various forms. For example, the pattern of growth for *Homo erectus* local to frontozygomatic junction is most like *Pan,* but growth local to zygomaxillare superior more closely resembles the pattern seen in modern *Homo sapiens.* Patterns of growth for the osseous structures surrounding the nasal complex (1, 2, 13) are unlike either of the contemporary species. The pattern of growth of the anterior palate in *Homo erectus* (10, 11, 9) is most like *Pan*'s, while aspects of the growth pattern of the posterior palate and zygoma (3, 5, 16, 15) bear a closer resemblance to the growth pattern characterized for modern *Homo sapiens.*

SEXUAL DIMORPHISM OF HOMO ERECTUS GROWTH. Comparison of FDMs for female (KNM-WT 15000 compared with KNM-ER 3733) and simulated male (KNM-WT compared with either of the simulated male *Homo erectus* faces) growth in *Homo erectus*

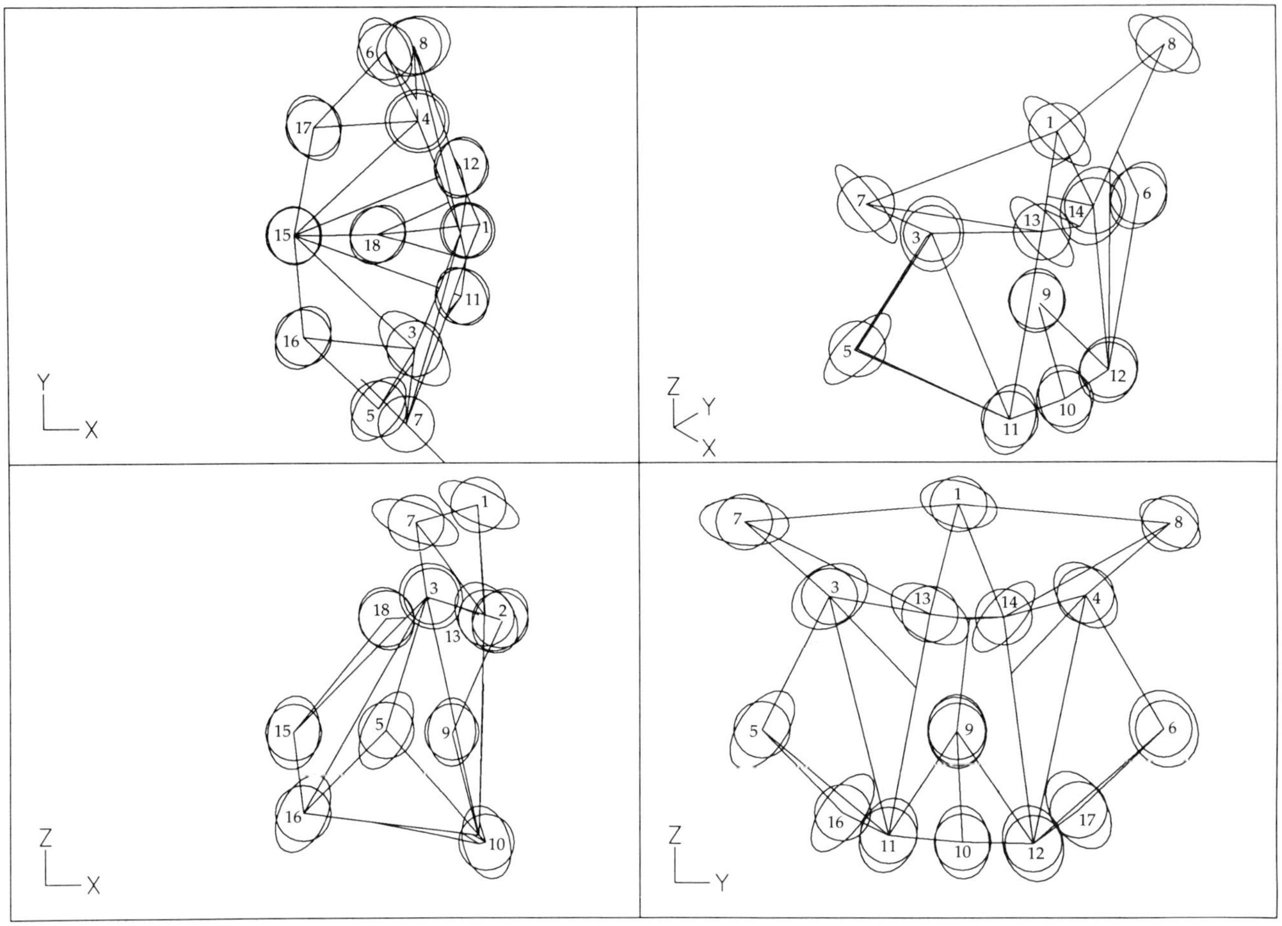

FIGURE 16.8.
FESA graphic of male HOMO ERECTUS *adolescent facial growth pattern produced by comparing KNM-WT 15000 with an adult male* HOMO ERECTUS *face simulated using the pattern of sexual dimorphism of the Eskimo face.*

are similar in terms of the ordering of linear distances from minimum to maximum change. The values of the ratios at the smaller end of the matrices are similar (about .84 to .90). However, the maximum ratio for the FDM describing female growth is 10% less than the maximum ratio for the FDMs describing male growth.

Subtle differences in growth directions at various landmarks can be seen by comparing Figure 16.7 with Figures 16.8 and 16.9. The greatest directional differences between male and female growth patterns are apparent local to landmarks 1, 7, and 8. Notwithstanding these details, overall patterns of growth in *Homo erectus* appear to be similar between the sexes.

Discussion

The most striking observation from this comparative analysis of growth in *Pan, Homo sapiens,* and *Homo erectus* is the overall similarity in adolescent facial growth patterns between sexes within species and between species. The separation of facial growth into an upper facial component, oriented primarily along an anteroinferior axis, and a lower facial component, oriented superoinferiorly, may be suggestive of a more generalized primate pattern of facial growth (Cheverud and Richtsmeier, 1986; Corner and Richtsmeier, 1991, 1992; Grausz, 1992).

We remind the reader that we are looking specifically at growth patterns responsible for producing an adult morphology from an adolescent morphology. Many of the morphological criteria that contribute to primate species designation are

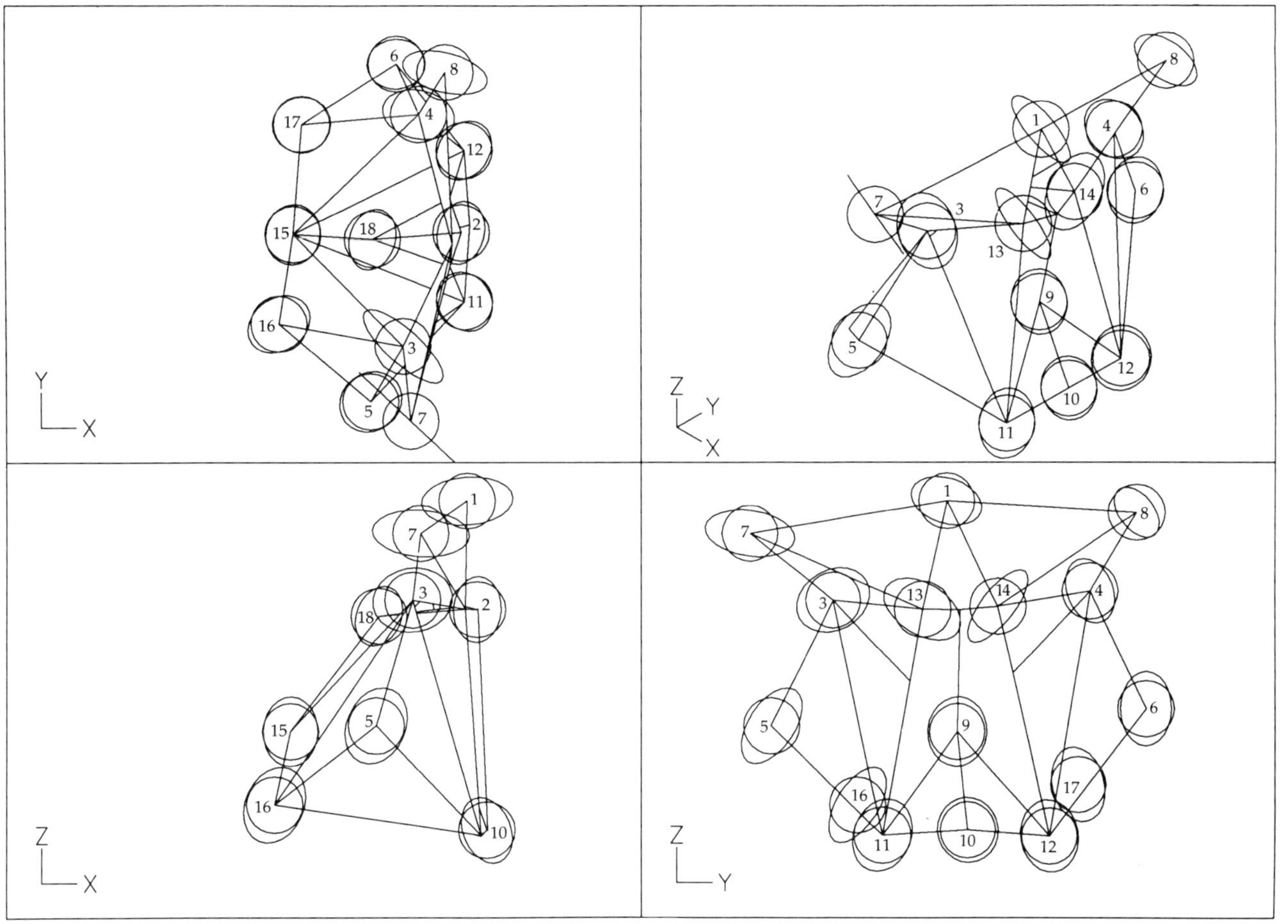

FIGURE 16.9.
FESA graphic of male HOMO ERECTUS *adolescent facial growth pattern produced by comparing KNM-WT 15000 with an adult male* HOMO ERECTUS *face simulated using the pattern of sexual dimorphism of the chimpanzee face.*

present at birth and are certainly well established by adolescence. There may be aspects of adolescent facial growth patterns that are shared among all primates, but at this time we have no systematic understanding of the similarities and the potential differences. Differences among these patterns may simply be variations that amplify certain sexual characteristics or functioning complexes. Perhaps some aspects of primate growth and development that are truly species-specific occur early in development or prenatally, even during the embryonic period. If this is the case, then adolescent growth patterns serve only to maintain or amplify an already established morphology during postnatal expansion of the form (see Richtsmeier et al., 1993, for an additional consideration).

Although local differences in growth were noted, we are more impressed by the similarities among these various growth patterns. We define *growth pattern* in a geometric sense, as the changing arrangement of skeletal structure through time. This includes magnitude and direction of change at individual landmarks but is based more heavily on the configuration of the collection of these local patterns: the relationship of growth at each landmark to all others, and to the whole. We emphasize that growth patterns need not be identical between species and sexes in order to suggest a shared developmental plan, but only similar.

We propose that our results are evidence of a generalized (perhaps primitive) pattern of adolescent facial growth in hominoids and that the diagnostic criteria of facial morphologies (i.e., the arrangement of characters that make an organism recognizably male, female, or a member of a particular species) are established early in development, perhaps prenatally. This would mean that taxa-specific features of growth patterns would also appear early in development. There may exist an underlying, generalized facial growth pattern shared by all primates. We expect morphological complexes that distinguish, for example, *Pan* from *Homo* to have underlying, unique developmental patterns. However, there may be aspects of the growth patterns that cross phylogenetic boundaries. If described adequately, postnatal facial growth patterns could be used as characters for phylogenetic analyses (see de Queiroz, 1985). Deviations from a shared pattern might be interpreted as derived characters, adjustments to particular life history or ecological circumstances with a genetic basis.

Acknowledgments

We thank David Hunt of the Smithsonian Institution for assistance in retrieval of specimens from the Division of Anthropology skeletal collection, Smithsonian Institution National Museum of Natural History, and Doug Ubelaker for access to the collections. Bruce Latimer kindly made specimens of immature *Pan troglodytes* from the Cleveland Museum of Natural History available for study. Subhash Lele and Leslie Marcus provided assistance in programming the solution to determining three-dimensional coordinates from an over-determined set of linear distances. Elaine Kasmer drew Figure 16.1. We thank Brian Corner and Hannah Grausz for reading and commenting on previous versions of this manuscript. This work was made possible in part by grants from the Whitaker Foundation, the March of Dimes Birth Defects Foundation, the John Simon Guggenheim Memorial Foundation, and the John D. and Catherine T. MacArthur Foundation.

C H A P T E R 1 7

Perspectives on the Nariokotome Discovery

ALAN WALKER

Analysis of the Nariokotome skeleton and the new insights that it has given us in interpreting other, more fragmentary, specimens have led to several revisions of our view of human evolution. We have found few instances in our investigations where these early hominids differed substantially from modern humans, but those we have found testify to differences that are probably all ultimately and indirectly related to differences in higher mental function. The discovery of this 1.5-million-year-old skeleton has changed the narrative of our own evolution and illuminated yet other areas of our ignorance.

The nearly complete skeleton KNM-WT 15000 found at Nariokotome III by Kamoya Kimeu is one of the truly great discoveries in paleoanthropology. Eugene Dubois found the first *Homo erectus* in 1891, but up until 1971 only isolated parts of skeletons had been found in Asia and Africa. Meave Leakey found a fragmentary partial skeleton (KNM-ER 803) near Ileret, Lake Turkana, in 1971 and, although we did not know it at the time, this specimen was the first to include associated dental and postcranial material of *H. erectus.* The next year Kamoya Kimeu discovered the first associated bones of a single diseased individual (KNM-ER 1808); this specimen was more complete, but the long bones were covered with a surface layer of bone owing to a bout of periostitis. Although it was an interesting case for paleopathology (Walker et al., 1982), the fact that the individual had suffered the disease made comparative studies very difficult.

The discovery in 1984, then, of KNM-WT 15000 was the climax of nearly a century of diligent search. What we have found out already about this individual and his population's characteristics, about the life history strategies of this species and what they have to tell us about our own evolution, shows us that the search was worth it. I hope my colleagues and I have laid the groundwork for many future studies of this skeleton.

Although the chapters in this volume cover a wide field of topics, there remain a few issues that have not been addressed. One wonders about the life history strategy of early African *H. erectus* populations, and in what sorts of environments they lived. What morphological features characterized the species? And what can this skeleton tell us about the evolution of our own lineage? The comments that follow are often only speculative, but they might lead other researchers to invent ways of answering some unresolved important questions.

THE SIZE AND WEIGHT OF EARLY AFRICAN *HOMO ERECTUS*

When we discovered the first limb bones of KNM-WT 15000 in 1984 we were surprised by their size. They were long for an individual whose second molars had just erupted and whose upper milk canines were still in place. When still in camp at Nariokotome, we thought that we must have found a rare, tall individual. But we found out on our return to Nairobi that we had underestimated the size of the incomplete bones that had been collected previously and that all known early African *H. erectus* individuals were tall. The results of the detailed study of body size and body shape (Ruff and Walker, this volume) have confirmed our initial assessment (Brown et al., 1985).

Six *H. erectus* individuals from East Africa have parts complete enough to predict their stature and body weight. These six probably include both males and females. The Nariokotome skeleton is almost certainly male on the basis of skull and pelvic characteristics. Walker at al. (1982) and Leakey and Walker (1985) thought that KNM-ER 1808 was a female, but McHenry (1991) thought otherwise. In Chapter 10, this volume, Walker and Ruff have carried out computerized shape analyses on hip bones, but they were unable to apply that analysis to KNM-ER 1808 because the sciatic notch is not preserved in its entirety. McHenry contends that the sciatic notch is asymmetric and thus more similar to males, although he agrees that it appears to have been broad. We could match the remaining curvature of the notch in the fossil to either male or female, depending on how we oriented the specimen. We still maintain that the brow ridges, the depth of the mandible, and the size of the temporal make it more likely that the individual was female. The partial skeleton KNM-ER 803 is smaller in stature and in tooth dimensions and may be female, and the Olduvai femur and hip bone, OH 28, is definitely female, according to visual examination by Day (1971) and to Walker and Ruff (Chapter 10, this volume) by computerized shape analysis.

The average stature is 170 cm (158–185 cm) and the average weight is 58 kg (51–68 kg) for this population of early *H. erectus.* Any comparisons with modern humans can only be illustrative, for modern *H. sapiens* is an extremely diverse, polytypic species. But for the sake of illustration, we can look at the distributions of size and shape in various human populations around the world. Martin (1928) compiled data about body height and weight from many sources. For height he

TABLE 17.1.

World distribution, by average male height in cm, of modern human populations (data from Martin, 1928).

	SMALL (140.8–159.9)	MEDIUM (160–169.9)	LARGE (170–200)
Africa	8	25	12
Europe	4	51	13
Asia	35	43	4
Oceania	9	10	4
New World	10	19	11
Total number of populations	66	148	44

broke these populations down into 3 groups—small, medium, and large. The small group had 66, the medium group had 148, and the tall group had 44 populations.

As can be seen from Table 17.1, the early African *H. erectus* population, with a mean stature of 170, would be in the tallest 17% of modern human populations, even if we make the comparison only with modern males. The average female heights in these modern populations are about 10 cm shorter, so if the *H. erectus* sample included some females the population would be among an even smaller percentage of the tallest modern human populations. A smaller amount of body weight data is available for modern populations, but according to the data in Martin (1928) the body weights estimated for the early African *H. erectus* sample would match those of the middle-sized modern human populations. This is because the morphometric model used in the weight estimation allows for the elongated body shape that we know KNM-WT 15000 had.

Confirmation of this relatively large body size comes from a set of footprints found on the east side of Lake Turkana (Behrensmeyer and Laporte, 1981). These footprints were apparently made by a *Homo erectus* individual walking across a sandy mudflat covered by shallow water. The prints are in sediments that are about the same age as those of the site of Nariokotome III (NK3). Feibel and Brown (Chapter 3, this volume) show that the sediments containing the prints are at virtually the same stratigraphic level as the footprint tuff at site NK3. Behrensmeyer and Laporte estimated, from the length of the footprints, a stature of between 1.6 and 1.7 meters for the individual. This estimate can be refined by using Tuttle's (1987) data on stature and foot length data in over 50 modern human populations. The regression equation is:

$$\text{Stature} = 5.366 \times \text{Foot length} + 28.97$$
$$(r^2 = .879,\ N = 52,\ p < .001)$$

This gives a calculated height of 1.685 meters for the individual and is very close to the average for stature of 1.7 meters calculated by Ruff and Walker (Chapter 11, this volume) from the bones of 6 individuals.

With good stature and body weight estimates on what is admittedly a small sample of early African *H. erectus,* together with knowledge of their body proportions, we can begin to answer some important questions about human evolution. The first is the problem of climatic adaptation and the second is the interpretation of the reduction of stature that seems to have taken place in many parts of the world with the advent of agriculture. We can also look into several aspects of life history strategy because body weight is correlated very strongly with many of them (Harvey, 1990).

Body proportions and thermoregulation

In an earlier publication (Leakey and Walker, 1989), we used Trinkaus's regression method (Trinkaus, 1981) to determine what mean annual temperature the KNM-WT 15000 individual was adapted to. We calculated values of 18.5 and 25.3 °C using the preliminary brachial and crural indices that we had estimated then. We now have better estimates of the radial length (see Ruff and Walker, Chapter 11, this volume) and have also found that there is an error in the published regression equations. The correct equation for the hindlimb is:

$$\text{Mean annual temperature} = 3.328 \times \text{Crural index} - 263.602$$

This gives a mean annual temperature of 29.2 °C using our crural index of 88. Using our new brachial index of 79.9 in Trinkaus's (1981) other equation, we calculate a mean annual temperature of 30.8 °C. These values are extremely close to one another and confirm the findings presented in Chapter 11 that KNM-WT 15000 was adapted to a climate with extremely high mean annual temperatures. It is true, as Ruff and Walker pointed out in Chapter 11, that ratios can be misleading if used over a wide range, but these two new temperature values are consistent with each other and with the results of other ways of looking at body proportions.

Feibel et al. (1991) report that the Lake Turkana region has had the same climate as today since about 1.5 Ma. They base this on many lines of evidence, from soil profiles to animal diversity, and point to the fresh and unweathered state of 1.5 Ma volcanic outflows as a clear sign of continuous aridity since then. If, as seems likely, the general climate of the Lake Turkana region then was much the same as it is today, and our deductions about the climatic adaptations of early *H. erectus* populations are correct (Ruff, 1991), then these early humans were shaped the way they were for efficient cooling. People cool off by sweating, not by panting or a combination of panting and sweating as do nearly all other mammals. A human needs a prodigious amount of water to keep cool in a hot climate—as all of us who have worked at Lake Turkana know! All this points to the conclusion that early *H. erectus* was under high radiant heat stress and was very dependent on water (see also Wheeler, 1991).

It is possible that early African *H. erectus* were salt-dependent too, for by relying only on sweating for cooling, people also lose a large amount of salt. Panting animals reingest most of their salt (Schmidt-Nielsen, 1975) and this enables them to cool themselves without depleting their salt resouces. Many animal and human populations ingest salt directly by geophagy, and it is possible that early African *H. erectus* did so too. Another source of salt might have been animal blood and animal tissue, for it is virtually certain that these early humans were eating meat (Bunn, 1981; Potts and Shipman, 1981). And finally, Carrier (1984) put forward an argument based on energetic principles that early hominids had a physiology that allowed endurance running at speeds which forced prey to avoid them in an inefficient way. This, he hypothesized, made them effective predators who could run down prey without succumbing to overheating. What we have learned from the Nariokotome skeleton can be taken to support Carrier.

Body Size and the Origins of Agriculture

It used to be thought that agriculture brought health and happiness wherever it had developed, but lately there has been an increasing realization that agriculture can bring with it increased risk of disease, greater susceptibility to starvation, and poor nutritional status for its practitioners. The most useful recent compilation of data and ideas is Cohen and Armelagos (1984). As Cohen and Armelagos say in their summation, "The data point fairly clearly to a decline in health associated with the origin of agriculture." Some of this data is evidence in several populations at different times and in different parts of the world of a reduction in stature. None of the contributors to the Cohen and Armelagos volume were dealing with the early paleolithic record and, of course, most concentrated their attention on the transition to agriculture. It is not surprising, then, that the beginnings of hunter-gathering practices were not considered. Cohen and Armelagos point out that the health trends before the adoption of agriculture are of great importance in interpreting trends subsquent to the start of it. But, they said, "Unfortunately, these trends are particularly hard to unravel in the absence of good early ('Paleo') hunter-gatherers."

The new evidence for stature in early *H. erectus* is of great help in looking at the problem of what agriculture does to people. It is most revealing to learn that populations of what might be the earliest hunter-gatherer humans were taller than most modern human ones. These hominids were living in equatorial Africa in extremely hot climates without the benefit of domesticated stock upon which the present-day people depend for survival. They also lived, we assume, in low population densities. Yet their tall statures support the conclusion that people living a hunter-gatherer existence might be healthier, better fed, and less subject to famine.

Antecedent populations in the same area of

Africa were not as tall as the early *H. erectus* ones (McHenry, 1991; Ruff and Walker, Chapter 11, this volume) if the small amount of data can be trusted. McHenry (1991) calculated almost exactly the same estimated statures that we do for the 3 individuals from the antecedent populations. His three averaged 146.6 cm while Ruff and Walker give the average as 147.0. Thus the earlier individuals seem to have been about 23 cm (about 14%) shorter on average than *H. erectus*. McHenry (1988) found the weight of these three early individuals to average 49.6 kg, while Ruff and Walker, using a different model, calculate the average as 46 kg. Thus the early sample could have averaged as much as 12 kg (about 20%) less than the early *H. erectus*. If this small sample gives a true indication of what was going on about 2 million years ago, then it would appear that both stature and body weight increased at the same time that a hunter-gatherer life-style was being established for the first time.

Five femurs about half a million years old have been recovered from Zhoukoudian (Weidenreich, 1941). C. B. Ruff has kindly calculated for me the body weights using his climate model of human proportions (Ruff, 1991). He reports an average of a little over 56 kg. This is very close to the mean of 58 kg calculated for the early African *H. erectus* population (Ruff and Walker, Chapter 11, this volume). On this slim basis, then, it would appear that there were not any major changes in body weight in the evolving *H. erectus* lineage for over 1 Ma.

An Example of Stasis or of Gradual Evolution?

Rightmire and Wolpoff carried on a short exchange of papers on whether *Homo erectus* was an example of stasis in evolution or whether gradual evolution had taken place. Rightmire (1981) began with a regression analysis of four traits in a sample of *H. erectus* which included a few specimens, such as Olduvai Hominid 13 and Petralona, that are not usually thought of as belonging to this species. He found that the slopes of the regressions of traits versus geological time were not significantly different from zero when he dropped contentious specimens such as these from his analysis. He concluded that no significant trends can be observed and that stasis had occurred for over a million years before rapid evolutionary change began toward the end of the Middle Pleistocene. Wolpoff's (1984) reply was based on an enlarged sample of *H. erectus* which he termed "conservative." He divided the sample into three "low-resolution" time spans and tested the differences between the means by Student's t tests. He found that there were significant changes in a number of skull and tooth features. Rightmire later (1986) said that Wolpoff had not composed his samples in a biologically reasonable way and, further, that some specimens included in Wolpoff's sample couldn't be measured properly. Wolpoff then (1986) dropped the offending samples and was still able to point to significant differences among his time samples.

It is worth pointing out that the two methods are very different and that, perhaps, the gradualist and punctuationist viewpoints have been forced into caricatures of extremes. Even ardent gradualists admit that there is little to no change in most systems of the body for most species for most of the time (Rose and Bown, 1992). And even those who promote the idea of punctuated equilibrium would have to admit to small, fluctuating changes through time as opposed to absolutely no change at all. The most telling point in the debate over change in the *H. erectus* lineage, as far as the issues go concerning KNM-WT 15000, is the composition of Wolpoff's early category. His sample for the Lower Pleistocene includes specimens from the earliest African ones, about 1.8 million years old, to later ones from Sangiran in Java. It seems more than likely that no hominids in Asia are much older than a million years. Three authorities (J. DeVos, J. Franzen, and Wu Rukang) agree that the earliest *H. erectus* populations in East and Southeast Asia might be no older than 700,000 years (in Wolpoff and Nkini, 1985). Thus Wolpoff's early category might cover over a million years of evolution. Any changes over that time period must, then, be relatively small. For example, the *Homo erectus* crania from Sangiran and Trinil in Java have an average cranial capacity of 931 cc with a range from 815 to 1059 cc. The average cranial capacity of early African *Homo erectus* from about 1.7–1.2 Ma is 907 cc with a range from 804 to 1067 cc (Begun and Walker, Chapter 14, this

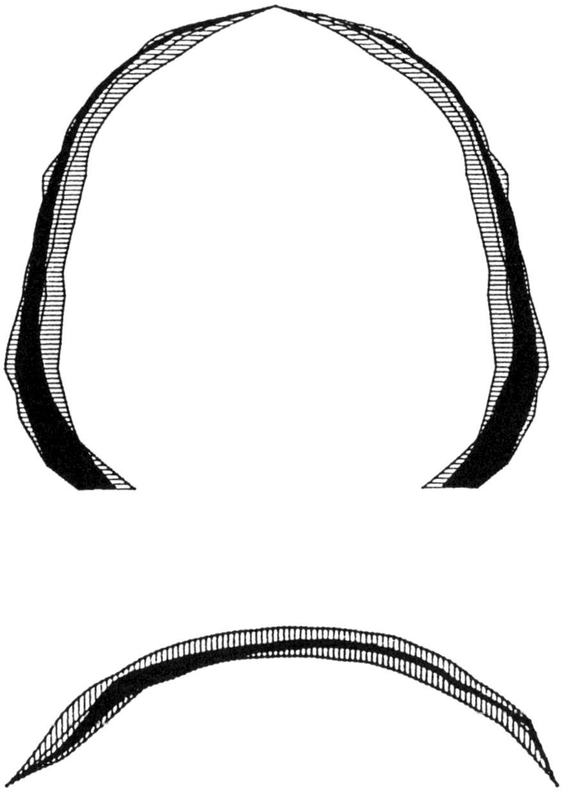

FIGURE 17.1.
Computer-generated shape envelopes for coronal and sagittal sections taken through HOMO ERECTUS *crania. In cross-hatching are shown later crania from Indonesia and China, in solid black the two early crania KNM-ER 3733 and 3883.*

volume). The shapes of the early and late skulls are also not very different.

Figure 17.1 shows computer shape envelopes generated in the same manner as for the sciatic notches by Walker and Ruff, Chapter 9, this volume, for early and late *Homo erectus.* Sagittal and coronal shape envelopes for later specimens from Indonesia and Java are shown in cross-hatching on the sagittal and coronal profiles. In solid black is the envelope generated for the two early specimens from Kenya, KNM-ER 3733 and 3883. As can be seen, this envelope falls within that of the later specimens.

In Chapter 9 (this volume), Brown and Walker show that the teeth from KNM-WT 15000 are very like those described from 500,000-year-old Chinese *H. erectus* and that, as with stature and brain size, there is relatively little change over the million-year span. In any case, though there are some demonstrated small changes through the lineage, these appear dwarfed by the sudden changes in morphology that took place at about 1.8 million years ago and then again with the appearance of anatomically modern people.

Brain weight and body weight in KNM-WT 15000

The Nariokotome skeleton is complete enough for us to estimate both the brain weight and body weight in the same individual. This cannot be accomplished for any other early hominid. Whole skeletons were hardly ever left for sedimentary burial, it seems, until the time humans began to bury the dead. Although partial skeletons of early hominids have been found before, none has a cranium preserved well enough to measure a cranial capacity.

In an influential paper on size and scaling in human evolution, Pilbeam and Gould (1974) looked at the scaling of the brain and the teeth in human evolution. They demonstrated that the African great apes were, in respect to the way the cranial capacity scaled to body weight, scaled versions of each other. They also showed that the three species of *Australopithecus* known then were also, in the same way, scaled versions of each other: the larger species were no more encephalized than the smaller but showed the usual mammalian relationship of brain size to body size. The difference between the ape and australopithecine samples did not lie in the slopes of the power functions calculated for them but in the intercepts. The hominids were more encephalized—the intercept of the line calculated to fit the hominid points was higher, but it had the same slope as the line calculated to fit the ape data. Thus both the African apes and the australopithecines were seen (in their relationship of these two features) to be series of animals differing only in size and not in design.

Pilbeam and Gould came to a different result when they applied this method to the human lineage. They used the sequence *Australopithecus africanus, Homo habilis, Homo erectus, Homo sapiens* to represent the human lineage. They found that the slope of the regression line calculated to fit the data for brain weight and body weight of this lineage was very different from the slopes of the lines for australopithecines and apes. The slope was 1.73, far greater than the other two slopes of .339 and .329 and far and away steeper than any slope representing functional equivalence. Pilbeam and Gould took this to show quite clearly that the brain had increased in size with marked positive allometry during our own evolution. Thus they concluded that "all australopithecines had brains equally expanded beyond the ape grade" but that that in our lineage brain volume had increased dramatically. One reason *Homo habilis* should be in our genus, they added, was that it was the first species whose brain had increased beyond the australopithecine level.

We now have a chance to reevaluate this work. Another species of *Australopithecus* has been found since 1974 and we have better estimates for body weight and brain weight, especially for early *H. erectus.* The slope of the power function calculated for the australopithecines is not changed significantly by the addition of brain and body weight estimates for *A. afarensis.* The estimates for *H. habilis* and *H. erectus* do not now fit on the regression line for the human lineage calculated by Pilbeam and Gould. This is because the 68 kg adult body weight estimate for KNM-WT 15000 is much greater than the 53 kg that Pilbeam and Gould used for *H. erectus,* and the cranial capacity of the specimen is less than the value of 1050 cc that they used. The point representing *H. erectus* would still be very far from the line which Pilbeam and Gould calculated for the human lineage even if we were to use mean values for the early African *H. erectus* populations rather than the values for KNM-WT 15000 itself. This is because the body weight estimates are higher (58 kg) and the cranial capacity estimates lower (mean for four capacities = 907 cc). A line through the *H. habilis* and *H. erectus* points has a slope very close to regression lines calculated for the apes and australopithecines. This is especially remarkable because there are only two points. The straight line which Pilbeam and Gould could place through the points representing the human lineage can no longer be drawn. This is because the average body weight of early *H. erectus* is as big as those found in some modern populations with very big people, but the average cranial capacity of early *H. erectus* is much smaller than modern human cranial capacities. These authors would also have seen a more complicated picture had they used several human populations in their analysis. Their values for *H. sapiens* was a combined average of the data for male and female Australian aboriginal people. Figure 17.2 shows the data for several human populations, and these data, too, are scattered along a line with a slope very similar to the three others.

Figure 17.2 suggests an alternative view of brain and body scaling in human evolution. It seems there might be one scaling pattern for all hominoid groups rather than a separate pattern for members of the human lineage fundamentally different from those of the ape and australopithecine ones. We can postulate four allometric or scaling *grades* for the hominoid pattern: one each for African apes, australopithecines, *H. habilis–H. erectus,* and *H. sapiens.* Each of these groups scales in a similar way, which means that the animals in the group are similar in form but different in size. The difference between the groups is the level of encephalization. This view seems to be more likely, to my way of thinking, than the idea that there was a different set of relationships for our lineage alone.

Evolution from one grade to the next could have come about in several ways. Martin (1983) gives a theoretical discussion of them. Grade shifts require changes that result in vertical readjustments on the graph. It is theoretically possible that in this one respect, for instance, the chimpanzees could have given rise to australopithecines had their bodies but not their brains decreased in size. It is also possible that brain size could have increased while body size remained constant. And

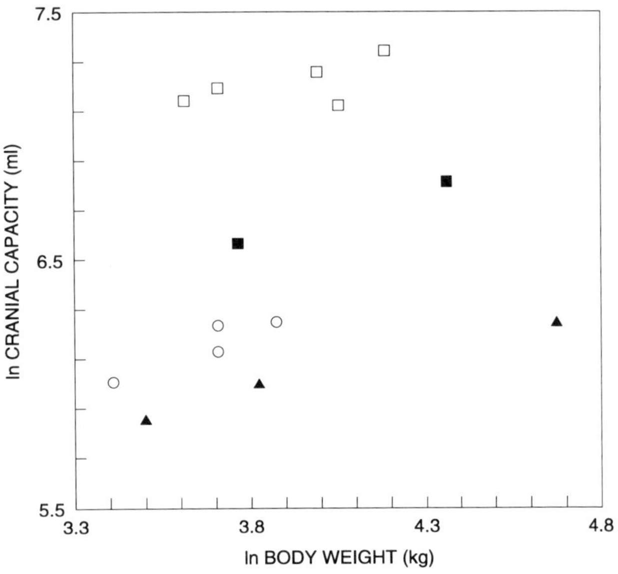

FIGURE 17.2.
Plot of estimates of body weights and brain weights (in logarithms) of four hominoid groups. Triangles: *African apes (data from Pilbeam and Gould, 1974);* circles: AUSTRALOPITHECUS *species;* filled squares: HOMO HABILIS *and* HOMO ERECTUS *(the point at upper right based on KNM-WT 15000 adult estimate);* open squares: HOMO SAPIENS *(population means taken from Martin, 1928). Regression lines have not been drawn because the data points are so few, but the trends are apparent.*

lastly it is possible that both body size and brain size increased. In the case of the three postulated scaling grades of hominids, it seems not to have been possible for the first to have happened, because the lowest brain values in the two later grades are bigger than the highest in the preceding ones. Correlated increases in both body and brain size could have happened and dwarfing could have happened with brain size increase in the evolution of *H. sapiens* from *H. erectus.* If, as some believe (e.g., Andrews, 1984), Asian *H. erectus* were not on the lineage to modern humans, the same result applies. It is likely that most modern human populations have reduced body size relative to our early ancestors.

THE TAXONOMY OF EARLY AFRICAN *HOMO ERECTUS*

Leakey and Walker gave a brief account of the problem of taxonomy of early African *H. erectus* in Chapter 1 (this volume). When KNM-ER 3773 was first found it was considered an early African representative of this species. It was earlier even than Olduvai Hominid 9, which has been almost unanimously considered to belong in *Homo erectus* (its finder, Louis Leakey, was an exception to the consensus—see Leakey, 1964). In Chapter 6 (this volume), Walker and Leakey have given an account of all early African *H. erectus* specimens. I hope that this review will encourage people making taxonomic and phylogenetic assessments to take into consideration the high degree of variability to be found in the fossil sample.

The 1980s saw the rise of Hennegian cladistics as a common tool in taxonomy and phylogenetics in all branches of biology. There have been several cladistic studies of human taxonomy and phylogeny, but the one that I find most illuminating is that of Bilsborough and Wood (1986). They used two alternative methods to examine the questions of what *Homo erectus* really is and what were its origins and fate. The first was shape analysis, an analysis of a series of linear measurements using multivariate statistical procedures. The second was a cladistic analysis based upon the states of a selected series of characters.

The results of this study are illuminating because they give such different answers. The phenetic (or shape) analysis showed that early *H. erectus* from Africa are very close to late *H. erectus* from Asia but that there was a "distinct break" between the antecedent African fossils and early *H. erectus.*

The results of the cladistic study show major differences between some early *H. erectus* specimens from Africa and those from Asia. These differences are great enough to exclude two important early crania from Koobi Fora from the species *H. erectus,* because they lack the unique derived features of the Asian forms. The cladistic analysis links these early crania with those early *Homo* which immediately antedate them. The finding of no shared derived features with which one can link the early and late *H. erectus* contrasts strongly with the clear shape similarities. And yet the analysis links early *H. erectus* with specimens that do not look like them. This paradoxical result is illuminating in that it shows clearly that the two methods are unlikely to yield congruent answers when one is dealing with such closely related organisms.

Some practitioners of cladistics stick to a strict taxonomy that relies on splitting points. Changes of whatever magnitude within a species lineage cannot be used to define as species any isolated part of the lineage. Wiley (1978) and Hull (1979), for instance, insist that, because an evolutionary species is in itself a lineage, it can give rise to other species only by branching. If this position is accepted, then all species of *Homo* back to the branching point from an ancestral species would be, by definition, in the same species. That species would be *Homo sapiens* for reasons of priority.

Rose and Bown (1993) give a clear exposition on the demerits of this sort of taxonomic scheme. As field paleontologists dealing with a very dense fossil record, they find such a scheme of little practical use, for, in the first and last members of their species, lineages might be composed of very different sorts of animals. I feel that taxonomy should be an aid to describing the real world and not a barrier. The real reason that cladistic treatments of very closely related organisms are difficult is the variability both within and between populations. When comparing closely related organisms it is difficult to find features that are discrete enough that all members of the populations have them. The use of continuous variables (or indices made from them) in cladistic analyses has grave problems—see Kennedy (1991) for a list of a few of them—and, further, makes it even more difficult to imagine that these "traits" are anything that would be genetically controlled in a straightforward way without being correlated very strongly with each other.

If evolution is the subject of interest, rather than taxonomy, the use of cladistics to sort out fossils in the genus *Homo* is frustrating. Evolution acts, as we have known since Darwin, upon the variability present in populations. By ignoring variability, in the attempt to make cleanly defined traits, much of the information in the fossil record can be lost. As Hopson (1989) says, "Although I am an advocate of cladistics, I find that the application of its methods to the dense fossil record of Cenozoic mammals, in which differences among closely related taxa often have a complex distribution, carries with it the danger of oversimplifying such distributions in the interest of clearcut results."

Several recent attempts to apply these methods have included *H. erectus* as an object of cladistic scrutiny (e.g., Andrews, 1984; Bilsborough and Wood, 1989; Chamberlain and Wood, 1987; Stringer, 1984, 1987; Wood, 1984, 1991). Kennedy's (1991) appraisal of supposed autapomorphic traits of *Homo erectus* is instructive in evaluating them. She looked at twelve possible traits that were said in several studies to be autapomorphic for the species and found that neither the qualitative discrete traits nor the quantitative traits were autapomorphic. She concluded that "*H. erectus*, as presently defined and using the methodology of phylogenetic systematics, cannot be considered a valid species." Brauer (1990) also attacked the idea that the Asian sample had a unique set of derived features. He looked at seven characters that Andrews (1984) believed were present in the Asian sample but not in the early African one. Brauer found that all seven could be found in the early African *erectus* grade sample. Most tellingly, Stringer, whose analyses in 1984 led him to conclude that the African fossils could be included in *H. erectus* only if a grade concept were being used, had by 1987 decided that they could be properly included in the *erectus* clade.

Finally, Turner and Chamberlain (1989) argue that cladistic analyses that attempt to define species using apomorphic character states may prove to be unworkable unless the characters themselves can be related to the fertilization system of the species. Here they follow Paterson (1985) by using his "recognition concept," which states that only those characters of sexual organisms that are related to mating and fertilization (or the bases for them) will necessarily be correlated with speciation itself. These authors do not attempt to give any details of *H. erectus* morphology

TABLE 17.2.

Thickness (mm) of frontals at or near bregma.

KNM-ER 730	8.5
KNM-ER 1466	10.0
KNM-ER 3883	8.0
KNM-ER 3892	9.2
KNM-WT 15000	7.3
KNM-ER 3733	(damaged here)
OH 9*	9.5

*Estimated from diagram of CT scan in Maier and Nkini, 1984.

that could be related to the specific mate recognition system of Paterson. Rather, they point out that most, if not all, of the autapomorphic characters used in prior cladistic analyses are hardly likely to have influenced early hominid sexual behavior. They believe that there is no consistent correlation between a speciation event and a change in morphology and that, when due recognition is given to the level of variation in early African *H. erectus* grade hominids, there is no good reason not to include them in the species.

This last point about variation seems very important, and without intending to belabor the issue, I shall give yet another example. It is often said that the crania of the early African sample have thinner vaults than later crania. Table 17.2 gives frontal bone thickness measured at bregma or as near as possible to bregma in the early African sample for comparison with the later Asian ones. Even counting KNM-WT 15000, which is a juvenile, and not counting Olduvai Hominid 9, which is from somewhat younger deposits, the mean for the early sample is 8.6 mm with a range of 7.3 to 10.0 mm. If Olduvai Hominid 9 is included, the mean is 8.75 mm. The mean and ranges for vault thickness near bregma for Zhoukoudian, according to Weidenreich (1943), is 8.78 mm with a range of 7.0 to 9.9 mm. Thus the vault thicknesses in these two samples are very nearly equal.

With all this in mind, we decided for the purposes of this volume to assign KNM-WT 15000 to *H. erectus.* The following alternatives to this assignment may or may not be testable in the future.

(1) For those who see speciation as occurring only by splitting, KNM-WT 15000 could be called one of three things. If there has been no split of the lineage leading to *Homo sapiens* since the early Pleistocene, it will be part of our species *by definition.* If there has been a split of the lineage, then KNM-WT 15000 might either be placed in the species *Homo erectus* or be declared a new species, depending on the distribution of derived traits.

(2) For those who believe that species can arise anagenetically, then time-successive species within the genus *Homo* might be definable. If it becomes apparent that the early African sample is the same species as the later Asian ones, then it might be placed in *Homo erectus,* but if there are sufficient unique differences then it might be accommodated in a new species.

In either case, if a new name is needed for the species in which the early African sample belongs, then that name is, as suggested by Wood (1991, 1992), *H. ergaster.* This point is taken up in the next section.

THE STATUS OF OTHER FOSSILS

With a fairly complete skeleton of an early African *H. erectus,* it is now possible to know which other fossils belong in this species and which do not. The discovery of KNM-WT 15000, together with the recently discovered association of a mandibular fragment with a small *Australopithecus boisei* skeleton, means that we can sort most postcranial bones from the later part of the Turkana Basin succession into one or the other species. The task of sorting is still difficult for the period up to 1.7 Ma because it appears that some of these earlier postcranial bones could either be small *H. habilis* or *A. boisei.* In Chapters 6 and 7 (this volume), Walker and Leakey have pointed out which other fossils from East and West Turkana can be certainly assigned to *H. erectus.* Table 17.3 gives a list of these.

The most important, as far as taxonomy

goes, is KNM-ER 992, the type of *Homo ergaster.* A nearly complete mandible, KNM-ER 992 was discovered in 1971. Leakey et al. (1978) stated that this specimen belonged to the same taxon as cranium KNM-ER 1813 and that there were affinities with *Australopithecus africanus* from South Africa. Groves and Mazak (1975) made the mandible the type specimen of *Homo ergaster* but, curiously, did not attempt to make a differential diagnosis between this specimen and those attributed to *H. erectus.* Wood (1976) concluded from a metrical study of the mandible that it could be allocated to *H. erectus.* Howell (1978) also placed it in that species.

Until KNM-WT 15000 was found, however, it was still difficult to establish the identity of this jaw. This was because only one of the mandibles from Lake Turkana was associated with an *erectus* grade cranium. This specimen (KNM-ER 730) also was an old adult with anterior alveolar resorption and with the three remaining teeth worn to the gum. Comparisons could be made with cranium KNM-ER 1813, which is not of *erectus* grade. Groves (1989) regards this comparison favorably and includes both KNM-ER 992 and 1813 in his *H. ergaster.* He made statements about the shape of the whole mandible that are not accurate. As noted in the original description (Leakey and Wood, 1973), the break between the two halves is not a simple one, and some crushing has occurred. Leakey and Wood wrote: "Breakage and displacement of bone at the anterior end of the body, especially on the left side, precludes accurate articulation of the hemi-mandibles, and requires that comments on the shape of the tooth row and dimensions of the whole mandible must await expert reassembly of the damaged area." This has not yet been done. The photograph in Groves (1989, figure 7.4) of the casts pushed together at the anterior break is misleading.

We can now compare the mandible of the Nariokotome specimen with the type of *H. ergaster.* When allowance is made for the juvenile state of the former and for the interstitial and occlusal wear of the latter (Brown and Walker, Chapter 8, this volume), the similarities are striking, even down to details of the subocclusal morphology as revealed by X-ray. If, as Wood (1991) suggests, the early African *erectus*-grade hominids need their own taxon, then *H. ergaster* is the appropriate name for it. Two other mandibular specimens now known to be from this sample have been included here in Figures 17.3 and 17.4. These are KNM-ER 820 and KNM-ER 730. The first is a juvenile younger even than KNM-WT 15000 (see Smith, Chapter 9, this volume) and the second is an old adult, older in terms of tooth-wear than KNM-ER 992. These four mandibles have been arranged in the figures in order of individual age, and they make a very impressive age series showing devel-

TABLE 17.3.

Early *Homo erectus* specimens from Lake Turkana.

SPECIMEN NUMBER	BODY PART	STRATIGRAPHIC MEMBER
KNM-ER 164	Cranial fragment, vertebrae, phalanges	KBS
KNM-ER 730	Partial skull	KBS
KNM-ER 731	Mandible fragment	Okote
KNM-ER 736	Femur	Okote
KNM-ER 737	Femur	Okote
KNM-ER 741	Tibia	Okote
KNM-ER 803	Partial skeleton	Okote
KNM-ER 806	Isolated teeth	Okote
KNM-ER 807	Maxilla fragment	KBS
KNM-ER 808	Isolated teeth	Okote
KNM-ER 820	Juvenile mandible	Okote
KNM-ER 992	Mandible	Okote
KNM-ER 1466	Frontal fragment	Okote
KNM-ER 1808	Partial skeleton	KBS
KNM-ER 1821	Parietal fragment	KBS
KNM-ER 2592	Parietal fragment	Okote
KNM-ER 3733	Cranium	KBS
KNM-ER 3883	Cranium	Okote
KNM-ER 3892	Frontal fragment	Okote
KNM-ER 5428	Talus	KBS
KNM-WT 15000	Skeleton	Natoo
KNM-WT 16001	Cranial fragments	Natoo
KNM-WT 19700	Tibial fragment	Nachukui

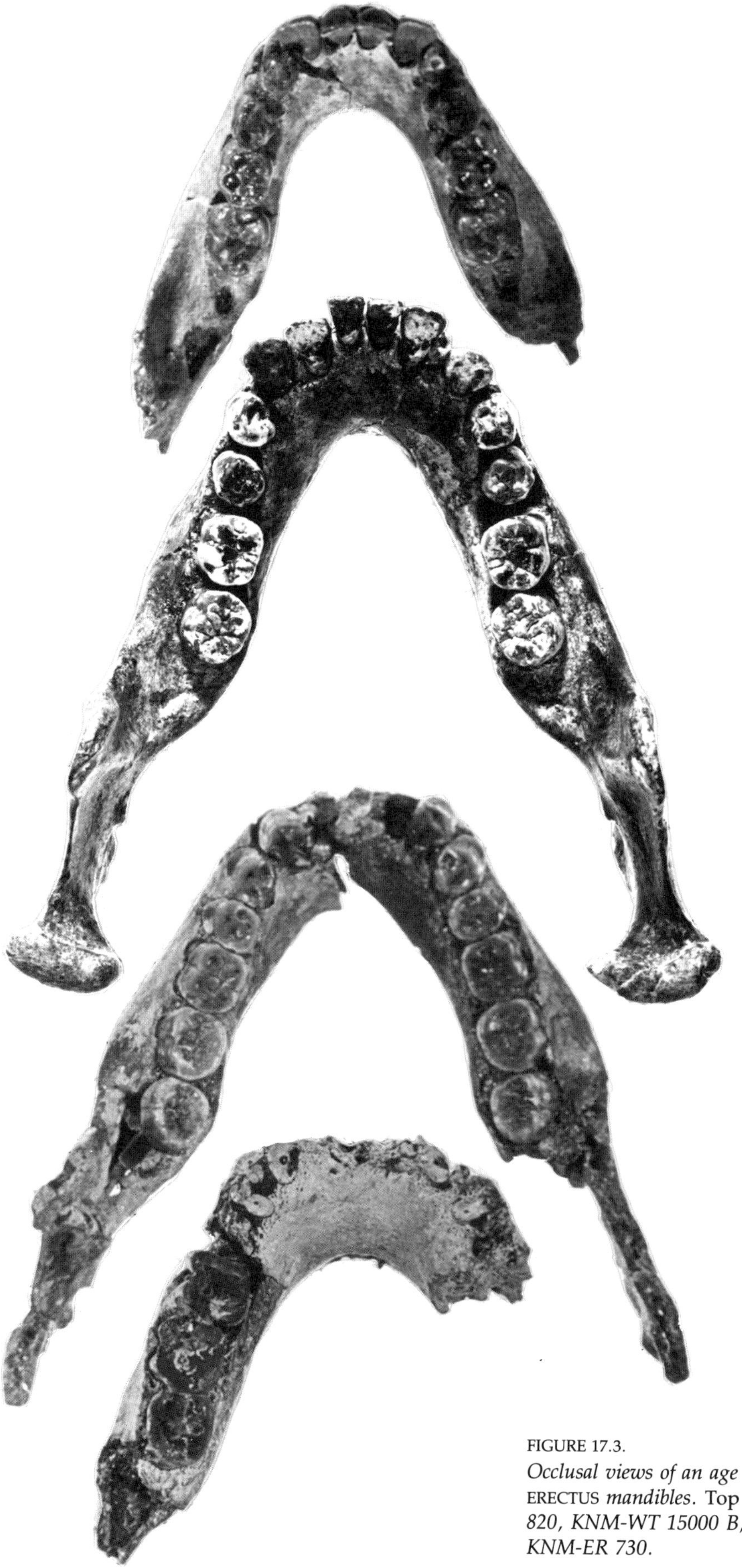

FIGURE 17.3.
Occlusal views of an age series of early HOMO ERECTUS *mandibles*. Top to bottom: *KNM-ER 820, KNM-WT 15000 B, KNM-ER 992, and KNM-ER 730.*

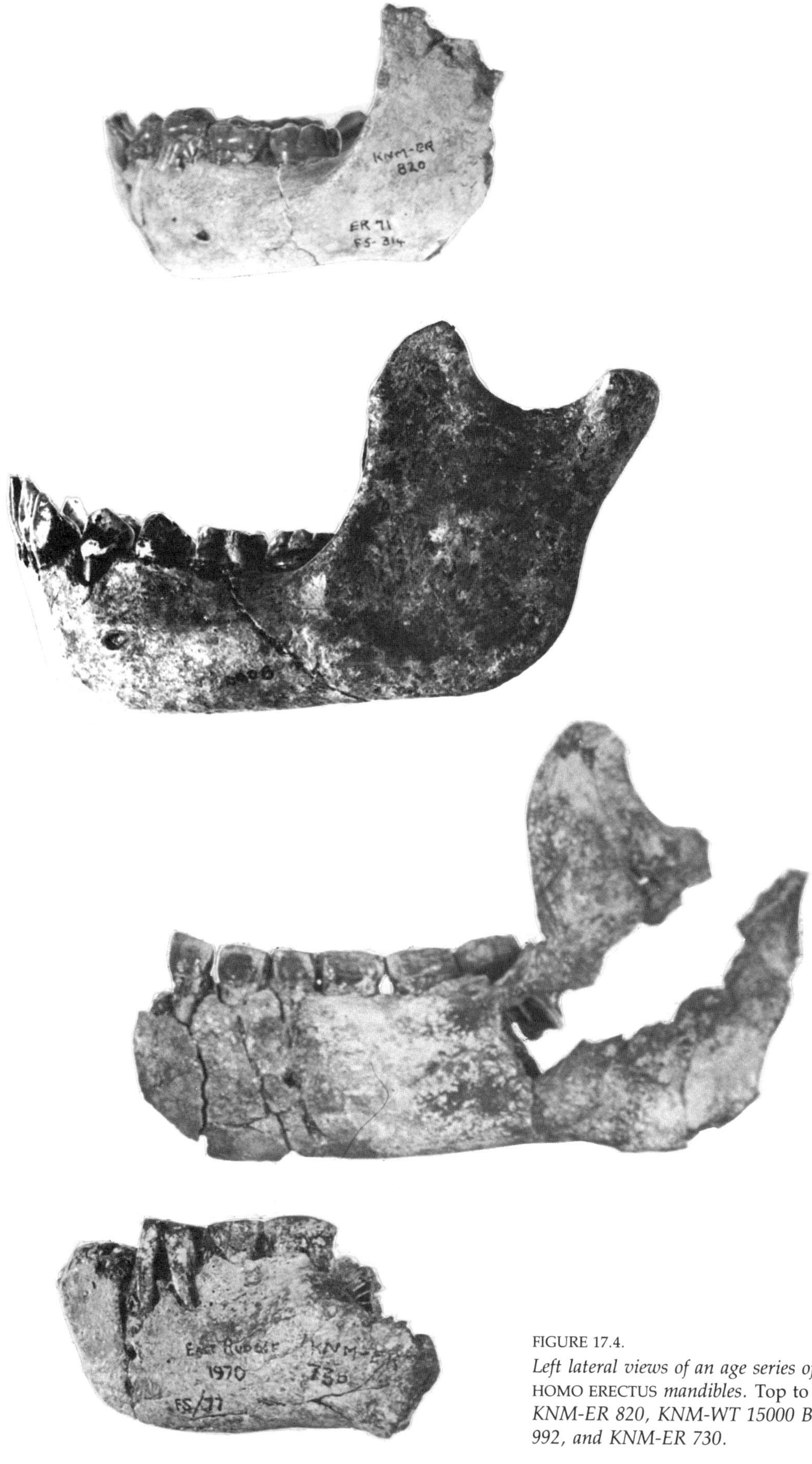

FIGURE 17.4.
Left lateral views of an age series of early HOMO ERECTUS *mandibles.* Top to bottom: *KNM-ER 820, KNM-WT 15000 B, KNM-ER 992, and KNM-ER 730.*

opment of the body, mesial drift of the teeth, and, in the case of KNM-WT 15000 and KNM-ER 992, the development of the ramus.

THE FUNCTIONAL MOPHOLOGY OF EARLY AFRICAN *HOMO ERECTUS*

Many aspects of early African *H. erectus* morphology have been covered in the accounts presented in this volume. Here I review a few that deserve special mention.

The overall similarity between early *H. erectus* skeletons and those of modern humans is striking. As we have shown, the limb proportions are those found in modern human populations who inhabit extremely hot, dry places. The facial skeleton seems remarkably modern in many aspects, although the paranasal sinus development is advanced for an individual of that dental age and the face as a whole is more prognathic. The upper limb skeleton is unremarkable from a modern human perspective. The thorax had already developed the unique human features of being barrel-shaped and having a ventrally invaginated thoracic column and ribs that hang downwards and forwards. The vertebrae show that the lordotic and kyphotic curvatures typical of modern humans were already in place. Even the six lumbar vertebrae which seem to be typical of early hominids can be found in a small percentage of modern people. The major exceptions to a modern morphology are in the size and shape of the braincase, the size of the thoracic neural canal, the narrowness of the pelvis, the long femoral necks, the marked medullary cavity stenosis of the limb bones, and the markedly concave anterior borders of the fibulae.

These exceptions have been dealt with in the various chapters concerning them, but a few further remarks are necessary. Cranial capacities of early African *H. erectus* individuals were small by modern human standards. As an illustration, the capacity for an average individual that stood 1.7 meters tall and weighed nearly 60 kg was about 900 cc. As a crude illustration, this is about the cranial capacity of a human child between the ages of three and four whose stature might be around 1 meter and whose weight around 15 kg. There is, of course, no obvious correlation between mental abilities and brain size in people, but the gross difference in relative adult brain size between *H. erectus* and *H. sapiens* must point to differences in behavior. This difference should be enough to caution workers against using much of the behavior of modern people as a model for the behavior of these hominids 1.5 million years ago.

Despite this relative small brain size, however, the uniquely human trait of having secondarily altricial infants was already established among early hominids, as the extremely narrow pelvis of these specimens shows (Walker and Ruff, Chapter 10, this volume). Selection for larger brains was already operating, then, despite stabilizing selection, which kept pelvises narrow and femoral necks long either for climatic or locomotor reasons. The abductor mechanism of the pelvis can be made more efficient by either decreasing the distance between the hip joints (and thereby narrowing the birth canal) or lengthening the femoral neck or both. *H. erectus* individuals had both narrow birth canals and long femoral necks and thus a more mechanically efficient abductor mechanism than modern *H. sapiens.* It is a matter of some importance to trace the subsequent development of the human condition and find reasons for the compromise of the abductor mechanism. This will require the finding of more complete fossil material, but a speculation can be offered here.

Stabilizing selection acting on the pelvis together with positive selection for larger brain size had led to secondarily altricial infants in *H. erectus* by 1.5 Ma. Modern humans could have evolved wide birth canals and kept primitively long femoral necks only if climatic conditions allowed a wide body build. This type of build is seen, in fact, among neandertals who had adapted to extreme cold and who had very wide pelvises and still retained relatively long femoral necks. A reasonable suggestion would be that modern humans had their origin in equatorial latitudes and that climatic stabilizing selection maintained the primitively narrow body condition, but that positive selection pressure for larger brains led to a compromise in the efficiency of the abductor mechanism. This led to wide birth canals and short femoral necks. An African origin of modern humans has been postulated for some time now (Stringer, 1990), and although present tree-making algorithms do not allow the mitochondrial DNA data to be interpreted as supporting or not an African origin for modern humans (Templeton, 1992), other genetic data hint at this conclusion. The very early modern human fossils from Qafzeh (Vandermeersch, 1969) also are considered to be African in affinities and are earlier in age than some neandertals in their region (Stringer, 1990).

The limb bones of early hominids are more robustly built than those of modern *H. sapiens*. Ruff et al. (1992) have demonstrated that "robusticity" (structural strength of bone relative to body mass) declines exponentially in the genus *Homo* from the early Pleistocene to the present day. There was little change in robusticity with the species *H. erectus*, however, because the decline is exponential. These changes probably reflect decreased mechanical loading of the skeleton and may not have any strong genetic basis. Early African *H. erectus* were, therefore, probably very active and placed high stresses on their postcranial skeletons.

THE ORIGINS OF SPEECH AND LANGUAGE

There are two schools of thought on the origin of human language. One believes that language of some sort appeared early in the human lineage, by the beginning of the Pleistocene at least. The other believes that human language and its use of speech appeared very late in our evolution, perhaps as late as 40,000 years ago. Several different sorts of evidence are used by both schools, but both use evidence from the morphology of the fossils and the nature of the archeological record. The morphological evidence involves the central nervous system and the pharyngeal airway. The primary evidence for speech noted by the first school is the size of the brain together with the presence of asymmetries and certain details on the endocranial surface. The second school holds as evidence of speech the bony parts from which it might be possible to make a reconstruction of the pharyngeal resonators. Other evidence includes the complexity of the archeological record, language capabilities of other primates, and the sequence of language development in human children. Noble and Davidson (1991) give a recent review of some of the arguments in this complex field. I shall deal here only with the evidence that can be gathered from the morphology of the fossils.

It has been thought for some time now that evidence of defined speech areas can be found on the endocasts of early hominids from at least 2.0 Ma. The two areas that might be expected to leave impressions are Broca's and Wernicke's (see Chapter 14, this volume). Tobias (1983) has claimed the presence of the former in hominids as early as *Australopithecus africanus* and the latter in the earliest members of the genus *Homo*. Tobias (1983, 1991), following Holloway (1969, 1983), links the development of language with stone tool complexity. As he puts it (1991, p. 836), "the complexity which archeologists have shown in, or inferred from, the lifestyle of *H. habilis* seems to the author to mark the point at which adequate and efficient transmission of cultural practices and advances to the offspring required at least rudimentary language."

Holloway (1983) gives the clearest account of what can and what cannot be learned about language from looking at endocasts of fossils. He suggests that four traits can be of use, but concludes that paleoneurological evidence can only corroborate or correlate with other evidence. He considers that the archeological record is the only "true" evidence for early human cognitive behavior. The four traits are brain size and evidence of well-defined areas for the motor control of speech, for Wernicke's area, and for cerebral asymmetries. Many advances in human and primate brain research have been made since Holloway's seminal paper (1969), and a summary of those that pertain to the origins of language can be made here.

Brain size

As Holloway himself pointed out, brain size by itself cannot help one decide whether an early hominid could speak or not, even though some mimumum size must be required. Tobias (1983; 1991) calls attention to the coincidence of clear speech areas appearing at the same time as brain enlargment and encephalization became "obtrusive" and stone tools appear in the geological record. As Begun and I have pointed out (Chapter 14, this volume), the evidence now suggests that *H. erectus* individuals were larger than had previously been suspected and that they are about as encephalized as *H. habilis.* If there was language in these two species, as Tobias thinks, then there seems to have been no corresponding increase in encephalization with changes in stone tool technology or during the dramatic spread of *H. erectus* over the Old World.

Broca's area

As Begun and I pointed out (Chapter 14, this volume), there now is good evidence from physiological studies on living people that the cortical area of the brain thought by Broca to be a motor area for speech is more likely to be associated with motor programming rather than with speech function. Tobias (1983) thought that the traditionally defined Broca's area was present in *A. africanus* but that Wernicke's area was not. This does not make much intuitive sense, since it is difficult to imagine how downstream parts of a command system could be selected for without the corresponding upstream parts. Holloway, in fact, pointed out (1983, p. 111) that Broca's caps could be found on chimpanzee endocasts. It could be argued, I suppose, that a language area deep to Broca's cap could be responsible for its development, but that is not very convincing.

Wernicke's area

This is the most difficult area to assess. Most of the endocasts I know are distorted somewhat. That of KNM-WT 15000, for instance, is pushed over to the left and so features on the left side appear fuller. Tobias (1991, p. 729) says that the mark of the inferior parietal lobule shows strong development on all four of the Olduvai *H. habilis* endocasts. The following remarks can be made about the four endocasts. Olduvai Hominid 24 was squashed flat when found. Despite almost heroic efforts by R. J. Clarke to undo this distortion, he could not unflatten vault bones. Olduvai Hominid 16 has mainly fragments of vault bone, and the orientation of these pieces probably will never be completely settled. Olduvai Hominid 7 has most of the left and a piece of the right parietal. And Olduvai Hominid 13 has only the top part of the left parietal. In all of these cases it is very difficult to decide whether or not there were asymmetries in the configurations of the inferior parietal lobule impressions or even, I suspect, whether the impression itself is there. It is even more difficult to imagine how any such differences could be measured. Asymmetries of the sylvian sulcus impression are not necessarily indications of a well-developed speech area, as LeMay (1975) has proposed. This is because a longer left sylvian fissure is found in chimpanzees and *Macaca* (Falk et al., 1986).

Cerebral asymmetries

It has been known for some time that human brains are often asymmetrical and also, since the work of R. W. Sperry, that in humans the left hemisphere is usually dominant in speech (see Williams et al., 1989, pp. 1064–1068, for a very clear review). It has been shown recently that it is the linguistic and not the symbolic or motor properties of language that are lateralized in the left hemisphere (Corina et al., 1992). It is also now known, however, that some Old World monkeys and chimpanzees are left hemisphere dominant for perception of meaningful (species-specific) vocalizations and that their brains are asymmetrical in ways similar to those of humans (Falk et al., 1990), including in sylvian fissure asymmetry (Falk et al., 1986). As Falk et al. (1990) conclude, the neurological substrate for brain lateralization may have been present during early anthropoid evolution. Thus care must be taken not to equate brain asymmetries necessarily with speech, even though in most humans the left hemisphere is the verbally skillful one.

There are definite associations between certain cerebral asymmetries and handedness in humans (LeMay, 1976). In KNM-WT 15000 the endocast shows a wider right frontal, right frontopetalia, and equal occipital petalia (Chapter 14, this volume). In this the fossil is most like modern male right-handers. There are also asymmetries in the postcranial skeleton of this specimen. The right ulna is longer than the left and the clavicles have asymmetrical ligament pits and muscle scars (see Chapter 7, this volume). Schultz (1937) reviewed the asymmetries in lengths of long bones and found that right upper limb bones were usually longer in humans. He did not measure ulnas, but he found that the right radius was the longer in nearly 70% of cases and the left longer in 18%. The distributions are much more even in anthropoids. The fact that the clavicles of KNM-WT 15000 are asymmetrical and that the right ulna is longer suggests, together with the endocranial evidence, that this youth was right-handed. Since both brain asymmetries (LeMay, 1976) and limb asymmetries (Ruff and Jones, 1981) are known to be associated with handedness, it is likely that they are part of the same lateralization process. It is also likely that these combined asymmetries in the fossil skeleton show that this youth was right-handed for tool use. Toth (1985) has found evidence that early hominids at Lake Turkana had a preference for right-handedness in stone tool making. It does not necessarily follow, as I hope I have shown above, that he could speak because he was right-handed.

The Acheulean stone technologies which are associated with these early *H. erectus* populations require cores to be processed by reduction from large flakes. Sometimes such large flakes can be found already detached from their source rock, but in many instances these have to be removed from boulders or outcrops. Practically the only means of removing large flakes that was available to these early hominids was the throwing of one boulder against another. Even if the spheroids and bifaces often found on these sites were not throwing objects, throwing must have been involved in flake removal. Darlington (1957, 1975) hypothesized that throwing was an important part of hunting behavior in early hominids and that selection for throwing efficiency would have brought about changes in those parts of the central and peripheral nervous system that were required in the complex act of "correlation of three moving objects (the animal, the object thrown, and another animal)." He suggests that such changes would then be preadaptive for later, more complex human faculties. This idea has been developed more fully by Calvin (1983) and others who see hammering and throwing as important skills which would have been selected for in early hominids and which would also lead to selection for handedness and increased circuitry in those parts of the central nervous system responsible for accurate throwing and the feedback required for it. Calvin himself thinks that the Acheulean bifaces were specialized objects purposefully made for throwing, but it is worth pointing out that at Olorgesalie, where bifaces abound in some places, only flakes struck from bifaces and not the bifaces themselves are found at butchery sites and, further, that edge wear is restricted to flakes (Rick Potts, personal communication). Perhaps the lateralization and the development of Broca's caps in early *H. erectus* is a reflection of throwing and hammering ability rather than of speech. Noble and Davidson (1991) hypothesize that selection for one-handed throwing could provide the neural mechanism for pointing, since pointing can be thought of as aiming without throwing.

The spinal cord

KNM-WT 15000 has many of its vertebrae preserved, enough for Latimer and Ward (Chapter 12, this volume) to show that he had all the spinal features associated with the fully human form of upright bipedalism. The vertebrae also allowed MacLarnon (Chapter 15, this volume) to examine whether or not the spinal canal was built in the same way as it is in humans. She found that the overall shape and proportions of the canal were most similar to humans among primates, and many of the features she examined were only found in humans. There was one important feature, however, in which *H. erectus* differed from modern humans—the relative size of the thoracic vertebral canal. In this feature MacLarnon found that KNM-WT 15000 was like other primates and not like humans, who have larger relative thoracic canals than all other primates. She concluded that *H. erectus* did not have the typically human increased bulk of grey matter in the thoracic cord nor enlarged thoracic spinal nerves. She also gave two possible explanations for this. The first is that improvements in the control of bipedalism might be responsible for the later enlargements in humans. This alternative I find unlikely since hominids had been bipedal for more than 2 million years by the time the Nariokotome boy was born. And, further, all parts of the locomotor system and body proportions in KNM-WT 15000 and the other fossils from that population seem to be completely human, even down to conforming to Allen's rule (Chapter 11, this volume). The second alternative is that the enlargement might be related to speech and that more control of the intercostal muscles and abdominal wall muscles is needed in modern people than was needed in *H. erectus.* If speech had not developed in *H. erectus,* then increased control over a normal primate thoracic cage and abdomen would not have been needed and the thoracic cord enlargement would not have been present. William Calvin (personal communication) also points out that the enlargement could have stabilized the thorax and abdomen for throwing.

The upper respiratory tract

A great deal of interest and controversy has been generated since Leiberman and Crelin (1971) first concluded that neandertals could not speak. They based their analysis on studies of the upper pharynx and other resonators in the respiratory tract of humans and apes and on a reconstruction of the same in neandertals. Laitman and his colleagues, following this lead, have looked at the basicranial anatomy of early hominids (Laitman, 1985) and concluded that early African *H. erectus* (as exemplified by KNM-ER 3773) are the earliest hominids to have a cranial base that is derived relative to the basic mammalian condition. Laitman suggested further that the larynx may have been somewhat lowered in these populations and that the human type of upper respiratory tract might have begun to develop then. Laitman (personal communication) does not himself think that selection for laryngeal descent is related to the onset of human speech. Rather, he thinks a purely respiratory function would be selected for first.

Fransiscus and Trinkaus (1988) reviewed another part of the early hominid repiratory tract, the nose and nasal cavity, and concluded that early African *H. erectus* nasal morphology reflected the increased physiological demands of life in an arid environment. They point out that the change from an ape-like, flat, non-projecting nose to an anteriorly positioned external nose as in modern humans came first with these African populations. They argue that this new nasal structure allowed a greater volume of inhaled air to be moisturized with conserved exhaled moisture. They also pointed out that the appearance of the human type of nose in early African *H. erectus* followed the development of limbs that show a full adaptation to terrestrial bipedalism; it also coincided with the development of the Acheulean lithic complex and preceded the spread of the species out of Africa. The observations and ideas of Laitman and colleagues and Fransiscus and Trinkaus can be seen to converge, then, on the point that greater respiratory volume served as an adaptation for prolonged active bouts in arid environments. They might provide reasons for the initial respiratory adaptations from which efficient pharyngeal resonators could have developed, but they do not lend direct support to the presence of speech 1.5 million years ago.

Conclusion

The reader is referred to Noble and Davidson (1991) for a recent critical review of the origins of language. They reason that since the constituent signs of language are used symbolically, the earliest signs of language will be the archeological traces of symbols. They find no evidence of symbolic signs prior to 32,000 Ma and conclude that human language is a very late evolutionary event. Their review did not cover the morphology of early hominids, but from the brief review given above I feel that those who have done so and then claimed a very early origin for human language and speech are standing on ground that is steadily crumbling away. The reading of complex behaviors into the archeological record is all too easy and practically untestable. The notion that language is needed for cultural transmission if the inferred life-style and tools appear complex (Tobias, 1983) is testable, however. An experiment could be devised, for instance, whereby people are asked to make tools by following spoken instructions or by following an example. I suspect that it would be easier to make an Acheulean biface after watching someone else make one than to attempt the same by listening to instructions, let alone by reading an instruction manual! What such an experiment would tell us about hominid behavior 1.5 million years ago, though, is another matter. Potts (1991) has recently observed and measured a large sample of the Oldowan stone tool "culture" that impressed Tobias with its complexity and has concluded that standardized categories of broken stones cannot be found. Potts finds complete overlap in counts of scars, lengths of edges, and suchlike on cores and a complete spectrum of flake size and shape.

THE ORIGINS OF *HOMO ERECTUS*

Most authorities agree that the early African members of the genus *Homo* such as KNM-WT 15000 are from the early part of the human lineage, even if they do not agree on the name for them. But the origins of these populations are still a matter of dispute. With the exception of Groves (1989), who sees four species of the genus *Homo* about 2 million years ago, most workers fall into two groups. The first group thinks that there was only one non-*boisei* species at this time and that this species contained the populations antecedent to early African *H. erectus*. The second thinks that there were two non-*boisei* species at this time. The problem is not an easy one to solve, because there are about as many hypodigms of *H. habilis* as there are workers in the field. Whole crania can be placed in different species or even genera by different people. *H. habilis* (Leakey, Tobias, and Napier, 1964) was based on a broken, subadult mandible and some skull fragments that are probably associated with it and some postcranial pieces that are possibly associated with it. Tobias (1991) has now completed his analyses of the *H. habilis* fossils from Olduvai, but there are now many more fossils attributed to this species than there are from just that one site. These have been found in South Africa, Kenya, and Ethiopia. The variation to be seen in this species has increased with the sample. For some people (e.g., Howell, 1978; Johanson, 1989; McHenry, 1991; Tobias, 1991; Wolpoff, 1980) there is only one, variable species. For others (e.g., Leakey and Walker, 1978; Rightmire, 1990; Stringer, 1986; Walker, 1991; Wood, 1991), the level of variability suggests that there is more than one species present. At present there is no way of resolving this issue. We need more and better fossil hominids from about 2 Ma. We can, however, give two alternative accounts from these two perspectives on the number of early *Homo* species about 2 million years ago.

If there was only one non-*boisei* hominid species antecedent to *H. erectus*, then the following can be said about the transition. *H. habilis* in this account would be a very variable, sexually dimorphic species with body size dimorphism approaching that of species of *Australopithecus*. Females were short, perhaps only a meter tall, and had humerofemoral proportions more like those of chimpanzees than humans (Johanson et al., 1987; Johanson, 1989). They were relatively megadont and had skulls which retained many features reminiscent of *Australopithecus* (Walker, 1976), including the flat pyriform aperture and nasal region. This species made a punctuational transition to *H. erectus* about 1.8 million years ago (OH 62 is about 1.8 Ma). The individuals of the new species were larger and displayed less sexual dimorphism. (That is, the females were as large as the males in the previous species). Brain size increased along with body size and limb proportions became essentially human. These physical changes coincided with major changes in lithic technology. This is the hypothesis put forward by the discoverers and describers of Olduvai Hominid 62 (Johanson et al., 1987) in the telegraphic form usually found in *Nature* articles and expanded a little in Johanson (1989).

The second account involves two non-*boisei*

hominids co-existing about 2 million years ago. One of these species that are presently lumped together as one in *H. habilis* was ancestral to *H. erectus* and is represented by specimens such as KNM-ER 1470, 1471, 1590, and 3228. It is not, however, a very well-known taxon. The other species is better represented in both South and East Africa and includes specimens like Olduvai Hominids 24 and 62 and KNM-ER 1813. This hypothesis was put forward by Leakey and Walker (1978) and does not necessarily involve a punctuational event.

The testing of these two competing hypotheses is critical to our understanding of the origin of the genus *Homo.* The resolution of this problem should determine whether or not our lineage evolved in a punctuated evolutionary manner and whether or not major external stimulants such as global climatic shifts affected the course of our own evolution. Whether we can recognize any correspondence between morphological changes and major faunal changes or the appearance of new lithic technologies depends on our ability to sort out the phylogeny of the earliest members of the genus *Homo.* We have been able in the past to find new evidence to test hypotheses (e.g., Leakey and Walker, 1976), so it should be a matter of urgency that we plan field expeditions to look for the more complete fossils that will solve this problem.

A Note on Accession Numbers

The Preparation and Casting of the Specimen

References • Acknowledgments

Contributors • Index

A NOTE ON ACCESSION NUMBERS

Several colleagues have asked us why, after years of low accession numbers (such as KNM-ER 1470), the Kenya National Museum should suddenly label specimens with numbers like KNM-WT 15000 and 17000. Under the old accessions system, each fossil collected from a site was given a KNM prefix, a site prefix (ER for East Rudolf, for example, or WT for West Turkana), and a number. Naturally, with fossils being collected from sites all over Kenya, these numbers were not unique. When L. L. Jacobs was placed in charge of the Palaeontology Department at the museum, he instituted the more common system requiring unique specimen numbers. From then on each specimen was entered in the accession catalogue in the order of collection. The site prefixes were retained for ease of use. Since nearly 15,000 fossils were in the museum catalogue in August 1984, we looked ahead for a suitable round number for the exceptional specimen from Nariokotome. It became KNM-WT 15000.

Observers will note that when, a year later, another important specimen was discovered at West Turkana, about 2,000 more fossils had been found and accessioned; the new find therefore became KNM-WT 17000.

KNM-WT 15000 has many parts, and so suffixes were added for each piece. The cranium is KNM-WT 15000 A, the mandible B, and so on. After the alphabet was exhausted, the suffixes were AA, AB, AC, and so on. At the finish the list ended in CA. There is no obvious order to the postcranial pieces, since their suffixes were given as the specimens were found and taken from the Nariokotome site to Nairobi.

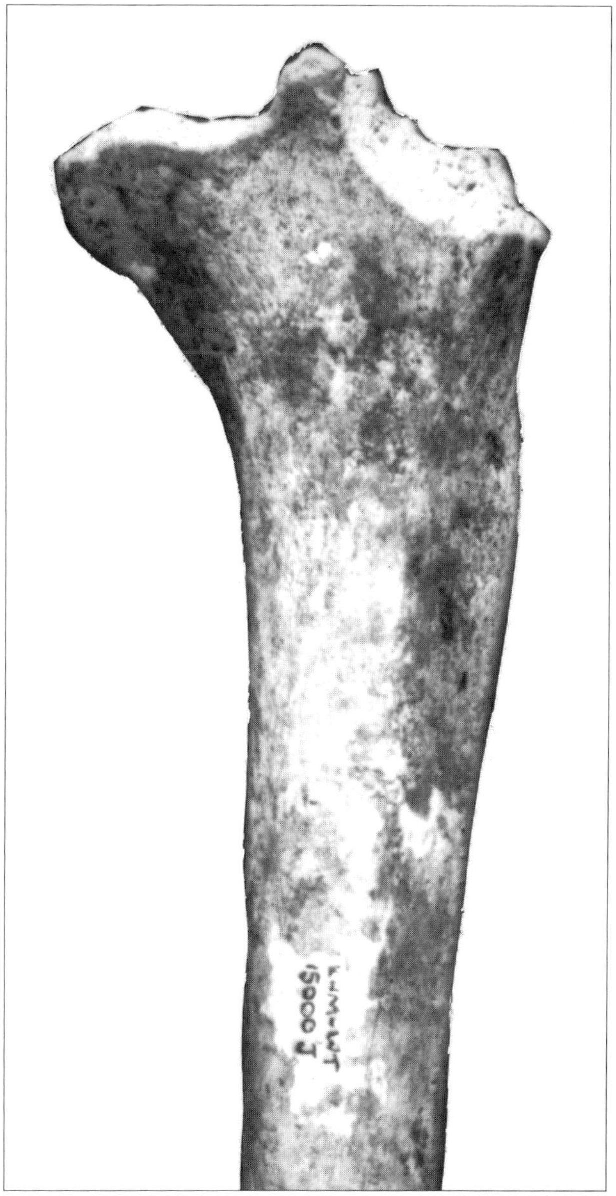

THE PREPARATION AND CASTING OF THE SPECIMEN

The task of removing the crust of calcified sediment from the bones of KNM-WT 15000 was carried out for the most part by Joseph Mutaba and Christopher Kiarie, with the occasional assistance of one of us (A.W.). The matrix was removed using pneumatic Airscribes (Chicago Pneumatic Corp.) under Zeiss binocular microscopes. A very few bones were too thin and fragile to stand the vibration of the Airscribes, so they were cleaned using diamond dental bits in a turbo dental drill. The tooth sockets were cleaned using both Airscribes and dental drills. The bones are relatively hard, by and large, but a plastic preservative (vinyl acetate solution) was used to make them stronger.

The task of casting the skeleton was left to Simon Kasinga and Alphonse Katula. Silastic molds of RTV (Dow Corning Corp.) were made of the specimen at several stages during its preparation. These master molds are housed in the National Museums of Kenya. Single-sided reference casts were first made from the master molds so that the specimens would not have to be subjected to molding again.

The individual cranial bones were all molded separately. The individual fragments of pelvis, together with all isolated vertebral fragments, were also molded separately. In addition, molds were made of the reconstructed cranium and its endocast. Jennie Smith of the Cleveland Museum of Natural History also made molds of the reconstructed pelvis. For those primarily interested in instruction, and especially for those interested in mounting the casts as a standing specimen in an exhibit, it is preferable by far to purchase the reconstructions. This is because there is always slight distortion in replicating bones and there is no guarantee that the individual pieces will be able to be glued together easily to make an undistorted cranium. The pelvic fragments also require a great deal of careful reconstruction to make a pelvis suitable for mounting, especially when one considers that only parts of the individual sacral elements were found and all were unfused (see Chapter 10).

Casts of all parts, including the reconstructed cranium and pelvis, are available from the National Museums of Kenya. Information and prices can be had from the Casting Director, National Museums of Kenya, P.O. Box 40568, Nairobi, Kenya.

Black-and-white photographs of individual bones (as in Chapters 6–8), as well as of the mounted skeleton illustrated in the color section, are available from the National Museums of Kenya, and color slides of the skeleton and the reconstructed skull and mandible are also available. Enquiries should be sent to the Curator of Palaeoanthropology at the address given above.

REFERENCES

1. Introduction

Andrews, P. J. 1984. An alternative interpretation of the characters used to define *Homo erectus. Courier Forschungsinstitut Senckenberg* 69: 167–175.

Brown, F. H., and C. S. Feibel. 1991. Stratigraphy, depositional environments, and paleogeography of the Koobi Fora Formation. In *Koobi Fora Research Project,* vol. 3, ed. J. M. Harris, pp. 1–30. Oxford: Clarendon Press.

Brown, F., J. Harris, R. Leakey, and A. Walker. 1985. Early *Homo erectus* from west Lake Turkana, Kenya. *Nature* 316: 788–792.

Champion, A. M. 1937. Physiography of the region to the west and southwest of Lake Rudolf. *Geographical Journal* 89: 97–118.

Cohen, M. N., and G. J. Armelagos. 1984. *Paleopathology at the Origins of Agriculture.* Orlando: Academic Press.

Conroy, G. C. 1990. *Primate Evolution.* New York: Norton.

Coppens, Y. 1983. Les hominidés du Pliocène et du Pléistocène d'Afrique orientale et leur environment. In *Morphologie évolutive, morphogenèse du crâne et origine de l'homme,* ed. M. Sakka, pp. 185–194. 8th International Primatological Congress. Paris: Editions du Centre national de la recherche scientifique.

Coppens, Y., F. C. Howell, G. Ll. Isaac, and R. E. F. Leakey. 1976. Preface. In *Earliest Man and Environments in the Lake Rudolf Basin,* ed. Y. Coppens, F. C. Howell, G. Ll. Isaac, and R. E. F. Leakey. Chicago: University of Chicago Press.

Day, M. H. 1986. *Guide to Fossil Man.* Chicago: University of Chicago Press.

Feibel, C. S., J. M. Harris, and F. H. Brown. 1991. Paleoenvironmental context for the late Neogene of the Turkana Basin. In *Koobi Fora Research Project,* vol. 3, ed. J. M. Harris, pp. 321–370. Oxford: Clarendon Press, Oxford.

Groves, C. P. 1989. *A Theory of Human and Primate Evolution.* Oxford: Clarendon Press.

Harris, J. M. 1983. *Koobi Fora Research Project,* vol. 2. Oxford: Clarendon Press.

——— 1986. The fossil boy. *Terra* 24: 21–24.

——— 1991. *Koobi Fora Research Project,* vol. 3. Oxford: Clarendon Press.

Harris, J. M., and F. H. Brown. 1985. New hominid locality west of Lake Turkana, Kenya. *National Geographic Research* 1: 289–297.

Harris, J. M., F. H. Brown, and M. G. Leakey. 1988b. Stratigraphy and paleontology of Pliocene and Pleistocene localities west of Lake Turkana, Kenya. *Contributions in Science* 399: 1–128.

Harris, J. M., F. H. Brown, M. G. Leakey, A. C. Walker, and R. E. Leakey. 1988a. Pliocene and Pleistocene hominid-bearing sites from west of Lake Turkana, Kenya. *Science* 239: 27–33.

Howell, F. C. 1978. Hominidae. In *Evolution of African Mammals,* ed. H. B. S. Cooke and V. J. Maglio, pp. 154–248. Cambridge: Harvard University Press.

Howell, F. C., and Y. Coppens. 1983. Introduction. *Annales, Musée Royale de l'Afrique Centrale, Tervuren* 8: 1–5.

Leakey, M. D. 1971. *Olduvai Gorge,* vol. 3. Cambridge: Cambridge University Press.

Leakey, M. G., and R. E. Leakey, eds. *Koobi Fora Research Project,* vol. 1. Oxford: Clarendon Press.

Leakey, R. E. F., and A. C. Walker. 1976. *Australopithecus, Homo erectus* and the single species hypothesis. *Nature* 261: 572–574.

——— 1985. *Homo erectus* unearthed. *National Geographic* 168: 624–629.

——— 1989. Early *Homo erectus* from west Lake Turkana, Kenya. In *Hominidae: Proceedings of the Second International Congress of Human Paleontology,* ed. G. Giacobini, pp. 209–215. Milan: Jaca Book.

Lewin, R. 1987. *Bones of Contention.* New York: Simon and Schuster.

Rightmire, P. G. 1979. Cranial remains of *Homo erectus* from Beds II and IV, Olduvai Gorge, Tanzania. *American Journal of Physical Anthropology* 53: 99–115.

——— 1991. *The Evolution of Homo erectus.* Cambridge: Cambridge University Press.

Ruff, C. B. 1991. Climate and body shape in hominid evolution. *Journal of Human Evolution* 21: 81–105.

Stringer, C. B. 1984. The definition of *Homo erectus* and the existence of the species in Africa and Europe. *Courier Forschungsinstitut Senckenberg* 69: 131–143.

——— 1987. A numerical cladistic analysis of the genus *Homo. Journal of Human Evolution* 16: 135–146.

Tobias, P. V. T. 1978. The earliest Transvaal members of the genus *Homo* with another look at some problems of hominid taxonomy and systematics. *Zeitschrift für Morphologie und Anthropologie* 69: 225–265.

——— 1991. *Olduvai Gorge,* vol. 4. Cambridge: Cambridge University Press.

Wolpoff, M. H. 1980. *Paleoanthropology.* New York: Knopf.

Wood, B. A. 1976. Remains attributable to *Homo* in the East Rudolf succession. In *Earliest Man and Environments in the Lake Rudolf Basin,* ed. Y. Coppens, F. C. Howell, G. Ll. Isaac, and R. E. F. Leakey, pp. 490–506. Chicago: University of Chicago Press.

——— 1991. *Koobi Fora Research Project,* vol. 4. Oxford: Clarendon Press.

2. Geologic Setting and Age

Brown, F. H., and T. E. Cerling. 1982. Stratigraphic significance of the Tulu Bor Tuff. *Nature* 299: 212–215.

Brown, F. H., and C. S. Feibel. 1985. Stratigraphic notes on the Okote Tuff Complex at Koobi Fora. *Nature* 316:794–797.

——— 1986. Revision of lithostratigraphic nomenclature in the Koobi Fora region, Kenya. *Journal of the Geological Society* (London) 143: 297–310.

Brown, F. H., I. McDougall, T. Davies, and R. Maier. 1985. An integrated Plio-Pleistocene chronology of the Turkana Basin. In *Paleoanthropology: The Hard Evidence,* ed. E. Delson, pp. 82–90. New York: Alan R. Liss, Inc.

Cerling, T. E., and F. H. Brown. 1982. Tuffaceous marker horizons in the Koobi Fora region and the Lower Omo Valley. *Nature* 299: 216–221.

Feibel, C. S. 1988. Paleoenvironments from the Koobi Fora Formation, Turkana Basin, northern Kenya. Ph.D. dissertation, University of Utah.

Feibel, C. S., F. H. Brown, and I. McDougall. 1989. Stratigraphic context of fossil hominids from the Omo Group deposits, northern Turkana Basin, Kenya and Ethiopia. *American Journal of Physical Anthropology* 78: 595–622.

Haileab, B., and F. H. Brown. 1992. Turkana Basin–Middle Awash Valley correlations and the age of the Sagantole and Hadar Formations. *Journal of Human Evolution* 22: 453–468.

Harris, J. M., and F. H. Brown. 1985. New hominid locality west of Lake Turkana, Kenya. *National Geographic Research* 1: 289–297.

Harris, J. M., F. H. Brown, and M. G. Leakey. 1988. Geology and paleontology of Plio-Pleistocene localities west of Lake Turkana, Kenya. *Contributions in Science* 399: 1–128.

McDougall, I. 1981. $^{40}Ar/^{39}Ar$ spectra from the KBS Tuff, Koobi Fora Formation. *Nature* 294: 120–124.

——— 1985. K-Ar and $^{40}Ar/^{39}Ar$ dating of the hominid-bearing Pliocene-Pleistocene sequence at Koobi Fora, Lake Turkana, northern Kenya. *Geological Society of America Bulletin* 96: 159–175.

McDougall, I., T. Davies, R. Maier, and R. Rudowski. 1985. Age of the Okote Tuff Complex at Koobi Fora, Kenya. *Nature* 316: 792–794.

McDougall, I., R. Maier, P. Sutherland-Hawkes, and A. J. W. Gleadow. 1980. K-Ar age estimate for the KBS Tuff, East Turkana, Kenya. *Nature* 284: 230–234.

3. Microstratigraphy and Paleoenvironments

Behrensmeyer, A. K., and L. F. Laporte. 1981. Footprints of a Pleistocene hominid in northern Kenya. *Nature* 289: 167–169.

Brown, F. H., and C. S. Feibel. 1985. Stratigraphic notes on the Okote Tuff Complex at Koobi Fora. *Nature* 316: 794–797.

——— 1986. Revision of lithostratigraphic nomenclature in the Koobi Fora region, Kenya. *Journal of the Geological Society* (London) 143: 297–310.

——— 1988. "Robust" hominids and Plio-Pleistocene paleogeography of the Turkana Basin, Kenya and Ethiopia. In *Evolutionary History of the "Robust" Australopithecines,* ed. F. E. Grine, pp. 325–341. New York: Aldine de Gruyter.

Brown, F., J. Harris, R. Leakey, and A. Walker. 1985. Early *Homo erectus* skeleton from west Lake Turkana, Kenya. *Nature* 316: 788–792.

Carbonel, P., and J. P. Peypouquet. 1983. Ostracoda as indicators of ionic concentrations and dynamic variations: methodology (Lake Bogoria, Kenya). In *Applications of Ostracoda,* ed. R. F. Maddocks, pp. 264–276. Houston: University of Houston Geoscience.

de Heinzelin, J. 1983. *The Omo Group.* Musée Royale de l'Afrique Centrale, Tervuren, Belgique, Annales, Série in 8°, Sciences Géologiques, No. 85.

Feibel, C. S. 1983. Stratigraphy and paleoenvironments of the Koobi Fora Formation along the western Koobi Fora Ridge, East Turkana, Kenya. M.S. thesis, Iowa State University.

——— 1987. Fossil fish nests from the Koobi Fora Formation (Plio-Pleistocene) of northern Kenya. *Journal of Paleontology* 61: 130–134.

——— 1988. Paleoenvironments from the Koobi Fora Formation, Turkana Basin, northern Kenya. Ph.D. dissertation, University of Utah.

——— In preparation. Neogene lake phases of the Turkana Basin, Kenya and Ethiopia.

Feibel, C. S. In review. Freshwater stingrays from the Plio-Pleistocene of the Turkana Basin, Kenya and Ethiopia. Submitted to *Lethaia.*

Feibel, C. S., F. H. Brown, and I. McDougall. 1989. Stratigraphic context of fossil hominids from the Omo Group deposits, northern Turkana Basin, Kenya and Ethiopia. *American Journal of Physical Anthropology* 78: 595–622.

Feibel, C. S., J. M. Harris, and F. H. Brown. 1991. Neogene paleoenvironments of the Turkana Basin. In *Koobi Fora Research Project,* vol. 3, ed. J. M. Harris, pp. 321–370. Oxford: Clarendon Press.

Ferguson, A. J. D., and B. J. Harbott. 1982. Geographical, physical and chemical aspects of Lake Turkana. In *Lake Turkana, a Report on the Findings of the Lake Turkana Project,* ed. A. J. Hopson, pp. 1–107. London: Overseas Development Administration.

Haileab, B. 1988. Characterization of tephra from the Shungura Formation, southwestern Ethiopia. M.S. thesis, University of Utah.

Haileab, B., and F. H. Brown. 1992. Turkana Basin–Middle Awash Valley correlations and the age of the Sagantole and Hadar Formations. *Journal of Human Evolution* 22: 453–468.

Harris, J. M., F. H. Brown, and M. G. Leakey. 1988. Geology and paleontology of Plio-Pleistocene localities west of Lake Turkana, Kenya. *Contributions in Science* 399: 1–128.

McDougall, I., T. Davies, R. Maier, and R. Rudowski. 1985. Age of the Okote Tuff Complex at Koobi Fora, Kenya. *Nature* 316: 792–794.

Munsell Color Co. 1975. *Munsell Soil Color Charts.* Baltimore: Munsell Color Co.

Neale, J. W. 1988. Ostracods and palaeosalinity reconstruction. In *Ostracoda in the Earth Sciences,* ed. P. DeDecker, J. P. Colin, and J. P. Peypouquet, pp. 125–155. Amsterdam: Elsevier.

Schwartz, H. L. 1983. Paleoecology of late Cenozoic fishes from the Turkana basin, northern Kenya. Ph.D. dissertation, University of California, Santa Cruz.

Vesey-Fitzgerald, D. 1963. Central African grasslands. *Journal of Ecology* 51: 243–247.

——— 1970. The origin and distribution of valley grasslands in East Africa. *Journal of Ecology* 58: 51–75.

Williams, S. K. 1989. Fossil sponges from the Turkana Basin, northern Kenya and southern Ethiopia, East Africa. B.S. thesis, University of Utah.

Yuretich, R. F., and T. E. Cerling. 1983. Hydrogeochemistry of Lake Turkana, Kenya: Mass balance and mineral reactions in an alkaline lake. *Geochimica et Cosmochimica Acta* 47: 1099–1109.

4. Taphonomy

Behrensmeyer, A. K. 1978. Taphonomic and ecologic information from bone weathering. *Paleobiology* 4: 150–162.

Brown, D. S. 1980. *Freshwater Snails of Africa and Their Medical Importance.* London: Taylor and Francis.

Brown, F., J. Harris, R. Leakey, and A. Walker. 1985. Early *Homo erectus* skeleton from west Lake Turkana, Kenya. *Nature* 316: 788–792.

Leakey, R. E., and A. Walker. 1985. *Homo erectus* unearthed. *National Geographic* 168: 624–629.

Shipman, P. 1981. *Life History of a Fossil.* Cambridge: Harvard University Press.

Voorhies, M. R. 1969. Taphonomy and population dynamics of an Early Pliocene vertebrate fauna, Knox County, Nebraska. *Contributions to Geology,* Special Paper No. 1: 1–69.

5. THE FAUNAL CONTEXT

Brown, D. S. 1980. *Freshwater Snails of Africa and Their Medical Importance.* London: Taylor and Francis.

Coryndon, S. 1976. Fossil Hippopotamidae from Plio-Pleistocene successions of the Rudolf Basin. In *Earliest Man and Environments in the Lake Rudolf Basin,* ed. Y. Coppens, F. C. Howell, G. Ll. Isaac, and R. E. F. Leakey, pp. 238–250. Chicago: University of Chicago Press.

Feibel, C. S. 1988. Paleoenvironments of the Koobi Fora Formation, Turkana Basin, northern Kenya. Ph.D. dissertation, University of Utah.

Fitzsimons, V. F. 1943. *The Lizards of South Africa.* Pretoria: Transvaal Museum.

Gentry, A. W., and A. Gentry. 1978. Fossil Bovidae (Mammalia) of Olduvai Gorge, Tanzania: Parts I-II. *Bulletin of the British Museum (Natural History): Geology* 29: 289–446; 30: 1–30.

Harris, J. M. 1985. Fossil ungulate faunas from Koobi Fora. In *L'environnement des Hominidés au Plio-Pléistocène,* ed. Y. Coppens, pp. 151–162. Paris: Masson.

Harris, J. M., F. H. Brown, and M. G. Leakey. 1988. Stratigraphy and paleontology of Pliocene and Pleistocene localities west of Lake Turkana, Kenya. *Contributions in Science* 399: 1–128.

Harris, J. M., F. H. Brown, M. G. Leakey, A. C. Walker, and R. E. Leakey. 1988. Pliocene and Pleistocene hominid-bearing sites from west of Lake Turkana, Kenya. *Science* 239: 27–33.

Kingdon, J. 1982. *East African Mammals,* vol. 3, part C: *Bovids.* London: Academic Press.

Shoen, A. 1971. The effect of heat stress and water deprivation on the environmental physiology of the bushbuck, the reedbuck, and the Uganda kob. *East African Agricultural and Forestry Journal* 37: 1–7.

6. THE SKULL

Day, M. H., and R. E. F. Leakey. 1973. New evidence of the genus *Homo* from East Rudolf, Kenya (I). *American Journal of Physical Anthropology* 39: 341–354.

——— 1974. New evidence of the genus *Homo* from East Rudolf, Kenya (III). *American Journal of Physical Anthropology* 41: 367–380.

Day, M. H., R. E. F. Leakey, A. C. Walker, and B. A. Wood. 1976. New hominids from East Rudolf, Kenya (I). *American Journal of Physical Anthropology* 42:461–475.

Groves, C. P., and V. Mazak. 1975. An approach to the taxonomy of the Hominidae: Gracile Villafranchian hominids of Africa. *Casopsis pro Mineralogii a Geologii* 20: 225–247.

Holloway, R. L. 1975. Early hominid endocasts: Volumes, morphology and significance for hominid evolution. In *Primate Functional Morphology and Evolution,* ed. R. H. Tuttle, pp. 393–415. The Hague: Mouton.

Leakey, R. E. F., and A. C. Walker. 1976. *Australopithecus, Homo erectus* and the single species hypothesis. *Nature* 261: 572–574.

——— 1985 Further hominids from the Plio-Pleistocene of Koobi Fora, Kenya. *American Journal of Physical Anthropology* 67: 135–163.

Leakey, R. E. F., and B. A. Wood. 1973. New evidence of the genus *Homo* from East Rudolf, Kenya (II). *American Journal of Physical Anthropology* 39: 355–368.

Maier, W., and A. Nkini. 1984. Olduvai Hominid 9: New results of investigation. *Courier Forschungsinstitut Senckenberg* 69: 123–130.

Walker, A. C., R. E. F. Leakey, and M. R. Zimmerman. 1982. A possible case of hypervitaminosis A in *Homo erectus. Nature* 296: 248–250.

Weidenreich, F. 1943. The skull of *Sinanthropus pekinensis. Palaeontologia Sinica,* new series D, 10: 1–291.

Wolpoff, M. H. 1984. Evolution in *Homo erectus:* The question of stasis. *Paleobiology* 10: 389–406.

7. THE POSTCRANIAL BONES

Brown, F., J. Harris, R. Leakey, and A. Walker. 1985. Early *Homo erectus* skeleton from west Lake Turkana, Kenya. *Nature* 316: 788–792.

Day, M. H. 1971. Postcranial remains of *Homo erectus* from Bed IV, Olduvai Gorge, Tanzania. *Nature* 232: 383–387.

——— 1986. *Catalogue of Fossil Man.* Chicago: University of Chicago Press.

Day, M. H., and R. E. F. Leakey, R.E.F. 1973 New evidence of the genus *Homo* from East Rudolf, Kenya (I). *American Journal of Physical Anthropology* 39: 341–354.

——— 1974. New evidence of the genus *Homo* from East Rudolf, Kenya (III). *American Journal of Physical Anthropology* 41: 367–380.

Day, M. H., R. E. F. Leakey, A. C. Walker, and B. A. Wood. 1975. New hominids from East Turkana, Kenya. *American Journal of Physical Anthropology* 45: 369–436.

Feibel, C. S., F. H. Brown, and I. McDougall. 1989. Stratigraphic context of fossil hominids from the Omo Group Deposits: Northern Turkana Basin, Kenya and Ethiopia. *American Journal of Physical Anthropology* 78: 595–622.

Hay, R. L. 1976. *Geology of the Olduvai Gorge: A Study of Sedimentation in a Semiarid Basin.* Berkeley: University of California Press.

Howell, F. C. 1978. Hominidae. In *Evolution of African Mammals,* ed. V. J. Magio and H. B. S. Cooke, pp. 154–248. Cambridge: Harvard University Press.

Kennedy, G. E. 1983. A morphometric and taxonomic assessment of a hominine femur from the Lower Member, Koobi Fora, Lake Turkana. *American Journal of Physical Anthropology* 61: 429–436.

Leakey, R. E. F., M. G. Leakey, and A. K. Behrensmeyer. 1972. The hominid catalogue. In *Koobi Fora Reseach Project,* vol. 1, ed. R. E. F. Leakey and M. G. Leakey, pp. 86–182. Oxford: Clarendon Press.

Leakey, R. E. F., and A. C. Walker. 1985. Further hominids from the Plio-Pleistocene of Koobi Fora, Kenya. *American Journal of Physical Anthropology* 67:135–163.

Robinson, J. T. 1972. *Early Hominid Posture and Locomotion.* Chicago: University of Chicago Press.

Rose, M. D. 1984. A hominine hip bone, KNM-ER 3228, from East Lake Turkana, Kenya. *American Journal of Physical Anthropology* 63: 371–378.

Solan, R., and M. H. Day. 1992. The Kapthurin ulna. *Journal of Human Evolution,* 22: 307–313.

Schultz, A. H., and W. L. Straus. 1945. The numbers of vertebrae in primates. *Proceedings of the American Philosophical Society* 89: 601–626.

Susman, R. L. 1988. Hand of *Paranthropus robustus* from Member 1, Swartkrans: Fossil evidence for tool behavior. *Science* 240: 781–784.

Trinkaus, E. 1984. Does KNM-ER 1481A establish *Homo erectus* at 2.0 myr BP? *American Journal of Physical Anthropology* 64: 137–139.

Walker, A., M. R. Zimmerman, and R. E. F. Leakey. 1982. A possible case of hypervitaminosis A in *Homo erectus. Nature* 296: 248–250.

Weidenreich, F. 1941. The extremity bones of *Sinanthropus pekinensis. Palaeontologia Sinica,* new series D, 5: 1–150.

8. THE DENTITION

Andrews, P. 1984. An alternative interpretation of the characters used to define *Homo erectus. Courier Forschungsinstitut Senckenberg* 69: 167–175.

Bilsborough, A., and B. Wood. 1986. The nature, origin and fate of *Homo erectus.* In *Major Topics in Primate and Human Evolution,* ed. B. Wood, L. Martin, and P. Andrews, pp. 295–316. New York: Cambridge University Press.

Bromage, T. G., and M. C. Dean. 1985. Re-evaluation of the age at death of immature fossil hominids. *Nature* 317: 525–527.

Day, M. H., and R. E. F. Leakey. 1973. New evidence of the genus *Homo* from East Rudolf, Kenya (I). *American Journal of Physical Anthropology* 39: 341–354.

——— 1974. New evidence of the genus *Homo* from East Rudolf, Kenya (III). *American Journal of Physical Anthropology* 41: 367–380.

Day, M. H., R. E. F. Leakey, A. C. Walker, and B. A. Wood. 1976. New hominids from East Turkana, Kenya. *American Journal of Physical Anthropology* 45: 369–436.

Dean, M. C. 1987a. Growth layers and incremental markings in hard tissues; a review of the literature and some preliminary observations about enamel structure in *Paranthropus boisei. Journal of Human Evolution* 16: 157–172.

——— 1987b. The dental developmental status of six East African juvenile fossil hominids. *Journal of Human Evolution* 16: 197–213.

Groves, C. P. 1989. *A Theory of Human and Primate Evolution.* Oxford: Clarendon Press.

Groves, C. P., and V. Mazak. 1975. An approach to the taxonomy of the Hominidae: Gracile Villafranchian hominids of Africa. *Casopis pro Mineralogii a Geologii* 20: 225–247.

Johanson, D. C., F. T. Masao, G. G. Eck, T. D. White, R. C. Walter, W. H. Kimbel, B. Asfaw, P. Manega, P. Ndessokia, and G. Suwa. 1987. New partial skeleton of *Homo habilis* from Olduvai Gorge, Tanzania. *Nature* 327: 205–209.

Kovacs, I. 1971. A systematic description of dental roots. In *Dental Morphology and Evolution,* ed. A. A. Dahlburg, pp. 211–256. Chicago: University of Chicago Press.

Leakey, R. E. F., and A. C. Walker. 1976. *Australopithecus, Homo erectus* and the single species hypothesis. *Nature* 261: 572–574.

——— 1985. Further hominids from the Plio-Pleistocene of Koobi Fora, Kenya. *American Journal of Physical Anthropology* 67: 135–163.

Leakey, R. E. F., and B. Wood. 1973. New evidence of the genus *Homo* from East Rudolf, Kenya (II). *American Journal of Physical Anthropology* 39: 355–368.

——— 1974. New evidence of the genus *Homo* from East Rudolf, Kenya (IV). *American Journal of Physical Anthropology* 41: 237–244.

Rightmire, G. P. 1984. Comparisons of *Homo erectus* from Africa and Southeast Asia. *Courier Forschungsinstitut Senckenberg* 69: 83–98.

——— 1990. *The Evolution of Homo erectus.* New York: Cambridge University Press.

Skelton, R. R., H. M. McHenry, and G. M. Drawhorn. 1986. Phylogenetic analysis of early hominids. *Current Anthropology* 27: 21–43.

Smith, B. H. 1986. Dental development in *Australopithecus* and early *Homo. Nature* 323: 327–330.

Stringer, C. B. 1984. The definition of *Homo erectus* and the existence of the species in Africa and Europe. *Courier Forschungsinstitut Senckenberg,* 69: 131–143.

——— 1986. The credibility of *Homo habilis.* In *Major Topics in Primate and Human Evolution,* ed. B. Wood, L. Martin, and P. Andrews, pp. 266–294. New York: Cambridge University Press.

Teaford, M. F., and O. J. Oyen. 1989. In vivo and in vitro turnover in dental microwear. *American Journal of Physical Anthropology* 80: 447–460.

Tobias, P. V. 1991. *Olduvai Gorge,* vol. 4. Cambridge: Cambridge University Press.

Walker, A. C., and R. E. F. Leakey. 1978. The hominids of East Turkana. *Scientific American* 239: 54–66.

Weidenreich, F. 1937. The dentition of *Sinanthropus pekinensis:* A comparative odontography of the hominids. *Palaeontologica Sinica* 101: 1–180 and plates I–XXXVI.

Wolpoff, M. H. 1980. *Paleoanthropology.* New York: Knopf.

Wood, B. 1984. The origin of *Homo erectus. Courier Forschungsinstitut Senckenberg* 69: 99–111.

——— 1985. Early *Homo* in Kenya, and its systematic relationships. In *Ancestors: The Hard Evidence,* ed. E. Delson, pp. 206–214. New York: Alan R. Liss, Inc.

——— 1991. Early *Homo:* How many species? *American Journal of Physical Anthropology,* supplement 12: 186.

——— 1992. *Koobi Fora Research Project,* vol. 4. Oxford: Clarendon Press.

Wood, B. A., S. A. Abbott, and H. Uytterschaut. 1988. Analysis of the dental morphology of Plio-Pleistocene hominids, IV. Mandibular postcanine root morphology. *Journal of Anatomy* 156: 107–139.

9. THE PHYSIOLOGICAL AGE OF KNM-WT 15000

Abbie, A. A. 1967. Skinfold thickness in Australian aborigines. *Archaeology and Physical Anthropology in Oceania* 2: 207–209.

Acheson, R. M. 1966. Maturation of the skeleton. In *Human Development,* ed. F. Falkner, pp. 465–502. Philadelphia: Saunders.

Adler, P. 1959. Der Geschlechtsunterschied im Zahnwechsel. *Deutsche Zahn-, Mund- und Kieferheilkunde* 31: 20–33.

Alvarez, J.O., and J. M. Navia. 1989. Nutritional status, tooth eruption, and dental caries: A review. *American Journal of Clinical Nutrition* 49: 417–426.

Anderson, D. L., G. W. Thompson, and F. Popovich. 1976. Age of attainment of mineralization stages of the permanent dentition. *Journal of Forensic Sciences* 21: 191–200.

Barrett, M. J., T. Brown, and K. M. Cellier. 1964. Tooth eruption sequence in a tribe of central Australian aborigines. *American Journal of Physical Anthropology* 22: 79–89.

Beynon, A. D., and B. A. Wood. 1987. Patterns and rates of enamel growth in the molar teeth of early hominids. *Nature* 326: 493–496.

Bogin, B. 1988. *Patterns of Human Growth.* Cambridge: Cambridge University Press.

——— 1990. The evolution of human childhood. *BioScience* 40: 16–25.

Bromage, T. G. 1987. The biological and chronological maturation of early hominids. *Journal of Human Evolution* 16: 257–272.

Bromage, T. G., and M. C. Dean. 1985. Re-evaluation of the

age at death of immature fossil hominids. *Nature* 317: 525–527.

Brown, F., J. Harris, R. Leakey, and A. Walker. 1985. Early *Homo erectus* skeleton from west Lake Turkana, Kenya. *Nature* 316: 788–792.

Brown, T. 1978. Tooth emergence in Australian aboriginals. *Annals of Humam Biology* 5: 41–54.

Brown, T., and M. J. Barrett. 1972. Growth in central Australian aborigines: Weight. *Medical Journal of Australia* 2: 999–1002.

Brown, T., M. J. Barrett, and K. C. Grave. 1971. Facial growth and skeletal maturation at adolescence. *Danish Dental Journal* 75: 1211–1222.

Chertkow, S. 1980. Tooth mineralization as an indicator of the pubertal growth spurt. *American Journal of Orthodontics* 77: 79–91.

Cheverud, J. M. 1981. Epiphyseal union and dental eruption in *Macaca mulatta. American Journal of Physical Anthropology* 56: 157–167.

Crossner, C. G., and L. Mansfeld. 1983. Determination of dental age in adopted non-European children. *Swedish Dental Journal* 7: 1–10.

Dahlberg, A. A., and R. M. Menegaz-Bock. 1958. Emergence of the permanent teeth in Pima Indian children. *Journal of Dental Research* 37: 1123–1140.

Dean, M. C. 1987a. The dental developmental status of six east African juvenile fossil hominids. *Journal of Human Evolution* 16: 197–213.

——— 1987b. Growth layers and incremental markings in hard tissues; a review of the literature and some preliminary observations about enamel structure in *Paranthropus boisei. Journal of Human Evolution* 16: 157–172.

Dean, M. C., and B.A. Wood. 1981. Developing pongid dentition and its use for ageing individual crania in comparative cross-sectional growth studies. *Folia Primatologica* 36: 111–127.

Dobzhansky, T. 1962. *Mankind Evolving*. New Haven: Yale University Press.

Eveleth, P. 1986. Population differences in growth: Environmental and genetic factors. In *Human Growth: A Comprehensive Treatise*, vol. 3, 2d ed., ed. F. Falkner and J. M. Tanner, pp. 221–239. New York: Plenum.

Falkner, F., and J. M. Tanner. 1986. *Human Growth: A Comprehensive Treatise*, vols. 1–3, 2d ed. New York: Plenum.

Fanning, E. A., and C. F. A. Moorrees. 1969. A comparison of permanent mandibular molar formation in Australian aborigines and caucasoids. *Archives of Oral Biology* 14: 999–1006.

Feldesman, M. R., J. G. Kleckner, and J. K. Lundy. 1990. Femur/stature ratio and estimates of stature in mid- and late-Pleistocene fossil hominids. *American Journal of Physical Anthropology* 83: 359–372.

Feldesman, M. R., and J. K. Lundy. 1988. Stature estimates for some African Plio-Pleistocene fossil hominids. *Journal of Human Evolution* 17: 583–596.

Fisher, D. C. 1987. Mastodont procurement by Paleoindians of the Great Lakes Region: Hunting or scavenging? In *The Evolution of Human Hunting*, ed. M. H. Nitecki and D. V. Nitecki, pp. 309–421. New York: Plenum.

Friedlaender, J. S., and H. L. Bailit. 1969. Eruption times of the deciduous and permanent teeth of natives on Bougainville Island, Territory of New Guinea: A study of racial variation. *Human Biology* 41: 51–65.

Garn, S.M. 1966. Body size and its implications. In *Review of Child Research Development*, ed. L. W. Hoffman and M. L. Hoffman, pp. 529–561. New York: Russell Sage Foundation.

Garn, S. M., A. B. Lewis, and R. M. Blizzard. 1965a. Endocrine factors in dental development. *Journal of Dental Research* 44: 243–248.

Garn, S. M., A. B. Lewis, and R. S. Kerewsky. 1965b. Genetic, nutritional, and maturational correlates of dental development. *Journal of Dental Research* 44: 228–242.

Gavan, J. A. 1953. Growth and development of the chimpanzee; a longitudinal and comparative study. *Human Biology* 25: 93–143.

Gavan, J. A., and D. R. Swindler. 1966. Growth rates and phylogeny in primates. *American Journal of Physical Anthropology* 24: 181–190.

Goldstein, H. 1979. *The Design and Analysis of Longitudinal Studies*. London: Academic Press.

Greulich, W. W., and S. I. Pyle. 1950. *Atlas of Skeletal Development of the Hand and Wrist*. Stanford: Stanford University Press.

Grøn, A. 1962. Prediction of tooth emergence. *Journal of Dental Research* 41: 573–585.

Gustafson, G., and G. Koch. 1974. Age estimation up to 16 years of age. *Odontologisk Revy* 25: 297–306.

Hagg, U., and L. Matsson. 1985. Dental maturity as an indicator of chronological age: The accuracy and precision of three methods based on dental development. *European Journal of Orthodontics* 7: 25–34.

Harvey, P. H., and T. H. Clutton-Brock. 1985. Life history variation in primates. *Evolution* 39: 559–581.

Helm, S. 1969. Secular trend in tooth eruption: A comparative study of Danish school children of 1913 and 1965. *Archives of Oral Biology* 14: 1177–1191.

——— 1990. Relationship between dental and skeletal maturation in Danish schoolchildren. *Scandinavian Journal of Dental Research* 98: 313–317.

Houpt, M. I., S. Adu-Aryee, and R. M. Grainger. 1967. Eruption times of permanent teeth in the Brong Ahafo region of Ghana. *American Journal of Orthodontics* 53: 95–99.

Hurme, V. O., and G. van Wagenen. 1953. Basic data on the emergence of deciduous teeth in the monkey *(Macaca mulatta). Proceedings of the American Philosophical Society* 97: 291–315.

——— 1961. Basic data on the emergence of permanent teeth in the rhesus monkey *(Macaca mulata). Proceedings of the American Philosophical Society* 105: 105–140.

Jaswal, S. 1983. Age and sequence of permanent-tooth emergence among Khasis. *American Journal of Physical Anthropology* 62: 177–186.

Kerley, E. R. 1966. Skeletal age changes in the chimpanzee. *Tulane Studies in Zoology* 13: 71–80.

Kimura, M. 1977. *A Bibliography of Age Determination of Mammals (with Special Emphasis on the Cetacea)*. Available on request from the National Oceanic and Atmospheric Administration, National Marine Fisheries Service, Southwest Fisheries Center, La Jolla, California 92038.

Knorr, D., F. Bidlingmaier, O. Butenandt, H. Fendel, and R. Ehrt-Wehle. 1974. Plasma testosterone in male puberty. *Acta Endocrinologica* 75: 181–194.

Kraemer, H. C., J. R. Horvat, C. Doering, and P. R. McGinnis. 1982. Male chimpanzee development focusing on adolescence: Integration of behavioral with physiological changes. *Primates* 23: 393–405.

Krogman, W. M. 1962. *The Human Skeleton in Forensic Medicine*. Springfield, Ill.: Charles C. Thomas.

Krumholt, L., B. Roed-Peterson, and J. J. Pindborg. 1971. Eruption times of the permanent teeth in 622 Ugandan children. *Archives of Oral Biology* 16: 1281–1288.

Laird, A. K. 1967. Evolution of the human growth curve. *Growth* 31: 345–355.

Lee, M. M. C., W. D. Low, and K. S. F. Chang. 1965. Eruption of the permanent dentition of southern Chinese children in Hong Kong. *Archives of Oral Biology* 10: 849–861.

Leinonen, A., B. Wasz-Höckert, and P. Vuorinen. 1972. Usefulness of the dental age obtained by orthopantomography as an indicator of the physical age. *Finnish Dental Society, Proceedings* 68: 235–242.

Lewis, A. B., and S. M. Garn. 1960. The relationship between tooth formation and other maturational factors. *Angle Orthodontist* 30: 70–77.

Little, M. A., K. Galvin, and M. Mugambi. 1983. Cross-sectional growth of nomadic Turkana pastoralists. *Human Biology* 55: 811–830.

Little, M. A., and B. R. Johnson. 1987. Mixed-longitudinal growth of nomadic Turkana pastoralists. *Human Biology* 59: 695–707.

MacKay, D. H., and W. J. Martin. 1952. Dentition and physique of Bantu children. *Journal of Tropical Medicine* 55: 265–275.

Mann, A. E. 1975. *Paleodemographic Aspects of the South African Australopithecines.* Philadelphia: University of Pennsylvania Press.

Mann, A. E., M. Lampl, and J. Monge. 1987. Maturational patterns in early hominids. *Nature* 328: 673–674.

Martin, D. E., R. B. Swenson, and D. C. Collins. 1977. Correlation of serum testosterone levels with age in male chimpanzees. *Steroids* 29: 471–481.

Martin, R. D. 1983. *Human Brain Evolution in an Ecological Context: Fifty-Second James Arthur Lecture on the Evolution of the Human Brain.* New York: American Museum of Natural History.

Mayhall, J. T., P. L. Belier, and M. F. Mayhall. 1978. Canadian Eskimo permanent tooth emergence timing. *American Journal of Physical Anthropology* 49: 211–216.

Moorrees, C. F. A., E. A. Fanning, and E. E. Hunt. 1963. Age variation of formation stages for ten permanent teeth. *Journal of Dental Research* 42: 1490–1502.

Murchison, M. A., A. J. Riopelle, and D. W. Owsley. 1988. Dental development in protein-deprived infant rhesus monkeys. *Human Biology* 60: 383–394.

Neville, A. C. 1967. Daily growth layers in animals and plants. *Biological Review* 42: 421–441.

Nissen, H. W., and A. H. Riesen. 1964. The eruption of the permanent dentition of chimpanzee. *American Journal of Physical Anthropology* 22: 285–294.

Niswander, J. D., and C. Sujaku. 1965. Permanent tooth eruption in children with major physical defect and disease. *Journal of Dentistry for Children* 32: 266–268.

Nolla, C. M. 1952. The development of permanent teeth. M.S. thesis. Ann Arbor: University of Michigan.

Owsley, D. W., and R. L. Jantz. 1983. Formation of the permanent dentition in Arikara Indians: Timing differences that affect dental age assessments. *American Journal of Physical Anthropology* 61: 467–471.

Preece, M. A. 1988. Prediction of adult height: Methods and problems. *Acta Paediatrica Scandinavica,* supplement 347: 4–11.

Roberts, D. F. 1960. Effects of race and climate on human growth as exemplified by studies on African children. In *Human Growth,* ed. J. M. Tanner, pp. 59–72. Oxford: Pergamon.

Rose, R. M., I. S. Bernstein, T. P. Gordon, and J. G. Lindsley. 1978. Changes in testosterone and behavior during adolescence in the male rhesus monkey. *Psychosomatic Medicine* 40: 60–70.

Sacher, G. A. 1959. Relation of lifespan to brain weight and body weight in mammals. In *CIBA Foundation Colloquia on Ageing,* vol. 5: *The Lifespan of Animals,* ed. G. E. W. Wolstenholme and M. O'Connor, pp. 115–133. London: Churchill.

——— 1975. Maturation and longevity in relation to cranial capacity in hominid evolution. In *Primate Functional Morphology and Evolution,* ed. R. H. Tuttle, pp. 417–441. The Hague: Mouton.

——— 1978. Longevity, aging, and death: An evolutionary perspective. *Gerontologist* 18: 112–119.

Sacher, G. A., and E. F. Staffeldt. 1974. Relation of gestation time to brain weight for placental mammals. *American Naturalist* 108: 593–616.

Sarich, V. M. 1968. The origins of the hominids: An immunological approach. In *Perspectives on Human Evolution,* ed. S. L. Washburn and P. C. Jay, pp. 94–121. New York: Holt, Rinehart, and Winston.

Schultz, A. H. 1940. Growth and development of the chimpanzee. *Contributions to Embryology* 28: 1–63.

——— 1956a. *The Life of Primates.* New York: Universe.

——— 1956b. Postembryonic age changes. In *Primatologia,* vol. 1, ed. H. Hofer, A. H. Schultz, and D. Starck, pp. 887–964. Basel: Karger.

——— 1960. Age changes in primates and their modification in man. In *Human Growth,* ed. J. M. Tanner, pp. 1–20. Oxford: Pergamon.

Shigehara, N. 1980. Epiphyseal union, tooth eruption, and sexual maturation in the common tree shrew, with reference to its systematic problem. *Primates* 21: 1–19.

Singh, S. P. 1980. Eruption of permanent teeth in Gaddi Rajput males of Dhaula Dhar Range of the Himalayas. *Zeitschrift für Morphologie und Anthropologie* 70: 295–301.

Smith, B. H. 1986. Dental development in *Australopithecus* and early *Homo. Nature* 323: 327–330.

——— 1989a. Dental development as a measure of life history in primates. *Evolution* 43: 683–688.

——— 1989b. Growth and development and its significance for early hominid behaviour. *Ossa* 14: 63–96.

——— 1990. The cost of a large brain [comment on: Brain evolution in *Homo:* the "radiator" theory, by Dean Falk]. *Behavioral and Brain Sciences* 13: 365–366.

——— 1991a. Age of weaning approximates age of emergence of the first permanent molar in nonhuman primates. *American Journal of Physical Anthropology,* supplement 12: 163–164 (abstract).

——— 1991b. Dental development and the evolution of life history in Hominidae. *American Journal of Physical Anthropology* 86: 157–174.

——— 1991c. Standards of human tooth formation and dental age assessment. In *Advances in Dental Anthropology,* ed. M. A. Kelley and C. S. Larsen, pp. 143–168. New York: Wiley-Liss.

Smith, B. H., and S. M. Garn. 1987. Polymorphisms in eruption sequence of permanent teeth in American children. *American Journal of Physical Anthropology* 74: 289–303.

Tanner, J. M. 1962. *Growth at Adolescence,* 2d ed. Oxford: Blackwell Scientific Publications.

Tanner, J. M., M. E. Wilson, and C. G. Rudman. 1990. Pubertal growth spurt in the female rhesus monkey: Relation to menarche and skeletal maturation. *American Journal of Human Biology* 2: 101–106.

Tonge, C. H., and R. A. McCance. 1973. Normal development of the jaws and teeth in pigs, and the delay and malocclusion produced by calorie deficiencies. *Journal of Anatomy* 115: 1–22.

Tooby, J., and I. DeVore. 1987. The reconstruction of hominid behavioral evolution through strategic modeling. In *The Evolution of Human Behavior: Primate Models,* ed. W. G. Kinzey, pp. 183–238. Albany: State University of New York Press.

Watts, E. S. 1986a. Skeletal development. In *Comparative Primate Biology,* vol. 3: *Reproduction and Development,* ed. W. R. Dukelow and J. Erwin, pp. 415–439. New York: Alan R. Liss.

——— 1986b. Evolution of the human growth curve. In *Human Growth: A Comprehensive Treatise,* vol. 1, 2d ed., ed. F. Falkner and J. M. Tanner, pp. 153–166. New York: Plenum.

Watts, E. S., and J. A. Gavan. 1982. Postnatal growth of non-

human primates: The problem of the adolescent spurt. *Human Biology* 54: 53–70.
Weidenreich, F. 1936. The mandibles of *Sinanthropus pekinensis:* A comparative study. *Palaeontologia Sinica,* series D, vol. 7, fasc. 3.
——— 1937. The dentition of *Sinanthropus pekinensis:* A comparative odontography of the hominids. *Palaeontologia Sinica,* n.s. 1: 120–180.

10. The Reconstruction of the Pelvis

Brown, F., J. Harris, R. Leakey, and A. Walker. 1985. Early *Homo erectus* skeleton from west Lake Turkana, Kenya. *Nature* 316: 788–792.
Dobbing, J., and J. Sands. 1978. Head circumference, biparietal diameter and brain growth in fetal and postnatal life. *Early Human Development* 2: 81–87.
Gould, S. J. 1977. *Ontogeny and Phylogeny.* Cambridge: Harvard University Press.
——— 1985. The most compelling pelvis since Elvis. *Discover,* December 1985, 54–58.
Levi, S., and F. Erbsman. 1974. Antenatal growth from the nineteenth week. *American Journal of Obstetrics and Gynecology* 121: 262–268.
Martin, R. D. 1983. *Human Brain Evolution in an Ecological Context: Fifty-Second James Arthur Lecture on the Evolution of the Human Brain.* New York: American Museum of Natural History.
Portmann, A. 1941. Die Tragseiten der Primaten und die Dauer der Schwangerschaft beim Menschen: Ein Problem der vergleichenden Biologie. *Revue Suisse de Zoologie* 48: 511–518.
Rose, M. D. 1984. A hominine hip bone, KNM-ER 3228, from East Lake Turkana, Kenya. *American Journal of Physical Anthropology* 63: 371–378.
Tague, R. G. 1989. Variation in pelvic size between males and females. *American Journal of Physical Anthropology* 80: 59–71.
Tobias, P. V. 1971. *The Brain in Hominid Evolution.* New York: Columbia University Press.
Walker, A. 1984. Extinction in hominid evolution. In *Extinctions,* ed. M. H. Nitecki, pp. 119–152. Chicago: University of Chicago Press.

11. Body Size and Body Shape

Abbie, A. A. 1967. Skinfold thickness in Australian Aborigines. *Archaeology and Physical Anthropology in Oceania* 2: 207–219.
Allbrook, D. 1961. The estimation of stature in British and East African males. *Journal of Forensic Medicine* 8: 15–28.
Anderson, M., W. T. Green, and M. B. Messner. 1963. Growth and predictions of growth in the lower extremities. *Journal of Bone and Joint Surgery* 45A: 1–14.
Bazett, H. C. 1968. The regulation of body temperatures. In *Physiology of Heat Regulation and the Science of Clothing,* ed. L. H. Newburgh, pp. 109–192. New York: Hafner.
Brown, F., J. Harris, R. Leakey, and A. Walker. 1985. Early *Homo erectus* skeleton from West Lake Turkana, Kenya. *Nature* 316: 788–792.
Calder, W. A., III. 1984. *Size, Function, and Life History.* Cambridge: Cambridge University Press.
Eveleth, P. B. 1966. The effects of climate on growth. *Annals of the New York Academy of Sciences* 134: 750–759.
Eveleth, P. B., and J. M. Tanner. 1976. *Worldwide Variation in Human Growth.* Cambridge: Cambridge University Press.
Feldesman, M. R., J. G. Kleckner, and J. K. Lundy. 1990. The femur/stature ratio and estimates of stature in mid- and late-Pleistocene fossil hominids. *American Journal of Physical Anthropology* 83: 359–372.
Feldesman, M. R., and J. K. Lundy. 1988. Stature estimates for some African Plio-Pleistocene fossil hominids. *Journal of Human Evolution* 17: 583–596.
Flecker, H. 1932/33. Roentgenographic observations of the times of appearance of the epiphyses and their fusion with the diaphyses. *Journal of Anatomy* 67: 118–164.
Foley, R. A., and P. C. Lee. 1989. Finite social space, evolutionary pathways, and reconstructing hominid behavior. *Science* 243: 901–906.
Franciscus, R. G., and T. W. Holliday. In press. Hindlimb skeletal allometry in Plio-Pleistocene hominids with special reference to AL 288-1 ("Lucy"). *Bulletin et Mémoires de la Société de l'Anthropologie de Paris.*
Geissmann, T. 1986. Estimation of australopithecine stature from long bones: A.L.288–1 as a test case. *Folia Primatologica* 47: 119–127.
Gindhart, P. S. 1973. Growth standards for the tibia and radius in children aged one month through eighteen years. *American Journal of Physical Anthropology* 39: 41–48.
Green, W. T., G. M. Wyatt, and M. Anderson. 1946. Orthoroentgenography as a method of measuring the bones of the lower extremities. *Journal of Bone and Joint Surgery* 28: 60–65.
Harvey, R. G. 1974. An anthropometric survey of growth and physique of the populations of Karkar Island and Lufa subdistrict, New Guinea. *Philosophical Transactions of the Royal Society of London* B268: 279–292.
Hautvast, J. 1971. Physical growth and menarcheal age in Tanzanian schoolchildren and adults. *Human Biology* 43: 421–443.
Hesse, R., W. C. Allee, and K. P. Schmidt. 1951. *Ecological Animal Geography.* New York: Wiley.
Hiernaux, J. 1968. *La Diversité Humaine en Afrique Subsaharienne: Recherches Biologiques.* Bruxelles: Institut de Sociologie, Université Libre de Bruxelles.
——— 1975. *The People of Africa.* New York: Scribner.
Johanson, D. C., F. T. Masao, G. G. Eck, T. D. White, R. C. Walter, W. H. Kimbel, B. Asfaw, and P. Manega. 1987. New partial skeleton of *Homo habilis* from Olduvai Gorge, Tanzania. *Nature* 327: 205–209.
Jungers, W. L. 1982. Lucy's limbs: Skeletal allometry and locomotion in *Australopithecus afarensis. Nature* 297: 676–678.
——— 1988a. Lucy's length: Stature reconstruction in *Australopithecus afarensis* (A.L. 288–1) with implications for other small-bodied hominids. *American Journal of Physical Anthropology* 76: 227–231.
——— 1988b. New estimates of body size in australopithecines. In *Evolutionary History of the "Robust" Australopithecines,* ed. F. E. Grine, pp. 115–125. New York: Aldine de Gruyter.
Krogman, W. M. 1962. *The Human Skeleton in Forensic Medicine.* Springfield, Ill.: Charles C. Thomas.
Leakey, R. E., and A. Walker. 1989. Early *Homo erectus* from West Lake Turkana, Kenya. In *Hominidae: Proceedings of the 2nd International Congress of Human Paleontology,* ed. G. Giacobini, pp. 209–215. Milan: Jaca Book.
Little, M. A., K. Galvin, and M. Mugambi. 1983. Cross-sectional growth of nomadic Turkana pastoralists. *Human Biology* 55: 811–830.
Little, M. A., and B. R. Johnson. 1987. Mixed-longitudinal growth of nomadic Turkana pastoralists. *Human Biology* 59: 695–707.
Maresh, M. M. 1970. Measurements from roentgenograms, heart size, long bone lengths, bone, muscles and fat widths, skeletal maturation. In *Human Growth and Development,* ed. R. W. McCammon, pp. 155–200. Springfield, Ill.: Charles C. Thomas.
Martin, R. 1928. *Lehrbuch der Anthropologie.* Jena: Fischer.
Mayr, E. 1963. *Animal Species and Evolution.* Cambridge: Harvard University Press.

McHenry, H. M. 1975. Fossil hominid body weight and brain size. *Nature* 254: 686–688.

——— 1976. Early hominid body weight and encephalization. *American Journal of Physical Anthropology* 45: 77–84.

——— 1988. New estimates of body weight in early hominids and their significance to encephalization and megadontia in "robust" australopithecines. In *Evolutionary History of the "Robust" Australopithecines*, ed. F. E. Grine, pp. 133–148. New York: Aldine de Gruyter.

——— 1991a. Sexual dimorphism in *Australopithecus afarensis. Journal of Human Evolution* 20: 21–32.

——— 1991b. Femoral lengths and stature in Plio-Pleistocene hominids. *American Journal of Physical Anthropology* 85: 149–158.

Modi, J. P. 1957. *Medical Jurisprudence and Toxicology*, 12th ed. Bombay: Tripathy Private, Ltd.

Oleksiak, D. A. 1986. The estimation of body weight for Neanderthals and early anatomically modern humans. *American Journal of Physical Anthropology* 69:248.

Parizkova, J. 1973. Body composition and lipid metabolism. *Nutrition Society, Proceedings* 32: 181–186.

Rak, Y., and B. Arensburg. 1987. Kebara 2 neanderthal pelvis: First look at a complete inlet. *American Journal of Physical Anthropology* 73: 227–231.

Rightmire, G. P. 1986. Body size and encephalization in *Homo erectus. Anthropos* 23: 139–149.

Roberts, D. F. 1978. *Climate and Human Variability*, 2d ed. Menlo Park: Cummings.

Roberts, D. F., and D. R. Bainbridge. 1963. Nilotic physique. *American Journal of Physical Anthropology* 21: 341–370.

Robinson, J. T. 1972. *Early Hominid Posture and Locomotion*. Chicago: University of Chicago Press.

Robinson, S. 1968. Physiological adjustments to heat. In *Physiology of Heat Regulation*, ed. L. H. Newburgh, pp. 193–231. New York: Hafner.

Ruff, C. B. 1988. Hindlimb articular surface allometry in Hominoidea and *Macaca*, with comparisons to diaphyseal scaling. *Journal of Human Evolution* 17: 687–714.

——— 1991. Climate, body size and body shape in hominid evolution. *Journal of Human Evolution* 21: 81–105.

Ruff, C. B., W. W. Scott, and A. Y.-C. Liu. 1991. Scaling of proximal femoral dimensions to body mass in a living human sample. *Am. J. Phys. Anthro.* 86: 397–413.

Ruff, C. B., E. Trinkaus, A. Walker, and C. S. Larson. 1993. Postcranial robusticity in *Homo*. 1: Temporal trends and mechanical interpretation. *Am. J. Phys. Anthro.* 91: 21–53.

Ruff, C. B., A. C. Walker, and M. F. Teaford. 1989. Body mass, sexual dimorphism and femoral proportions of *Proconsul* from Rusinga and Mfangano Islands, Kenya. *Journal of Human Evolution* 18: 515–536.

Schmidt-Nielson, K. 1984. *Scaling: Why Is Animal Size So Important?* Cambridge: Cambridge University Press.

Schreider, E. 1964. Ecological rules, body-heat regulation, and human evolution. *Evolution* 18: 1–9.

Schultz, A. H. 1937. Proportions, variability and asymmetries of the long bones of the limbs and the clavicles in man and apes. *Human Biology* 9: 281–328.

Shipman, P., and A. Walker. 1989. The costs of being a predator. *Journal of Human Evolution* 18: 373–392.

Stinson, S., and A. R. Frisancho. 1978. Body proportions of Highland and Lowland Peruvian Quecha children. *Human Biology* 50: 57–68.

Tague, R. G. 1989. Variation in pelvic size between males and females. *American Journal of Physical Anthropology* 80: 59–71.

Tague, R. G., and C. O. Lovejoy. 1986. The obstetric pelvis of A.L. 288–1 (Lucy). *Journal of Human Evolution* 15: 237–255.

Tanner, J. M., T. Hayashi, M. A. Preece, and N. Cameron. 1982. Increase in length of leg relative to trunk in Japanese children and adults from 1957 to 1977: Comparison with British and with Japanese Americans. *Annals of Human Biology* 9: 411–423.

Tobias, P. V. 1985. The negative secular trend. *Journal of Human Evolution* 14: 347–356.

Trinkaus, E. 1981. Neanderthal limb proportions and cold adaptation. In *Aspects of Human Evolution*, ed. C. B. Stringer, pp. 187–224. London: Taylor and Francis.

Trotter, M., and G. C. Gleser. 1952. Estimation of stature from long bones of American whites and Negroes. *American Journal of Physical Anthropology* 10: 463–514.

——— 1958. A re-evaluation of estimation of stature based on measurements of stature taken during life and of long bones after death. *Am. J. Phys. Anthro.* 16: 79–123.

Tupman, G. S. 1962. A study of bone growth in normal children and its relationship to skeletal maturation. *Journal of Bone and Joint Surgery* 44B: 42–67.

Y'Edynak, G. 1976. Long bone growth in Western Eskimo and Aleut skeletons. *Am. J. Phys. Anthro.* 45: 569–574.

Zar, J. H. 1984. *Biostatistical Analysis*. Englewood Cliffs: Prentice-Hall.

12. THE THORACIC AND LUMBAR VERTEBRAE

Bick, E. M., and J. W. Copel. 1950. Longitudinal growth of the human vertebra. *Journal of Bone and Joint Surgery* 32A: 803–814.

Brain, C. K. 1981. *The Hunters or the Hunted?* Chicago. University of Chicago Press.

Brown, F., J. Harris, R. Leakey and A. Walker. 1985. Early *Homo erectus* skeleton from west Lake Turkana, Kenya. *Nature* 316: 788–792.

Cihak, R. 1970. Variations of lumbosacral joints and their morphogenesis. *Acta Universitatis Carolinae Medica* 16: 145–165.

Cunningham, D. J. 1886. "Cunningham Memoirs": Lumbar curve in man and the apes with an account of the topographical anatomy of the chimpanzee and orangutan. *Royal Irish Academy*, No. II.

Digiovanni, B. F., P. V. Scoles, and B. M. Latimer. 1989. Anterior extension of the thoracic vertebral bodies in Scheuermann's kyphosis: An anatomic study. *Spine* 14: 712–716.

Dommisse, G. F. 1975. Morphological aspects of the lumbar spine and lumbosacral region. *Orthopedic Clinics of North America* 6: 163–175.

Huxley, T. H. 1863. *Man's Place in Nature*. New York: D. Appleton and Company.

Jungers, W. L. 1985. Body size and scaling of limb proportions in primates. In *Size and Scaling in Primate Biology*, ed. W. L. Jungers, pp. 345–381. New York: Plenum Press.

Keith, A. 1902. The extent to which the posterior segments of the body have been transmuted and suppressed in the evolution of man and allied primates. *Journal of Anatomy* 37: 18–40.

Latimer, B., J. C. Ohman, and C. O. Lovejoy. 1987. The talocrural joint in African hominids: Implications for *Australopithecus afarensis*. *American Journal of Physical Anthropology* 74: 155–176.

Leakey, M. G., and R. E. Leakey, eds. 1978. *Koobi Fora Research Project*, vol. 1. Oxford: Clarendon Press.

Lovejoy, C. O. 1988. Evolution of human walking. *Scientific American* 259: 118–125.

Mensforth, R. P. 1985. Relative tibia long bone growth in the Libben and BT-5 Prehistoric skeletal populations. *American Journal of Physical Anthropology* 68: 247–262.

Noback, C. R., and J. K. Harting. 1971. *Spinal Cord (Spinal Medulla)*. Primatologia, No. II. Basel: Karger.

Roaf, R. 1971. The growth of the spinal articular process and their clinical significance. In *Scoliosis and Growth*, ed. P. A. Zorab, pp. 92–97. Baltimore: Williams and Wilkins.

Robinson, J. T. 1972. *Early Hominid Posture and Locomotion*. Chicago: University of Chicago Press.

Schultz, A. H. 1930. The skeleton of the trunk and limbs of higher primates. *Human Biology* 2: 303–438.

——— 1953. The relative thickness of the long bones and the vertebrae in primates. *American Journal of Physical Anthropology* 11: 277–311.

Scoles, P. V., A. E. Linton, B. Latimer, M. E. Levy, and B.F. Digiovanni. 1988. Vertebral body and posterior element morphology: The normal spine in middle life. *Spine* 13: 1082–1086.

Scoles, P. V., B. M. Latimer, B. F. DiGiovanni, E. Vargo, S. Bauza, and L. M. Jellema. 1991. Vertebral alterations in Scheuermann's kyphosis. *Spine* 16: 509–515.

Tague, R. G., and C. O. Lovejoy. 1986. The obstetric pelvis of A.L. 288–1 (Lucy). *Journal of Human Evolution* 15: 237–255.

Tanner, J. M., R. H. Whitehouse, N. Cameron, W. A. Marshall, M. J. R. Healey, and H. Goldstein. 1990. *Assessment of Skeletal Maturity and Prediction of Adult Height (TW2 Method)*. New York: Academic Press.

Tanner, J. M., R. H. Whitehouse and M. Takaishi. 1966. Standards from birth to maturity for height, weight, height velocity and weight velocity: British Children. *Archives of Disease in Childhood* 41: 454–71, 613–35.

Tulsi, R. S. 1971. Growth of the human vertebral column—An osteological study. *Acta Anatomica* 79: 570–580.

Veldhuizen, A. G., P. Baasand, and P. J. Webb. 1986. Observations on the growth of the adolescent spine. *Journal of Bone and Joint Surgery* 68B: 724–728.

Ward, C. V. 1991. Functional anatomy of the lower back and pelvis of the Miocene hominid *Proconsul nyanzae* from Mfangano Island, Kenya. Ph.D. dissertation, The Johns Hopkins University.

Ward, C. V., A. Walker, M. F. Teaford, and I. Odhiambo. 1993. A partial skeleton of *Proconsul nyanzae* from Mfangano Island, Kenya. *American Journal of Physical Anthropology* 90: 77–111.

Williams, P. L., and R. Warwick. 1980. *Gray's Anatomy*, 36th ed. Philadelphia: W. B. Saunders Company.

13. THE RIB CAGE

Brain, C. K. 1981. *The Hunters or the Hunted?* Chicago: University of Chicago Press.

Brown, F., J. Harris, R. Leakey, and A. Walker. 1985. Early *Homo erectus* skeleton from West Lake Turkana, Kenya. *Nature* 316: 788–792.

Gehr, P., et. al. 1981. Design of the Mammalian Respiratory System, V. Scaling morphometric pulmonary diffusing capacity to body mass: Wild and domestic mammals. *Respiratory Physiology* 44: 61–86.

Jenkins, F. A. 1970. Anatomy and function of expanded ribs in certain edentates and primates. *Journal of Mammology* 51: 288–301.

Jordanoglou, J. 1970. Vector analysis of rib movement. *Respiration Physiology* 10: 109–120.

Jungers, W. L. 1988. New estimates of body size in Australopithecines. In *Evolutionary History of the "Robust" Australopithecines*, ed. F. E. Grine, pp. 115–125. New York: Aldine de Gruyter.

Keith, A. 1903. The nature and anatomy of Enteroptosis (Glenard's disease). *Lancet*, March 7, 1903, pp. 631–640.

——— 1923. Man's posture: Its evolution and disorders. *The British Medical Journal*, 1:451–454, 545–548, 587–590, 624–626, 669–672.

Latimer, B., and J.C. Ohman (in preparation).

Lovejoy, C. O. 1988. Evolution of human walking. *Scientific American*, 259: 118–125.

Ohman, J. C. 1984. Morphological variation in the cervicothoracic region of the vertebral column in modern humans, extant apes, fossil hominids and with reference to Mammalia in general. M.A. thesis, Kent State University.

Ohman, J. C. 1986. The first rib of Hominids. *American Journal of Physical Anthropology* 70: 209–229.

Ruff, C. B. 1991. Climate and body shape in hominid evolution. *Journal of Human Evolution* 21: 81–105.

Schultz, A. H. 1926. Fetal growth of Man and other Primates. *Quarterly Review of Biology* 1: 465–521.

Schultz, A. H. 1950. Origins of the human stock: The specializations of man and his place in the Catarrhine primates. *Cold Spring Harbor Symposia on Quantitative Biology*, 15: 37–53.

——— 1961. Vertebral column and thorax. *Primatologica* 4: 1–66.

——— 1963. Age changes, sex differences and variability as factors in the classification of primates. In *Classification and Human Evolution*, S. L. Washburn, pp. 85–115. Viking Fund Publications in Anthropology, No. 37. Chicago: Aldine.

——— 1969. The skeleton of the chimpanzee. In *The Chimpanzee*, vol. 1, ed. G. H. Bourne, pp. 50–103. Basel: Karger.

Scoles, P. V., et al. 1991. Vertebral alterations in Scheuermann's kyphosis. *Spine* 16: 509–515.

Tague, R. G., and C. O. Lovejoy. 1986. The obstetric pelvis of A.L. 288–1 (Lucy). *Journal of Human Evolution* 15: 237–255.

Todd, T. W. 1912. The decent of the shoulder after birth. *Anatomischer Anzeiger* 41: 385–397.

14. THE ENDOCAST

Bauchot, R. 1978. Encephalization in vertebrates: A new mode of calculation for allometry coefficients and isopondral indices. *Brain, Behavior and Evolution* 15: 1–18.

Bauchot, R., and H. Stephan. 1969. Encéphalisation et niveau évolutif chez les simiens. *Mammalia* 33: 225–275.

Browder, J., and H. A. Kaplan. 1976. *Cerebral Dural Sinuses and Their Tributaries*. Springfield, Ill.: Charles C. Thomas.

Brown, F., J. Harris, R. Leakey, and A. Walker. 1985. Early *Homo erectus* from west Lake Turkana, Kenya. *Nature* 316: 788–792.

Caramazza, A., and A. E. Hillis. 1991. Lexical organization of nouns and verbs in the brain. *Nature* 349: 788–790.

Chamberlain, A. T., and B. A. Wood. 1987. Early hominid phylogeny. *Journal of Human Evolution* 16: 119–133.

Conroy, G. C. 1982. A study of cerebral vascular evolution in primates. In *Primate Brain Evolution: Methods and Concepts*, ed. E. Armstrong and D. Falk, pp. 247–262. New York: Academic Press.

Dart, R. A. 1925. *Australopithecus africanus:* The man-ape of South Africa. *Nature* 115: 195–199.

Day, M. H. 1971. Postcranial remains of *Homo erectus* from Bed IV, Olduvai Gorge, Tanzania. *Nature* 232: 383–387.

——— 1986. *Guide to Fossil Man*. Chicago: University of Chicago Press.

Deacon, T. W. 1990. Problems of ontogeny and phylogeny in brain-size evolution. *International Journal of Primatology* 11: 237–282.

Falk, D. 1980. Hominid brain evolution: The approach from paleoneurology. *Yearbook of Physical Anthropology* 23: 93–107.

——— 1983. The Taung endocast: A reply to Holloway. *American Journal of Physical Anthropology* 60: 479–489.

——— 1987. Hominid paleoneurology. *Annual Review of Anthropology* 16: 13–30.

Feibel, C. S., F. H. Brown, and I. McDougall. 1989. Stratigraphic context of fossil hominids from the Omo group deposits: Northern Turkana basin, Kenya and Ethiopia. *American Journal of Physical Anthropology* 78: 595–622.

Gould, S. J. 1977. *Ontogeny and Phylogeny.* Cambridge: Harvard University Press.

Harvey, P. H. 1988. Allometric analysis and brain size. In *Intelligence and Evolutionary Biology,* ed. H. J. Jerison and I. Jerison, pp. 199–210. Berlin: Springer-Verlag.

Hay, R. L. 1976. *Geology of the Olduvai Gorge: A Study of the Sedimentation in a Semiarid Basin.* Berkeley: University of California Press.

Hofman, M. A. 1983. Encephalization in hominids: Evidence for the model of punctuationalism. *Brain, Behavior and Evolution* 22: 102–117.

Holloway, R. L. 1980. Within species brain-body weight variability: A reexamination of the Danish data and other primate species. *American Journal of Physical Anthropology* 53: 109–121.

——— 1982. *Homo erectus* brain endocasts: Volumetric and morphological observations, with some comments on cerebral asymmetries. *Congres Internationale de Paleontologie Humaine, 1er Congress,* pp. 355–369.

——— 1983. Human paleontological evidence relevant to language behavior. *Human Neurobiology* 2: 105–114.

Holloway, R. L., and M. C. De La Coste-Lareymondie. 1982. Brain endocast asymmetry in pongids and hominids: Some preliminary findings on the paleontology of cerebral dominance. *American Journal of Physical Anthropology* 58: 101–110.

Jerison, H. J. 1973. *The Evolution of Brain and Intelligence.* New York: Academic Press.

Jungers, W. L. 1988. New estimates of body size in australopithecines. In *The Evolution of the "Robust" Australopithecines,* ed. F. E. Grine, pp. 115–125. New York: Aldine de Gruyter.

Kimbel, W. L. 1984. Variation in the pattern of cranial venous sinuses and hominid phylogeny. *American Journal of Physical Anthropology* 63: 243–263.

Leakey, M. D., R. J. Clarke, and L. S. B. Leakey. 1971. New hominid skull from Bed I, Olduvai Gorge, Tanzania. *Nature* 232: 308–312.

LeMay, M. 1976. Morphological cerebral asymmetries of modern man, fossil man and nonhuman primates. In *Origins and Evolution of Language and Speech,* ed. S. R. Harnard, H. D. Steklis, and J. Lancaster, pp. 349–366. *Annals of the New York Academy of Sciences,* vol. 280.

Leutenegger, W. 1982. Encephalization and obstetrics in primates with particular reference to human evolution. In *Primate Brain Evolution: Methods and Concepts,* ed. E. Armstrong and D. Falk, pp. 85–95. New York: Plenum Press.

Liu, T., and M. Ding. 1984. A tentative chronological correlation of early human fossil horizons in China with loess-deep sea records. *Acta Anthropologica Sinica* 3: 93–101.

Mann, A. E. 1975. *Paleodemographic Aspects of the South African Australopithecines.* Philadelphia: University of Pennsylvania.

Martin, R. 1928. *Lehrbuch der Anthropologie.* Jena: Fischer.

Martin, R. D. 1981. Relative brain size and basal metabolic rate in terrestrial vertebrates. *Nature* 293: 57–60.

——— 1983. *Human Brain Evolution in an Ecological Context: Fifty-Second James Arthur Lecture on the Evolution of the Human Brain.* New York: American Museum of Natural History.

——— 1989. *Primate Origins and Evolution: A Phyletic Reconstruction.* Princeton: Princeton University Press.

Martin, R. D. and P. H. Harvey 1985. Brain size allometry: ontogeny and phylogeny. In *Size and Scaling in Primate Biology,* ed. W. L. Jungers, pp. 147–174. New York: Plenum Press.

McHenry, H. M. 1982. The pattern of human evolution: Studies on bipedalism, mastication and encephalization. *Annual Review of Anthropology* 11: 151–173.

——— 1988. New estimates of body weight in early hominids and their significance to encephalization and megadontia in robust australopithecines. In *The Evolution of the "Robust" Australopithecines,* ed. F. E. Grine, pp. 133–148. New York: Aldine de Gruyter.

Pagel, M. D., and P. H. Harvey. 1988. The taxon-level problem in the evolution of mammalian brain size: Facts and artefacts. *American Naturalist* 132: 344–359.

Passingham, R. E. 1985. Rates of brain development in mammals including man. *Brain, Behavior and Evolution* 26: 167–175.

Peterson, S. E., P. T. Fox, M. I. Posner, M. Mintun, and M. E. Raichle. 1988. Positron emission tomographic studies of the cortical anatomy of single-word processing. *Nature* 331: 585–589.

——— 1989. Positron emission tomographic studies of the processing of single words. *Journal of Cognitive Neuroscience* 1: 163–170.

Pilbeam, D., and S. J. Gould. 1974. Size and scaling in human evolution. *Science* 186: 892–901.

Pope, G. G., and J. E. Cronin. 1984. The Asian Hominidae. *Journal of Human Evolution* 13: 377–396.

Portmann, A. 1941. Die Tragzeiten der Primaten und die Dauer der Schwangerschaft beim Menschen: Ein Problem der vergleichenden Biologie. *Revue Suisse de Zoologie* 48: 511–518.

Radinsky, L. 1982. Some cautionary notes on making inferences about relative brain size. In *Primate Brain Evolution: Methods and Concepts,* ed. E. Armstrong and D. Falk, pp. 29–38. New York: Academic Press.

Rightmire, G. P. 1985. The tempo of change in the evolution of mid-Pleistocene *Homo.* In *Ancestors: The Hard Evidence,* ed. E. Delson, pp. 255–264. New York: Alan R. Liss.

——— 1986. Body size and encephalization in *Homo erectus. Anthropos* 23: 139–150.

——— 1990. *The Evolution of Homo erectus.* Cambridge: Cambridge University Press.

Ruff, C. 1990. Body mass and hindlimb bone cross-sectional and articular dimensions in anthropoid primates. In *Body Size and Mammalian Paleobiology: Estimation and Biological Implications,* ed. J. Damuth and B.J. MacFadden, pp. 119–149. Cambridge: Cambridge University Press.

——— 1991. Climate, body size and body shape in hominid evolution. *Journal of Human Evolution* 21: 81–105.

Saban, R. 1986. Veines méningées et hominisation. *Anthropos* 23: 15–33.

Smith, G. E. 1908. Right-handedness. *British Medical Journal* 1908: 596–597.

Smith, J. A., and B. M. Latimer. 1989. A method for making three-dimensional reproductions of bones and fossils. *Kirtlandia* 44: 3–16.

Tobias, P. V. 1967. *Olduvai Gorge,* vol. 1. Cambridge: Cambridge University Press.

——— 1971. *The Brain in Hominid Evolution.* New York: Columbia University Press.

——— 1975. Brain evolution in the Hominoidea. In *Primate Functional Morphology and Evolution,* ed. R. H. Tuttle, pp. 353–392. The Hague: Mouton.

——— 1981. The emergence of man in Africa and beyond. *Philosophical Transactions of the Royal Society* 292: 43–66.

——— 1983. Recent advances in the evolution of the hominids with especial reference to brain and speech. In *Recent Advances in the Evolution of Primates,* ed. C. Chagas, pp. 85–140. Citta del Vaticano: Pontificia Academia Scientiarum.

Walker, A. 1980. Remains attributable to *Australopithecus* in the East Rudolf succession. In *Stratigraphy, Evolution and Palaeoecology in the Lake Rudolf Basin,* ed. Y. Coppens, F. C. Howell, G. Ll. Isaac, and R. E. Leakey, PP. 484–489. Chicago: Chicago University Press.

Williams, P. L., R. Warwick, M. Dyson, and L. H. Bannister. 1989. *Gray's Anatomy.* Edinburgh: Churchill Livingstone.

Weidenreich, F. 1941. The extremity bones of *Sinanthropus pekinensis. Palaeontologica Sinica,* New Series D, 5: 1–150.

——— 1943. The skull of *Sinanthropus pekinensis:* A comparative study on a primitive hominid skull. *Palaeontologica Sinica,* new series D, 10: 1–484.

Wu, R. 1985. New Chinese *Homo erectus* and recent work at Zhoukoudian. In *Ancestors: The Hard Evidence,* ed. E. Delson, pp. 245–248. New York: Alan R. Liss.

15. The Vertebral Canal

Amonoo-Kuofi, H. S. 1982. Maximum and minimum interpedicular distances in normal Nigerians. *Journal of Anatomy* 135: 225–233.

Amonoo-Kuofi, H. S., P. J. Patel, and J. A. Fatani. 1990. Transverse diameter of the lumbar spinal canal in normal adult Saudis. *Acta Anatomica* 137: 124–128.

Ariëns Kappers, C. U., G. C. Huber, and E. C. Crosby. 1936. *The Comparative Anatomy of the Nervous System of Vertebrates, Including Man.* New York: Macmillan.

Barr, M. L. 1979. *The Human Nervous System,* 3d ed. New York: Harper and Row.

Berry, J. L., J. M. Moran, W. S. Berg, and A. D. Steffee. 1987. A morphometric study of human lumbar and selected thoracic vertebrae. *Spine* 12: 362–367.

Bhalla, S. P., S. K. Lall, J. S. Sodhi, and O. N. Negi. 1977. Comparative values of intervertebral space, spinal canal and intervertebral foramina in normal and in cases of cervical spondylosis in Indian population. *Indian Journal of Radiology* 31: 170–175.

Brown, F., J. Harris, R. Leakey, and A. Walker. 1985. Early *Homo erectus* skeleton from west Lake Turkana, Kenya. *Nature* 316: 788–792.

Burrows, E. H. 1963. The sagittal diameter of the spinal canal in cervical spondylosis. *Clinical Radiology* 14: 77–86.

Donaldson, H. H., and D. J. Davis. 1903. A description of charts showing the areas of the cross-sections of the human spinal cord at the level of each spinal nerve. *Journal of Comparative Neurology* 13: 19–40.

Donaldson, H. H., and S. Hatai. 1911. A comparison of the Norway rat with the albino rat in respect to body length, brain weight, spinal cord weight and the percentage of water in both the brain and the spinal cord. *Journal of Comparative Neurology* 21: 417–458.

Eisenstein, S. 1976. Measurements of the lumbar spinal canal in 2 racial groups. *Clinical Orthopaedics* 115: 42–46.

——— 1977. The morphometry and pathological anatomy of the lumbar spine in South African Negroes and Caucasoids with specific reference to spinal stenosis. *Journal of Bone and Joint Surgery* 59B: 173–180.

Gupta, S. K., R. C. Roy, and A. Srivastava. 1982. Sagittal diameter of the cervical canal in normal Indian adults. *Clinical Radiology* 33: 681–685.

Haworth, J. B., and G. W. Keillor. 1962. Use of transparencies in evaluating the width of the spinal canal in infants, children, and adults. *Radiology* 79: 109–114.

Hinck, V. C., W. M. Clark, and C. E. Hopkins. 1966. Normal interpediculate distances (minimum and maximum) in children and adults. *American Journal of Roentgenology* 97: 141–153.

Hinck, V. C., C. E. Hopkins, and W. M. Clark. 1965. Sagittal diameter of the lumbar spinal canal in children and adults. *Radiology* 85: 929–937.

Hinck, V. C., C. E. Hopkins, and B. S. Savara. 1962. Sagittal diameter of the cervical spinal canal in children. *Radiology* 79: 97–108.

Hopf, A., and C.-P. Claussen. 1971. Comparative studies on the fresh weights of the brains and spinal cords of *Theropithecus gelada, Papio hamadryas* and *Cercopithecus aethiops. Proceedings of the Third International Congress of Primatology, Zurich 1970,* vol 1., pp. 115–121. Basel: Karger.

Huizinga, J., J. A. Heiden, and P. J. Vinken. 1951. The human lumbar vertebral canal: A biometric study. *Koninklijke Nederlandse Akademie van Wetenschappen, Proceedings* C55: 22–33.

Krompecher, St., and J. Lipak. 1966. A simple method for determining cerebralization, brain weight and intelligence. *Journal of Comparative Neurology* 127: 113–120.

Larsen, J. L. 1981. The lumbar spinal canal in children. Part II: The interpedicular distance and its relation to the sagittal diameter and transverse pedicular width. *European Journal of Radiology* 1: 312–321.

Larsen, J. L., and D. Smith. 1981. The lumbar spinal canal in children. Part I: The sagittal diameter. *European Journal of Radiology* 1: 163–170.

Latimer, H. B. 1950. The weights of the brain and of its parts, and the weight and length of the spinal cord in the adult male guinea pig. *Journal of Comparative Neurology* 93: 37–51.

Latimer, H. B., and P.B. Sawin. 1955. Morphogenetic studies of the rabbit. XIII. The influence of the dwarf gene upon organ size and variability in race X. *Anatomical Record* 123: 447–466.

——— 1957. Morphogenetic studies of the rabbit. XIX. Organ size in relation to body size in large race III and in small race X. *Anatomical Record* 129: 457–472.

Leakey, R. E., and A. Walker. 1989. Early *Homo erectus* from west Lake Turkana, Kenya. In *Hominidae: Proceedings of the Second International Congress of Human Paleontology,* ed. G. Giacobini, pp. 209–215. Milan: Jaca Book.

MacLarnon, A. M. 1987. Size relationships of the spinal cord and associated skeleton in primates. PhD. thesis, University of London.

——— 1989. Applications of the Reflex instruments in quantitative morphology. *Folia Primatologica* 53: 33–49.

Markuske, H. 1977. Sagittal diameter measurements of the bony cervical spinal canal in children. *Pediatric Radiology* 6: 129–131.

Martin, R. D. 1981. Relative brain size and basal metabolic rate in terrestrial vertebrates. *Nature* 293: 57–60.

——— 1983. *Human Brain Evolution in an Ecological Context: Fifty-Second James Arthur Lecture on the Evolution of the Human Brain.* New York: American Museum of Natural History.

McHenry, H. M. 1988. New estimates of body weight in early hominids and their significance to encephalization and megadontia in "robust" australopithecines. In *Evolutionary History of the "Robust" Australopithecines,* ed. F. E. Grine, pp. 133–148. New York: Aldine de Gruyter.

Payne, E. E., and J. D. Spillane. 1957. The cervical spine: An anatomico-pathological study of seventy specimens (using a special technique) with particular reference to the problem of cervical spondylosis. *Brain* 80: 571–596.

Piera, V., A. Rodríguez, A. Cobos, R. Hernández, and P. Cobos. 1988. Morphology of the lumbar vertebral canal. *Acta Anatomica* 131: 35–40.

Porter, R. W., and D. Pavitt. 1987. The vertebral canal: I. Nutrition and development, an archaeological study. *Spine* 12: 901–906.

Postacchini, F., M. Ripani, and S. Carpano. 1983. Morphometry of the lumbar vertebrae. *Clinical Orthopaedics* 172: 296–303.

Ridgway, S. H., N. J. Flanigan, and J. G. McCormick. 1966. Brain-spinal cord ratios in porpoises: Possible correlations with intelligence and ecology. *Psychonomic Science* 6:491–492.

Romanes, G. D., ed. 1972. *Cunningham's Textbook of Anatomy,* 11th ed. London: Oxford University Press.

——— ed. 1984. *Cunningham's Manual of Practical Anatomy,* combined 14th ed. London: Oxford University Press.

Sand, P. G. 1970. The human lumbosacral vertebral column: An osteometric study. Universitetsforlaget Oslo.

Schultz, A.H. 1961. Vertebral column and thorax. In *Primatologia,* vol. 4(5), ed. H. Hofer, A. H. Schultz and D. Starck, pp. 1–66. Basel: Karger.

Stanley, J. H., S. I. Schabel, G. D. Frey, and G. D. Hungerford. 1986. Quantitative analysis of the cervical spinal canal by computed tomography. *Neuroradiology* 28: 139–143.

Twomey, L., and J. Taylor. 1988. Age changes in the lumbar spinal and intervertebral canals. *Paraplegia* 26: 238–49.
Wholey, M. H., A. J. Bruwer, and H. L. Baker. 1958. The lateral roentgenogram of the neck (with comments on the atlanto-odontoid-basion relationship). *Radiology* 71: 350–356.
Wolf, B. S., M. Khilnani, and L. L. Malis. 1956. The sagittal diameter of bony cervical canal and its significance. *Journal of Mount Sinai Hospital* 23: 283–292.
Yousefzadeh, D. K., G. Y. El-Khoury, and W. L. Smith. 1982. Normal sagittal diameter and variation in the pediatric cervical spine. *Radiology* 144: 319–325.

16. A MORPHOMETRIC STUDY OF FACIAL GROWTH

Atchley, W. R. 1987. Developmental quantitative genetics and the evolution of ontogenies. *Evolution* 41(2): 316–330.
Atchley, W. R., and B. K. Hall. 1991. A model for development and evolution of complex morphological structures. *Biological Reviews* 66: 101–157.
Bonner, J. T., ed. 1982. *Evolution and Development.* Dahlem Workshop Reports. Life Sciences Research Report 22. Berlin: Springer-Verlag.
Cheverud, J. M., and J. T. Richtsmeier. 1986. Finite-element scaling applied to sexual dimorphism in rhesus macaque *(Macaca mulatta)* facial growth. *Systematic Zoology* 35(3): 381–399.
Corner, B. D., and J. T. Richtsmeier. 1991. Morphometric analysis of craniofacial growth in *Cebus apella. American Journal of Physical Anthropology* 84(3): 323–342.
——— 1992. Cranial growth in the squirrel monkey *(Saimiri sciureus):* A quantitative analysis using three-dimensional coordinate data. *American Journal of Physical Anthropology* 87: 67–82.
——— 1993. Cranial growth and growth dimorphism in *Ateles geoffroyi. American Journal of Physical Anthropology,* in press.
Corner, B. D., S. Lele, and J. T. Richtsmeier. 1992. Measurement error of three-dimensional landmark data. *Quantitative Anthropology* 3(4): 347–359.
de Queiroz, K. 1985. The ontogenetic method for determining character polarity and its relevance to phylogenetic systematics. *Systematic Zoology* 34(3): 280–299.
Garn, S. M., and K. Koski. 1957. Tooth eruption sequence in fossil and recent man. *Nature* 180: 442–443.
Gould, S. J. 1966. Allometry and size in ontogeny and phylogeny. *Biological Reviews* 41: 587–640.
——— 1977. *Ontogeny and Phylogeny.* Cambridge: Harvard University Press.
——— 1988. The uses of heterochrony. In *Heterochrony in Evolution,* ed. M. L. McKinney, pp. 1–16. New York: Plenum Press.
Gower, J. 1975. Generalized procrustes analysis. *Psychometrika* 40: 33–50.
Grausz, H. M. 1992. Growth of the human perinatal craniofacial skeleton characterized in three dimensions. Ph.D. thesis, The Johns Hopkins University.
Leakey, R. E., and A. Walker. 1989. Early *Homo erectus* from West Lake Turkana, Kenya. In *Hominidae: Proceedings of the Second International Congress of Human Paleontology,* ed. G. Giacobini, pp. 209–215. Milan: Jaca Book.
Lele, S. 1991. Some comments on coordinate free and scale invariant methods in morphometrics. *American Journal of Physical Anthropology* 85: 407–417.
Lele, S., and J. T. Richtsmeier. 1991. Euclidean distance matrix analysis: A coordinate free approach for comparing biological shapes using landmark data. *American Journal of Physical Anthropology* 86: 415–428.
——— 1992. On comparing biological shapes: Detection of influential landmarks. *American Journal of Physical Anthropology* 87: 49–66.
Lewis, J. L., W. D. Lew, and J. R. Zimmerman. 1980. A nonhomogeneous anthropometric scaling method based on finite element principles. *Biomechanics* 13: 815–824.
Lozanoff, S., and V. M. Diewert. 1986. Measuring histological form change with finite element methods: An application using Diazo–Oxo–Norleucine (DON)–treated rats. *American Journal of Anatomy* 177: 187–201.
Martin, R. D. 1983. *Human Brain Evolution in an Ecological Context: Fifty-Second James Arthur Lecture on the Evolution of the Human Brain.* New York: American Museum of Natural History.
McKinney, M. L., ed. 1988. *Heterochrony in Evolution.* New York: Plenum Press.
Morris, G. R. 1991. FIESCA: Graphic and analytical software for finite element scaling analysis in biological research. Department of Civil Engineering, G. W. C. Whiting School of Engineering, The Johns Hopkins University.
Raff, R. A., and A. C. Raff. 1987. *Development as an Evolutionary Process.* New York: Alan R. Liss.
Richtsmeier, J. T. 1990. Quantitative analysis of three-dimensional data. *SPIE Proceedings* 1380: 49–58.
——— 1991. Implications of the use of varying element types in finite-element analysis of craniofacial growth. *American Journal of Physical Anthropology,* supplement 12: 151.
Richtsmeier, J. T., and J. M. Cheverud. 1986. Finite-element scaling analysis of human craniofacial growth. *Journal of Craniofacial Genetics and Developmental Biology* 6(3): 289–323.
Richtsmeier, J. T., and S. Lele. 1990. Analysis of craniofacial growth in Crouzon syndrome using landmark data. *Journal of Craniofacial Genetics and Developmental Biology* 10: 39–62.
Richtsmeier, J. T., J. M. Cheverud, and G. R. Morris. 1990a. Comparison of three-dimensional form. *Journal of Anatomy* 170: 221–223.
Richtsmeier, J. T., G. R. Morris, J. L. Marsh, and M. W. Vannier. 1990b. The biological implications of varying element design in finite-element scaling analyses of growth. *IEEE Proceedings* 12(1): 387–388.
Richtsmeier, J. T., J. M. Cheverud, B. C. Corner, S. M. Danahey, and S. Lele. 1993. Sexual dimorphism of ontogeny in the crab eating macaque *(Macaca fascicularis). Journal of Human Evolution,* in press.
Schultz, A. H. 1962. Metric age changes and sex differences in primate skulls. *Yearbook of Physical Anthropology* 10: 129–154.
Shea, B. T. 1983. Allometry and heterochrony in the African apes. *American Journal of Physical Anthropology* 62(3): 275–290.
——— 1989. Heterochrony in human evolution: The case for neoteny reconsidered. *Yearbook of Physical Anthropology* 32: 69–101.
Skalak, R., G. Dasgupta, M. Moss, E. Otten, P. Dullemeijer, and H. Vilmann. 1982. Analytical description of growth. *Journal of Theoretical Biology* 94: 555–577.
Slatkin, M. 1987. Quantitative genetics of heterochrony. *Evolution* 41(4): 799–811.
Strauss, R. E. 1990. Heterochronic variation in the developmental timing of cranial ossifications in poeciliid fishes (Cyprinodontiformes). *Evolution* 44(6): 1558–1567.
Watts, E. S. 1985. Adolescent growth and development of monkeys, apes and humans. In *Nonhuman Primate Models for Human Growth and Development,* ed. E. S. Watts, pp. 41–65. New York: Alan R. Liss.
Watts, E. S., and J. A. Gavan. 1982. Postnatal growth of nonhuman primates: The problem of the adolescent spurt. *Human Biology* 54: 53–70.
Wood, B. A., Y. Li, and C. Willoughby. 1991. Intraspecific variation and sexual dimorphism in cranial and dental var-

iables among higher primates and their bearing on the hominid fossil record. *Journal of Anatomy* 174: 185–205.

17. Perspectives on the Nariokotone Discovery

Adelola, A., K. R. Kattan, and F. N. Silverman. 1975. Thickness of the normal skull in the American blacks and whites. *American Journal of Physical Anthropology* 43: 23–30.

Andrews, P. J. 1984. An alternative interpretation of the characters used to define *Homo erectus. Courier Forschungsinstitut Senckenberg* 69: 167–175.

Behrensmeyer, A. K., and L. F. Laporte. 1981. Footprints of a Pleistocene hominid in northern Kenya. *Nature* 289: 167–169.

Bilsborough, A., and B. A. Wood. 1986. The origin and fate of *Homo erectus*. In *Major Topics in Primate and Human Evolution*, ed. B. A. Wood, L. Martin, and P. Andrews, pp. 295–316. Cambridge: Cambridge University Press.

Brauer, G. 1990. The occurrence of some controversial *Homo erectus* cranial features in the Zoukoudian and East African hominids. *Acta Anthropologica Sinica* 9: 352–358.

Brown, F., J. Harris, R. Leakey, and A. Walker. 1985. Early *Homo erectus* from west Lake Turkana, Kenya. *Nature* 316: 788–792.

Bunn, H. T. 1981. Archaeological evidence for meat-eating by Plio-Pleistocene hominids from Koobi Fora and Olduvai Gorge. *Nature* 291: 574–577.

Calvin, W. H. 1983. A stone's throw and its launch window: Timing precision and its implications for language and hominid brains. *Journal of Theoretical Biology* 104: 121–135.

Carrier, D. R. 1984. The energetic paradox of human running and hominid evolution. *Current Anthropology* 25: 483–495.

Chamberlain, A. T., and B. A. Wood. 1987. Early hominid phylogeny. *Journal of Human Evolution* 16: 119–133.

Cohen, M. N., and G. J. Armelagos. 1984. *Paleopathology at the Origins of Agriculture.* Orlando: Academic Press.

Corina, D. P., J. Vaid, and U. Bellugi. 1992. The linguistic basis of left hemisphere lateralization. *Science* 255: 1258–1260.

Darlington, P. R. 1957 *Zoogeography.* New York: Wiley.

——— 1975. Group selection, altruism, reinforcement, and throwing in human evolution. *Proceedings of the National Academy of Sciences* 72: 3748–3752.

Day, M. H. 1971. Postcranial remains of *Homo erectus* from Bed IV, Olduvai Gorge, Tanzania. *Nature* 221: 383–387.

Falk, D., J. Cheverud, M. W. Vannier, and G. C. Conroy. 1986. Advanced computer graphics technology reveals cortical asymmetry in endocasts of rhesus monkeys. *Folia Primatologica* 46: 98–103.

Feibel, C., J. M. Harris, and F. H. Brown. 1991. Palaeoenvironmental context for the late Neogene of the Turkana Basin. In *Koobi Fora Research Project,* vol. 3, ed. J. M. Harris. Oxford: Clarendon Press.

Fransiscus, R. G., and E. T. Trinkaus. 1988. Nasal morphology and the emergence of *Homo erectus. American Journal of Physical Anthropology* 75: 517–527.

Groves, C. P. 1989. *A Theory of Human and Primate Evolution.* Oxford: Clarendon Press.

Groves, C. P., and V. Mazek. 1975. An approach to the taxonomy of the Hominidae: Gracile Villafrancian hominids of Africa. *Casopis pro Mineralogii a Geologii* 20: 225–247.

Harvey, P. H. 1990. Life-history variation: Size and mortality patterns. In *Primate Life History and Evolution,* ed. C. J. DeRousseau, pp. 81–88. New York: Wiley-Liss.

Holloway, R. L. 1969. Culture: A *human* domain. *Current Anthropology* 10: 395–407.

——— 1983. Human paleontological evidence relevant to language behavior. *Human Neurobiology* 2: 105–114.

Hopson, J. A. 1989. Leonard Burton Radinsky (1937–1985). In *The Evolution of the Perissodactyls,* ed. D. R. Prothero and R. M. Schoch, pp. 3–12. Oxford: Oxford University Press.

Howell, F. C. 1978. Hominidae. In *Evolution of African Mammals,* ed. H. B. S. Cooke and V. J. Maglio, pp. 154–248. Cambridge: Harvard University Press.

Hull, D. L. 1979. The limits of cladism. *Systematic Zoology* 28: 416–440.

Johanson, D. C. 1989. A partial *Homo habilis* skeleton from Olduvai Gorge, Tanzania: A summary of preliminary results. In *Hominidae: Proceedings of the Second International Congress of Human Paleontology,* ed. G. Giacobini, pp. 155–166. Milan: Jaca Book.

Johanson, D. C., F. T. Masao, G. E. Eck, T. D. White, R. C. Walter, W. H. Kimbel, B. Asfaw, P. Manega, P. Ndessokia, and G. Suwa. 1987. New partial skeleton of *Homo habilis* from Olduvai Gorge, Tanzania. *Nature* 327: 205–209.

Kennedy, G. E. 1991. On the autopomorphic traits of *Homo erectus. Journal of Human Evolution* 20: 375–412.

Laitman, J. T. 1985. Evolution of the hominid upper respiratory tract: The fossil evidence. In *Hominid Evolution:* Past, Present and Future, ed. P. V. Tobias, pp. 281–286. New York: Alan R. Liss.

Leakey, L. S. B. 1964. Very early African Hominidae and their ecological setting. In *African Ecology and Human Evolution,* ed. F. C. Howell and F. Bourliere, pp. 448–457. London: Methuen.

Leakey, L. S. B., P. V. T. Tobias, and J. R. Napier. 1964. A new species of the genus *Homo* from Olduvai Gorge. *Nature* 202: 7–9.

Leakey, R. E., M. G. Leakey, and A. K. Behrensmeyer. 1978. The hominid catalogue. In *Koobi Fora Research Project,* vol. 1, ed. M. G. Leakey and R. E. Leakey, pp. 86–182. Oxford: Clarendon Press.

Leakey, R. E., and A. C. Walker. 1976. *Australopithecus, Homo erectus* and the single species hypothesis. *Nature* 261: 572–574.

——— 1978. The hominids of East Turkana. *Scientific American* 239: 54–66.

——— 1985. Further hominids from the Plio-Pleistocene of Koobi-Fora, Kenya. *American Journal of Physical Anthropology* 67: 135–163.

——— 1989. Early *Homo erectus* skeleton from west Lake Turkana, Kenya. In *Hominidae: Proceedings of the Second International Congress of Human Paleontology,* ed. G. Giacobini, pp. 209–215. Milan: Jaca Book.

Leakey, R. E. F., and B. A. Wood. 1973. New evidence for the genus *Homo* from East Rudolf, Kenya (II). *American Journal of Physical Anthropology* 39: 355–368.

Leiberman, P. E., and E. S. Crelin. 1971. On the speech of Neanderthal man. *Linguistic Inquiry* 2: 203–222.

LeMay, M. 1976. Morphological cerebral asymmetries of modern man, fossil man, and nonhuman primate. *Annals of the New York Academy of Sciences* 280: 349–366.

Maier, W., and A. Nkini. 1984. Olduvai Hominid 9: New results of investigation. *Courier Forschungsinstitut Senckenberg* 69: 123–130.

Martin, R. 1928. *Lehrbuch der Anthropologie.* Jena: Fischer.

Martin, R. D. 1983. *Human Brain Evolution in an Ecological Context: Fifty-Second James Arthur Lecture on the Human Brain.* New York: American Museum of Natural History.

McHenry, H. M. 1988. New estimates of body weight in early hominids and their significance to encephalization and megadontia in "robust" australopithecines. In *Evolutionary History of the "Robust" Australopithecines,* ed. F. E. Grine, pp. 133–148. New York: Aldine de Gruyter.

——— 1991. Femoral lengths and stature in Plio-Pleistocene hominids. *American Journal of Physical Anthropology* 85: 149–158.

Noble, W., and I. Davidson. 1991. The evolutionary emergence of modern human behaviour: Language and its archeology. *Man* 26: 223–254.

Paterson, H. E. H. 1986. Environment and species. *South African Journal of Science* 82: 62–65.

Pilbeam, D. R., and S. J. Gould. 1974. Size and scaling in human evolution. *Science* 186: 892–901.

Potts, R. 1991. Why the Oldowan? *Journal of Anthropological Research* 47: 153–176.

Potts, R., and P. Shipman. 1981. Cutmarks made by stone tools on bones from Olduvai Gorge, Tanzania. *Nature* 291: 577–580.

Rightmire, P. G. 1981. Patterns in the evolution of *Homo erectus*. *Paleobiology* 7: 241–246.

——— 1990. *The Evolution of Homo erectus.* Cambridge: Cambridge University Press.

——— 1986. Stasis in *Homo erectus* defended. *Paleobiology* 12: 324–325.

Rose, K. D., and T. M. Bown. 1993. Species concepts and species recognition in Eocene primates. In *Species, Species Concepts and Primate Evolution,* ed. L. Martin and W. Kimbel. New York: Plenum.

Ruff, C. B. 1991. Climate and body shape in hominid evolution. *Journal of Human Evolution* 21: 81–105.

Ruff, C. B., and H. H. Jones. 1981. Bilateral asymmetry in cortical bone of the humerus and tibia—Sex and age factors. *Human Biology* 53: 69–86.

Ruff, C. B., E. Trinkaus, A. Walker, and C. S. Larson. 1993. Postcranial robusticity in *Homo,* I: Temporal trends and mechanical interpretation. *American Journal of Physical Anthropology* 91: 21–53.

Schmidt-Nielsen, K. 1975. *Animal Physiology.* Cambridge: Cambridge Univerisity Press.

Schultz, A. H. 1937. Proportions, variability and asymmetries of the long bones of the limbs and the clavicles in man and apes. *Human Biology* 9: 281–328.

Stringer, C. B. 1984. The definition of *Homo erectus* and the existence of the species in Africa and Europe. *Courier Forschungsinstitut Senckenberg* 69: 131–143.

——— 1987. A numerical cladistic analysis of the genus *Homo.* *Journal of Human Evolution* 16: 135–146.

——— 1990. The emergence of modern humans. *Scientific American* 264: 98–97.

Templeton, A. 1992. Human origins and analysis of mitochondrial DNA. *Science* 255: 737.

Tobias, P. V. 1988. Recent advances in the evolution of the hominids with especial reference to brain and speech. *Pontifical Academy of Sciences: Scripta Varia* 50: 85–140.

——— 1991. *Olduvai Gorge,* vol. 4. Cambridge: Cambridge University Press.

Toth, N. 1985. Archaeological evidence for preferential right-handedness in the lower and middle Pleistocene, and its possible implications. *Journal of Human Evolution* 14: 607–614.

Trinkaus, E. 1981. Neanderthal limb proportions and cold adaptation. In *Aspects of Human Evolution,* ed. C. B. Stringer, pp. 187–224. London: Taylor and Francis.

Turner, A., and A. Chamberlain. 1989. Speciation, morphological change and the status of African *Homo erectus. Journal of Human Evolution* 18: 115–130.

Tuttle, R. H. 1987. Kinesiological inferences and evolutionary implications from Laetoli bipedal trails G-1, G-2/3, and A. In *Laetoli: A Pliocene Site in Northern Tanzania,* ed. M. D. Leakey and J. M. Harris, pp. 503–523. Oxford: Clarendon Press.

Vandermeersch, B. 1969. Les nouveaux découvertes de restes humains dans les couches Levalloiso-Mousteriennes du gisement de Qafzeh (Israel). *Comptes Rendus, Académie des Sciences, Paris* 268: 2562–2565.

Walker, A. 1976. Remains attributable to *Australopithecus* in the East Rudolf succession. In *Earliest Man and Environments in the Lake Rudolf Basin,* ed. Y. Coppens, F. C. Howell, G. Ll. Isaac, and R. E. F. Leakey, pp. 484–489. Chicago: University of Chicago Press.

——— 1991. The origin of the genus *Homo.* In *Evolution of Life,* ed. S. Osawa and T. Honjo, pp. 379–389. Tokyo: Spinger-Verlag.

Walker, A., M. R. Zimmerman, and R. E. F. Leakey. 1982. A possible case of hypervitaminosis A in *Homo erectus. Nature* 296: 248–250.

Weidenreich, F. 1943. The skull of *Sinanthropus pekinensis. Palaeontologia Sinica,* new series D, 10: 1–291.

Wheeler, P. E. 1991. The thermoregulatory advantages of hominid bipedalism in open equatorial environments: The contribution of increased convective heat loss and cutaneous evaporative cooling. *Journal Human Evolution* 21: 107–115.

Wiley, E. O. 1978. The evolutionary species concept reconsidered. *Systematic Zoology* 27: 17–26.

Williams, P. L., R. Warwick, M. Dyson, and L. H. Bannister, eds. 1989. *Gray's Anatomy,* 37th ed. Edinburgh: Churchill Livingstone.

Wolpoff, M. H. 1980. *Paleoanthropology.* New York: Knopf.

——— 1984. Evolution in *Homo erectus:* The question of stasis. *Paleobiology* 10: 389–406.

——— 1986. Stasis in the interpretation of evolution of *Homo erectus:* A reply to Rightmire. *Paleobiology* 12: 325 –328.

Wolpoff, M. H., and A. Nkini. 1985. Early and early middle Pleistocene hominids in Asia and Africa. In *Ancestors,* ed. E. Delson, pp. 202–205. New York: Alan R. Liss.

Wood, B. A. 1976. Remains attributable to *Homo* in the East Rudolf succession. In *Earliest Man and Environments in the Lake Rudolf Basin,* ed. Y. Coppens, F. C. Howell, G. Ll. Isaac, and R. E. F. Leakey, pp. 484–489. Chicago: University of Chicago Press.

——— 1984. The origins of *Homo erectus. Courier Forschungsinstitut Senckenberg* 69: 99–111.

——— 1991. *Koobi Fora Research Project,* vol. 4. Oxford: Clarendon Press.

——— 1992. Origin and evolution of the genus *Homo. Nature* 355: 783–790.

ACKNOWLEDGMENTS

Funding for the expeditions which led to the discovery and recovery of KNM-WT 15000 and thousands of other fossils was provided by the National Museums of Kenya and the generous support of the National Geographic Society, Washington, D.C. Any royalties from the sale of this volume will go to the Kenya National Museums for the express purpose of helping paleontological research and the training of young Kenyan paleontologists.

We thank the Government of Kenya and the Governors of the National Museums of Kenya for permission to carry out the research. A.W. benefited from fellowships from the John Simon Guggenheim Memorial Foundation and the John D. and Catherine T. MacArthur Foundation.

Kamoya Kimeu is to be congratulated on his discovery and thanked for his leadership, friendship, and hard work on these expeditions. To all the members of his team, "The Hominid Gang," go our sincere thanks. The team changed during the years, but at various times it included Joseph Mutaba, Nzube Mutiwa, Bernard Ngeneyo, Wambua Mangao, Benson Kyongo, Musa Kyeva, Christopher Kiarie, and Mwongela Muoka. In camp and on the road, Daniel Mutinda, Omari Hassan, Solomon s/o Derekich, and Kabai Nd'unda provided invaluable assistance. The children of the Nariokotome Primary School gave us help and more.

Several people visited the excavations, helped us, and left with blisters from digging in the hard rock. In particular we wish to thank Louise Leakey, Samira Leakey, Meave Leakey, Mary Leakey, Pat Shipman, Audrone Biknevicius, Mark Teaford, Fred Grine, Harvey Schreter, Phyllis Schreter, Mary Muungu, Emma Mbua, Alfreda Ibui, Alice Maundu, Isiah Odhiambo, Cheryl Taylor, Delta Willis, Virginia Morrell, William Anyonge, Nancy Reynolds, Anne Cooksie, David Brill, Tom Smith, Rick Potts, Ellery Ingal, and Teresa Nganga. Our colleagues Frank Brown, Craig Feibel, and John Harris deserve our great thanks for their major contributions and for adding their own mix of genius and personality to these memorable expeditions. We also express our sincere thanks to the contributors to this volume and ask them to forgive our nagging.

In Nairobi Emma Mbua and Meave Leakey gave us invaluable curatorial assistance and help at all times. Joseph Mutaba and Christopher Kiarie prepared practically all of the specimen, and Simon Kasinga and Alphonse Katula cast it. Nina Mudida, John Kimengich, and Charles Kibi helped with faunal identifications. In Baltimore Tom Urquhart printed the plates. Barbara Brown helped enormously with the final collation of the manuscripts and with the illustrations. Linda Kurant and Calotta Welch helped with secretarial matters. Peter Ungar and Kathy Rafferty helped in checking details.

Finally, we thank our many past and present editors and their assistants at Harvard University Press, especially Howard Boyer, who helped us start, and Michael Fisher, who helped us finish. Kate Schmit deserves the most thanks from us, for her help in making sense out of an extremely complicated manuscript. We're also grateful to Sally Bindari of Designworks for the elegant look she has given this volume.

CONTRIBUTORS

DAVID BEGUN
Department of Anthropology
University of Toronto
100 St. George Street
Toronto, Ontario M5S 1A1
Canada

BARBARA BROWN
Department of Anthropology
Yale University
P.O. Box 2114 Yale Station
New Haven, Connecticut 06520

FRANCIS H. BROWN
Department of Geology and Geophysics
University of Utah
Salt Lake City, Utah 84112

CRAIG S. FEIBEL
Department of Geology and Geophysics
University of Utah
Salt Lake City, Utah 84112

JOHN M. HARRIS
Earth Sciences Division
Natural History Museum of Los Angeles County
900 Exposition Boulevard
Los Angeles, California 90007

LYMAN M. JELLEMA
Department of Physical Anthropology
The Cleveland Museum of Natural History
1 Wade Oval Drive
University Circle
Cleveland, Ohio 44106

BRUCE LATIMER
Department of Physical Anthropology
The Cleveland Museum of Natural History
1 Wade Oval Drive
University Circle
Cleveland, Ohio 44106

MEAVE LEAKEY
Division of Palaeontology
National Museums of Kenya
P.O. Box 40658
Nairobi, Kenya

CONTRIBUTORS

RICHARD LEAKEY
Kenya Wildlife Service
P.O. Box 40241
Nairobi, Kenya

ANN MACLARNON
Department of Biological and Chemical Sciences
Roehampton Institute
Whitelands College
West Hill
London SW15 3SN, United Kingdom

IAN MCDOUGALL
Research School of Earth Sciences
The Australian National University
G.P.O. Box 4
Canberra, ACT 2601, Australia

JOAN T. RICHTSMEIER
Department of Cell Biology and Anatomy
The Johns Hopkins University
School of Medicine
725 North Wolfe Street
Baltimore, Maryland 21205

CHRISTOPHER B. RUFF
Department of Cell Biology and Anatomy
The Johns Hopkins University
School of Medicine
725 North Wolfe Street
Baltimore, Maryland 21205

B. HOLLY SMITH
Museum of Anthropology
University of Michigan
Ann Arbor, Michigan 48109

ALAN WALKER
Department of Cell Biology and Anatomy
The Johns Hopkins University
School of Medicine
725 North Wolfe Street
Baltimore, Maryland 21205

CAROL V. WARD
Department of Anthropology
The University of Missouri
200 Swallow Hall
Columbia, Missouri 65211

INDEX

A

B

C

I

J

K

L

M

N

O

P

R

S

T

U

V

W

Z

THIS VOLUME WAS
designed by Sally Bindari/Designworks.
The text was composed by DEKR Corporation
in 10-point Palatino,
a typeface created in 1948 by Hermann Zapf.
Text files and computer graphs were prepared by
Technologies 'N Typography.
The halftone and text negatives were made by
Jay's Publishers Services.
The color insert was printed by the Henry Sawyer Company.
The text was printed and bound by Vail Ballou Press.
The paper, 70-pound Dulcet, was manufactured by
Monadnock Paper Mills
and supplied by Pratt Paper Company.